# Series and Products in the Development of Mathematics

## Volume 1

This is the first volume of a two-volume work that traces the development of series and products from 1380 to 2000 by presenting and explaining the interconnected concepts and results of hundreds of unsung as well as celebrated mathematicians. Some chapters deal with the work of primarily one mathematician on a pivotal topic, and other chapters chronicle the progress over time of a given topic. This updated second edition of *Sources in the Development of Mathematics* adds extensive context, detail, and primary source material, with many sections rewritten to more clearly reveal the significance of key developments and arguments. Volume 1, accessible even to advanced undergraduate students, discusses the development of the methods in series and products that do not employ complex analytic methods or sophisticated machinery. Volume 2 treats more recent work, including de Branges's solution of Bieberbach's conjecture, and requires more advanced mathematical knowledge.

RANJAN ROY (1947–2020) was the Ralph C. Huffer Professor of Mathematics and Astronomy at Beloit College, where he was a faculty member for 38 years. Roy published papers and reviews on Riemann surfaces, differential equations, fluid mechanics, Kleinian groups, and the development of mathematics. He was an award-winning educator, having received the Allendoerfer Prize, the Wisconsin MAA teaching award, and the MAA Haimo Award for Distinguished Mathematics Teaching and was twice named Teacher of the Year at Beloit College. He coauthored *Special Functions* (2001) with George Andrews and Richard Askey and coauthored chapters in the *NIST Handbook of Mathematical Functions* (2010); he also authored *Elliptic and Modular Functions from Gauss to Dedekind to Hecke* (2017) and the first edition of this book, *Sources in the Development of Mathematics* (2011).

Ranjan Roy 1948–2020

# Series and Products in the Development of Mathematics

## Second Edition

### Volume 1

RANJAN ROY

*Beloit College*

CAMBRIDGE
UNIVERSITY PRESS

# CAMBRIDGE
## UNIVERSITY PRESS

University Printing House, Cambridge CB2 8BS, United Kingdom

One Liberty Plaza, 20th Floor, New York, NY 10006, USA

477 Williamstown Road, Port Melbourne, VIC 3207, Australia

314-321, 3rd Floor, Plot 3, Splendor Forum, Jasola District Centre, New Delhi - 110025, India

103 Penang Road, #05-06/07, Visioncrest Commercial, Singapore 238467

Cambridge University Press is part of the University of Cambridge.

It furthers the University's mission by disseminating knowledge in the pursuit of
education, learning and research at the highest international levels of excellence.

www.cambridge.org
Information on this title: www.cambridge.org/9781108709453
DOI: 10.1017/9781108627702

First edition © Ranjan Roy 2011
Second edition © Ranjan Roy 2021

First published as *Sources in the Development of Mathematics*, 2011
Second edition 2021

*A catalogue record for this publication is available from the British Library*

ISBN 2-Volume Set 978-1-108-70943-9 Paperback
ISBN Volume 1 978-1-108-70945-3 Paperback
ISBN Volume 2 978-1-108-70937-8 Paperback

# Contents

# Contents of Volume 2

# Preface

*Sources in the Development of Mathematics: Series and Products from the Fifteenth to the Twenty-first Century*, my book of 2011, was intended for an audience of graduate students or beyond. However, since much of its mathematics lies at the foundations of the undergraduate mathematics curriculum, I decided to use portions of my book as the text for an advanced undergraduate course. I was very pleased to find that my curious and diligent students, of varied levels of mathematical talent, could understand a good bit of the material and get insight into mathematics they had already studied as well as topics with which they were unfamiliar. Of course, the students could profitably study such topics from good textbooks. But I observed that when they read original proofs, perhaps with gaps or with slightly opaque arguments, students gained very valuable insight into the process of mathematical thinking and intuition. Moreover, the study of the steps, often over long periods of time, by which earlier mathematicians refined and clarified their arguments revealed to my students the essential points at the crux of those results, points that may be more difficult to discern in later streamlined presentations. As they worked to understand the material, my students witnessed the difficulty and beauty of original mathematical work, and this was a source of great enjoyment to many of them. I have now thrice taught this course, with extremely positive student response.

In order for my students to follow the foundational mathematical arguments in *Sources*, I was often required to provide additional material, material actually contained in the original works of the mathematicians being studied. I therefore decided to expand my book, as a second edition in two volumes, to make it more accessible to readers, from novices to accomplished mathematicians. This second edition contains about 250 pages of new material, including more details within the original proofs, elaborations and further developments of results, and additional results that may give the reader a better perspective. Furthermore, to give the material greater focus, I have limited this second edition to the topics of series and products, areas that today permeate both applied and pure mathematics; the second edition is thus entitled *Series and Products in the Development of Mathematics*.

This first volume of my work discusses the development of the fundamental though powerful and essential methods in series and products that do not employ complex analytic methods or sophisticated machinery such as Fourier transforms. Much of this material would be accessible, perhaps with guidance, to advanced undergraduate students. The second volume deals with more recent work and requires considerable mathematical background. For example, in volume 2, I discuss Weil's 1949 paper on solutions of equations in finite fields and de Branges's conquest of the Bieberbach conjecture. Each volume contains the same complete bibliography.

The exercises at the end of the chapters present many additional original results and may be studied simply for the supplementary theorems they contain. The exercises are accompanied by references to the original works, as an aid to further research. Readers may attempt to prove the results in the problems and, by use of the references, compare their own solutions with the originals. Moreover, many of the exercises can be tackled by methods similar to those given in the text, so that some exercises can be realistically assigned to a class as homework. I assigned many exercises to my classes, and found that the students enjoyed and benefited from their efforts to find solutions. Thus, the exercises may be useful as problems to be solved, and also for the results they present.

Detailed study of original mathematical works provides a point of entry into the minds of the creators of powerful theories, and thus into the theories themselves. But tracing the discovery and evolution of mathematical ideas and theorems entails the examination of many, many papers, letters, notes, and monographs. For example, in this work I have discussed the work of more than three hundred mathematicians, including arguments and theorems contained in approximately one hundred works and letters of Euler alone. Locating, studying, and grasping the interconnections among such original works and results is a ponderous, complex, and rewarding effort. In this second edition, I have added numerous footnotes and almost five hundred works to the bibliography. My hope is that the detailed footnotes and the expanded bibliography, containing both original works and works of distinguished expositors and historians of mathematics, may encourage and facilitate the efforts of those who wish to search out and study the original sources of our inherited mathematical wealth.

I first wish to thank my wife, who typeset and edited this work, made innumerable corrections and refinements to the text, and devotedly assisted me with translations and locating references. I am also very grateful to NFN Kalyan for his encouragement and for creating the eloquent artwork for the cover of these volumes. I greatly appreciate Maitreyi Lagunas's unflagging support and interest. I thank Bruce Atwood who cheerfully constructed the nice diagrams contained in this work, and Paul Campbell who generously provided expert technical support and advice. I am grateful to my student Shambhavi Upadhyaya, who has an unusual ability to proofread very accurately, for spending so much time giving useful suggestions for improvement. I am indebted to my students whose questions and enthusiasm helped me refine this second edition. I also thank the very capable librarians at Beloit College, especially Chris Nelson and Cindy Cooley. Finally, I wish to acknowledge the inspiration provided me by my friend, the late Dick Askey.

# 1

## Power Series in Fifteenth-Century Kerala

### 1.1 Preliminary Remarks

More than two and a half centuries before Isaac Newton discovered the sine and cosine series and James Gregory the arctan series, the Indian astronomer and mathematician Madhava (c. 1340–c. 1425) gave expressions for $\sin x$, $\cos x$, and $\arctan x$ as infinite power series.[1] Madhava's work may have been motivated by his studies in astronomy, since he concentrated mainly on the trigonometric functions. There appears to be no connection between the work of Madhava's school and that of Newton and other European mathematicians. In spite of this, the Keralese and European mathematicians shared some similar methods and results. Both were fascinated with transformation of series, though they used very different methods.

The mathematician-astronomers of medieval Kerala lived, worked, and taught in large family compounds called illams. Madhava, believed to have been the founder of the school, worked in the Bakulavihara illam in the town of Sangamagrama, a few miles north of Cochin. He was an Emprantiri Brahmin, then considered socially inferior to the dominant Namputiri (or Nambudri) Brahmin. This position does not appear to have curtailed his teaching activities; his most distinguished pupil was Paramesvara, a Namputiri Brahmin. No mathematical works of Madhava have been found, though three of his short treatises on astronomy are extant. The most important of these describes how to accurately determine the position of the moon at any time of the day. Other surviving mathematical works of the Kerala school attribute many very significant results to Madhava. Although his algebraic notation was almost primitive, Madhava's mathematical skill allowed him to carry out highly original and difficult research.

Paramesvara (c.1380–c.1460), Madhava's pupil, was from Asvattagram, about thirty-five miles northeast of Madhava's home town. He belonged to the Vatasreni illam, a famous center for astronomy and mathematics. He made a series of observations of the eclipses of the sun and the moon between 1395 and 1432 and composed several astronomical texts, the last of which was written in the 1450s,

---

[1] Newton (1959–1960) vol. 2, pp. 20–47, especially p. 36; Turnbull (1939) p. 170; Jyesthadeva et al. (2008).

near the end of his life. Sankara Variyar attributed to Paramesvara a formula for the radius of a circle in terms of the sides of an inscribed quadrilateral. See Exercise 4. Paramesvara's son, Damodara, was the teacher of Jyesthadeva (c. 1500–c. 1570) whose works survive and give us all the surviving proofs of this school. Damodara was also the teacher of Nilakantha (c. 1450–c. 1550) who composed the famous treatise called the *Tantrasangraha* (c. 1500), a digest of the mathematical and astronomical knowledge of his time. His works allow us determine his approximate dates because, in his *Aryabhatyabhasya*, Nilakantha refers to his observation of solar eclipses in 1467 and 1501. Nilakantha made several efforts to establish new parameters for the mean motions of the planets and vigorously defended the necessity of continually correcting astronomical parameters on the basis of observation. Sankara Variyar (c. 1500–1560) was his student.

The surviving texts containing results on infinite series are Nilakantha's *Tantrasangraha*, a commentary on it by Sankara Variyar called *Yuktidipika*, the *Yuktibhasa* by Jyesthadeva and the *Kriyakramakari*, started by Variyar and completed by his student Mahisamangalam Narayana. In addition, there is a text called *Karanapaddhati* of Putumana Somayaji, thought by some to have been written around 1700. However, the four translators of this work present an argument that Somayaji was a junior contemporary of Nilakantha and composed his work between 1532 and 1566. All these works are in Sanskrit except the *Yuktibhasa*, written in Malayalam, the language of Kerala. These works, especially the *Yuktibhasa* that gives detailed arguments, provide a summary of major results on series discovered by these original mathematicians of the indistinct past:

A. Series expansions for arctangent, sine, and cosine:

(1) $\theta = \tan\theta - \frac{\tan^3\theta}{3} + \frac{\tan^5\theta}{5} - \cdots$ ,

(2) $\sin\theta = \theta - \frac{\theta^3}{3!} + \frac{\theta^5}{5!} - \cdots$ ,

(3) $\cos\theta = 1 - \frac{\theta^2}{2!} + \frac{\theta^4}{4!} - \cdots$ ,

(4) $\sin^2\theta = \theta^2 - \frac{\theta^4}{2^2-\frac{2}{2}} + \frac{\theta^6}{(2^2-\frac{2}{2})(3^2-\frac{3}{2})} - \frac{\theta^8}{(2^2-\frac{2}{2})(3^2-\frac{3}{2})(4^2-\frac{4}{2})} + \cdots$ .

In the proofs of the formulas contained in (A), the range of $\theta$ for the first series was $0 \le \theta \le \frac{\pi}{4}$ and for the second and third was $0 \le \theta \le \frac{\pi}{2}$. Although the series for sine and cosine converge for all real values, the concept of periodicity of the trigonometric functions was discovered much later.

B. Series for $\pi$:

(1) $\frac{\pi}{4} \approx 1 - \frac{1}{3} + \frac{1}{5} - \cdots \mp \frac{1}{n} \pm f_i(n+1)$,   $i = 1, 2, 3$, where

$$f_1(n) = \frac{1}{2n}, \quad f_2(n) = \frac{n}{2(n^2+1)},$$

and

$$f_3(n) = \frac{n^2+4}{2n(n^2+5)};$$

(2) $\frac{\pi}{4} = \frac{3}{4} + \frac{1}{3^3-3} - \frac{1}{5^3-5} + \frac{1}{7^3-7} - \cdots$ ;

(3) $\frac{\pi}{4} = \frac{4}{1^5+4\cdot1} - \frac{4}{3^5+4\cdot3} + \frac{4}{5^5+4\cdot5} - \cdots$ ;

(4) $\frac{\pi}{2\sqrt{3}} = 1 - \frac{1}{3\cdot3} + \frac{1}{5\cdot3^2} - \frac{1}{7\cdot3^3} + \cdots$ ;

(5) $\frac{\pi}{6} = \frac{1}{2} + \frac{1}{(2\cdot2^2-1)^2-2^2} + \frac{1}{(2\cdot4^2-1)^2-4^2} + \frac{1}{(2\cdot6^2-1)^2-6^2} + \cdots$ ;

(6) $\frac{\pi-2}{4} \approx \frac{1}{2^2-1} - \frac{1}{4^2-1} + \frac{1}{6^2-1} - \cdots \mp \frac{1}{n^2-1} \pm \frac{1}{2((n+1)^2+2)}$ ;

(7) $\frac{\pi}{8} = \frac{1}{2^2-1} + \frac{1}{6^2-1} + \frac{1}{10^2-1} + \cdots$ ;

(8) $\frac{\pi}{8} = \frac{1}{2} - \frac{1}{4^2-1} - \frac{1}{8^2-1} - \frac{1}{12^2-1} - \cdots$ .

These results were stated in verse form. Thus, the series for sine was described:[2]

The arc is to be repeatedly multiplied by the square of itself and is to be divided [in order] by the square of each even number increased by itself and multiplied by the square of the radius. The arc and the terms obtained by these repeated operations are to be placed in sequence in a column, and any last term is to be subtracted from the next above, the remainder from the term then next above, and so on, to obtain the jya (sine) of the arc.

So if $r$ is the radius and $s$ the arc, then the successive terms of the repeated operations mentioned in the description are given by

$$s \cdot \frac{s^2}{(2^2+2)r^2}, \quad s \cdot \frac{s^2}{(2^2+2)r^2} \cdot \frac{s^2}{(4^2+4)r^2}, \ldots$$

and the equation is

$$y = s - s \cdot \frac{s^2}{(2^2+2)r^2} + s \cdot \frac{s^2}{(2^2+2)r^2} \cdot \frac{s^2}{(4^2+4)r^2} - \cdots,$$

where $y = r \sin \frac{s}{r}$.

Nilakantha's *Aryabhatyabhasya* attributes the sine series to Madhava. The *Kriyakramakari* attributes to Madhava the first two cases of (B.1), the arctangent series, and series (B.4); note that (B.4) can be derived from the arctangent by taking $\theta = \frac{\pi}{6}$. The extant manuscripts do not appear to attribute the other series to a particular person. The *Yuktidipika* gives series (B.6), including the remainder; it is possible that this series is due to Sankara Variyar, the author of the work. Series (B.7) and (B.8) are mentioned in the *Yuktibhasa* and are easily transformable into series (A.1) with $\theta = \frac{\pi}{4}$. We can safely conclude that the power series for arctangent, sine, and cosine were obtained by Madhava.

The series for $\sin^2 \theta$, (A.4), follows directly from the series for $\cos \theta$ by an application of the double angle formula, $\sin^2 \theta = \frac{1}{2}(1 - \cos 2\theta)$. The series for $\frac{\pi}{4}$, (B.1), has several points of interest. When $n \to \infty$, it is simply the series discovered by Leibniz in 1673, that he communicated to Newton.[3] However, this series is not useful for computational purposes because it converges extremely slowly. To make it more effective in this respect, Madhava added a rational approximation for the

[2] Rajagopal and Rangachari (1977) p. 96.
[3] Newton (1959–1960) vol II, pp. 57–71, especially p. 67.

remainder after $n$ terms. We present Jyesthadeva's derivation for the expressions $f_1(n)$ and $f_2(n)$ in (B.1) later in this chapter. However, if we set

$$\frac{\pi}{4} = 1 - \frac{1}{3} + \frac{1}{5} - \cdots \mp \frac{1}{n} \pm f(n+1), \tag{1.1}$$

then the remainder $f(n)$ has the continued fraction expansion

$$f(n+1) = \frac{1}{2} \cdot \frac{1}{n+} \frac{1^2}{n+} \frac{2^2}{n+} \frac{3^2}{n+} \cdots, \tag{1.2}$$

where $f(n+1)$ satisfies the functional relation

$$f(n+1) + f(n-1) = \frac{1}{n}. \tag{1.3}$$

The first three convergents of this continued fraction are

$$\frac{1}{2n} = f_1(n+1), \quad \frac{n}{2(n^2+1)} = f_2(n+1), \quad \text{and} \quad \frac{1}{2}\frac{n^2+4}{n(n^2+5)} = f_3(n+1). \tag{1.4}$$

Although this continued fraction is not mentioned in any extant works of the Kerala school, their approximants indicate that they must have known it, at least implicitly. In fact, continued fractions appear in much earlier Indian works. The *Lilavati* of Bhaskara (c. 1150) used continued fractions to solve first-order Diophantine equations and Variyar's *Kriyakramakari* was a commentary on Bhaskara's book.

The approximation in equation (B.6) is similar to that in (B.1) and gives further evidence that the Kerala mathematicians saw a connection between series and continued fractions. If we write

$$\frac{\pi - 2}{4} = \frac{1}{2^2 - 1} - \frac{1}{4^2 - 1} + \frac{1}{6^2 - 1} - \cdots \pm \frac{1}{n^2 - 1} \pm g(n+1), \tag{1.5}$$

then

$$g(n) = \frac{1}{2n} \cdot \frac{1}{n+} \frac{1 \cdot 2}{n+} \frac{2 \cdot 3}{n+} \frac{3 \cdot 4}{n+} \cdots \tag{1.6}$$

and

$$g_1(n) = \frac{1}{2n}, \quad g_2(n) = \frac{1}{2(n^2+2)}. \tag{1.7}$$

Newton, who was very interested in the numerical aspects of series, also found the $f_1(n) = \frac{1}{2n}$ approximation when he saw Leibniz's series. He wrote in a letter in 1676[4] to Henry Oldenburg:

> By the series of Leibniz also if half the term in the last place be added and some other like device be employed, the computation can be carried to many figures.

---

[4] Newton (1959–1960) vol. 2, pp. 110–149, especially p. 140.

Though the accomplishments of Madhava and his followers are quite impressive, the members of the school do not appear to have had any interaction with people outside of the very small region where they lived and worked. By the end of the sixteenth century, the school ceased to produce any further original works. Thus, there appears to be no continuity between the ideas of the Kerala scholars and those outside India or even from other parts of India.

## 1.2 Transformation of Series

The series in equations (B.2) and (B.3) are transformations of

$$\sum_{k=1}^{\infty} \frac{(-1)^{k-1}}{k}$$

by means of the rational approximations for the remainder. To understand this transformation in modern notation, observe:

$$\frac{\pi}{4} = (1 - f_1(2)) - \left(\frac{1}{3} - f_1(2) - f_1(4)\right) + \left(\frac{1}{5} - f_1(4) - f_1(6)\right) - \cdots. \quad (1.8)$$

The $(n+1)$th term in this series is

$$\frac{1}{2n+1} - f_1(2n) - f_1(2n+2) = \frac{1}{2n+1} - \frac{1}{4n} - \frac{1}{4(n+1)} = \frac{-1}{(2n+1)^3 - (2n+1)}. \quad (1.9)$$

Thus, we arrive at equation (B.2). Equation (B.3) is similarly obtained:

$$\frac{\pi}{4} = (1 - f_2(2)) - \left(\frac{1}{3} - f_2(2) - f_2(4)\right) + \left(\frac{1}{5} - f_2(4) - f_2(6)\right) - \cdots, \quad (1.10)$$

and here the $(n+1)$th term is

$$\frac{1}{2n+1} - \frac{n}{(2n)^2 + 1} - \frac{n+1}{(2n+2)^2 + 1} = \frac{4}{(2n+1)^5 + 4(2n+1)}. \quad (1.11)$$

Clearly, the $n$th partial sums of these two transformed series can be written as

$$s_i(n) = 1 - \frac{1}{3} + \frac{1}{5} - \frac{1}{7} + \cdots \mp \frac{1}{2n-1} \pm f_i(2n), \quad i = 1, 2. \quad (1.12)$$

Since series (1.8) and (1.10) are alternating, and the absolute values of the terms are decreasing, it follows that

$$\frac{1}{(2n+1)^3 - (2n+1)} - \frac{1}{(2n+3)^3 - (2n+3)} < \left|\frac{\pi}{4} - s_1(n)\right|$$

$$< \frac{1}{(2n+1)^3 - (2n+1)}. \quad (1.13)$$

Also

$$\frac{4}{(2n+1)^5 + 4(2n+1)} - \frac{4}{(2n+3)^5 + 4(2n+3)} < \left| \frac{\pi}{4} - s_2(n) \right|$$

$$< \frac{4}{(2n+1)^5 + 4(2n+1)}. \tag{1.14}$$

Thus, taking fifty terms of $1 - \frac{1}{3} + \frac{1}{5} - \cdots$ and using the approximation $f_2(n)$, the last inequality shows that the error in the value of $\pi$ becomes less than $4 \times 10^{-10}$. The Leibniz series with fifty terms is normally accurate in computing $\pi$ up to only one decimal place; by contrast, the Keralese method of rational approximation of the remainder produces numerically useful results.

## 1.3   Jyesthadeva on Sums of Powers

The Sanskrit texts of the Kerala school with few exceptions contain merely the statements of results without derivations. It is therefore extremely fortunate that Jyesthadeva's Malayalam text *Yuktibhasa*, containing the methods for obtaining the formulas, has survived. Sankara Variyar's *Yuktidipika* is a modified Sanskrit version of the *Yuktibhasa*. It seems that the *Yuktibhasa* was the text used by Jyesthadeva's students at his illam. From this, one may surmise that Variyar, a student of Nilakantha, also studied with Jyesthadeva whose illam was very close to that of Nilakantha.

A basic result used by the Kerala school in the derivation of their series is that

$$\lim_{n \to \infty} \frac{1}{n^{k+1}} \sum_{j=1}^{n} j^k = \frac{1}{k+1}. \tag{1.15}$$

Jyesthadeva gave an inductive proof of this result.[5] He noted

$$S_n^{(1)} = n + (n-1) + (n-2) + \cdots + 1$$
$$= n \cdot n - (1 + 2 + \cdots + (n-1))$$
$$= n^2 - S_{n-1}^{(1)}.$$

He then observed that for large $n$, $S_{n-1}^{(1)} \approx S_n^{(1)}$ and hence

$$S_n^{(1)} \approx n^2 - S_n^{(1)} \quad \text{or} \quad S_n^{(1)} \approx \frac{1}{2} n^2. \tag{1.16}$$

Now

$$S_n^{(2)} = n^2 + (n-1)^2 + \cdots + 1^2$$

---

[5] Jyesthadeva et al. (2008) pp. 192–196.

and

$$nS_n^{(1)} = n(n + (n-1) + \cdots + 1),$$

so that

$$
\begin{aligned}
nS_n^{(1)} - S_n^{(2)} &= 1 \cdot (n-1) + 2(n-2) + 3(n-3) + \cdots + (n-1) \cdot 1 \\
&= (n-1) + (n-2) + (n-3) + \cdots + 1 \\
&\quad + (n-2) + (n-3) + \cdots + 1 \\
&\quad\quad + (n-3) + \cdots + 1 \\
&\quad\quad\quad \cdots\cdots\cdots \\
&= S_{n-1}^{(1)} + S_{n-2}^{(1)} + S_{n-3}^{(1)} + \cdots + S_1^{(1)}.
\end{aligned}
\tag{1.17}
$$

By using (1.16) in (1.17), Jyesthadeva had, for large $n$,

$$
\begin{aligned}
nS_n^{(1)} - S_n^{(2)} &\approx \frac{1}{2}(n-1)^2 + \frac{1}{2}(n-2)^2 + \frac{1}{2}(n-3)^2 + \cdots + \frac{1}{2} \cdot 1^2 \\
&\approx \frac{1}{2} S_n^{(2)}.
\end{aligned}
\tag{1.18}
$$

Note that for large $n$, $S_n^{(2)} \approx S_{n-1}^{(2)}$ was used to obtain (1.18). Again, by applying (1.16) in (1.18), he had

$$S_n^{(2)} \approx \frac{1}{3} n^3. \tag{1.19}$$

Next

$$
\begin{aligned}
nS_n^{(2)} - S_n^{(3)} &= 1 \cdot (n-1)^2 + 2 \cdot (n-2)^2 + \cdots + (n-1) \cdot 1^2 \\
&= S_{n-1}^{(2)} + S_{n-2}^{(2)} + \cdots + S_1^{(2)}.
\end{aligned}
$$

Applying equation (1.19) yielded

$$
\begin{aligned}
nS_n^{(2)} - S_n^{(3)} &\approx \frac{1}{3}(n-1)^3 + \cdots + \frac{1}{3} \cdot 1^3 \\
&\approx \frac{1}{3} S_n^{(3)}
\end{aligned}
$$

or

$$S_n^{(3)} \approx \frac{1}{4} n^4.$$

Jyesthadeva next presented the general principle of summation, that we may express within our notation as:

Let

$$S_n^{(k)} = n^k + (n-1)^k + \cdots + 1^k$$

and suppose that $S_n^{(k-1)}$ has been estimated to be

$$S_n^{(k-1)} \approx \frac{1}{k} n^k.$$

Then

$$
\begin{aligned}
n S_n^{(k-1)} - S_n^{(k)} &= S_{n-1}^{(k-1)} + S_{n-2}^{(k-1)} + \cdots + S_1^{(k-1)} \\
&\approx \frac{1}{k} \left( (n-1)^k + (n-2)^k + \cdots + 1^k \right) \\
&\approx \frac{1}{k} S_n^{(k)};
\end{aligned}
\tag{1.20}
$$

hence

$$S_n^{(k)} \approx \frac{1}{k+1} n^{k+1} \tag{1.21}$$

and this proved (1.15).

As noted, Jyesthadeva did not write formulas in the symbolic form we have used. Rather, he gave verbal descriptions of his relations and formulas. His application of induction is very clearly executed. He writes that the case $k = 1$ implies the case $k = 2$, that in turn implies the case $k = 3$; in the same manner, a higher value of $k$ will imply the next value, and so on.[6] Also note that formula (1.20) was known to al-Haytham (965–1039) for $k = 1$ to $k = 4$, but who most probably knew that it could be generalized, though he did not do it explicitly.[7]

## 1.4    Arctangent Series in the *Yuktibhasa*

The derivation of the arctangent series,[8] as given by Jyesthadeva, boils down to the integration of $\frac{1}{1+x^2}$, as do the methods of Gregory and Leibniz.

In Figure 1.1, $AC$ is a quarter circle of radius one with center $O$; $OABC$ is a square. The side $AB$ is divided into $n$ equal parts of length $\delta$ so that $n\delta = 1$ and $P_{k-1}P_k = \delta$. $EF$ and $P_{k-1}D$ are perpendicular to $OP_k$. Now, the triangles $OEF$ and $OP_{k-1}D$ are similar, implying that

$$\frac{EF}{OE} = \frac{P_{k-1}D}{OP_{k-1}} \quad \text{or} \quad EF = \frac{P_{k-1}D}{OP_{k-1}}.$$

The similarity of the triangles $P_{k-1}P_kD$ and $OAP_k$ gives

$$\frac{P_{k-1}P_k}{OP_k} = \frac{P_{k-1}D}{OA} \quad \text{or} \quad P_{k-1}D = \frac{P_{k-1}P_k}{OP_k}.$$

---

[6] ibid. pp. 65–66.
[7] See Katz (1995) p. 125.
[8] Jyesthadeva et al. (2008) pp. 183–191.

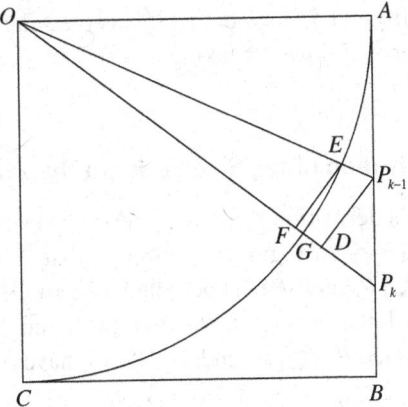

Figure 1.1 Rectifying a circle by the arctangent series.

Thus,

$$EF = \frac{P_{k-1}P_k}{OP_{k-1}OP_k} \simeq \frac{P_{k-1}P_k}{OP_k^2} = \frac{P_{k-1}P_k}{1+AP_k^2} = \frac{\delta}{1+k^2\delta^2}.$$

Now

$$\text{arc } EG \simeq EF \simeq \frac{\delta}{1+k^2\delta^2},$$

and if we write $AP_k = x = \tan\theta$, where $\theta = A\widehat{O}P_k$, then

$$\arctan x = \lim_{k\to\infty} \sum_{j=1}^{k} \frac{\delta}{1+j^2\delta^2}. \qquad (1.22)$$

To compute this limit, Jyesthadeva expanded $\frac{1}{1+j^2\delta^2}$ as a geometric series. He derived the series by an iterative procedure:

$$\frac{1}{1+x} = 1 - x\left(\frac{1}{1+x}\right) = 1 - x\left(1 - x\left(\frac{1}{1+x}\right)\right).$$

Thus, (1.22) is converted to

$$\arctan x = \lim_{k\to\infty}\left(\delta\sum_{j=1}^{k} 1 - \delta^3\sum_{j=1}^{k} j^2 + \delta^5\sum_{j=1}^{k} j^4 - \cdots\right)$$

$$= \lim_{k\to\infty}\left(\frac{x}{k}\sum_{j=1}^{k} 1 - \frac{x^3}{k^3}\sum_{j=1}^{k} j^2 + \frac{x^5}{k^5}\sum_{j=1}^{k} j^4 - \cdots\right)$$

$$= x - \frac{x^3}{3} + \frac{x^5}{5} - \cdots.$$

The last step follows from (1.15). Note that this is the Madhava–Gregory series for arctan $x$ and the series for $\frac{\pi}{4}$ follows by taking $x = 1$.

## 1.5 Derivation of the Sine Series in the *Yuktibhasa*

Once again, Jyesthadeva's derivation of the sine series has similarities with Leibniz's derivation of the cosine series. In Figure 1.2, suppose that $A\widehat{O}P = \theta$, $OP = R$, $P$ is the midpoint of the arc $P_{-1}P_1$, and $PQ$ is perpendicular to $OA$, where $O$ is the origin of the coordinate system. Let $P = (x, y)$, $P_1 = (x_1, y_1)$, and $P_{-1} = (x_{-1}, y_{-1})$. From the similarity of the triangles $P_{-1}Q_1P_1$ and $OPQ$, we have

$$\frac{P_{-1}P_1}{OP} = \frac{x_{-1} - x_1}{y} = \frac{y_1 - y_{-1}}{x}. \tag{1.23}$$

Jyesthadeva took an arc, $P_{-1}P = R\frac{\Delta\theta}{2} = \frac{\Delta s}{2}$, small enough that he could set it equal to the line segment $P_{-1}P$; we can then write (1.23) as

$$\cos\left(\theta + \frac{\Delta\theta}{2}\right) - \cos\left(\theta - \frac{\Delta\theta}{2}\right) = -\sin\theta\,\Delta\theta \tag{1.24}$$

and

$$\sin\left(\theta + \frac{\Delta\theta}{2}\right) - \sin\left(\theta - \frac{\Delta\theta}{2}\right) = \cos\theta\,\Delta\theta. \tag{1.25}$$

In fact, in his *Siddhanta Siromani*, Bhaskara[9] had stated (1.25) and proved it in the same way; he applied it to the discussion of the instantaneous motion of planets. Interestingly, in the 1650s, Pascal[10] used a very similar argument to show that $\int \cos\theta\,d\theta = \sin\theta$ and $\int \sin\theta\,d\theta = -\cos\theta$.

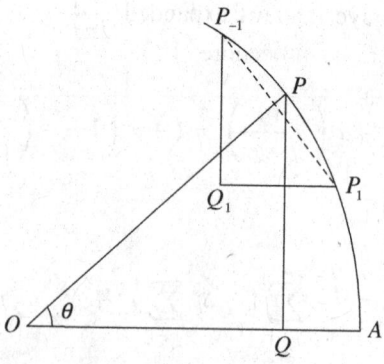

Figure 1.2 Derivation of the sine series.

[9] Bhaskara (2010).
[10] Struik (1969) vol. 2, p. 239.

From (1.24) and (1.25) Jyesthadeva derived the result, given in modern notation:

$$\sin\theta - \theta = -\int_0^\theta \int_0^t \sin u \, du \, dt = -\int_0^\theta (1 - \cos t)dt. \qquad (1.26)$$

We also note that Leibniz found the series for cosine using a similar method of repeated integration.[11] In Jyesthadeva, the integrals are replaced by sums and double integrals by sums of sums. The series is then obtained by using successive polynomial approximations for $\sin\theta$. For example, when the first approximation $\sin u \approx u$ is used in the right-hand side of (1.26), the result is

$$\sin\theta - \theta \sim -\frac{\theta^3}{3!} \quad \text{or} \quad \sin\theta \sim \theta - \frac{\theta^3}{3!}.$$

When this approximation is employed in (1.26), we obtain

$$\sin\theta - \theta \sim -\frac{\theta^3}{3!} + \frac{\theta^5}{5!}.$$

Briefly, Jyesthadeva arrived at the sums approximating (1.26) by first dividing $AP$ into $n$ equal parts using division points $P_1, P_2, \ldots, P_{n-1}$. Denote the midpoint of the arc $P_{k-1}P_k$ as $P_{k-\frac{1}{2}}$. Then by (1.23) and using $\Delta s = R\Delta\theta$

$$x_{k+\frac{1}{2}} - x_{k-\frac{1}{2}} = -\frac{\Delta s}{R} y_k, \quad k = 1, 2, \ldots, n - 1. \qquad (1.27)$$

We also have

$$(y_{k+1} - y_k) - (y_k - y_{k-1}) = \frac{\Delta s}{R}\left(x_{k+\frac{1}{2}} - x_{k-\frac{1}{2}}\right), \quad k = 1, 2, \ldots, n - 1 \qquad (1.28)$$

or

$$y_{k+1} - 2y_k + y_{k-1} = -\left(\frac{\Delta s}{R}\right)^2 y_k, \quad k = 1, 2, \ldots, n - 1. \qquad (1.29)$$

Now in (1.29), start with $k = n - 1$ and multiply the equations by $1, 2, \ldots, n - 1$ respectively and sum up the resulting equations. We then have

$$y_n - ny_1 = -\left(\frac{\Delta s}{R}\right)^2 (y_{n-1} + 2y_{n-2} + \cdots (n-1)y_1)$$

$$= -\left(\frac{\Delta s}{R}\right)^2 (y_1 + (y_1 + y_2) + \cdots + (y_1 + y_2 + \cdots y_{n-1})), \qquad (1.30)$$

the result corresponding to (1.26). To obtain the successive polynomial approximations, Jyesthadeva had to work with sums of powers of integers; in order to deal with these sums, he applied the same lemma (1.15) he had used for the arctangent series.

---

[11] Newton (1959–1960) vol. 2, p. 74.

Observe that (1.30) involves a sum of sums; in fact, Jyesthadeva's work has a section devoted to the topic of repeated summation.[12] Since this is a topic we shall see repeatedly in this book, we explain how Jyesthadeva dealt with it. Denote the sum of the first $n$ natural numbers by

$$V_n^{(1)} = n + n - 1 + \cdots + 2 + 1,$$

a sum whose value was given by Jyesthadeva as

$$V_n^{(1)} = \frac{n(n+1)}{1 \cdot 2}.$$

He remarked that this result was easier to understand by making a diagram with a shaded square in the first row, two shaded squares on the second row, three in the third row, and so on, with the last row containing as many shaded squares as terms. A diagram with $n = 4$:

With $n$ rows of squares, one may see that the result is $\frac{1}{2}n(n+1)$.

Jyesthadeva next noted, with a brief geometric argument, that the second summation was

$$
\begin{aligned}
V_n^{(2)} &= V_n^{(1)} + V_{n-1}^{(1)} + \cdots + V_1^{(1)} \\
&= \frac{n(n+1)}{2} + \frac{(n-1)n}{2} + \cdots + \frac{1 \cdot 2}{2} \\
&= \frac{n(n+1)(n+2)}{1 \cdot 2 \cdot 3},
\end{aligned}
\tag{1.31}
$$

writing that the successive summations continued in the same manner. Thus, in modern notation we have

$$
\begin{aligned}
V_n^{(k)} &= V_n^{(k-1)} + V_{n-1}^{(k-1)} + \cdots + V_1^{(k-1)} \\
&= \frac{n(n+1)(n+2)\cdots(n+k)}{1 \cdot 2 \cdot 3 \cdots (k+1)}.
\end{aligned}
\tag{1.32}
$$

We note that (1.32) is an important formula; it occurs in the earlier *Ganita Kaumudi* of Naryana Pandita (c. 1350). It is also given in Zhu Shijie's 1303 *Siyuan Yujian*.[13] This formula was rediscovered by several European mathematicians in the

---

[12] Jyesthadeva et al. (2009) pp. 226–228.
[13] Hoe (2007) p. 400.

seventeenth century. We note a method of proof given by Nicole[14] and others. Taking (1.32) as the definition of $V_n^{(k)}$, we have

$$
\begin{aligned}
V_j^{(k)} - V_{j-1}^{(k)} &= \frac{j(j+1)\cdots(j+k-1)(j+k)}{1\cdot 2\cdots k\cdot(k+1)} - \frac{(j-1)j\cdots(j+k-1)}{1\cdot 2\cdots k\cdot(k+1)} \\
&= \frac{j(j+1)\cdots(j+k-1)}{1\cdot 2\cdots k\cdot(k+1)}\cdot\left(j+k-(j-1)\right) \\
&= \frac{j(j+1)\cdots(j+k-1)}{1\cdot 2\cdots k} = V_j^{(k-1)}, \quad j = 1, 2, \cdots, n. \quad (1.33)
\end{aligned}
$$

By a repeated application of (1.33), one obtains

$$
\begin{aligned}
V_n^{(k-1)} &+ V_{n-1}^{(k-1)} + \cdots + V_2^{(k-1)} \\
&= \left(V_n^{(k)} - V_{n-1}^{(k)}\right) + \left(V_{n-1}^{(k)} - V_{n-2}^{(k)}\right) + \cdots + \left(V_2^{(k)} - V_1^{(k)}\right) \\
&= V_n^{(k)} - V_1^{(k)} \quad (1.34)
\end{aligned}
$$

and since $V_1^{(k)} = V_1^{(k-1)} = 1$, the result is proved. Jyesthadeva observed that for large $n$,

$$
V_n^{(k)} \approx \frac{n^{k+1}}{1\cdot 2\cdots(k+1)}. \quad (1.35)
$$

We now pick up the thread of Jyesthadeva's derivation of the sine and cosine series; in the course of this derivation, he employed the versine function vers $\theta$, defined by versine $\theta = 1 - \cos\theta$. See equation (1.26). Thus, referring to Figure 1.2,

$$
AQ \equiv z = R\, \text{vers}\,\theta = R - R\cos\theta.
$$

With $x$ denoting the $R$ cos and $y$ the $R$ sin values in (1.27) through (1.30), we can see that $z_k = R - x_k$; Jyesthadeva could take (1.27) as

$$
z_{k+\frac{1}{2}} - z_{k-\frac{1}{2}} = \frac{\Delta s}{R}\, y_k, \quad k = 1, 2, \ldots, n-1. \quad (1.36)
$$

Adding the $n - 1$ equations produced

$$
z_{n-\frac{1}{2}} - z_{\frac{1}{2}} = \frac{\Delta s}{R}\,(y_{n-1} + y_{n-2} + \cdots + y_1).
$$

Taking $n$ to be very large, he could replace $z_{n-\frac{1}{2}}$ by $z_n$ and $z_{\frac{1}{2}}$ by zero to obtain

$$
z_n = \frac{\Delta s}{R}\,(y_{n-1} + y_{n-2} + \cdots + y_1) = \frac{s}{nR}\sum_{j=1}^{n-1} y_j. \quad (1.37)
$$

---

[14] Nicole (1717).

Now for very large $n$

$$y_j \approx \frac{js}{n} \qquad (1.38)$$

so that

$$z_n \approx \frac{s}{nR} \left( (n-1)\frac{s}{n} + (n-2)\frac{s}{n} + \cdots + \frac{s}{n} \right),$$

but

$$\frac{s}{nR} \left( (n-1)\frac{s}{n} + (n-2)\frac{s}{n} + \cdots + \frac{s}{n} \right) = \frac{1}{R} \left( \frac{s}{n} \right)^2 (n-1+n-2+\cdots+1)$$

so, taking $k = 1$ in (1.35),

$$z_n \approx \frac{1}{R} \frac{s^2}{2}. \qquad (1.39)$$

Next, by (1.38) and (1.39), and taking $k = 2$ in (1.35), for (1.30) Jyesthadeva had

$$y_n \approx s - \frac{1}{R^2} \left( \frac{s}{n} \right)^2 (1+2+\cdots+n-1) + (1+2+\cdots+n-2) + \cdots)$$

$$\approx s - \frac{1}{R^2} \frac{s^3}{1 \cdot 2 \cdot 3}; \qquad (1.40)$$

here note that

$$y_n = R \sin \theta. \qquad (1.41)$$

Equation (1.40) gave Jyesthadeva a value of $y_j$ better than that in (1.38); he employed this to obtain an improved approximation for $z_n$. Thus, he substituted

$$y_j \approx \frac{js}{n} - \frac{1}{R^2} \frac{\left( \frac{js}{n} \right)^3}{1 \cdot 2 \cdot 3}$$

in (1.37) to find that

$$z_n \approx \frac{1}{R} \left( \frac{s}{n} \right)^2 (n-1+n-2+\cdots)$$

$$- \frac{s}{nR} \cdot \frac{1}{k^2} \cdot \left( \frac{s}{n} \right)^3 \frac{1}{1 \cdot 2 \cdot 3} ((n-1)^3 + (n-2)^3 + \cdots)$$

$$\approx \frac{1}{R} \frac{s^2}{1 \cdot 2} - \frac{1}{R^3} \frac{s^4}{1 \cdot 2 \cdot 3 \cdot 4}. \qquad (1.42)$$

Finally, employing (1.40) in (1.30) yielded Jyesthadeva the result

$$y_n = R \sin \theta \approx s - \frac{1}{R^2} \frac{s^3}{1 \cdot 2 \cdot 3} + \frac{1}{R^4} \frac{s^5}{1 \cdot 2 \cdot 3 \cdot 4 \cdot 5}. \qquad (1.43)$$

Repeating this process infinitely often would give the result

$$\sin\theta = \theta - \frac{\theta^3}{3!} + \frac{\theta^5}{5!} - \frac{\theta^7}{7!} + \cdots, \tag{1.44}$$

completing Jyesthadeva's derivation for the series for $\sin\theta$; clearly, the series for $\cos\theta$ was an immediate consequence.

Jyesthadeva's also noted that by applying (1.35) to (1.34), for large $n$ one would obtain

$$\frac{n^k}{k!} + \frac{(n-1)^k}{k!} + \cdots + \frac{1^k}{k!} \approx \frac{n^{k+1}}{(k+1)!}$$

or

$$S_n^{(k)} \approx \frac{n^{k+1}}{k+1}. \tag{1.45}$$

He thus gave a second method of proving (1.21) and thus (1.15),[15] but by using $V_n^{(k)}$, now called pyramidal or tetrahedral numbers. As we shall see in Chapter 2, this method of proof of (1.45) was rediscovered by Fermat.

The Kerala mathematicians, and indeed Fermat as well, must demand our admiration for their ability to perform such elaborate and intricate analytic calculations with only a very rudimentary notation at their disposal. We would do well to keep in mind the advantage afforded us by the mathematical power of our modern symbolic notation.

## 1.6  Continued Fractions

The noted twelfth-century Indian mathematician Bhaskara, who lived and worked in the area now known as Karnataka, used continued fractions in his c. 1150 *Lilavati*. The Kerala school was certainly familiar with Bhaskara's work, since they commented on it. It is therefore possible that they were aware of the specific continued fractions (1.2) and (1.6) for the error terms, even though they mentioned only the first few convergents of these fractions.

The *Yuktibhasa* indicated a method[16] by which the first two approximations for $f(n+1)$ in (1.4) could be derived. Jyesthadeva observed that if the correction in (1.1) was performed after the term $\frac{1}{n-2}$, then

$$\frac{\pi}{4} = 1 - \frac{1}{3} + \frac{1}{5} - \cdots \pm \frac{1}{n-2} \mp f(n-1). \tag{1.46}$$

---

[15] Jyesthadeva (2008) pp. 98–99, 227.
[16] ibid. pp. 201–205.

Subtracting (1.46) from (1.1) would yield (1.3). Now Jyesthadeva noted that $f(n+1) = \frac{1}{2n}$ was merely an approximation, because it implied that $f(n-1) = \frac{1}{2n-4}$, so that equation (1.3) would not be satisfied. But he next argued that it would be possible to bring the values of both $f(n+1)$ and $f(n-1)$ close to $\frac{1}{2n}$ by taking

$$f(n-1) \approx \frac{1}{2n-2} \quad \text{and} \quad f(n+1) \approx \frac{1}{2n+2}.$$

Subtracting series (1.46) from (1.1) gave a measure of the error, $er(n)$, involved in choosing these values:

$$er(n) = f(n+1) + f(n-1) - \frac{1}{n}$$

$$= \frac{1}{2n+2} + \frac{1}{2n-2} - \frac{1}{n}$$

$$= \frac{2n^2 - 2n}{4n^3 - 4n} + \frac{2n^3 + 2n}{4n^3 - 4n} - \frac{4n^2 - 4}{4n^3 - 4n} = \frac{1}{n^3 - n}, \qquad (1.47)$$

that is essentially the same as the result obtained in (1.9). Jyesthadeva's next step was to point out that the error given by (1.47) was positive, but adding 1 to the denominator of $f(n+1)$, so that $f(n+1) \approx \frac{1}{2n+3}$, would make the error negative:

$$er(n) = \frac{1}{2n+3} + \frac{1}{2n-1} - \frac{1}{n}$$

$$= \frac{2n^2 - n}{4n^3 + 4n^2 - 3n} + \frac{2n^2 + 3n}{4n^3 + 4n^2 - 3n} - \frac{4n^2 + 4n - 3}{4n^3 + 4n^2 - 3n}$$

$$= \frac{-2n + 3}{4n^3 + 4n^2 - 3n}. \qquad (1.48)$$

However, the error in (1.48) was not an improvement over (1.47) because of the term $-2n$ in the numerator. In order to improve upon (1.48), Jyesthadeva observed that a quantity less than 1 must be added to the denominators of $f(n+1)$ and $f(n-1)$; he remarked that adding 1 to the denominators of $f(n+1)$ and $f(n-1)$ had introduced an extra factor of $2n$ in the numerator. Thus, if the denominator of $f(n+1)$ in (1.47) were changed to $2n + 2 + \frac{1}{2n+2}$, then the contribution $2n$ of the error terms would become $\frac{2n}{2n+2} \approx 1$. But the term $-\frac{1}{n}$ added $-4n - 1$ in the numerator. By changing the denominator of $f(n+1)$ to $2n + 2 + \frac{4}{2n+2}$, the contribution of $-\frac{1}{n}$ would essentially amount to $-1$, and would cancel with $+1$ from the error terms $f(n+1) + f(n-1)$. Thus, reasoned Jyesthadeva, he should take

$$f(n+1) = \frac{1}{2n + 2 + \frac{4}{2n+2}}.$$

The corresponding $er(n)$ would then be given as

$$
\begin{aligned}
er(n) &= \frac{2(n+1)}{4((n+1)^2+1)} + \frac{2(n-1)}{4((n-1)^2+1)} - \frac{1}{n} \\
&= \frac{n+1}{2((n^2+2)+2n)} + \frac{n-1}{2((n^2+2)-2n)} - \frac{1}{n} \\
&= \frac{(n+1)(n^2+2-2n) + (n-1)(n^2+2+2n)}{2(n^4+4)} - \frac{2(n^4+4)}{n(n^4+4)} \\
&= \frac{-4}{n^5+4n}.
\end{aligned}
\tag{1.49}
$$

Note that (1.49) is the same as (1.11) and leads to the series (B.3).[17]
Jyesthadeva also gave a derivation of (B.6) by taking $f(n+1) = \frac{1}{2n}$, so that
$f(n-1) = \frac{1}{2n-4}$ and

$$
\begin{aligned}
f(n+1) + f(n-1) - \frac{1}{n} &= \frac{1}{2n} + \frac{1}{2n-4} - \frac{1}{n} \\
&= \frac{1}{(n-1)^2 - 1}.
\end{aligned}
$$

He next derived a third-order correction:

$$
\begin{aligned}
f(n+1) &= \cfrac{1}{2n+2+\cfrac{4}{2n+2+\frac{16}{2n+2}}} \\
&= \frac{\left(\frac{n+1}{2}\right)^2 + 1}{((n+1)^2+5)\frac{n+1}{2}}.
\end{aligned}
$$

There is a method for finding a continued fraction for $f(n)$ that goes back to Wallis;[18] Whiteside's reworked form of this method is quite clear:[19] Start with the functional equation (1.4) for $f(n)$,

$$
f(n+1) + f(n-1) = \frac{1}{n}.
\tag{1.50}
$$

It is obvious that a first approximation for $f(n)$ is given by $f(n) \approx \frac{1}{2n}$. As a first step toward the continued fraction for $f(n)$, set

$$
f(n) = \frac{1}{2r_n^{(0)}} \quad \text{and} \quad r_n^{(0)} = n + \frac{1}{r_n^{(1)}}.
\tag{1.51}
$$

---

[17] ibid. p. 205.
[18] Wallis (2004) pp. 167–174.
[19] Whiteside (1961b) p. 212.

It follows from (1.50) that $r_n^{(0)}$ satisfies

$$\left(2r_{n+1}^{(0)} - n\right)\left(2r_{n-1}^{(0)} - n\right) = n^2. \tag{1.52}$$

From (1.51)

$$2r_{n+1}^{(0)} - n = n + 2 + \frac{2}{r_{n+1}^{(1)}},$$

and a similar relation holds for $r_{n-1}^{(0)}$. When these values are substituted in (1.52), some calculation gives us

$$\left(2r_{n+1}^{(1)} - (n-2)\right)\left(2r_{n-1}^{(1)} - (n+2)\right) = n^2. \tag{1.53}$$

Once again, $r_n^{(1)} \approx n$. So assume $r_n^{(1)} = n + \frac{1}{s_n^{(2)}}$ and substitute in (1.53) to get, after simplification,

$$16s_{n-1}^{(2)}s_{n-1}^{(2)} - 2(n+4)s_{n+1}^{(2)} - 2(n-4)s_{n-1}^{(2)} - 4 = 0. \tag{1.54}$$

To obtain an equation such as (1.52) or (1.53), multiply (1.54) by 4, set

$$s_n^{(2)} = \frac{r_n^{(2)}}{2^2},$$

and add $n^2$ to both sides to get

$$\left(2r_{n+1}^{(2)} - (n-4)\right)\left(2r_{n-1}^{(2)} - (n+4)\right) = n^2. \tag{1.55}$$

We then have $r_n^{(1)} = n + \frac{2^2}{r_n^{(2)}}$. A similar calculation shows that

$$r_n^{(2)} = n + \frac{3^2}{r_n^{(3)}}$$

satisfies the equation

$$\left(2r_{n+1}^{(3)} - (n-6)\right)\left(2r_{n-1}^{(3)} - (n+6)\right) = n^2. \tag{1.56}$$

It can be shown inductively that if

$$r_n^{(k-1)} = n + \frac{k^2}{r_n^{(k)}}$$

and

$$\left(2r_{n+1}^{(k-1)} - (n - 2(k-1))\right)\left(2r_{n-1}^{(k-1)} - (n + 2(k-1))\right) = n^2,$$

then

$$\left(2r_{n+1}^{(k)} - (n - 2k)\right)\left(2r_{n-1}^{(k)} - (n + 2k)\right) = n^2.$$

It follows that $f(n)$ has the continued fraction expansion (1.2). In a similar way, we may obtain the continued fraction (1.6) for $g(n)$ if we start with the functional relation

$$g(n - 1) + g(n + 1) = \frac{1}{n^2 - 1}.$$

It may be instructive to consider another method, first published by Gauss,[20] for finding the continued fractions of the Kerala school, also a method for obtaining the successive convergents. There is certainly no clear indication that this method was discovered before Gauss did so in his work on approximate quadrature, presented in 1814, published in 1815. Start with a series of the form

$$f(n) = \frac{a_1}{n} + \frac{a_2}{n^2} + \frac{a_3}{n^3} + \cdots . \tag{1.57}$$

Note that it is always possible to associate a continued fraction with (1.57) by applying successive division. Write (1.57) as

$$f(n) = \frac{1}{\frac{n}{a_1} - \frac{a_2}{a_1^2} + t_1(n)}. \tag{1.58}$$

From this we can see that

$$t_1(n) = \frac{b_1}{n} + \frac{b_2}{n^2} + \frac{b_3}{n^3} + \cdots ,$$

a series of the same kind as (1.57). So the process can be continued, and the result is a continued fraction for $f(n)$. To find the numbers $a_1, a_2, a_3, a_4, \ldots$, substitute (1.57) in (1.50). The first four values are

$$a_1 = \frac{1}{2}, \, a_2 = 0, \, a_3 = -\frac{1}{2}, \, a_4 = 0.$$

From these values, we obtain the second convergent of the continued fraction for $f(n)$ by applying the process described in (1.58):

$$\frac{1}{2n + \frac{2}{n}} = \frac{n}{2(n^2 + 1)}.$$

By also using $a_5 = \frac{5}{2}$ and $a_6 = 0$, we obtain the third convergent: [21]

$$\frac{n^2 + 4}{2n(n^2 + 5)}.$$

---

[20] Gauss (1815).
[21] See Srinivasiengar (1967) pp. 149–151.

One problem that arises in the computation of $a_1$, $a_2$, $a_3$, etc., is finding the series expansions of $\frac{1}{(n+1)^2}$, $\frac{1}{(n-1)^2}$, $\frac{1}{(n+1)^3}$, etc. Although this may appear to require knowledge of the binomial theorem for negative integer powers, observe that the series may be obtained by repeatedly multiplying the geometric series by itself. In our Chapter 4, we see that Newton verified the correctness of his binomial theorem by multiplying series.

## 1.7   Exercises

(1) Prove that if $C$ is the circumference and $D$ the diameter of a circle, then

$$C = 3D + 6D \left( \frac{1}{(2 \cdot 2^2 - 1)^2 - 2^2} + \frac{1}{(2 \cdot 4^2 - 1)^2 - 4^2} \right.$$
$$\left. + \frac{1}{(2 \cdot 6^2 - 1)^2 - 6^2} + \cdots \right).$$

This result, equivalent to (B.5), is easily derived from series (B.2); it is contained in the *Karanapaddhati*, by an unknown author from the Putumana illam in Sivapur, Kerala. The result is described: "Six times the diameter is divided separately by the square of twice the squares of even integers minus one, diminished by the squares of the even integers themselves. The sum of the resulting quotients increased by thrice the diameter is the circumference." See Bag (1966) or Pai et al. (2018) pp. 150–151. Also see Srinivasiengar (1967) p. 149.

(2) Compute

$$1 - \frac{1}{3} + \frac{1}{5} - \cdots + \frac{1}{149} - f_3(150)$$

where $f_3$ is defined in (1.4). This gives $\pi$ correct to eleven decimal places. In one of his astronomical works, Madhava gave a value of $\pi$: "For a circle of diameter $9 \times 10^{11}$ units, the circumference is 2,827,433,38,233 units." This gives the approximate value of $\pi$ as 3.14159265359, correct to eleven decimal places. The *Sadratnamala* by Sankara Verman of unknown date gives $\pi$ to seventeen decimal places. See Parameswaran (1983) p. 194, and Srinivasiengar (1967).

(3) Prove al-Haytham's formula (1.20).

(4) This exercise outlines the proof of Paramesvara's formula for the radius of the circle circumscribing a cyclic quadrilateral, as given in the *Kriyakramakari*. First, prove that the product of the flank sides of any triangle divided by the diameter of its circumscribed circle is equal to the altitude of the triangle. This result follows from a rule given by Brahmagupta (c. 628) in an astronomical work, the *Brahmasphutasiddhanta*.

Next, prove that the area of the cyclic quadrilateral is given by

$$A = \sqrt{s(s-a)(s-b)(s-c)} \quad \text{where} \quad s = \frac{a+b+c+d}{2}$$

and $a, b, c, d$ are the lengths of the sides of the quadrilateral. This was also stated by Brahmagupta. The *Yuktibhasa* contains a complete proof. See also Kichenassamy (2010), who convincingly argues that Brahmagupta had a proof and reconstructs it from indications in *Brahmasphutasiddhanta*.

Then, let $ABCD'$ be the quadrilateral obtained from $ABCD$ by interchanging the sides $AD$ and $CD$, so that $AD' = CD = c$ and $CD' = AD = d$. Show that if $x, y, z$ denote the three diagonals $AC, BD, BD'$, respectively, then

$$yz = ab + cd, \, zx = bc + da, \, xy = ca + bd.$$

This is, of course, Ptolemy's theorem. Ptolemy's formula is equivalent to the addition formula for the sine function; his *Almagest*, containing this relation, is heavily indebted to the *Chords in a Circle* of Hipparchus. Bhaskara defined the three diagonals in his *Lilavati*. See Boyer and Merzbach (1991) and Maor (1998) pp. 87–94. Finally, prove that the radius of the circle circumscribing the cyclic quadrilateral is

$$r = \sqrt{\frac{(ad+bc)(ac+bd)(ab+cd)}{(b+c+d-a)(c+d+a-b)(d+a+b-c)(a+b+d-d)}}.$$

This is Paramesvara's formula, sometimes attributed to S. A. J. L'Huillier, who published it in 1782. See Gupta (1977).

(5) Use al-Haytham's formula (1.20) to obtain

$$\sum_{k=1}^{n} k^3 = \frac{1}{4}n^4 + \frac{1}{2}n^3 + \frac{1}{4}n^2$$

and

$$\sum_{k=1}^{n} k^4 = \frac{1}{5}n^5 + \frac{1}{2}n^4 + \frac{1}{3}n^3 - \frac{1}{30}n.$$

(6) Derive (1.6) by the methods described in this chapter.
(7) Prove formulas (B.8) and (B.9), given in Section 1.1.

## 1.8 Notes on the Literature

It seems that the work of Madhava and his followers on series became known outside India only when a British civil servant and Indologist, Charles M. Whish, wrote a paper on the subject, posthumously published in the *Transactions of the Royal Asiatic Society of Great Britain and Ireland* in 1835. This journal was founded by British Indologists in the early 1830s, though Sir William Jones had first conceived the idea about fifty years earlier. Unfortunately, Whish's paper had little impact. Interest in the Kerala school was renewed in the twentieth century by the efforts of C. Rajagopal and his associates, who published several papers on the topic. See Rajagopal (1949), and Rajagopal and Aiyar (1951), Rajagopal and Venkataraman (1949), and Rajagopal and Rangachari (1977), and Rajagopal and Rangachari (1986).

The *Yuktibhasa* of Jyesthadeva and the *Tantrasangraha* of Nilakantha have recently been published with commentaries in English by Sarma: Nilakantha (1977) and Jyesthadeva (2008). Jyesthadeva (2008) was published after Sarma's death, with additional notes by Ramasubramanian, Srinivasa, and Sriram. This two-volume translation with extensive and informative commentary contains both the mathematical and astronomical portions; the original Malayalam text extends to 300 pages. Sarma (1972) also discusses the Kerala school, but from the astronomical point of view. Biographical information on the members of the Kerala school, as well as numerous other ancient and medieval Indian astronomers and mathematicians, can be found in David Pingree's five-volume work (1970–1994).

Readers who wish to read more on the Indian work on series, but with modern notation, may consult Roy (1990), Katz (1995), and Bressoud (2002). These papers are conveniently available in Anderson, Katz, and Wilson (2004). Also see the papers by Parameswaran (1983) on Madhava, Bag (1966) on the *Karanapaddhati*, Gupta (1977) on Paramesvara's rule for radius of the cyclic quadrilateral, and Sarma and Hariharan (1991) on the *Yuktibhasa*. Plofker (2009) presents a scholarly, detailed, and readable discussion of Kerala mathematics, with several excerpts on $\pi$ translated from Sankara Variyar's *Kriyakramakari*. She also presents the derivation of the sine series with translations from the *Yuktidipika* and describes Takao Hayashi's suggested reconstruction of Madhava's remainder term results. In order to derive the continued fraction, Hayashi and his collaborators have compared the values of partial sums of Madhava's series for $\pi$ with the then-known rational approximations for $\pi$. Van Brummelen (2009), on the history of trigonometry, discusses the contributions of the Kerala school and relates them to the astronomical work of medieval India. In the context of the development of astronomy, Van Brummelen (2009) presents the *Yuktibhasa* derivation of the sine series. This accessible presentation is very helpful, since the mathematics of the Kerala school was largely motivated by an interest in astronomy.

# 2

## Sums of Powers of Integers

### 2.1 Preliminary Remarks

In his work on spirals, and on conoids and spheroids, Archimedes gave proofs of results equivalent to the formulas for the sum of the first $n$ integers and for the sum of the squares of those integers; he represented these magnitudes by lines. In the proof of his proposition 11 of his study on spirals,[1] he considered an ascending arithmetic progression of magnitudes $A_1, A_2, \ldots, A_n$, whose common difference was the least term $A_1$. He placed the lines representing these magnitudes in order and parallel, and extended each line so that it would be equal to $A_n$. Thus, he had

$$A_1 + A_{n-1} = A_2 + A_{n-2} = \cdots = A_{n-1} + A_1 = A_n$$

and so

$$(A_1 + A_2 + \cdots + A_{n-1}) + A_n = \frac{n-1}{2}A_n + A_n = \frac{n+1}{2}A_n. \qquad (2.1)$$

Taking $A_1$ to be a unit $= 1$, (2.1) then reduces to

$$1 + 2 + \cdots + (n-1) + n = \frac{n(n+1)}{2}. \qquad (2.2)$$

For the sum of squares of integers, Archimedes stated as the tenth proposition of his work on spirals, and also as a lemma to the second proposition of his work on conoids and spheroids:[2]

If $A_1, A_2, \ldots, A_n$ be $n$ lines forming an ascending arithmetic progression in which the common difference is equal to the least term $A_1$, then

$$(n+1)A_n^2 + A_1(A_1 + A_2 + A_3 + \cdots + A_n) = 3(A_1^2 + A_2^2 + \cdots + A_n^2). \qquad (2.3)$$

---

[1] Archimedes and Heath (1953) pp. 105 and 163.
[2] ibid. pp. 107–109.

If we take $A_1 = 1$, then we can write (2.3) as

$$3(1^2 + 2^2 + \cdots + n^2) = (n+1)n^2 + \frac{n(n+1)}{2}$$

$$= \frac{(n+1)n}{2}(2n+1)$$

or

$$1^2 + 2^2 + \cdots + n^2 = \frac{n(n+1)(2n+1)}{6}. \tag{2.4}$$

Archimedes applied these formulas to area and volume problems. As mentioned in Chapter 1, the mathematician and physicist al-Haytham (965–1040) extended Archimedes' results to cubes and fourth powers by a method that could be extended to higher powers as well. In Chapter 1, we also discussed Jyesthadeva's two methods for deriving the results for higher powers. One of these was based on al-Haytham's procedure, given in general form in relation (1.20). Jyesthadeva's purpose was to arrive at the result that for a positive integer $k$

$$S_n^{(k)} \sim \frac{n^{k+1}}{k+1} \quad \text{as } n \to \infty, \tag{2.5}$$

where $S_n^{(k)} = 1^k + 2^k + \cdots + n^k$. Note that for the sequences of positive numbers $\{a_n\}$ and $\{b_n\}$, we write

$$a_n \sim b_n \tag{2.6}$$

if

$$\lim_{n \to \infty} \frac{a_n}{b_n} = 1. \tag{2.7}$$

Observe that Jyesthadeva, following Madhava, was mainly interested in the result

$$\lim_{n \to \infty} \frac{S_n^{(k)}}{n^{k+1}} = \frac{1}{k+1}, \tag{2.8}$$

equivalent to

$$\int_0^1 x^k \, dx = \frac{1}{k+1}. \tag{2.9}$$

The work of Archimedes, al-Haytham, and Jyesthadeva showed that $S_n^{(k)}$ could be expressed as a polynomial in $n$ of degree $k + 1$ for $k = 1, 2, 3, \ldots$. However, these mathematicians were primarily interested in the highest power term of the polynomial, because in modern terms they wished to prove (2.8).

By contrast, Johann Faulhaber (1580–1635) brought an algebraic and number theoretic approach to study of the sums of powers of integers. He found a recursive method for expressing $S_n^{(k)}$ as a polynomial and determined some properties of the coefficients. Faulhaber was also motivated by a fascination with figurate numbers. In Chapter 1, we denoted the $n$th term of the $k$th sequence of figurate numbers by $V_n^{(k)}$ and noted formula (1.32).

The one-dimensional figurate numbers are merely the consecutive positive integers $1, 2, 3, \ldots, n$. The two-dimensional figurate numbers are the triangular numbers, where the $n$th triangular number is the sum of the first $n$ consecutive numbers:

$$1, 1 + 2 = 3, 1 + 2 + 3 = 6, 1 + 2 + 3 + 4 = 10, \ldots, \frac{n(n+1)}{2}.$$

The three-dimensional figurate numbers are the pyramidal or tetrahedral numbers such that the $n$th pyramidal number is the sum of the first $n$ triangular numbers:

$$1, 4, 10, 20, \ldots, \frac{n(n+1)(n+2)}{6}, \ldots.$$

These formulas in modern notation can be written as

$$\sum_{k=1}^{n} \binom{k}{1} = \binom{n+1}{2}, \tag{2.10}$$

$$\sum_{k=1}^{n} \binom{k+1}{2} = \binom{n+2}{3}. \tag{2.11}$$

When written this way, it is clear that the figurate numbers are related to the number of combinations of $k$ things chosen from $m$ different things, for appropriate $m$ and $k$. It appears that the connection between figurate numbers and combinations was recognized by Narayana Pandita whose *Ganita Kaumudi* of c. 1356 makes this explicit in chapter 13, sutra 67, example 30.

Narayana Pandita also algebraically extended the figurate numbers by taking sums of sums of sequences. So the sequence after the tetrahedral numbers would be

$$1, 1 + 4 = 5, 1 + 4 + 10 = 15, 1 + 4 + 10 + 20 = 35, \ldots.$$

Some earlier mathematicians may have refrained from doing this because they did not conceive of dimensionality beyond three as meaningful. In effect, Narayana had the formula

$$\sum_{k=1}^{n} \binom{k+p-1}{p} = \binom{n+p}{p+1}, \quad p = 1, 2, 3, \ldots. \tag{2.12}$$

Note that Narayana's notation did not allow him to state formula (2.12) in general. He showed it true for small values of $k$ and $p$ and indicated that the process could be continued. From Narayana's reference in his chapter 13, sutra 39, to Indian mathematicians of an earlier era, we surmise that they may also have been aware of formula (2.12). Again, Zhu Shijie, in his 1303 *Siyuan Yujian*, gave this formula for

$p = 1$ to $p = 5$. From the manner in which Zhu Shijie discusses it,[3] we may conclude that he was aware that the formula would hold for all positive integers $p$.

The work of the English mathematicians Thomas Harriot and Henry Briggs on problems related to interpolation shows that they also understood formula (2.12). The German algebraist and arithmetician Johann Faulhaber also independently discovered (2.12), but his motivation was an interest in numbers and in particular the figurate numbers. However, his results do not seem to have influenced Fermat, Harriot, or Briggs.[4] Note that Fermat used (2.12) to prove (2.8), just as Jyesthadeva had done.[5]

Faulhaber (1580–1635) was born in Ulm, Germany, and learned the weaving trade from his father. His love of computation led him to study mathematics. His knowledge of Latin was not very good, so in the course of his studies, he laboriously translated several mathematical texts, ancient and modern, into German. He founded a school for engineers in the early 1600s and wrote treatises on arithmetical questions. Faulhaber gave an algorithm for expressing $S_n^{(p)}$ as a polynomial in $n$; though he worked with Bernoulli numbers, he failed to note their significance. It was not the practice in Faulhaber's time to give proofs of algorithms. Two centuries later, in a paper on the Euler–Maclaurin formula, Jacobi provided proofs of some of Faulhaber's formulas.

Around 1700, Jakob Bernoulli gave a simple method for computing the polynomial in $n$ for $S_n^{(k)}$. Bernoulli numbers, a sequence of rational numbers, play a significant role in the determination of this polynomial. Bernoulli's interest in the summation of finite and infinite series was connected with his study of probability theory. Jakob Bernoulli (1654–1705) was the eldest in an illustrious scientific and mathematical family, including his brother Johann, nephews Niklaus I, Niklaus II, Daniel, and Johann II. In 1676, Bernoulli received a degree in theology from the University of Basel, intending to go into the ministry. He then traveled in Europe, coming into contact with the Dutch mathematician Hudde and members of the Royal Society. These experiences aroused his interest in science and mathematics. In the 1680s, he taught himself mathematics by reading short treatments by Leibniz on differentiation and integration; he then taught this subject to his younger brother Johann. One of the first mathematicians to grasp Leibniz's calculus, Jakob Bernoulli proceeded to apply it to fundamental problems in mechanics and to differential equations. The study of Huygens's treatise on games of chance led Bernoulli to a study of probability theory, on which he wrote the first known full-length text, *Ars Conjectandi*. From 1687 until his death, Bernoulli happily served as professor of mathematics at Basel, in spite of a salary more meager than he would have received as a clergyman. This professorship was occupied by a member of the Bernoulli family for one hundred years.

Although he spent many years on the problems contained in his probability treatise, Jakob Bernoulli never completed it. It appears that he wished to include several problems arising out of "civil, moral, and economic matters," i.e., applications to practical situations. For example, even in the year of his death, he repeated his

---

[3] Hoe (2007) pp. 383–390.
[4] Edwards (2002) pp. 10–15.
[5] ibid. p. 88.

earlier request to Leibniz for a hard-to-find copy of Jan de Witt's work[6] on annuities and life expectancy. *Ars Conjectandi* was posthumously published in 1713 by Jakob Bernoulli's son Niklaus with a foreword by his nephew Niklaus I. Publication was delayed when Jakob's immediate family, fearing academic dishonesty, refused to hand over the manuscript to Johann or to Niklaus I.

Seki was another independent discoverer of the Bernoulli numbers. The Japanese mathematician Seki Takakazu was probably born in 1642 and we do not know who his mathematical mentor might have been. However, we know that he studied the thirteenth century works of the Chinese mathematicians Yang Hui and Zhu Shijie. From *Yang Hui's Methods of Computation* (*Yang Hui Suanfa*) of 1275, Seki read of the method for solving algebraic equations with numerical coefficients; this is now known as Horner's method, or the Horner-Ruffini method. Seki made refinements to this method and also discovered Newton's method. Seki made original contributions to the theory of determinants. His collected papers are accompanied by a helpful English summary of his mathematical accomplishments.[7]

## 2.2  Johann Faulhaber

In 1631, Johann Faulhaber published *Academia Algebrae* in which he listed the formulas for

$$S_n^{(k)} = 1^k + 2^k + \cdots + n^k, \ \ k = 1, 3, 5, \ldots, 17,$$

given as polynomials in $N = \frac{n(n+1)}{2}$, but without indication of a derivation or motivation. Donald Knuth found these formulas striking and useful, and he noted them in his 1993 paper on Faulhaber.[8] Note that for odd $k$, if $S_n^{(k)}$ is a polynomial in $N$, then $S_n^{(k)}$ is a polynomial in $n$, but the converse does not hold. Following Knuth's exposition, we note that Faulhaber wrote his formulas in the form

$$\sum_{i=1}^{n} i^{2k+1} = b_0 N^{k+1} - b_1 N^k + b_2 N^{k-1} - \cdots + (-1)^{k-1} b_{k-1} N^2, \ \ k \geq 1, \quad (2.13)$$

where $b_0, b_1, b_2, \cdots$ represented positive rational numbers; when $k = 0$, then $\sum_{i=1}^{n} i = N$. Faulhaber also observed that

$$b_0 = \frac{2^k}{k+1} \quad \text{and} \quad b_{k-2} = 4b_{k-1}. \qquad (2.14)$$

Faulhaber also derived formulas for some even powers of consecutive integers. He noticed that (2.13) would hold if and only if

---

[6] Bernoulli and Sylla (2006) p. 46.
[7] Seki and Hirayama et al. (1974).
[8] Knuth (1993).

$$\sum_{i=1}^{n} i^{2k} = \frac{2n+1}{2(2k+1)}\left((k+1)b_0 N^k - kb_1 N^{k-1}\right.$$

$$\left. + (k-1)b_2 N^{k-2} - \cdots + (-1)^{k-1} 2b_{k-1} N\right). \quad (2.15)$$

In 1834, Jacobi proved formulas (2.13) through (2.15), using a method that clearly would have been unfamiliar to Faulhaber.[9] Knuth has given an interesting reconstruction of the methods Faulhaber might well have employed. A. W. F. Edwards[10] gave a technique for finding the sum of odd powers as a polynomial in $N$ by use of the sums of lower order odd powers. In brief, first use the binomial theorem to expand

$$x^k (x+1)^k - x^k (x-1)^k = 2\left(\binom{k}{1} x^{2k-1} + \binom{k}{3} x^{2k-3} + \cdots\right).$$

Next, successively set $x = n, n-1, n-2, \ldots, 1$ and add the corresponding formulas to obtain

$$(n(n+1))^k = 2\left(\binom{k}{1} \sum_{i=1}^{n} i^{2k-1} + \binom{k}{3} \sum_{i=1}^{n} i^{2k-3} + \cdots\right). \quad (2.16)$$

Thus, from (2.16) one may ascertain the sum of the $2k-1$ powers of integers when the sums of the odd powers of integers up to the exponent $2k-3$ are known.

## 2.3 Fermat

Fermat rediscovered Narayana's formula (2.12) in approximately 1635, although he apparently never wrote down a proof. Rather, in the margin of his copy of Diophantus's *On Polygonal Numbers*, he wrote that he had discovered this proposition, calling it "beautiful."[11] In a 1636 letter to Roberval, Fermat stated the result:[12]

> The last number multiplied by the next larger number is double the collateral triangle; the last number multiplied by the triangle of the next larger is three times the collateral pyramid; the last number multiplied by the pyramid of the next larger is four times the collateral triangulo-triangle; and so on to infinity by this uniform method.

The first line of Fermat's description becomes $n(n+1) = 2\sum_{k=1}^{n} k$; the second line becomes $\frac{n(n+1)(n+2)}{2} = 3\sum_{k=1}^{n} \frac{k(k+1)}{2}$, and so on. Fermat wrote that these results had helped him determine the sums of powers of integers and that these in turn gave him the areas under the curves $y = x^k$, $k = 1, 2, 3 \ldots$. The example of the curve $y = x^4$, as given to Roberval,[13] does not indicate a general method. He wrote,

---

[9] Jacobi (1834).
[10] Edwards (1986).
[11] Boyer (1943).
[12] ibid. p. 238.
[13] See Mahoney (1973) p. 230.

If you multiply four times the greatest number increased by two by the square of the triangle of numbers, and from the product you subtract the sum of the squares of the individual numbers, five times the sum of the fourth powers will result.

His description, in our symbolic form, can be given as

$$5 \sum_{i=1}^{n} i^4 = (4n + 2) \frac{n^2(n+1)^2}{4} - \sum_{i=1}^{n} i^2.$$

In general, Fermat possibly had the following procedure in mind: Write (2.12) in the form

$$\sum_{k=1}^{n} k(k+1) \cdots (k+p-1) = \frac{1}{p+1} \big( n(n+1) \cdots (n+p) \big). \tag{2.17}$$

Note that the product in the summation may be written as a polynomial in $k$:

$$k^p + A_1 k^{p-1} + A_2 k^{p-2} + \cdots + A_{p-1} k, \tag{2.18}$$

and the right-hand side of (2.17) may be written as a polynomial in $n$:

$$\frac{1}{p+1} \big( n^{p+1} + B_1 n^p + B_2 n^{p-1} + \cdots + B_p n \big). \tag{2.19}$$

As discussed in Section 10.7, the coefficients $A_1, \ldots, A_{p-1}$ and $B_1, \ldots, B_p$ are Stirling numbers, but this fact is not required to solve the problem of finding the sums of powers of integers. Observe that by using (2.18) and (2.19) in (2.17) we obtain

$$\begin{aligned} & S_n^{(p)} + A_1 S_n^{(p-1)} + A_2 S_n^{(p-2)} + \cdots + A_{p-1} S_n^{(1)} \\ &= \frac{1}{p+1} \big( n^{p+1} + B_1 n^p + \cdots + B_p n \big). \end{aligned} \tag{2.20}$$

Note that $S_n^{(1)} = \frac{1}{2} n^2 + \frac{1}{2} n$ is a polynomial in $n$ of degree 2 and the coefficient of the highest power is $\frac{1}{2}$. We inductively assume that for $k = 2, 3, \ldots, p-1$, $S_n^{(k)}$ is a polynomial in $n$ of degree $k+1$ and the coefficient of the highest power is $\frac{1}{k+1}$, so that (2.20) would imply that $S_n^{(p)}$ is a polynomial of degree $p+1$, with the coefficient of the highest power of $n$ being $\frac{1}{p+1}$.

This result would be sufficient to show that

$$\int_0^a x^p \, dx = \frac{a^{p+1}}{p+1}$$

and verifies Fermat's remark on the determination of the area under curves of the form $y = x^p$. Fermat worked out the case $p = 4$ in his letters to Roberval and Mersenne. He clearly thought that the cases $p = 1, 2, 3$ had been found in antiquity, but that he had presented a new case. He was apparently unaware[14] of the work of al-Haytham (965–1040), who had explicitly determined $S_n^{(4)}$, or of the work of the Kerala mathematicians or, indeed, of the work of Faulhaber who had found $S_n^{(p)}$ through $p = 17$.

## 2.4 Pascal

Pascal also gave a proof that $S_n^{(k)} = \frac{1}{k+1} n^{k+1}+$ terms of lower power, but his proof differed from that of Fermat. Pascal treated this problem in the second part of his monograph on the arithmetical triangle, published posthumously,[15] where he showed that

$$(n+1)^{k+1} - 1 = \binom{k}{1} S_n^{(k)} + \binom{k}{2} S_n^{(k-1)} + \cdots + \binom{k}{k} S_n^{(1)}. \qquad (2.21)$$

In fact, Pascal proved a more general formula covering the sums of powers of any arithmetic progression,[16] rather than the particular case $1, 2, \ldots, n$. The case (2.21) will serve quite nicely, however, to illustrate Pascal's approach. To verify (2.21) by Pascal's method, start with the binomial theorem for positive integer exponents

$$(x+1)^{k+1} - x^{k+1} = \sum_{j=1}^{k+1} \binom{k+1}{j} x^{k+1-j}. \qquad (2.22)$$

In (2.22), successively set $x = n, n-1, \ldots, 2, 1$ to arrive at

$$(n+1)^{k+1} - n^{k+1} = \sum_{j=1}^{k+1} \binom{k+1}{j} n^{k+1-j}$$

$$(n)^{k+1} - (n-1)^{k+1} = \sum_{j=1}^{k+1} \binom{k+1}{j} (n-1)^{k+1-j}$$

$$(n-1)^{k+1} - (n-2)^{k+1} = \sum_{j=1}^{k+1} \binom{k+1}{j} (n-2)^{k+1-j}$$

$$\cdots \qquad \cdots \qquad \cdots \qquad \cdots$$

$$3^{k+1} - 2^{k+1} = \sum_{j=1}^{k+1} \binom{k+1}{j} 2^{k+1-j}$$

$$2^{k+1} - 1^{k+1} = \sum_{j=1}^{k+1} \binom{k+1}{j} 1^{k+1-j}.$$

[14] ibid. p. 231, footnote 35.
[15] Pascal (1665).
[16] Boyer (1943) provides an English translation of this formula on p. 239.

Add the $n$ equations. The left-hand side reduces to $(n+1)^{k+1} - 1$ after cancellation, and the right-hand side sums to

$$\sum_{j=1}^{k+1} \binom{k+1}{j} S_n^{(k+1-j)},$$

implying that if $S_n^{(j)}$, $j = 0, 1, \ldots, k-1$, are known, then $S_n^{(k)}$ can be determined.

## 2.5 Seki and Jakob Bernoulli on Bernoulli Numbers

Seki Takakazu (1643–1708) and Jakob Bernoulli (1651–1705) discovered the sequence of rational numbers now designated Bernoulli numbers at around the same time. Seki's contribution to this topic was published posthumously in 1712 in volume one of his *Katsuyō Sampō*;[17] these books have not been translated into English, but the editors have given useful comments on and some summaries of the contents. In the following year, Bernoulli's work was also published posthumously in his book on probability, *Ars Conjectandi*.[18] Especially since Seki's book appeared earlier, it has been suggested that Bernoulli numbers be renamed "Seki–Bernoulli numbers."[19]

Seki began by expressing in Japanese sentences the formulas:

$$S_n^{(1)} = 1 + 2 + \cdots + n = \frac{1}{2}(n + n^2),$$

$$S_n^{(2)} = 1^2 + 2^2 + \cdots + n^2 = \frac{1}{6}(n + 3n^2 + 2n^3),$$

$$S_n^{(3)} = 1^3 + 2^3 + \cdots + n^3 = \frac{1}{4}(n^2 + 2n^3 + n^4),$$

$$S_n^{(4)} = 1^4 + 2^4 + \cdots + n^4 = \frac{1}{30}(-n + 10n^3 + 15n^4 + 6n^5),$$

up to $S_n^{11}$, without explaining how he arrived at these formulas.

The editors of Seki's collected papers think it probable that he used a method explained earlier in his book. According to this method, Seki would have first assumed, for example, that $S_n^{(2)}$ was a polynomial of degree 3:[20]

$$1^2 + 2^2 + \cdots + n^2 = an + bn^2 + cn^3.$$

Note that because the sum of the terms with $n = 0$ would be 0, there is no constant term. Seki would then have taken $n = 1, 2, 3$ to obtain the three linear equations

[17] Seki and Hirayama et al. (1974).
[18] Bernoulli and Sylla (2006).
[19] Arakawa et al. (2014), especially p. 3.
[20] Seki and Hirayama et al. (1974) pp. 40–43.

$$a + b + c = 1$$
$$2a + 4b + 8c = 5$$
$$3 + 9b + 27c = 14$$

and would have solved them to find $a = \frac{1}{6}, b = \frac{1}{2}, c = \frac{1}{3}$. Note that Seki's approach here reflects his extensive work in linear equations and determinants.

After giving the statements describing the formulas for $S_n^{(1)}, S_n^{(2)}, \ldots, S_n^{(11)}$, Seki presented the method for deriving the general formula. He expanded

$$(1 + x)^n - 1 \quad \text{for} \quad n = 1, 2, 3, \ldots$$

to get

$$1 + x - 1 = x, (1 + x)^2 - 1 = 2x + x^2, (1 + x)^3 - 1 = 3x + 3x^2 + x^3, \ldots ;$$

he then introduced numbers, denoted here by $K_0, K_1, K_2$, as solutions of the equations

$$1 = K_0$$
$$2 = 1 + 2K_1$$
$$3 = 1 + 3K_1 + 3K_2$$
$$4 = 1 + 4K_1 + 6K_2 + 4K_3$$
$$\cdots \quad \cdots \quad \cdots \quad \cdots .$$

$$(2.23)$$

Observe that the integers on the right-hand side of the equations (2.23) are binomial coefficients. For example, the integers $1, 3, 3$ in the third row are the coefficients in the expansion of $(1 + x)^3 - 1$. Similarly, the integers in the fourth row are $1, 4, 6, 4$, the coefficients in the expansion of $(1 + x)^4 - 1$. Seki calculated some values of $K_i$ and then presented the following values in order:

$$1, \frac{1}{2}, \frac{1}{6}, 0, -\frac{1}{30}, 0, \frac{1}{42}, 0, -\frac{1}{30}, 0, \frac{5}{66}, 0, \ldots . \qquad (2.24)$$

Thus, the fifth number in the list, $-\frac{1}{30}$, is $K_4$, and so on. Seki made a table that shows that

$$S_n^{(1)} = \frac{1}{2}(n^2 + 2K_1 n),$$

$$S_n^{(2)} = \frac{1}{3}(n^3 + 3K_1 n^2 + 3K_2 n),$$

$$S_n^{(3)} = \frac{1}{4}(n^4 + 4K_1 n^3 + 6K_2 n^2 + 4K_3 n)$$

$$= \frac{1}{4}(n^4 + 2n^3 + n^2 + 0n).$$

Seki gave no proof of these results and also apparently gave no clue as to how he came upon them. Observe that the sequence (2.24) is the sequence of Bernoulli numbers, with the difference that we now take the second number to be $-\frac{1}{2}$. With the Bernoulli numbers, defined in the following section, are denoted by $B_k$, we see that $B_n = K_n$, except that $B_1 = K_1 - 1$. Note that for Jakob Bernoulli, too, $K_1 = \frac{1}{2}$.

## 2.6 Jakob Bernoulli's Polynomials

The second part of Bernoulli's great probability treatise contains results on permutations and combinations. He rigorously worked out the connection between binomial coefficients and figurate numbers. He thought that he was the first to do this, but Pascal anticipated him in 1654, as did others in earlier centuries. Bernoulli also rediscovered the formula (2.12) and applied it to the problem of finding the sums of powers of integers. Herein he made his enduring discovery of the role played by the sequence of rational numbers now named after him. Bernoulli found a pattern in the coefficients of the polynomials for $S_n^{(p)}$ that had been missed by so outstanding an arithmetician as Faulhaber.

Bernoulli began by explicitly expressing $S_n^{(p)}$ for $p = 1, 2, \ldots, 10$ as polynomials in $n$:[21]

Sums of Powers

$$\int n = \frac{1}{2}nn + \frac{1}{2}n.$$

$$\int n^2 = \frac{1}{3}n^3 + \frac{1}{2}nn + \frac{1}{6}n.$$

$$\int n^3 = \frac{1}{4}n^4 + \frac{1}{2}n^3 + \frac{1}{4}nn.$$

$$\int n^4 = \frac{1}{5}n^5 + \frac{1}{2}n^4 + \frac{1}{3}n^3 * -\frac{1}{30}n.$$

$$\int n^5 = \frac{1}{6}n^6 + \frac{1}{2}n^5 + \frac{5}{12}n^4 * -\frac{1}{12}nn.$$

$$\int n^6 = \frac{1}{7}n^7 + \frac{1}{2}n^6 + \frac{1}{2}n^5 * -\frac{1}{6}n^3 + \frac{1}{42}n.$$

$$\int n^7 = \frac{1}{8}n^8 + \frac{1}{2}n^7 + \frac{7}{12}n^6 * -\frac{7}{24}n^4 * \frac{1}{12}nn.$$

$$\int n^8 = \frac{1}{9}n^9 + \frac{1}{2}n^8 + \frac{2}{3}n^7 * -\frac{7}{15}n^5 * +\frac{2}{9}n^3 * -\frac{1}{30}n.$$

$$\int n^9 = \frac{1}{10}n^{10} + \frac{1}{2}n^9 + \frac{3}{4}n^8 * -\frac{7}{10}n^6 * +\frac{1}{2}n^4 * -\frac{1}{12}nn.$$

$$\int n^{10} = \frac{1}{11}n^{11} + \frac{1}{2}n^{10} + \frac{5}{6}n^9 * -1n^7 * 1n^5 * -\frac{1}{2}n^3 * +\frac{5}{66}n.$$

[21] Hellegouarch (2002) p. 52 for Schneps's translation; Bernoulli and Sylla (2006) pp. 215–216 also has a translation.

A. W. F. Edwards has noted that the last term in the polynomial for $\int n^9$ should be $-\frac{3}{20}n^2$, rather than $-\frac{1}{12}n^2$.[22]

Bernoulli went on:[23]

Any one who carefully observed the symmetry properties of this table will easily be able to continue it. If we let $c$ denote an arbitrary exponent, we have

$$\int n^c = \frac{1}{c+1}n^{c+1} + \frac{1}{2}n^c + \frac{c}{2}An^{c-1} + \frac{c(c-1)(c-2)}{2\cdot3\cdot4}Bn^{c-3}$$
$$+ \frac{c(c-1)(c-2)(c-3)(c-4)}{2\cdot3\cdot4\cdot5\cdot6}Cn^{c-5}$$
$$+ \frac{c(c-1)(c-2)(c-3)(c-4)(c-5)(c-6)}{2\cdot3\cdot4\cdot5\cdot6\cdot7\cdot8}Dn^{c-7} + \cdots, \quad (2.25)$$

the exponents of $n$ decreasing by 2 until $n$ or $nn$ is reached. The capitals $A, B, C, D$, etc. denote, in order, the last terms in the expressions of $\int nn$, $\int n^4$, $\int n^6$, $\int n^8$ etc. namely

$$A = \frac{1}{6}, \; B = -\frac{1}{30}, \; C = \frac{1}{42}, \; D = -\frac{1}{30}.$$

But these coefficients are so established that each of the coefficients along with the others of its order adds up to one. Thus $D = -\frac{1}{30}$, since

$$\frac{1}{9} + \frac{1}{2} + \frac{2}{3} - \frac{7}{15} + \frac{2}{9} + D = 1.$$

Using this table, it took me less than a quarter of an hour to compute the tenth powers of the first 1000 integers; the result is

91, 409, 924, 241, 424, 243, 424, 241, 924, 242, 500.

This example shows the uselessness of the book *Arithmetica Infinitorum* by Ismael Bullialdus, which is entirely devoted to a tremendously large computation of the sums of the six first powers – less than what I have accomplished in a single page.

If we denote $A, B, C, D, \ldots$ by $B_2, B_4, B_6, B_8, \ldots$ respectively, then equation (2.25) can be written as

$$1^c + 2^c + \cdots + n^c = \frac{1}{c+1}n^{c+1} + \frac{1}{2}n^c + \frac{c}{2}B_2n^{c-1} + \frac{c(c-1)(c-2)}{2\cdot3\cdot4}B_4n^{c-3} + \cdots,$$
$$(2.26)$$

where $c$ is a positive integer. Observe that Bernoulli was able to find a recurrence relation for the Bernoulli numbers by setting $n = 1$ in (2.26), obtaining

$$\frac{1}{c+1} + \frac{1}{2} + \binom{c}{1}\frac{B_2}{2} + \binom{c}{3}\frac{B_4}{4} + \binom{c}{5}\frac{B_6}{6} + \cdots = 1; \quad (2.27)$$

this equation can be used to determine $B_2, B_4, \ldots$. Thus, take $c = 2$ to find $B_2 = \frac{1}{6}$; take $c = 4$ to obtain

[22] Edwards (2002) p. 128.
[23] Hellegouarch (2002) pp. 52–53. Bernoulli and Sylla (2006) also has a nice translation.

$$\frac{1}{5} + \frac{1}{2} + \frac{1}{3} + 4\frac{B_4}{4} = 1 \quad \text{or} \quad B_4 = -\frac{1}{30}.$$

Observe that the coefficients of $n^{c-2}, n^{c-4}, \ldots$ in Bernoulli's (2.26) are all apparently zero, though this was not explained by Bernoulli. Later in this chapter, we shall give Euler's argument for this point.

Today, we would sum up to $(n-1)^c$ on the left-hand side of (2.26); we would therefore subtract $n^c$ from each side of the equation and also insert the coefficients of the missing powers of $n$:

$$1^c + 2^c + \cdots + (n-1)^c = \frac{1}{c+1}\left(n^{c+1} + \binom{c+1}{1}B_1 n^c + \binom{c+1}{2}B_2 n^{c-1} \right.$$
$$\left. + \binom{c+1}{3}B_3 n^{c-2} + \binom{c+1}{4}B_4 n^{c-3} + \cdots + B_c n\right),$$

$$(2.28)$$

where $B_1 = -\frac{1}{2}$ and $B_3, B_5, \ldots$ are all zero. Now if $B_{c+1}$ is added to the polynomial in $n$ in parentheses on the right-hand side of (2.28), the resulting polynomial would be the Bernoulli polynomial of degree $c+1$, denoted as $B_{c+1}(n)$; we could then write (2.28) as

$$\sum_{i=1}^{n-1} i^c = \frac{1}{c+1}(B_{c+1}(n) - B_{c+1}). \qquad (2.29)$$

Bernoulli left it to the reader to use "the symmetry properties of this table" to figure out how he obtained his general formula for the sums of powers. A little earlier in his book, Bernoulli presented a "table of combinations or figurate numbers" and analyzed it columnwise.[24] Apply this idea to Bernoulli's table on sums of powers. The progression in the first column is easy to understand. Now in the second column, factor out $\frac{1}{2}$ to obtain the progression $1, 1, 1, \ldots$; these form the first column of Bernoulli's table of figurate numbers. Next factor out the first number in the third column, $\frac{1}{6}$, to obtain the sequence $1, \frac{3}{2}, 2, \frac{5}{2}, 3, \ldots$, and this turns out to be $\frac{1}{2}$ of the sequence $2, 3, 4, 5, 6, \ldots$ appearing in the second column of the figurate numbers table. The fourth column of the sums of powers table consists of only zeros but factor out $-\frac{1}{30}$ from the fifth column to obtain the progression $1, \frac{5}{2}, 5, \frac{35}{4}, 14, \ldots$. This last sequence is equal to $\frac{1}{4}$ of the fourth column in the figurate numbers table. These observations clarify Bernoulli's comment on the "symmetry properties of the table." Note that the Bernoulli numbers are formed by the sequence of coefficients of $n$ in the polynomial expansions of the various sums of powers. Today, however, we take the first Bernoulli number to be $-\frac{1}{2}$ rather than $\frac{1}{2}$ so that the signs alternate.

It is possible, in fact, to derive the Bernoulli–Seki formula for $S_n^{(m)}$ from Pascal's identity (2.21), as Boyer has shown.[25] To see this, write the polynomial for $S_n^{(m)}$ as

---

[24] See Bernoulli and Sylla (2006) p. 206.
[25] Boyer (1943) pp. 242–243.

$$S_n^{(m)} = \frac{n^{m+1}}{m+1} + A_1(m)n^m + A_2(m)n^{m-1} + A_3(m)n^{m-3} + A_4(m)n^{m-4} + \cdots,$$

$$(2.30)$$

where $A_1(m), A_2(m), A_3(m), \ldots$ are functions of $m$. Thus, (2.21) takes the form

$$n^{m+1} + \binom{m+1}{1}n^m + \binom{m+1}{2}n^{m-1} + \binom{m+1}{3}n^{m-2} + \cdots + \binom{m+1}{m}n$$

$$= \binom{m+1}{1}\left(\frac{n^{m+1}}{m+1} + A_1(m)n^m + A_2(m)n^{m-1} + \cdots\right)$$

$$+ \binom{m+1}{2}\left(\frac{n^m}{m} + A_1(m-1)n^{m-1} + A_2(m-1)n^{m-2} + \cdots\right)$$

$$+ \binom{m+1}{3}\left(\frac{n^{m-1}}{m-1} + A_1(m-2)n^{m-2} + A_2(m-2)n^{m-3} + \cdots\right)$$

$$+ \cdots.$$

Equate the coefficients of $n^m$ on each side to obtain

$$\binom{m+1}{1} = \binom{m+1}{1}A_1(m) + \binom{m+1}{2}\cdot\frac{1}{m}.$$

Thus, $A_1(m) = \frac{1}{2}$, yielding $A_1(m-1) = A_1(m-2) = \cdots = \frac{1}{2}$. Similarly, equate the coefficients of $n^{m-1}$ to obtain

$$\binom{m+1}{2} = \binom{m+1}{1}A_2(m) + \binom{m+1}{2}\frac{1}{2} + \binom{m+1}{3}\frac{1}{m-1}.$$

Solve for $A_2(m)$:

$$A_2(m) = \frac{1}{6}\cdot\frac{m}{1\cdot 2} \quad \text{and thus} \quad A_2(m-1) = \frac{1}{6}\cdot\frac{m-1}{1\cdot 2} \quad \text{and so on.}$$

Equating the coefficients of $n^{m-2}$ produces

$$\binom{m+1}{3} = \binom{m+1}{1}A_3(m) + \binom{m+1}{2}\frac{1}{6}\cdot\frac{m-1}{1\cdot 2}$$

$$+ \binom{m+1}{3}\frac{1}{2} + \binom{m+1}{4}\frac{1}{m-2}.$$

Therefore, $A_3(m) = 0$. By the same method, equating the coefficients of $n^{m-3}$, we can determine that

$$A_4(m) = -\frac{1}{30}\cdot\frac{m(m-1)(m-2)}{1\cdot 2\cdot 3\cdot 4}.$$

Substituting these values into (2.30), we arrive at Bernoulli's formula (2.25).

It appears that Pascal did not give the coefficients of the powers of $n$, other than $n^{m+1}$. Seki and Bernoulli seem to have arrived at their formula for the sums of powers of integers by incomplete induction and we have no indication of their proofs. Bernoulli was apparently unaware of Pascal's book on the arithmetical triangle;[26] had he seen it, he might have derived the proof by using (2.30).

## 2.7 Euler

Both Seki and Bernoulli defined the Bernoulli–Seki numbers by means of the equations

$$K_0 = 1, \quad K_1 = \frac{1}{2},$$

and

$$\frac{1}{m+1}\left(K_0 + \binom{m+1}{1}K_1 + \binom{m+1}{2}K_2 + \cdots + \binom{m+1}{m}K_m\right) = 1, \quad (2.31)$$

where $m = 2, 3, \ldots$. Recall that $K_i$ here represents the modern $i$th Bernoulli number $B_i$, except that $B_1 = K_1 - 1$.

First, note that the generating function for a sequence $a_1, a_2, a_3, \ldots$ is defined as the function

$$f(x) = 1 + a_1 x + a_2 x^2 + a_3 x^3 + \cdots;$$

the exponential generating function is then defined for $a_1, a_2, a_3, \ldots$ as the function

$$F(x) = 1 + a_1 \frac{x}{1!} + a_2 \frac{x^2}{2!} + a_3 \frac{x^3}{3!} + \cdots.$$

Euler encountered the same relations as in (2.31) when he discovered the Euler–Maclaurin formula, given in a paper of 1739, published in 1750, "De Seriebus Quibusdam Considerationes."[27] In this paper, that also contains his asymptotic series for $f(x) - f(x+h) + f(x+2h) - \cdots$, now known as Boole's formula, he found the generating function for $C_0, C_1, C_2, \ldots$, defined by the equations

$$C_0 = 0$$

and

$$C_m = \frac{C_{m-1}}{2!} - \frac{C_{m-2}}{3!} + \frac{C_{m-3}}{4!} - \cdots + \frac{(-1)^{m-2}C_1}{m!} + \frac{(-1)^{m-1}}{(m+1)!}, \quad m = 1, 2, 3, \ldots.$$

---

[26] Bernoulli and Sylla (2006) p. 99.
[27] Eu. I-14 pp. 407–462. E 130 § 27.

Euler observed that these relations could be obtained by equating the coefficients of the powers of $z$ in the equation

$$(1 + C_1 z + C_2 z^2 + \cdots) \left( 1 - \frac{z}{1 \cdot 2} + \frac{z^2}{1 \cdot 2 \cdot 3} - \frac{z^3}{1 \cdot 2 \cdot 3 \cdot 4} + \cdots \right) = 1.$$

Euler thus found that the generating function for the numbers in question was

$$1 + C_1 z + C_2 z^2 + \cdots = \frac{1}{1 - \frac{z}{1 \cdot 2} + \frac{z^2}{1 \cdot 2 \cdot 3} - \frac{z^3}{1 \cdot 2 \cdot 3 \cdot 4} + \cdots}$$

$$= \frac{z}{1 - e^{-z}} = \frac{z e^z}{e^z - 1}. \tag{2.32}$$

But by equating the coefficients of the powers of $z$ in the relation obtained from (2.32):

$$(1 + C_1 z + C_2 z^2 + \cdots) \left( z + \frac{z^2}{2!} + \frac{z^3}{3!} + \cdots \right) = z + \frac{z^2}{1!} + \frac{z^3}{2!} + \frac{z^4}{3!} + \cdots,$$

one obtains the relations in (2.31). Thus, the sequence $\frac{K_i}{i!}$ is the same as the sequence $C_i$ for $i \geq 1$. Note that the odd Bernoulli–Seki numbers $K_{2k+1} = B_{2k+1}$ for $k \geq 1$ can be shown to be zero. As Euler noted in sections 24–28 of his "De numero memorabili in summatione progressiones harmonicae naturalis occurrente,"[28] since $K_1 = \frac{1}{2}$, it follows from (2.32) that

$$1 + \frac{K_2}{2!} z^2 + \frac{K_3}{3!} z^3 + \frac{K_4}{4!} z^4 + \cdots = \frac{z e^z}{e^z - 1} - \frac{z}{2}, \tag{2.33}$$

whose right-hand side is an even function. Therefore, the coefficients of the odd powers of $z$ in (2.33) must all be zero. This fact also follows from Faulhaber's equation (2.13). Observe that equation (2.13) shows that $S_n^{(2m+1)}$ is divisible by $N^2 = \frac{n^2(n+1)^2}{4}$ for $m \geq 1$, so that $S_n^{(2m+1)}$ is divisible by $n^2$, and thus that the coefficient of $n$ in the polynomial for $S_n^{(2m+1)}$ is zero when $m \geq 1$. Now the coefficient of $n$ is $K_{2m+1} = B_{2m+1}$; it follows that $K_{2m+1} = B_{2m+1} = 0$ for $m \geq 1$.

From this point onward, it makes sense to use only the expression $B_j$ and omit the use of $K_j$ to denote the $j$th Bernoulli number. Thus, the exponential generating function of the sequence of $B_1, B_2, B_3, B_4, \ldots$ by (2.33) would be given by[29]

$$1 + \sum_{n=1}^{\infty} B_n \frac{x^n}{n!} = \frac{x e^x}{e^x - 1} - x = \frac{x}{e^x - 1}. \tag{2.34}$$

Euler also found two different proofs that $(-1)^{n-1} B_{2n} \geq 0$ for $n = 1, 2, 3, \ldots$. One depended on the formula

---

[28] Eu. 1-15 pp. 569–603. E 583 § 24–28.
[29] ibid. § 24.

$$\sum_{n=1}^{\infty} \frac{1}{n^{2k}} = \frac{(-1)^{k-1} 2^{2k-1} \pi^{2k} B_{2k}}{(2k)!}. \tag{2.35}$$

Note that the left-hand side of (2.35) is positive, showing that $(-1)^{k-1} B_{2k}$ must be positive; in Chapter 16 we discuss how this formula may be proved, in connection with Euler's and Spence's work.

The second proof used the generating function: In sections 26–29 of his "De numero memorabili," he observed that since $B_{2m+1} = 0$ for $m \geq 1$, (2.33) implied that

$$f(z) \equiv 1 + \frac{B_2}{2!} z^2 + \frac{B_4}{4!} z^4 + \frac{B_6}{6!} z^6 + \cdots = \frac{z e^z}{e^z - 1} - \frac{z}{2} = \frac{z}{e^z - 1} + \frac{z}{2}. \tag{2.36}$$

Therefore

$$\begin{aligned}
zf'(z) &= \frac{z}{2} + \frac{z}{e^z - 1} - \frac{z^2 e^z}{(e^z - 1)^2} \\
&= \frac{z}{2} + \frac{z}{e^z - 1} - \frac{z^2}{e^z - 1} - \frac{z^2}{(e^z - 1)^2} \\
&= f - f^2 + \frac{z^2}{4}.
\end{aligned} \tag{2.37}$$

Substituting the series for $f$ into the relation (2.37) and equating the coefficient of $z^{2n}$, for $n > 1$, on each side, Euler had

$$2n \frac{B_{2n}}{(2n)!} = \frac{B_{2n}}{(2n)!} - \sum_{m=0}^{n} \frac{B_{2m}}{(2m)!} \cdot \frac{B_{2n-2m}}{(2n-2m)!}. \tag{2.38}$$

In fact, Euler explicitly wrote down the coefficients for $n = 1, 2, 3, 4, 5, 6$ and then wrote "etc." Observe that (2.38) implies

$$(2n + 1) B_{2n} = - \sum_{m=1}^{n-1} \binom{2n}{2m} B_{2m} B_{2n-2m},$$

from which we can inductively deduce that $(-1)^{n-1} B_{2n} > 0$: It is clear that $(-1)^0 B_2 = \frac{1}{6} > 0$. Supposing the result true up to $n - 1$, one has

$$(-1)^{n-1} (2n + 1) B_{2n} = \sum_{m=1}^{n-1} \binom{2n}{2m} (-1)^{m-1} B_{2m} (-1)^{n-m-1} B_{2(n-m)} > 0.$$

Naturally, in the mathematical approach of his time, Euler did not write this argument in the form presented here. Instead, he showed that

$$B_2 > 0 \Longrightarrow -B_4 > 0 \Longrightarrow B_6 > 0 \Longrightarrow -B_8 > 0 \Longrightarrow B_{10} > 0 \Longrightarrow -B_{12} > 0,$$

at which point he wrote "etc."

## 2.8   Lacroix's Proof of Bernoulli's Formula

In the first part of his 1755 differential calculus book *Institutiones Calculi Differentialis*, Euler suggested deriving Bernoulli's formula, (2.25), by means of finite differences, but he did not provide any details. In the second part of his book, Euler derived (2.25) using the Euler–Maclaurin summation formula. Sylvestre F. Lacroix (1765–1843), in the third volume of his important text on calculus, summarized Euler's ideas on finite differences and then indicated how they could be worked into a proof of (2.25).

Lacroix investigated partial differential equations under the tutelage of Gaspard Monge but did not pursue mathematical research. Rather, at the urging of Condorcet, he decided that his broad knowledge of eighteenth-century mathematics should be put to use in the writing of elementary and advanced mathematics textbooks. These books were widely popular, going into numerous editions and translations. I here summarize Lacroix's treatment of Bernoulli's formula (2.25).

Sylvestre Lacroix essentially redefined the first Bernoulli–Seki number to be $-\frac{1}{2}$ instead of $\frac{1}{2}$ by writing the sum $S_{n-1}^{(m)}$ as a polynomial in $n$. To see his approach, consider

$$S_n^{(m)} = \frac{1}{p+1}n^{m+1} + \frac{1}{2}n^m + \cdots ;$$

subtract $n^m$ from each side to get

$$S_{n-1}^{(m)} = \frac{1}{p+1}\left(n^{m+1} - \binom{m+1}{1}\frac{1}{2}n^m \right.$$

$$\left. + \binom{m+1}{2}B_2 n^{m-1} + \binom{m+1}{3}B_3 n^{m-2} + \cdots \right). \quad (2.39)$$

Now let $B_1 = -\frac{1}{2}$ to obtain the modern definition of the Bernoulli (or Bernoulli–Seki) numbers as the sequence $B_0 = 1, B_1 = -\frac{1}{2}, B_2 = \frac{1}{6}, \ldots$ as defined by the equations

$$B_0 + \binom{m+1}{1}B_1 + \binom{m+1}{2}B_2 + \binom{m+1}{3}B_3 +$$

$$\cdots + \binom{m+1}{m}B_m = 0, \quad m = 1, 2, 3, \ldots . \quad (2.40)$$

Equations (2.40) are obtained upon taking $n = 1$ in (2.39), because $S_{n-1}^{(m)} = 0$ for $n = 1$. Also observe that the $B_1, B_2, B_3, \ldots$ are uniquely defined by (2.40). On pages 69 and 70 of his *Traité des différences et des séries*, published in 1800, Lacroix gave a proof[30] of Bernoulli's result (2.39).

---

[30] Lacroix (1800) pp. 69–70.

Using Lacroix's notation, let $\sum x^m$ denote the sum $\sum_{k=1}^{x-1} k^m \equiv S(x)$. Note that, unlike Euler and Bernoulli, Lacroix took the sum up to $x - 1$, rather than $x$. With this in mind, he had

$$S(x + 1) - S(x) = x^m.$$

Now Lacroix did not use subscripts, as in $A_k$, but we use the modern notation: Assume

$$\sum x^m = \sum_{k=0}^{m+1} A_k x^{m+1-k}.$$

Then

$$x^m = \sum_{k=0}^{m+1} A_k \left( (x + 1)^{m+1-k} - x^{m+1-k} \right) = \sum_{k=0}^{m+1} A_k \left( \binom{m+1-k}{1} x^{m-k} \right.$$

$$\left. + \binom{m+1-k}{2} x^{m-k-1} + \cdots + \binom{m+1-k}{m-k} x + 1 \right).$$

Equate the powers of $x$ to get

$$A_0 = \frac{1}{m+1}; \quad A_1 = -A_0 \frac{(m+1)}{2} = -\frac{1}{2}; \quad A_2 = -A_0 \frac{(m+1)m}{2 \cdot 3} - A_1 \frac{m}{2} = \frac{1}{6} \cdot \frac{m}{2};$$

$$A_3 = -A_0 \frac{(m+1)(m(m-1))}{2 \cdot 3 \cdot 4} - A_1 \frac{m(m-1)}{2 \cdot 3} - A_2 \frac{(m-1)}{2} = 0; \text{ etc.}$$

Lacroix wrote that from these equations one could successively obtain the values of the coefficients $A_k$, and he explicitly gave the values of $A_k, k = 0, 1, \ldots, 20$.

Although he did not work out the general case, it is easy to do: Write $A_s = a_s \binom{m+1}{s}$ and equate the coefficient of $x^{m-k}$ to get

$$\binom{m+1}{0}\binom{m+1}{k+1} a_0 + \binom{m+1}{1}\binom{m}{k} a_1 + \binom{m+1}{2}\binom{m-1}{k-1} a_2 +$$

$$\cdots + \binom{m+1}{k}\binom{m+1-k}{1} a_k = 0.$$

Divide by $\binom{m+1}{k+1}$ to find that

$$\binom{k+1}{0} a_0 + \binom{k+1}{1} a_1 + \binom{k+1}{2} a_2 + \cdots + \binom{k+1}{k} a_k = 0.$$

Observe that since $a_0 = 1$, the equations defining $a_1, a_2, a_3, \ldots$ are identical to (2.40), so that $a_1 = B_1$, $a_2 = B_2$, $a_3 = B_3$, and so on. In this manner, Lacroix has demonstrated the Bernoulli–Seki formula.

### 2.9    Jacobi on Faulhaber

In an 1834 issue of *Crelle's Journal*, Jacobi published an important paper giving
a rigorous derivation of the Euler–Maclaurin summation formula with remainder
term.[31] We discuss this paper in Chapter 20. Near the end of this paper, however,
Jacobi gave a very brief treatment of Faulhaber's work on sums of powers of
integers, but without mentioning Faulhaber. I saw a copy of Faulhaber's book[32] in the
Cambridge University Library. It is stated on the title page that that book had belonged
to Jacobi; on the previous blank page "J. F. Pfaff" is written. It appears that Jacobi may
have perhaps acquired the book after the death of Pfaff in 1825. It is thus possible that
Jacobi was in possession of Faulhaber's book before writing his 1834 paper.

Jacobi wrote the formulas for $\sum_{i=1}^{n} i^{2k-1}$ in powers of $u = n(n+1)$, rather than
in powers of $N = \frac{n(n+1)}{2}$. He also observed that

$$\frac{1}{2k+1} \frac{d \sum_{i=1}^{n} i^{2k+1}}{dn} = \sum_{i=1}^{n} i^{2k}; \qquad (2.41)$$

this follows immediately from Bernoulli's formula (2.26) or (2.28), since $B_{2k+1} = 0$.
Moreover, he noted that

$$\frac{1}{2k} \frac{d \sum_{i=1}^{n} i^{2k}}{dn} = \sum_{i=1}^{n} i^{2k-1} - B_{2k}. \qquad (2.42)$$

Jacobi began the last part of his paper by explicitly writing the sums for $k = 2, 3, \ldots, 7$:

$$\sum_{x=1}^{n} x^3 = \frac{1}{4} u^2,$$

$$\sum_{x=1}^{n} x^5 = \frac{1}{6} u^2 \left( u - \frac{1}{2} \right),$$

$$\sum_{x=1}^{n} x^7 = \frac{1}{8} u^2 \left( u^2 - \frac{4}{3} u + \frac{2}{3} \right),$$

$$\cdots,$$

$$\sum_{x=1}^{n} x^{13} = \frac{1}{14} u^2 \left( u^5 - \frac{35}{6} u^4 + \frac{287}{15} u^3 - \frac{118}{3} u^2 + \frac{691}{15} u - \frac{691}{30} \right).$$

To obtain a general form of the sum $\sum_{i=1}^{n} i^{2k-1}$, Jacobi set

$$\sum_{i=1}^{n} i^{2k-3} = \frac{1}{2k-2} \left( u^{k-1} - a_1 u^{k-2} + a_2 u^{k-3} - \cdots + (-1)^{k-1} a_{k-3} u^2 \right) \qquad (2.43)$$

---

[31] Jacobi (1969) vol. 6, pp. 64–75.
[32] Faulhaber (1631).

and

$$\sum_{i=1}^{n} i^{2k-1} = \frac{1}{2k}\left(u^k - b_1 u^{k-1} + b_2 u^{k-2} - \cdots + (-1)^k b_{k-2} u^2\right). \tag{2.44}$$

He then stated that the coefficients $a_i$, $i = 1, \ldots, k-3$ and $b_j$, $j = 1, \ldots, k-2$ would satisfy the relations

$$2k(2k-1)a_1 = (2k-2)(2k-3)b_1 - k(k-1),$$

$$2k(2k-1)a_2 = (2k-4)(2k-5)b_2 - (k-1)(k-2)b_1,$$

$$2k(2k-1)a_3 = (2k-6)(2k-7)b_3 - (k-1)(k-3)b_2,$$

$$\cdots \quad \cdots \quad \cdots \quad \cdots \quad \cdots$$

$$2k(2k-1)a_{k-3} = 5 \cdot 6 b_{k-3} - 3 \cdot 4 b_{k-4},$$

$$0 = 3 \cdot 4 b_{k-2} - 2 \cdot 3 b_{k-3}. \tag{2.45}$$

Observe that these relations show that if $a_i > 0$, $i = 1, 2, \ldots, k-3$, then $b_j > 0$ for $j = 1, 2, \ldots, k-2$. The relation (2.45) shows that $b_{k-3} = 2b_{k-2}$. Taking note of (2.41), we have completed the proof of Faulhaber's statements (2.13), (2.14), and (2.15). Note that (2.15) is essentially the derivative of (2.13) with respect to $n$.

Though Jacobi did not provide details of the proof of the relations between $a_i$ and $b_i$, they are straightforward. They follow from an application of (2.41) through (2.44) and observing that since $u = n^2 + n$, $\frac{du}{dn} = 2n + 1$. Taking the second derivative of (2.44) with respect to $n$ produces

$$(2k-2)\sum_{i=1}^{n} i^{2k-3} + B_{2k-2}$$

$$= \frac{2}{2k(2k-1)}\left(ku^{k-1} - (k-1)b_1 u^{k-2} + (k-2)b_2 u^{k-3} - \cdots + (-1)^k 2b_{k-2}u\right)$$

$$+ \frac{(2n+1)^2}{2k(2k-1)} \times \left(k(k-1)u^{k-2} - (k-1)(k-2)b_1 u^{k-3}\right.$$

$$+ (k-2)(k-3)b_2 u^{k-4} - \cdots + (-1)^k 2b_{k-2}\Big)$$

$$= u^{k-1} - a_1 u^{k-2} + a_2 u^{k-3} - \cdots + (-1)^{k-1} a_{k-3} u^2 + B_{2k-2}. \tag{2.46}$$

Now note that $(2n+1)^2 = 4u + 1$; apply this in (2.46) and equate the coefficients of the respective powers of $u$ to obtain Jacobi's relations among the $a_i$ and $b_i$.

## 2.10 Jacobi and Raabe on Bernoulli Polynomials

Observe that equations (2.41) and (2.42) are not clearly stated: $n$ represents a positive integer in the sum $\sum_{i=1}^{n}$, while it is a real variable in the expression $\frac{d}{dn}$. In this section we will employ a very simple result on polynomials, used by Jacobi, to show how this inconsistency can in fact be ignored. First let $n$ denote an integer variable and $x$ denote a real variable. Thus, since for any integer $k \geq 0$, Bernoulli's formula would state that

$$\sum_{i=1}^{n-1} i^k = \frac{1}{k+1} \left( n^{k+1} + \binom{k+1}{1} B_1 n^k + \binom{k+1}{2} B_2 n^{k-1} + \cdots \right.$$

$$\left. + \binom{k+1}{2} B_{k-1} n^2 + B_k n \right). \tag{2.47}$$

Now define $B_k(x)$, the $k$th Bernoulli polynomial, by

$$B_k(x) = \frac{d}{dx} \left( \frac{1}{k+1} \left( x^{k+1} + \binom{k+1}{1} B_1 x^k + \cdots + \binom{k+1}{2} B_{k-1} x^2 + B_k x \right) \right)$$

$$= x^k + \binom{k}{1} B_1 x^{k-1} + \binom{k}{2} B_2 x^{k-2} + \cdots + \binom{k}{1} B_{k-1} x + B_k. \tag{2.48}$$

Recall from (2.24) that

$$B_1 = -\frac{1}{2}, \quad B_2 = \frac{1}{6}, \quad B_3 = 0, \quad B_4 = -\frac{1}{30}, \quad \ldots;$$

hence, all nonzero terms following after the term $\binom{k}{1} B_1 x^{k-1}$ must be of the form

$$\binom{k}{2s} B_{2s} x^{k-2s}.$$

Observe that, as we have seen in (2.29), (2.47) may be written as

$$\sum_{i=1}^{n-1} i^k = \frac{B_{k+1}(n) - B_{k+1}}{k+1}, \quad k \geq 1 \tag{2.49}$$

and that

$$\frac{d}{dx} B_k(x) = k B_{k-1}(x), \quad k \geq 0. \tag{2.50}$$

The general theorem on polynomials, used by Jacobi, is not difficult to prove and in fact could have been provided in the seventeenth century. Cauchy[33] stated it thus: Suppose $P(x)$ and $Q(x)$ are polynomials of degree at most $n$, and for at least $n + 1$ distinct complex numbers $x_1, x_2, \ldots, x_{n+1}$, $P(x_i) = Q(x_i)$ with $i = 1, 2, \ldots, n + 1$; then $P(x) = Q(x)$. To verify this, Cauchy first showed that if a polynomial vanished at $x = x_1, x_2, \ldots, x_{n+1}$, then it was divisible by $(x - x_1)(x - x_2) \cdots (x - x_{n+1})$. He next assumed that $P(x)$ was not identical with $Q(x)$ and pointed out that when they were not identical, $P(x) - Q(x)$ was a nonzero polynomial of degree at most $n$ and it vanished at $x = x_1, \ldots, x_{n+1}$. Hence, it was divisible by $(x-x_1)(x-x_2) \cdots (x-x_{n+1})$,

---

[33] Cauchy (1989) § 4.1.

a polynomial of degree $n + 1$. This was absurd; therefore, $P(x) = Q(x)$. Thus, $x$ can be replaced by $n$ as in equations (2.41) and (2.42).

The author has not found any application of this principle before the nineteenth century, and it in spite of Cauchy's work, was used only infrequently until about 1850. For example, in a paper of 1845 on the $q$-extension of the binomial theorem, Eisenstein used this principle on polynomials, but then gave an explicit statement of it in a footnote, apparently reflecting his impression that the idea was not well-known. Joseph Raabe did not use this principle in his 1848 work, *Die Jacob Bernoullische Function*,[34] although it would have been natural to do so and would have greatly shortened his proof of a result on Bernoulli polynomials.

Interestingly, however, the mathematical basis for the proof of this theorem was actually available in the seventeenth century and would seem to have been readily achievable by Descartes or Newton. In his 1821 work *Analyse algébrique*,[35] Cauchy proved a theorem whose attempted proof by Euler had had a gap; Cauchy completed the proof by use of this theorem on polynomials. We discuss this result in Chapter 4.

As an application of this general theorem on polynomials, consider Jacobi's proof of the formula[36]

$$B_k(1 - x) = (-1)^k B_k(x). \tag{2.51}$$

Although Jacobi proved (2.51) only for the case in which $k$ is even, the proof actually extends to all cases. Jacobi first noted that (2.49) would imply

$$B_k(n + 1) = k \sum_{i=1}^{n} i^{k-1} + B_k = k \sum_{i=1}^{n-1} i^{k-1} + B_k + kn^{k-1}$$

$$= B_k(n) + kn^{k-1}, \quad \text{for} \quad n = 1, 2, 3, \ldots.$$

Thus, we can state that

$$B_k(x + 1) - B_k(x) = kx^{k-1}, \tag{2.52}$$

because it holds true for an infinite number of values $x = 1, 2, 3, \ldots$. Now when $k$ is even, then every term in $B_k(x)$ must be even with the exception of $-\frac{1}{2}x^{k-1}$; when $k$ is odd, the reverse is the case. Therefore

$$(-1)^{k-1} B_k(-x) + B_k(x) = -kx^{k-1}. \tag{2.53}$$

Adding (2.52) and (2.53) then produces

$$B_k(x + 1) + (-1)^{k-1} B_k(-x) = 0. \tag{2.54}$$

Changing $x$ to $-x$ in (2.54), Jacobi obtained (2.51).

---

[34] Raabe (1848).
[35] Cauchy (1989).
[36] Jacobi (1834) or Jacobi (1969) vol. 6, pp. 64–75. See equations (16) through (22) and the remark given after equation (19).

In section 4 of his 1834 paper on the Euler–Maclaurin formula, Jacobi gave a method of finding the generating function of the Bernoulli polynomials. He actually gave the generating function for only the even-degree polynomials, but his method clearly applies in general. He first employed (2.49) and $B_1(n) = n + B_1$ from (2.48) to obtain

$$\sum_{k=0}^{\infty} B_k(n) \frac{t^k}{k!} = 1 + B_1(n)t + \sum_{k=2}^{\infty} \left( k \sum_{j=1}^{n-1} j^{k-1} + B_k \right) \frac{t^k}{k!}$$

$$= 1 + (n + B_1)t + \sum_{k=1}^{\infty} t \sum_{j=1}^{n-1} \frac{(jt)^k}{k!} + \sum_{k=2}^{\infty} B_k \frac{t^k}{k!}$$

$$= 1 + \sum_{k=1}^{\infty} B_k \frac{t^k}{k!} + nt + t \sum_{j=1}^{n-1} (e^{jt} - 1)$$

$$= \frac{t}{e^t - 1} + t + \sum_{j=1}^{n-1} t e^{jt}$$

$$= \frac{t}{e^t - 1} + t \left( 1 + \frac{e^t (e^{(n-1)t} - 1)}{e^t - 1} \right)$$

$$= \frac{t}{e^t - 1} + \frac{t e^{nt} - 1}{e^t - 1} = \frac{t e^{nt}}{e^t - 1}.$$

Jacobi thus obtained the generating function for Bernoulli polynomials in the integer variable $n$. Now take $x$ to be a real or complex variable and suppose, as Jacobi himself could have, that

$$\frac{t e^{xt}}{e^t - 1} = \sum_{k=0}^{\infty} C_k(x) \frac{t^k}{k!}.$$

Clearly, $C_k(x)$ is a polynomial of degree $k$ and $C_k(n) = B_k(n)$ for all positive integers $n$. Hence $C_k(x) = B_k(x)$ and the generating function for the Bernoulli polynomials is

$$\sum_{k=0}^{\infty} B_k(x) \frac{t^k}{k!} = \frac{t e^{xt}}{e^t - 1}. \tag{2.55}$$

Joseph Raabe called $\frac{1}{k}(B_k(x) - B_k)$ the "Bernoullische Function" in his 1848 book of that name. In this book he proved some interesting properties of $B_k(x)$, including what is now called the multiplication formula for Bernoulli polynomials:

$$B_k(nx) = n^{k-1} \sum_{j=0}^{n-1} B_k \left( x + \frac{j}{n} \right), \quad k \geq 0. \tag{2.56}$$

Raabe first proved (2.56) for the case in which $k$ is an odd integer; when $k = 1$ the result is immediate. Raabe observed that when $k$ is odd and $1 \leq j \leq n - 1$, (2.51) would imply that

$$B_k \left( 1 - \frac{j}{n} \right) + B_k \left( \frac{j}{n} \right) = 0, \quad j = 1, 2, \ldots, n - 1. \tag{2.57}$$

Adding the $n - 1$ formulas in (2.57), Raabe concluded that

$$\sum_{j=1}^{n} B_k \left( \frac{j}{n} \right) = 0. \tag{2.58}$$

He next noted that, with $r$ a positive integer, a repeated application of (2.52) would imply

$$B_k(x + r) = B_k(x) + k \left( x^{k-1} + (x + 1)^{k-1} + \cdots + (x + r - 1)^{k-1} \right). \tag{2.59}$$

Setting $x = \frac{j}{n}$, $j = 1, 2, \ldots, n - 1$ in (2.59) and adding the $n - 1$ equations, Raabe obtained

$$\sum_{j=0}^{n-1} B_k \left( r + \frac{j}{n} \right) = \sum_{j=1}^{n-1} B_k \left( \frac{j}{n} \right) + B_k(r) + k \sum_{j=1}^{n-1} \sum_{s=0}^{r-1} \left( \frac{sn + j}{n} \right)^{k-1}. \tag{2.60}$$

Since Raabe took $k$ as odd, $B_k = 0$ and (2.29) produced

$$B_k(r) = k \left( 1^{k-1} + 2^{k-1} + \cdots + (r - 1)^{k-1} \right)$$
$$= k \left( \left( \frac{n}{n} \right)^{k-1} + \left( \frac{2n}{n} \right)^{k-1} + \cdots + \left( \frac{n(r - 1)}{n} \right)^{k-1} \right).$$

Thus, (2.58) implied that the right-hand side of (2.60) could be rewritten as

$$\frac{k}{n^{k-1}} \left( 1^{k-1} + 2^{k-1} + \cdots + (nr - 1)^{k-1} \right) = \frac{1}{n^{k-1}} B_k(nr). \tag{2.61}$$

This actually completes the proof of (2.56) for $k$ odd, since both sides of the equation are polynomials and the equation holds for an infinite number of integers. But Raabe did not draw this conclusion at this point. After (2.61), he gave a lengthy argument to prove that $r$ could be taken to be a positive rational number and then went on to show how to extend the result to negative rationals. Finally, he applied the idea of continuity to further extend his result to all real numbers. However, we omit the details of this part of Raabe's reasoning because, in fact, he had already completed the proof of (2.56) for $k$ odd when he demonstrated that it was true for an arbitrary positive integer $x = r$.

To prove (2.56) for even $k \geq 0$, take the derivative of (2.56) for odd $k$ and then use (2.50) to conclude that (2.56) is true for all $k \geq 0$. Now Raabe did not give a proof for *k* even in this manner, but presented a much more elaborate argument. Recall that for Raabe the *k*th Bernoullische Function was

$$C_k(x) = \frac{1}{k}(B_k(x) - B_k),$$

implying that while

$$B_{2k+1}(x) = (2k+1)C_{2k+1}(x),$$

he had

$$B_{2k}(x) = 2kC_{2k}(x) + B_{2k},$$

leading him to a more involved argument for *k* even.


## 2.11  Ramanujan's Recurrence Relations for Bernoulli Numbers

Jakob Bernoulli's recurrence relation (2.27) is not too practical to use in computational situations; to find $B_{2n}$ we require the values of $B_2, B_4, \ldots, B_{2n-2}$. Taking Bernoulli's recurrence relation as having a gap of 2, Ramanujan's relations have gaps of $4, 6, 8, 10, 12, 14$ and these are clearly much more efficient for computations. For example, given a formula with gaps of 6 with $B_2$ known, one can immediately obtain $B_8$ and then $B_{14}$, and so on. We give one such example from Ramanujan's paper, "Some properties of Bernoulli numbers." Bruce Berndt, editor of Ramanujan's notebooks, has commented on this paper: "It is fitting that Ramanujan's first paper is on Bernoulli numbers, for he clearly loved them. They permeate much of the work in his notebooks."[37] Ramanujan stated his result:[38]

Suppose *n* is an odd integer; then

$$\binom{n}{3}|B_{n-3}| + \binom{n}{9}|B_{n-9}| + \binom{n}{15}|B_{n-15}| + \cdots = 0, \qquad (2.62)$$

where the constant term is

$$(-1)^{\frac{n-1}{6}}\frac{n}{6}, \quad (-1)^{\frac{n+1}{6}}\frac{n}{3}, \quad \text{or} \quad (-1)^{\frac{n-3}{6}}\frac{n-3}{3},$$

depending on whether *n* is of the form $6k + 1$, $6k + 5$, or $6k + 3$.

Observe here that since

$$\cot x = \frac{i(e^{2ix} + 1)}{e^{2ix} - 1}, \qquad (2.63)$$

---

[37] Ramanujan (2000) p. 357.
[38] Ramanujan (1911) p. 222.

(2.34) implies that we can state

$$x \cot x = 1 + \sum_{k=1}^{\infty} (-1)^k B_{2k} \frac{(2x)^{2k}}{(2k)!}. \tag{2.64}$$

We will prove (2.62) later in this section. For now, observe that the series (2.64) has only even powers with odd powers missing. Thus, we have a series with gaps of 2 between the powers of $x$. In order to obtain relations among Bernoulli numbers with gaps of 6, we must use (2.64) to construct series with further gaps of 3 among the even powers of $x$. This method for constructing such series was published in 1759 by the self-taught English mathematician Thomas Simpson. We discuss this general method in Section 13.3 and here describe two particular cases of this method using square roots and cube roots of unity.

Suppose

$$f(x) = a_0 + a_1 x + a_2 x^2 + a_3 x^3 + \cdots.$$

Then

$$f(-x) = a_0 - a_1 x + a_2 x^2 - a_3 x^3 + \cdots,$$

$$\frac{1}{2}(f(x) + f(-x)) = a_0 + a_2 x^2 + a_4 x^4 + \cdots$$

and

$$\frac{1}{2}(f(x) - f(-x)) = a_1 x + a_3 x^3 + a_5 x^5 + \cdots.$$

Thus, Simpson's suggested method has allowed us, from a given series, to produce two series with gaps of 2 by using the square roots $\pm 1$ of unity. In a similar manner, we can use cube roots of unity to produce gaps of 3. Denote the cube roots of unity by $1, \omega, \omega^2$, where $\omega = \frac{-1+\sqrt{-3}}{2}$ and note that

$$1 + \omega + \omega^2 = 0 \quad \text{and} \quad \omega^3 = 1. \tag{2.65}$$

Next, let us consider how to produce gaps of 6 from (2.64). First, to avoid the alternating signs in (2.64), we take the absolute values of $B_{2k}$ and use $(-1)^{k-1} B_{2k} = |B_{2k}|$ to obtain

$$-\frac{x}{2} \cot \frac{x}{2} = -1 + |B_2| \frac{x^2}{2!} + |B_4| \frac{x^4}{4!} + |B_6| \frac{x^6}{6!} + \cdots. \tag{2.66}$$

Changing $x$ to $\omega x$ and then $x$ to $\omega^2 x$ in (2.66) we arrive at the two equations

$$\frac{-\omega x}{2} \cot \frac{\omega x}{2} = -1 + \omega^2 |B_2| \frac{x^2}{2!} + \omega |B_4| \frac{x^4}{4!} + |B_6| \frac{x^6}{6!} + \cdots \tag{2.67}$$

$$-\frac{\omega^2 x}{2} \cot \frac{\omega^2 x}{2} = -1 + \omega |B_2| \frac{x^2}{2!} + \omega^2 |B_4| \frac{x^4}{4!} + |B_6| \frac{x^6}{6!} + \cdots . \qquad (2.68)$$

Upon adding (2.66), (2.67), and (2.68), while applying (2.65), we find that

$$-\frac{x}{2} \left( \cot \frac{x}{2} + \omega \cot \frac{\omega x}{2} + \omega^2 \cot \frac{\omega^2 x}{2} \right) = 3 \left( -1 + |B_6| \frac{x^6}{6!} + |B_{12}| \frac{x^{12}}{12!} + \cdots \right).$$
$$(2.69)$$

If we multiply (2.66), (2.67), (2.68) by $1, \omega, \omega^2$ respectively and then add the results, we arrive at

$$-\frac{x}{2} \left( \cot \frac{x}{2} + \omega^2 \cot \frac{\omega x}{2} + \omega \cot \frac{\omega^2 x}{2} \right)$$
$$= 3 \left( |B_2| \frac{x^2}{2!} + |B_8| \frac{x^8}{8!} + |B_{14}| \frac{x^{14}}{14!} + \cdots \right). \qquad (2.70)$$

Again, multiplying (2.66), (2.67), (2.68) by $2, 2\omega^2, 2\omega$ respectively, and then adding the resulting equations, we obtain

$$-x \left( \cot \frac{x}{2} + \cot \frac{\omega x}{2} + \cot \frac{\omega^2 x}{2} \right) = 6 \left( |B_4| \frac{x^4}{4!} + |B_{10}| \frac{x^{10}}{10!} + |B_{16}| \frac{x^{16}}{16!} + \cdots \right).$$
$$(2.71)$$

Ramanujan discovered recurrence relations for $|B_{2k}|$ with gaps of 6 by finding other expressions for the left-hand sides of (2.69), (2.70), and (2.71).[39] He observed that

If $1, \omega, \omega^2$ be the three cube roots of unity, then

$$4 \sin x \sin \omega x \sin \omega^2 x = -(\sin 2x + \sin 2\omega x + \sin 2\omega^2 x), \qquad (2.72)$$

as may be easily verified.

One way to verify (2.72) is to make use of the two trigonometric identities that follow from the addition formula for the sine and cosine functions:

$$2 \sin A \sin B = \cos(A - B) - \cos(A + B), \qquad (2.73)$$
$$2 \sin A \cos B = \sin(A + B) + \sin(A - B). \qquad (2.74)$$

Now by (2.73) and (2.65)

$$2 \sin x \sin \omega x = \cos(1 - \omega)x - \cos(1 + \omega)x = \cos(1 - \omega)x - \cos \omega^2 x;$$

multiply by $2 \sin \omega^2 x$ and apply (2.74) to obtain

<hr />

[39] Ramanujan (1911) pp. 221–222.

$$4 \sin x \, \sin \omega x \, \sin \omega^2 x = 2 \sin \omega^2 x \, \cos(1 - \omega)x - 2 \sin \omega^2 x \, \cos \omega^2 x$$
$$= \sin(1 - \omega + \omega^2)x + \sin(\omega + \omega^2 - 1)x - \sin 2\omega^2 x - \sin 0$$
$$= - \sin 2\omega x - \sin 2x - \sin 2\omega^2 x.$$

This completes the proof of (2.72); Ramanujan next took its logarithmic derivative to get

$$\cot x + \omega \cot \omega x + \omega^2 \cot x \omega^2 = \frac{2(\cos 2x + \omega \cos 2x\omega + \omega^2 \cos 2x\omega^2)}{\sin 2x + \sin 2\omega x + \sin 2\omega^2 x}.$$

Then, writing $\frac{x}{2}$ for $x$ and multiplying by $\frac{x}{2}$, he found another expression for the left-hand side of (2.69):

$$-\frac{x}{2}\left(\cot \frac{x}{2} + \omega \cot \frac{\omega x}{2} + \omega^2 \cot \frac{\omega^2 x}{2}\right) = -\frac{x(\cos x + \omega \cos \omega x + \omega^2 \cos \omega^2 x)}{\sin x + \sin \omega x + \sin \omega^2 x}. \tag{2.75}$$

Applying the power series expansions of $\sin x$ and $\cos x$, given in chapter 1, he could express the right-hand side of (2.62) as a quotient of two power series with gaps of 6. Combining these with (2.69), Ramanujan arrived at

$$3\left(-1 + |B_6| \frac{x^6}{6!} + |B_{12}| \frac{x^{12}}{12!} + \cdots\right) = \frac{\frac{x^2}{2!} - \frac{x^8}{8!} + \frac{x^{14}}{14!} - \cdots}{\frac{x^3}{3!} - \frac{x^9}{9!} + \frac{x^{15}}{15!} - \cdots}. \tag{2.76}$$

To obtain a similar formula for (2.70), he noted that

$$\cot \frac{\omega x}{2} - \cot \frac{\omega^2 x}{2} = \frac{\cos \omega^2 x - \cos \omega x}{2 \sin \frac{x}{2} \sin \frac{\omega x}{2} \sin \frac{\omega^2 x}{2}} = \frac{2(\cos \omega x - \cos \omega^2 x)}{\sin x + \sin \omega x + \sin \omega^2 x}. \tag{2.77}$$

For the next step, Ramanujan multiplied (2.77) by $-\frac{x}{2}(\omega^2 - \omega)$ and added this result to (2.75), arriving at

$$-\frac{x}{2}\left(\cot \frac{x}{2} + \omega^2 \cot \frac{\omega x}{2} + \omega \cot \frac{\omega^2 x}{2}\right) = \frac{-x(\cos x + \omega^2 \cos \omega x + \omega \cos \omega^2 x)}{\sin x + \sin \omega x + \sin \omega^2 x}. \tag{2.78}$$

This led to the relation

$$3\left(|B_2| \frac{x^2}{2!} + |B_8| \frac{x^8}{8!} + |B_{14}| \frac{x^{14}}{14!} + \cdots\right) = x \frac{\frac{x^4}{4!} - \frac{x^{10}}{10!} + \frac{x^{16}}{16!} - \cdots}{\frac{x^3}{3!} - \frac{x^9}{9!} + \frac{x^{15}}{15!} - \cdots}. \tag{2.79}$$

Ramanujan then wrote,

Similarly,

$$-x\left(\cot \frac{x}{2} + \cot \frac{\omega x}{2} + \cot \frac{\omega^2 x}{2}\right) = \frac{x(\cos x + \cos \omega x + \cos \omega^2 x - 3)}{\sin x + \sin \omega x + \sin \omega^2 x} \tag{2.80}$$

and therefore

$$6\left(|B_4|\frac{x^4}{4!} + |B_{10}|\frac{x^{10}}{10!} + |B_{16}|\frac{x^{16}}{16!} + \cdots\right) = x\frac{\frac{x^6}{6!} - \frac{x^{12}}{12!} + \frac{x^{18}}{18!} - \cdots}{\frac{x^3}{3!} - \frac{x^9}{9!} + \frac{x^{15}}{15!} - \cdots}. \qquad (2.81)$$

To prove this as Ramanujan did, write the left-hand side of (2.80) as

$$\frac{-x\left(\cos\frac{x}{2}\sin\frac{\omega x}{2}\sin\frac{\omega^2 x}{2} + \cos\frac{\omega x}{2}\sin\frac{\omega}{2}\sin\frac{\omega^2 x}{2} + \cos\frac{\omega^2 x}{2}\sin\frac{x}{2}\sin\frac{\omega x}{2}\right)}{\sin\frac{x}{2}\sin\frac{\omega x}{2}\sin\frac{\omega^2}{2}}. \qquad (2.82)$$

Now apply the identities (2.73) and (2.74) to the first term in the numerator of (2.82); obtain the other two terms in the numerator by changing $x$ to $\omega x$ and $x$ to $\omega^2 x$ respectively. For the denominator, use (2.72).

Ramanujan multiplied each of the equations (2.76), (2.79), (2.81) by the denominator on the right-hand side of each and then equated the coefficients of $x^n$, arriving at the result stated in (2.62). He then used this recurrence relation to calculate the absolute values of the Bernoulli numbers up through $B_{40}$; we show how he found $B_6, B_{12}, B_{18}$ in this manner. He employed the case for which the constant term was $(-1)^{\frac{n-3}{6}}\frac{n-3}{3}$. Now, since $B_6$ is positive, $n = 9$ yields

$$\binom{9}{3}B_6 - 2 = 0 \quad \text{or} \quad B_6 = \frac{2}{84} = \frac{1}{42}.$$

With $n = 15$, we get

$$-\binom{15}{3}B_{12} - \binom{15}{9}B_6 + 4 = 0;$$

divide by $\binom{15}{3}$ to find that

$$B_{12} = -11\,B_6 + \frac{4}{455} = -\frac{11}{42} + \frac{4}{455} = -\frac{691}{2730}.$$

After taking $n = 21$ and dividing the resulting equation by $\binom{21}{3}$, we come to

$$B_{18} + 221\,B_{12} + \frac{204}{5}B_6 = \frac{3}{665} \quad \text{or} \quad B_{18} = \frac{43867}{798}.$$

As Wagstaff's paper[40] has pointed out, many of Ramanujan's results on Bernoulli numbers were anticipated.

---

[40] Wagstaff (1981).

## 2.12   Notes on the Literature

Seki's collected works were first published in 1974 in Japanese. However, the editors very helpfully added an English summary of Seki's main achievements, including the results on Bernoulli numbers. In her 2006 translation of Bernoulli's *Ars Conjectandi,* Edith Sylla has given a preface and an excellent 126-page introduction.

# 3

# Infinite Product of Wallis

## 3.1 Preliminary Remarks

In 1655, John Wallis produced the following very important infinite product:

$$\frac{4}{\pi} = \frac{3}{2} \cdot \frac{3}{4} \cdot \frac{5}{4} \cdot \frac{5}{6} \cdots . \tag{3.1}$$

This result appeared in his *Arithmetica Infinitorum*, published in 1656.[1] The passage of 350 years has not diminished the beauty and significance of Wallis's result, the culmination of a series of remarkable mathematical insights and audacious guesses; his book exercised great influence on the early mathematical work of Newton and Euler. We note that in 1593 François Viète gave the only earlier example of an infinite product, a calculation of the value of $\pi$ by inscribing regular polygons in a circle.[2] His formula can be written as

$$\frac{2}{\pi} = \frac{\sqrt{2}}{2} \cdot \frac{\sqrt{2 + \sqrt{2}}}{2} \cdot \frac{\sqrt{2 + \sqrt{2 + \sqrt{2}}}}{2} \cdots .$$

John Wallis (1616–1703) apparently received little mathematical training at school or at Emmanuel College, Cambridge. He taught himself elementary arithmetic from textbooks belonging to his younger brother, who was going into a trade. It was only during the English Civil War (1642–1648) that Wallis's mathematical inclinations began to be evident as he decoded letters for Parliament. The code operated by replacing letters with numerical values. Wallis gained a feeling for numerical relationships through this experience, and he applied it to his mathematical researches for the *Arithmetica Infinitorum*. In fact, the manner in which he presented and analyzed the mathematical data in his book is reminiscent of the way in which he decoded messages.

It was probably around 1646 that Wallis began delving more deeply into mathematics, by studying the famous *Clavis Mathematicae* by William Oughtred (1574–1660),

---

[1] Wallis (1656).
[2] Viète (1593) p. 30, second leaf.

inventor of the slide rule. First published in 1631 and composed for the instruction of the son of the Earl of Arundel, this book was widely studied and exerted a tremendous influence on seventeenth-century English mathematics. A second edition in English and then in Latin appeared in 1647 and 1648. The second edition was among the first mathematical texts studied by Newton as a student in 1664. In the 1690s Newton recommended that the book be reprinted for the new generation of students of mathematics. The *Clavis* introduced Wallis to algebraic notation and to the method of applying algebra to geometric problems in the manner developed by Viète in the 1590s.

In 1649, Wallis was appointed to the Savilian Chair of Geometry at Oxford. The valuable service Wallis had provided to the winning side in the Civil War helped him attain this post. The Savilian Chair, endowed by Sir Henry Savile in 1619 to promote development of mathematics in England, was the second endowed mathematics chair in England; the first was founded in 1597 at Gresham College, London. With the rapid advancement of mathematics after 1550, it had become clear that university instruction in mathematics was essential, especially since this subject was proving useful in navigation and military matters. In fact, Italy and France had already established a number of mathematics professorships.

At the time of his appointment, Wallis knew little more than the contents of the *Clavis*. But the professorship gave him access to the Savile Library with its fine collection of mathematics books. Wallis was most influenced by Frans van Schooten's 1649 Latin translation of Descartes's *La Géométrie* and Evangelista Torricelli's *Opera Geometrica* of 1644. Although Oughtred and Viète had employed algebra in the study of geometry, Descartes took the process to a higher level by reducing the study of curves to algebraic equations by means of coordinate systems. At around the same time, Pierre Fermat (1607–1666) also made this major step, but his expositions on this and other topics were unfortunately published only posthumously. Wallis's first book, *De Sectionibus Conicis*, written in 1652 and published in 1656, was clearly inspired by Descartes. He obtained properties of conic sections algebraically, making extensive use of the symbolic algebra developed by Harriot and Descartes. Wallis defined the parabola, hyperbola, and ellipse by means of algebraic equations; he remarked that "It is no more necessary that a parabola is the section of a cone by a plane parallel to a side than that a circle is a section of a cone by a plane parallel to the base, or that a triangle is a section through the vertex."[3]

Wallis learned of Bonaventura Cavalieri's method of indivisibles from Torricelli; Wallis regarded his own *Arithmetica Infinitorum* as a continuation of Cavalieri, an accurate assessment. Wallis spent a fair amount of his book computing the area under $y = x^m$ when $m$ was a positive integer, using an arithmetical approach, as contrasted with Torricelli's geometrical method. Wallis then extended the result to the case $m = \frac{1}{n}$ where $n$ was a positive integer, by observing that the curve $y = x^{\frac{1}{n}}$ was identical to $x = y^n$ when seen from the $y$-axis. Now when the area under $y = x^n$ on the interval $(0, 1)$ was added to the area under $x = y^n$, taken on the same interval on

---

[3] Wallis and Stedall (2004) p. xiii.

the $y$-axis, the result was a square of area 1. But since Wallis had already found the area under $y = x^n$ to be $\frac{1}{n+1}$, the area under $y = x^{\frac{1}{n}}$ turned out to be[4]

$$1 - \frac{1}{n+1} = \frac{n}{n+1} = \frac{1}{\frac{1}{n}+1}.$$
(3.2)

Wallis then jumped to the conclusion that the area under $y = x^{\frac{m}{n}}$ over the unit interval, where $m$ and $n$ were positive integers, was[5]

$$\frac{1}{\frac{m}{n}+1}.$$
(3.3)

In the *Arithmetica*, Wallis's aim was to obtain the arithmetical quadrature of the circle. In modern term this means that he wished to evaluate the integral $\int_0^1 (1-x^2)^{\frac{1}{2}}dx$ using numerical calculations. Since (3.3) gave the value of $\int_0^1 x^{\frac{m}{n}} dx$, Wallis's plan of attack was to compute $\int_0^1 (1 - x^{\frac{1}{p}})^q dx$ for positive integer values of $p$ and $q$ and then interpolate the values of the integral for fractional $p$ and $q$. The area of the quarter circle was obtained when $p = q = \frac{1}{2}$. To compute $\int_0^1 (1 - x^{\frac{1}{p}})^q dx$ for integer $q$, Wallis expanded the integrand and integrated term by term. For example, for $q = 3$ one has (in modern notation)

$$\int_0^1 (1-x^{\frac{1}{p}})^3 dx = \int_0^1 (1 - 3x^{\frac{1}{p}} + 3x^{\frac{2}{p}} - x^{\frac{3}{p}})dx = 1 - \frac{3}{\frac{1}{p}+1} + \frac{3}{\frac{2}{p}+1} - \frac{1}{\frac{3}{p}+1}.$$

In proposition 131, he tabulated thirty-six values of these integrals (or areas) for $1 \le p, q \le 6$, of which we present the reciprocals:

|   | 2 | 3 | 4 | 5 | 6 | 7 |
|---|---|---|---|---|---|---|
|   | 3 | 6 | 10 | 15 | 21 | 28 |
|   | 4 | 10 | 20 | 35 | 56 | 84 |
|   | 5 | 15 | 35 | 70 | 126 | 210 |
|   | 6 | 21 | 56 | 126 | 252 | 462 |
|   | 7 | 28 | 84 | 210 | 462 | 924 |

Here the rows are given by $p$ and the columns by $q$. Wallis observed that these were figurate numbers. For example, the second row/column consisted of triangular numbers, the third row/column of pyramidal numbers, and so on. It was already known (though Wallis may have rediscovered this) that these numbers could be expressed as ratios of two products. Thus, as discussed in our Section 2.1, the numbers in the $p$th row were given by

$$\frac{(q+1)(q+2)\cdots(q+p)}{p!}.$$

---

[4] ibid. propositions 54–57.
[5] ibid. proposition 59.

Therefore, if

$$w(p,q) = \frac{1}{\int_0^1 (1 - x^{\frac{1}{p}})^q dx},$$

then Wallis had

$$w(p,q) = \frac{(q + 1)(q + 2) \cdots (q + p)}{p!}. \tag{3.4}$$

Wallis then assumed that the formula continued to hold when $q$ was a half integer. Of course, $p$ could not be taken to be a half integer since neither the denominator nor the numerator would have meaning in that case. However, for $p = \frac{1}{2}$ the integral would be $\int_0^1 (1 - x^2)^q dx$; this could be easily computed when $q$ was an integer. So Wallis had a row corresponding to $p = \frac{1}{2}$, and in proposition 168 he got the values of $w(\frac{1}{2}, q)$ for $q = 0, 1, 2, 3, \ldots$ as

$$1, \frac{3}{2} = \frac{\frac{1}{2} + 1}{1!}, \frac{15}{8} = \frac{(\frac{1}{2} + 1)(\frac{1}{2} + 2)}{2!}, \frac{105}{48} = \frac{(\frac{1}{2} + 1)(\frac{1}{2} + 2)(\frac{1}{2} + 3)}{3!}, \ldots$$

To find $w(\frac{1}{2}, q)$ when $q$ was a half integer, he observed that (3.4) implied (in our notation)

$$w(p, q + 1) = w(p, q) \frac{p + q + 1}{q + 1}. \tag{3.5}$$

From this relation, he could get the value of $w(\frac{1}{2}, \frac{1}{2} + n)$, for integer $n$, in terms of $w(\frac{1}{2}, \frac{1}{2})$. So if $A$ denoted $w(\frac{1}{2}, \frac{1}{2}) = \frac{4}{\pi}$, then proposition 189 stated that the row corresponding to $p = \frac{1}{2}$ and $q = -\frac{1}{2}, 0, \frac{1}{2}, 1, \frac{3}{2}, 2, \frac{5}{2}, \cdots$ would be

$$\frac{1}{2}A, \quad 1, \quad A, \quad \frac{3}{2}, \quad \frac{4}{3}A, \quad \frac{3 \times 5}{2 \times 4}, \quad \frac{4 \times 6}{3 \times 5}A, \quad \frac{3 \times 5 \times 7}{2 \times 4 \times 6}, \quad \cdots. \tag{3.6}$$

Wallis understood that (3.5) provided the rule for forming the subsequence of the first, third, fifth, ... terms and the subsequence of the second, fourth, sixth, ... terms, but he was initially unable to see how the two sequences were related. Wallis's research was stalled at this stage in the spring of 1652. He consulted a number of his mathematical friends at Oxford including Christopher Wren, the famous architect, but none could help him. Three years later, he informed Oughtred of the progress he had made and where he was still stymied, ending his letter[6] with the request, "wherein if you can do me the favour to help me out; it will be a very great satisfaction to me, and (if I do not delude myself) of more use than at the first view it may seem to be." Apparently, Oughtred could provide no assistance and eventually in the spring of 1655, Wallis requested help from Brouncker, who sent back an infinite continued fraction to solve the problem. It is likely that Brouncker's solution inspired Wallis

---

[6] Wallis and Stedall (2004) p. xviii. For the full letter, see Rigaud (1841) vol. I, pp. 85–86.

to discover his own very different one, though some speculate that Wallis made his discovery independently.

William Brouncker (c. 1620–1684) may have studied at Oxford around 1636, though he told his friend John Aubrey that he was "of no university."[7] However, Brouncker was very proficient in languages as well as mathematics. He did all his surviving mathematical work in association with Wallis, with the exception of his series for $\ln 2$. In addition to the continued fraction for $\pi$, he wrote a short piece on the rectification of the semicubical parabola $y = x^{\frac{3}{2}}$, probably after seeing William Neil's work. He also gave a method for solving Fermat's problem of finding integer solutions of $x^2 - Ny^2 = 1$ for a given positive integer $N$. This solution can also be described in terms of continued fractions, but when Wallis wrote up Brouncker's method, he did not use that form. A letter of 1669 from Collins to James Gregory,[8] suggests that Brouncker found the series for $(1 - x^2)^{\frac{1}{2}}$ independently of Newton. Indeed, Charles II chose Brouncker as the inaugural President of the Royal Society, a post he held from 1662 to 1677. The Society's *Philosophical Transactions* was founded during his tenure; the April 1668 issue contained his proof of the formula[9]

$$\ln 2 = \frac{1}{1 \cdot 2} + \frac{1}{3 \cdot 4} + \frac{1}{5 \cdot 6} + \cdots . \tag{3.7}$$

Brouncker provided no explanation of how he obtained his very intriguing result on the continued fraction for $\pi$ and in his book, Wallis presented only a sketch of Brouncker's argument. In the course of this discussion, Wallis included a short account of a few fundamental results on continued fractions, including the recurrence relations satisfied by the numerators and denominators of the successive convergents of a continued fraction. Brouncker's result, as well as Wallis's exposition of it, suggests connections between continued fractions and series, products, integrals, and rational approximations. It is surprising to note that, although Huygens and Cotes gave isolated results, no mathematician before Euler made a systematic study of continued fractions. Wallis's book had a tremendous impact on Euler who, at the age of 22, used it as his starting point for his theory of gamma and beta functions. At about the same time, Euler began his investigations into continued fractions, as indicated by a 1731 letter from Euler to his friend Goldbach.[10] He explained how he had applied continued fractions to solve a Riccati equation. Shortly after that, he began researching the relation between continued fractions and infinite series, infinite products, and integrals. It is a remarkable fact that when Euler chanced upon a mathematical avenue or by-path, such as those suggested by Wallis, he explored it with vigor and almost always found numerous results of interest and value.

---

[7] Stedall (2000) p. 295.
[8] Turnbull (1939).
[9] Brouncker (1668).
[10] Fuss (1968) pp. 56–59.

## 3.2   Wallis's Infinite Product for π

Although he did not give an explicit definition, the concept of a logarithmically convex
sequence is crucial for understanding Wallis's derivation of the product for $\pi$. Wallis,
Newton, and Euler all made use of this idea. A sequence of positive numbers $\{a_n\}$ is
called logarithmically convex if

$$\ln a_n \le \frac{1}{2}(\ln a_{n-1} + \ln a_{n+1}), \quad n = 1, 2, 3, \ldots \qquad (3.8)$$

or

$$a_n^2 \le a_{n-1}a_{n+1}, \quad n = 1, 2, 3, \ldots. \qquad (3.9)$$

Now a sequence $\{a_n\}$ is logarithmically concave if

$$a_n^2 \ge a_{n-1}a_{n+1}, \quad n = 1, 2, 3, \ldots. \qquad (3.10)$$

In addition, a positive function $f$ on an interval $(a,b)$ is called logarithmically
convex if $f$ is continuous and if for every pair of points $x_1, x_2 \in (a,b)$

$$\ln f\left(\frac{x_1 + x_2}{2}\right) \le \frac{1}{2}(\ln f(x_1) + \ln f(x_2)).$$

Wallis, we may recall, was searching for a rule capable of describing (3.6) in some
form. He eventually arrived at the deep insight that the sequence of the reciprocals
was logarithmically convex. This allowed him to express the first term of the sequence
as an infinite product. To reach his insight, Wallis first denoted the numbers in the
sequence (3.6) by the letters $\alpha$, $a$, $\beta$, $b$, $\gamma$, $c$, $\delta$, $d$ etc. He observed in proposition 191
that the ratios

$$\frac{\beta}{\alpha} = \frac{2}{1}, \frac{b}{a} = \frac{3}{2}, \frac{\gamma}{\beta} = \frac{4}{3}, \frac{c}{b} = \frac{5}{4}, \frac{\delta}{\gamma} = \frac{6}{5}, \frac{d}{c} = \frac{7}{6}, \text{etc.}$$

were decreasing. He then assumed the same for the ratios $\frac{a}{\alpha}, \frac{\beta}{a}, \frac{b}{\beta}, \frac{\gamma}{b}$, etc. This meant
that $a^2 > \alpha\beta, \beta^2 > ab, b^2 > \beta\gamma$, and so on. So, if we denote three consecutive
members of (3.6) by $a_{n-1}$, $a_n$, $a_{n+1}$, we must have

$$a_n^2 > a_{n-1}a_{n+1}, \ a_0 = \frac{1}{2}A. \qquad (3.11)$$

Since this indicates logarithmic concavity, the reciprocals must be logarithmically
convex. Wallis wrote down the first few of these inequalities explicitly. Thus, $c^2 > \gamma\delta$
and $\delta^2 > cd$ gave him

$$A < \frac{3 \times 3 \times 5 \times 5}{2 \times 4 \times 4 \times 6}\sqrt{\frac{6}{5}},$$

$$A > \frac{3 \times 3 \times 5 \times 5}{2 \times 4 \times 4 \times 6} \sqrt{\frac{7}{6}}.$$

In general, we can write these inequalities as

$$\frac{3 \times 3 \times \cdots \times 2n - 1 \times 2n - 1}{2 \times 4 \times \cdots \times 2n - 2 \times 2n} \sqrt{\frac{2n + 1}{2n}} < A$$

$$< \frac{3 \times 3 \times \cdots \times 2n - 1 \times 2n - 1}{2 \times 4 \times \cdots \times 2n - 2 \times 2n} \sqrt{\frac{2n}{2n - 1}}.$$

Studying the pattern evident in only the first few cases of these two inequalities, Wallis concluded that

$$A = \frac{4}{\pi} = \frac{3 \times 3 \times 5 \times 5 \times 7 \times 7 \times \cdots}{2 \times 4 \times 4 \times 6 \times 6 \times 8 \times \cdots}. \tag{3.12}$$

Newton studied Wallis as a student in the winter of 1664–65 and made notes in a notebook now held by the University Library, Cambridge. Here Newton observed[11] that Wallis's proof of (3.12) could be simplified, writing in his notebook, "Thus Wallis doth it, but it may bee [*sic*] done thus." He noted that the sequence (3.6) was increasing, though he did not explain why. Observe, however, that the terms of the sequence are the reciprocals of the integrals

$$\int_0^1 (1 - x^2)^m dx, \ m = -\frac{1}{2}, 0, \frac{1}{2}, 1, \ldots. \tag{3.13}$$

The integrand decreases as $m$ increases and hence so does the integral. Therefore, we see that

$$\frac{3 \times 5 \times \cdots \times 2n - 1}{2 \times 4 \times \cdots \times 2n - 2} < \frac{4 \times 6 \times \cdots \times 2n}{3 \times 5 \times \cdots \times 2n - 1} A < \frac{3 \times 5 \times \cdots \times 2n - 1 \times 2n + 1}{2 \times 4 \times \cdots \times 2n - 2 \times 2n}.$$

And these two inequalities together imply (3.12). Newton's argument certainly shortened the proof of Wallis. But Wallis's use of (3.11) gave a deep insight into the connection between interpolation of factorials and logarithmic convexity. Note that the inequality (3.11) implies the logarithmic convexity of the sequence $\frac{1}{a_n}$. This was fully understood only in the 1920s, when Bohr and Mollerup showed that logarithmic convexity was one of the defining properties of the gamma function, by which the factorial is interpolated; in this connection, see our Chapter 17. Thus, as Bourbaki also commented,[12] Wallis's methods are very similar to those used today in the theory of the gamma function. It is possible that by 1890 the Dutch mathematician T. J. Stieltjes had also gained an understanding of the significance of logarithmic convexity as it related to the gamma function.

To understand more clearly the meaning of the logarithmic convexity of the sequence in (3.13), observe that $m - \frac{1}{2}, m, m + \frac{1}{2}$ are three successive values of $m$,

---

[11] Newton (1967–1981) vol. 1, p. 103.
[12] Bourbaki (1994) p. 187.

where the least value of $m$ is 0. If we denote the integral in the equation (3.13) by $W_m$, then the logarithmic convexity of the sequence in that equation entails that

$$W_m^2 < W_{m-\frac{1}{2}} W_{m+\frac{1}{2}} \tag{3.14}$$

and this is the inequality being assumed by Wallis.

## 3.3  Brouncker and Infinite Continued Fractions

Indian mathematicians between 700 and 1500 discussed finite continued fractions.[13] We have noted that it is possible that the Kerala school also had a conception of infinite continued fractions. It seems, however, that the first explicit discussions of infinite continued fractions appeared in the works of two professors of mathematics at the University of Bologna: Rafael Bombelli (1526–1572) and Pietro Antonio Cataldi (1548–1626). In 1572, Bombelli described a method for computing $\sqrt{13}$,[14] amounting to the continued fraction expansion

$$\sqrt{13} = 3 + \frac{4}{6+} \frac{4}{6+} \cdots ;$$

observe that in this notation, the left-hand side denotes the continued fraction

$$3 + \cfrac{4}{6 + \cfrac{4}{6+\cdots}}.$$

Though Cataldi's work appeared later than that of Bombelli, he may fairly be regarded as the creator of the theory of infinite continued fractions. He explained how to expand the square root of a number in terms of fractions in such a way as to clearly show that an infinite continued fraction must result.[15] Moreover, he introduced a modern notation for continued fractions, also used by Wallis. Cataldi also gave the recurrence relations satisfied by the successive convergents of the continued fraction representation of a quadratic irrational. Finally, he showed that the convergents were successively larger and smaller than the continued fraction and that they converged to it.

Brouncker utilized continued fractions to present an ingenious solution to Wallis's longstanding problem of finding the law of formation of the sequence (3.6). He stated that the continued fraction

$$\phi(n) = n + \frac{1^2}{2n+} \frac{3^2}{2n+} \frac{5^2}{2n+} \cdots, \quad n = 0, 1, 2, 3, \ldots \tag{3.15}$$

---

[13]  See Brezinski (1991) chapter 1.

[14]  ibid. pp. 62–64 gives excerpts from the first and second editions of Bombelli's algebra book, where he described the method for finding the continued fraction for $\sqrt{13}$.

[15]  ibid. pp. 65–70.

had the two properties:[16]

$$\phi(n-1)\phi(n+1) = n^2, \quad n = 0, 1, 2, \ldots \qquad (3.16)$$

and

$$\phi(1) = \frac{4}{\pi} \equiv A. \qquad (3.17)$$

It follows from these properties that the $m$th term of the sequence (3.6), starting at 1 rather than at $\frac{A}{2}$, is given by

$$\frac{A}{2} \cdot \frac{2}{\phi(1)} \cdot \frac{4}{\phi(3)} \cdots \frac{2m}{\phi(2m-1)}, \quad m = 1, 2, 3, \ldots. \qquad (3.18)$$

If we take the empty product in (3.18) to be 1, then for $m = -1$, we also get the term $\frac{A}{2}$ in (3.6).

Wallis was able to prove (3.17) from his formula (3.12) combined with (3.16).[17] We note briefly that by (3.16),

$$\phi(1) = \frac{2^2}{\phi(3)} = \frac{2^2}{4^2}\phi(5) = \frac{2^2}{4^2} \cdot \frac{6^2}{\phi(7)} = \frac{2^2 \cdot 6^2 \cdots (4m-2)^2}{4^2 \cdot 8^2 \cdots (4m)^2}\phi(4m+1)$$

$$= \frac{1}{2} \cdot \frac{3^2 \cdot 5^2 \cdots (2m-1)^2}{2 \cdot 4^2 \cdots (2m-1)^2 2m} \cdot \frac{\phi(4m+1)}{2m}. \qquad (3.19)$$

Now by (3.15), $n < \phi(n) < n+1$, and, therefore,

$$\frac{4m+1}{2m} < \frac{\phi(4m+1)}{2m} < \frac{4m+2}{2m}.$$

If we let $m \to \infty$ in (3.19), then these inequalities and Wallis's formula imply that $\phi(1) = \frac{4}{\pi}$. Wallis did not give a complete proof of (3.16), but one may reconstruct his thought from the arguments he gave. He wrote that Brouncker had noticed that the product of two consecutive odd or even numbers was one less than a square, since $(n-1)(n+1) = n^2 - 1$. He then asked by what fraction the factors should be increased so that one obtained $n^2$ rather than $n^2 - 1$. We may say that he looked for a function $\phi(n)$ such that

$$\phi(n-1)\phi(n+1) = n^2.$$

Since $\phi(n) = n$ gives $n^2 - 1$, we take

$$\phi(n) = n + \frac{\alpha_1}{\phi_1(n)}, \qquad (3.20)$$

---

[16] See also Eu. I-14 pp. 291–349. E123, § 15.

[17] See Wallis's commentary on proposition 191 in Wallis and Stedall (2004) pp. 168–178. See also Stedall (2000) pp. 300–305.

where $\alpha_1$ is a constant to be determined.[18] Substituting in (3.16), we get

$$-\phi_1(n-1)\phi_1(n+1) + \alpha_1(n+1)\phi_1(n+1) + \alpha_1(n-1)\phi_1(n-1) + \alpha_1^2 = 0. \tag{3.21}$$

The symmetry of (3.16) is preserved if we take $\alpha_1 = 1$, for then (3.21) can be written as

$$(\phi_1(n-1) - (n+1))(\phi_1(n+1) - (n-1)) = n^2. \tag{3.22}$$

Now let

$$\phi_1(n) = 2n + \frac{\alpha_2}{\phi_2(n)}, \tag{3.23}$$

so that (3.22) simplifies to

$$-9\phi_2(n-1)\phi_2(n+1) + \alpha_2(n+3)\phi_2(n+1) + \alpha_2(n-3)\phi_2(n-1) + \alpha_2^2 = 0. \tag{3.24}$$

If we take $\alpha_2 = 3^2$, then we get an equation similar to (3.22):

$$(\phi_2(n-1) - (n+3))(\phi_2(n+1) - (n-3)) = n^2.$$

So set

$$\phi_2(n) = 2n + \frac{\alpha_3}{\phi_3(n)},$$

and it turns out that $\alpha_3 = 5^2$. One can continue in this way to get the continued fraction expansion (3.15).

Wallis's contribution to the theory of continued fractions was to note the recurrence relations for the convergents of a general continued fraction.[19] Take a continued fraction

$$C = b_0 + \frac{a_1}{b_1+} \frac{a_2}{b_2+} \cdots , \tag{3.25}$$

and set the $n$th convergent (or approximant) of the continued fraction to be

$$C \equiv \frac{P_n}{Q_n} \equiv b_0 + \frac{a_1}{b_1+} \frac{a_2}{b_2+} \cdots \frac{a_n}{b_n}, \quad n = 1,2,3,\dots \tag{3.26}$$

with $P_0 = b_0, P_{-1} = 1, Q_0 = 1$, and $Q_{-1} = 0$. Then Wallis's recurrence relations for the numerators and denominators $P_n, Q_n$ of the convergents can be written as

$$P_n = b_n P_{n-1} + a_n P_{n-2}, \tag{3.27}$$
$$Q_n = b_n Q_{n-1} + a_n Q_{n-2}. \tag{3.28}$$

[18] See Whiteside (1961b) pp. 211–212.
[19] Wallis and Stedall (2004) p. 176.

Wallis wrote the continued fraction (3.25) with $b_0 = 0$ as

$$\cfrac{a}{\alpha - \cfrac{b}{\beta - \cfrac{c}{\gamma - \cfrac{d}{\sigma - \cfrac{e}{\epsilon}}}}} \quad \text{etc.}$$

and gave the first four convergents. He stated the rules (3.27) and (3.28) in words and showed how it worked by an example. He remarked that these results allowed one to compute the convergents by starting at the beginning of the fraction rather than from the end. The twelfth-century Indian mathematician Bhaskara, in his *Lilavati* (1150), also gave the rules (3.27) and (3.28).[20] Since he considered continued fractions of only rational numbers, the value of $a_k$ was 1.

## 3.4  Méray and Stieltjes: The Probability Integral

In his 1730 work on the interpolation of the sequence of factorials,[21] Euler noted the relation between an integral and Wallis's infinite product:

$$\int_0^1 \left( \ln \frac{1}{x} \right)^{\frac{1}{2}} dx = \lim_{n \to \infty} \frac{4}{3} \cdot \frac{6}{5} \cdots \frac{2n}{2n-1} \cdot \frac{1}{\sqrt{n+1}}$$

$$= \frac{\sqrt{\pi}}{2}. \tag{3.29}$$

We remark that Euler did not use the idea of a limit; he wrote the infinite product instead. However, a change of variables and integration by parts produces the probability integral:

$$\int_0^\infty e^{-t^2} dt = \frac{\sqrt{\pi}}{2}. \tag{3.30}$$

Although Euler's 1730 paper did not contain a proof of (3.29) or (3.30), several proofs were well-known by the time Charles Méray published his 1888 paper,[22] "Valeur de l'intégrale définie $\int_0^\infty e^{-x^2} dx$ déduite de la formule de Wallis." In his paper, Méray wrote that the integral (3.30) played a considerable role in the theory of least errors and that he had presented in his paper an interesting proof with the advantage of complete rigor. His paper showed how the integral under discussion could be directly expressed in terms of Wallis's product; this may well have been a new development. Thus, his derivation is worth studying for its own sake, and because Méray's work has been somewhat overlooked by the mathematical community. His 1869 paper,[23] "Remarques sur la nature des quantités définies par la condition

---

[20] Brezinski (1991) pp. 32–33.
[21] Eu. I-14 pp. 1–24. E 19. Also see our Chapter 17.
[22] Méray (1888).
[23] Méray (1869).

de servir de limites à des variables données," gave the first published version of a theory of real numbers. Now Weierstrass had presented his theory of real numbers in his Berlin lectures in the 1860s and in 1858 Dedekind had worked out his theory, published in 1872, using Dedekind cuts,[24] although these were not published until later. And Méray's 1869 paper was ignored in France because during that period, interest in this subject seems to have been limited to Germany. Upon reading Méray's 1888 paper on the probability integral, Stieltjes produced a simplified argument that is also of interest.

Méray started with the integral

$$I_n = \int_0^\infty x^n e^{-x^2} dx \qquad (3.31)$$

with $n \geq 0$ an integer. Writing $x^n e^{-x^2} = xe^{-x^2}x^{n-1}$ and integrating by parts yielded

$$\int x^n e^{-x^2} dx = -\frac{1}{2}e^{-x^2}x^{n-1} + \frac{n-1}{2}\int x^{n-2}e^{-x^2} dx, \qquad (3.32)$$

or

$$I_n = \frac{n-1}{2}I_{n-2}. \qquad (3.33)$$

Thus when $n$ was even, say $n = 2m$,

$$I_{2m} = \frac{2m-1}{2} \cdot \frac{2m-3}{2} \cdot \cdots \cdot \frac{1}{2}I_0, \qquad (3.34)$$

and with $n$ odd, say $n = 2m + 1$,

$$I_{2m+1} = m! \, I_1 = \frac{m!}{2}. \qquad (3.35)$$

With these values in hand, Méray made the substitution in integral (3.31)

$$x = \left(\frac{n-1}{2}y\right)^{\frac{1}{2}}$$

to obtain

$$I_n = \frac{1}{2}\left(\frac{n-1}{2}\right)^{\frac{n+1}{2}} e^{-\frac{n-1}{2}} T_n \qquad (3.36)$$

where

$$T_n = \int_0^\infty (e\, y\, e^{-y})^{\frac{n-1}{2}} dy. \qquad (3.37)$$

[24] Dedekind (1872). For a translation into English, see Dedekind (1963).

He next observed that $e\,y\,e^{-y}$ was increasing for $y < 1$, decreasing for $y > 1$, and equal to 1 when $y = 1$. This implied that $T_n$ would decrease when $n$ increased; thus

$$T_{2m+1} < T_{2m} < T_{2m-1}. \tag{3.38}$$

He wrote $T_n$ in terms of $I_n$, using (3.36), so that (3.38), combined with (3.34) and (3.35) gave him

$$\left(\frac{2m}{2}\right)^{-\frac{2m+2}{2}} e^{\frac{2m}{2}} \frac{m!}{2} < \left(\frac{2m-1}{2}\right)^{-\frac{2m+1}{2}} e^{\frac{2m-1}{2}} \frac{2m-1}{2} \cdots \frac{1}{2} I_0$$

$$< \left(\frac{2m-2}{2}\right)^{-\frac{2m}{2}} e^{\frac{2m-2}{2}} \frac{(m-1)!}{2}.$$

Divide across by $m^{-m} e^m \frac{(m-1)!}{2}$ to obtain the inequalities

$$1 < \left(1 - \frac{1}{2m}\right)^{-m-\frac{1}{2}} e^{-\frac{1}{2}} \frac{1}{\sqrt{m}} \frac{(2m-1)(2m-3)\cdots 3\cdot 1}{\sqrt{m}(2m-2)\cdots 4\cdot 2} I_0$$

$$< \left(1 - \frac{1}{m}\right)^{-m} e^{-1}.$$

Méray then let $m \to \infty$, noting for example that

$$e^{-\frac{1}{2}} \lim_{m\to\infty} \left(1 - \frac{1}{2m}\right)^{-m-\frac{1}{2}} = 1,$$

and using Wallis's formula to arrive at

$$I_0^2 = \frac{1}{2} \lim_{m\to\infty} \frac{2^2 \cdot 4^2 \cdots (2m-2)^2}{3^2 \cdot 5^2 \cdots (2m-1)^2} \cdot (2m)$$

$$= \frac{1}{2} \cdot \frac{\pi}{2}.$$

Stieltjes wrote in an 1890 paper on the same topic[25] that Méray's proof became simpler when one noted that the sequence $I_n$ was logarithmically convex, a fact for which he gave a very easy proof:

He observed that for an arbitrary real number $x$,

$$I_{n+1} + 2x I_n + x^2 I_{n-1} = \int_0^\infty u^{n-1}(u + x)^2 e^{-u^2}\, du > 0,$$

equivalent to

$$(x I_{n-1} + I_n)^2 > I_n^2 - I_{n-1} I_{n+1},$$

25 Stieltjes (1890).

so that Stieltjes could conclude that

$$I_n^2 < I_{n-1}I_{n+1}. \tag{3.39}$$

To see this, simply take $x = -\frac{I_n}{I_{n-1}}$. From (3.33) and (3.39), Stieltjes found

$$I_n^2 < \frac{n}{2} I_{n-1}^2. \tag{3.40}$$

Inequalities (3.39) and (3.40) produced the two inequalities

$$I_{2k}^2 > \frac{2}{2k+1} I_{2k+1}^2 \quad \text{and} \quad I_{2k}^2 < I_{2k-1}I_{2k+1}.$$

Therefore by (3.35), Stieltjes had

$$I_{2k}^2 > \frac{(1 \cdot 2 \cdot 3 \cdots k)^2}{4k+2} \quad \text{and} \quad I_{2k}^2 < \frac{(1 \cdot 2 \cdot 3 \cdots k)^2}{4k}$$

or

$$I_{2k}^2 = \frac{(1 \cdot 2 \cdot 3 \cdots k)^2}{4k+2}(1+\epsilon), \quad 0 < \epsilon < \frac{1}{2k}.$$

At this point, Stieltjes used (3.34) to conclude that

$$2I_0^2 = \frac{(2 \cdot 4 \cdot 6 \cdots 2k)^2}{(1 \cdot 3 \cdot 5 \cdots (2k-1))^2(2k+1)}(1+\epsilon),$$

$$2I_o^2 = \frac{\pi}{2}, \quad I_0 = \frac{\sqrt{\pi}}{2}.$$

This was clearly a more direct route to Méray's result. Now Stieltjes's argument may be used to prove Wallis's conjectured inequality (3.14):

$$W_{m+\frac{1}{2}} + 2x W_m + x^2 W_{m-\frac{1}{2}} = \int_0^1 (1-t^2)^{m-\frac{1}{2}}(x^2 + 2x(1-t^2)^{\frac{1}{2}} + (1-t^2))dt$$

$$= \int_0^1 (1-t^2)^{m-\frac{1}{2}}(x + (1-t^2)^{\frac{1}{2}})^2 dt > 0,$$

implying Wallis's inequality:

$$W_m^2 < W_{m-\frac{1}{2}}W_{m+\frac{1}{2}}.$$

## 3.5 Euler: Series and Continued Fractions

Wallis's discussion of Brouncker's continued fractions convinced Euler of their importance in analysis. Quite early in his career, he found a connection with the

Riccati equation[26] and saw the necessity of relating continued fractions with series, products, and definite integrals. In this way, Euler succeeded in fleshing out the methods of which Wallis and Brouncker had only given key examples.

Euler presented his general theorems on the conversion of series to continued fractions in such a way that the $n$th partial sum of the series and the $n$th convergent of the continued fraction were identical. Euler's first paper on this topic, of 1737,[27] treated this topic somewhat briefly but the second one, of 1739,[28] was more detailed. It explicitly stated the formulas for obtaining the corresponding series starting with a given continued fraction and, conversely, for obtaining the continued fraction from the given series.

We follow Euler's approach from the first book in which he treated this topic, *Introductio in analysin infinitorum*;[29] he discusses continued fractions in his chapter 18. He wrote a continued fraction in the form

$$a + \cfrac{\alpha}{b + \cfrac{\beta}{c + \cfrac{\gamma}{d + \cfrac{\delta}{e + \text{etc.}}}}} \qquad (3.41)$$

Using subscripts to clarify Euler's expressions for the modern reader, we replace $a, b, c, d, \ldots$ by $b_0, b_1, b_2, b_3, \ldots$ and $\alpha, \beta, \gamma, \delta, \ldots$ by $a_1, a_2, a_3, a_4, \ldots$ so that in modern notation, Euler's (3.41) would be written as

$$b_0 + \frac{a_1}{b_1+} \frac{a_2}{b_2+} \frac{a_3}{b_3+} \cdots . \qquad (3.42)$$

He first observed that the successive fractions would be

$$\frac{b_0}{1},$$

$$b_0 + \frac{a_1}{b_1} = \frac{b_0 b_1 + a_1}{b_1},$$

$$b_0 + \cfrac{a_1}{b_1 + \frac{a_2}{b_2}} = \frac{b_0 b_1 b_2 + b_0 a_2 + a_1 b_2}{b_1 b_2 + a_2},$$

$$b_0 + \cfrac{a_1}{b_1 + \cfrac{a_2}{b_2 + \frac{a_3}{b_3}}} = \frac{b_0 b_1 b_2 b_3 + b_0 b_3 a_2 + b_2 b_3 a_1 + b_0 b_1 a_3 + a_1 a_3}{b_1 b_2 b_3 + b_3 a_2 + b_1 a_3},$$

$$\cdots\cdots .$$

For brevity, we may denote the successive fractions by

$$\frac{A_0}{B_0}, \ \frac{A_1}{B_1}, \ \frac{A_2}{B_2}, \ \frac{A_3}{B_3}, \ \frac{A_4}{B_4}, \cdots .$$

[26] Fuss (1969) vol. 1, pp. 56–59.
[27] Eu. I-14 pp. 187–216. E 71.
[28] Eu. I-14 pp. 291–349. E 123.
[29] Euler (1988) provides a translation into English.

Euler placed the fraction $\frac{1}{0}$ as the first fraction of his list of successive fractions so that the recurrence relations for forming the numerators and denominators would hold when $n = 1$. Thus he had:

$$\frac{1}{0}, \frac{A_0}{B_0}, \frac{A_1}{B_1}, \frac{A_2}{B_2}, \frac{A_3}{B_3}, \frac{A_4}{B_4}, \ldots \tag{3.43}$$

and he could state these rules:

$$A_n = b_n A_{n-1} + a_n A_{n-2}, \quad n = 1, 2, 3, \ldots \tag{3.44}$$
$$B_n = b_n B_{n-1} + a_n B_{n-2}, \quad n = 1, 2, 3, \ldots, \tag{3.45}$$

where $A_{-1} = 1$ and $B_{-1} = 0$. Observe that $A_0 = b_0$ and $B_0 = 1$.

To obtain the series corresponding to a given continued fraction, Euler considered the differences of the successive fractions:[30]

$$\frac{A_1}{B_1} - \frac{A_0}{B_0}, \quad \frac{A_2}{B_2} - \frac{A_1}{B_1}, \quad \ldots .$$

Using (3.44) and (3.45), he found

$$\begin{aligned}
\frac{A_n}{B_n} - \frac{A_{n-1}}{B_{n-1}} &= \frac{A_n B_{n-1} - A_{n-1} B_n}{B_n B_{n-1}} \\
&= \frac{(b_n A_{n-1} + a_n A_{n-2}) B_{n-1} - A_{n-1}(b_n B_{n-1} + a_n B_{n-2})}{B_n B_{n-1}} \\
&= \frac{a_n(A_{n-2} B_{n-1} - A_{n-1} B_{n-2})}{B_n B_{n-1}} \\
&= \frac{-a_n(A_{n-1} B_{n-2} - A_{n-2} B_{n-1})}{B_n B_{n-1}} \\
&= \frac{a_n a_{n-1}(A_{n-2} B_{n-3} - A_{n-3} B_{n-2})}{B_n B_{n-1}} \\
&= \frac{(-1)^{n-1} a_n a_{n-1} a_{n-2} \cdots a_1}{B_n B_{n-1}}.
\end{aligned}$$

Euler noted that the successive partial fractions (3.43) were given by

$$\frac{A_0}{B_0}, \quad \frac{A_0}{B_0} + \left(\frac{A_1}{B_1} - \frac{A_0}{B_0}\right) = \frac{A_1}{B_1}, \quad \frac{A_0}{B_0} + \left(\frac{A_1}{B_1} - \frac{A_0}{B_0}\right) + \left(\frac{A_2}{B_2} - \frac{A_1}{B_1}\right) = \frac{A_2}{B_2}, \quad \ldots .$$

The $n$th partial fraction would thus be given by the series

$$\begin{aligned}
\frac{A_n}{B_n} &= \frac{A_0}{B_0} + \left(\frac{A_1}{B_1} - \frac{A_0}{B_0}\right) + \cdots + \left(\frac{A_n}{B_n} - \frac{A_{n-1}}{B_{n-1}}\right) \\
&= \frac{A_0}{B_0} + \frac{a_1}{B_0 B_1} - \frac{a_1 a_2}{B_1 B_2} + \frac{a_1 a_2 a_3}{B_2 B_3} + \cdots + (-1)^{n-1} \frac{a_1 a_2 \cdots a_n}{B_{n-1} B_n}.
\end{aligned}$$

---

[30] Euler (1988) pp. 306–308.

Following Euler, consider the case $A_0 = b_0 = 0$, so that $\frac{A_n}{B_n}$ could be expressed as an alternating series:

$$\frac{A_n}{B_n} = \frac{a_1}{B_0 B_1} - \frac{a_1 a_2}{B_1 B_2} + \frac{a_1 a_2 a_3}{B_2 B_3} - \cdots + (-1)^{n-1} \frac{a_1 a_2 \cdots a_n}{B_{n-1} B_n}.$$

Now, assuming the existence of $\lim_{n \to \infty} \frac{A_n}{B_n}$, he had the infinite continued fraction (3.42), $b_0 = 0$, expressed as a series:

$$\frac{a_1}{B_0 B_1} - \frac{a_1 a_2}{B_1 B_2} + \frac{a_1 a_2 a_3}{B_2 B_3} - \cdots . \tag{3.46}$$

Euler next showed how to obtain a continued fraction from a given series;[31] denote the series by

$$C_1 - C_2 + C_3 - C_4 + C_5 - \cdots . \tag{3.47}$$

Comparing (3.47) with (3.46), he found that

$$C_1 = \frac{a_1}{B_0 B_1}, \quad C_2 = \frac{a_1 a_2}{B_1 B_2}, \quad C_3 = \frac{a_1 a_2 a_3}{B_2 B_3}, \quad \ldots, \quad C_n = \frac{a_1 a_2 \cdots a_n}{B_{n-1} B_n}. \tag{3.48}$$

Euler observed that (3.48) gave him

$$\frac{C_n}{C_{n-1}} = \frac{a_n B_{n-2}}{B_n}, \quad n = 2, 3, 4, \ldots ; \tag{3.49}$$

subtracting each side of (3.49) from (3.44) and then using (3.45), he obtained

$$C_{n-1} - C_n = \frac{C_{n-1}(B_n - a_n B_{n-2})}{B_n} = \frac{C_{n-1} B_{n-1} b_n}{B_n}, \quad n = 2, 3, 4, \ldots . \tag{3.50}$$

Taking the product of the differences in (3.50) yielded

$$(C_{n-1} - C_n)(C_n - C_{n+1}) = \frac{C_{n-1} B_{n-1} b_n}{B_n} \cdot \frac{C_n B_n b_{n+1}}{B_{n+1}}$$

$$= \frac{C_{n-1} C_n B_{n-1} b_n b_{n+1}}{B_{n+1}}, \quad n = 2, 3, 4, \ldots .$$

Thus

$$\frac{B_{n+1}}{B_{n-1}} = \frac{C_{n-1} C_n b_n b_{n+1}}{(C_{n-1} - C_n)(C_n - C_{n+1})}, \quad n = 2, 3, 4, \ldots . \tag{3.51}$$

Finally, from (3.48), (3.49), and (3.51) he concluded

$$a_1 = C_1 b_1, \quad a_2 = \frac{C_2 b_1 b_2}{C_1 - C_2},$$

[31] ibid. pp. 308–313.

$$a_3 = \frac{C_1 C_3 b_2 b_3}{(C_1 - C_2)(C_2 - C_3)}, \quad a_4 = \frac{C_2 C_4 b_3 b_4}{(C_2 - C_3)(C_3 - C_4)}, \quad \dots \qquad (3.52)$$

Euler then observed that, since the numerators of the continued fraction, that is $a_1, a_2, a_3, \dots$, were known, the values of the denominators $b_1, b_2, b_3, \dots$ could be arbitrarily chosen. That is, if $C_1, C_2, C_3, \dots$ were integers, then $b_1, b_2, b_3, \dots$ could be so selected that $a_1, a_2, a_3, \dots$ would turn out to be integers. Thus, taking the values of $b_1, b_2, b_3, b_4, \dots$ to be $1, C_1 - C_2, C_2 - C_3, C_3 - C_4, \dots$ respectively, then

$$a_1 = C_1, \quad a_2 = C_2, \quad a_3 = C_1 C_3, \quad a_4 = C_2 C_4, \dots.$$

The continued fraction corresponding to (3.47) would then be, in modern notation:

$$\frac{C_1}{1+} \ \frac{C_2}{C_1 - C_2+} \ \frac{C_1 C_3}{C_2 - C_3+} \ \frac{C_2 C_4}{C_3 - C_4+} \ \dots \ .$$

On the other hand, Euler observed, if the series were

$$\frac{1}{C_1} - \frac{1}{C_2} + \frac{1}{C_3} - \frac{1}{C_4} + \cdots , \qquad (3.53)$$

then the equations in (3.52) could be written as

$$a_1 = \frac{b_1}{C_1}, \quad a_2 = \frac{C_1 b_1 b_2}{C_2 - C_1}, \quad a_3 = \frac{C_2^2 b_2 b_3}{(C_2 - C_1)(C_3 - C_2)},$$

$$a_4 = \frac{C_3^2 b_3 b_4}{(C_3 - C_2)(C_4 - C_3)}, \quad \dots \ .$$

He could then take

$$b_1 = C_1, \quad b_2 = C_2 - C_1, \quad b_3 = C_3 - C_2, \quad b_4 = C_4 - C_3, \quad \cdots$$

to see that

$$a_1 = 1, \quad a_2 = C_1^2, \quad a_3 = C_2^2, \quad a_4 = C_3^2 \quad \cdots$$

Series (3.53) could thus be converted into the continued fraction

$$\frac{1}{C_1+} \ \frac{C_1^2}{C_2 - C_1+} \ \frac{C_2^2}{C_3 - C_2+} \ \frac{C_3^2}{C_4 - C_3+} \ \dots \ . \qquad (3.54)$$

As an example of the series in (3.53), Euler considered

$$\log 2 = \int_0^1 \frac{1}{1+x} \, dx = \int_0^1 (1 - x + x^2 - x^3 + \cdots) \, dx$$

$$= 1 - \frac{1}{2} + \frac{1}{3} - \frac{1}{4} + \cdots .$$

Since $C_1 = 1, C_2 = 2, C_3 = 3, C_4 = 4$, the corresponding continued fraction was

$$\log 2 = \frac{1}{1+} \frac{1}{1+} \frac{4}{1+} \frac{9}{1+} \frac{16}{1+} \cdots.$$

Then again,

$$\frac{\pi}{4} = \int_0^1 \frac{1}{1+x^2} dx = 1 - \frac{1}{3} + \frac{1}{5} - \frac{1}{7} + \cdots,$$

so that $C_1 = 1, C_2 = 3, C_3 = 5, C_4 = 7, \ldots$ and

$$\frac{\pi}{4} = \frac{1}{1+} \frac{1}{2+} \frac{9}{2+} \frac{25}{2+} \frac{49}{2+} \cdots;$$

note that Brouncker discovered the reciprocal of this result.

More generally, Euler observed that

$$\int_0^1 \frac{x^{n-1}}{1+x^m} dx = \int_0^1 x^{n-1}(1 - x^m + x^{2m} + \cdots) \, dx$$

$$= \frac{1}{n} - \frac{1}{m+n} + \frac{1}{2m+n} - \frac{1}{3m+n} + \cdots$$

$$= \frac{1}{n+} \frac{n^2}{m+} \frac{(m+n)^2}{m+} \frac{(2m+n)^2}{m+} \frac{(3m+n)^2}{m+} \cdots, \qquad (3.55)$$

where the last step follows from (3.54).


## 3.6  Euler: Riccati's Equation and Continued Fractions

Euler found continued fractions for $e$, its square and cube roots, and other related numbers. In his first paper on the topic,[32] "De Fractionibus Continuis Dissertatio" written in 1737 and published in 1744, he explained that he had initially found these expansions by studying the patterns in the continued fractions for the rational approximations of these numbers. It was only later that he attempted to prove the results. In the process, he discovered a connection with the Riccati equation and he employed this to establish his formulas.[33] It is interesting that Euler gave the main theorem of this paper in his 1731 letter to Goldbach.[34] For $e$ he had the expansion

$$e = 2 + \frac{1}{1+} \frac{1}{2+} \frac{1}{1+} \frac{1}{1+} \frac{1}{4+} \frac{1}{1+} \frac{1}{1+} \frac{1}{6+} \cdots, \qquad (3.56)$$

obtained by taking the approximation $e = 2.71828182845904$ and applying the division algorithm. Cotes had earlier given this expansion by applying the same

[32] Eu. 1-14 pp. 187–216. E 71, § 21–22.
[33] See our Section 14.9.
[34] Fuss (1968) pp. 57–59.

procedure.[35] He used the continued fraction (3.56) to obtain rational approximations for $e$, noting that the successive convergents were alternately bigger or smaller than $e$. To find a continued fraction for $\pi$, take the approximation $\pi \approx 3.1416$. Observe that

$$.1416 = \frac{1416}{100000} = \frac{177}{1250} = \frac{1}{\frac{1250}{177}} = \frac{1}{7 + \frac{11}{177}}$$

$$= \frac{1}{7 + \frac{1}{\frac{177}{11}}} = \frac{1}{7 + \frac{1}{16 + \frac{1}{11}}}$$

$$= \frac{1}{7+} \frac{1}{16+} \frac{1}{11}.$$

Similar to Cotes's method, Euler took $\sqrt{e} = 1.6487212707$ and found

$$\sqrt{e} = 1 + \frac{1}{1+} \frac{1}{1+} \frac{1}{1+} \frac{1}{5+} \frac{1}{1+} \frac{1}{1+} \frac{1}{9+} \frac{1}{1+} \frac{1}{1+} \frac{1}{13+} \cdots . \tag{3.57}$$

Then again

$$\frac{e^{\frac{1}{3}} - 1}{2} = 0.1978062125 = \frac{1}{5+} \frac{1}{18+} \frac{1}{30+} \frac{1}{42+} \frac{1}{54+} \cdots \tag{3.58}$$

and

$$\frac{e+1}{e-1} = 2 + \frac{1}{6+} \frac{1}{10+} \frac{1}{14+} \frac{1}{18+} \frac{1}{22+} \frac{1}{26+} \cdots . \tag{3.59}$$

He observed that in (3.56) and (3.57), the arithmetic progressions of the denominators $2, 4, 6, \ldots$ and $1, 5, 9, 13, \ldots$ were interrupted by consecutive 1's, whereas in (3.58) and (3.59) they were not. He showed how to convert the interrupted progressions into non-interrupted progressions. When he applied this procedure to (3.56) and (3.57), he got

$$e = 2 + \frac{1}{1+} \frac{2}{5+} \frac{1}{10+} \frac{1}{14+} \frac{1}{18+} \frac{1}{22+} \frac{1}{26+} \cdots \tag{3.60}$$

and

$$\sqrt{e} = 1 + \frac{2}{3+} \frac{1}{12+} \frac{1}{20+} \frac{1}{28+} \cdots . \tag{3.61}$$

Euler then noted that he had not really proved any of these expansions and that it was only probable that the arithmetic progressions continued in the manner indicated. He wrote that after some exertion he had found a rigorous though peculiar proof that related the problem to differential equations. He stated without proof the theorem[36] that if

---

[35] Cotes (1714) p. 11. An English translation of this paper is available in appendix 1 of Gowing (1983).
[36] See E 71 § 28.

$$q = \cfrac{1}{p+} \cfrac{1}{\frac{3}{p}+} \cfrac{1}{\frac{5}{p}+} \cdots \cfrac{1}{\frac{2n-1}{p}+} \cfrac{1}{\frac{1}{x^{\frac{2n}{2n+1}} y}}, \tag{3.62}$$

where $p = (2n+1)x^{\frac{1}{2n+1}}$, then $y$ satisfied the differential equation

$$dy + y^2 \, dx = x^{-\frac{4n}{2n+1}} \, dx. \tag{3.63}$$

Euler's expression for $q$ also contained a parameter $a$ but this can be taken to be equal to 1 without loss of generality.

It is possible to give an inductive proof of this theorem and it is very likely that Euler had discovered that argument. Note that when $n = 0$ and when $n = 1$, (3.62) takes the form

$$q = y \quad \text{and} \quad q = \cfrac{1}{p+} \cfrac{1}{\frac{1}{x^{\frac{2}{3}} y}}. \tag{3.64}$$

The corresponding differential equations would be

$$dy + y^2 dx = dx, \tag{3.65}$$

$$dy + y^2 dx = x^{-\frac{4}{3}} dx. \tag{3.66}$$

In this way, the solution of (3.66) required the solution of (3.65). More generally, the solution of the Riccati equation (3.63) depended on that of (3.65). However, Euler easily solved (3.65) by observing that it was equivalent to

$$\frac{dy}{1-y^2} = dx \quad \text{or} \quad \frac{1}{2} \ln \frac{1+y}{1-y} = x.$$

Since $x = p$ and $y = q$ for $n = 0$, Euler wrote the solution as

$$q = \frac{e^{2p}+1}{e^{2p}-1}, \tag{3.67}$$

or, in modern terms, $q = \coth p$. Euler observed that when $n$ was an infinite number in (3.62), then

$$q = \cfrac{1}{p+} \cfrac{1}{\frac{3}{p}+} \cfrac{1}{\frac{5}{p}+} \cfrac{1}{\frac{7}{p}+} \cdots. \tag{3.68}$$

The result (3.68) is now called Lambert's continued fraction, although Euler found it earlier. Now, since $e^{2p} = 1 + \frac{2}{q-1}$, Euler saw that

$$e^{2p} = 1 + \cfrac{2}{\frac{1-p}{p}+} \cfrac{1}{\frac{3}{p}+} \cfrac{1}{\frac{5}{p}+} \cfrac{1}{\frac{7}{p}+} \cdots,$$

or

$$e^{\frac{1}{s}} = 1 + \cfrac{2}{2s-1+}\ \cfrac{1}{6s+}\ \cfrac{1}{10s+}\ \cfrac{1}{14s+}\cdots. \tag{3.69}$$

He then noted that (3.69) would in fact produce all those continued fractions he had obtained experimentally by using rational approximations.

## 3.7 Exercises

(1) Prove that

$$s + \cfrac{1}{s+}\ \cfrac{4}{s+}\ \cfrac{9}{s+}\ \cfrac{16}{s+}\cdots = \left(2\int_0^1 \frac{x^s dx}{1+x^2}\right)^{-1}.$$

See Eu. I-14 pp. 292–297.

(2) Show that

$$a + \cfrac{1}{m+}\ \cfrac{1}{n+}\ \cfrac{1}{b+}\ \cfrac{1}{m+}\ \cfrac{1}{n+}\ \cfrac{1}{c+}\ \cfrac{1}{m+}\ \cfrac{1}{n+}\ \cfrac{1}{d+}\cdots$$

$$= \frac{1}{mn+1}\left((mn+1)a + n + \cfrac{1}{(mn+1)b+m+n+}\ \cfrac{1}{(mn+1)c+m+n+}\cdots\right).$$

See Euler (1985) p. 313, and Eu. I-14 p. 205.

(3) Show that

$$a + \cfrac{1}{m+}\ \cfrac{1}{n+}\ \cfrac{1}{p+}\ \cfrac{1}{q+}\ \cfrac{1}{b+}\ \cfrac{1}{m+}\cdots$$

$$= \frac{1}{p}\left(P + npq + n + q + \cfrac{1}{Pb+Q+}\ \cfrac{1}{Pc+Q+}\ \cfrac{1}{Pd+q+}\cdots\right),$$

where $P = mnpq + mn + mq + pq$ and $Q = mnp + npq + m + n + p + q$.
See Euler (1985) p. 318, and Eu. I-14 p. 208.

(4) Show that

$$\sqrt{2} = 1 + \cfrac{1}{25+}\ \cfrac{1}{2+}\ \cfrac{1}{2+}\cdots$$

$$\sqrt{3} = 1 + \cfrac{1}{1+}\ \cfrac{1}{2+}\ \cfrac{1}{1+}\ \cfrac{1}{2+}\cdots.$$

See Euler (1985) pp. 307–308 and Eu. I-14 p. 200.

(5) Show that if $x = a + \cfrac{1}{b+}\ \cfrac{1}{b+}\cdots$, then $x = a - \frac{b}{2} + \sqrt{1 + \frac{b^2}{4}}$. See Euler (1985) p. 308, and Eu. I-14 p. 201.

(6) Show that if $x = a + \dfrac{1}{a_1+} \dfrac{1}{a_2+} \cdots \dfrac{1}{a_n+} \dfrac{1}{a_1+} \cdots$, that is, if $x$ is a periodic continued fraction, then $x$ satisfies a quadratic equation. See Eu. I-14 p. 203. Euler stated the result in words, as opposed to symbolically.

(7) Show that

$$\tan \frac{\pi x}{4} = \frac{x}{1+} \frac{1 - x^2}{2+} \frac{3 - x^2}{2+} \frac{5 - x^2}{2+} \cdots.$$

See Stieltjes's letter to Hermite of March 4, 1891, in Baillaud and Bourget (1905) p. 157.

## 3.8   Notes on the Literature

The introduction to Stedall's excellent English translation of Wallis's 1656 *Arithmetica Infinitorum*, Wallis and Stedall (2004), discusses the evolution of Wallis's ideas and the influence of his book on his contemporaries and mathematical heirs. The article by Stedall in Grattan-Guinness (2005) may also be helpful, especially for its insight into how Wallis's work influenced Newton. The fruitful collaboration of Wallis and Brouncker is also the subject of two interesting notes by Stedall (2000). In his Cambridge thesis, Whiteside (1961b) reconstructed Wallis's attempt to recreate the continued fraction formula communicated to him without proof by Brouncker. This thesis is an informative and perceptive resource on seventeenth-century mathematics. Brezinski (1991) is a very useful book; it contains excerpts from original works accompanied by interesting historical commentary.

Surprisingly, a translation into English of Euler's *De Fractionibus Continuis, Dissertatio* appeared in the applied mathematics journal *Mathematical Systems Theory* (1985); the editors requested this translation, since they thought Euler's discussion of Riccati's equation could be useful to their readers. Khrushchev (2008) contains an English translation of Euler's *De Fractionibus Continuis, Observationes*. Khrushchev gives a systematic and well-organized summary of the work on continued fractions by Wallis, Brouncker, Huygens, the Bernoullis, Euler, Lagrange, Gauss, Chebyshev, Stieltjes, and others. Khrushchev illustrates the process by which the ideas of earlier researchers in continued fractions have evolved into important modern theories, such as that of orthogonal polynomials.

# 4

---

# *The Binomial Theorem*

## 4.1 Preliminary Remarks

The discovery of the binomial theorem for general exponents exerted a tremendous impact on the development of analysis, especially the theory of power series. It also led to an understanding that an exponential function was defined by the property $f(a + b) = f(a)f(b)$. The binomial theorem was pivotal not only in the initial discovery of series for other important functions but also in the eventual consolidation of the foundations of analysis as a whole. The development of the theorem is particularly fascinating because it was independently found by both Newton and Gregory; because of the various approaches to its proof, including one by Euler; and because the validation of these proofs elicited the efforts of the best mathematicians of the nineteenth century.

The binomial theorem for a positive integer exponent $n$ states that

$$(a + b)^n = a^n + A_1^n a^{n-1} b + A_2^n a^{n-2} b^2 + \cdots + A_{n-1}^n a b^{n-1} + b^n, \qquad (4.1)$$

where the coefficients $A_k^n$ satisfy the additive rule

$$A_k^n = A_{k-1}^{n-1} + A_k^{n-1}, \qquad (4.2)$$

and the multiplicative rule

$$A_k^n = \frac{n(n-1)\cdots(n-k+1)}{1 \cdot 2 \cdots k}, \qquad (4.3)$$

where it is understood that $A_0^n = 1$. We here use a notation unusual today, because the notation $\binom{n}{k}$ or $C_k^n$, or $C_{n,k}$, may be suggestive of recent developments, whereas we wish to understand how these coefficients developed over time. Now we note that the additive rule (4.2) is not difficult to obtain. In terms of the notation used in (4.1), we can write

77

$$(a + b)^{n-1} = a^{n-1} + A_1^{n-1} a^{n-2} b + \cdots + A_{k-1}^{n-1} a^{n-k} b^{k-1}$$
$$+ A_k^{n-1} a^{n-k-1} b^k + \cdots + b^{n-1}.$$

Multiplying both sides by $a + b$, we have

$$(a + b)\left(a^{n-1} + \cdots + A_{k-1}^{n-1} a^{n-k} b^{k-1} + A_k^{n-1} a^{n-k-1} b^k + \cdots + b^{n-1}\right)$$
$$= a^n + \cdots + A_k^n a^{n-k} b^k + \cdots + b^n.$$

Equating the coefficients of $a^{n-k} b^k$ on each side, we obtain (4.2). The multiplicative rule is somewhat more difficult to obtain. Observe that

$$(a + b)^n = (a + b)(a + b) \cdots (a + b),$$

where there are $n$ factors $(a+b)$. To find the coefficient of $a^{n-k} b^k$, note that for $a^{n-k} b^k$ we must take $b$ from $k$ of the factors $a + b$; the remaining $n - k$ factors $a + b$ contribute $n - k$ of the $a$'s. Thus, we see that the coefficient of $a^{n-k} b^k$ represents the number of ways $k$ $b$'s can be chosen from $n$ factors $a + b$. This number is given by the right-hand side of (4.3).

The binomial theorem has a complicated history. Some of its components can be traced back to the third or second century BCE, to Pingala's Sanskrit *Chandas sutra*[1] (also *Chhandas sutra*) or *Prosody aphorisms*. Pingala most probably lived in the third century BCE. In the eighth and final chapter of his work, he dealt with the construction of tables (or *prastāra*) of all possible sequences of $n$ syllables, each syllable being either short (*laghu* = l) or long (*guru* = g). We call a sequence of $n$ syllables a meter.

Pingala's rule for the construction of a table (or *prastāra*) of all possible meters of $n$ syllables was to place $n$ g's, or $n$ long syllables, in the first row. After $k$ rows, the $k+1$th row would be constructed by entering g's in each position until the first g appeared directly above in the preceding row, when $l$ (representing a short syllable) would be entered. The $k + 1$th row would then be completed by making entries identical with the ones directly above. We present a table of the possible combinations of long and short syllables for meters of three syllables, and a corresponding table in which we set $g = 0$ and $l = 1$.

| g g g | 0 0 0 |
|-------|-------|
| l g g | 1 0 0 |
| g l g | 0 1 0 |
| l l g | 1 1 0 |
| g g l | 0 0 1 |
| l g l | 1 0 1 |
| g l l | 0 1 1 |
| l l l | 1 1 1. |

---

[1] Sridharan (2005) especially pp. 47–59.

Pingala gave a rule for the number of the row in which a given syllable sequence, a given meter, would be found. He also gave a rule predicting the exact $n$-syllable sequence to be found in a given row. If $g$ is set to be 0 and $l$ to be 1, as in our table for words of three syllables, then Pingala's rules amount to writing the numbers in binary notation. Thus, the fifth row reads $g$ $g$ $l$ because

$$5 - 1 = 4 = 0 \cdot 1 + 0 \cdot 2 + 1 \cdot 2^2$$

and conversely.

Pingala proceeded to raise and answer questions on combinations: how many meters of a given length have one long syllable, two long syllables, and so on? What is the total number of meters of a given length? Although the sutras containing the answers to such questions appear to us somewhat obscure, later prosodists have elaborated on these, clarifying them for us. In his tenth-century *Mritasanjivani*, a commentary on Pingala's *Chandas sutra*, Halāyudha explains the answer to the question concerning the number of long syllables contained in meters of a given length:[2]

> Draw a square. Beginning at half of the square, draw two other similar squares below it, below the two, three other squares, and so on. By putting one in the first square, the marking should be started. In the two squares of the second line, put 1 in each. In the third line put 1 in the two squares at the ends and in the middle square the sum of the digits in the two squares lying above it. In the fourth line put one in the two squares at the ends. In the middle ones put the sum total of the digits in the two squares above each. Proceed on in this way. Of these the second line gives the combinations with one syllable.... The third line gives the combinations with two syllables and etc.

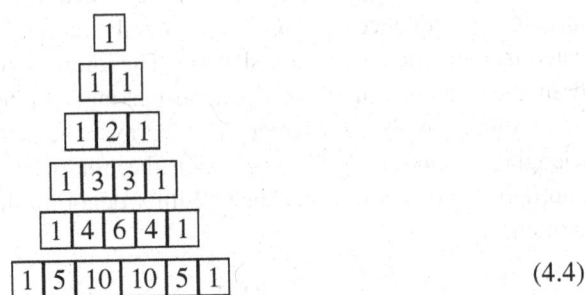

$$(4.4)$$

Thus, (4.4) gives six lines of the *Meru Prastāra*, the Sanskrit name for Pascal's triangle. Using this table, Pingala could determine, for instance, that the number of meters of length $n$ was $2^n$. The sixth-century mathematician Varāhamihira also clearly described the procedure for constructing this triangle. Observe that the method for constructing the *Meru Prāstara* yields the additive rule for binomial coefficients. Another commentator from the tenth-century, Bhattopala, gave the additive rule and also the multiplicative rule:[3] "Putting down (the figures [numbers]) once in the reverse

[2] Chakravarti (1932) p. 83.
[3] ibid. p. 85.

order, put them below again in the direct order. (In finding the final result) multiply the numbers in the process, i.e., from left to right and divide by the corresponding numbers below." Thus, the rule specifies that we write the numbers as

$$n \quad (n-1) \quad (n-2) \quad \cdots \quad 1$$
$$1 \quad 2 \quad 3 \quad \cdots \quad n$$

so, to find the number of meters of $n$ syllables and with 3 *guru* or long syllables, one would write

$$n \cdot (n-1) \cdot (n-2) \div 1 \cdot 2 \cdot 3.$$

Since each row of Pascal's triangle is created by adding two consecutive numbers from the previous row with just one from the first row, we can deduce that Pingala understood the additive property of the binomial coefficients. Before Bhattopala, the multiplicative property was presented by Mahavira around 850 A.D. and may have been known by Indian mathematicians before that, since Mahavira was heir to a thousand-year line of Indians of the Jaina tradition researching combinatorial questions.[4]

In 628, the Indian mathematician Brahmagupta had explicitly stated the binomial formula (4.1) for $n = 3$ and immediately applied this to find the cube root of a given number[5]. It appears that Brahmagupta would have been able to write down the formula for higher values of $n$, based on Pingala's rule, and certainly Mahavira could have worked out the formula in general.

In chapter 13 of his 1356 work, *Ganita-Kaumudi*, Narayana Pandita gave the binomial theorem. The topic of this chapter was combinatorial problems and Narayana begins the chapter thus: "For the pleasure of mathematicians, [I] now describe briefly *aṅka-pāśa* [sequences of numbers or combinatorics] where bad, wicked and intoxicated mathematicians' vanity shatters. The knowledge of *aṅka-pāśa* is very useful in dramatics, prosody, medicine, garland-making, architecture, and mathematics."[6]

In sutras 36–39 of chapter 13,[7] Narayana extends the idea of *meru* (Pascal's triangle) to *sumeru*, a table in which the binomial coefficients were multiplied by appropriate powers of a constant. With $s$ denoting this constant, a *sumeru* could be written as

$$
\begin{array}{lllll}
1 \\
s & 1 \\
s^2 & 2s & 1 \\
s^3 & 3s^2 & 3s & 1 \\
s^4 & 4s^3 & 6s^2 & 4s & 1.
\end{array}
\tag{4.5}
$$

[4] See Datta (1929).
[5] Brahmagupta (1817) p. 279.
[6] Narayana Pandita (2001) p. 23.
[7] ibid. p. 33.

Thus, adding the quantities in a given row of (4.5) would yield an expression for a power of $(s + 1)$, with the first row giving the 0th power, the second row giving first power, and so on.

After describing this construction, Narayana specified that this method for the formation of *sumeru* had been given by the learned mathematicians. In sutra 67,[8] Narayana wrote that the sums of the rows of the *sumeru* were in geometric progression. Thus, in our notation, the sum of the rows would be $(s + 1)^n$, $n = 0, 1, 2, 3, 4 \ldots$ and in this way, Narayana stated the binomial theorem.

Now in the 1261 work of Yang Hui, *A detailed analysis of mathematical methods in the nine chapters and their reclassifications*,[9] Pascal's triangle is presented up through the sixth row, or by counting the zeroth row as the first, up through the seventh row. However, Yang Hui explained that his Pascal's triangle had appeared earlier,[10] in the work of the eleventh-century mathematician Jia Xian, and that Jia Xian had applied it to calculate the roots of numbers up through the fifth root by a method he had devised,[11] called "the method for extracting roots by iterated multiplication." This method could be applied not only to equations of the form $x^n = N$, but also to general polynomial equations $f(x) = 0$. Jia Xian's method is akin to Horner's method for finding approximate solutions of algebraic equations. Jia Xian's works, unfortunately, appear to have been lost and his contributions are now known through the attributions of Yang Hui. Clearly, the extraction of the fourth and fifth roots would require the binomial theorem for $n = 4$ and $n = 5$; thus, Jia Xian appears to have been aware of the binomial theorem. In addition, Pascal's triangle through the eighth row may be found in Zhu Shijie's 1303 book, *Siyuan Yujian* or *Jade Mirror of the Four Unknowns*.[12]

The algebraist al-Karaji (953–1029) apparently lived in Baghdad during his most productive period.[13] In his book *Al-bahir*, al-Samawal (c. 1130–1180) attributed the additive law of binomial coefficients as well as the expansion of $(a + b)^n$ to al-Karaji. In addition, al-Samawal showed how the expansion for $(a + b)^2$ implied the expansion for $(a + b)^3$, which in turn implied that of $(a + b)^4$. The argument he used was a type of induction also used by Euler and Lagrange; al-Samawal attributed the discovery of this type of argument to al-Karaji. In his surviving work *al-Fakhre*, al-Karaji gave the expansion for $(a + b)^3$ and in his *al-Badi*, he presented expansions for $(a - b)^3$ and $(a + b)^4$. Al-Kashi, who died in 1429, gave Pascal's triangle up through the ninth power; he was aware of the additive and multiplicative rules for the binomial coefficients.

In fact, in a lost work of around 1100, the noted poet Omar Khayyam apparently presented a method for finding the fourth, fifth, and higher roots of a given number.[14] He wrote, "I have composed a book demonstrating the soundness of these methods

---

[8] ibid. p. 56.
[9] Li and Du (1987) p. 110.
[10] ibid. p. 122.
[11] ibid. pp. 117–121.
[12] Hoe (2007).
[13] Rashed (1970).
[14] Yadegari (1980) p. 402.

leading to the discovery of required values and I have added methods for the solution of various other types.... I refer to the extraction of the sides of the square of a square the square of a cube, and the cube of a cube, etc. ... all of which is new. These proofs are arithmetical." These results would certainly suggest knowledge of the binomial theorem.

Al-Zanjani (d. 1262) gave a method for finding $(a + b + c + \cdots)^n$:[15]

> We have concerned ourselves with the expression consisting of two terms because those which consist of three, four, or more terms are nothing but special cases of the two terms. Don't you see that if you want to find the cube of a three term expression you combine two of them into one? That is, combine the first two and raise it to the third power. Also raise the third term itself into a cube. Multiply the third term by the square of the sum of the first two thrice. Then multiply the sum of the first two by the first two by the third thrice. The sum is the final answer to the original one. Follow the procedure for all the other powers.

More than a hundred years before Pascal wrote his 1654 treatise on his triangle, Pascal's triangle was surely known in Europe. In 1527, the German mathematician Petrus Apianus published his *Arithmetic*, and on its title page he gave Pascal's triangle up through the ninth power. This triangle apparently became a part of received knowledge after the Italian mathematician Niccolò Tartaglia wrote his *General Trattato* of 1556. Nevertheless, it certainly appears that Newton was not familiar with the binomial theorem for positive integral exponents at the time he discovered his theorem of rational exponents.

Newton discovered the general binomial theorem in the winter of 1664–65,[16] while he was still a student at Cambridge. He was motivated by this discovery to develop his "method of infinite series" and apply it to several important problems. Indeed, the binomial theorem played a basic role in his approach to such topics as algebraic equations in two variables and differential equations. James Gregory independently found this theorem between 1668 and 1670, and it formed an important part of his original work on infinite series.[17]

Newton discussed particular cases of his theorem in two papers written in 1669 and 1671. However, the first explicit statement of the general theorem for rational exponents appeared in a letter from Newton to Oldenburg, dated June 13, 1676. This letter was a response to an inquiry from Leibniz, who had learned of Newton's series for arcsin $x$ and sin $x$ from the Danish mathematician Georg Mohr. Newton's letter also introduced his new notation for exponents, as he explained:[18]

> These are the foundation of these reductions: but extractions of roots are much shortened by this theorem,

$$(P + PQ)^{\frac{m}{n}} = P^{\frac{m}{n}} + \frac{m}{n}AQ + \frac{m-n}{2n}BQ + \frac{m-2n}{3n}CQ + \frac{m-3n}{4n}DQ + \text{etc.}$$

> where $P + PQ$ signifies the quantity whose root or even any power, or the root of a power, is to be found: $P$ signifies the first term of that quantity, $Q$ the remaining terms divided by the

---

[15] ibid. p. 404.
[16] Newton (1967–1981) vol. 1, pp. 104–108.
[17] Turnbull (1939) p. 131.
[18] Newton (1959–1960) vol. 2, pp. 32 and 42.

first, and $\frac{m}{n}$ the numerical index of the power of $P + PQ$, whether that power is integral or (so to speak) fractional, whether positive or negative. For as analysts, instead of $aa, aaa$, etc., are accustomed to write $a^2, a^3$, etc., so instead of $\sqrt{a}, \sqrt{a^3}, \sqrt{c} : a^5$, etc. I write $a^{\frac{1}{2}}, a^{\frac{3}{2}}, a^{\frac{5}{3}}$, and instead of $\frac{1}{a}, \frac{1}{aa}, \frac{1}{a^3}$, I write $a^{-1}, a^{-2}, a^{-3}$. And so for

$$\frac{aa}{\sqrt{c} : (a^3 + bbx)}$$

I write $aa(a^3 + bbx)^{-\frac{1}{3}}$, and for $\frac{aab}{\sqrt{c:(a^3+bbx)(a^3+bbx)}}$ I write $aab(a^3 + bbx)^{-\frac{2}{3}} \cdots$.

In Newton's formula, $A$ denotes the first term, $B$ the second term, and so on, such notation being common at that time. Note also that $\sqrt{c} : x$ stands for the cube root of $x$.

Intrigued by Newton's groundbreaking work, Leibniz responded with some of his own discoveries on series and requested details about the origin and derivation of Newton's results, especially the binomial theorem.[19] Newton wrote a lengthy reply amounting, to nineteen printed pages in his letter of October 24, 1676,[20] again through Oldenburg. Newton explained that in 1664–1665, he was inspired by Wallis's *Arithmetica Infinitorum* to consider the integral $\int_0^x (1 - t^2)^{\frac{1}{2}} \, dt$ and to expand the integrand. He looked at the absolute values of the coefficients of the polynomials

$$(1 - x^2)^0 = 1, \ (1 - x^2)^1 = 1 - x^2, \ (1 - x^2)^2 = 1 - 2x^2 + x^4,$$

$$(1 - x^2)^3 = 1 - 3x^2 + 3x^4 - x^6, \ (1 - x^2)^4 = 1 - 4x^2 + 6x^4 - 4x^6 + x^8, \ldots$$

and asked how the (absolute) values of the first two coefficients of any of these polynomials could produce the remaining coefficients:[21]

found that on putting $m$ for the second figure [coefficient], the rest could be produced by a continual multiplication of the terms of this series,

$$\frac{m - 0}{1} \times \frac{m - 1}{2} \times \frac{m - 2}{3} \times \frac{m - 3}{4} \times \frac{m - 4}{5}, \text{ etc.}$$

For example, let $m = 4$, and $4 \times \frac{1}{2}(m - 1)$, that is 6 will be the third term, and $6 \times \frac{1}{3}(m - 2)$, that is 4 the fourth, and $4 \times \frac{1}{4}(m - 3)$, that is 1 the fifth, and $1 \times \frac{1}{5}(m - 4)$, that is 0 is the sixth, at which term in this case the series stops. According, ..., for the circle, ..., I put $m = \frac{1}{2}$ and the terms arising were

$$\frac{1}{2} \times \frac{\frac{1}{2} - 1}{2} \text{ or } -\frac{1}{8}, \ -\frac{1}{8} \times \frac{\frac{1}{2} - 2}{3} \text{ or } +\frac{1}{16}, \ \frac{1}{16} \times \frac{\frac{1}{2} - 3}{4} \text{ or } -\frac{5}{128},$$

and so to infinity.

Thus, Newton learned how to generate the binomial series when the exponent was any number $m$ and, by taking $m = \frac{1}{2}$, he obtained the expansion

[19] ibid. pp. 57–71.
[20] ibid. pp. 110–161.
[21] ibid. pp. 130–131.

$$(1 - x^2)^{\frac{1}{2}} = 1 - \frac{1}{2}x^2 - \frac{1}{8}x^4 - \frac{1}{16}x^6 - \frac{5}{128}x^8 - \cdots$$

from which he derived the value of the integral $\int_0^x (1 - t^2)^{\frac{1}{2}} \, dt$ as an infinite series.

It is curious that Newton was unaware of the work of Briggs, Pascal, and others on the multiplicative formula for binomial coefficients. It seems that the mathematical texts Newton studied as a student did not contain the multiplicative formula. In fact, Wallis wrote in 1685 that he had not known this formula when he wrote his *Artithmetica Infinitorum*.[22] This is surprising because this work included the multiplicative expression for figurate numbers, intimately connected with binomial coefficients. In any case, Wallis's book was apparently sufficiently suggestive for Newton to make his discovery about $C_{n,k}$ for integral $n$ and then extend it to fractional $n$ by following Wallis once again. Newton attempted to verify his theorem by the interpolation methods he had learned from Wallis[23] but he soon found more satisfactory techniques, described in his letter of October 24, 1676.[24]

For in order to test these processes, I multiplied

$$1 - \frac{1}{2}x^2 - \frac{1}{8}x^4 - \frac{1}{16}x^6, \text{ etc.} \tag{4.6}$$

into itself; and it became $1 - x^2$, the remaining terms vanishing by the continuation of the series to infinity. And even so $1 - \frac{1}{3}x^2 - \frac{1}{9}x^4 - \frac{5}{81}x^6$, etc. multiplied twice into itself also produced $1 - x^2$. And as this was not only sure proof of these conclusions so too it guided me to try whether, conversely, these series, which it thus affirmed to be roots of the quantity $1 - x^2$, might not be extracted out of it in an arithmetical manner. And the matter turned out well. This was the form of the working in square roots.

$$1 - x^2(1 - \frac{1}{2}x^2 - \frac{1}{8}x^4 - \frac{1}{16}x^6, \text{ etc.}$$

$$1$$

$$\overline{0 - x^2}$$

$$-x^2 + \frac{1}{4}x^4$$

$$\overline{\phantom{-x^2+}-\frac{1}{4}x^4}$$

$$-\frac{1}{4}x^4 + \frac{1}{8}x^6 + \frac{1}{64}x^8$$

$$\overline{0 \phantom{..} \cdot -\frac{1}{8}x^6 - \frac{1}{64}x^8.}$$

After getting this clear I have quite given up the interpolation of series, and have made use of these operations only, as giving more natural foundations.

Newton realized that all the algebraic operations could be applied to infinite series and that series could be viewed as the algebraic analogs of infinite decimals. Just as

[22] Wallis (1685) pp. 318–320.
[23] Newton (1967–1981) vol. 1, pp. 106–107.
[24] Newton (1959–1960) vol. 2, pp. 131–132.

the latter appear when division and root extraction are performed on integers, infinite series result when these operations are performed on polynomials. In the preceding quote, Newton explained that when he applied the square root algorithm to $1 - x^2$, the result was the series for $(1 - x^2)^{\frac{1}{2}}$. For division, Newton gave the example of the geometric series

$$\frac{1}{d+e} = \frac{1}{d} - \frac{e}{d^2} + \frac{e^2}{d^3} - \frac{e^3}{d^4} + \text{etc.} \qquad (4.7)$$

In searching for the proof of the binomial theorem, Newton looked no further than a few cases, and he verified these by multiplication. We shall see that this method is the basis for one proof of the binomial theorem, due to Euler.[25]

James Gregory first revealed his discovery of the binomial theorem in a letter to his longtime correspondent, John Collins. First, on March 24, 1670, Collins wrote to Gregory, mentioning some mysterious work done by Newton:[26] "Mr Newtone of Cambridge sent the following series for finding the Area of a Zone of a Circle to Mr. Dary, to compare with the said Dary's approaches, putting $R$ the radius and $B$ the parallell [*sic*] distance of a Chord from the Diameter the Area of the space or Zone betweene [*sic*] them is $= 2RB - \frac{B^3}{3R} - \frac{B^5}{20R^3} - \frac{B^7}{56R^5} - \frac{5B^9}{576R^7}$." This area is given by the integral $2 \int_0^B \sqrt{R^2 - x^2}\, dx$. We note that Newton obtained the series by expanding the integrand as a binomial series and then doing term-by-term integration.

Gregory then formulated the binomial theorem in a letter to Collins of November 30, 1670,[27] stated as the solution of a problem: Use the numbers $b, b+d$ and the values of their logarithms, $e$ and $e + c$, respectively, to find the number whose logarithm is $e + a$. Gregory wrote that the desired number was given by the series

$$b + \frac{a}{c}d + \frac{a(a-c)}{c \cdot 2c}\frac{d^2}{b} + \frac{a(a-c)(a-2c)}{c \cdot 2c \cdot 3c}\frac{d^3}{b^2} + \text{etc.} \qquad (4.8)$$

Since

$$\ln\left(b\left(1 + \frac{d}{b}\right)^{\frac{a}{c}}\right) = \ln b + \left(\frac{a}{c}\right)\left(\ln(b+d) - \ln b\right) \qquad (4.9)$$

$$= e + \left(\frac{a}{c}\right)(e + c - e) = e + a,$$

we see that the series is the binomial expansion of $b\left(a + \frac{d}{b}\right)^{\frac{a}{c}}$. In this letter, Gregory also stated his general interpolation formula; the manner in which he stated this formula and the binomial theorem suggests that the latter was derived from the former.

In spite of the fact that he had already found the binomial theorem, it was not until December 1670 that Gregory could perceive the origin of Newton's series. Gregory explained in a letter of December 19 to Collins, that he had derived numerous

[25] Eu. I-15 pp. 207–216. E 465.
[26] Turnbull (1939) p. 89.
[27] ibid. pp. 118–137.

series for the circle and had mistakenly expected Newton's to be a corollary of at least one of them. He added, "I admire much my own dulness [*sic*], that in such a considerable time, I had not taken notice"[28] that Newton's series followed from a binomial expansion.

Note that the interpolation formula can be written as

$$f(x) = f(0) + x\Delta f(0) + \frac{x(x-1)}{2!}\Delta^2 f(0) + \frac{x(x-1)(x-2)}{3!}\Delta^3 f(0) + \cdots ;$$

in this connection, see Section 9.3.

To derive the binomial theorem, take

$$f(x) = b\left(1 + \frac{d}{b}\right)^x,$$

so that

$$\Delta f(x) = f(x+1) - f(x) = b\left(1 + \frac{d}{b}\right)^x \frac{d}{b},$$

$$\Delta^2 f(x) = \Delta f(x+1) - \Delta f(x) = b\left(1 + \frac{d}{b}\right)^x \frac{d^2}{b^2},$$

$$\Delta^3 f(x) = b\left(1 + \frac{d}{b}\right)^x \frac{d^3}{b^3},$$

$$\cdots\cdots\cdots .$$

Thus,

$$\Delta f(0) = d, \ \Delta^2 f(0) = \frac{d^2}{b}, \ \Delta^3 f(0) = \frac{d^3}{b^2}, \ \ldots,$$

and we get Gregory's series (4.8) by taking $x = \frac{a}{c}$ in the interpolation formula.

Interestingly, Gregory's derivation is logically more sound than Newton's original argument by a Wallis interpolation. Yet the two derivations both involve interpolation with respect to the exponent. In spite of their highly imaginative and useful work in this area, both Newton and Gregory failed to give well-founded derivations of the binomial series. Eighteenth-century mathematicians made very interesting attempts to fill this gap, but it took until the nineteenth century to find a completely rigorous derivation.

In the eighteenth century, it was generally known that the binomial expansion for $f(x) = (1+x)^\alpha$ could be obtained as the series solution of the equation

$$(1+x)\frac{dy}{dx} = \alpha y. \tag{4.10}$$

28  ibid. p. 148.

Of course, whether a series could be differentiated term by term and whether a differential equation had a series solution were not then seen as problems. The point that bothered the English mathematician John Landen (1719–1790) was that the proof used derivatives (fluxions) to obtain a result in algebra. In his 1758 *Discourse Concerning the Residual Analysis*,[29] Landen applied the algebraic identity

$$\frac{x^{\frac{m}{n}} - v^{\frac{m}{n}}}{x - v} = x^{\frac{m}{n}-1} \times \frac{1 + q + q^2 + q^3 + \cdots + q^{m-1}}{1 + q^{\frac{m}{n}} + q^{\frac{2m}{n}} + q^{\frac{3m}{n}} + \cdots + q^{(n-1)\frac{m}{n}}}, \tag{4.11}$$

where $q = \frac{v}{x}$ and $m$ and $n$ were integers, to avoid differentiation.

Euler took a different approach, presenting a proof using Newton's idea of multiplication of series.[30] He showed that if

$$f(m) = 1 + \frac{m}{1}x + \frac{m}{1} \cdot \frac{m-1}{2}x^2 + \text{etc., then}$$

$$f(m+n) = f(m) \cdot f(n). \tag{4.12}$$

His proof consisted in demonstrating that the coefficients of $x^k$ on both sides of equation (4.12) were the same. This was sufficient to derive the binomial theorem for rational exponents, except that he did not address convergence questions, particularly in the case of the product of two series. We must note that seventeenth- and eighteenth-century mathematicians had more or less clear ideas of convergence of series, but only occasionally did they apply these ideas to the series arising in their work. As examples of rigor, Grégoire St. Vincent (1584–1667) gave an entirely rigorous treatment of the geometric series in his *Opus Geometricum* of 1647. Twenty years later, Wallis discussed the logarithmic series[31] with a careful analysis of the remainder term, obtainable from the remainder in a geometric series. Leibniz gave a clear account of an alternating series in which the terms decrease to zero; he wrote to Jakob Hermann about this in a letter of June 26, 1705,[32] although he had done this work two decades earlier. Later, on January 10, 1714, in a letter to Johann Bernoulli,[33] Leibniz discussed alternating series, among other topics. In addition, he included his alternating series theorem as proposition 49 of his *De quadratura arithmetica circuli ellipseos et hyperbolae*, the first full and accurate publication of which appeared only in the twentieth century, thanks to the diligent efforts of the Leibniz scholar Eberhard Knobloch.[34]

Some gems from the eighteenth century include Stirling's 1719 criterion for convergence[35] based on second differences of the terms of a series (though it required an amendment) and Maclaurin's statement and proof of the integral test in his 1742 *Treatise of Fluxions*. Moreover, d'Alembert made some comments on the convergence

[29] Landen (1958) pp. 5–7.
[30] Eu. I-15 pp. 207–216. E 465.
[31] Wallis (1668).
[32] Leibniz (1971) vol. 4, pp. 272–275, especially p. 273.
[33] Bernoulli and Leibniz (1745) vol. 2, pp. 329–331, especially p. 330.
[34] Leibniz and Knobloch (1993) p. 115.
[35] Stirling (1719), especially pp. 1067–1070.

of series from which the ratio test can be developed.[36] Also of note is Edward Waring's 1776 statement without proof of the ratio test, now known as Raabe's test, from which Waring derived the convergence of the series $\sum_{n=1}^{\infty} \frac{1}{n^p}$;[37] in this connection, see our Section 23.6. We mention that Gauss greatly extended this ratio test in his famous work on hypergeometric series.

To extend Euler's proof to all real exponents, it was necessary to give a precise definition of continuity. Bernard Bolzano (1781–1848) and A. L. Cauchy (1789–1856) independently accomplished this. Bolzano was a professor of theology at Prague; his main interests were in philosophy and mathematics. He defined continuity in an 1817 paper[38] on the intermediate value theorem: A function $f(x)$ varies according to the law of continuity for all values of $x$ inside or outside certain limits if the difference $f(x+w) - f(x)$ can be made smaller than any given quantity, provided $w$ can be taken as small as we please. Bolzano's definition leaves little to be desired.

Cauchy emphasized rigor in analysis from the very beginning of his teaching career at the École Polytechnique. His published lectures from 1821 and 1823 discussed the concepts of limits, continuity, and convergence. His 1821 lectures *Analyse algébrique* gave his definition of continuity, not quite as good as Bolzano's: The function $f(x)$ will be a continuous function of the variable $x$ between two assigned bounds if, for each value of $x$ between these bounds, the numerical value of the differences $f(x + \alpha) - f(x)$ decrease indefinitely with $\alpha$.

Cauchy derived the continuity of the series $f(m)$ in (4.12) from the erroneous result that if every term of an infinite series is continuous and the series is convergent, then the series is continuous. In fact, in his 1826 paper on the binomial theorem,[39] Abel noted that Cauchy's theorem on the continuity of a series admits of exception. For example, in a footnote Abel wrote that

$$\sin\phi - \frac{1}{2}\sin 2\phi + \frac{1}{3}\sin 3\phi - \cdots \tag{4.13}$$

was discontinuous for every value $(2m + 1)\pi$ of $\phi$, where $m$ is a whole number. Abel then proceeded to state and prove his famous continuity theorem for power series, using the method of summation by parts. This method had been known for over a century, but Abel was the first to apply it to problems of convergence of series. Dirichlet profited from Abel's paper and used these ideas very effectively in his study of $L$-series less than a decade later. Interestingly, Abel gleaned ideas of mathematical rigor from Cauchy's lectures, obtained from his friend Crelle's library; Abel's paper appeared in Crelle's newly founded journal.

The concept of uniform convergence was implied in Abel's continuity theorem, but its explicit formulation came later. First, C. Gudermann observed in an 1838 paper[40] on modular functions, published in Crelle's journal, that he had obtained a certain series having the same convergence rate for all values of the variable. A year later,

---

[36] See Grabiner (1981) pp. 60–64.
[37] See Gonzalez-Velasco (2011) p. 391.
[38] Bolzano (1980) contains an English translation of this paper by S. B. Russ.
[39] Abel (2007) pp. 105–138, especially p. 111, footnote 3.
[40] Gudermann (1838) pp. 251–252.

K. Weierstrass was the only student in Gudermann's course on modular functions. Weierstrass introduced the term uniform convergence, understood its importance, and gave its definition in an 1841 paper "Zur Theorie der Potenzreihen,"[41] submitted as part of his examination for teaching certification. Gudermann declared, "The candidate hereby enters by birthright into the ranks of discoverers crowned with glory."[42] Unfortunately, the paper was not published until 1894. During the winter of 1859–60, Weierstrass lectured at the University of Berlin on the foundations of analysis, but it took some time before his ideas spread to other European countries and to America. In a letter of 1881 to his former student Hermann A. Schwarz, Weierstrass observed that people in France were finally grasping the importance of the idea of uniform convergence. Finally, in the last two decades of the nineteenth century, textbooks containing Weierstrassian ideas appeared in several languages. The British mathematical physicist G. G. Stokes[43] (1819–1903) and Dirichlet's student P. Seidel[44] (1821–1896), also wrote on concepts related to uniform convergence, though their papers did not have much influence.

Interestingly, Cauchy wrote a paper in 1853[45] acknowledging his mistake on continuity, noting that it was easy to rectify. He then proceeded to work with uniform convergence without naming the concept, so it is not clear whether he fully realized the wider significance of the idea.

## 4.2 Landen's Derivation of the Binomial Theorem

In 1758, John Landen presented the standard eighteenth-century derivation of the binomial theorem,[46] in which one assumes the series expansion

$$(1+x)^{\frac{m}{n}} = 1 + ax + bx^2 + cx^3 + dx^4 + \text{ etc.} \tag{4.14}$$

Now take the derivative of each side to get

$$\left(\frac{m}{n}\right)(1+x)^{\frac{m}{n}-1} = a + 2bx + 3cx^2 + 4dx^3 + \text{ etc.} \tag{4.15}$$

Multiply the last equation by $1+x$ to see that

$$\left(\frac{m}{n}\right)(1 + ax + bx^2 + cx^3 + \text{ etc.}) = (1+x)(a + 2bx + 3cx^2 + 4dx^3 + \text{ etc.}).$$

Equate coefficients to obtain

$$a = \frac{m}{n}, \quad 2b + a = \frac{m}{n}a, \quad 3c + 2b = \frac{m}{n}b, \quad 4d + 3c = \frac{m}{n}c, \ldots$$

[41] Weierstrass (1894–1927) vol. 1, pp. 67–74.
[42] Klein (1979) p. 263.
[43] Stokes (1849).
[44] Seidel (1847).
[45] Cauchy (1853).
[46] Landen (1758) pp. 5–7.

so that

$$b = \frac{\frac{m}{n}\left(\frac{m}{n}-1\right)}{2!},$$

$$c = \frac{\frac{m}{n}\left(\frac{m}{n}-1\right)\left(\frac{m}{n}-2\right)}{3!},$$

$$d = \frac{\frac{m}{n}\left(\frac{m}{n}-1\right)\left(\frac{m}{n}-2\right)\left(\frac{m}{n}-3\right)}{4!}, \ldots,$$

proving the binomial theorem. Landen also gave an alternative method, avoiding differentiation, by starting with (4.14) and applying (4.11) to get

$$\frac{(1+x)^{\frac{m}{n}} - (1+y)^{\frac{m}{n}}}{x-y} = (1+x)^{\frac{m}{n}-1} \times \frac{1 + \frac{1+y}{1+x} + \cdots + \left(\frac{1+y}{1+x}\right)^{m-1}}{1 + \left(\frac{1+y}{1+x}\right)^{\frac{m}{n}} + \cdots + \left(\frac{1+y}{1+x}\right)^{(n-1)\frac{m}{n}}}$$

$$= a + b(x+y) + c(x^2 + xy + y^2) + d(x^3 + x^2y + xy^2 + y^3) + \text{etc.}$$

He then observed that the last equation is an algebraic identity true for all values of $y$ and so that he could take $y = x$ to obtain (4.15).

Almost seven decades later, Abel objected to differentiation in this context, not because he perceived the binomial series as algebraic but, as he wrote from Berlin to his friend and former teacher Holmboe, he thought it impermissible to apply operations on infinite series as if they were finite.[47] He noted that it had not been proved that the derivative of an infinite series could be obtained by taking the derivative of each term, and that there were numerous counterexamples. For example, he observed, the sum of the series (4.13) was $\frac{\phi}{2}$ in the interval $-\pi < \phi < \pi$. Taking derivatives gave

$$\frac{1}{2} = \cos\phi - \cos 2\phi + \cos 3\phi - \text{etc.},$$

a clearly false result, because the series was divergent. In 1841, Weierstrass finally addressed Abel's concerns when he developed the theorems for differentiation and integration of series.[48]

## 4.3 Euler: Binomial Theorem for Rational Exponents

In 1773, Euler presented his paper, published in 1775, "Demonstratio theorematis Newtoniani"[49] to the Petersburg Academy, giving a proof of the binomial theorem.

[47] Abel (2007) pp. 482–487, translation by Horowitz.
[48] Weierstrass (1894–1927) vol. 1, pp. 67–74.
[49] Eu. 1-15 pp. 207–216. E 465.

In the first section of this paper, Euler wrote that this theorem stood as a foundation for the whole of higher analysis, so that a rigorous proof was clearly called for. He explained that to avoid circularity, he wished to give a demonstration not using differentiation, since he had used the binomial theorem in his differential calculus book of 1755 to show that the derivative of $x^n$ is $nx^{n-1}$.[50]

Eighteenth-century mathematicians used differentiation in some form to find the binomial expansion. They also assumed that $(1 + x)^n$, where $n$ was any real number, was expandable as an infinite series; for example, Landen had made this assumption. In this paper, Euler avoided this difficulty.

Moreover, in 1763, the German scientist Franz Aepinus published an inductive proof for positive integral exponents in the Petersburg Academy journal.[51] Euler thought that the argument, while ingenious, was quite obscure.

Euler started by observing that since $(a + b)^n = a^n \left(1 + \frac{b}{a}\right)^n$, it was sufficient to obtain the expansion of $(1 + x)^n$. He set

$$[m] = 1 + \frac{m}{1}x + \frac{m}{1} \cdot \frac{m-1}{2}x^2 + \text{etc.}, \qquad (4.16)$$

with the aim of proving that $[m] = (1 + x)^m$ when $m$ was a fraction. Note that he already knew that the result was true when $m$ was a positive integer. The important step in his proof was to show that $[m] \cdot [n] = [m + n]$. He wrote

$$[n] = 1 + \frac{n}{1}x + \frac{n}{1} \cdot \frac{n-1}{2}x^2 + \text{etc.},$$

so that

$$[m] \cdot [n] = 1 + \frac{m}{1}x + \frac{m}{1} \cdot \frac{m-1}{2}x^2 + \text{etc.}$$
$$+ \frac{n}{1}x + \frac{m}{1} \cdot \frac{n}{1}x^2 + \text{etc.}$$
$$+ \frac{n}{1} \cdot \frac{n-1}{2}x^2 + \text{etc.}$$

Thus, the product had the form $1 + Ax + Bx^2 + Cx^3 + \text{etc.}$, where

$$A = m + n, \quad B = \frac{nn - n}{2} + mn + \frac{m+n-1}{2} = \frac{m+n}{1} \cdot \frac{m+n-1}{2}.$$

Euler then observed that it was very laborious to compute $C, D, E$, etc. by this method. To see in modern terms what was involved, write the coefficient of $x^k$ in $[m]$ as $\binom{m}{k}$, that is,

$$\binom{m}{k} = \frac{m(m-1)\cdots(m-k+1)}{k!}. \qquad (4.17)$$

[50] Eu. 1-10₁, chapter 5. E 212.
[51] See Eu. I-15 p. 208, footnote 2, for a reference to Aepinus.

Thus, the coefficients are $A = \binom{m+n}{1}$, $B = \binom{m+n}{2}$. So we expect, and Euler had to show, that the coefficient of $x^k$ in $[m] \cdot [n]$, that is,

$$\binom{m}{k}\binom{n}{0} + \binom{m}{k-1}\binom{n}{1} + \cdots + \binom{m}{0}\binom{n}{k} \qquad (4.18)$$

is equal to

$$\binom{m+n}{k} = \frac{(m+n)(m+n-1)\cdots(m+n-k+1)}{k!}. \qquad (4.19)$$

He noted that it was safe to conclude that his method showed that the expressions for the coefficients $A$, $B$, $C$, etc., depended upon $m$ and $n$ but did not require $m$ and $n$ to be integers. He then observed that when $m$ was an integer,

$$[m] = \binom{m}{0} + \binom{m}{1}x + \binom{m}{2}x^2 + \cdots + \binom{m}{m}x^m = (1+x)^m.$$

Thus, when $m$ and $n$ were positive integers,

$$[m] \cdot [n] = (1+x)^m \cdot (1+x)^n = (1+x)^{m+n} = [m+n], \qquad (4.20)$$

so that the coefficient of $x^k$ was given by (4.19). Hence, Euler argued, (4.19) gave the expression for the coefficient of $x^k$ for any real $m$ and $n$. He was in effect arguing that if the equation

$$\binom{m}{k}\binom{n}{0} + \binom{m}{k-1}\binom{n}{1} + \cdots + \binom{m}{0}\binom{n}{k} = \binom{m+n}{k}, \qquad (4.21)$$

where $\binom{m}{k}$ was defined by (4.17), held true for all positive integers $m$ and $n$, then it must also hold for any pair of real numbers $m$ and $n$. Though his argument was sufficiently persuasive to his contemporaries, including Legendre, and although he applied the same argument in other situations, we observe that it is quite incomplete, an early example of Peacock's principle of permanence of equivalent forms. We return to this topic later in this section. To continue with Euler's proof of the binomial theorem, we suppose with Euler that he had proved that

$$[m] \cdot [n] = [m+n] \qquad (4.22)$$

was true for all real $m$ and $n$. Euler could then deduce the binomial theorem for rational exponents. He supposed $m = \frac{p}{q}$ where $p$ and $q$ were positive integers. Then by (4.20)

$$(1+x)^p = [p] = \left[\frac{p}{q} + \frac{p}{q} + \cdots + \frac{p}{q}\right] = \left[\frac{p}{q}\right] \cdot \left[\frac{p}{q}\right] \cdots \left[\frac{p}{q}\right] = \left[\frac{p}{q}\right]^q,$$

or

$$\left[\frac{p}{q}\right] = (1+x)^{\frac{p}{q}}, \tag{4.23}$$

proving the theorem for positive rational exponents. Euler extended this result to negative rational exponents by noting that

$$[m] \cdot [-m] = [m - m] = [0] = 1$$

and therefore

$$\left[\frac{p}{q}\right] \cdot \left[-\frac{p}{q}\right] = 1.$$

By (4.23), this meant that $(1+x)^{\frac{p}{q}} \left[-\frac{p}{q}\right] = 1$ and thus $\left[-\frac{p}{q}\right] = (1+x)^{-\frac{p}{q}}$. Euler did not discuss convergence questions here. He certainly knew that the series for $[m]$ converged when $|x| < 1$, but apparently he had not given thought to the more subtle questions related to the convergence of the products of infinite series. Cauchy, Abel, and Dirichlet eventually addressed such issues.

The gap in Euler's reasoning, a full proof for (4.21), was addressed by Cauchy in his 1821 *Analyse Algébrique*.[52] In section 4.1 of this work, he first proved that if two polynomials of one variable and of degree $n - 1$ were equal at $n$ different values, then they were identical. This proof is contained in our Section 2.10. In the next section of his book, Cauchy extended this result to polynomials in many variables. For polynomials in two variables $x$ and $y$, the theorem can be stated: If $P(x, y)$ and $Q(x, y)$ are two polynomials of degree $n - 1$ in $x$ and $y$ that become equal when $x$ takes one of the values $x_0, x_1, \dots, x_{n-1}$, and $y$ takes one of the values $y_0, y_1, \dots, y_{n-1}$, then the two polynomials are identical. Cauchy proved this result by arguing that $P(x_0, y)$ and $Q(x_0, y)$ were two polynomials of degree $n - 1$ in one variable, $y$. He reasoned that since these polynomials were equal for $n$ values, $y = y_0, y_1, \dots, y_{n-1}$, they were then identical. Similarly, for $i = 1, 2, \dots, n - 1$,

$$P(x_i, y) \equiv Q(x_i, y).$$

Now for any real value of $y$, $P(x, y) = Q(x, y)$, for $x = x_0, x_1, \dots, x_{n-1}$. Thus, Cauchy could conclude that $P(x, y) \equiv Q(x, y)$.

Cauchy went on to apply this theorem to prove (4.21), observing that both sides were polynomials of degree $k$ in two variables $m$ and $n$; furthermore, since these polynomials were equal for infinitely many integer values of $m$ and $n$, they were identical.

As we discuss in Chapter 17, Euler once again saw no need to validate his move from integers to all real numbers in the case where the function was not a polynomial.

---

[52] Cauchy (1989).

## 4.4  Cauchy: Proof of the Binomial Theorem for Real Exponents

In his lectures at the École Polytechnique, Cauchy attempted to put the work of his predecessors on a more solid foundation,[53] although his students did not appreciate his efforts and apparently complained about it. In order to make Euler's work of the previous section more rigorous, Cauchy had to define a continuous function, and he also needed to work out the definitions of convergence, absolute convergence, and the product of infinite series.

Cauchy considered the infinite series

$$\mu_0 + \mu_1 + \mu_2 + \cdots$$

by setting the $n$th partial sum

$$s_n = \mu_0 + \mu_1 + \mu_2 + \cdots + \mu_{n-1}.$$

Cauchy then stated that if $s_n$ approached a fixed limit as $n$ increased indefinitely, then the infinite series would converge; otherwise, it diverged. The limit of $s_n$ was said to be $s$ if for any small number $\epsilon$, the value of $s_n$ fell between $s - \epsilon$ and $s + \epsilon$, for large enough $n$.

He stated and proved the ratio test for convergence of a series and deduced that the binomial series for $|x| < 1$ converged. He also defined what is known as the Cauchy product of two series

$$u_0 + u_1 + u_2 + \cdots \quad \text{and} \quad v_0 + v_1 + v_2 + \cdots$$

as

$$u_0 v_0 + (u_0 v_1 + u_1 v_0) + (u_0 v_2 + u_1 v_1 + u_2 v_0) + \cdots$$

$$+ (u_0 v_n + u_1 v_{n-1} + \cdots + u_n v_0) + \cdots.$$

Here note that if we define the degree of a term $u_i v_j$ to be $i + j$, then $u_0 v_0$ would be of degree 0, $u_0 v_1 + u_1 v_0$ would contain all the terms of degree 1, $u_0 v_2 + u_1 v_1 + u_2 v_0$ would contain all the terms of degree 2, and so on.

Cauchy then proved that if the two series were absolutely convergent and converged to $s$ and $s'$, then the product series converged to $s'' = ss'$. For the case in which all $u$ and $v$ were positive, Cauchy observed that

$$s_{m+1} s'_{m+1} < s''_n < s_n s'_n, \qquad (4.24)$$

where $m = \frac{n-1}{2}$ for all $n$ odd and $m = \frac{n-2}{2}$ for $n$ even. To quickly check the validity of these two inequalities in (4.24), take the case when $n$ is even and equal to $2m + 2$. In that case, $s''_n$ contains every term of degree less than or equal to $2m + 1$, while $s_{m+1} s'_{m+1}$ cannot contain any term of degree greater than $2m$. The same approach applies for $n$ odd.

---

[53] Cauchy (1989) § 4.3 and Note 6.

The two inequalities (4.24) implied the theorem for series with positive terms, taking the limit as $m \to \infty$. When the terms $u_0, u_1, u_2, \ldots$ and $v_0, v_1, v_2 \ldots$ were positive as well as negative, Cauchy first observed that

$$s_n s'_n - s''_n = u_{n-1} v_{n-1} + (u_{n-1} v_{n-2} + u_{n-2} v_{n-1}) + \cdots + (u_{n-1} v_1 + \cdots + u_1 v_{n-1}).$$

He denoted the absolute values of the $u$ and $v$ by $P_0, P_1, P_2, \ldots$ and $P'_0, P''_1, P'_2, \ldots$ respectively and remarked that, from the result on the convergence of the product of series with all positive terms,

$$P_{n-1} P'_{n-1} + (P_{n-1} P'_{n-2} + P_{n-2} P'_{n-1}) + \cdots + (P_{n-1} P'_1 + \cdots + P_1 P'_{n-1})$$

tended to zero as $n \to \infty$. Since this expression was an upper bound for $|s_n s'_n - s''_n|$, it followed that $s_n s'_n - s''_n \to 0$ as $n \to \infty$, proving the result.

With these theorems in hand, Cauchy could close the gaps in Euler's proof of the binomial theorem. The binomial series $[m]$ and $[n]$ in (4.16) were absolutely convergent for $|x| < 1$. Hence by Euler's argument and Cauchy's result on products of series, $[m] \cdot [n] = [m + n]$. Finally, if $[m]$ was a continuous function of $m$, and if for integers $p$ and $q$, $[\frac{p}{q}] = (1 + x)^{\frac{p}{q}}$, then $[m] = (1 + x)^m$ for all real exponents $m$. Recall that Cauchy's proof of the continuity of $[m]$ was inadequate, and the gap was filled by Abel.

Cauchy gave another proof of the binomial theorem as a corollary of Taylor's or Maclaurin's theorem. It was well known in the eighteenth century that the binomial theorem could be formally derived from Maclaurin's theorem, but Cauchy was the first to understand how an analysis of the remainder term could be applied to obtain a rigorous proof of the binomial theorem for real exponents. Note that Cauchy gave the remainder in two different forms, as discussed in our Chapter 11:

$$f(x) = f(0) + \frac{x}{1} f'(0) + \frac{x^2}{1 \cdot 2} f''(0) + \cdots + \frac{x^{n-1}}{1 \cdot 2 \cdots \cdot (n-1)} f^{(n-1)}(0) + R_n,$$

where

$$R_n = \frac{x^n}{1 \cdot 2 \cdots n} f^{(n)}(\theta x), \quad 0 < \theta < 1,$$

or

$$R_n = \frac{x^n}{1 \cdot 2 \cdots \cdot (n-1)} (1 - \theta_1)^{n-1} f^{(n)}(\theta_1 x), \quad 0 < \theta_1 < 1.$$

When $f(x) = (1 + x)^\mu$, he had

$$f^{(k)}(x) = \mu(\mu - 1) \cdots (\mu - k + 1)(1 + x)^{\mu - k}$$

and hence

$$\frac{f^{(k)}(0)}{k!} = \frac{\mu(\mu - 1) \cdots (\mu - k + 1)}{k!},$$

---

so the binomial series was obtained. To determine the values of $x$ for which the series equaled $(1 + x)^{\mu}$, it was necessary to find the values of $x$ for which $R_n \to 0$ as $n \to \infty$. Taking $m$ large enough that $|\frac{\mu}{m}| < 1$, Cauchy noted[54] that

$$\mu_{n-1} \equiv \frac{\mu(\mu - 1) \cdots (\mu - n + 1)}{1 \cdot 2 \cdot 3 \cdots n} x^{n-1}$$

$$= \frac{\mu(\mu - 1) \cdots (\mu - m + 1)}{1 \cdot 2 \cdot 3 \cdots m} x^{m-1} \cdot \left(1 - \frac{\mu + 1}{m + 1}\right) \cdots \left(1 - \frac{\mu + 1}{n}\right) (-x)^{n-m},$$

and hence for $|x| < 1, \mu_{n-1} \to 0$ as $n \to \infty$. Now the first form of the remainder would be

$$R_n = \frac{\mu(\mu - 1) \cdots (\mu - n + 1)}{1 \cdot 2 \cdot 3 \cdots n} x^n (1 + \theta x)^{\mu - n}$$

$$= \mu_{n-1} \cdot x(1 + \theta x)^{\mu} \left(\frac{1}{1 + \theta x}\right)^n.$$

The factor $\left(\frac{1}{1+\theta x}\right)^n$ was bounded only for positive $x$, and he could deduce that $R_n \to 0$ as $n \to \infty$ for $0 \leq x < 1$. Cauchy needed the second form of the remainder to be able to deduce the binomial theorem for $|x| < 1$. Using the second form, Cauchy had

$$R_n = \mu_{n-1} \cdot x(1 + \theta_1 x)^{\mu - 1} \left(\frac{1 - \theta_1}{1 + \theta_1 x}\right)^{n-1}.$$

Clearly, $\left(\frac{1-\theta_1}{1+\theta_1 x}\right)^n$ was bounded for $|x| < 1$ and $R_n \to 0$ as $n \to \infty$ for $|x| < 1$.

## 4.5   Abel's Theorem on Continuity

Abel's continuity theorem was a response to Cauchy's 1821 result requiring uniform convergence, a concept developed later by Weierstrass. Implicitly accounting for uniform convergence in his result, Abel proved a theorem yielding the binomial theorem for real exponents. Though Abel found a mistake in Cauchy's work, he acknowledged his indebtedness to Cauchy and wrote in his paper that every analyst who loved rigor in mathematics should study Cauchy's *Cours d'analyse*; we note that this work is also called *Analyse algébrique*. Abel's basic result on power series is now called Abel's continuity theorem; in modern terms, it states that if an infinite series $\sum a_n$ converges, then the series $\sum a_n x^n$ converges uniformly for $0 \leq x \leq 1$ and also tends to $\sum a_n$ as $x$ tends to $1^-$. In 1897, Alfred Tauber (1866–1942) proved a conditional converse,[55] leading to the extensive Tauberian theory, developed and named by Hardy and Littlewood .

[54] Cauchy (1829) pp. 85–86, 102–105.
[55] Tauber (1897).

In his January 1826 letter to Holmboe, Abel discussed his continuity theorem:

Let $a_0 + a_1 + a_2 + a_3 + a_4+$ etc. be any infinite Series and thus You know that a very useful Manner of adding up this Series is to seek the sum of the following: $a_0 + a_1 x + a_2 x^2 + a_3 x^3 + \cdots$ and then later, put $x = 1$ in the Results. This is correct; but it seems to me that one cannot accept it without Proof.

Abel applied his theorem[56] to show that if $A$ and $B$ were convergent infinite series and their Cauchy product $C$ was convergent, then $AB = C$. Abel's continuity theorem was based on a lemma using summation by parts: If $t_0, t_1, \ldots, t_m, \ldots$ denoted a sequence of arbitrary quantities, and if the quantity $p_m = t_0 + t_1 + \cdots + t_m$ was less than a definite quantity $\delta$, then

$$r = \epsilon_0 t_0 + \epsilon_1 t_1 + \cdots + \epsilon_m t_m < \delta \epsilon_0, \tag{4.25}$$

where $\epsilon_0, \epsilon_1, \epsilon_2, \ldots$ were positive decreasing quantities. To prove this result, Abel noted that

$$r = \epsilon_0 p_0 + \epsilon_1(p_1 - p_0) + \epsilon_2(p_2 - p_1) + \cdots + \epsilon_m(p_m - p_{m-1})$$
$$= p_0(\epsilon_0 - \epsilon_1) + p_1(\epsilon_1 - \epsilon_2) + \cdots + p_{m-1}(\epsilon_{m-1} - \epsilon_m) + p_m \epsilon_m \tag{4.26}$$
$$< \delta(\epsilon_0 - \epsilon_1 + \epsilon_1 - \epsilon_2 + \cdots + \epsilon_{m-1} - \epsilon_m + \epsilon_m) = \delta \epsilon_0.$$

The last step in this proof was valid because $\epsilon_0 - \epsilon_1, \epsilon_2 - \epsilon_1, \ldots$ were positive. Next, for the continuity theorem, Abel wrote that if the series

$$f(\alpha) = v_0 + v_1 \alpha + v_2 \alpha^2 + \cdots + v_m \alpha^m + \cdots$$

converged for $\alpha = \delta$, it would also converge for every smaller value of $\alpha$; likewise, $f(\alpha - \beta)$, for continually decreasing values of $\beta$, would come arbitrarily close to the limit $f(\alpha)$, with $\alpha$ equal to or smaller than $\delta$. To prove this, Abel let

$$v_0 + v_1 \alpha + \cdots + v_{m-1}\alpha^{m-1} = \phi(\alpha),$$

$$v_m \alpha^m + v_{m+1}\alpha^{m+1} + \cdots = \psi(\alpha).$$

Then

$$\psi(\alpha) = \left(\frac{\alpha}{\delta}\right)^m \cdot v_m \delta^m + \left(\frac{\alpha}{\delta}\right)^{m+1} + \cdots < \left(\frac{\alpha}{\delta}\right)^m p,$$

where p was the maximum of

$$v_m \delta^m, \, v_m \delta^m + v_{m+1}\delta^{m+1}, \, v_m \delta^m + v_{m+1}\delta^{m+1} + v_{m+2}\delta^{m+2}, \ldots.$$

---

[56] Abel (2007) pp. 107–112.

Note that the inequality followed from his lemma proving (4.25). Then for $0 \leq \alpha \leq \delta$, $m$ could be chosen large enough so that $\psi(\alpha) = \omega$. We observe that Abel used the symbol $\omega$ to denote an arbitrarily small quantity. Next,

$$f(\alpha) = \phi(\alpha) + \psi(\alpha),$$

and hence

$$f(\alpha) - f(\alpha - \beta) = \phi(\alpha) - \phi(\alpha - \beta) + \omega.$$

Since $\phi(\alpha)$ was a polynomial, $\beta$ could be taken small enough that $\phi(\alpha) - \phi(\alpha - \beta) = \omega$ and hence $f(\alpha) - f(\alpha - \beta) = \omega$, proving the theorem.

To address the defect in Cauchy's proof of the binomial theorem, Abel stated and proved the theorem: Let $v_0 + v_1 \delta + v_2 \delta^2 + \cdots$ be a convergent series, in which $v_0, v_1, v_2, \ldots$ are continuous functions of a variable quantity $x$ between $x = a$ and $x = b$; then the series $f(x) = v_0 + v_1 \alpha + v_2 \alpha^2 + \cdots$, where $\alpha < \delta$, will be convergent and a continuous function of $x$ between the same limits. As in the proof of the previous theorem, Abel set

$$v_0 + v_1 \alpha + \cdots + v_{m-1} \alpha^{m-1} = \psi(x) \quad \text{and} \quad v_m \alpha^m + v_{m+1} \alpha^{m+1} + \cdots = \phi(x).$$

Then

$$\psi(x) = \left(\frac{\alpha}{\delta}\right)^m v_m \delta^m + \left(\frac{\alpha}{\delta}\right)^{m+1} v_{m+1} \delta^{m+1} + \left(\frac{\alpha}{\delta}\right)^{m+2} v_{m+2} \delta^{m+2} + \cdots.$$

By the summation by parts lemma, if $\theta(x)$ denoted the largest of the quantities

$$v_m \delta^m, v_m \delta^m + v_{m+1} \delta^{m+1}, v_m \delta^m + v_{m+1} \delta^{m+1} + v_{m+2} \delta^{m+2}, \ldots,$$

then $\psi(x) < \left(\frac{\alpha}{\delta}\right)^m \theta(x)$. Thus, for $m$ large enough, $\psi(x) = \omega$ and $f(x) = \phi(x) + \omega$, where $w$ was less than any assignable quantity. Similarly,

$$f(x) - f(x - \beta) = \phi(x) - \phi(x - \beta) + \omega.$$

Since $\phi(x)$ was a finite sum of continuous functions, it was also continuous and hence $\phi(x) - \phi(x - \beta) = \omega$. Therefore, $f(x) - f(x - \beta) = \omega$, which meant that $f(x)$ was continuous. It was here that Abel pointed out in a footnote that Cauchy's theorem on an infinite sum of continuous functions had some exceptions. But Abel succeeded in filling the gap, so that the proof of the binomial theorem for real exponents was complete.

Abel went on[57] to prove the binomial theorem for a complex variable $x$ and complex exponent. He finally stated his result as

---

[57] ibid. p. 125.

$$1 + \frac{m+ni}{1}(a+bi) + \frac{(m+ni)(m-1+ni)}{1 \cdot 2}(a+bi)^2$$

$$+ \frac{(m+ni)(m-1+ni)(m-2+ni)}{1 \cdot 2 \cdot 3}(a+bi)^3 + \cdots$$

$$+ \frac{(m+ni)(m-1+ni)\cdots(m-\mu+1+ni)}{1 \cdot 2 \cdot 3 \cdots \mu}(a+bi)^\mu + \cdots$$

$$= \left[ \cos\left( m \arctan \frac{b}{1+a} + \frac{1}{2}n \ln\left[(a+a)^2 + b^2\right]\right) \right.$$

$$\left. + i \sin\left( m \arctan \frac{b}{1+a} + \frac{1}{2}n \ln\left[(1+a)^2 + b^2\right]\right) \right]$$

$$\times \left[(1+a)^2 + b^2\right]^{\frac{m}{2}} e^{-n \arctan \frac{b}{1+a}}.$$

Note that Abel wrote $\sqrt{-1}$ for $i$ and log for ln; the right-hand side was the principal value of $(1+a+bi)^{m+ni}$.

Liouville found Abel's proof of the continuity theorem difficult to understand and asked Dirichlet to explain it. Dirichlet presented a proof on the spot; Liouville then used it in his lectures at the Collège de France. After Dirichlet's death, Liouville published the proof in honor of his friend.[58] He stated and proved the theorem:

If the series $a_0 + a_1 + a_2 + \cdots$ converges to $A$, then

$$\lim_{p \to 1^-} \sum_{n=0}^{\infty} a_n p^n = A.$$

Let $\delta_n = a_0 + a_1 + \cdots + a_n$ and $0 < p < 1$. Then

$$s = a_0 + a_1 p + a_2 p^2 + \cdots + a_n p^n + \cdots$$

$$= \delta_0 + (\delta_1 - \delta_0)p + (\delta_2 - \delta_1)p^2 + \cdots + (\delta_n - \delta_{n-1})p^n + \cdots$$

$$= (1-p)(\delta_0 + \delta_1 p + \delta_2 p^2 + \cdots + \delta_n p^n + \cdots). \tag{4.27}$$

Note that the last equation, (4.27), is valid because the first $n+1$ terms of the two series differ by $\delta_n p^{n+1}$, and this tends to zero as $n \to \infty$. Next, break up the last series into two parts:

$$S(p) = (1-p)(\delta_0 + \delta_1 p + \cdots + \delta_{n-1}p^{n-1}) + (1-p)(\delta_n p^n + \delta_{n+1}p^{n+1} + \cdots).$$

Let $P_n$ be a number between the maximum and minimum values of the sequence

$$\delta_n, \delta_{n+1}, \delta_{n+2}, \cdots$$

such that the second series is equal to

$$(1-p)P_n(p^n + p^{n+1} + p^{n+2} + \cdots) = P_n p^n.$$

---

[58] Dirichlet (1863).

Clearly, $P_n \to A$ as $n \to \infty$. So if we let $p \to 1$ and then let $n \to \infty$, the finite series tends to zero and the other series tends to $A$ and the theorem is proved.

## 4.6   Harkness and Morley's Proof of the Binomial Theorem

Weierstrass promulgated his fundamental ideas primarily through his teaching. Thus, it was left to others to write up and disseminate these ideas. For example, in 1898, J. Harkness of Bryn Mawr and F. Morley of Haverford College wrote *Introduction to the Theory of Analytic Functions*. They explained:[59] "we recognized that readers approaching the subject for the first time could not fail to be hampered by the non-existence in English of any text-book giving a consecutive and elementary account of the fundamental concepts and processes employed in the theory of functions." In his delightful article, *A Mathematical Education*, the great English analyst J. E. Littlewood mentioned that his study of Harkness and Morley's book was one of the brighter spots in his education up to the time he took his Tripos examination in 1905.[60]

Harkness and Morley's proof of the binomial theorem is different from the other proofs presented here, and it considers the general case where the variable and exponent are both complex numbers. The proof depends on a theorem attributed to Weierstrass:[61] Let $u_q$, $q = 0, 1, 2, \ldots$ be series in powers of $x$:

$$u_q = a_{q0} + a_{q1}x + a_{q2}x^2 + \cdots + a_{qn}x^n + \cdots .$$

Given that the separate series $u_q$ and the collective series $\sum u_q$ converge within the circle $(R)$ and that the series $\sum u_q$ converges uniformly along every circle $(R_1)$ where $R_1 < R$, then within the circle $(R)$ we have

$$\sum_{q=0}^{\infty} u_q = \sum_{n=0}^{\infty} a_n x^n,$$

where $a_n$ is the sum of the coefficients of $x^n$ in the series of $us$. Now consider the function $(1 - a)^{-x} = \exp(-x \log(1 - a))$, where $a$ and $x$ are complex with $|a| < 1$ and where log takes its principal value.[62] Then

$$(1 - a)^{-x} = 1 + u + \frac{u^2}{2!} + \frac{u^3}{3!} + \cdots$$

where

$$u = -x \log(1 - a) = xa + \frac{xa^2}{2} + \frac{xa^3}{3} + \cdots .$$

[59]  Harkness and Morley (1898) p. v.
[60]  Littlewood (1986) pp. 80–93, especially p. 82.
[61]  ibid. p. 134.
[62]  ibid. p. 169.

It is clear that the series in $u$ is absolutely and uniformly convergent in every circle $|u| \leq R$, while the series for $\frac{u^n}{n!}$ in powers of $a$ is absolutely and uniformly convergent in $|a| \leq 1 - \delta$ for every $\delta > 0$. Therefore, by Weierstrass's theorem,

$$\sum_{n=0}^{\infty} \frac{u^n}{n!} = \sum_{n=0}^{\infty} \frac{x_n a^n}{n!}$$

for $|a| < 1$. It remains to find $x_n$. Note that only $u, u^2, \ldots, u^n$ contribute to the expression. So

$$x_n = x^n + \cdots + (n-1)! \, x$$

is a polynomial of degree $n$ in $x$. When $x = 0, -1, -2, \ldots, -n+1$, we have $(1-a)^{-x} = (1-a)^m$ where $m = 0, 1, 2, \ldots, n-1$. For these values of $m$, the coefficient of $a^n$ in $(1-a)^m$ is zero. Hence, $x_n = 0$ for $x = 0, -1, -2, \ldots, -n+1$, and we have

$$x_n = x(x+1)(x+2) \cdots (x+n-1)$$

and thus

$$(1-a)^{-x} = \sum_{n=0}^{\infty} \frac{x(x+1) \cdots (x+n-1)}{n!} a^n.$$

## 4.7  Exercises

(1) Following Newton, apply the procedure for finding the square root of a number to the algebraic expression $1 - x^2$ and show that you get the series

$$1 - \frac{1}{2}x^2 - \frac{1}{8}x^4 - \frac{1}{16}x^6 - \cdots .$$

(2) Apply the Gregory–Newton difference formula to the function $f(\alpha) = (1 + x)^\alpha$ and show that you get the binomial series.

(3) Prove that the Cauchy product of the series $\sum_{n=0}^{\infty} \frac{(-1)^n}{\sqrt{n+1}}$ with itself diverges. Cauchy gave this example in his *Analyse algébrique*, chapter 6.

(4) Cauchy stated the ratio test in chapter 6 of his *Analyse algébrique*: If for $n$ increasing and positive, the ratio $\frac{u_{n+1}}{u_n}$ converges to a fixed limit $k$, then the series $u_0 + u_1 + u_2 + \cdots$ converges when $k < 1$ and diverges when $k > 1$. Prove this theorem.

(5) Cauchy's *Analyse algébrique*, chapter 6, also gave the condensation test: A series of positive and decreasing terms $u_0 + u_1 + u_2 + \cdots$ converges if and only if $u_0 + 2u_1 + 4u_3 + 8u_7 + \cdots$ converges. Prove this theorem, and use it to prove that $\sum_{n=1}^{\infty} \frac{1}{n^\mu}$ converges for $\mu > 1$ and diverges for $\mu \leq 1$. Cauchy gave this application. Earlier, the fourteenth-century French theologian and scientific thinker N. Oresme used the condensation test in the case $\mu = 1$.

(6) Prove Abel's theorem on products, that if $c_n = a_0 b_n + a_1 b_{n-1} + \cdots + a_n b_0$, and $A = \sum_{n=0}^{\infty} a_n$, $B = \sum_{n=0}^{\infty} b_n$, $C = \sum_{n=0}^{\infty} c_n$ are all convergent, then $AB = C$. See Abel (1965) vol. 1, p. 226.

(7) Prove F. Mertens's extension of Cauchy's product theorem: If $A$ is absolutely convergent and $B$ conditionally convergent, then $AB = C$. See Mertens (1875).

(8) Prove that if $A$ and $B$ are convergent, and $|a_n| \le \frac{k}{n}$, $|b_n| \le \frac{k}{n}$ for all $n$, then $C$ is convergent. This theorem is due to G. H. Hardy; see Hardy (1966–1979) vol. 6, pp. 414–416.

(9) Follow Abel in proving that $\sum_{k=2}^{\infty} \frac{1}{k \ln k}$ diverges. Use $\ln(1+x) < x$ to show that that $\ln \ln(1+n) < \ln \ln n + \frac{1}{n \ln n}$. Conclude that $\ln \ln(1+n) < \ln \ln 2 + \sum_{k=1}^{n} \frac{1}{k \ln k}$. See Abel (1965) vol. 1, pp. 399–400.

(10) Show that if $A_n = a_1 + a_2 + \cdots + a_n$, $|A_n|$ is bounded for all $n$, and $\sum_{k+1}^{n} |b_{k+1} - b_k|$ is bounded for all $n$, and $b_n \to 0$ as $n \to \infty$, then $\sum_{n=1}^{\infty} a_n b_n$ converges. Dedekind stated this theorem in Supplement IX to Dirichlet's lectures on number theory. See Dirichlet and Dedekind (1999) pp. 261–264.

(11) (a) Suppose $\phi$ is a real valued continuous function and $\phi(x + y) = \phi(x) + \phi(y)$. Show that $\phi(x) = ax$ for some constant $a$.

(b) Suppose $\phi$ is a real valued continuous function and $\phi(x+y) = \phi(x)\phi(y)$. Show that $\phi(x) = A^x$ for some positive constant $A$.

(c) Suppose $\phi$ is a real valued continuous function and for $x > 0, y > 0$, $\phi(xy) = \phi(x) + \phi(y)$. Show that $\phi(x) = a \ln x$.

(d) Suppose $\phi$ is a real valued continuous function and for $x > 0, y > 0$, $\phi(xy) = \phi(x)\phi(y)$. Show that $\phi(x) = x^a$ for $x > 0$.

(e) Suppose $\phi$ is a real valued continuous function and $\phi(y + x) + \phi(y - x) = 2\phi(x)\phi(y)$. Show that if $0 \le \phi(x) \le 1$ and $\phi$ is not constant, then $\phi(x) = \cos(ax)$ where $a$ is a constant. If $\phi(x) \ge 1$, then there exists a positive constant $A$ such that $\phi(x) = A^x$. The solutions of these five problems take up all of chapter 5 of Cauchy's *Analyse algébrique* of 1821.

(12) Let

$$\phi(x) = 1 + \frac{x}{1} + \frac{x^2}{1 \cdot 2} + \frac{x^3}{1 \cdot 2 \cdot 3} + \cdots .$$

Show that $\phi(x + y) = \phi(x) \cdot \phi(y)$ and hence that $\phi(x) = (\phi(1))^x = e^x$. See Cauchy (1989) pp. 168–169.

(13) Prove the binomial theorem: Let

$$f_\alpha(x) = \sum_{k=0}^{\infty} \frac{\alpha(\alpha + 1) \cdots (\alpha + k - 1)}{k!} x^k .$$

Show that $\frac{d}{dx} f_\alpha(x) = \alpha f_{\alpha+1}(x)$ and $f_{\alpha+1}(x) - f_\alpha(x) = x f_{\alpha+1}(x)$. Deduce that $f_\alpha(x) = (1 - x)^{-\alpha}$. Gauss did not explicitly give this proof, but the first two steps are very special cases of results in his paper on hypergeometric functions.

(14) With the development of set theory and the language of sets by Cantor, Dedekind and others in the period 1870–1900, mathematicians could ask whether the equation $\phi(x + y) = \phi(x) + \phi(y)$ had solutions other than $\phi(x) = ax$, $a$ a constant, and, if so, whether a condition weaker than continuity would imply $\phi(x) = ax$. Hilbert's student Georg Hamel (1877–1954) used Zermelo's result, that the set of real numbers can be well-ordered, to obtain a basis $B$ for the vector space of real numbers over the field of rational numbers. Thus, $B$ has the property that for every real number $x$, $x = r_1\alpha_1 + r_2\alpha_2 + \cdots + r_n\alpha_n$ for some $n$, rationals $r_1, \ldots, r_n$ and basis elements $\alpha_1, \ldots, \alpha_n$. Use a Hamel basis to show that $\phi(x + y) = \phi(x) + \phi(y)$ has solutions different from $\phi(x) = ax$, where $a$ is a constant. See G. Hamel (1905).

(15) Prove that if $\phi$ is measurable and satisfies $\phi(x + y) = \phi(x) + \phi(y)$, then $\phi(x) = ax$. This was proved by M. Fréchet in 1913. For a simple proof, see Kac (1979) pp. 64–65.

## 4.8  Notes on the Literature

In 1712, Newton cited his two letters to Leibniz and the letters of James Gregory to Collins to establish his priority over Leibniz in the invention of the calculus. Whiteside discusses this at length in volume 8 of Newton's works, pp. 469–632. In the context of the calculus controversy, it was argued by Newton that Gregory did his work on series after seeing Newton's series on the area of a zone of a circle. Consequently, Gregory's work was perhaps relegated to a secondary position, though his discoveries were independent. The 1939 publication of the Gregory memorial volume has helped to reestablish Gregory's position as one of the greatest mathematicians of the seventeenth century.

Landen's 43-page booklet was a contribution to the mathematical tendency of that time, to employ algebra to avoid infinitesimals and fluxions. Also part of this tradition, Hutton (1812) discusses the binomial theorem, expressing appreciation for Landen's proof. Hutton's three-volume work is entertaining reading, with articles on building bridges, experiments with gunpowder, histories of trigonometric and logarithmic tables, and a long history of algebra.

Cauchy started teaching analysis at the École Polytechique in 1817. He divided his course into two parts, the first dealing with infinite series of real and complex variables and the second with differential and integral calculus. Following eighteenth-century usage, he called the first part algebraic analysis. These lectures were published in 1821 with the title *Analyse algébrique*. Bradley and Sandifer (2009) present an English translation with useful notes. This was the first textbook dealing fairly rigorously with the basic concepts of infinite series: limits, convergence, and continuity.

Abel's paper on the binomial theorem appeared in the first issue of A. L. Crelle's *Journal für die reine und angewandte Mathematik* in 1826. It is the oldest mathematical journal still being published today. A large majority of Abel's papers were published in this journal, helping this journal quickly secure a high standing in the mathematics community. There are two editions of Abel's collected papers. Abel's college teacher B. M. Holmboe (1795–1850) published the first in 1837. Unfortunately, Holmboe could not include Abel's great paper, "Mémoire pour une propriété générale d'une classe très-étendue des fonctions transcendantes," presented by Abel to the French Academy in 1826. This manuscript was lost and found several times before it was finally published in Paris in 1841. The manuscript was again lost, possibly stolen by G. Libri; in 1952 a portion was recovered in Florence by the Norwegian mathematician V. Brun. In 2000, Andrea Del Cantina discovered the remaining parts, with the exception of four pages. Abel's "Mémoire" was included in the second and larger edition of his collected work, edited by L. Sylow and S. Lie, published in 1881. Abel (1965) is a reprint of this edition. For an English translation of Abel's papers on analysis, see Abel (2007); see the footnote on p. 111 concerning Abel's example of a series of continuous functions converging to a discontinuous function.

# 5

## The Rectification of Curves

### 5.1 Preliminary Remarks

Up until the seventeenth century, geometry was pursued along the Greek model. Thus, second-order algebraic curves were studied as conic sections, though higher-order curves were also considered. Algebraic relationships among geometric quantities were considered, but algebraic equations were not used to describe geometric objects. In the course of his attempts during the late 1620s to recreate the lost work of Apollonius, it occurred to Fermat that geometry could be studied analytically by expressing curves in terms of algebraic equations. Conic sections are defined by second-degree equations in two variables, but this new perspective expanded geometry to include curves of any degree. Fermat's work in algebraic geometry was not published in his lifetime, so its influence was not great. But during the 1620s, René Descartes (1596–1650) developed his conception of algebraic geometry and his seminal work, *La Géométrie*, was published in 1637. Note that Fermat's work on algebraic geometry, *Ad locos planos et solidos isagoge*, written in Viète's notation, was published in 1679, though he most probably wrote it before Descartes's 1637 book.[1] The variety of new curves made possible by this new perspective, combined with the development of the differential method, spurred the efforts to discover a general method for determining the length of an arc. In the late 1650s, Hendrik van Heuraet (1634–c. 1660) and William Neil(e) (1637–1670) gave a solution to this problem by reducing it to the problem of finding the area under a related curve. In this and other areas, Descartes's new approach to geometry served as a guiding backdrop.

Descartes's early training in mathematics included the study of the classical texts of the fourth-century Greek mathematician Pappus, the *Arithmetica* of Diophantus, and the contemporary algebra of Peter Roth and Christoph Clavius. Descartes's meeting with Johann Faulhaber in the winter 1619–20 also contributed to his understanding of algebra. However, from the very beginning, Descartes was determined to develop and follow his own methods, and this eventually led him to a symbolic algebra whose notation was very similar to the one we now use.

---

[1] Struik (1987) p. 99.

In the first part of his *Géométrie*, Descartes explained how geometric curves could be reduced to polynomial equations in two variables. He did not consistently use what are now called the Cartesian orthogonal axes but chose the angle of his axes to suit the problem. On the subject of the rectification of curves, Descartes wrote, "geometry should not include lines that are like strings, in that they are sometimes straight and sometimes curved, since the ratios between straight and curved lines are not known, and I believe cannot be discovered by human minds."[2] But in a November 1638 letter to Mersenne, Descartes nevertheless discussed the rectification of the logarithmic spiral, a curve he defined as making a constant angle with the radius vector at each point. To understand this apparent contradiction in his thinking, note that Descartes made a distinction between geometric and mechanical curves. Geometric curves were defined by algebraic equations; mechanical curves would today be termed transcendental. So when Descartes referred to unrectifiable curves, he meant the geometric ones. He maintained that the study of geometry should be restricted to algebraic curves, for which algebraic methods should be used. Thus, for constructing tangents and normals to curves, he used algebraic methods, as opposed to the infinitesimal methods of Fermat. Descartes considered Fermat's methods inappropriate for geometric (for us, algebraic) curves. And, since the length of an arc could not be found by algebraic methods, he stated that the length of geometric curves could not be obtained.[3] He allowed, however, that the lengths of mechanical (transcendental) curves could be determined by infinitesimal methods. In fact, in 1638 Descartes succeeded in rectifying the equiangular (or logarithmic) spiral,[4] and he was therefore one of the earliest mathematicians to find the length of a noncircular arc. Soon after this, Torricelli also rectified this spiral. We note that by approximately 1614, Harriot had completed his work on this spiral, in connection with his researches related to navigation, but he did not publish his results.[5]

Frans van Schooten (1615–1660) played an important role in the solution of the rectification problem. A Dutch mathematician of considerable ability, he was certainly one of the great teachers of mathematics. He attracted a number of talented young students to mathematics, even when their primary interests lay in other disciplines. Van Schooten studied at the University of Leiden where his father, Frans van Schooten the Elder, was a professor of mathematics. The younger van Schooten received a thorough grounding in the Dutch mathematical tradition. In 1635, he met Descartes, and by the summer of 1637 he had seen his *Géométrie*, though he did not immediately understand it. In 1646 he inherited his father's professorship and in 1647, van Schooten published a Latin translation of Descartes's book with his own commentary. This translation made Descartes's ideas accessible to many more mathematicians and simultaneously helped build van Schooten's reputation. A second edition of this translation appeared in 1659. The work was about a thousand pages long, ten times longer than *Géométrie*;

---

[2] Descartes, Smith, and Latham (1954) p. 91.
[3] Hofmann (1990) vol. 2, p. 132.
[4] Hofmann (1974) p. 103.
[5] Pepper (1968).

it included several significant contributions by van Schooten's students: Christiaan Huygens, Jan Hudde, Jan de Witt, and Hendrik van Heuraet.[6]

Van Heuraet entered the University of Leiden as a medical student. He was inspired to study the rectification problem by Huygens's 1657 discovery that the arc length of a parabola could be measured by the quadrature of an equilateral hyperbola. In modern terms, this means that the arc length of $y = x^2$ can be computed by the integral $\int (1 + 4x^2)^{\frac{1}{2}} dx$. Sometime in 1658, van Heuraet solved the general problem; he communicated his work to van Schooten in a letter dated January 13, 1659.[7] In the course of applying his method to the semicubical parabola $y^2 = ax^3$, he used a rule of Hudde concerning multiple roots of polynomials.

Jan Hudde (1628–1704) studied law at Leiden around 1648 and later served as burgomaster of Amsterdam for 21 years. He stated his rule in the article *De Maximis et Minimis*, communicated to van Schooten in a letter of February 26, 1658.[8] His rule provided a method for determining maxima and minima of functions and a simplification of Descartes's method for finding the normal to an algebraic curve. In unpublished work from 1656,[9] Hudde used the logarithmic series in the form $x + \frac{x^3}{3} + \frac{x^5}{5} + \cdots$. Note that N. Mercator published this series in 1668[10] and Newton's unpublished results on this topic date from 1665.

Around the same time as van Heuraet, the English mathematician William Neil gave a method for rectifying the semicubical parabola; this method could also be generalized to other curves. Wallis included Neil's work in his *Tractatus duo* of 1659.[11] The methods of Neil and van Heuraet were lacking in rigor, but Pierre Fermat very soon filled the gap. He showed in his *Comparatio Curvarum Linearum* of 1660 that a monotonically increasing curve will have a length. James Gregory, apparently independently of Fermat, also found a rigorous proof of this fact, technically better than Fermat's, using the same basic idea. Gregory's proof appeared in his *Geometriae Pars Universalis* of 1668. The inspiration behind both Fermat and Gregory was Christopher Wren's rectification of the cycloid in 1659.

## 5.2 Descartes's Method of Finding the Normal

Descartes's conception of geometry may be summarized by his own remarks from his *Geometry*, given just before he described his method of finding the normal:[12]

Finally, all other properties of curves depend only on the angles which these curves make with other lines. But the angle formed by two intersecting curves can be as easily measured as the angle between two straight lines, provided that a straight line can be drawn making right angles with one of these curves at its point of intersection with the other. This is my reason for believing

6 See Bissel (1987) for more details on the contributions of these Dutch mathematicians.
7 van Schooten (1659) vol. I, pp. 517–520.
8 ibid. pp. 507–516.
9 Hofmann (2003) p. 39. Leibniz (1920) p. 123.
10 Mercator (1668).
11 Wallis (1659) pp. 90–94.
12 Descartes, Smith, and Latham (1954) p. 95.

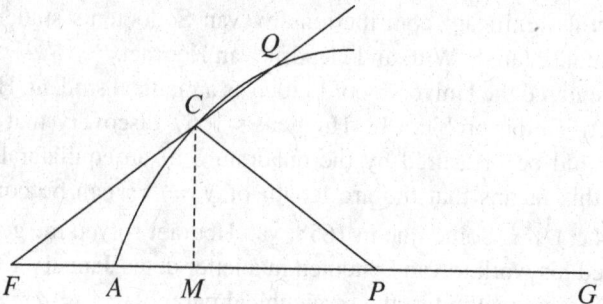

Figure 5.1 Descartes's construction of a normal.

that I shall have given here a sufficient introduction to the study of curves when I have given a general method of drawing a straight line making right angles with a curve at an arbitrarily chosen point upon it. And I dare say that this is not only the most useful and most general problem in geometry that I know, but even that I ever have desired to know.

In Figure 5.1, as given by Descartes, suppose $y = f(x)$ is the curve $ACQ$ and $C$ is a point on the curve. Also suppose that $CF$ is the tangent at $C$ that intersects the $x$-axis at $F$. Let $CP$ be the normal at $C$, meaning that $FCP$ is a ninety-degree angle, with $P$ a point on the $x$-axis. Then $FM$ and $PM$ are called, respectively, the subtangent and the subnormal at $C$. Observe that if the lengths of the subtangent and subnormal are known, then it is easy to draw or construct the tangent and the normal to the curve at $C$. Descartes gave a method for constructing the normal at a point on a curve $y = f(x)$ where $(f(x))^2$ was a polynomial. He assumed that the center $P$ of the circle tangent to the curve at $C$ had coordinates $(v, 0)$. The equation of the circle of radius $s$ was thus

$$y^2 + (x - v)^2 = s^2. \tag{5.1}$$

In general, a circle would cut the curve at $C$ and at another point. These two points would coalesce into one point when the circle was tangent to the curve. So Descartes noted that when the variable $y$ was eliminated from the two equations (5.1) and $y = (f(x))^2$, then the polynomial

$$y = (f(x))^2 + (x - v)^2 - s^2$$

would have a factor $(x - e)^2$ for some $e$, or (with $g(x)$ some polynomial)

$$(f(x))^2 + (x - v)^2 - s^2 = (x - e)^2 g(x).$$

By equating the coefficients of the powers of $x$, he was able to find the value of $e$ in terms of $v$ and thus the value of the subnormal $v - e$. As his words indicate, Descartes felt that this method solved the fundamental problem of geometry.

Let $CA$ in Figure 5.1 be an algebraic curve. Note that in the original book, Descartes interchanged $x$ and $y$. But we will suppose $CP$ is normal to the curve at $C$ and let

$AM = x$ and $CM = y$. The problem is to find $v = AP$. Descartes took the $x$-axis such that the center $P$ of the required circle fell on the axis. The equation of the circle was $(x - v)^2 + y^2 = s^2$, where $CP = s$ was the radius. He used this equation to eliminate $y$ from the equation of the curve, obtaining an expression in $x$ with a double root when the circle was tangent to the curve. He could then write this expression such that it had a factor $(x - e)^2$. Descartes then found the required result by equating the powers of $x$. He explained his method by applying it to some examples such as the ellipse $\frac{r}{q}x^2 - rx + y^2 = 0$, in which case he used the equation of the circle to eliminate $y$ and obtain the equation

$$x^2 + \frac{qr - 2qv}{q - r}x + \frac{q(v^2 - s^2)}{q - r} = 0.$$

He set the left-hand side as $(x - e)^2$ and equated coefficients of powers of $x$ to find the necessary result:

$$-2e = \frac{q(r - 2v)}{q - r}, \quad \text{or} \quad v = \frac{r}{2} + \frac{q - r}{q}e = \frac{r}{2} + \frac{q - r}{q}x.$$

Claude Rabuel's 1730 commentary[13] on Descartes's *Géométrie* gave the example of the parabola $y^2 = rx$. In this case,

$$x^2 + (r - 2v)y + v^2 - s^2 = 0$$

must have a double root. The resulting equations, when the left-hand side is set equal to $(x - e)^2$, are

$$r - 2v = -2e, \quad v^2 - s^2 = e^2.$$

So $v = \frac{r}{2} + e$, and this implies that $v = \frac{r}{2} + x$, since $x = e$.

It is easy to see that Descartes's method would become cumbersome for curves of higher degree. After equating coefficients, one would end up with a large number of equations. It is for this reason that Hudde searched for a simpler approach.

## 5.3 Hudde's Rule for a Double Root

In his letter of February 1658 to van Schooten, Hudde gave a rule to determine conditions for a polynomial to have a double root:[14]

> If in an equation two roots are equal and this is multiplied by an arbitrary arithmetical progression, naturally the first term of the equation by the first term of the progression, the second term of the equation by the second term of the progression and so on: I say that the product will be an equation in which one of the afore-mentioned roots will be found.

[13] Rabuel (1730) p. 314.
[14] Grootendorst and van Maanen (1982) p. 107.

Hudde gave a proof for a fifth-degree polynomial,[15] and it works in general. In modern notation, suppose $b$ is a double root of $f(x) = 0$. Then

$$f(x) = (x - b)^2(c_0 + c_1 x + c_2 x^2 + \cdots + c_{n-2} x^{n-2})$$

$$= \sum_{k=0}^{n-2} c_k(x^{k+2} - 2bx^{k+1} + b^2 x^k).$$

If this equation is multiplied term-wise by an arithmetic progression $p + qk$ where $p$ and $q$ are integers and $k = 0, 1, \ldots, n - 2$, then we arrive at the polynomial

$$g(x) = \sum_{k=0}^{n-2} c_k((p + q(k + 2))x^{k+2} - 2b((p + q(k + 1))x^{k+1} + b^2(p + qk)x^k))$$

$$= \sum_{k=0}^{n-2} c_k(p + qk)x^k(x^2 - 2bx + b^2) + \sum_{k=0}^{n-2} c_k 2q(x^{k+2} - bx^{k+1})$$

$$= (x - b)^2 \sum_{k=0}^{n-2}(p + qk)x^k + 2q(x - b) \sum_{k=0}^{n-2} c_k x^{k+1}.$$

Clearly, $x = b$ is a root of $g(x)$. For a modern proof of Hudde's rule, observe that $g(x) = pf(x) + qxf'(x)$, where $f'(x)$ is the derivative of $f(x)$. By writing $f(x) = (x - a)^2 h(x)$, we get $f'(x) = 2(x - a)h(x) + (x - a)^2 h'(x)$. Thus, if $x = a$ is a double root of $f(x)$, then $x = a$ is also a root of $f'(x)$ and conversely.

Hudde's rule greatly reduced the computation required for Descartes's method, especially if the arithmetic progression was chosen judiciously. Van Schooten's book gave several examples of the application of Hudde's rule, and van Heuraet used it in his work on rectification.

## 5.4   Van Heuraet's Letter on Rectification

We present van Heuraet's 1659 rectification method largely in his own terms with some modification.[16] Concerning Figure 5.2, van Heuraet wrote, in the classical style, "$CM$ is to $CQ$ as $\Sigma$ to $MI$," where $\Sigma$ denoted a fixed line segment. We describe this relationship as $\frac{CQ}{CM} = MI$, eliminating the reference to $\Sigma$. Unlike van Heuraet, we now view $\Sigma$ as a number and set it equal to 1. Van Heuraet set out to find the length of the curve $ACE$. $CQ$ was normal to the curve, and $CN$ was the tangent at $C$. The point $I$ was determined so that $\frac{CQ}{CM} = MI$, where $MI$ was perpendicular to $AQ$. The locus of the point $I$ then determined the curve $GIL$, and the area under this curve rectified $ACE$. Once again, recall van Heuraet's perspective: He wrote that the area under the curve $GIL$ equaled the area of the rectangle with one side as $\Sigma$ and

---

[15] ibid. pp. 107–108.
[16] For the original letter and English translation, see ibid. pp. 95–105.

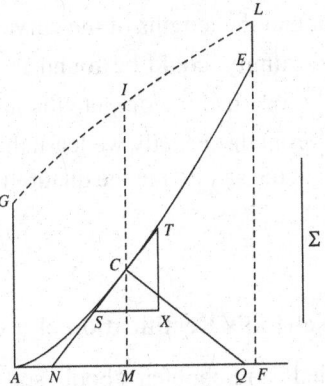

Figure 5.2  Van Heuraet's diagram.

the other side equal to the length of the curve $ACE$. To prove this, he observed that the similarity of the triangles $STX$ and $CMQ$ gave

$$\frac{ST}{SX} = \frac{CQ}{CM} = MI, \quad \left(= \frac{MI}{\Sigma}\right), \quad \text{or} \quad ST = MI \cdot SX.$$

Thus, the length of $ST$ was the area of the rectangle of base $SX$ and height $MI$. The lengths of the tangents taken at successive points along $AE$ approximated the length of the curve $ACE$; when the number of points was increased to infinity, the length of the curve equaled the area under $GIL$.

Van Heuraet then explained how the result might be applied to the semicubical parabola $y^2 = \frac{x^3}{a}$, where $AM = x$ and $MC = y$. He let $AQ = s$, $CQ = v$. Then

$$s^2 - 2sx + x^2 + \frac{x^3}{a} = v^2. \tag{5.2}$$

Following Descartes, van Heuraet noted that there were two equal roots of the equation implied by the simultaneous equations $y^2 = \frac{x^3}{a}$ and $(s-x)^2 + y^2 = v^2$. So he multiplied equation (5.2), according to Hudde's method, by 0, 1, 2, 3, 0 to get

$$-2sx + 2x^2 + \frac{3x^3}{a} = 0 \quad \text{or} \quad s = x + \frac{3x^2}{2a}.$$

Thus

$$MI = \frac{CQ}{CM} = \frac{v}{y} = \left(1 + \frac{9}{4a}x\right)^{\frac{1}{2}},$$

and the area under the curve $GIL$ could be expressed as

$$\frac{8a}{27}\left(1 + \frac{9}{4a}x\right)^{\frac{3}{2}} - \frac{8a}{27}.$$

Van Heuraet then pointed out that the lengths of the curves defined by $y^4 = x^{\frac{5}{a}}$, $y^6 = x^{\frac{7}{a}}$, $y^8 = x^{\frac{9}{a}}$, and so on to infinity, could be found in a similar way. However, in the case of a parabola $y = \frac{x^2}{a}$, one had to compute the area under the hyperbola $y = \sqrt{4x^2 + a^2}$. He concluded, "From this exactly we learn that the length of the parabolic curve cannot be found unless at the same time the quadrature of the hyperbola is found and vice versa."

## 5.5  Newton's Rectification of a Curve

Isaac Newton carefully studied van Schooten's book, so he understood van Heuraet's method. Newton worked out a simpler method of rectification, based on his conception of a curve as a dynamic entity, or as a moving point. In his 1671 treatise, *Of the Method of Fluxions and Infinite Series*,[17] Newton treated arc length by the approach he developed in his 1666 tract on calculus (or fluxions). Referring to Figure 5.3, he explained his derivation in the text presented by the 1737 editor, modified to include Newton's later "dot" notation:

> The Fluxion of the Length is discovered by putting it equal to the square root of the sum of the squares of the Fluxion of the Absciss and of the Ordinate. For let $RN$ be the perpendicular Ordinate moving upon the Absciss $MN$. And let $QR$ be the proposed Curve, at which $RN$ is terminated. Then calling $MN = s$, $NR = t$, and $QR = v$, and their Fluxions $\dot{s}$, $\dot{t}$, and $\dot{v}$, respectively; conceive the line $NR$ to move into the place $nr$ infinitely near the former, and letting fall $Rs$ perpendicularly to $nr$; then $Rs$, $sr$ and $Rr$, will be contemporaneous moments of the lines $MN$, $NR$, and $QR$, by the accession of which they become $Mn$, $nr$, and $Qr$; but as these are to each other as the Fluxions of the same lines, and because of the Rectangle $Rsr$, it will be $\sqrt{\overline{Rs}^2 + \overline{sr}^2} = Rr$, or $\sqrt{\dot{s}^2 + \dot{t}^2} = \dot{v}$.

Later in the treatise, he added that one may take $\dot{s} = 1$. This gives exactly the formula we have in textbooks now. It is interesting that some of his examples still appear in modern textbooks. For example, Newton considered the equation $y = \frac{z^3}{aa} + \frac{aa}{12z}$, with

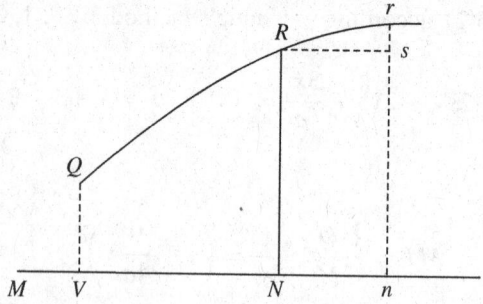

Figure 5.3 Newton's rectification of a curve.

[17] Newton (1964–1967) vol. I, pp. 173–174.

$a$ a constant. Taking $\dot{z} = 1$, he had $\dot{y} = \frac{3zz}{aa} - \frac{aa}{12zz}$ and $\sqrt{1 + \dot{y}^2} = \frac{3zz}{aa} + \frac{aa}{12zz}$. Thus, the arc length was given by $\frac{z^3}{aa} - \frac{aa}{12z}$. Newton also went on to find the constant of integration.

## 5.6 Leibniz's Derivation of the Arc Length

In 1673, Leibniz became interested in problems related to arc length;[18] following Pascal, he considered the characteristic triangle of a curve, as shown in Figure 5.4. Note that the characteristic triangle has sides of lengths $dx$, $dy$, and $ds$. After reading van Heurat's work on arc length, Leibniz attempted to find the arc length of a parabola by means of an infinite series. In a later, undated manuscript, probably from around 1680,[19] he noted the length of the infinitesimal arc as $\sqrt{dx^2 + dy^2}$. Thus,

$$ds = \sqrt{dx^2 + dy^2}.$$

By 1680, Leibniz had developed his notation and his ideas on integration so that he could write

$$s = \int \sqrt{(dx)^2 + (dy)^2} = \int \sqrt{1 + \left(\frac{dy}{dx}\right)^2} \, dx.$$

## 5.7 Exercises

(1) Find the lengths, between two arbitrary points, of curves defined by the equations: $y = \frac{2(a^2 + z^2)^{\frac{3}{2}}}{3a^2}$; $ay^2 = z^3$; $ay^4 = z^5$; $ay^6 = z^7$; $ay^{2n} = z^{2n+1}$; $ay^{2n-1} = z^{2n}$; $y = (a^2 + bz^2)^{\frac{1}{2}}$. See Newton (1964–1967) vol. 1, pp. 181–187. Note that in $y = (a^2 + bz^2)^{\frac{1}{2}}$, Newton expressed the arc length as an infinite series.

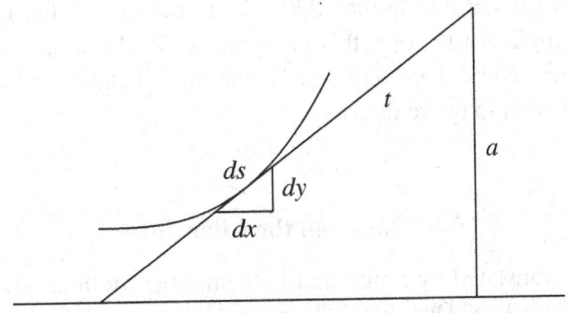

Figure 5.4 Leibniz on arc length.

18 See Probst (2015) p. 118.
19 Leibniz (1920) pp. 139–141.

(2) Find the length of any part of the equiangular or logarithmic spiral. Note that this spiral in polar coordinates is defined by $r = ae^{\theta \cot \alpha}$ where $a$ and $\alpha$ are constants.

(3) Find the length of any arc of the Archimedean spiral defined by $r = a\theta$. Stone (1730) worked out the examples in Exercises 2 and 3. Edmund Stone (1700–1768) was the son of the gardener of the Duke of Argyll. He taught himself mathematics, Latin, and French and translated mathematical works from these languages into English. Elected to the Royal Society in 1725, Stone was financially supported by the Duke whose death in 1743 left him destitute. See Pierpoint (1997).

(4) Take a triangle $ABC$ whose sides $AB = x$, $AC = a$, $BC = \frac{x^2}{a}$, and the perpendicular (altitude) $CD = \frac{a^2}{x}$ form a geometric progression. Show that

$$x = a\sqrt{\frac{1}{2} + \frac{\sqrt{5}}{2}}.$$

See Newton (1964–1967) p. 63. This exercise is taken from Newton's book on algebra, illustrating that algebra could be used to solve geometric problems. Viète had earlier used algebra in this way. Note, however, that this use of algebra is different from the algebraic geometry of Descartes and Fermat, in which algebraic equations are used from the outset to define curves.

(5) In a triangle $ABC$, let $AC = a$, $BC = b$, $AB = x$. Let $CD = c$ bisect the angle at $C$. Show that $x = (a + b)\sqrt{\frac{ab - cc}{ab}}$. See Simpson (1800) p. 261. This is an example of algebra being used in the service of geometrically defined problems.

(6) Suppose $ABC$ is a triangle such that the length of the bisectors of the angles $B$ and $C$ are equal. Use the result of the previous exercise to prove that the triangle is isosceles. In 1840, this theorem, now known as the Lehmus-Steiner theorem or the internal bisectors problem, was suggested as a problem by D. C. Lehmus (1780–1863) to the great Swiss geometer, Jacob Steiner (1796–1867). As a high school student during the 1930s, A. K. Mustafy rediscovered Simpson's result and applied it to solve this problem. A. S. Mittal has pointed out to me that if the trisectors or $n$-sectors are equal, the triangle must be isosceles; the reader may wish to prove this also.

## 5.8 Notes on the Literature

Descartes (1954), translated by Smith and Latham, gives both an English translation and the original French of Descartes's book on geometry. Rabuel's 1730 commentary on Descartes's *Geometrie* is very helpful because presents many examples of Descartes's method. Edmund Stone, mentioned in Exercise 3, wrote a 1730 book on calculus, with two parts: a translation of G. l'Hôpital's 1696 differential calculus text and a treatment of the integral calculus. Guicciardini's (1989) discussion of other eighteenth-century British calculus textbooks provides much interesting information on those books and their authors. For example, Guicciardini suggests that Bishop (George) Berkeley may have used Stone's presentation as mathematical background for his 1734 work, containing his philosophical objection to the concept of the infinitesimal. Pierpoint (1997) gives Stone's dates as 1700–1768, as opposed to others who give 1695 as his date of birth.

# 6

# *Inequalities*

## 6.1 Preliminary Remarks

In his 1928 presidential address to the London Mathematical Society,[1] G. H. Hardy observed, "A thorough mastery of elementary inequalities is to-day one of the first necessary qualifications for research in the theory of functions." He also recalled, "I think that it was Harald Bohr who remarked to me that 'all analysts spend half their time hunting through the literature for inequalities which they want to use and cannot prove.'" It is surprising, however, that the history of one of the most basic inequalities, the arithmetic and geometric means inequality (AMGM), is tied up with the theory of algebraic equations. Inequalities connected with the symmetric functions of the roots of an equation were used to determine the number of that equation's positive and negative roots. In 1665–1666, Newton laid down the foundation in this area when, in order to determine the bounds on the number of positive and negative roots of equations, he stated a far-reaching generalization of Descartes's rule of signs.

The arithmetic and geometric means inequality states that if there are $n$ nonnegative numbers $a_1, a_2, \ldots, a_n$ and there are $n$ positive numbers $q_1, q_2, \ldots, q_n$, such that $\sum_{i=1}^{n} q_i = 1$, then

$$a_1^{q_1} a_2^{q_2} \cdots a_n^{q_n} \leq \sum_{i=1}^{n} q_i a_i, \tag{6.1}$$

where equality holds only when all the $a_i$ are equal. The theory of equations has an interesting connection with the case for which $q_i = \frac{1}{n}$:

$$(a_1 a_2 \cdots a_n)^{\frac{1}{n}} \leq \sum_{i=1}^{n} \frac{a_i}{n}. \tag{6.2}$$

---

[1] Hardy (1929).

116

Note that when $n = 2$, the AMGM is simply another form of

$$(a_1 - a_2)^2 \geq 0; \tag{6.3}$$

this case can probably be attributed to Euclid. The nature of the relationship between the AMGM and algebraic equations is clear from (6.2). To see this, suppose that $a_1, a_2, \ldots, a_n$ are the roots of

$$x^n - A_1 x^{n-1} + A_2 x^{n-2} + \cdots + (-1)^n A_n = 0.$$

Then (6.2) is identical with the inequality $A_n^{\frac{1}{n}} \leq \frac{A_1}{n}$.

The three-dimensional case of the AMGM was first stated and proved by Thomas Harriot (c. 1560–1621) to analyze the roots of a cubic. Before this, François Viète (1540–1603) gave the condition under which a cubic could have distinct positive roots; Harriot then noted that Viète's condition was insufficient and that one also required $A_3^{\frac{1}{3}} < \frac{A_1}{3}$. A substantial portion of Harriot's algebraic work arose from his attempts to improve on both the notation and the results of the algebraist Viète. One of Harriot's innovations was to make algebra completely symbolic. Much of his work dates from about 1594 to the early 1600s, but his book on algebra, *Artis Analyticae Praxis*, was published in 1631, ten years after his death. And even this book omitted significant portions of Harriot's original text and, in places, changed the order of presentation so that the text lost its clarity. Harriot set up a new, convenient notation for inequality relations, and it is still in use, although Harriot's inequality symbol was very huge. William Oughtred, whose work was done later, independently introduced a different notation for inequality. One can see this cumbersome notation in the early manuscripts of Newton, but it soon fell into disuse. Harriot, with his effective notation, showed that one could carry out algebraic operations without using explanatory sentences or words and he demonstrated the superiority of his notation by rewriting Viète's expressions. The Harriot scholar J. Stedall points out[2] some key examples of this: Where Viète wrote: If to $\frac{A \text{ plane}}{B}$ there should be added $\frac{Z \text{ squared}}{G}$, the sum will be $G$ times $A$ plane $+ \frac{B \text{ times } Z \text{ squared}}{B \text{ times } G}$; Harriot wrote: $\frac{ac}{b} + \frac{zz}{g} = \frac{acg+bzz}{bg}$. Viète, under the influence of the Greek mathematicians, wrote $A$ plane, meaning that $A$ was a two-dimensional object; Harriot instead had $ac$. Then again, Viète described his example of antithesis:

> *A squared* minus *D plane* is supposed equal to *G squared* minus *B* times *A*. I say that *A squared* plus *B* times *A* is equal to *G squared* plus *D plane* and that by this transposition and under opposite signs of conjuction the equation is not changed.

Harriot's streamlined notation gives us:

> Suppose $aa - dc = gg - ba$. I say that $aa + ba = gg + dc$ by antithesis.

Descartes made similar advances in notation and in some places went beyond Harriot. For example, Descartes wrote $a^3$ or $a^4$, in place of $aaa$ or $aaaa$, retaining

---

[2] Harriot and Stedall (2003) pp. 8–11.

*aa* for $a^2$. Moreover, Descartes also stated a rule giving an upper bound for the number of positive (negative) roots of an equation. Its extension by Newton contributed further to the discovery of some important inequalities. In his 1637 *La Géométrie* Descartes stated his rule:[3] "We can determine also the number of true [positive] and false [negative] roots that any equation can have as follows: An equation can have as many true roots as it contains changes of sign from + to − or from − to +; and as many false roots as the number of times two + signs or two − signs are found in succession." Thus, in Descartes's example, $x^4 - 4x^3 - 19x^2 + 106x - 120 = 0$, the term $+x^4$ followed by the term $-4x^3$, then $-19x^2$ followed by $106x$ and, finally, $106x$ followed by $-120$ net a total of three changes of sign. According to Descartes's recipe, there can therefore be three positive roots. In fact, the roots are 2, 3, 4, and −5. The negative root is indicated by one repeated sign: $-4x^3$ followed by $-19x^2$. A rudimentary form of this rule of signs can be found in the earlier work of Faulhaber and Roth. Indeed, Manders has suggested[4] with some justification that Faulhaber and Descartes may have collaborated in analyzing the work of Roth.

Since Descartes's rule gave an upper bound for the number of positive roots and for the number of negative roots, it also determined a lower bound for the number of complex roots. Newton gave an extension of this rule, yielding a more accurate lower bound for many cases. This extension also directly connected this problem with certain inequalities satisfied by the coefficients of the given polynomial. Such inequalities included the AMGM. Newton's rule involved the consideration of the sequence of polynomials quadratic in the coefficients of the original polynomial. Newton stated this rule,[5] called by Sylvester "Newton's incomplete rule," in his *Arithmetica Universalis*, in the section "Of the Nature of the Roots of an Equation."

But you may know almost by this rule how many roots are impossible.

*Make a series of fractions, whose denominators are numbers in this progression 1, 2, 3, 4, 5, &c. going on to the number which shall be the same as that of the dimensions of the equation; and the numerators the same series of numbers in a contrary order. Divide each of the latter fractions by each of the former. Place the fractions that come out over the middle terms of the equation. And under any of the middle terms, if its square, multiplied into the fraction standing over its head, is greater than the rectangle of the terms on both sides, place the sign +; but if it be less, the sign −. But under the first and last term place the sign +. And there will be as many impossible roots as there are changes in the series of the under-written signs from + to −, and − to +.*

He made the following remarks for the case in which two or more successive terms of the polynomial were zero:[6]

Where two or more terms are wanting together, under the first of the deficient terms you must write the sign −, under the second sign +, under the third the sign −, and so on, always varying the signs, except that under the last of such deficient terms you must always place +, when the terms next on both sides the deficient terms have contrary signs. As in the equations

$$x^5 + ax^4 * * * + a^5 = 0, \quad \text{and} \quad x^5 + ax^4 * * * - a^5 = 0;$$
$$+ \quad\quad + \,-\,+\,- \,+ \quad\quad\quad\quad\quad + \quad\quad + \,-\,+\,+ \quad +$$

---

[3] Descartes (1954) p. 160.
[4] Manders (2006).
[5] Newton (1964–1967) vol. 2, pp. 103–105.
[6] ibid.

the first whereof has four, and the latter two impossible roots. Thus also the equation,

$$x^7 - 2x^6 + 3x^5 - 2x^4 + x^3 * * -3 = 0$$
$$+ \quad - \quad + \quad - \quad + - \quad +$$

has six impossible roots.

To understand Newton's incomplete rule, we use modern notation. Let the polynomial be

$$a_0x^n + a_1x^{n-1} + a_2x^{n-2} + \cdots + a_{n-1}x + a_n.$$

In the sequence of fractions $\frac{n}{1}, \frac{n-1}{2}, \frac{n-2}{3}, \ldots, \ldots, \frac{1}{n}$, divide the second term by the first, the third by the second and so on to get $\frac{n-1}{2n}, \frac{2(n-2)}{3(n-1)}, \frac{3(n-3)}{4(n-2)}, \ldots$. So $\frac{n-1}{2n}$ is placed over $a_1$, $\frac{2(n-2)}{3(n-1)}$ over $a_2$ and so on. If $\frac{(n-1)}{2n}a_1^2 > a_0a_2$, then a + sign is placed under $a_1$ and a − sign if the inequality is reversed. Similarly, if $\frac{2(n-2)}{3(n-1)}a_2^2 > a_1a_3$, then place a + sign under $a_2$ and so on. These inequalities take a simpler form if we follow J. J. Sylvester's notation from his 1865 paper[7] in which Newton's rule was proved for the first time, almost two hundred years after it was discovered. Write the polynomial as

$$f(x) = p_0x^n + np_1x^{n-1} + \frac{1}{2}n(n-1)p_2x^{n-2} + \cdots + np_{n-1}x + p_n. \qquad (6.4)$$

The inequalities become $p_1^2 > p_0p_2$ or $p_1^2 - p_0p_2 > 0$, $p_2^2 - p_1p_3 > 0$, and so on. Thus, Newton's sequence of signs is obtained from the sequence of numbers

$$A_0 = p_0^2, \ A_1 = p_1^2 - p_0p_2, \ A_2 = p_2^2 - p_1p_3, \ldots,$$
$$A_{n-1} = p_{n-1}^2 - p_{n-2}p_n, \ A_n = p_n^2.$$

$A_0$ and $A_n$ are always positive, while a plus sign is written under $a_k$ if

$$A_k = p_k^2 - p_{k-1}p_{k+1} > 0 \qquad (6.5)$$

and a minus sign if

$$A_k = p_k^2 - p_{k-1}p_{k+1} < 0. \qquad (6.6)$$

Newton gave several examples of his method, including the polynomial equation $x^4 - 6xx - 3x - 2 = 0$. Here $a_1 = 0$ and the signs of $A_1, A_1, A_2, A_3, A_4$ came out to be + + + − +. So, Newton wrote that there were [at least] two "impossible roots." In a 1728 paper[8] appearing in the *Philosophical Transactions*, the Scottish mathematician George Campbell published an incomplete proof of the incomplete rule (Newton's rule for complex roots); his efforts were sufficient to obtain the AMGM.

[7] Sylvester (1973) vol. 2, p. 498.
[8] Campbell (1728).

A little later, in 1729, Colin Maclaurin published a paper[9] in the same journal proving similar results. A priority dispute arose in this context, although it is generally recognized that Maclaurin's work was independent. Both Campbell and Maclaurin use the idea that the derivative of a polynomial with only real roots also had only real roots. Note that, in fact, this follows from Rolle's theorem, published in 1691,[10] though neither Campbell nor Maclaurin referred to Rolle. Campbell wrote that the derivative result was well-known to algebraists "and is easily made evident by the method of the maxima and minima." He began his paper by stating the condition under which a quadratic would have complex roots. He then showed that a general polynomial would have complex roots if, after repeated differentiation, it produced a quadratic with complex roots. Extremely little is known of Campbell's life; he was elected to the Royal Society on the strength of his paper in the *Transactions*.

In his 1729 paper, Maclaurin stated and proved, among other results, his inequalities:

$$p_1 \geq p_2^{\frac{1}{2}} \geq p_3^{\frac{1}{3}} \geq \cdots \geq p_n^{\frac{1}{n}}, \tag{6.7}$$

where the $p_k$ were all positive and defined by (6.4) with $p_0 = 1$. Most of Maclaurin's work in algebra arose out of his efforts to prove Newton's unproven statements and he presented them in his *Treatise of Algebra*, unfortunately published only posthumously in 1748.

Later, especially in the nineteenth century, the arithmetic and geometric means and related inequalities became objects of study on their own merits; they were then stated and proved independent of their use in analyzing the roots of algebraic equations. In the 1820s, Cauchy gave an inductive proof of AMGM in his lectures at the École Polytechnique. He started with $n = 2$ and then proved it for all powers of 2. He then obtained the result for all positive integers by a proof containing an interesting trick. In 1906, Jensen discovered that Cauchy's method could be generalized to convex functions, a fruitful concept he discovered and named, though it is implicitly contained in the work of Otto Hölder. To understand Jensen's motivation, observe that the two-dimensional case of (6.1) could be written as

$$e^{x_1} + e^{x_2} \geq 2e^{\frac{x_1+x_2}{2}}.$$

This led Jensen to define a convex function on an interval $[a,b]$ as a continuous function satisfying

$$\phi(x_1) + \phi(x_2) \geq 2\phi\left(\frac{x_1 + x_2}{2}\right) \tag{6.8}$$

for all pairs of numbers $x_1, x_2$ in $[a,b]$. Cauchy's proof could be applied in this situation without any change and Jensen was able to prove that for any $n$ numbers $x_1, x_2, \ldots, x_n$ in $[a,b]$,

---

[9] Maclaurin (1729).
[10] Rolle (1691).

$$\phi\left(\frac{x_1 + x_2 + \cdots + x_n}{n}\right) \le \frac{\phi(x_1) + \phi(x_2) + \cdots + \phi(x_n)}{n}. \tag{6.9}$$

As we shall see later in this chapter, Jensen did not require continuity up to this point, but he needed it for the generalization to (6.1).

Johan Jensen (1859–1925) was a largely self-taught Danish mathematician. He studied in an engineering college where he took courses in mathematics and physics. To support himself, he took a job in a Copenhagen telephone company in 1881. His energy and intelligence soon got him a high position in the company where he remained for the rest of his life. The rapid early development of telephone technology in Denmark was mainly due to Jensen. His spare time, however, was devoted to the study of mathematics; the function theorist Weierstrass, also self-taught to a great extent, was his hero. Jensen himself made a significant contribution to the theory of complex analytic functions, laying the foundation for Nevanlinna's theory of meromorphic functions of the 1920s. Jensen wrote his generalization of (6.1) in the form[11]

$$\phi\left(\frac{\sum a_\mu x_\mu}{a}\right) \le \frac{\sum a_\mu \phi(x_\mu)}{a}, \tag{6.10}$$

where $a = \sum a_\mu$ and $a_\mu > 0$. He took $\phi(x) = x^p, p > 1, x > 0$ to obtain the important inequality named after Hölder, one form of which states that for $p, b_\mu,$ and $c_\mu$ all positive, if $\frac{1}{p} + \frac{1}{q} = 1$, then

$$\sum b_\mu c_\mu \le \left(\sum b_\mu^p\right)^{\frac{1}{p}} \left(\sum c_\mu^q\right)^{\frac{1}{q}}. \tag{6.11}$$

Interestingly, in 1888, L. J. Rogers was the first to state and prove the inequality (6.11);[12] he first proved (6.1) and then derived several corollaries, including the Hölder inequality. A year later, Hölder gave the generalization (6.10),[13] except that he took $\phi(x)$ to be differentiable, with $\phi'(x)$ increasing. It is not difficult to prove that such functions are convex in Jensen's sense. Hölder noted that his work was based on that of Rogers, and Jensen also credited Rogers.

The case $p = q = 2$ of (6.11) is called the Cauchy–Schwarz inequality. Cauchy derived it[14] from an identity with the form, here given for three dimensions:

$$(ax + by + cz)^2 + (ay - bx)^2 + (az - cx)^2 + (bz - cy)^2$$
$$= (a^2 + b^2 + c^2)(x^2 + y^2 + z^2). \tag{6.12}$$

It is clear that the identity implies

$$\sum ax \le \left(\sum a^2\right)^{\frac{1}{2}} \left(\sum x^2\right)^{\frac{1}{2}} \tag{6.13}$$

[11] Jensen (1906).
[12] Rogers (1888).
[13] Hölder (1889).
[14] Cauchy et al. (2007) p. 304.

and that equality would hold if $ay = bx, az = cx$, and $bz = cy$, that is, $\frac{a}{x} = \frac{b}{y} = \frac{c}{z}$. It is evident that identity (6.12) can be extended to any number of variables. Eighteenth-century mathematicians such as Euler and Lagrange applied this and other identities involving sums of squares to physics problems, number theory and other areas.

In 1885, Hermann Schwarz (1843–1921) gave the integral analog of the inequality with which his name is now associated.[15] But this analog was actually presented as early as 1859[16] by Viktor Bunyakovski (1804–1859), a Russian mathematician with an interest in probability theory who had studied with Cauchy in Paris. Though Bunyakovski made no claim to this result, it is sometimes called the Cauchy–Schwarz–Bunyakovski inequality. Bunyakovski was very familiar with Laplace's work in probability theory, a subject in which he did his best work and for which he worked out a Russian terminology, introducing many terms which have became standard in that language.

We briefly mention other related results, without detailed definitions. One of the earliest applications of the infinite form of the Cauchy–Schwarz and Hölder inequalities was in functional analysis, dealing with infinite series and integrals. For example, in a pioneering paper of 1906,[17] David Hilbert defined $l^2$ spaces consisting of sequences of complex numbers $\{a_n\}$ such that the sum of the squares of absolute values converged. The infinite form of the Cauchy–Schwarz inequality may be employed to show that an inner product can be defined on $l^2$. In a paper of 1910,[18] the Hungarian mathematician Frigyes Riesz (1880–1956) generalized Hilbert's work. Dieudonné called this paper "second only in importance for the development of Functional Analysis to Hilbert's 1906 paper."[19] Riesz kept well abreast of the work of Hilbert, Erhard Schmidt, Ernst Hellinger, Otto Toeplitz, Ernst Fischer, Henri Lebesgue, Jacques Hadamard, and Maurice Fréchet. With such inspiration, Riesz was able to define and develop the theory of $l^p$ and $L^p$ spaces. By using Minkowski's inequality, he proved that these were vector spaces; employing the Hölder inequality, he showed that $l^q$ and $L^q$ were duals of $l^p$ and $L^p$, where $q = \frac{(p-1)}{p}$ and $p > 1$. In a proof very different from Minkowski's proof related to the geometry of numbers, Riesz demonstrated that Minkowski's inequality could be obtained from Hölder's. Thus, inequalities originating in the study of algebraic equations eventually led to inequalities now fundamental to analysis.

## 6.2  Harriot's Proof of the Arithmetic and Geometric Means Inequality

Harriot proved the AMGM only in the cases of two and three dimensions, but his motivation, notation, and mode of presentation are worthy of note. Harriot began by proving the inequality for dimension 2:[20]

[15] Schwarz (1885).
[16] Bunyakovski (1859).
[17] Hilbert (1906).
[18] Riesz (1910).
[19] Dieudonné (1981) p. 124.
[20] Harriot and Stedall (2003) p. 195.

**Lemma I** Suppose $b, a, \frac{aa}{b}$ are in continued proportion and suppose $b > a$. I say that $b + \frac{aa}{b} >$ $2a$ that is $bb + aa > 2ab$ so $bb - ba > ba - aa$ that is

$$\left. \begin{array}{c} b - a \\ b \end{array} \right| > \left. \begin{array}{c} b - a \\ a \end{array} \right. \tag{6.14}$$

so $b > a$ and this is so. Therefore the lemma is true.

Note that the expression on the left-hand side of (6.14) was Harriot's notation for $(b - a)b$. Harriot used this lemma to analyze the different forms taken by a cubic with one positive root. He proved the three-dimensional case in connection with a result of Viète. In his *De Numerosa Potestatum Resolutione*, Viète discussed a condition for a cubic to have three distinct roots:[21] "A cubic affected negatively by a quadratic term and positively by a linear term is ambiguous [has distinct roots] when three times the square of one-third the linear coefficient [of the square term] is greater than the plane coefficient [of the first power]." Viète's example was $x^3 - 6x^2 + 11x = 6$. Here $3\left(\frac{6}{3}\right)^2 > 11$ and the roots were $1, 2, 3$. Harriot commented that Viète's statement required an amendment; in order to get three positive roots, he required that "the cube of a third of the coefficient of the square term is greater than the given constant." This would yield the three-dimensional case of the inequality. Harriot went on to give an example, showing why Viète's condition was inadequate. He noted that $aaa - 6aa +$ $11a = 12$ had only one positive root (namely, 4) even though Viète's condition was satisfied. In a similar way, he amended Viète's remarks for the case of equal roots.

Harriot stated and proved additional lemmas, of which we give two; he gave the comment, "But what need is there for verbose precepts, when with the formulae from our reduction, it is possible to show all the roots directly, not only for these cases, but for any other case you like. However, if a demonstration of these precepts is required, we adjoin the three following lemmas."[22]

$$\left. \begin{array}{c} 3, \frac{b+c+d}{3} \\ \frac{b+c+d}{3} \end{array} \right| > bc + cd + bd \quad \text{and} \quad \left. \begin{array}{c} \frac{b+c+d}{3} \\ \frac{b+c+d}{3} \\ \frac{b+c+d}{3} \end{array} \right| > bcd.$$

These inequalities are particular cases of (6.7) and can be written as

$$3\left(\frac{b+c+d}{3}\right)^2 > bc + cd + bd \; ; \quad \left(\frac{b+c+d}{3}\right)^3 > bcd.$$

As we noted previously, Descartes made advances over Harriot in terms of notation, though he continued to write $aa$ instead of $a^2$; in fact, this practice continued well into the nineteenth century as one may see in the work of Gauss, Riemann, and others. The notation for the fractional or irrational power was introduced by Newton in his earliest mathematical work.

[21] Viète (1983) p. 360.
[22] ibid. pp. 233–234.

## 6.3  Maclaurin's Inequalities

Maclaurin's novel proof of the arithmetic and geometric means inequality is worth studying, though it used an unproved assumption on the existence of a maximum. The proof consists of two steps, lemmas $V$ and $VI$, contained in his 1729 paper on algebraic equations.[23]

> Lemma V    Let the given line $AB$ be divided anywhere in $P$ and the rectangles of the parts $AP$ and $PB$ will be a maximum when the parts are equal.

In algebraic symbols, Maclaurin's lemma would be stated: If $AB = a$ and $AP = x$, then $x(a - x)$ is maximized when $x = \frac{a}{2}$ for $x$ in the interval $0 \le x \le a$. Maclaurin wrote that this followed from Euclid's *Elements*. He then stated and proved the following generalization:

> Lemma $VI$    If the line $AB$ is divided into any numbers of parts $AC, CD, DE, EB$, the product of all those parts multiplied into one another will be a maximum when the parts are equal among themselves.

$$\overline{\qquad A \quad\ C \quad\ D \qquad E \qquad e \qquad\ B \qquad}$$

> For let the point $D$ be where you will, it is manifest that if $DB$ be bisected in $E$, the product $AC \times CD \times DE \times EB$ will be greater than $AC \times CD \times De \times eB$, because $DE \times EB$ is greater than $De \times eB$; and for the same reason $CE$ must be bisected in $C$ and $D$; and consequently all the parts $AC, CD, DE, EB$ must be equal among themselves, that their product may be a maximum.

In other words, Maclaurin argued that if $\alpha_1, \alpha_2, \dots, \alpha_n$ are positive quantities not all equal to each other and their sum $\sum \alpha_i = A$ is a constant, then there exist $\alpha_1', \alpha_2', \dots, \alpha_n'$ with $\sum \alpha_i' = A$ and $\alpha_1' \alpha_2' \cdots \alpha_n' > \alpha_1 \alpha_2 \cdots \alpha_n$. Thus, if a maximum value of the product exists, then it must occur when all the $\alpha$ are equal. Maclaurin assumed that such a maximum must exist; proving this would boil down to showing that the continuous function of $n - 1$ variables

$$\alpha_1 \alpha_2 \cdots \alpha_{n-1} \left( A - \alpha_1 - \alpha_2 - \cdots - \alpha_{n-1} \right)$$

has a maximum in the closed domain $\alpha_1 \ge 0, \alpha_2 \ge 0, \dots, \alpha_{n-1} \ge 0$, $\alpha_1 + \alpha_2 + \cdots + \alpha_{n-1} \le A$. It was common for eighteenth-century mathematicians to assume the existence of such a maximum. Lagrange did this extensively in his derivation of the Taylor theorem with remainder. The inequality for the arithmetic and geometric means follows from these lemmas. We see that if all values of $\alpha_i$ are equal, then $\alpha_i = \frac{A}{n}$ and we can conclude that

$$\alpha_1 \alpha_2 \cdots \alpha_n \le \left( \frac{A}{n} \right)^n = \left( \frac{\alpha_1 + \alpha_2 + \cdots + \alpha_n}{n} \right)^n.$$

Moreover, equality holds if and only if all the $\alpha_i$ are identical.

---

[23] Maclaurin (1729).

## 6.4 Comments on Newton's and Maclaurin's Inequalities

The purpose of Newton's inequalities (6.5) and (6.6) was to determine the number of complex roots of an algebraic equation

$$p_0 x^n + \binom{n}{1} p_1 x^{n-1} + \binom{n}{2} p_2 x^{n-2} + \cdots + \binom{n}{n-1} p_{n-1} x + p_n = 0. \quad (6.15)$$

Newton's inequalities have interesting implications in the case in which all the roots of the equation are real and not all identical. In such a case, the inequalities are given by

$$A_k = p_k^2 - p_{k-1} p_{k+1} > 0, \quad k = 1, 2, \ldots, n-1. \quad (6.16)$$

We reproduce the essential ideas due to George Campbell[24] and Colin Maclaurin[25] to show that if all the roots of an algebraic equation

$$a_0 x^n + a_1 x^{n-1} + \cdots + a_n = 0, \quad a_i = \binom{n}{i} p_i \quad (6.17)$$

are real and not all identical, then the inequalities (6.16) hold true. Campbell and Maclaurin applied Rolle's theorem: If the value of a polynomial $p(x)$ is zero at two points, then $p'(x)$ has value zero at a point in between those points. They also used the fact that if a polynomial $p(x)$ has a zero of order $m \geq 1$ at $x_1$, then $p'(x)$ has a zero of order $m - 1$ at $x_1$. Note that $x_1$ is a zero of order $m$ for $p(x)$ if $p(x) = (x - x_1)^m q(x)$ and $q(x_1) \neq 0$, that is, if $(x - x_1)^m \mid p(x)$ but $(x - x_1)^{m+1} \nmid p(x)$.

Suppose that $x = 0$ is not a root of (6.17). Then $a_n \neq 0$. Now let $\alpha_1, \alpha_2, \ldots, \alpha_n$ be the roots of (6.17). Campbell and Maclaurin then considered the equation whose roots were $\frac{1}{\alpha_1}, \frac{1}{\alpha_2}, \ldots, \frac{1}{\alpha_n}$ and the derivatives of such an equation, that is, the equation

$$a_n y^n + a_{n-1} y^{n-1} + \cdots + a_0 = 0. \quad (6.18)$$

In order to deal with both (6.17) and (6.18) simultaneously, we consider the equation in two variables, $x$ and $y$,

$$a_0 x^n + a_1 x^{n-1} y + a_2 x^{n-2} y^2 + \cdots + a_n y^n = 0, \quad (6.19)$$

and take the partial derivatives of (6.19) with respect to $x$ as well as $y$. Supposing all the roots $\frac{x}{y}$ of (6.19) to be real, we obtain an equation after several partial derivatives with respect to $x$ as well as $y$. It follows from Rolle's theorem and its implications that, if this new equation (obtained after several differentiations) has a root $\alpha$ of order $m > 1$, then $\alpha$ must be a root of order $m + 1$ of the equation from which the new equation is obtained directly by a single differentiation.

---

[24] Campbell (1728).
[25] Maclaurin (1729).

If $a_n \neq 0$ is assumed, then $\frac{x}{y} = 0$ is not a root of (6.19) and this implies that $\frac{x}{y} = 0$ cannot be a multiple root of any equation obtained by partial differentiation of (6.19) with respect to $x$ and $y$. Now take $m - 1$ derivatives of (6.19) with respect to $y$, followed by $n - m - 1$ derivatives with respect to $x$; one arrives at the result

$$(m-1)! \frac{(n-m+1)!}{2!} a_{m-1} x^2 + \frac{m!}{1!} \frac{(n-m)!}{1!} a_m \, xy$$
$$+ \frac{(m+1)!}{2!} (n-m-1)! \, a_{m+1} \, y^2 = 0. \qquad (6.20)$$

Divide (6.20) by $\frac{(n-m)!}{2!}$ and set $p_k = \frac{a_k}{\binom{n}{k}}$ to get

$$p_{m-1} x^2 + 2 p_m \, xy + p_{m+1} \, y^2 = 0. \qquad (6.21)$$

Now note that $p_m$ and $p_{m+1}$ cannot both be zero, because in that case the derived equation (6.20) would have zero as a multiple root. Hence the quadratic in (6.21) is not identically zero and it has real roots. This implies that

$$p_{m-1} \, p_{m+1} \leq p_m^2, \qquad (6.22)$$

where equality holds in case the roots of (6.21) are equal. The considerations on Rolle's theorem show that the roots of (6.21) are equal when all the roots of (6.19) are identical, but we have assumed this is not true. Hence

$$p_{m-1} \, p_{m+1} < p_m^2 \qquad (6.23)$$

and Newton's identities hold.

As we have seen, Maclaurin did not derive his refinement of the AMGM inequality (6.7) directly from Newton's inequalities. However, the intimate relationship between the two sets of inequalities is quite interesting. Observe that, with $p_0 = 1$ and

$$p_k^2 - p_{k-1} \, p_{k+1} > 0, \qquad (6.24)$$

when the points $(k, \ln p_k)$ are plotted, the straight lines joining consecutive points have decreasing slopes. Observe this by writing $y_k = \ln p_k$ and noting that (6.24) can be rewritten as[26]

$$y_{k+1} - y_k < y_k - y_{k-1}. \qquad (6.25)$$

Now in Figure 6.1, $y_k - y_{k-1}$ is the slope of the line joining the points $(k-1, y_{k-1})$ and $(k, y_k)$. Figure 6.1 then shows the plot of the points; note that the lines joining the consecutive points have decreasing slopes. The Maclaurin inequalities

$$p_{k-1}^{\frac{1}{k-1}} > p_k^{\frac{1}{k}}, \quad k = 2, 3, \ldots, n$$

[26] See Steele (2004) p. 180.

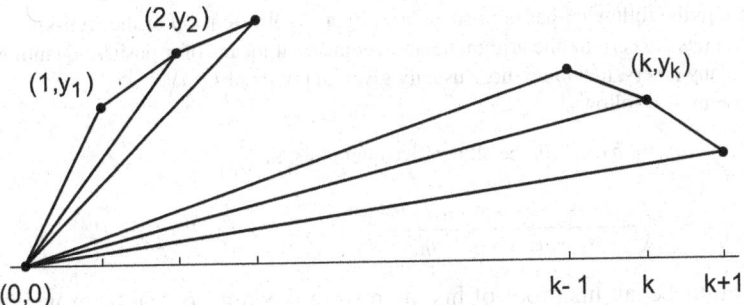

Figure 6.1  Maclaurin's inequality.

are equivalent to

$$\frac{1}{k-1} y_{k-1} > \frac{1}{k} y_k, \quad k = 2, 3, \ldots, n.$$

Clearly, the slopes of the lines joining the origin to $(k, y_k)$, with $k = 1, 2, 3, \ldots, n$, are decreasing. Although this is obvious from Figure 6.1, a rigorous proof can be given by induction. Since $p_0 = 1$, the case $k = 2$ holds true because the inequality $y_1 > \frac{1}{2} y_2$ is equivalent to the inequality $p_2 \, p_0 < p_1^2$. Now suppose the result true for $k < n$, that is

$$\frac{1}{k} y_k < \frac{1}{k-1} y_{k-1} \quad \text{when} \quad k < n. \tag{6.26}$$

Next, note that (6.26) implies that $y_{k-1} > \frac{k-1}{k} y_k$ so that

$$y_{k+1} - y_k < y_k - y_{k-1}$$
$$< y_k - \frac{k-1}{k} y_k = \frac{1}{k} y_k$$

or

$$y_{k+1} < y_k + \frac{1}{k} y_k = \frac{k+1}{k} y_k,$$

completing the inductive proof of Maclaurin's inequalities.

## 6.5  Rogers

L. J. Rogers began his 1888 paper,[27] "An extension of a certain theorem in inequalities," with the remark,

[27] Rogers (1888).

I propose in the following pages to show how, by a slight extension of the well-known theorem in inequalities concerning the arithmetic and geometrical means of $n$ positive quantities, we can deduce many others, including those usually given in text-books. The theorem is as follows:

If $a_1, a_2, \ldots, a_n, b_1, b_2, \ldots, b_n$ be all positive quantities, then

$$\left( \frac{a_1 b_1 + a_2 b_2 + \cdots + a_n b_n}{a_1 + a_2 + \cdots + a_n} \right)^{a_1 + a_2 + \cdots + a_n} \geq b_1^{a_1} b_2^{a_2} \cdots b_n^{a_n}. \tag{6.27}$$

Rogers thus began his proof of his interesting theorem (6.27), from which he was able to deduce many standard inequalities. He first let $a_1, a_2, \ldots, a_n$ be positive integers so that the inequality (6.27) was reduced to the known result (6.2). He next took $a_1, a_2, \ldots, a_n$ to represent rational numbers and let $N$ denote the least common multiple of the denominators of $a_1, a_2, \ldots, a_n$ so that $Na_1 = A_1$, $Na_2 = A_2$, $\ldots$, $Na_n = A_n$, with $A_1, A_2, \ldots, A_n$ integers. He noted that since the inequality (6.27) was true for integers $A_1, A_2, \ldots, A_n$, it followed that

$$\left( \frac{A_1 b_1 + A_2 b_2 + \cdots + A_n b_n}{A_1 + A_2 + \cdots + A_n} \right)^{A_1 + A_2 + \cdots + A_n} \geq b_1^{A_1} b_2^{A_2} \cdots b_n^{A_n}.$$

Taking the positive $N$th root of each side then produced the required result. He finally took $a_1, a_2, \ldots, a_n$ to be irrational numbers and applied the continuity argument. He gave the argument, "Then we may substitute for each of these quantities fractions, which may differ from them by less than any assigned quantities, and since the theorem is true for the substituted fractions, we may assume it is also true for the given incommensurables." This proved the inequality (6.27) completely.

Rogers derived a new inequality from (6.27): For $m > r > t > 0$,

$$\left( \sum_{i=1}^{n} a_i b_i^r \right)^{m-t} \leq \left( \sum_{i=1}^{n} a_i b_i^m \right)^{r-t} \left( \sum_{i=1}^{n} a_i b_i^t \right)^{m-r}. \tag{6.28}$$

To prove (6.28), he denoted $S_k = \sum_{i=1}^{n} a_i^k$ and first showed that

$$S_r^{m-t} \leq S_m^{r-t} S_t^{m-r} \tag{6.29}$$

by replacing $a_i$ by $a_i^r$ and $b_i$ by $a_i^{m-r}$ in (6.27) to obtain

$$\left( \frac{S_m}{S_r} \right)^{S_r} \geq \left( a_1^{a_1^r} a_2^{a_2^r} \cdots \right)^{m-r}. \tag{6.30}$$

Next, Rogers replaced $b_i$ by $a_i^{t-r}$ to get

$$\left( \frac{S_t}{S_r} \right)^{S_r} \geq \left( a_1^{a_1^r} a_2^{a_2^r} \cdots \right)^{-(r-t)},$$

$$\left( \frac{S_r}{S_t} \right)^{S_r} \leq \left( a_1^{a_1^r} a_2^{a_2^r} \cdots \right)^{(r-t)}. \tag{6.31}$$

Combining (6.30) and (6.31) yielded

$$\left(\frac{S_m}{S_r}\right)^{\frac{S_r}{m-r}} \geq \left(\frac{S_r}{S_t}\right)^{\frac{S_r}{r-t}} ;$$

and by taking the $S_r$th root of each side and rearranging terms, Rogers arrived at (6.29) and, taking $a_i = 1$, rewrote it as

$$\left(\sum_{i=1}^{n} b_i^r\right)^{m-t} \leq \left(\sum_{i=1}^{n} b_i^m\right)^{r-t} \left(\sum_{i=1}^{n} b_i^t\right)^{m-r}. \qquad (6.32)$$

Rogers wrote that (6.32) implied (6.28) in the same way that the case for integers, which he called the classical case, implied (6.27). To see this, first take $a_i$, $i = 1, 2, \ldots n$ to be rational and reduce to the integer case. Finally, take $a_i$ to be irrational and apply the continuity argument.

To put these inequalities in more standard form, take the $(m-t)$th root of each side of (6.28) to get

$$\sum_{i=1}^{n} a_i b_i^r \leq \left(\sum_{i=1}^{n} a_i b_i^m\right)^{\frac{r-t}{m-t}} \left(\sum_{i=1}^{n} a_i b_i^t\right)^{\frac{m-r}{m-t}}. \qquad (6.33)$$

Let $\frac{1}{p} = \frac{r-t}{m-t}$ and $\frac{1}{q} = \frac{m-r}{m-t}$ so that $\frac{1}{p} + \frac{1}{q} = 1$. Also let $A_i^p = a_i b_i^m$ and $B_i^q = a_i b_i^t$ so that $A_i B_i = a_i b_i^r$.

Now, with $p > 0$ and $q > 0$ (6.33) can be rewritten as

$$\sum_{i=1}^{n} A_i B_i \leq \left(\sum_{i=1}^{n} A_i^p\right)^{\frac{1}{p}} \left(\sum_{i=1}^{n} B_i^q\right)^{\frac{1}{q}}, \quad \frac{1}{p} + \frac{1}{q} = 1. \qquad (6.34)$$

The integral form of (6.28) was also given by Rogers. Note that the integral form of (6.34) would be: Supposing $f$ and $g$ to be positive, integrable functions in $[a, b]$, then

$$\int_a^b f(x)g(x)\,dx \leq \left(\int_a^b f^p(x)\,dx\right)^{\frac{1}{p}} \left(\int_a^b g^q(x)\,dx\right)^{\frac{1}{q}}. \qquad (6.35)$$

Inequalities (6.34) and (6.35) are now called Hölder's inequalities. Observe that the integral form (6.35) can be obtained from (6.34) by writing the integrals as the limits of sums.

Rogers applied (6.35) to derive inequalities for the gamma and beta integrals; note these integrals are defined

$$\Gamma(s) = \int_0^{\infty} x^{s-1} e^{-x}\,dx$$

and

$$B(s,t) = \int_0^1 x^{s-1}(1-x)^{t-1}\,dx.$$

The inequalities for these integrals were found by Rogers as:

$$\big(\Gamma(r)\big)^{m-t} \le \big(\Gamma(m)\big)^{r-t}\big(\Gamma(t)\big)^{m-r}, \quad m > r > t \qquad (6.36)$$

and

$$B(l,r)^{m-t} \le B(l,m)^{r-t}\,B(l,t)^{m-r}, \quad m > r > t; \qquad (6.37)$$

note that they follow immediately from the integral form of (6.32), equivalent to (6.35).

## 6.6  Hölder

In his 1889 paper,[28] "Ueber einen Mittelwerthabsatz," Otto Hölder wrote that he had a general theorem from which Rogers's inequalities could be derived and that this theorem had connections to the principles of differential calculus. In order to state the Rogers's theorem, we define the concept of a convex function: A function $f(x)$ on $[a,b]$ is called convex if the line joining any pair of points on the curve $y = f(x)$ lies above the curve as shown in Figure 6.2.

Let $t$ be any point in $[a,b]$. The curve $PSQ$ is given by $y = f(x)$ and the coordinates of $P$, $S$, and $Q$ are respectively $(a,f(a))$, $(t,f(t))$, and $(b,f(b))$. Note that $t$ can be written uniquely as $t = (1-\alpha)a + \alpha b$, where $0 \le \alpha \le 1$. By means

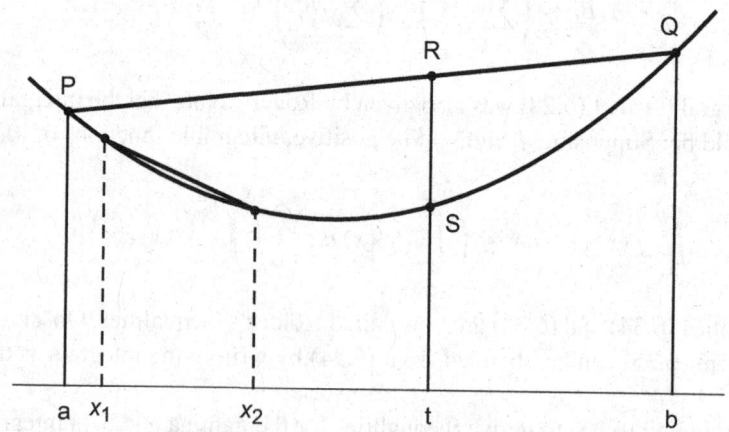

Figure 6.2  A convex function.

28  Hölder (1889).

of similar triangles, one can show that the coordinates of the point $R$ must be $(t, (1 - \alpha) f(a) + \alpha f(b))$. The fact that $R$ lies above the point $S$ can thus be expressed by the inequality

$$f\big((1 - \alpha)a + \alpha b\big) \leq (1 - \alpha) f(a) + \alpha f(b).$$

Thus, a function $f$ is convex on $[a, b]$ if, for every pair of points $x_1, x_2 \in [a, b]$, we have

$$f\big((1 - \alpha)x_1 + \alpha x_2\big) \leq (1 - \alpha) f(x_1) + \alpha f(x_2), \quad 0 \leq \alpha \leq 1. \tag{6.38}$$

We may therefore take the inequality (6.38) as the definition of a convex function, a definition equivalent to Jensen's, given in equation (6.8) with accompanying condition.

Next, suppose that $x_1 < x_2 < x_3 < \cdots < x_n$ are $n$ points in the interval $[a, b]$ where $f(x)$ is convex. Let $\alpha_1, \alpha_2, \ldots, \alpha_n$ represent $n$ positive numbers whose sum is one, that is, $\alpha_1 + \alpha_3 + \cdots + \alpha_n = 1$. In that case,

$$f(\alpha_1 x_1 + \alpha_2 x_2 + \cdots + \alpha_n x_n) \leq \alpha_1 f(x_1) + \alpha_2 f(x_2) + \cdots + \alpha_n f(x_n) = 1. \tag{6.39}$$

Hölder proved the lemma (6.39) by induction; clearly, the result is true for $n = 2$. Now assume the result true for $n - 1$. That means that, given $n - 1$ positive numbers such that $\beta_1 + \beta_2 + \cdots + \beta_{n-1} = 1$ and given $y_1 < y_2 < \cdots < y_{n-1}$ points in $[a, b]$, one has

$$f(\beta_1 y_1 + \beta_2 y_2 + \cdots + \beta_{n-1} y_{n-1}) \leq \beta_1 f(y_1) + \cdots + \beta_{n-1} f(y_{n-1}). \tag{6.40}$$

Let

$$s_{n-1} = \frac{\alpha_1 x_1 + \alpha_2 x_2 + \cdots + \alpha_{n-1} x_{n-1}}{\alpha_1 + \alpha_2 + \cdots + \alpha_{n-1}};$$

the inductive hypothesis then implies that

$$f(s_{n-1}) \leq \frac{\alpha_1 f(x_1) + \alpha_2 f(x_2) + \cdots + \alpha_{n-1} f(x_{n-1})}{\alpha_1 + \alpha_2 + \cdots + \alpha_{n-1}}.$$

Now we can see that

$$s_n = \alpha_1 x_1 + \cdots + \alpha_{n-1} x_{n-1} + \alpha_n x_n = (\alpha_1 + \cdots + \alpha_{n-1}) s_{n-1} + \alpha_n x_n$$

so that

$$f(s_n) \leq (\alpha_1 + \cdots + \alpha_{n-1}) f(s_{n-1}) + \alpha_n f(x_n)$$
$$\leq \alpha_1 f(x_1) + \cdots + \alpha_n f(x_n).$$

We can now state Hölder's basic theorem: Suppose that $f(x)$ is twice differentiable in $(a, b)$. Then if $f'(x)$ is increasing in $(a, b)$ [or if $f''(x) \geq 0$ in $(a, b)$], then $f$ is convex in $(a, b)$.

To prove this, Hölder set $\check{x}_1, x_2 \in (a, b)$ and let $x_1 < s < x_2$ so that there must exist positive numbers $a_1, a_2$ such that

$$s = \frac{a_1 x_1 + a_2 x_2}{a_1 + a_2}.$$

He applied the mean value theorem to $f(x)$ on the interval $[x_1, s]$ to obtain

$$f(s) - f(x_1) = f'(c_1)(s - x_1)$$

$$= f'(c_1) \frac{a_2(x_2 - x_1)}{a_1 + a_2}, \quad x_1 < c_1 < s, \tag{6.41}$$

and next applied the mean value theorem to $f(x)$ on $[s, x_2]$ to arrive at

$$f(x_2) - f(s) = f'(c_2)(x_2 - s)$$

$$= f'(c_2) \frac{a_1(x_2 - x_1)}{a_1 + a_2}, \quad s < c_2 < x_2. \tag{6.42}$$

Hölder multiplied equation (6.41) by $a_1$ and (6.42) by $a_2$ and then subtracted the resulting first equation from the second to get

$$a_1 f(x_1) + a_2 f(x_2) - (a_1 + a_2) f(s) = \left(f'(c_2) - f'(c_1)\right) \frac{a_1 a_2 (x_2 - x_1)}{a_1 + a_2}. \tag{6.43}$$

Since $c_2 > c_1$ and since $f'$ was increasing, the right-hand side of (6.43) would be positive and Hölder could conclude that

$$f(s) = f\left(\frac{a_1 x_1 + a_2 x_2}{a_1 + a_2}\right) \leq \frac{a_1}{a_1 + a_2} f(x_1) + \frac{a_2}{a_1 + a_2} f(x_2), \tag{6.44}$$

completing the proof that $f$ was convex.

From (6.44), an inductive proof could be given that if $f$ were convex in an interval and $x_1, x_2, \ldots, x_n$ were points in that interval, with $a_1, a_2, \ldots, a_n$ positive numbers, then

$$f\left(\frac{a_1 x_1 + a_2 x_2 + \cdots + a_n x_n}{a_1 + a_2 + \cdots a_n}\right) \leq \sum_{i=1}^{n} \frac{a_i}{a_1 + \cdots + a_n} f(x_i). \tag{6.45}$$

Hölder observed that $f(x) = e^x$ was a convex function, because $f''(x) = e^x > 0$. Hence (6.45) implied that

$$\exp \sum_{i=1}^{n} \frac{a_i x_i}{a_1 + \cdots + a_n} \leq \sum_{i=1}^{n} \frac{a_i e^{x_i}}{a_1 + \cdots + a_n}. \tag{6.46}$$

Now note that, given positive numbers $b_i$, $i = 1, 2, \ldots, n$, there exist numbers $x_i$ such that $e^{x_i} = b_i$. When Hölder substituted these values in (6.46), the result was Rogers's inequality (6.27).

To derive Rogers's second inequality (6.28), Hölder applied Taylor's theorem with remainder:

$$f(x_\nu) = f(\sigma) + (x_\nu - \sigma) f'(\sigma) + \frac{1}{2}(x_\nu - \sigma)^2 f''(\sigma_\nu), \qquad (6.47)$$

where $\sigma_\nu$ lies between $\sigma$ and $x_\nu$. He multiplied (6.47) by $a_\nu$ and added for all $\nu = 1, 2, \ldots, n$ to obtain

$$\sum_{\nu=1}^{n} a_\nu f(x_\nu) = f(\sigma) \sum_{\nu=1}^{n} a_\nu + f'(\sigma) \sum_{\nu=1}^{n} a_\nu(x_\nu - \sigma) + \frac{1}{2} \sum_{\nu=1}^{n}(x_\nu - \sigma)^2 f''(\sigma_\nu). \qquad (6.48)$$

He assumed $f$ to be convex so that $f''(\sigma_\nu) \geq 0$ and thus

$$\sum_{\nu=1}^{n} (x_\nu - \sigma)^2 f''(\sigma_\nu) \geq 0.$$

He then took

$$\sigma = \frac{\sum_{\nu=1}^{n} a_\nu x_\nu}{\sum_{\nu=1}^{n} a_\nu}$$

so that the term $\sum_{\nu=1}^{n} a_\nu(x_\nu - \sigma)$ in (6.48) became zero and (6.48) produced the inequality

$$\sum_{\nu=1}^{n} a_\nu f(x_\nu) \geq \sum_{\nu=1}^{n} a_\nu f\left(\frac{\sum a_\nu x_\nu}{\sum a_\nu}\right). \qquad (6.49)$$

To obtain (6.28), Hölder took $f(x) = x^m$, $m > 1$, so that $f''(x) = m(m-1) x^{m-2} > 0$ for $x > 0$, showing that (6.49) was true for $f(x) = x^m$, $m > 1$. Thus, (6.49) implied that

$$\left(\sum_{\nu=1}^{n} a_\nu\right)^{m-1} \sum_{\nu=1}^{n} a_\nu x_\nu^m \geq \left(\sum_{\nu=1}^{n} a_\nu x_\nu\right)^m. \qquad (6.50)$$

It is not hard to show, using a suitable change in variables, that (6.50) is equivalent to (6.28) and thus to what is now known as Hölder's inequality: For $p, q > 0$ and $\frac{1}{p} + \frac{1}{q} = 1$,

$$\sum_{\nu=1}^{n} a_\nu b_\nu \leq \left(\sum_{\nu=1}^{n} a_\nu^p\right)^{\frac{1}{p}} \left(\sum_{\nu=1}^{n} b_\nu^q\right)^{\frac{1}{q}}$$

or

$$\int_a^b f(x) g(x)\, dx \leq \left(\int_a^b f^p(x)\, dx\right)^{\frac{1}{p}} \left(\int_a^b g^q(x)\, dx\right)^{\frac{1}{q}},$$

where $f$ and $g$ are nonnegative in $(a, b)$.

Let us now interpret Rogers's inequality (6.36), noting that it can be written for $0 \le \alpha \le 1$ as

$$\log \Gamma\big(\alpha x + (1 - \alpha)y\big) \le \alpha \log \Gamma(x) + (1 - \alpha) \log \Gamma(y), \quad x, y > 0.$$

This means that $\log \Gamma(x)$ is a convex function for $x > 0$. Thus, $\Gamma(x)$ would be called logarithmically convex and, as we see in Section 17.14, this is one of the defining properties of $\Gamma(x)$, the others properties being the obvious ones: $\Gamma(1) = 1$ and $\Gamma(x + 1) = x \Gamma(x)$.

## 6.7  Jensen's Inequality

Jensen proved (6.9)[29] by following Cauchy's proof of (6.2) in detail. From the definition of convexity (6.8), he deduced that

$$\phi(x_1) + \phi(x_2) + \phi(x_3) + \phi(x_4) \ge 2\phi\left(\frac{x_1 + x_2}{2}\right) + 2\phi\left(\frac{x_1 + x_2}{2}\right)$$
$$\ge 4\phi\left(\frac{x_1 + x_2 + x_3 + x_4}{4}\right).$$

He showed by an inductive argument that

$$\sum_{\nu=1}^{2^m} \phi(x_\nu) \ge 2^m \phi\left(2^{-m} \sum_{\nu=1}^{2^m} x_\nu\right).$$

This proved the inequality for the case in which the number of $x$s was a power of two. To prove the theorem for any number of $x$s, Jensen, still following Cauchy, applied Cauchy's ingenious idea: For any positive integer $n$, choose $m$ so that $2^m > n$ and set

$$x_{n+1} = x_{n+2} = \cdots = x_{2^m} = \frac{x_1 + x_2 + \cdots + x_n}{n}.$$

Then

$$\sum_{\nu=1}^{n} \phi(x_\nu) + (2^m - n)\phi\left(\frac{1}{n}\sum_{\nu=1}^{n} x_\nu\right) \ge 2^m \phi\left(\frac{1}{n}\sum_{\nu=1}^{n} x_\nu\right)$$

or

$$\phi\left(\frac{1}{n}\sum_{\nu=1}^{n} x_\nu\right) \le \frac{1}{n}\sum_{\nu=1}^{n} \phi(x_\nu).$$

[29] Jensen (1906).

Jensen then used the continuity of $\phi$ to get the more general inequality (6.10). He supposed $a_1, a_3, \ldots, a_m$ to be $m$ positive numbers with sum $a$, as in (6.10). He chose sequences of positive integers $n_1, n_2, \ldots, n_m$ with $n_1 + n_2 + \cdots + n_m = n$ such that

$$\lim_{n\to\infty} \frac{n_1}{n} = \frac{a_1}{a}, \quad \lim_{n\to\infty} \frac{n_2}{n} = \frac{a_2}{a}, \quad \cdots, \quad \lim_{n\to\infty} \frac{n_{m-1}}{n} = \frac{a_{m-1}}{a}.$$

Consequently, he could write

$$\lim_{n\to\infty} \frac{n_m}{n} = \frac{a_m}{a}.$$

Now (6.9) implied that

$$\phi\left(\frac{n_1 x_1 + n_2 x_2 + \cdots + n_m x_m}{n}\right) \le \frac{n_1}{n}\phi(x_1) + \frac{n_2}{n}\phi(x_2) + \cdots + \frac{n_m}{n}\phi(x_m);$$

from this Jensen got (6.10) by letting $n \to \infty$ and using the continuity of $\phi$. Jensen also gave an integral analog of this inequality. He supposed that $a(x)$ and $f(x)$ were integrable on $(0,1)$ and $a(x)$ was positive; $\phi(x)$ was assumed to be convex and continuous in the interval $(g_0, g_1)$, where $g_0$ and $g_1$ were, respectively, the inferior and superior limits of $f(x)$ in $(0,1)$. Then he had

$$\phi\left(\frac{\sum_{\nu=1}^{n} a\left(\frac{\nu}{n}\right) f\left(\frac{\nu}{n}\right) \frac{1}{n}}{\sum_{\nu=1}^{n} a\left(\frac{\nu}{n}\right) \frac{1}{n}}\right) \le \frac{\sum_{\nu=1}^{n} a\left(\frac{\nu}{n}\right) \phi\left(f\left(\frac{\nu}{n}\right)\right) \frac{1}{n}}{\sum_{\nu=1}^{n} a\left(\frac{\nu}{n}\right) \frac{1}{n}}.$$

By letting $n \to \infty$, he found

$$\phi\left(\frac{\int_0^1 a(x) f(x)\, dx}{\int_0^1 a(x)\, dx}\right) \le \frac{\int_0^1 a(x)\phi\left(f(x)\right)\, dx}{\int_0^1 a(x)\, dx}.$$

## 6.8 Riesz's Proof of Minkowski's Inequality

Riesz's derivations[30] of Hölder's and Minkowski's inequalities were contained in his letter to Leonida Tonelli of February 5, 1928. Although Riesz had worked out these ideas almost two decades earlier and had presented them in papers, his object in this letter was to present proofs of the inequalities without any mention of the applications. These proofs are essentially the same as our standard derivations of all these inequalities. Here we use the concept of a measurable set $E$, but $E$ can also be replaced with an interval $(a, b)$. Stating that he did this work around 1910, Riesz started with

$$A^\alpha B^{1-\alpha} \le \alpha A + (1-\alpha)B, \quad 0 < \alpha < 1, \quad A \ge 0, \quad B \ge 0.$$

---

[30] Riesz (1960).

This followed immediately from the convexity of the exponential function, but Riesz gave a simpler proof. After this proof, he supposed $f(x)$ and $g(x)$ were nonnegative functions defined on a measurable set $E$ such that

$$\int_E f^p\, dx = \int_E g^{\frac{p}{p-1}}\, dx = 1, \quad p > 1.$$

He then took $A = f^p$, $B = g^{\frac{p}{p-1}}$, $\alpha = \frac{1}{p}$ to get

$$fg \le \frac{1}{p} f^p + \frac{p-1}{p} g^{\frac{p}{p-1}};$$

thus

$$\int_E fg\, dx \le \frac{1}{p} + \frac{p-1}{p} = 1.$$

For general $f$ and $g$, he replaced $f$ and $g$ by $\dfrac{|f|}{\left|\int_E |f|^p\, dx\right|^{\frac{1}{p}}}$ and $\dfrac{|g|}{\left|\int_E |g|^{\frac{p}{p-1}}\, dx\right|^{\frac{p-1}{p}}}$,

respectively, to obtain

$$\int_E |fg|\, dx \le \left|\int_E |f|^p\, dx\right|^{\frac{1}{p}} \left|\int_E |g|^{\frac{p}{p-1}}\, dx\right|^{\frac{p-1}{p}}.$$

He next cleverly observed that

$$\int_E (f+g)^p\, dx = \int_E f(f+g)^{p-1}\, dx + \int_E g(f+g)^{p-1}\, dx.$$

With $f \ge 0$ and $g \ge 0$, he had

$$\int_E (f+g)^p\, dx \le \left(\int_E f^p\, dx\right)^{\frac{1}{p}} \left(\int_E (f+g)^p\, dx\right)^{\frac{p-1}{p}}$$
$$+ \left(\int_E g^p\, dx\right)^{\frac{1}{p}} \left(\int_E (f+g)^p\, dx\right)^{\frac{p-1}{p}}.$$

Dividing across by $\left(\int_E (f+g)^p\, dx\right)^{\frac{p-1}{p}}$, he could obtain

$$\left(\int_E (f+g)^p\, dx\right)^{\frac{1}{p}} \le \left(\int_E f^p\, dx\right)^{\frac{1}{p}} + \left(\int_E g^p\, dx\right)^{\frac{1}{p}},$$

and this was Minkowski's inequality, stated and used by Minkowski for sums in geometry of numbers.

## 6.9 Exercises

(1) Let $s_n = u_0 + u_1 + \cdots + u_n$, where the terms $u_i$ are positive.

(a) Show that $\ln \frac{s_n}{s_{n-1}} < \frac{u_n}{s_{n-1}}$.

(b) Deduce that

$$\frac{u_1}{s_0} + \frac{u_2}{s_1} + \cdots + \frac{u_n}{s_{n-1}} > \ln s_n - \ln s_0.$$

(c) Prove that if $\sum_{n=1}^{\infty} u_n$ is divergent, then $\sum_{n=1}^{\alpha} \frac{u_n}{s_n^\alpha}$ is divergent for $\alpha \leq 1$.

(d) Show that when $\alpha > 0$,

$$s_{n-1}^{-\alpha} - s_n^{-\alpha} = (s_n - u_n)^{-\alpha} - s_n^{-\alpha} > s_n^{-\alpha} + \alpha s_n^{-\alpha-1} u_n - s_n^{-\alpha} = \alpha \cdot \frac{u_n}{s_n^{1+\alpha}}.$$

(e) Deduce that if $\sum_{n=1}^{\infty} u_n$ is divergent, then $\sum_{n=1}^{\infty} \frac{u_n}{s_n^{1+\alpha}}$ is convergent for $\alpha > 0$. See Abel (1965) vol. 2, pp. 197–98.

(2) Suppose $p > 1$ and $a_i > 0$. Suppose that the series $L = \sum_{i=1}^{\infty} a_i x_i$ converges for every system of positive numbers $x_i$ ($i = 1, 2, \ldots$) such that $\sum_{i=1}^{\infty} x_i^p = 1$.

Use Abel's result in Exercise 1 to prove that $\sum_{i=1}^{\infty} a_i^{\frac{p}{p-1}}$ is convergent and that

$$L \leq \left( \sum_{i=1}^{\infty} a_i^{\frac{p}{p-1}} \right)^{\frac{p-1}{p}}. \text{ See Landau (1907c)}.$$

(3) Prove that if $h$ is measurable and $\int_a^b |f(x)h(x)|\, dx$ exists for all functions $f \in L^p(a,b)$, then $h \in L^{\frac{p}{p-1}}(a,b)$. See Riesz (1960) vol. 1, pp. 449–451.

(4) Show that

$$\sum_{m=1}^{\infty} \sum_{n=1}^{\infty} \frac{a_m b_n}{m+n} < 2\pi \left( \sum_{m=1}^{\infty} a_m^2 \right)^{\frac{1}{2}} \left( \sum_{n=1}^{\infty} b_n^2 \right)^{\frac{1}{2}}.$$

Hilbert presented this result in his lectures on integral equations. It was first published in 1908 in Hermann Weyl's doctoral dissertation. I. Schur proved that the constant $2\pi$ could be replaced by $\pi$. See Steele (2004).

(5) Where $p_1, p_2, \ldots, p_n$ are real, let

$$f(x,y) = (x + \alpha_1 y)(x + \alpha_2 y) \cdots (x + \alpha_n y)$$

$$= x^n + n p_1 x^{n-1} y + \binom{n}{2} p_2 x^{n-2} y^2 + \cdots + \binom{n}{n} p_n y^n.$$

(a) Derive the quadratic polynomial obtained by first taking the $r$th derivative of $f(x,y)$ with respect to $y$ and then the $(n - r - 2)$nd derivative with respect to $x$ of $f_y^{(r)}(x,y)$.

(b) Use the quadratic polynomial to show that if all $\alpha_1, \alpha_2, \ldots, \alpha_n$ are real, then $p_{r+1}^2 \geq p_r p_{r+2}$. This is in effect the argument George Campbell gave to show that if $p_{r+1}^2 < p_r p_{r+2}$ for some $r$, then $f(x,1)$ has at least two complex roots. In fact, he did not use the variable $y$; instead, he applied the lemma he stated and proved:

> Whatever be the number of impossible roots in the equation
>
> $$x^n - Bx^{n-1} + Cx^{n-2} - Dx^{n-3} + \cdots \pm dx^3 \mp cx^2 \pm bx \mp A = 0,$$
>
> there are just as many in the equation
>
> $$Ax^n - bx^{n-1} + cx^{n-2} - dx^{n-3} + \cdots \pm Dx^3 \mp Cx^2 \pm Bx \mp 1 = 0.$$
>
> For the roots of the last equation are the reciprocals of those of the first as is evident from common algebra.

This lemma is also contained in Newton's *Arithmetica Universalis*. Newton explained that the equation for the reciprocals of the roots of $f(x)$ was given by $x^n f\left(\frac{1}{x}\right) = 0$.

(6) Suppose that $\alpha_1, \alpha_2, \ldots, \alpha_n$ in Exercise 5 are positive. Show that

$$p_2 (p_1 p_3)^2 (p_2 p_4)^3 \cdots (p_{k-1} p_{k+1})^k < p_1^2 p_2^4 p_3^6 \cdots p_k^{2k}.$$

Deduce Maclaurin's inequality (6.7) that $p_{k+1}^{\frac{1}{k+1}} < p_k^{\frac{1}{k}}$. See Hardy, Littlewood, and Pólya (1967).

(7) Fourier's proof of Descartes's rule of signs: Suppose that the coefficients of the given polynomial have the following signs:

$$+ \; + \; - \; + \; - \; - \; - \; + \; + \; - \; + \; - \; .$$

Multiply this polynomial by $x - p$ where $p$ is positive. The result is

$$
\begin{array}{c}
+ \; + \; - \; + \; - \; - \; - \; + \; + \; - \; + \; - \\
\underline{\; - \; - \; + \; - \; + \; + \; + \; - \; - \; + \; - \; +} \\
+ \; \pm \; - \; + \; - \; \mp \; \mp \; + \; \pm \; - \; + \; - \; +
\end{array}.
$$

The ambiguous sign $\pm$ appears whenever there are two terms with different signs to be added. Show that in general the ambiguous sign appears whenever $+$ follows $+$ or $-$ follows $-$. Next show that the number of sign variations is not diminished by choosing either of the ambiguous signs. Also prove that there is always one variation added at the end, whether or not the original polynomial ends with a variation, as in our example. Show by induction that these facts, taken together, demonstrate Descartes's rule. Descartes indicated no proof for his rule. In 1728, J. A. von Segner gave a proof and in 1741 the French Jesuit priest J. de Gua de Malves gave a similar proof, apparently independently. We remark that de Gua also wrote a short history of algebra in which he emphasized French contributions to algebra at the expense of the English, in order to counter Wallis's 1685 history, emphasizing the opposite.

Fourier presented the method described in this exercise in his lectures at the École Polytechnique, soon after its inauguration in November 1794. In 1789, Fourier communicated a paper on the theory of equations to the Académie des Sciences in Paris but due to the outbreak of the French Revolution the paper was lost. In the late 1790s, Fourier's interests turned to problems of heat conduction; it was not until around 1820 that he returned to the theory of equations. His book on equations was published posthumously in 1831.

(8) Gauss's proof of Descartes's rule: With his extraordinary mathematical insight, Gauss saw the essence of Fourier's argument and presented it in a general form. He supposed

$$x^{n+1} + A_1 x^n + A_2 x^{n-1} + \cdots + A_{n+1}$$
$$= (x - p)(x^n + a_1 x^{n-1} + a_3 x^{n-2} + \cdots + a_n),$$

and that the sign changes occurred at $a_{k_1}, a_{k_2}, \ldots, a_{k_s}$. Show that $A_{k_j} = a_{k_j} - p a_{k_{j-1}}$ and that this in turn implies that the signs of $A_{k_j}$ and $a_{k_j}$ are identical for $j = 1, 2, \ldots, s$. Deduce also that there is an odd number of sign changes between $A_{k_{i-1}}$ and $A_{k_i}$. Conclude, by induction, that the number of sign changes is an upper bound for the number of positive roots and that the two differ by an even number. Gauss published this result in 1828 in the newly founded *Crelle's Journal*. Note that Gauss did not use subscripts; we use them for convenience. See Gauss (1863–1927) vol. 3, pp. 67–70.

(9) Fourier's extension of Descartes's rule gives an upper bound on the number of real roots of a polynomial $f(x)$ of degree $n$ in an interval $(a, b)$. Suppose $r$ is the number of real roots in $(a, b)$, $m$ is the number of sign changes in the sequence

$$f(x), f'(x), f''(x), \ldots, f^{(n)}(x)$$

when $x = a$, and $k$ is the number of sign changes when $x = b$. Prove that then $(m - k) - r = 2p$, where $p$ is a nonnegative integer. Descartes's rule follows when $a = 0$ and $b = \infty$. In his 1831 book, Fourier gave a very leisurely account of this theorem with numerous examples.

(10) Ferdinand François Budan's (1761–1840) extension of Descartes's rule: With the notation as in the previous exercise, suppose that $m$ is the number of sign changes in coefficients of powers of $x$ in $f(x + a)$, and that $k$ is the corresponding number in $f(x + b)$. Then, $r \leq m - k$. Prove this theorem and also prove that it follows from Fourier's theorem. Budan was born in Haiti and was a physician by training. In 1807, he wrote a pamphlet on his theorem; then in 1811 he presented a paper to the Paris Academy. Lagrange and Legendre recommended it be published, but the Academy's journal was not printed until 1827, partly due to political problems. With the appearance of Fourier's papers in 1818 and 1820, Budan felt compelled to republish his pamphlet with the paper as an appendix. In response, Fourier pointed out that he had lectured on this theorem in the 1790s, as some of his students were willing

to testify. Some of Fourier's lecture notes from this period have survived; they contain a discussion of algebraic equations, in particular Descartes's rule, but they do not discuss Fourier's more general theorem. See the monograph, Budan (1822).

(11) Let $f_0(x) = f(x)$ and $f_1(x) = f'(x)$. Apply the Euclidean algorithm to $f_0$ and $f_1$, but take the negatives of the remainders. Thus,

$$f_0(x) = q_1(x)f_1(x) - f_2(x),$$
$$f_1(x) = q_1(x)f_2(x) - f_3(x),$$

............................

$$f_{m-2}(x) = q_{m-1}(x)f_{m-1}(x) - f_m(x).$$

Consider the sequence $f_0(x), f_1(x), \ldots, f_m(x)$. Prove that the difference between the number of changes of sign in the sequence when $x = a$ is substituted and the number when $x = b$ is substituted gives the actual number of real roots in the interval $(a, b)$. Charles Sturm (1803–1855) published this theorem in 1829. Sturm was a great friend of Liouville; they jointly founded the spectral theory of second order differential equations. He also worked as an assistant to Fourier, who helped him in various ways. See Sturm (1829).

(12) Let $F(x) = Ax^p + \cdots + Mx^r + Nx^s + \cdots + Rx^u$, and let the powers of $x$ run in increasing (or decreasing) order. Let $m$ be the number of variations of signs of the coefficients and let $\alpha$ be an arbitrary real number. Prove that the number of positive roots of $xF'(x) - \alpha F(x) = 0$ is one less than the number of positive roots of $F(x) = 0$. Prove also that if $\alpha$ lies between $r$ and $s$, then the number of sign variations in the coefficients of $xF' - \alpha F$ is the same as the number of sign variations in the sequence $A, \ldots, M, -N, \ldots, -R$; in other words, $m - 1$. From this, deduce Descartes's rule and prove that the equation

$$x^3 - x^2 + x^{\frac{1}{3}} + x^{\frac{1}{7}} - 1 = 0$$

has at most three positive roots and no negative roots. These results were given by Laguerre in 1883. See Laguerre (1972) vol. 1, pp. 1–3.

(13) Prove de Gua's observation that when $2m$ successive terms of an equation have 0 as coefficient, the equation has $2m$ complex roots; if $2m + 1$ successive terms are 0, the equation has $2m + 2$, or $2m$ complex roots, depending on whether the two terms, between which the missing terms occur, have like or unlike signs. See Burnside and Panton (1960) vol. 1, chapter 10.

(14) In his book on the theory of equations, Robert Murphy took $f(x) = x^3 - 6x^2 + 8x + 40$ to illustrate Sturm's theorem in Exercise 11. Carry out the details. See Murphy (1839) p. 25.

(15) Suppose $f(x)$ is a polynomial of degree $n$. Prove Newton's rule that if $f(a), f'(a), \ldots, f^{(n)}(a)$ are all positive, then all the real roots of $f(x) = 0$ are less than $a$. Newton gave this rule in his *Arithmetica Universalis* in the section "Of the Limits of Equations."

(16) Following Fourier, let $f(x) = x^5 - 3x^4 - 24x^3 + 95x^2 - 46x - 101$. Consider the sequence $f^V(x), f^{IV}(x), \ldots, f'(x), f(x)$ and find the number of sign variations when $x = -10, x = -1, x = 0, x = 1$, and $x = 10$. What does your analysis show about the real roots of $f(x)$? Now apply Sturm's method to this polynomial. The tediousness of this computation explains why one might wish to rely on Fourier's procedure.

(17) Let

$$f_0(x) = A_0 x^m + A_1 x^{m-1} + A_2 x^{m-2} + \cdots + A_{m-1} x + A_m.$$

Set $f_m(x) = A_0$, and $f_i(x) = x f_{i+1}(x) + A_{m-i}, i = m-1, m-2, \ldots, 0$. Prove that the number of variations of sign in $f_m(a), f_{m-1}(a), \ldots, f_0(a), a > 0$, is an upper bound for the number of roots of $f_0(x)$ greater than $a$; show that the two numbers differ by an even number. This result is due to Laguerre. See Laguerre (1972) vol. 1, p. 73.

(18) After his examples of the incomplete rule, Newton moved on to state what has become known as Newton's complete rule for complex roots. In 1865, J. J. Sylvester offered a description of this rule:

Let $fx = 0$ be an algebraical equation of degree $n$. Suppose

$$fx = a_0 x^n + n a_1 x^{n-1} + \frac{1}{2}(n-1) a_2 x^{n-2} + \cdots + n a_{n-1} x + a_n;$$

$a_0, a_1, a_2, \ldots, a_n$ may be termed the simple elements of $fx$. Suppose

$$A_0 = a_0^2, \quad A_1 = a_1^2 - a_0 a_2,$$
$$A_2 = a_2^2 - a_1 a_3, \ldots A_{n-1} = a_{n-1}^2 - a_{n-2} a_n, \quad A_n = a_n^2;$$

$A_0, A_1, A_2, \ldots, A_n$ may be termed the quadratic elements of $fx$. $a_r, a_{r+1}$ is a succession of simple elements, and $A_r, A_{r+1}$ of quadratic elements.

$$\left.\begin{array}{l} a_r \\ A_r \end{array}\right\} \text{ is an } \textit{associated} \text{ couple of elements;}$$

$$\left.\begin{array}{ll} a_r & a_{r+1} \\ A_r & A_{r+1} \end{array}\right\} \text{ is an associated couple of } \textit{successions.}$$

A succession may contain a permanence or a variation of signs, and will be termed for brevity a permanence or a variation, as the case may be. Each succession in an associated couple may be respectively a *permanence* or a *variation*. Thus an associated couple may consist of two permanences or two variations, or a superior permanence and inferior variation, or an inferior permanence and superior variation; these may be denoted respectively by the symbols $pP, vV, pV, vP$, and termed *double* permanences, *double* variations, permanence variations, variation permanences. The meaning of the simple symbols $p, v, P, V$ speaks for itself.

Newton's rule in its complete form may be stated as follows: On writing the complete series of quadratic under the complete series of simple elements of $fx$ in their natural order, the number of double permanences in the associated series, or pair of progressions so formed, is a superior limit to the number of negative roots, and the number of variation permanences in the same is a superior limit to the number of positive roots in $fx$. Thus the number of negative roots = or < $\sum pP$ ..., positive roots = or < $\sum vP$. This is the Complete Rule as given in other terms by Newton. The rule for negative roots is deducible from that for positive, by changing $x$ into $-x$. As a corollary, the total number of real roots = or < $\sum pP + \sum vP$, that is = or < $\sum P$. Hence, the number of imaginary roots

$$= \text{or} > n - \sum P, \text{ that is } = \text{or} > \sum V.$$

This is Newton's incomplete rule, or *first part* of complete rule, the rule as stated by every author whom the lecturer has consulted except Newton himself.

Read Sylvester's proof of this rule. Though Newton did not write down a proof, Sylvester writes in another paper of the same year, "On my mind the internal evidence is now forcible that Newton was in possession of a proof of this theorem (a point which he has left in doubt and which has often been called into question), and that, by singular good fortune, whilst I have been enabled to unriddle the secret which has baffled the efforts of mathematicians to discover during the last two centuries, I have struck into the very path which Newton himself followed to arrive at his conclusions." See Sylvester (1973) vol. 2, pp. 494 and 498–513. See also Acosta (2003).

## 6.10   Notes on the Literature

Newton's *Arithmetica Universalis*, written in 1683, contains his account of the undergraduate algebra course he taught at Cambridge in the 1670s. This was partly based on Newton's extensive notes on N. Mercator's Latin translation of Gerard Kinckhuysen's 1661 algebra text in Dutch. The later parts of the *Arithmetica* present Newton's own researches in algebra, carried out in the 1660s. This work was first published in 1707, in Latin; Newton was reluctant to have it published, perhaps because the first portion depended much on Kinckhuysen. An English translation appeared in 1720, motivating Newton to make a few changes and corrections and publish a new Latin version in 1721. In 1722, the English translation was republished with the same minor changes. Whiteside published the 1722 version in vol. 2 of Newton (1964–67). Harriot and Stedall (2003) presents Harriot's original text on algebra for the first time, although in English. The 1631 book published as Harriot's algebra was in fact a mutilated and somewhat confused version. Stedall's introduction explains this unfortunate occurrence.

A good source for references to early work on inequalities is Hardy, Littlewood, and Pólya (1967), though they omit Campbell. See Grattan-Guinness (1972) for an interesting historical account of Fourier's work on algebraic equations and Fourier series. Dieudonné (1981) is an excellent history of functional analysis and covers the period 1900–1950, from Hilbert and Riesz to Grothendieck. For functional analysis after 1950, see the comprehensive history of Pietsch (2007).

# 7

## The Calculus of Newton and Leibniz

### 7.1 Preliminary Remarks

Newton was a student at Cambridge University from 1661 to 1665, but he does not appear to have undertaken a study of mathematics until 1663. According to de Moivre, Newton purchased an astrology book in the summer of 1663; in order to understand the trigonometry and diagrams in the book, he took up a study of Euclid. Soon after that, he read Oughtred's *Clavis* and then Descartes's *Géométrie* in van Schooten's Latin translation. By the middle of 1664, Newton became interested in astronomy; he studied the work of Galileo and made notes and observations on planetary positions. This in turn required a deeper study of mathematics and Newton's earliest mathematical notes date from the summer of 1664. On July 4, 1699, Newton wrote in his 1664–65 annotations on Wallis's work that a little before Christmas 1664 he bought van Schooten's commentaries and a Latin translation of Descartes. He also wrote that he borrowed Wallis's *Arithmetica Infinitorum* and other works. In fact, his meditations on van Schooten and Wallis during the winter of 1664–65 resulted in his discovery of his method of infinite series and of the calculus.

Following the methods of van Schooten's commentaries, Newton devoted intense study to problems related to the construction of the subnormal, subtangent, and the radius of curvature at a point on a given curve. Newton's analyses of these problems gradually led him to discover a general differentiation procedure based on the concept of a small quantity, denoted by $o$, that ultimately vanished. Later in life, Newton wrote that he received a hint of this method of Fermat from the second volume of van Schooten's commentaries, although this gave only a brief summary based on P. Hérigone's 1642 outline of Fermat's method of finding the maximum or minimum of a function. Newton found the derivative, just as Fermat had, by expanding $f(x+o) = f(x) + of'(x) + O(o^2)$. Newton realized that the derivative was a powerful tool for the analyses of the subtangent, subnormal, and curvature and by the middle of 1665 he had worked out the standard algorithms for derivatives in general. Wallis's work motivated Newton to research the integration of rational and algebraic functions. Newton combined this with a study of van Heuraet's rectification of curves; in the summer of 1665, he began to understand the inverse method of tangents, that is, the connection between derivatives and integrals.

Newton left Cambridge in summer 1665 due to the plague, and returned to his home in Lincolnshire for two years. This gave him the opportunity to organize his thoughts on calculus and several other subjects. He gave up the idea of infinitesimal increments and adopted the concepts of fluents and fluxions as the new foundation for calculus. Fluents were flowing quantities; their finite instantaneous speeds were called fluxions, for which he later used the dot notation, such as $\dot{x}$, where $x$ was the fluent. From this point of view, Newton regarded it as obvious that the fluxion of the area generated by the ordinate $y$ along the $x$-axis would be $y$ itself. In other words, the derivative of the area function was the ordinate. In the fall of 1665, Newton ran into trouble with an uncritical application of the parallelogram of forces method, but he soon realized his mistake and by the spring of 1666 he was able to apply the method to an analysis of inflection points. Note that in 1640, the French mathematician G. Roberval warned that a curve could be viewed as the result of a moving point, but that there were pitfalls to using the parallelogram of forces method to find the tangent. What was the origin of Newton's conception of a curve as a moving point? A half century later, Newton wrote that, though his memory was unclear, he might have learned of a curve as a moving point from Barrow. Another possible source was Galileo but Newton did not mention him in this connection. In any case, Newton organized his concentrated research on calculus into a short thirty-page essay without title; he later referred to it as the October 1666 tract, published only in 1967 in the first volume of Whiteside's edition of Newton's mathematical papers.

In 1671, Newton wrote up the results of his researches on calculus and infinite series as a textbook on methods of solving problems on tangents, curvature, inflection, areas, volumes, and arc length. The portions of this work on infinite series were expanded from his 1669 work *De Analysi*. Whiteside designated the 1671 book as *De Methodis Serierum et Fluxionum* because Newton once referred to it this way, but Newton's original title is unknown because the first page of the original manuscript was lost. English translations of 1736 and 37 were given the title *The Method of Fluxions and Infinite Series*. Unfortunately, Newton was unable to publish this work in the 1670s, though he made several attempts. At that time, the market for advanced mathematics texts was not good; the publisher of Barrow's lectures on geometry, for example, went bankrupt. The controversy with Leibniz, causing wasted time and effort, would have been avoided had Newton succeeded in publishing his work.

Newton's *De Methodis* dealt with fluxions analytically, but it was never actually completed; in some places he merely listed the topics for discussion. However, when he revised the text in the winter of 1671–72, Newton added a section on the geometry of fluxions, developed axiomatically; he later called this the synthetic method of fluxions. Note that in the *Principia* Newton employed his insightful geometric approach. The Nobel Prize winner and *Principia* expert S. Chandrasekhar commented on the mode of Newton's proofs:[1]

[1] Wali (1991) pp. 242–243.

I first constructed the proofs for myself. Then I compared my proofs with those of Newton. The experience was a sobering one. Each time, I was left in sheer wonder at the elegance, the careful arrangement, the imperial style, the incredible originality, and above all the astonishing lightness of Newton's proofs; and each time I felt like a schoolboy admonished by his master.

As Newton was completing his researches on the calculus and infinite series, Gottfried Leibniz (1646–1716) was starting his mathematical studies. He studied law at the University of Leipzig but received his degree from the University of Altdorf, Nuremberg in 1666. At that time, he conceived the idea of reducing all reasoning to a symbolic computation, although he had not yet studied much mathematics. Leibniz's mathematical education started with his meeting with Huygens in 1672, at whose suggestion he studied Pascal and then went on to read Grégoire St. Vincent's *Opus Geometricum* and other mathematical works.

From the beginning, Leibniz searched for a general formalism, or symbolic method, capable of handling infinitesimal problems in a unified way. In a paper of 1673, Leibniz began to denote geometric quantities associated with a curve, such as the tangent, normal, subtangent and subnormal, as functions. He began to set up tables of specific curves and their associated functions in order to determine the relations among these quantities. Thus, he raised the question of determining the curve, given some property of the tangent line. In 1673, Leibniz came to the conclusion that this problem, the inverse tangent problem, was reducible to the problem of quadratures. By the end of the 1670s, Leibniz had independently worked out his differential and integral calculus. In 1684, his first paper on differentiation appeared, and in 1686 his first paper on integration was published. The notation of Leibniz, including the differential and integral signs, gave insight into the processes and operations being performed. The Bernoulli brothers were among the first to learn and exploit the calculus of Leibniz and in the 1690s, they began to make contributions to the development of calculus in tandem with Leibniz.

In the May 1690 issue of the *Acta Eruditorum*, Jakob Bernoulli proposed the problem of finding the curve assumed by a chain/string hung freely from two fixed points, named a catenary by Leibniz. Leibniz was the first to solve the problem, announcing his construction without details in the July 1691 issue of the *Acta*. Johann Bernoulli (1667–1748) soon published a solution, in which he explained that he and his brother had been surprised that this everyday problem had not attracted anyone's attention. But in his paper, Leibniz wrote that the problem had been well known since Galileo had articulated it; moreover, Leibniz stated that he would refrain from publishing his solution by means of differential calculus, to give others a chance to work out a solution. Jakob had trouble with this question, since he initially thought the curve was a parabola, until Johann corrected him. According to a 1717 letter from Johann Bernoulli to Montmort, Jakob had initially assumed that the catenary was a parabola, as had Galileo. However, when Johann showed his brother the correct solution, Jakob was able to extend his brother's method, developing a general theory of flexible strings.[2]

---

[2] Spiess (1955) vol. 1, pp. 97–98.

In his 1638 work, that we translate as *Mathematical Discourses Concerning Two New Sciences*, Galileo suggested that the catenary was a parabola. In 1646, Christian Huygens (1629–1695) showed that it could not be a parabola.[3] In the 1690s, Huygens offered a geometric solution to the problem posed by Bernoulli, using classical methods of which he was a master. In their approach, Leibniz and Johann Bernoulli applied mechanical principles to determine the differential equation of the catenary, making use of the work of Pardies. The Jesuit priest Ignace-Gaston Pardies (1636–1673) published a 1673 work on theoretical mechanics, developing his original idea of tension along the string, a concept fully clarified by Jakob Bernoulli. Thus, Leibniz and Johann found the differential equation of the catenary: $\frac{dy}{dx} = \frac{s}{a}$ where $s$ was the length of the curve.[4] They showed that the solution of this differential equation was the integral

$$x = \int \frac{a\,dy}{\sqrt{y^2 - a^2}}.$$

In his 1691 paper, Leibniz presented a geometric figure and explained that the points on the catenary could be found from an exponential curve, called by Leibniz the logarithmic line. Details of this proof can be found in his letters to Huygens[5] and von Bodenhausen.[6] In modern notation, the solution would be expressed as $y = \frac{a}{2}(e^{\frac{x}{a}} + e^{-\frac{x}{a}})$. Johann Bernoulli also failed to publish details but presented two geometric constructions of the catenary, one using the area under a curve related to a hyperbola and the other using the length of an arc of a parabola. In the 1690s, this kind of solution would have been acceptable, because the coordinates of any point on the catenary were then described in terms of geometric quantities related to known curves such as the hyperbola and parabola. In modern terms, the area and length can be written as the integrals

$$\int \frac{dx}{\sqrt{(x-a)^2 - a^2}} \quad \text{and} \quad \int \sqrt{\frac{2a+x}{x}}\,dx.$$

In the 1690s, Leibniz and Johann Bernoulli were arriving at an understanding of an exponential function. In a letter of May 1694,[7] Bernoulli wrote to Leibniz that he had written a paper for the *Acta Eruditorum* in which he defined the exponential and the meaning and construction of $x^x$. He also mentioned that the area under $x^x$ over the interval $(0, 1)$ was given by the series

$$1 - \frac{1}{2^2} + \frac{1}{3^3} - \frac{1}{4^4} + \frac{1}{5^5} - \cdots . \tag{7.1}$$

[3] For an analysis of Galileo's work, see Truesdell (1960) pp. 43–47.
[4] For a summary of Huygens's paper and that of Leibniz and Bernoulli, see Truesdell (1960) pp. 64–75.
[5] For this letter and references to this and other letters to Huygens, see Truesdell (1960) p. 71.
[6] Leibniz (1971) vol. 7, pp. 370–372.
[7] Bernoulli and Leibniz (1745) vol. 1, pp. 5–9.

In his reply of June 1694,[8] Leibniz wrote that he had written to Huygens about these matters and he went on to explain that $y = x^x$ meant that $\ln y = x \ln x$ or

$$\int (dy : y) = x \int (dx : x). \tag{7.2}$$

Leibniz then gave the differential form of (7.2) as

$$dy : y = dx + dx \int (dx : x)$$

or

$$dy : dx = y \left( 1 + \int (dx : x) \right).$$

Bernoulli did not offer a proof of (7.1) in his paper, but he included a proof in his collected papers.[9] See Exercise 10 at the end of this chapter.

## 7.2 Newton's 1671 Calculus Text

The *De Methodis Serierum et Fluxionum*[10] of Newton began by considering the general problem, called Problem 1, of determining the relation of the fluxions, given the relations to one another of two flowing quantities. As an example, Newton took

$$x^3 - ax^2 + axy - y^3 = 0. \tag{7.3}$$

His rule for finding the fluxional equation was to first write the equation in decreasing powers of $x$, as in (7.3), and then multiply the terms by $\frac{3\dot{x}}{x}, \frac{2\dot{x}}{x}, \frac{\dot{x}}{x}$, and 0, respectively, to get

$$3\dot{x}x^2 - 2\dot{x}ax + \dot{x}ay. \tag{7.4}$$

Thus, if the term were $x^n y^m$, then it would be multiplied by $\frac{n\dot{x}}{x}$. He next wrote the equation in powers of $y$: $\quad -y^3 + axy + (x^3 - ax^2)$ and multiplied the terms by $-\frac{3\dot{y}}{y}, \frac{\dot{y}}{y}$ and 0 to obtain

$$-3\dot{y}y^2 + a\dot{y}x. \tag{7.5}$$

In order to obtain the equation expressing the relation between the fluxions $\dot{x}$ and $\dot{y}$ Newton added (7.4) and (7.5) and set the sum equal to zero:

$$3\dot{x}x^2 - 2a\dot{x}x + a\dot{x}y - 3\dot{y}y^2 + a\dot{y}x = 0.$$

[8] ibid. pp. 10–13.
[9] Bernoulli, Joh. (1968) vol. 3, pp. 380–381.
[10] Newton (1967–1981) vol. 3, pp. 32–353, especially pp. 74–83.

From this it followed that

$$\dot{x} : \dot{y} = (3y^2 - ax) : (3x^2 - 2ax + ay).$$

Newton also presented more examples, involving more complex expressions such as

$$\sqrt{a^2 - x^2} \quad \text{and} \quad \sqrt{ay + x^2}.$$

Explaining why his rule for finding fluxional (differential) equations worked, Newton pointed out that a fluent quantity $x$ with speed $\dot{x}$ would change by $\dot{x}o$ during the small interval of time $o$. So the fluent quantity $x$ would become $x + \dot{x}o$ at the end of that time interval. Hence, the quantities $x + \dot{x}o$ and $y + \dot{y}o$ would satisfy the same relation as $x$ and $y$, and when substituted in (7.3) gave him

$$(x^3 + 3\dot{x}ox^2 + 3\dot{x}^2o^2x + \dot{x}^3o^3) - (ax^2 + 2a\dot{x}ox + a\dot{x}^2o^2)$$
$$+ (axy + a\dot{x}oy + a\dot{y}ox + a\dot{x}\dot{y}o^2) - (y^3 + 3\dot{y}oy^2 + 3\dot{y}^2o^2y + \dot{y}^3o^3) = 0.$$
$$\tag{7.6}$$

After subtracting (7.3) from (7.6) and dividing by o, Newton had

$$3\dot{x}x^2 + 3\dot{x}^2ox + \dot{x}^3o^2 - 2a\dot{x}x - a\dot{x}^2o + a\dot{x}y + a\dot{y}x$$
$$+ a\dot{x}\dot{y}o - 3\dot{y}y^2 - 3\dot{y}^2oy - \dot{y}^3o^2 = 0.$$

Here Newton explained that quantities containing the factor $o$ could be neglected,[11] "since o is supposed to be infinitely small so that it be able to express the moments of quantities, terms which have it as a factor will be equivalent to nothing in respect of the others. I therefore cast them out and there remains

$$3\dot{x}x^2 - 2a\dot{x}x + a\dot{x}y + a\dot{y}x - 3\dot{y}y^2 = 0.\text{"}$$

Note that this amounts to the result of implicit differentiation with respect to a parameter. Actually, Newton here used the letters $m$ and $n$ for $\dot{x}$ and $\dot{y}$, respectively. He introduced the dot notation in the early 1690s. From this he had the slope

$$\dot{y} : \dot{x} = 3x^2 - 2ax + ay : 3y^2 - ax.\tag{7.7}$$

Observe that to construct the tangent, rather than work with slope, it is better to find the point where the tangent intersects the $x$-axis and join this point to the point of tangency on the curve. Now if $(x, y)$ is the point on the curve, then the length of the segment of the $x$-axis from $(x, 0)$ to the intersection with the tangent is given by the magnitude of $\frac{y}{\frac{dy}{dx}}$. In his discussion of the tangent, Newton computed this quantity to obtain

[11] ibid. p. 81.

$$\frac{3y^3 - axy}{3x^2 - 2ax + ay}.$$

In the section of his book on maxima and minima, Newton gave a method and then two examples and nine exercises to be solved using that method. He never completed this section of his book. To find a maximum or minimum, he explained, the derivative should be set equal to zero at an extreme value:[12]

> When a quantity is greatest or least, at that moment its flow neither increases nor decreases; for if it increases, that proves that it was less and will at once be greater than it now is, and conversely so if it decreases. Therefore seek its fluxion by Problem 1 [above] and set it equal to nothing.

In the first application of this principle, he sought the greatest value of $x$ in equation (7.3) by setting $\dot{x} = 0$ in the fluxional equation (7.4) to get

$$-3y^2 + ax = 0; \tag{7.8}$$

using this result in the original equation, one obtains the largest value of $x$. Newton remarked that equation (7.8) illustrated the "celebrated Rule of Hudde, that, to obtain the maxima or minima of the related quantity, the equation should lie ordered according to the dimensions of the correlate one and then multiplied by an arithmetical progression." He added that his method extended to expressions with surd quantities, whereas the earlier rules and techniques did not. As an example, he gave the problem of finding the greatest value of $y$ in the equation

$$x^3 - ay^2 + \frac{by^3}{a+y} - x^2\sqrt{ay + x^2} = 0.$$

Newton wrote that the equation for the fluxions of $x$ and $y$ would come out to be

$$3\dot{x}x^2 - 2a\dot{y}y + \frac{3ab\dot{y}y^2 + 2b\dot{y}y^3}{a^2 + 2ay + y^2} - \frac{4a\dot{x}xy + 6\dot{x}x^3 + a\dot{y}x^2}{2\sqrt{ay + x^2}} = 0.$$

He then observed that by hypothesis $\dot{y} = 0$ and hence, after substituting this in the equation and dividing by $\dot{x}x$,

$$3x - \frac{2ay + 3x^2}{\sqrt{(ay + x^2)}} = 0 \text{ or } 4ay + 3x^2 = 0.$$

Newton noted that this equation should be used to eliminate $x$ or $y$ from the original equation; the maximum would be obtained by solving the resulting cubic.

The next section of Newton's book discussed the problem of constructing tangents to curves and he mentioned seven problems solvable by the principles he explained. For example:[13]

---

[12] ibid. p. 117.
[13] ibid. p. 149.

(1) To find the point in a curve where the tangent is parallel to the base (or any other straight line given in position) or perpendicular to it or inclined to it at any given angle.

(2) To find the point where a tangent is most or least inclined to the base or to another straight line given in position – to find, in other words, the bound of contrary flexure. I have already displayed an example of this above in the conchoid.

By "the bound of contrary flexure" Newton meant the point of inflection and at this point, $\frac{d^2y}{dx^2} = 0$. In the example of the conchoid of Nichomedes, defined by

$$yx = (b + y)\sqrt{c^2 - y^2},$$

Newton actually minimized the $x$-intercept of the tangent given by

$$x - y\,\frac{dx}{dy}. \tag{7.9}$$

Whiteside has noted[14] that in 1653, Huygens determined the inflection points of this conchoid by this method, using Fermat's procedure to obtain the minimum value. Newton was most likely aware of Huygens's work and wanted to show that calculus algorithms could simplify Huygens's calculation. It should be noted that Huygens's criterion that the inflection points in general could be obtained by minimizing (7.9) is false, though it is true for the conchoid.

Newton was very interested in problems related to curvature and intended to devote several chapters to the topic, but many of these are barely outlines. However, he presented a procedure for finding radius of curvature, and we explain this later on. He also included sections on arc length and the area of surface of revolution.

## 7.3   Leibniz: Differential Calculus

Leibniz gave a very terse account of his differential calculus in his 1684 *Acta Eruditorum* paper,[15] starting with the basic rules for the differentials of geometric quantities (variables). Leibniz's approach was not to find the derivative of a function. As he conceived things, geometric quantities had differentials; when the quantities stood in a certain relationship to one another, then the differentials also satisfied certain relations. To determine these relations, for a constant $a$ and variable quantities $v, x, y$, etc., he stated the rules for the differentials $dv, dx, dy$, etc.:

$$da = 0, \quad d(ax) = a\,dx, \quad d(z - y + w + x) = dz - dy + dw + dx,$$

$$d(xv) = x\,dv + v\,dx, \quad d\frac{v}{y} = \pm\frac{v\,dy \mp y\,dv}{yy}.$$

---

[14] ibid. vol. 2, pp. 198–199, footnote 9.
[15] Leibniz (1684). For an English translation, see Struik (1969) pp. 272–280.

Concerning signs of differentials, Leibniz explained that if the ordinate $v$ increased, then $dv$ was positive, and when $v$ decreased, $dv$ was negative.

In only one paragraph, Leibniz described in terms of differentials: maxima and minima, concavity or convexity, and inflection points of curves. He explained that at a maximum or minimum for an ordinate $v$, $dv = 0$ since $v$ was neither increasing nor decreasing. For concavity, the difference of the differences $d\,dv$ had to be positive, and for convexity $d\,dv$ had to be negative. Note here that Leibniz took the definitions of concavity and convexity to be the reverse of our present definition. At a point of inflection $d\,dv = 0$. After this, Leibniz gave the rules for the differentials of powers and roots, that is

$$dx^a = ax^{a-1}\,dx, \quad \text{and} \quad d\sqrt[b]{x^a} = \frac{a}{b}\sqrt[b]{x^{a-b}}. \tag{7.10}$$

He wrote that with this differential calculus, he could solve problems dealing with tangents and with maxima and minima by a uniform technique, lacking in the earlier expositions. To demonstrate the power of his method, he found the tangent to a curve defined by a complicated algebraic relation between the variables $x$ and $y$. As another application of the differential calculus, he gave the derivation of Snell's law in optics, one of the standard examples in modern textbooks.

As a final example, Leibniz considered the problem that de Beaune proposed to Descartes in 1639. Florimond de Beaune (1601–1652) was a jurist who carefully studied Descartes's book on geometry. He observed that, though Descartes had given a method for finding the tangent to a curve, he had not indicated how to obtain the curve, given a property of the tangent. One of de Beaune's problems was to find the curve for which the subtangent was the same for each point of the curve. This problem translates to the differential equation $\frac{dy}{dx} = \frac{y}{a}$, where $a$ is a constant. It is well known that the solution is $\ln y = \frac{1}{a}x + c$ and Descartes came close to a solution.[16] In the course of his work, he obtained, without mentioning logarithms, particular cases of the inequality, written in modern notation as

$$\frac{1}{n} + \frac{1}{n+1} + \cdots + \frac{1}{m-1} > \ln\frac{m}{n} > \frac{1}{n+1} + \frac{1}{n+2} + \cdots + \frac{1}{m}.$$

To tackle this problem, Leibniz described the differential equation by saying that $y$ was to $a$ as $dy$ was to $dx$. He then noted that $dx$ could be chosen arbitrarily and hence could be taken to be a constant $b$. Then

$$dy = \frac{b}{a}y, \quad \text{or} \quad y = \frac{a}{b}\,dy.$$

He observed that this implied that if the $x$ formed an arithmetic progression, then the $y$ formed a geometric progression. Leibniz did not explain or prove this statement, but it is easy to check that if

[16] Descartes (1897–1905) vol. 2, pp. 510–519. For a complete discussion of Descartes's solutions, see Scriba (1961). See also Hofmann (1990) vol. 2, pp. 279–281.

$$y(x) = \frac{a}{b} \, dy(x), \quad \text{then}$$

$$y(x + dx) = y(x) + dy(x) = \left(1 + \frac{b}{a}\right) y(x).$$

Again,

$$y(x + 2\,dx) = y(x + dx) + dy(x + dx)$$

$$= y(x + dx) + \frac{b}{a} y(x + dx)$$

$$= \left(1 + \frac{b}{a}\right) y(x + dx)$$

$$= \left(1 + \frac{b}{a}\right)^2 y(x).$$

Similarly,

$$y(x + 3dx) = y(x + 2dx) + \frac{b}{a} \, dy(x + 3dx)$$

$$= \left(1 + \frac{b}{a}\right)^3 y(x),$$

and in general

$$y(x + ndx) = \left(1 + \frac{b}{a}\right)^n y(x).$$

This proves Leibniz's claim, since he took $dx$ to be a constant; thus, $x, x + dx, x + 2dx, \ldots$ is an arithmetic progression, and the values of $y$ at these points form a geometric progression. This idea illustrates the seventeenth-century understanding of logarithms. In fact, Leibniz was already suggesting that the logarithm be defined by means of the integral $\int \frac{dx}{x}$ and later on, he did so.

Leibniz could have integrated to get $\ln y = \int \frac{dy}{y} = \frac{1}{a} \int dx = \frac{x}{a}$, but in his 1684 paper, he did not use or discuss integration, though he had been aware of it for several years. Surprisingly, he gave a brief exposition of his ideas on integration in a review of John Craig's 1685 book on quadrature.[17] In the review,[18] Leibniz introduced the symbol $\int$ for the summation of infinitesimal quantities and gave an illustration of its power when used in conjunction with differentials. Leibniz also pointed out that the integral symbol could be used to represent transcendental quantities such as the arcsine or logarithm, in such a way that it revealed a property of the quantity.

In his 1684 paper on the differential calculus, Leibniz gave a derivation of Snell's law by applying Fermat's principle of least time.

[17] Craig (1685).
[18] Leibniz (1971) 3/II, pp. 226–235.

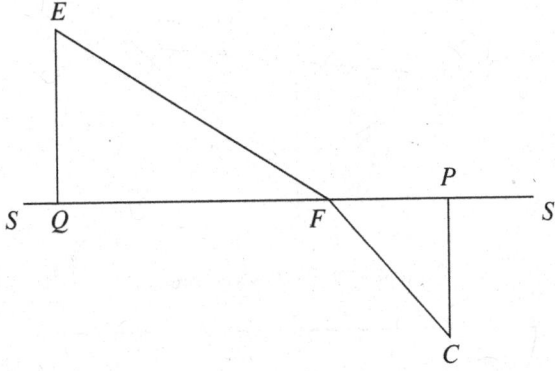

Figure 7.1 Leibniz's figure for derivation of Snell's law.

In Figure 7.1, with the lines $PC$ and $QE$ perpendicular to $SQ$, light traveled from point $C$ to point $E$ and the line $QP$ separated an upper medium of density $r$ from a lower medium of density $h$. Leibniz explained that density should be understood to be with respect to the resistance to a ray of light.

Let $QF = x$, $QP = p$, $CP = c$, and $EQ = e$. Then

$$FC = \sqrt{cc + pp - 2px + xx} = \text{(in short) } \sqrt{l},$$

$$EF = \sqrt{ee + xx} = \text{(in short) } \sqrt{m}.$$

Leibniz gave the quantity to be minimized when the densities were taken into account as $w = h\sqrt{l} + r\sqrt{m}$. He then argued that to minimize, set $dw = 0$, to obtain

$$0 = hdl : 2\sqrt{l} + rdm : 2\sqrt{m}.$$

Note that Leibniz specified that he would denote $\frac{x}{y}$ by $x : y$. He then observed that $dl = -2(p - x)$ and $dm = 2xdx$; hence he had Snell's law:

$$h(p - x) : \sqrt{l} = rx : \sqrt{m}.$$

## 7.4 Leibniz on the Catenary

Leibniz developed a theory of second- and higher-order differentials in order to apply differential calculus to geometry and mechanics. In his applications, including the catenary problem,[19] he often took one of the variables, say $y$, to be such that the second-order differential $ddy$ was 0 or, equivalently, that the first-order differential $dy$ was a constant. This amounted to taking $y$ to be the independent variable. To describe the curve of the catenary, Leibniz used Pardies's important mechanical principle,

---

[19] Leibniz (1971) vol. 5, pp. 243–247.

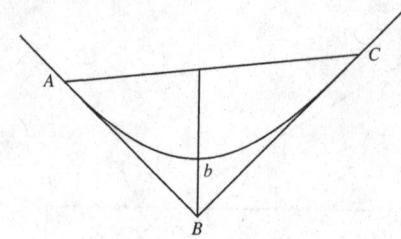

Figure 7.2 Pardies's theorem.

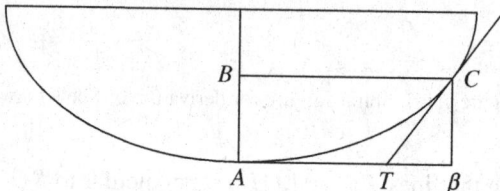

Figure 7.3 Leibniz's figure of catenary, made for Huygens.

dating from 1673, that for any portion $AC$ of the curve made by the string, the vertical line through the center of gravity of $AC$, and the tangents at $A$ and at $C$ intersected at one point[20] (Figure 7.2).

Leibniz's letters to Huygens and Bodenhausen offered the following details of his derivation of the catenary.[21] In Leibniz's Figure 7.3, $A$ is the lowest point of the catenary; $CT$ is the tangent at a point $C$ on the catenary; and $C\beta, AB$ are perpendicular to $A\beta$, the tangent at $A$. We follow Leibniz's notation and argument.[22] Let $AB = x, BC = y, T\beta = x\,dy : dx$ and $AT = y - x\,dy : dx$. Then, by Pardies's theorem, the $y$ coordinate of the center of gravity of the arc $AC$ of length $c$ is $\frac{1}{c}\int y\,dc$. Thus,

$$\frac{1}{c}\int y\,dc = y - x\,dy : dx. \qquad (7.11)$$

Now multiply both sides by $c$ and differentiate to get

$$y\,dc = c\,dy + y\,dc - x\overline{dy : dx}\,dc - c\,dy - cxd, \overline{dy : dx}. \qquad (7.12)$$

Note that Leibniz used a comma to separate the operator $d$ from the quantity $dy : dx$, where the line above the expression was used instead of parentheses. Upon simplification, obtain in Leibniz's notation

$$dc\,\overline{dy : dx} + cd, \overline{dy : dx} = 0. \qquad (7.13)$$

[20] See Truesdell (1960) pp. 50–53.
[21] For Leibniz's letters to Bodenhausen on the catenary, see Leibniz (1971) vol. 7, pp. 359–361 and pp. 370–372.
[22] See Truesdell (1960) pp. 71–72.

Suppose that $y$ increases uniformly, so that $dy$ is constant and $ddy = 0$. This implies by the quotient rule that

$$d, \overline{dy : dx} = -dy\,ddx : \overline{dx\,dx},$$

so that (7.13) is transformed into $dc\,dx - c\,ddx = 0$.

By differentiating $dx : c = dy : a$ we get the previous equation, indicating that this is the integral of that equation. Rewrite this integral as

$$a\,dx = c\,dy. \tag{7.14}$$

This is the differential equation of the catenary, and its differential is

$$a\,ddx = dc\,dy. \tag{7.15}$$

Following Leibniz, one may solve this equation by observing that in general, since $c$ denotes arc length,

$$dc\,dc = dy\,dy + dx\,dx. \tag{7.16}$$

Differentiate this, using $ddy = 0$ and (7.15), to obtain

$$dc\,ddc = dy\,ddy + dx\,ddx = dx\,ddx = dx\,dc\,\frac{dy}{a}.$$

By integration (Leibniz used the term summation), we arrive at $a\,dc = (x + b)dy$, where $b$ is a constant. Next set $z = x + b$ to rewrite, obtaining $a\,dc = z\,dy$. Combining this with $dc\,dc = dz\,dz + dy\,dy$, the result emerges as

$$aa\,dz\,dz + aa\,dy\,dy = zz\,dy\,dy. \tag{7.17}$$

Thus, as Leibniz wrote,

$$y = a\int \overline{dz : \sqrt{zz - aa}}, \tag{7.18}$$

or in modern notation

$$a\int \frac{dz}{\sqrt{z^2 - a^2}}$$

gives the area under the curve with ordinate $\frac{a}{\sqrt{z^2-a^2}}$. This integral can be computed in terms of the logarithm. Although we today would wish to evaluate the integral, and write it as the logarithm of a specific function, mathematicians of the seventeenth century were satisfied with a result expressed in terms of areas or arc lengths of known curves, so that from Leibniz's point of view, this result was sufficient to define the catenary.

We remark that the meaning of the exponential curve, or exponential function as we call it, and of the logarithmic function, was not clearly understood in the 1690s.

Leibniz's paper on the catenary[23] thus devoted some space to these curves. In his correspondence with Huygens, Leibniz was called upon to explain these concepts.[24] Again, l'Hôpital asked Johann Bernoulli the meaning of $m^n$, since the prevailing point of view saw the magnitudes $m$ and $n$ as represented by lines.[25] To clarify these matters, in 1697 Johann Bernoulli published a paper in the *Acta Eruditorum*, "Principia calculi exponentialium seu percurrentium."[26] In this paper, Bernoulli defined the logarithmica curve as one whose subtangent was a constant. Since the subtangent is given by

$$\frac{y}{\frac{dy}{dx}},$$

and taking the constant to be one, we essentially have

$$y = \exp(x)$$

and the logarithmica is the exponential curve. Bernoulli applied the logarithmica to define $y = x^x$ and showed how to construct this curve pointwise. In effect, he set $x^x = \exp(x \ln x)$. In this way, we see how Leibniz and the Bernoullis and their followers found it necessary to work with formulas instead of geometric objects.

## 7.5  Johann Bernoulli on the Catenary

In his 1691–1692 lectures on integral calculus, Bernoulli gave details to supplement the treatment of the catenary in his 1691 paper.[27] He first set down the mechanical principles required to obtain the fundamental equation.

In Figure 7.4, Bernoulli set $BG = x$, $GA = y$, $Ha = dy$, $HA = dx$, and $BA = s$. He then applied the laws of statics, and in effect Pardies's law, to obtain the differential

$$\frac{dx}{dy} = \frac{ks}{ka} = \frac{s}{a} \tag{7.19}$$

for some constant $a$. Bernoulli's solution, like that of Leibniz, amounted to showing that (7.19) was equivalent to

$$dy = \frac{a\,dx}{\sqrt{x^2 - a^2}}. \tag{7.20}$$

To show this, he wrote (7.19) as

$$dy = \frac{a\,dx}{s} \quad \text{or} \quad dy^2 = \frac{a^2}{s^2}dx^2.$$

[23] Leibniz (1971) vol. 5, pp. 243–247.
[24] See Bos (1996) and Truesdell (1960), pp. 64–72.
[25] Spiess (1955) vol. 1, p. 172.
[26] Joh. Bernoulli (1968) vol. I, pp. 179–187.
[27] Joh. Bernoulli (1742) vol. 3, pp. 404–406.

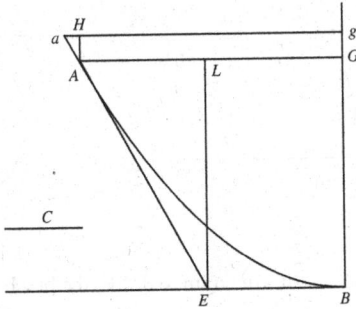

Figure 7.4 Bernoulli's diagram for a catenary.

Therefore,

$$ds^2 = dx^2 + dy^2 = \frac{s^2\,dx^2 + a^2\,dx^2}{s^2}$$

and

$$ds = \frac{dx\,\sqrt{s^2+a^2}}{s}, \quad \text{or} \quad dx = \frac{s\,ds}{\sqrt{s^2+a^2}}.$$

By integration,

$$x = \sqrt{s^2+a^2} \quad \text{or} \quad s = \sqrt{x^2-a^2}.$$

By differentiating, he got

$$ds = \frac{x\,dx}{\sqrt{x^2-a^2}} = \sqrt{dx^2+dy^2}.$$

Squaring this, he had

$$x^2\,dy^2 - a^2\,dy^2 = a^2\,dx^2,$$

equivalent to (7.20).

## 7.6  Johann Bernoulli: The Brachistochrone

In 1696, Bernoulli conceived of and solved the brachistochrone problem:[28] Given two points in a vertical plane but not vertically aligned, find the curve along which a point mass must fall under gravity, starting at one point and passing through the other in the shortest possible time. He argued that this mechanical problem was identical to an optical problem of the path taken by light moving from one point to another, following

---

[28] Bernoulli, Joh. (1697). For an English translation of the main points of the paper, see Struik (1969) pp. 392–396.

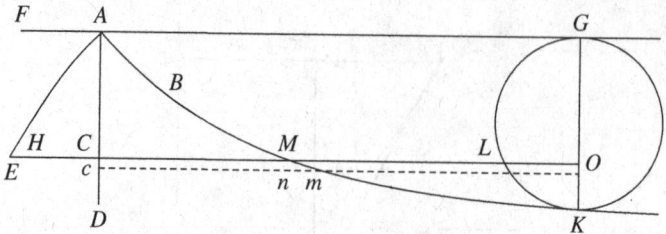

Figure 7.5 Bernoulli's diagram to derive the brachistochrone.

the curve of least time, passing through a medium whose ever-changing density is inversely proportional to the velocity of a falling body. As light passes continuously from one medium to another, the quantity $\frac{\sin \alpha}{v}$ remains constant, where $\alpha$ is the angle between the vertical and the direction of the path and $v$ is the velocity.

We change Bernoulli's notation slightly; he used $t$ for the velocity and interchanged the $x$ and $y$. So in Figure 7.5, let $AC = y, CM = x, mn = dx, Cc = dy, Mm = ds$, and $\alpha = \angle nMm$. Since $\sin \frac{\alpha}{c}$ is a constant, we have

$$\frac{dx}{ds} = \frac{v}{a}, \quad \text{or} \quad a^2 dx^2 = v^2 (dx^2 + dy^2),$$

where $a$ is a constant. Since for a falling body $v^2 = 2gy$, we get $dx = \sqrt{\frac{y}{c-y}}\, dy$, where $c = \frac{a^2}{2g}$. Now

$$dy \sqrt{\frac{y}{c-y}} = \frac{1}{2} \frac{c\, dy}{\sqrt{cy - y^2}} - \frac{1}{2} \frac{c\, dy - 2y\, dy}{\sqrt{cy - y^2}}.$$

Integrating this, obtain $CM = \text{arc } GL - LO$ and, since $MO = CO - \text{arc } GL + LO = \text{arc } LK + LO$, it follows that $ML = \text{arc } LK$. Thus, the curve is a cycloid.

Bernoulli was particularly proud of having linked mechanics with optics. In his brachistochrone paper he wrote,[29] "In this way I have solved at one stroke two important problems – an optical and a mechanical one – and have achieved more than I have demanded from others: I have shown that the two problems, taken from entirely separate fields of mathematics, have the same character." Bernoulli mentioned the link between geometrical optics and mechanics more than once in his works, but this concept was not developed until the 1820s when the Irish mathematician William Rowan Hamilton independently worked out the same idea.

## 7.7 Newton's Solution to the Brachistochrone

In 1696, although he had already solved the problem, Johann Bernoulli made a public challenge of the brachistochrone problem and another problem, perhaps directed at

[29] Struik (1969) p. 394.

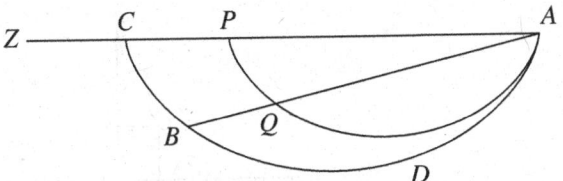

Figure 7.6 Newton's solution to the brachistochrone problem.

Newton. At that time, Newton was serving in London as Warden of the Mint, having given up mathematics. However, upon receiving these problems after a full day's work, he set upon them immediately and reportedly solved both problems within twelve hours. Whiteside commented that, although this was a marvelous feat, Newton was then out of practice, and thus he took hours instead of minutes.[30] We note that in 1685, Newton had addressed a problem mathematically similar to the brachistocrone, of the solid of revolution of least resistance in a uniform fluid; his solution was included in the *Principia*. In 1697, Newton published a short note solving both of Bernoulli's problems, with an accompanying diagram in the *Philosophical Transactions*, stating that the solution to Bernoulli's brachistochrone problem was a cycloid. Then in 1700, he wrote up the details, apparently for the purpose of explaining the solution to David Gregory, nephew of James Gregory. In his brief note in the *Transactions*, Newton gave Figure 7.6 and stated:

> From the given point $A$ draw the unbounded straight line $APCZ$ parallel to the horizontal and upon this same line describe both any cycloid $AQP$ whatever, meeting the straight line $AB$ (drawn and, if need be, extended) in the point $Q$, and then another cycloid $ADC$ whose base and height [as $AC : AP$] shall be to the previous one's base and height respectively as $AB$ to $AQ$. This most recent cycloid will then pass through the point $B$ and be the curve in which a heavy body shall, under the force of its own weight, most swiftly reach the point $B$ from the point $A$. As was to be found.

We summarize Newton's solution for David Gregory, based upon his diagram (Figure 7.7) and Whiteside's commentary.[31] Let $AB = x$, $BC = o = CD$, $BE = y(= y(x))$. By Taylor's expansion

$$CN = y(x + o) = y + \dot{y}o + \frac{1}{2}\ddot{y}o^2,$$
$$DG = y(x + 2o) = y + 2\dot{y}o + 2\ddot{y}o^2.$$

From this it follows that

$$HN = IK = \dot{y}o + \frac{1}{2}\ddot{y}o^2, \quad IG = \dot{y}o + \frac{3}{2}\ddot{y}o^2, \text{ and } LG = 2\dot{y}o + 2\ddot{y}o^2.$$

Define $p$ and $q$ by $FN = q$ and $GL = 2p$.

---

[30] Newton (1967–1981) vol. 8, pp. 72–73, footnotes 1 and 2.
[31] Newton (1967–1981) vol. 8, pp. 87–91.

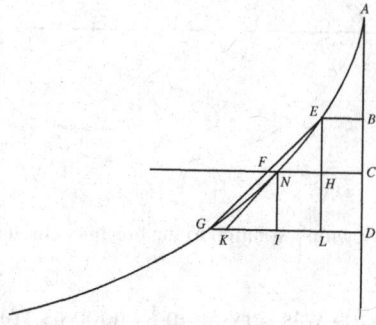

Figure 7.7 Newton's solution for David Gregory.

The time taken to travel from $E$ to $G$ is to be minimized as $q$ varies. The expression for time is given by

$$\frac{\sqrt{o^2 + (p-q)^2}}{\sqrt{x}} + \frac{\sqrt{o^2 + (p+q)^2}}{\sqrt{o+x}} = R + S,$$

where

$$R^2 = \frac{o^2 + (p-q)^2}{x} \quad \text{and} \quad S^2 = \frac{o^2 + (p+q)^2}{o+x}.$$

Taking the derivative with respect to $q$,

$$2R\dot{R} = \frac{-2p\dot{q} + 2q\dot{q}}{x} \quad \text{and} \quad 2S\dot{S} = \frac{2p\dot{q} + 2q\dot{q}}{x+o}.$$

So the condition for minimum time is that

$$\frac{-p\dot{q} + q\dot{q}}{Rx} + \frac{p\dot{q} + q\dot{q}}{S(x+o)} = 0,$$

or

$$\frac{(p-q)\sqrt{x}}{\sqrt{(p-q)^2 + o^2}} = \frac{(p+q)\sqrt{x+o}}{\sqrt{(p+q)^2 + o^2}}.$$

This condition implies that $\dfrac{p\sqrt{x}}{\sqrt{p^2 + o^2}}$ must be a constant and since $\dfrac{o}{p} = \dfrac{\dot{x}}{\dot{y}}$, we have

$$\frac{\sqrt{x}}{\sqrt{1 + (\frac{\dot{x}}{\dot{y}})^2}} = \text{constant} \quad \text{or} \quad 1 + \left(\frac{dx}{dy}\right)^2 = \frac{x}{c}.$$

Thus, $dy = \sqrt{\dfrac{x}{c-x}}\, dx$, and we have the differential equation of a cycloid.

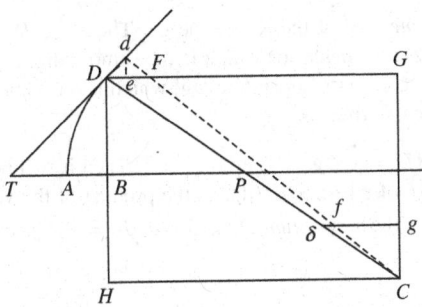

Figure 7.8 Newton's derivation of the radius of curvature.

## 7.8 Newton on the Radius of Curvature

From the time he studied van Schooten's book on Descartes's *Géométrie*, Newton was interested in the problem of finding the radius of curvature at a point on the curve. In 1664–1665, he grappled with this question, and after several attempts he found a solution. In the 1737 anonymous translation of his treatise on fluxions and series, he gave four items to consider in this connection, we present the second and third:[32]

> There are few Problems concerning Curves more elegant than This, or that give a greater Insight into their nature. In order to its resolution, I must premise the following general considerations.
>
> . . .
>
> II. If a Circle touches any Curve on its concave side in a given point, and its magnitude be such that no other Tangent Circle can be interscribed in the Angle of contact nearer that point, that Circle will be the same Curvature as the Curve is of in that point of contact. For that circle which comes between the curve and another Circle at the point of contact, varies less from the Curve and makes a nearer approach to its Curvature, than that other Circle does; and therefore that Circle approaches nearest to its Curvature, between which and the Curve no other Circle can intervene.
>
> III. Therefore the Center of Curvature at any point of a curve, is the Center of a Circle equally curved, and thus the Radius or Semi-diameter of Curvature is part of the perpendicular which is terminated at that Center.

After some discussion of properties of the center of curvature, he described one method for finding the radius of curvature by constructing normals at two infinitely close points, $D$ and $d$. The intersection of the normals gave the center $C$ of the circle of curvature and therefore $CD$ was the radius of curvature. Referring to Figure 7.8, he explained how to find $CD$.

> At any point $D$ of the Curve $AD$, let $DT$ be a Tangent, $DC$ a Perpendicular, and $C$ the Center of Curvature, as before. And let $AB$ be the Absciss, to which let $DB$ be applied at right angles, which $DC$ meets in $P$. Draw $DG$ parallel to $AB$, and $CG$ perpendicular to it, in which take $Cg$ of any given magnitude, and draw $g\delta$ perpendicular to it, which meets $DC$ in $\delta$. Then it will be $Cg : g\delta :: (TB : BD ::)$ as the Fluxion of the Absciss to the Fluxion of the Ordinate. Likewise imagine the point $D$ to move in the Curve an infinitely little distance $Dd$, and drawing $de$ perpendicular to $DG$, and $Cd$ perpendicular to the Curve, let $Cd$ meet $DG$ in $F$, and $\delta g$ in $f$. Then will $De$ be the *momentum* of the Absciss, $de$ the *momentum* of the Ordinate, and $\delta f$

32 Newton (1964–1967) vol. 1, pp. 81–85.

the contemporaneous *momentum* of the RightLine $g\delta$. Therefore $DF = De + \frac{de \times de}{De}$. Having therefore the ratios of these *momenta*, or which is the same thing, of their generating Fluxions, you will have the ratio of $GC$ to the given line $Cg$, which is the same as that of $DF$ to $\delta f$. And thence the point $C$ will be determined.

Therefore let $AB = x$, $BD = y$, $Cg = 1$, and $g\delta = z$. Then it will be $1 : z :: \dot{x} : \dot{y}$, or $z = \frac{\dot{y}}{\dot{x}}$. Now let the *momentum* $\delta f$ of $z$ be $\dot{z} \times o$, (that is the product of the velocity and of an infinitely small quantity $o$,) therefore the *momentum* $De = \dot{x} \times o$, $de = \dot{y} \times o$, and thence $DF = \dot{x}o + \frac{\dot{y}\dot{y}o}{\dot{x}}$. Therefore it is

$$Cg(1) : CG :: (\delta f : DF ::)\dot{z}o : \dot{x}o + \frac{\dot{y}\dot{y}o}{\dot{x}}, \text{ that is,}$$

$CG = \frac{\dot{x}\dot{x}+\dot{y}\dot{y}}{\dot{x}\dot{z}}$. And whereas we are at liberty to ascribe whatever velocity we please to the Fluxion of the Absciss, to which as to an equable Fluxion the rest may be referred, make $\dot{x} = 1$, and then $\dot{y} = z$, and $CG = \frac{1+zz}{\dot{z}}$; whence $GD = \frac{z+z^3}{\dot{z}}$; and $DC = \frac{\overline{1+zz}\sqrt{1+zz}}{\dot{z}}$.

## 7.9  Johann Bernoulli on the Radius of Curvature

In 1691, Guillaume l'Hôpital (1661–1704) met Johann Bernoulli, who informed him that he had found a formula for the radius of curvature. A keen student of mathematics, l'Hôpital was fascinated, and requested Bernoulli give him a course of lectures. In 1691–92, Bernoulli delivered these lectures, in which was necessarily included an elaboration of the calculus. L'Hôpital proceeded to write his famous differential calculus textbook, popular for a century. Bernoulli included his integral calculus lectures as *Lectiones Mathematicae* in vol. 3 of his *Opera Omnia*;[33] he mentions l'Hôpital in the subtitle of the lectures. The derivation of the formula for the radius of curvature is contained in Lecture 16 of this work.

In Figure 7.9, the lines $OD$ and $BD$ are radii normal to the curve, with $O$ and $B$ infinitely close, so that $BD$ is the radius of curvature. Bernoulli let $AE = x$, $EB = y$ with $BF = dx$, $FO = dy$. Then he could write

$$FC = \frac{dy^2}{dx} \left[ = \frac{dy}{dx}dy \right]$$

and therefore

$$BC = \frac{dx^2 + dy^2}{dx}.$$

Now $BF : FO = BE : EH$, so that

$$EH = y\frac{dy}{dx}, \quad BH = \frac{y\sqrt{dx^2 + dy^2}}{dx}, \quad AH = x + y\frac{dy}{dx};$$

[33] Bernoulli, Joh. (1742) vol. 3, pp. 187–193.

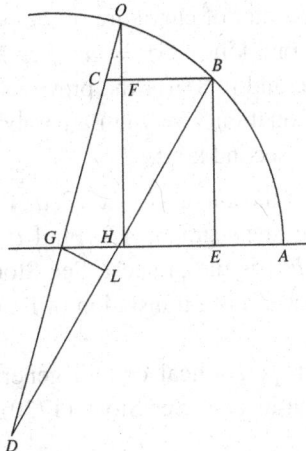

Figure 7.9 Bernoulli's figure for the radius of curvature.

taking $d^2x = 0$, the differential of $AH$ could be written as

$$HG = dx + \frac{dy^2 + yd^2y}{dx}.$$

Then because $BC : HG = BD : HD$, he had $(BC - HG) : BC = BH : BD$ and Bernoulli obtained the formula for the radius of curvature:

$$BD = \frac{(dx^2 + dy^2)\sqrt{dx^2 + dy^2}}{-dxd^2y}.$$

## 7.10 Exercises

(1) Let $b$ be a root of $y^3 + a^2y - 2a^3 = 0$. Show that if $y^3 + a^2y + axy - 2a^3 - x^3 = 0$ and $a^2 + 3b^2 = c^2$, then

$$y = b - \frac{abx}{c^2} + \frac{a^4bx^2}{c^6} + \frac{x^3}{c^2} + \frac{a^3b^3x^3}{c^8} - \frac{a^5bx^3}{c^8} + \frac{a^5bx^3}{c^{10}} + \cdots.$$

(2) Suppose $y^3 + y^2 + y - x^3 = 0$ where $x$ is known to be large. Show that

$$y = x - \frac{1}{3} - \frac{2}{9x} + \frac{7}{81x^2} + \frac{5}{81x^3} \quad \text{etc.}$$

See Newton (1964–1967) vol. 1, pp. 46–47 for the above two exercises.

(3) Show that if $\dot{y} = 3xy^{\frac{2}{3}} + y$, then $y^{\frac{1}{3}} = \frac{1}{2}x^2 + \frac{1}{15}x^3 + \frac{1}{216}x^4$ etc. See Newton (1964–1967) vol. 1, p. 63; see p. 66 for the next exercise.

(4) In a given triangle, find the dimensions of the greatest inscribed rectangle.

(5) Show that in the parabola $ax = yy$, the point at which the radius of curvature is of length $f$ is given by $x = -\frac{1}{4}a + \sqrt[3]{\frac{1}{16}af^2}$.

(6) Find the locus of the center of curvature of the parabola $x^3 = a^2 y$ and of the hyperbola (of the second kind) $xy^2 = a^3$. See Newton (1964–1967) vol. 1, p. 87 for this exercise and p. 85 for the previous exercise. Note that Newton called a polynomial equation $y = p(x)$ a parabola. Similarly, he called $y = \frac{a}{\sqrt{x}}$ a hyperbola of the second kind.

(7) Find the asymptotes of the curve $y^3 - x^3 = axy$. See Stone (1730) p. 19.

(8) Take a point $E$ on the line segment $AB$. Find $E$ such that the product of the square of $AE$ times $EB$ is the greatest. See Stone (1730) p. 58. Recall that this part of Stone's book was a translation of l'Hôpital's differential calculus book.

(9) Find the volume of a parabolical conoid generated by the rotation of the parabola $y^m = x$ about its axis. See Stone (1730) p. 121 of the appendix.

(10) Show that

$$\int x^3 \ln^3 x \, dx = \frac{1}{4}x^4 \ln^3 x - \frac{3}{4^2}x^4 \ln^2 x + \frac{3 \cdot 2}{4^3}x^4 \ln x - \frac{3 \cdot 2}{4^4}x^4.$$

More generally, find $\int x^m \ln^m x \, dx$. From this result, deduce that

$$\int_0^1 x^x \, dx = 1 - \frac{1}{2^2} + \frac{1}{3^3} - \frac{1}{4^4} + \frac{1}{5^5} - \frac{1}{6^6} + \text{etc.}$$

See Joh. Bernoulli (1968) vol. 3, pp. 380–381.

## 7.11   Notes on the Literature

Even though Newton was unable to publish his 1671 calculus treatise, the text was published several times, starting in the 1730s, in both Latin and English. However, Whiteside found that the translations were not completely adequate. Consequently, in vol. 3 of Newton (1967–1981), Whiteside presented his own translation accompanied by Newton's Latin text. Truesdell (1960) presents an interesting commentary on the work of Pardies, Leibniz, Huygens, and Jakob and Johann Bernoulli relating to the catenary. See especially pp. 64–75.

We have not dealt with Leibniz's higher differentials in any detail. An interesting account appears in Bos (1974). Euler showed that the complicated theory of higher differentials could be avoided by using dependent and independent variables. An English translation of Bernoulli's brachistochrone paper appears in Struik (1969), pp. 392–396. Simmons (1992) gives an entertaining account in modern terminology. The reader may enjoy seeing Knoebel, Laubenbacher, Lodder, and Pengelley (2007) for their discussion of Newton's derivation of the radius of curvature. They review this work of Newton in the context of the general notion of curvature and include a discussion of the ideas of Huygens, Euler, Gauss, and Riemann on the topic. Bourbaki (1994) contains a deep but concise summary of the development of the calculus; see pp. 166–198.

# 8

# De Analysi per Aequationes Infinitas

## 8.1  Preliminary Remarks

Newton's groundbreaking paper, revealing the power of infinite series to resolve intractable problems in algebra and calculus, was probably written in the summer of 1669. Before Newton, the only infinite power series to be studied in Europe, besides the infinite geometric series, was the logarithmic series, by J. Hudde and N. Mercator. In the winter of 1664–1665, inspired by Wallis's work on the area of a quadrant of a circle, Newton considered the more general problem of finding the area under $y = \sqrt{1 - t^2}$ on the interval $(0, x)$, for $x \leq 1$.[1] This question led Newton to make the extraordinary inquiry into the value of $(1 - t^2)^{\frac{1}{2}}$ in powers of $t$; Newton thus discovered the binomial theorem, first for exponent $\frac{1}{2}$ and soon for all rational exponents. He very quickly perceived the tremendous significance of this result, and more generally, the importance and usefulness of infinite series to analysis.

In this paper Newton resolved the general problem, at least in principle, of finding the area under a curve defined explicitly or implicitly. He showed that by means of infinite series the problem could be reduced to that of integrating $x^{\frac{m}{n}}$, where $m$ and $n$ were integers. If the equation was given explicitly as $y = f(x)$ with $f(x)$ a rational or algebraic function, then $f(x)$ could be expanded as an infinite series by the binomial theorem. The area under the curve could then be obtained after term-by-term integration. Among the examples he gave were the curves[2]

$$y = \frac{1}{1 + x^2}, \quad y = \frac{2x^{\frac{1}{2}} - x^{\frac{3}{2}}}{1 + x^{\frac{1}{2}} - 3x}, \quad y = \frac{\sqrt{1 + ax^2}}{\sqrt{1 - bx^2}}.$$

He wrote that the quadrature of the last example yielded the length of an elliptic arc.

The problem of integrating even these elementary functions would have been too difficult for the mathematicians before Newton, but his work on the integration of implicitly defined functions took algebra and analysis to a new level. In 1664,

---

[1] Newton (1967–1981) vol. 1, pp. 104–111.
[2] ibid. vol. 2, pp. 215–217.

Newton learned from the books of Viète and Oughtred how to solve algebraic equations $f(x) = 0$ by the method of successive approximation. One chose an approximate solution, and on that basis, one derived successively better ones. The Islamic mathematician Jamshid al-Kashi (1380–1429) had used a primitive form of this method to solve cubic equations and to compute roots of numbers, that is, to solve equations $x^p - N = 0$. With the concept of infinite series in hand, and the technical skill to work with it, Newton showed how to solve the equation $f(x, y) = 0$ in the form $y = g(x)$, where $g(x)$ was an infinite series.[3] Here he was consciously working with the analogy between decimals and infinite series. In his tract that has come to be known as *De methodus serierum et fluxionum*, he wrote,[4]

> Since the operations of computing in numbers and with variables are closely similar–indeed there appears to be no difference between them except in the characters by which quantities are denoted, definitely in the one case, indefinitely in the latter–, ... I am amazed that it has occurred to no one (if you except N. Mercator with his quadrature of the hyperbola) to fit the doctrine recently established for decimal numbers in similar fashion to variables, especially since the way is then open to more striking consequences. For since this doctrine in species has the same relationship to Algebra that the doctrine in decimal numbers has to common Arithmetic, its operations of Addition, Subtraction, Multiplication, Division, and Root-extraction may easily be learnt from the latter's provided the reader be skilled in each, both Arithmetic and Algebra, and appreciate the correspondence between decimal numbers and algebraic terms continued to infinity: namely, that to each single place in a decimal sequence decreasing continually to the right there corresponds a unique term in a variable array ordered according to the sequence of the dimensions of numerators or denominators continued in uniform progression to infinity (as you will see done in the sequel). And just as the advantage of decimals consists in this, that when all fractions and roots have been reduced to them they take on in a certain measure the nature of integers; so it is the advantage of infinite variable-sequences that classes of more complicated terms (such as fractions whose denominators are complex quantities, the roots of complex quantities and the roots of affected equations) may be reduced to the class of simple ones: that is, to infinite series of fractions having simple numerators and denominators and without the all but insuperable encumbrances which beset the others. I will first, consequently, show how to reduce other quantities to terms of this sort and then I will apply this Analysis to the resolution of problems.

Newton obtained higher and higher powers of $x$ by successive approximation. He then found the area under a curve defined implicitly as $f(x, y) = 0$ by integrating $g(x)$ term by term. He gave this method in the *De Analysi*, but he realized that one did not always obtain the solution $y = g(x)$ as a power series. In a longer treatise on calculus and infinite series of 1671, Newton gave examples where the solution around $x = 0$ was of the form $y = x^\alpha g(x)$, with $g(0) \neq 0$ and $\alpha$ a fraction.[5] He realized that for only certain values of $\alpha$ could $g(0)$ be determined; for those values, the functions $y = x^\alpha g(x)$ were solutions of $f(x, y) = 0$ in the neighborhood of $x = 0$. He devised a method now called Newton's polygon to determine the allowable values of $\alpha$. This method has important applications in algebraic geometry and analysis. Newton extended his method for solving $f(x, y) = 0$ to obtain the inverses of functions defined by infinite series. He knew that his formula, mentioned earlier, for the area of a sector

[3] ibid. p. 221, footnote 59.
[4] ibid. vol. 3, pp. 33 and 35.
[5] ibid. vol. 3, pp. 42–73.

of a circle was equivalent to the series for arcsine. By inversion he found the series for sine and from that the series for cosine. We have seen that Madhava earlier obtained the series for these functions by a different method.

Newton uncharacteristically wrote up his results on infinite series because by 1668 others were beginning to make similar discoveries. Mercator published a book, *Logarithmotechnia* in which he expanded $\frac{1}{1+x}$ as a series and integrated term by term to obtain his series for $\ln(1+x)$; in fact, Mercator gave the series for only some small values of $x$. The general series for $\ln(1+x)$ in powers of $x$, or $e$ in Wallis's notation, was given in Wallis's review of Mercator's book, contained in the 1668 volume of the *Philosophical Transactions*. In the spring of 1665, Newton had done exactly the same thing; after Mercator's publication, he realized that he would lose credit for his discoveries unless he made them known.

Newton submitted his paper to Isaac Barrow, then Lucasian Professor of Mathematics at Cambridge, who mentioned it to Collins in a letter of July 20, 1669, with the words:[6]

> A friend of mine here, that hath a very excellent genius to those things, brought me the other day some papers, wherein he hath sett downe methods of calculating the dimensions of magnitudes like that of Mr Mercator concerning the hyperbola, but very generall; as also of resolving aequations; which I suppose will please you; and I shall send you them by the next.

Barrow wrote Collins again on August 20, 1669:[7]

> I am glad my friends paper giveth you so much satisfaction. his name is Mr Newton; a fellow of our College, & very young (being but the second yeest [youngest] Master of Arts) but of an extraordinary genius & proficiency in these things. you may impart the papers if you please to my Ld Brounker [*sic*].

Collins made a complete copy of Newton's paper and communicated some of its results to his correspondents in Britain, France, and Italy. He and Barrow urged Newton to publish his paper as an appendix to Barrow's optical lectures but Newton resisted, perhaps because he had a much larger work in mind, finally written in 1671. This long but incomplete tract is referred to as *De Methodis Serierum et Fluxionum*, though its original title or whether it even had one is unclear, since the first page of the original manuscript is lost and mathematicians of Newton's own time referred to it by various titles. In the *De Methodis*, Newton showed how to find the derivative by implicit differentiation of the equation for the curve $f(x, y) = 0$. He applied this to problems on tangents, normals, and radii of curvature. Conversely, given a fluxional (differential) equation, he explained how it could be solved, particularly with infinite series. The equations he worked with here were algebraic differential equations.

It is interesting to note that when Newton wrote up his results on series, realizing that others were working on similar problems, this exercise gave him the opportunity to rethink his ideas and improve upon them. This happened to him several times. For example, in the spring of 1684, David Gregory, nephew of James, published a

---

[6] Newton (1959–60) vol. 1, pp. 13–14.
[7] ibid. pp. 14–15.

fifty-page tract *Exercitatio Geometrica de Dimensione Figurarum*,[8] discussing his uncle's results on infinite series related to the binomial theorem. He also promised to write a sequel with more results. This immediately spurred Newton to compose the *Matheseos Universalis Specimina*, in the first part of which he gave a brief history of his work on series and the results on this topic he had communicated to Collins and to Leibniz. He then went on to develop some new ideas on finite differences and series. The paper was not completed and in fact ended in the middle of a sentence. Very soon after this, he reorganized his ideas and presented them in a paper called *De Computo Serierum*. Here he left out the history but further clarified the new mathematical idea on series and differences, framed as the transformation formula now often called Euler's transformation. (See our Sections 10.1 and 10.2.) Unfortunately, Newton never published these papers. Similarly, in 1691 he wrote and rewrote the tract *De Quadratura Curvarum*, containing the first explicit statement of Taylor's theorem; he published only a portion of this work some years later. In this connection, see our Section 11.2.

## 8.2   Algebra of Infinite Series

Newton pointed out in his *De Analysi* that just as infinite decimals were needed to divide by numbers, extract roots of numbers, and solve equations with numerical coefficients, infinite series were needed to divide by polynomials, extract roots of algebraic expressions, and solve equations with algebraic coefficients. To illustrate division, Newton considered the equation $y = \frac{a^2}{b+x}$ and showed that the process led to the series[9]

$$y = \frac{a^2}{b} - \frac{a^2 x}{b^2} + \frac{a^2 x^2}{b^3} - \frac{a^2 x^3}{b^4} + \cdots . \tag{8.1}$$

From this he concluded that the area under the curve $y = \frac{a^2}{b+x}$ could be expressed as

$$\frac{a^2 x}{b} - \frac{a^2 x^2}{2b^2} + \frac{a^2 x^3}{3b^3} - \frac{a^2 x^4}{4b^4} + \cdots , \tag{8.2}$$

if the required area was taken over the interval $(0, x)$. If, however, the required area under $y = \frac{1}{1+x^2}$ was over $(x, \infty)$, then with $a = b = 1$, and $x$ replaced by $x^2$, he started with the series

$$y = \frac{1}{x^2 + 1} = x^{-2} - x^{-4} + x^{-6} - x^{-8} \quad \text{etc.} \tag{8.3}$$

The area was then given by

$$-x^{-1} + \frac{1}{3} x^{-3} - \frac{1}{5} x^{-5} + \frac{1}{7} x^{-7} \quad \text{etc.} \tag{8.4}$$

---

[8] Gregory (1684).
[9] Newton (1967–1981) vol. 2, pp. 213–215.

Newton noted that $x$ should be small in (8.2), but should be large in (8.4), though he did not specify how small or how large. At the end of the paper, he made some remarks on convergence. He observed that if $x = \frac{1}{2}$, then $x$ would be half of all of $x + x^2 + x^3 + x^4$ etc. and $x^2$ half of all of $x^2 + x^3 + x^4 + x^5$ etc. So if $x < \frac{1}{2}$, then $x$ would be more than half of all of $x + x^2 + x^3$ etc. and $x^2$ more than half of all of $x^2 + x^3 + x^4$ etc. He then extended the argument to the case $\frac{x}{b}$ where $b$ was a constant.

In his second example, Newton applied the algorithm for finding square roots of numbers to $\sqrt{(a^2 + x^2)}$, obtaining the infinite series

$$a + \frac{x^2}{2a} - \frac{x^4}{8a^3} + \frac{x^6}{16a^5} - \frac{5x^8}{128a^7} + \frac{7x^{10}}{256a^9} - \frac{21x^{12}}{1024a^{11}} \quad \text{etc.} \quad (8.5)$$

A little later in the paper, Newton explained his method of successive approximations to solve polynomial equations $f(x, y) = 0$. To illustrate the method, he first took an equation with constant coefficients:[10]

$$y^3 - 2y - 5 = 0. \quad (8.6)$$

An approximate solution would be 2, so he set $y = 2 + p$ to transform (8.6) to

$$p^3 + 6p^2 + 10p - 1 = 0. \quad (8.7)$$

He argued that since $p$ was small, the terms $p^3 + 6p^2$ could be neglected, though he noted that a better approximation would be obtained if only $p^3$ were neglected. Thus, he had $p = 0.1$, and he substituted $p = 0.1 + q$ in (8.7) to obtain

$$q^3 + 6.3q^2 + 11.23q + 0.061 = 0.$$

Newton linearized this equation to $11.23q + 0.061 = 0$, solved for $q$ to get $q = -0.0054$, set $q = -0.0054 + r$, and wrote that one could continue in this manner. In a similar way, he resolved the equation[11]

$$y^3 + a^2y - 2a^3 + axy - x^3 = 0 \quad (8.8)$$

for small values of $x$. He set $x = 0$ to obtain

$$y^3 + a^2y - 2a^3 = 0 \quad (8.9)$$

so that $y = a$ was a solution; he set $y = a + p$ in (8.8) and took the linear part of the equation to get $p = -\frac{1}{4}x$. In this manner, he had the series

$$y = a - \frac{1}{4}x + \frac{x^2}{64a} + \frac{131x^3}{512a^2} + \frac{509x^4}{16384a^3} \quad \text{etc.} \quad (8.10)$$

[10] ibid. pp. 219–221.
[11] ibid. pp. 223–227.

He then used the example

$$y^3 + axy + x^2 y - a^3 - 2x^3 = 0 \qquad (8.11)$$

to illustrate how to obtain a solution for large values of $x$. Here he started with the highest power terms in the equation (8.10) to get

$$y^3 + x^2 y - 2x^3 = 0.$$

Since $y = x$ was a solution of this, he set $y = x + p$ in (8.11) and proceeded as before.

Finally, Newton showed that similar methods could be applied when the equation had an infinite number of terms. The problem of interest was to solve $y = f(x)$, for $x$ in terms of $y$, where $f(x)$ was an infinite series. This gave him a series for the inverse function and in the *De Analysi*, he applied it to the cases where

$$f(x) = -\ln(1-x) = x + \frac{1}{2}x^2 + \frac{1}{2}x^3 + \cdots$$

and where

$$f(x) = \arcsin x = x + \frac{1}{6}x^3 + \frac{3}{40}x^5 + \frac{5}{112}x^7 + \cdots .$$

Thus, he found the series for the exponential and sine functions.[12]

Observe that Newton's method of successive approximations for finding the series for inverse functions is actually equivalent to the method of undetermined coefficients. One assumes that if $z$ denotes the series, say for $-\ln(1-x)$, then

$$x = a_0 + a_1 z + a_2 z^2 + \cdots ,$$

and the values of $a_0, a_1, a_2, \ldots$ are obtained by substituting back in the series and equating the coefficients of the powers of $z$ on both sides of the equation. Newton must have understood this because in his October 1676 letter to Oldenburg he wrote,[13]

Let the equation for the area of an hyperbola be proposed

$$z = x + \frac{1}{2}x^2 + \frac{1}{3}x^3 + \frac{1}{4}x^4 + \frac{1}{5}x^5, \quad \text{etc.}$$

and its terms being multiplied into themselves, there results

$$z^2 = x^2 + x^3 + \frac{11}{12}x^4 + \frac{5}{6}x^5, \quad \text{etc.,}$$

$$z^3 = x^3 + \frac{3}{2}x^4 + \frac{7}{4}x^5, \quad \text{etc.,}$$

$$z^4 = x^4 + 2x^5, \quad \text{etc.,}$$

$$z^5 = x^5, \quad \text{etc.}$$

[12] ibid. pp. 235–237.
[13] Newton (1959–1960) vol. 2, p. 146.

Now subtract $\frac{1}{2}z^2$ from $z$, and there remains

$$z - \frac{1}{2}z^2 = x - \frac{1}{6}x^3 - \frac{5}{24}x^4 - \frac{13}{60}x^5, \quad \text{etc.}$$

To this I add $\frac{1}{6}z^3$, and it becomes

$$z - \frac{1}{2}z^2 + \frac{1}{6}z^3 = x + \frac{1}{24}x^4 + \frac{3}{40}x^5, \quad \text{etc.}$$

I subtract $\frac{1}{24}z^4$ and there remains

$$z - \frac{1}{2}z^2 + \frac{1}{6}z^3 - \frac{1}{24}z^4 = x - \frac{1}{120}x^5, \quad \text{etc.}$$

I add $\frac{1}{120}z^5$ and it becomes

$$z - \frac{1}{2}z^2 + \frac{1}{6}z^3 - \frac{1}{24}z^4 + \frac{1}{120}z^5 = x$$

as nearly as possible; or

$$x = z - \frac{1}{2}z^2 + \frac{1}{6}z^3 - \frac{1}{24}z^4 + \frac{1}{120}z^5, \quad \text{etc.}$$

He then went on to state two general theorems:

Let $z = ay + by^2 + cy^3 + dy^4 + ey^5 + $ etc. Then conversely will

$$y = \frac{z}{a} - \frac{b}{a^3}z^2 + \frac{2b^2 - ac}{a^5}z^3 + \frac{5abc - 5b^3 - a^2d}{a^7}z^4$$
$$+ \frac{3a^2c^2 - 21ab^2c + 6a^2bd + 14b^4 - a^3e}{a^9}z^5 + \quad \text{etc.} \tag{8.12}$$

Let $z = ay + by^3 + cy^5 + dy^7 + ey^9 + $ etc. Then conversely will

$$y = \frac{z}{a} - \frac{b}{a^4}z^3 + \frac{3b^2 - ac}{a^7}z^5 + \frac{8abc - a^2d - 12b^3}{a^{10}}z^7$$
$$+ \frac{55b^4 - 55ab^2c + 10a^2bd + 5a^2c^2 - a^3e}{a^{13}}z^9 + \quad \text{etc.} \tag{8.13}$$

Newton observed that if he took $a = 1, b = \frac{1}{6}, c = \frac{3}{40}, d = \frac{5}{112}$, etc., in the second series, then the series for $\sin z$ would follow. Newton actually wrote the series for $r \sin z$, where $r$ was the radius of the circle, in powers of $\frac{z}{r}$ because $z$ was the length of an arc of the circle. Euler later eliminated the role of the radius and defined the trigonometric functions as we do now.

## 8.3 Newton's Polygon

In his *De Methodis*, Newton explained his method of solving $f(x, y) = 0$ by means of the Newton polygon, where the solution took the form $y = x^\alpha y_1$, with $\alpha$ rational and $y_1$ a power series in $x$.[14] To find the possible values of $\alpha$, he plotted the points $(b, a)$ for each term $cx^a y^b$ in $f(x, y)$, such that $b$ ran along the horizontal axis and $a$ along the vertical. He then took the lower portion of the convex hull of these points, consisting of the straight line(s) joining the vertical to the horizontal axis. Although it does not enclose an area, this lower portion is called the Newton polygon and the slope(s) of these line(s) gave him the values of $\alpha$. For example, if $m$ were the slope of a line in the polygon, then one value of $\alpha$ would be given by $-\frac{1}{m}$. These values of $\alpha$ permitted the evaluation of a nonzero value of $y_1(0)$. Note here that the lines joining the other pairs of points $(b, a)$ could allow for zero values of $y_1(0)$.

Newton considered the example

$$y^6 - 5xy^5 + \frac{x^3}{a}y^4 - 7a^2x^2y^2 + 6a^3x^3 + b^2x^4 = 0,$$

where he had the points $(6,0)$, $(5,1)$, $(4,3)$, $(2,2)$, $(0,3)$, and $(0,4)$. The line joining $(0,3)$, $(2,2)$, and $(6,0)$ formed the polygon and gave Newton the terms

$$y^6 - 7a^2x^2y^2 + 6a^3x^3;$$

setting these equal to zero, he obtained the lowest-order term in the expansion of $y$ as a series in $x$. The slope of the line in the polygon was $-\frac{1}{2}$ so in the case $y = cx^{\frac{1}{2}}$, the terms had the same power in $x$. Newton could then set $y = v\sqrt{ax}$ to reduce the equation to $v^6 - 7v^2 + 6 = 0$. He obtained $v = \pm 1$, $\pm\sqrt{2}$, $\pm\sqrt{-3}$ but rejected the complex roots. Thus, he had four possible initial values of $y$: $\pm\sqrt{ax}$, $\pm\sqrt{2ax}$. He wrote that all four expressions were acceptable initial values for $y$; by successive approximations, he went on to find more terms.

Newton's solutions of $f(x, y) = 0$ are the first known examples of the implicit function theorem. Significantly, though Newton did not give an existence proof, he presented an algorithm for deriving the solution. S. S. Abhyankar pointed out that this algorithm is applicable to existence proofs in analysis and can also produce the formal solutions required in algebraic geometry.[15]

## 8.4  Newton on Differential Equations

In his 1670–71 treatise *De Methodis Serierum et Fluxionum*, Newton discussed how to find the derivative from the equation $f(x, y) = 0$ and, conversely, how to find the relation between $x$ and $y$, given a first-order differential equation $f(x, y, \dot{x}, \dot{y}) = 0$. Note that in the 1690s, Newton began to use the dot notation to indicate a fluxion, or derivative. In his earliest work, including his work in the 1670s, he employed the letters $p$, $q$, or $m$, $n$.

---

[14] Newton (1967–1981) vol. 3, pp. 49–55.
[15] Abhyankar (1976) pp. 416–417.

To illustrate how series could be used to solve differential equations, Newton considered several examples,[16] including

$$\frac{\dot{y}}{\dot{x}} = 1 + \frac{y}{a-x},$$
$$\dot{y}^2 = \dot{x}\dot{y} + \dot{x}^2 x^2,$$

and

$$\dot{y}^3 + ax\dot{x}^2\dot{y} + a^2\dot{x}^2\dot{y} - \dot{x}^3 x^3 - 2\dot{x}^3 a^3 = 0.$$

He rewrote the first equation as

$$\frac{\dot{y}}{\dot{x}} = 1 + \frac{y}{a} + \frac{xy}{a^2} + \frac{x^2 y}{a^3} + \frac{x^3 y}{a^4} \quad \text{etc.}$$

and then showed how to obtain particular solutions of this equation by assuming a series solution. He rewrote the second equation as a quadratic in $\frac{\dot{y}}{\dot{x}}$ to get

$$\frac{\dot{y}^2}{\dot{x}^2} = \frac{\dot{y}}{\dot{x}} + x^2$$

and solved the quadratic algebraically to obtain

$$\frac{\dot{y}}{\dot{x}} = \frac{1}{2} \pm \sqrt{\frac{1}{4} + x^2}.$$

After expanding $\left(\frac{1}{4} + x^2\right)^{\frac{1}{2}}$ by the binomial theorem, Newton integrated the resulting infinite series term by term. He apparently did not observe that $\left(\frac{1}{4} + x^2\right)^{\frac{1}{2}}$ could be integrated directly in terms of the logarithm. Barrow gave the integral of $(a^2 + x^2)^{\frac{1}{2}}$ in his *Lectiones Geometricae*[17] of 1670 and Newton knew this work quite well. However, unlike Leibniz, Newton may not have been particularly interested in closed-form solutions.[18] Newton changed the third equation into a cubic in $\frac{\dot{y}}{\dot{x}}$:

$$\left(\frac{\dot{y}}{\dot{x}}\right)^3 + ax\left(\frac{\dot{y}}{\dot{x}}\right) + a^2\left(\frac{\dot{y}}{\dot{x}}\right) - x^3 - 2a^3 = 0,$$

the same cubic as in (8.8). So from (8.10) he saw that

$$\frac{\dot{y}}{\dot{x}} = a - \frac{x}{4} + \frac{x^2}{64a} + \frac{131x^3}{512a^2} \quad \text{etc.}$$

[16] Newton (1967–1981) vol. 3, pp. 83–91.

[17] Barrow (1916) pp. 160–161 and p. 185.

[18] Nevertheless, see Newton (1967–1981) vol. 3, pp. 199–209, where Newton discussed closed forms in terms of known curves.

and hence

$$y = ax - \frac{x^2}{8} + \frac{x^3}{192a} + \frac{131x^4}{2048a^2} \quad \text{etc.}$$

## 8.5  Newton's Earliest Work on Series

In some of the earliest material recorded in his mathematical notebooks, Newton raised the problem of finding $x$, given $\sin x$, observing that the problem was equivalent to finding the area of a segment of a circle. Newton did this work, inspired by Wallis's book, in winter 1664–65.[19]

In Figure 8.1, let $aec$ be a quarter of the circle of radius one with center $p$ and let $pq = x$. If the angle $ape = \theta$, then $x = \sin\theta$ and the area of the sector $ape$ would be $\frac{1}{2}\theta = \frac{1}{2}\arcsin x$. Newton's problem was to find an expression for the area given by $\frac{1}{2}\arcsin x$, when $x$ was known. He knew, from a study of Wallis's *Arithmetica Infinitorum*, that the area $aeqp$ was equal to the area under the curve $y = \sqrt{1 - t^2}$ over the interval $[0, x]$. The area of the triangle $peq$ was $\frac{1}{2}x\sqrt{1 - x^2}$. So his formula in modern notation would be given as

$$\text{area of sector } aep = \frac{1}{2}\arcsin x = \int_0^x \sqrt{1 - t^2}\, dt - \frac{1}{2}x\sqrt{1 - x^2}. \qquad (8.14)$$

In the course of this work, he discovered the binomial theorem. This gave him the result

$$(1 - x^2)^{\frac{1}{2}} = 1 - \frac{x^2}{2} - \frac{x^4}{8} - \frac{x^6}{16} - \frac{5x^8}{128} - \frac{7x^{10}}{256} - \frac{21x^{12}}{1024} \quad \text{etc.}$$

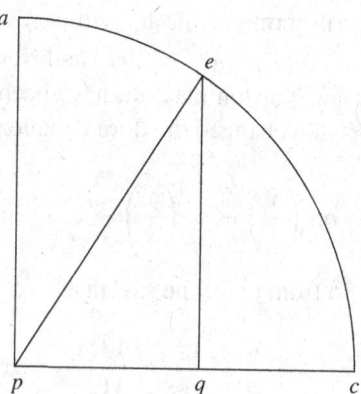

Figure 8.1  Newton's figure for derivation of the arcsine series.

[19] Newton (1967–1981) vol. I, pp. 104–110.

He substituted this series in the integral and integrated term by term to get the solution of his problem:

$$\arcsin x = x + \frac{x^3}{6} + \frac{3x^5}{40} + \frac{5x^7}{112} + \frac{35x^9}{1152} \quad \text{etc.} \tag{8.15}$$

In the later *De Analysi*, Newton found the series for arcsin $x$ by determining the arc length of the arc of a circle. In this case, he had to integrate $\frac{1}{\sqrt{1-t^2}}$. If this were combined with (8.14 ), the result would be

$$\int_0^x \frac{dt}{\sqrt{1-t^2}} = \arcsin x.$$

Thus, Newton was aware of the integral for arcsine as well as the formula

$$\int_0^x \sqrt{1-t^2}\, dt = \frac{1}{2}x\sqrt{1-x^2} + \frac{1}{2}\int_0^x \frac{dt}{\sqrt{1-t^2}}.$$

Modern textbooks usually derive the last formula by means of integration by parts. After 1666, Newton was effectively aware of substitution and integration by parts, but to obtain a result more simply, he often gave geometric arguments, similar to the preceding one; note, however, that he derived the series for arcsine in 1664–1665. Newton discovered another interesting formula from which the series for arcsine can be easily derived; the first mention of it occurs in his letter to Oldenburg dated June 13, 1676, in response to Leibniz:[20]

> If an arc is be taken in a given ratio to another arc, let $d$ be the diameter, $x$ the chord of the given arc, and the required arc be to that given arc as $n : 1$. Then the chord of the arc required will be
>
> $$nx + \frac{1-n^2}{2 \times 3d^2}x^2 A + \frac{9-n^2}{4 \times 5d^2}x^2 B + \frac{25-n^2}{6 \times 7d^2}x^2 C + \frac{36-n^2}{8 \times 9d^2}x^2 D + \frac{49-n^2}{10 \times 11d^2}x^2 E + \text{etc.}$$
>
> Here note that when $n$ is an odd number the series is no longer infinite and becomes the same as that which results by common algebra for multiplying the given angle by that number $n$.

As Newton explained in his letter, $A$ stood for the first term, $B$ for the second term, $C$ for the third term, etc. Observe that this formula is equivalent to

$$\sin n\theta = n\sin\theta - \frac{n(n^2-1)}{3!}\sin^3\theta + \frac{n(n^2-1)(n^2-3^2)}{5!}\sin^5\theta \\ - \frac{n(n^2-1)(n^2-3^2)(n^2-5^2)}{7!}\sin^7\theta + \cdots. \tag{8.16}$$

Note that the series for arcsine is obtained by dividing (8.16) by $n$ and letting $n$ tend to zero. The corresponding cosine series is given by

---

20 Newton (1959–1960) vol. 2, p. 36.

$$\cos n\theta = 1 - \frac{n^2}{2!}\sin^2\theta + \frac{n^2(n^2-2^2)}{4!}\sin^4\theta - \frac{n^2(n^2-2^2)(n^2-4^2)}{6!}\sin^6\theta + \cdots.$$

$$(8.17)$$

This series does not appear in the extant papers of Newton, although one may safely assume he must have known this result. Note that if we subtract 1 from both sides and then divide the equation by $n^2$, then we obtain the power series for $\arcsin^2 x$ when $n$ tends to zero. The result mentioned by Newton for $n$ odd, as obtainable from common algebra, may have been known to him since 1664 when he was a student.[21]

### 8.6   De Moivre on Newton's Formula for $\sin n\theta$

In 1698, de Moivre gave a derivation of Newton's series for $\sin n\theta$ in "A Method of Extracting the Root of an Infinite Equation" in the *Philosophical Transactions*.[22] Now de Moivre had already seen the method of undetermined coefficients used in this context, as presented in the letter of Newton for Leibniz. Note that British mathematicians of the 1690s were aware of these letters, since Wallis had published portions of them in his 1685 book on algebra[23] and had presented more complete accounts in the 1690s. In his paper, de Moivre considered the situation in which the series on the left side of the equation was in terms of a variable different from that on the right; one variable had to be determined in terms of the other. He stated the main result at the very beginning of his paper:

If $az + bzz + cz^3 + dz^4 + ez^5 + fz^6 + $ etc. $= gy + hyy + iy^3 + ky^4 + ly^5 + my^6 + $ etc., then $z$ will be

$$= \frac{g}{a}y + \frac{h-bAA}{a}y^2$$

$$+ \frac{i-2bAB-cA^3}{a}y^3$$

$$+ \frac{k-bBB-2bAC-3cAAB-dA^4}{a}y^4$$

$$+ \frac{l-2bBC-2bAD-3cABB-3cAAC-4dA^3B-eA^5}{a}y^5$$

$$+ \text{etc.}$$

Note that de Moivre also included the coefficient of $y^6$. Each capital letter denoted the coefficient of the preceding term. Thus, $A = \frac{g}{a}$ and $B = \frac{h-bAA}{a}$, and so on. His proof first assumed that $z$ had a series expansion $Ay + Byy + Cy^3 + Dy^4 + $ etc., then substituted this for each $z$ on the left side of the initial equation, and then equated coefficients of powers of $y$ to get $A$, $B$, $C$, $D$. To apply de Moivre's formula to get

---

[21] See, for example, Newton's annotations on Viète in Newton (1967–1981) vol. 1, pp. 78–83.
[22] de Moivre (1698).
[23] E.g. Wallis (1685) pp. 318–319 and pp. 330–346.

Newton's formula, recall that the latter involved the expansion of $z = \sin n\theta$ in powers of $y = \sin\theta$. Clearly, we can write

$$\arcsin z = n \arcsin y.$$

When the arcsines are replaced by their power series expansions, we have

$$z + \frac{z^3}{6} + \frac{3z^5}{40} + \frac{5z^7}{112} + \cdots = ny + \frac{ny^3}{6} + \frac{3ny^5}{40} + \frac{5ny^7}{112} + \cdots.$$

De Moivre applied his general result to this special equation to obtain

$$A = n, \quad B = 0, \quad C = -\frac{n(n^2 - 1)}{6}, \quad D = 0, \quad E = -\frac{n(n^2 - 1)(n^2 - 3^2)}{5!}, \ldots,$$

thereby completing a proof of Newton's formula.

## 8.7 Stirling's Proof of Newton's Formula

By studying Stirling's unpublished notebooks, Ian Tweddle discovered that Stirling gave yet another proof of Newton's formula, by means of differential equations.[24] This work was probably done before 1730. Stirling took variables $y = r\sin\theta$ and $v = r\sin n\theta$, where $n$ was any positive number. He used geometric considerations to define these variables and these required that $0 \le n\theta \le \frac{\pi}{2}$, but his proof is actually valid for $0 \le |\theta| \le \frac{\pi}{2}$ with $n$ any real number. Since $\theta = \arcsin\left(\frac{y}{r}\right)$, $\dot{\theta} = \frac{r\dot{y}}{\sqrt{r^2 - y^2}}$ and similarly $n\dot{\theta} = \frac{r\dot{v}}{\sqrt{r^2 - v^2}}$, Stirling obtained the fluxional equation

$$\frac{n\dot{y}}{\sqrt{rr - yy}} = \frac{\dot{v}}{\sqrt{rr - vv}}; \tag{8.18}$$

after squaring, he got

$$\frac{n^2\dot{y}^2}{rr - yy} = \frac{\dot{v}^2}{rr - vv}.$$

He cross multiplied to obtain

$$n^2 r^2 \dot{y}^2 - n^2 \dot{y}^2 v^2 = \dot{v}^2 r^2 - y^2 \dot{v}^2.$$

We note that until the middle of the nineteenth century, mathematicians sometimes wrote $xx$ and sometimes $x^2$. He took the fluxion (derivative) of this equation, assuming that $\dot{y}$ was uniform. This meant that $\ddot{y} = 0$. Thus, he had the equation

$$-n^2\dot{y}^2 v\dot{v} = \dot{v}\ddot{v}r^2 - \dot{v}\ddot{v}y^2 - y\dot{y}\dot{v}^2, \quad \text{or}$$
$$-n^2\dot{y}^2 v = \ddot{v}r^2 - \ddot{v}y^2 - y\dot{y}\dot{v}.$$

He next set $\dot{y} = 1$, without loss of generality, to obtain

$$\ddot{v}(r^2 - y^2) - y\dot{v} + n^2 v = 0. \tag{8.19}$$

Assuming a series solution, Stirling set

$$v = Ay + By^3 + Cy^5 + Dy^7 \text{ etc.} \tag{8.20}$$

After substituting (8.20) in (8.19), Stirling found the coefficients, completing the derivation:

$$B = \frac{1-n^2}{2 \cdot 3r^2} A, \quad C = \frac{9-n^2}{4 \cdot 5r^2} B, \quad D = \frac{25-n^2}{6 \cdot 7r^2} C, \text{ etc.}$$

In his 1730 *Methodus Differentialis*, proposition 15, Stirling briefly explained why he used a series of the form (8.20) to solve the differential equation. He set $v = Ay^m$ and substituted this expression in the differential equation to get

$$(m^2 - m)Ay^{m-2} + (n^2 - m^2)Ay^m = 0. \tag{8.21}$$

To obtain the lowest power of $y$ in the series solution, he then set $m^2 - m = 0$ to obtain $m = 0$ or $m = 1$. Thus, the lowest power of $y$ was either 0 or 1. By (8.21), the powers had to increase by two, so that either $v$ was given by the series (8.20), or else

$$v = A + By^2 + Cy^4 + Dy^6 + \cdots . \tag{8.22}$$

By using (8.22) in a similar way, he obtained the formula (8.17) for $\cos n\theta$.

Newton did not state the cosine formula, but he must have known it from his 1664 study of Viète's booklet on angular sections, written in 1591 but published in 1615 with proofs supplied by Alexander Anderson, an uncle of the great Scottish mathematician James Gregory. In this paper, Viète expressed in geometric terms the formulas for $\cos n\theta$ and $\sin n\theta$ in powers of $\cos\theta$, with $n$ an integer. He explicitly pointed out the appearance of the figurate numbers as coefficients of these polynomials. As a student, Newton made annotations on this work of Viète, though they do not indicate that he knew his formula (8.16) at that time.[25] The manner in which he wrote the coefficients of the powers of $\sin\theta$ in his letter for Leibniz suggests that he found the result after his discovery of the binomial theorem. It is even likely that he found the formula while reviewing his old notes before writing his first letter for Leibniz in June 1676.

The methods employed by de Moivre and Stirling to prove Newton's formulas were familiar to Newton in 1676. In fact, it is very likely that Newton had already found a proof. It is possible that he first came upon the formula for $\sin n\theta$ by interpolation, but he wrote in his letter for Leibniz that he had discarded interpolation as a method of proof. Since Newton was very cautious, he must have had an alternative derivation when he communicated it to Leibniz, though he gave no hint of what it was.

---

[25] See Newton (1967–1981) vol. 1, pp. 78–84.

In 1812, Gauss applied (8.16) to produce an unusual proof of Euler's gamma function formula $\Gamma(x)\Gamma(1-x) = \frac{\pi}{\sin \pi x}$.[26] He also briefly mentioned that he could prove (8.16) using transformations of hypergeometric functions. Yet another proof was given by Cauchy in his École Polytechnique lectures of 1821.[27] He also observed that the series for $\arcsin x$ could be derived by equating the coefficient of $n$ on both sides of (8.16). Similarly, by equating the coefficient of $n^2$ in (8.17), Cauchy obtained

$$\frac{1}{2}(\sin^{-1} x)^2 = \sum_{n=0}^{\infty} \frac{2^{2n}(n!)^2}{(2n+2)!} x^{2n+2}, \tag{8.23}$$

a result he attributed to J. de Stainville who published it in 1815. In November 1737, the particular case of the above series where $x = 1$ was discovered by Johann Bernoulli who communicated it to his former student Euler.[28] Euler responded by using differential equations to prove the more general formula (8.23).[29] As we shall see later, Bernoulli's method can be modified slightly to prove the general case. Even before Bernoulli and Euler, Takebe Katahiro published these two series in his 1722 *Yenri Tetsujutsu*.[30]

## 8.8 Zolotarev: Lagrange Inversion with Remainder

Newton's statement of his two theorems on the inversion of series suggests that he got them by using the method of undetermined coefficients, though his related work employs successive approximation. In 1769, Lagrange published a more interesting result now referred to as the Lagrange inversion formula. This work was done in connection with an application to celestial mechanics. Lagrange's formula stated that if $z = a + x\phi(z)$, then

$$F(z) = F(a) + x\phi(a)F'(a) + \frac{x^2}{1 \cdot 2}\frac{d}{da}(\phi^2(a)F'(a))$$
$$+ \frac{x^3}{1 \cdot 2 \cdot 3}\frac{d^2}{da^2}(\phi^3(a)F'(a)) + \cdots.$$

In support of this formula, Lagrange gave a complicated argument using divergent series.[31] The remainder term for the Lagrange series was apparently first determined by Robert Murphy in 1833;[32] then in 1861, A. Popoff rediscovered this

---

[26] Gauss (1813).
[27] Cauchy et al. (2009) p. 376.
[28] Eu. 4 A-2, p. 186.
[29] ibid. p. 196.
[30] Smith and Mikami (1914) pp. 148–149.
[31] See Johnson (2007) for interesting historical remarks and references. Johnson also fills out the details of Lagrange's sketchy proof from *Théorie des fonctions analytique*.
[32] Murphy (1833a).

remainder term.[33] In 1890, the mathematician J. J. Walker wrote a paper, "Of the Influence of Applied on the Progress of Pure Mathematics," and pointed out Murphy's priority:[34]

> The real novelty in Murphy's memoir was the expression given in the form of a definite integral– for the "error" involved in stopping at any given term of Lagrange's Series. This is quite different in form from Popoff's expression (C.R. [Comptes Rendus] 1861, pp. 795–8), which appears to be considered as the first attempt to sum up the remainder.

In 1876, Zolotarev gave a simple proof of the Lagrange series with remainder:[35]

$$F(z) = F(a) + \sum_{k=1}^{n} \frac{x^k}{k!} \frac{d^{k-1}}{da^{k-1}} \left( \phi^k(a) F'(a) \right)$$

$$+ \frac{1}{n!} \frac{d^n}{da^n} \left( \int_a^z (x\phi(u) + a - u)^n F'(u)\, du \right).$$

He proved this formula by setting

$$S_n = \int_a^z (x\phi(u) + a - u)^n F'(u)\, du$$

and observing that differentiation with respect to $a$ immediately yielded

$$\frac{dS_n}{da} = n S_{n-1} - x^n \phi^n(a) F'(a).$$

By setting $n = 1, 2, \ldots, n$, he arrived at the $n$ relations

$$S_0 = x\phi(a) F'(a) + \frac{dS_1}{da},$$

$$2S_1 = x^2 \phi^2(a) F'(a) + \frac{dS_2}{da},$$

$$\cdots$$

$$n S_{n-1} = x^n \phi^n(a) F'(a) + \frac{dS_n}{da}.$$

Noting that

$$S_0 = \int_a^z F'(u)\, du = F(z) - F(a),$$

and by substituting the $n$th equation into the $(n - 1)$th equation and continuing the process, he obtained the required formula.

---

[33] Popoff (1861).
[34] Walker (1890).
[35] Zolotarev (1876).

## 8.9 Exercises

(1) Show that Newton's series in (8.1) and (8.3) can be obtained by repeated division.

(2) Apply the method of finding square roots to the polynomial $a^2 + x^2$ to obtain Newton's series (8.5).

(3) Carry out Newton's procedure for successive approximation of a solution of (8.8) to obtain the series (8.10).

(4) Verify Newton's two theorems on the reversion of series, given in equations (8.12) and (8.13).

(5) Prove that for any real number $n$ and nonnegative integer $s$,

$$\binom{\frac{n}{2}}{s} + \binom{n}{2}\binom{\frac{n-2}{2}}{s-1} + \binom{n}{4}\binom{\frac{n-4}{2}}{s-2} + \cdots$$
$$= \frac{n^2(n^2 - 2^2)(n^2 - 4^2)\cdots(n^2 - (2s-2)^2)}{(2s)!};$$

$$\binom{n}{1}\binom{\frac{n-1}{2}}{s} + \binom{n}{3}\binom{\frac{n-3}{2}}{s-1} + \binom{n}{5}\binom{\frac{n-5}{2}}{s-2} + \cdots$$
$$= \frac{n(n^2 - 1^2)(n^2 - 3^2)\cdots(n^2 - (2s-1)^2)}{(2s+1)!}.$$

Cauchy used these identities without proof in his *Analyse algébrique* to prove Newton's formulas (8.16) and (8.17). A proof depends on the Vandermonde or Chu–Vandermonde identity; see Chapter 23.

(6) Show that for any real number $n$ and $|\theta| \le \frac{\pi}{4}$

$$\cos n\theta = \cos^n \theta - \frac{n(n-1)}{1 \cdot 2} \cos^{n-2}\theta \sin^2\theta$$
$$+ \frac{n(n-1)(n-2)(n-3)}{1 \cdot 2 \cdot 3 \cdot 4} \cos^{n-4}\theta \sin^4\theta - \cdots$$

and

$$\sin n\theta = \frac{n}{1} \cos^{n-1}\theta \sin\theta - \frac{n(n-1)(n-2)}{1 \cdot 2 \cdot 3} \cos^{n-3}\theta \sin^3\theta + \cdots.$$

Viète came close to stating these formulas. Cauchy pointed out in his *Analyse algébrique* that $|\theta| \le \frac{\pi}{4}$ was necessary to expand $(\cos\theta + i\sin\theta)^n$ by the binomial theorem when $n$ was not a positive integer.

(7) Replace $\cos^k \theta$ by $(1 - \sin^2\theta)^{\frac{k}{2}}$ in Exercise 6 and expand by the binomial theorem. Then deduce Newton's formulas (8.16) and (8.17) for $|\theta| \le \frac{\pi}{4}$.

(8) Prove that if $z = g(a + x\phi(z))$, then

$$f(z) = f(g(a)) + \sum_{k-1}^{n} \frac{x^k}{k!} \frac{d^{k-1}}{da^{k-1}}\left(\phi^k(g(a))\frac{d}{da}f(g(a))\right) + \frac{1}{n!}\frac{d^n}{da^n}I_n(a),$$

$$\text{where } I_n(a) = \int_a^{g^{-1}} (x\phi(g(t)) + a - t)^n f'(g(t))\, dt.$$

See Edwards (1954b) vol. I, pp. 373–374. An equivalent result was published by Emory McClintock (1881) pp. 96–97. McClintock (1840–1916), who served as president of the American Mathematical Society and was instrumental in the founding of the *Bulletin* and the *Transactions*, was an actuary by profession. See Johnson (2007) for historical remarks on the Lagrange series.

(9) Show that if $g(0) = 0$, then

$$g^{-1}(x) = x\left(\frac{x}{g(x)}\right)_{x=0} + \frac{x^2}{2!}\left(\frac{d}{dx}\left(\frac{x}{g(x)}\right)^2\right)_{x=0}$$
$$+ \frac{x^3}{3!}\left(\frac{d^2}{dx^2}\left(\frac{x}{g(x)}\right)^3\right)_{x=0} + \cdots.$$

See Edwards (1954a) p. 459.

(10) Show that Newton's differential equation $\frac{\dot y}{\dot x} = 1 + \frac{y}{a-x}$ can be written in the form

$$(a - x + y)(y' - 1) - yy' = 0,$$

where $y' = \frac{\dot y}{\dot x}$. Show that this can be directly integrated. This observation is due to Whiteside; see Newton (1967–81) vol. III, p. 101.

(11) Show that Newton's second differential equation $\frac{\ddot y^2}{\dot x^2} = \frac{\dot y}{\dot x} + x^2$ can be integrated in closed form in terms of the logarithmic function.

(12) Show that

$$\frac{\pi^2}{8} = 1 + \frac{1}{6} + \frac{1\cdot 4}{6\cdot 15} + \frac{1\cdot 4\cdot 9}{6\cdot 15\cdot 28} + \cdots.$$

This was proved by Tanzan Shokei in his 1728 *Yenri Hakki*.

(13) Show that

$$\frac{\pi^2}{9} = 1 + \frac{1^2}{3\cdot 4} + \frac{1^2\cdot 2^2}{3\cdot 4\cdot 5\cdot 6} + \frac{1^2\cdot 2^2\cdot 3^2}{3\cdot 4\cdot 5\cdot 6\cdot 7\cdot 8} + \cdots,$$
$$\frac{\pi^2}{3} = 1 + \frac{1^2}{4\cdot 6} + \frac{1^2\cdot 3^2}{4\cdot 6\cdot 8\cdot 10} + \frac{1^2\cdot 3^2\cdot 5^2}{4\cdot 6\cdot 8\cdot 10\cdot 12\cdot 14} + \cdots.$$

These were presented by Matsunaga Ryohitsu in his *Hoyen Sankyo* of 1738.

(14) Prove that

$$\frac{\pi}{4} = 1 - \frac{1}{2 \cdot 3} - \frac{1}{8 \cdot 5} - \frac{3}{48 \cdot 7} - \frac{15}{384 \cdot 9} - \frac{105}{3840 \cdot 11} - \cdots .$$

This was presented by Hasegawa Ko in his *Kyuseki Tsuko* of 1844.
   For Exercises 12–14, see Mikami (1974) pp. 213–215.

(15) Define

$$f(m) = 1 + m(i \sin \phi) + \frac{m^2}{2!}(i \sin \phi)^2 + \frac{m(m^2 - 1^2)}{3!}(i \sin \phi)^3$$

$$+ \frac{m^2(m^2 - 2^2)}{4!}(i \sin \phi)^4 + \cdots$$

where $m$ is real and $i = \sqrt{-1}$.

(a) Show that

$$f(m_1)f(m_2) = f(m_1 + m_2).$$

(b) Use the method of Exercise 7 to prove that Newton's formulas hold for $|\phi| \leq \frac{\pi}{2}$ when $n$ is a positive integer. Deduce that $f(m) = \cos m\phi + i \sin m\phi$ when $m$ is a positive integer.

(c) Show that $f(\frac{p}{q}) = \cos \frac{p\phi}{q} + i \sin \frac{p\phi}{q}$ when $p$ and $q$ are integers.

(d) Show that $f(m)$ is continuous and deduce Newton's formulas for $|\theta| \leq \frac{\pi}{2}$.

See Hobson (1957b) pp. 273–277.

## 8.10 Notes on the Literature

The *De Analysi* was first published by William Jones in 1711, over forty years after it was written. An English translation was published in 1745. This translation was reprinted in Newton (1964–1967). Whiteside's English translation is contained in Newton (1967–1981), vol. II. Newton wrote this paper so that he should not completely lose his priority in the discovery of the methods of infinite series; he circulated the manuscript privately to several people who were interested in the topic, but he did not want to publish it. For the purpose of publication, he wrote a much longer tract, *De Methodis Serierum et Fluxionum*, in 1671. Due to the difficulties of finding a publisher and other concerns, Newton did not complete the work or publish it. In 1736 John Colson published an English translation, soon retranslated by Castillione into Latin; in 1799 Samuel Horsley published the original Latin version. Whiteside remarked that a comparison of these two translations provides "an instructive check on the clarity and fluency of Newton's Latin style."

Newton used the material in the *De Methodis* to construct his two letters to Oldenburg for Leibniz in 1676. The second letter was quite long, and in it Newton gave a fairly complete account of his work on infinite series. In 1712, he included these letters in the *Commercium Epistolicum*, produced by a Royal Society committee headed by Newton to establish conclusively that he was the 'first inventor' of the calculus. The letters have been republished with English translations in the second volume of Newton's *Correspondence*, Newton (1959–1960).

It is remarkable that Newton's mathematical works were published in their entirety only 250 years after his death. Early attempts to accomplish this task were abandoned because his papers were in a state of disarray and were stored in several different locations. It was even assumed that all of Newton's significant results were already published. Thus, before Whiteside's monumental work, the world was unaware of a number of the results of Newton discussed in this book: transformation of series by finite differences, the first clear statement of Taylor's formula, and the expression of an iterated integral as a single integral. D. T. Whiteside (1932–2008) studied French and Latin at Bristol University; he was self-taught in mathematics. As a graduate student at Cambridge, he became deeply interested in the history of mathematics; his doctoral thesis on seventeenth-century mathematics became a classic. In the course of his studies, Whiteside asked to see the papers of Newton, still piled in boxes, and soon resolved to sort and edit them. Cambridge University Press, the world's oldest continually operating press, chartered by Henry VIII, published the eight handsome volumes between 1967 and 1982 from Whiteside's handwritten manuscript and hand-drawn diagrams, with facing pages giving the English and Latin. Whiteside's commentary and notes are extensive and invaluable. Whiteside executed this prodigious task in twenty years, with the excellent assistance of his thesis advisor Michael Hoskin and Adolf Prag, a teacher at Westminster School.

For a discussion of Takebe's work, see Mikami (1974) and Ogawa and Morimoto (2018). The formula for $(\arcsin x)^2$ was also discovered by Ming An-tu who was Manchurian by birth. It appeared in his *Ko-yuan Mi-lu Chieh-fa* of 1774, some years after his death. Ming had not completed the work before he died, and his son Hsin finished it. Ming An-tu's work on infinite series was inspired by the three infinite series communicated to Chinese mathematicians by the French Jesuit Pierre Jartoux in 1702. These were Newton's series for sine, cosine and arcsine.

Pierre Jartoux (1670–1720) was a French Jesuit missionary who entered China in 1701. He is said to have communicated either three or nine series for trigonometric functions to Chinese mathematicians. There is some doubt as to how much information he brought from Europe and how much the Chinese and Japanese mathematicians independently discovered. There is no doubt that he communicated the series for sine, cosine, and arcsine. But there is some question about the other six formulas, one of which is Takebe's series for $(\arcsin x)^2$. Though Jartoux's original notes are lost, Smith and Mikami (1914) suggested that the series for $(\arcsin x)^2$ was also introduced by Jartoux, who had been in correspondence with Leibniz. This appears to be unlikely. Jartoux was not a mathematician, and his correspondence with Leibniz was on an astronomical topic. If Jartoux knew the series for $(\arcsin x)^2$, he would have informed Leibniz, and perhaps others, because this would have been a new discovery. In fact, in 1737, when Euler and Bernoulli rediscovered this result and its particular case dealing with $\pi^2$, they regarded their formulas as original. And these mathematicians were very well aware of the works of all European mathematicians at that time. We may conclude that Takebe was the first to find the series for $(\arcsin x)^2$ and the corresponding series for $\pi^2$, while Ming's discovery was independent, though inspired by a knowledge of the series communicated by Jartoux.

# 9

---

# *Finite Differences: Interpolation and Quadrature*

## 9.1 Preliminary Remarks

The method of interpolation for the construction of tables of trigonometric functions has been used for over two thousand years. On this method, one may tabulate the values of a function $f(x)$ constructed from first principles (definitions) for $x = a$ and $x = a + h$, where $h$ is small, and then interpolate the values between $a$ and $a + h$, without further computation from first principles. For sufficiently small $h$, one may approximate the function $f(x)$ by a linear function on the interval $[a, a + h]$. This means that, in order to interpolate the values of the function in this interval, one may use the approximation $f(a + \lambda h) \approx f(a) + \lambda(f(a + h) - f(a))$, $0 \le \lambda \le 1$. In his *Almagest* of around 150 AD, Ptolemy applied linear interpolation to construct a table of lengths of chords of a circle as a function of the corresponding arcs. These are the oldest trigonometric tables in existence, though Hipparchus may well have constructed similar tables almost three centuries earlier.[1] In Ptolemy's table, the length of the chord was given as $2R \sin \theta$, where $R$ was the radius and $2\theta$ was the angle subtended by the arc. Later mathematicians in India, on the other hand, tabulated the half chord; when divided by the radius, this gives our sine.

From the beginning of the seventh century, Chinese and Indian mathematicians developed second-order interpolation, in which they employed second-degree polynomials in order to determine values between $a$ and $a + h$. To do this, they had to take second differences of the initially determined values. Let $\Delta$ denote the first difference:

$$\Delta f(a) = f(a + h) - f(a);$$

linear interpolation may then be performed by using

$$f(a + \lambda h) \approx f(a) + \lambda \Delta f(a),$$

---

[1] Chabert (1999) pp. 321–328.

or

$$f(a + \lambda h) = f(a) + \lambda \Delta f(a).\tag{9.1}$$

Around 600 AD, the Chinese astronomer Liu Zhuo used second-order differences to calculate the positions of the sun and moon. Suppose $f(a)$, $f(a + h)$, $f(a + 2h)$ are the observed values of the positions at equal time intervals $h$. Setting

$$\Delta_1 = f(a + h) - f(a), \quad \Delta_2 = f(a + 2h) - f(a + h),$$

for $0 \leq \lambda < 1$, Liu Zhuo's formula for the interpolated values was[2]

$$f(a + \lambda h) = f(a) + \frac{\lambda}{2}(\Delta_1 + \Delta_2) + \lambda(\Delta_1 - \Delta_2) - \frac{\lambda^2}{2}(\Delta_1 - \Delta_2).\tag{9.2}$$

The second difference appears in the terms containing $\Delta_1 - \Delta_2$. For $j$ a nonnegative integer, let $\Delta$ denote the first difference

$$\Delta f(a + jh) = f\big(a + (j + 1)h\big) - f(a + jh);$$

the second difference would then be given by

$$\begin{aligned}
\Delta^2 f(a + jh) &= \Delta\big(\Delta f(a + jh)\big) \\
&= \Delta f\big(a + (j + 1)h\big) - \Delta f(a + jh) \\
&= f\big(a + (j + 2)h\big) - f\big(a + (j + 1)h\big) \\
&\quad - f\big(a + (j + 1)h\big) + f(a + jh).
\end{aligned}$$

Thus one has

$$\begin{aligned}
\Delta_2 - \Delta_1 &= f(a + 2h) - f(a + h) - f(a + h) + f(a) \\
&= \Delta^2 f(a).
\end{aligned}$$

It is not difficult to show that Liu Zhuo's formula (9.2) is equivalent to

$$f(a + s) = f(a) + s\Delta f(a) + \frac{s(s - 1)}{2}\Delta^2 f(a), \quad s = \lambda h, \ 0 \leq \lambda < 1,\tag{9.3}$$

the Gregory–Newton interpolation formula, discussed in Section 9.3.

The astronomer Yi Xing (683–727) in 727 used an unequal intervals second difference formula to deal with the situation in which observations of the positions of the sun or moon were carried out at varying intervals.[3]

---

[2] Li Yan and Du Shiran (1987) pp. 89–91.
[3] Li Yan and Du Shiran (1987) p. 91.

In his 628 work, *Dhyanagrahopadesadhyaya*, the Indian mathematician and astronomer Brahmagupta (598–668) employed the formula,[4] here converted to modern notation:

$$f(a + \lambda h) = f(a) + \frac{h}{2}\big(\Delta f(a - h) + \Delta f(a)\big) + \frac{h^2}{2}\big(\Delta f(a) - \Delta f(a - h)\big), \quad (9.4)$$

the second-order Newton–Stirling interpolation formula, (9.17), given in Section 9.2. In his 665 *Khandakhadyaka*, Brahmagupta also described a general method of interpolation for unequal intervals, reducible to (9.4) when intervals were equal.

Parabolic or second-degree polynomial methods of interpolation were also employed by Islamic mathematicians such as al-Biruni (973–1048). Indian astronomers had constructed tables of sine and cosine values, multiplied by a radius. Having traveled and spent time in India, al-Biruni was aware of these tables and expanded them to include tangent and cotangent functions.

By the seventeenth century, the requirements of navigation and astronomy demanded finer tables of trigonometric and related functions; this led to the invention of the logarithm and better interpolation methods. Motivated by the needs of navigation, in 1611 or a little earlier, Thomas Harriot wrote a remarkable treatise, *De Numeris Triangularibus et inde de Progressionibus Arithmeticis: Magisteria Magna*, considering finite differences of third and higher order.[5] He gave the fifth-order interpolation formula, expressed in modern notation as

$$f(x) = f(0) + \binom{x}{1}\Delta f(0) + \binom{x}{2}\Delta^2 f(0) + \cdots + \binom{x}{5}\Delta^5 f(0), \quad (9.5)$$

where

$$\binom{x}{k} = \frac{x(x-1)(x-2)\cdots(x-k+1)}{k!} \quad (9.6)$$

were the binomial coefficients and $\Delta f(0) = f(1) - f(0)$, $\Delta^2 f(0) = \Delta(\Delta f(0)) = \Delta f(1) - \Delta f(0) = f(2) - f(1) - (f(1) - f(0))$, etc. In Harriot's work, $x$ took rational values and he used his formula to interpolate between unit values of the argument. He understood the values of (9.6) in terms of figurate numbers, instead of binomial coefficients, when $x$ was an integer.

Unfortunately, Harriot did not publish his work; some of his methods were rediscovered soon afterward by Henry Briggs (1561–1631). Briggs was the first professor of mathematics at Gresham College, London, as well as the first Savilian Professor at Oxford. In his *Arithmetica Logarithmica* of 1624, Briggs mentioned that the $n$th-order differences of the $n$th powers of integers were constants. According to Whiteside,[6] this work contained tables of logarithms obtained by second-order

---

[4] See Gupta (1969).

[5] Beery and Stedall (2009).

[6] Whiteside (1961b) p. 235. Whiteside remarks that Briggs apparently had some awareness of the later Gauss, Bessel, and Stirling interpolation formulas.

interpolation, that is, taking the first three terms on the right side of Harriot's formula (9.5). Observe that if the second differences are approximately identical, then the third and higher differences are approximately zero and can be neglected. More generally, if the $n$th differences are approximately constant, then $f(x)$ can be approximated by the polynomial of degree $n$ obtained by extending Harriot's formula (9.5) to $n$th differences.

Briggs also wrote *Trigonometria Britannica*, a book of trigonometric tables with a very long introduction giving details of his methods. Briggs's friend, Henry Gellibrand, had this work published in 1633, after Briggs's death. Unfortunately, the many users of these trigonometric tables may not have read the more important introduction in which Briggs gave some very interesting results, including the binomial theorem for exponent $\frac{1}{2}$.[7] However, the Scottish mathematician James Gregory studied Briggs's introduction and thereby learned interpolation methods. Thus, also making use of advances in algebraic notation, and employing N. Mercator's discovery of infinite series, Gregory obtained interpolation formulas containing up to an infinite number of terms. In an important letter to Collins, dated November 23, 1670, he communicated his formula,[8] given below in my translation, describing it as "both more easie and universal than either Briggs or Mercator's, and also performed without tables."

> I remember you did once desire of me my method of finding the proportional parts in tables, which is this: In figure 8 of my exercises [*Exercitationes Geometricae*], on the straight line $AI$ consider any segment $A\alpha$, to which there is a perpendicular $\alpha\gamma$, such that $\gamma$ lies on the curve $ABH$, the rest remaining the same; let there be an infinite series [sequence] $\frac{a}{c}, \frac{a-c}{2c}, \frac{a-2c}{3c}, \frac{a-3c}{4c}$, etc., and let the product of the first two terms of this series be $\frac{b}{c}$, of the first three terms $\frac{k}{c}$, of the first four terms $\frac{l}{c}$, of the first five terms $\frac{m}{c}$, etc., to infinity; the straight line $\alpha\gamma = \frac{ad}{c} + \frac{bf}{c} + \frac{kh}{c} + \frac{li}{c} +$ etc. to infinity.

Gregory defined $d$, $f$, $h$, $i$, etc., as the successive differences of the ordinates, at equal intervals $c$. He took $f(0) = 0$, so that $d = f(c) - f(0) = f(c)$, $f = f(2c) - 2f(c)$, etc. After inserting the values of $a$, $b$, $k$, $l$, ..., Gregory's formula can be written as

$$f(a) = \frac{a}{c}\Delta f(0) + \frac{a(a-c)}{2c^2}\Delta^2 f(0) + \frac{a(a-c)(a-2c)}{6c^3}\Delta^3 f(0)$$
$$+ \frac{a(a-c)(a-2c)(a-3c)}{24c^4}\Delta^4 f(0) + \cdots . \tag{9.7}$$

This result is now known as the Gregory–Newton forward difference formula, but it may also be called the Harriot–Briggs formula.

Newton's interest in finite differences and interpolation appears to have been a response to an appeal from one John Smith for help with the construction of tables

---

[7] ibid. pp. 233–234 and footnote 14 on p. 236. Also see Whiteside (1961a).
[8] Turnbull (1939) pp. 290–292.

of square, cube, and fourth roots of numbers. Collins broadcast this appeal; he wrote in a letter of November 23, 1674, to Gregory,[9] "We have one Mr. Smith here taking pains to afford us tables of the square and cube roots of all numbers from unit to 10000, which will much facilitate Cardan's rules." Smith was an accountant and compiler of tables whom Newton had helped five years earlier with the making of tables for the areas of segments of circles. Newton again assisted Smith, writing to him on May 8, 1675, giving details for the construction of tables of roots.[10] Newton explained to Smith that he should tabulate the roots of every hundredth number $n$. From these, he should construct the roots of every tenth number $n \pm 10$, $n \pm 20$, ... with a constant third difference and thence the roots of $n \pm 1$, $n \pm 2$, ... with a constant second difference. Newton also cautioned that all computations should be done to the tenth or eleventh decimal place so as to obtain a table accurate to eight places. Newton's ideas on finite difference interpolation developed quite rapidly after this.

Around October 1676, Newton started composing his (incomplete) "Regula Differentiarum" in which he presented the Newton–Stirling and Newton–Bessel formulas.[11] Perhaps he was not quite satisfied with this work; he next penned an untitled work on his general interpolation techniques,[12] a monograph fairly close to the final version that appeared in print in 1711 as the *Methodus Differentialis*.[13]

In a draft of his letter dated October 24, 1676,[14] intended for Leibniz through Oldenburg, secretary of the Royal Society of London, Newton set forth some of his insights on interpolation, including statements of his general formula and the Newton–Stirling formula. He later eliminated the portion of the letter on interpolation because he saw a copy of a letter from Leibniz to Oldenburg,[15] dated February 3, 1673, stating that Leibniz had independently discovered the Harriot–Briggs formula. Newton perhaps assumed from this that Leibniz may have made progress parallel to his own in the study of finite differences, though this was not the case.

In his *Principia* of 1687, in Lemma V of Book III, Newton published his general formula now called the divided difference formula, but he did not publish some of its corollaries, such as the Newton–Stirling and Newton–Bessel formulas. In 1708, Roger Cotes (1682–1716) independently found the latter formula and included both formulas in a paper, "Canonotechnia," published posthumously in his *Harmonia Mensurarum* of 1722.[16] In Whiteside's view[17] it might be more appropriate to refer to the Newton–Bessel formula as the Newton–Cotes formula, after its two independent creators.

Newton was the single most significant contributor to the theory of finite difference interpolation. Although many formulas in this subject are attributed jointly to Newton and some other mathematician, they are actually all due originally to Newton, with the

[9] Turnbull (1939) p. 291.
[10] Newton (1959–1960) vol. 1, pp. 342–344.
[11] Newton (1967–1981) vol. 4, pp. 36–51.
[12] ibid. pp. 54–69.
[13] See English translation in Newton (1964–1967) vol. 2, pp. 168–173.
[14] Newton (1959–1960) vol. 2, pp. 130–161.
[15] Gerhardt (1899) pp. 74–78.
[16] Cotes (1722). Also see the paper "De methodo differentiali Newtoniana," included in the same work.
[17] Newton (1967–1981) vol. 4, pp. 60–61, footnote 22.

exception of the Gregory–Newton formula, due to Harriot and Briggs. The secondary mathematicians usually made use of these formulas in their numerical work. As early as 1730, Stirling pointed this out[18] in his *Methodus Differentialis*: "After *Newton* several celebrated geometers have dealt with the description of the curve of parabolic type [defined by a polynomial] through any number of given points. But all their solutions are the same as those which have just been shown; indeed these differ scarcely from *Newton's* solutions." It is amusing that Stirling was subsequently honored by having his name attached to a formula he explicitly and modestly attributed to Newton. In his insightful work on the *Principia*, Chandrasekhar remarks, "It is a strange irony that in giving an account of Newton's published work of 1711, we have to hyphenate his name with Gauss, Stirling, and Bessel!"[19]

Newton's divided difference formula, in the notation of the French mathematicians A. M. Ampre (1775–1836) and A. L. Cauchy (1789–1857), was written as

$$f(x) = f(x_1) + (x - x_1)f(x_1, x_2) + (x - x_1)(x - x_2)f(x_1, x_2, x_3) + \cdots$$
$$+ (x - x_1) \cdots (x - x_{n-1})f(x_1, x_2, \ldots, x_n)$$
$$+ (x - x_1) \cdots (x - x_n)f(x_1, \ldots, x_n, x), \tag{9.8}$$

where

$$f(x_1, x_2) = \frac{f(x_1) - f(x_2)}{x_1 - x_2},$$

$$f(x_1, x_2, \ldots, x_k) = \frac{f(x_1, \ldots, x_{k-1}) - f(x_2, \ldots, x_k)}{x_1 - x_k}. \tag{9.9}$$

If we denote the last term in (9.8) by $R_n(x)$, and the remaining sum as $P_{n-1}(x)$, then $P_{n-1}(x)$ is a polynomial of degree $n - 1$ equal to $f(x_i)$ for $i = 1, 2, \ldots, n$. Note that this is true because $R_n(x_i) = 0$, $i = 1, \ldots, n$. Thus, $P_{n-1}(x)$ is the interpolating polynomial for a function $f(x)$ whose values are known at $x_1, x_2, \ldots, x_n$. In the 1770s, Lagrange and Waring gave a different expression for this polynomial, more convenient for many purposes, especially for numerical integration. The Waring–Lagrange interpolating polynomial is easy to obtain, yet it is interesting to see different proofs presented in the 1820s by Cauchy[20] and Jacobi.[21]

James Gregory used interpolating polynomials to approximately evaluate the area under a curve. He communicated his quadrature formula to Collins in the letter containing his interpolation formula, deriving it by integrating the interpolating polynomial, just as Newton did in his *Methodus Differentialis*. Newton derived his three-eighths rule by integrating the third-degree polynomial obtained by taking the first four terms of (9.5). He explained:[22]

---

[18] Stirling and Tweddle (2003) p. 122.
[19] Chandrasekhar (1995) p. 495.
[20] Cauchy (1989) Note V.
[21] Jacobi (1826).
[22] Newton (1964–1967) vol. 2, p. 172.

For example: If there are four ordinates at equal intervals, let $A$ be the sum of the first and fourth, $B$ the sum of the second and third, and $R$ the interval between the first and fourth; then the central ordinate will be $\frac{9B-A}{16}$ and the area between the first and fourth ordinates will be $\frac{A+3B}{8} R$.

In 1707, Cotes, unaware of Newton's then unpublished work in this area, composed a treatise on approximate quadrature. He wrote down formulas for areas when the number of ordinates was $3, 4, 5, \ldots, 11$. The coefficients became fairly large after six ordinates; for example, his formula for eight ordinates was[23]

$$\frac{751A + 3577B + 1323C + 2989D}{17280}R,$$

where $A$ was sum of the extreme ordinates, $B$ the sum of the ordinates closest to the extremes, $C$ the sum of the next ones, and $D$ the sum of the two in the middle. Cotes's paper, published posthumously in 1722, contained no proofs of his formulas.

Meanwhile, in 1719, Stirling published a paper in the *Philosophical Transactions*[24] on the same topic, presenting formulas for approximate areas for only the odd number of ordinates 3, 5, 7 and 9. He remarked that the approximations with an odd number of ordinates were more accurate than those with an even number. He did not prove this, though it is true. For example, it can be demonstrated that if $h = \frac{R}{n}$, where $4n$ is the number of ordinates, then the error will be $O(h^{n+2})$ for odd $n$ but $O(h^{n+1})$ for even $n$.[25]

The Newton–Cotes method of numerical integration was used for a century before Gauss developed a new approach, including a formula exact for any polynomial of degree $2n - 1$ or less when $n$ interpolation points were judiciously constructed. The Newton–Cotes formulas are exact only for polynomials of degree at most $n - 1$. Gauss's procedure will be discussed in Chapter 24. A drawback of the Newton–Cotes and Gauss formulas was that the coefficients of the ordinates were unequal. The Russian mathematician P. L. Chebyshev (1821–1894) observed that when the ordinates $f(x_i)$ were experimentally obtained, they were liable to errors. Assuming that the probability of error in each of the ordinates was the same, the linear combination of the ordinates with equal coefficients had the least probable error among all the linear combinations with a given fixed sum of coefficients. Chebyshev observed that a quadrature formula with equal coefficients might often be preferable. Chebyshev studied mathematics at Moscow University from 1837–1841. He was interested in building mechanical gadgets and some of his papers deal with the mathematics involved with these. Chebyshev was of the view that his job as a mathematician was to consider practical problems and to give solutions both theoretically satisfying and practically useful. He repeatedly professed this opinion in his lectures and advocated it in several papers; his work on numerical integration may be seen as an example of this perspective.

[23] Gowing (1983) p. 119.
[24] Stirling (1719).
[25] Milne-Thomson (1981) pp. 166–170.

In 1874, Chebyshev wrote a paper on quadrature with equal coefficients,[26] considering formulas of the type

$$\int_{-1}^{1} f(x)\phi(x)\,dx \equiv k(f(x_1) + f(x_2) + \cdots + f(x_n)), \qquad (9.10)$$

where $\phi(x)$ was the weight function and $k$ was the common coefficient of the ordinates. He found a method, exact for polynomials of degree less than $n$, for determining the interpolation points $x_1, x_2, \ldots, x_n$ and the constant $k$, such that they depended upon $\phi$ but not on $f$. He worked out the details with the weights given as $\phi(x) = 1$ and $\phi(x) = \frac{1}{\sqrt{1-x^2}}$. In particular, he showed that when $\phi(x) = 1$, then $k = \frac{2}{n}$ and $x_1, x_2, \ldots, x_n$ were the roots of that polynomial given by the polynomial portion of the expression

$$z^n e^{-\frac{n}{2\cdot 3z^2} - \frac{n}{4\cdot 5z^4} - \frac{n}{6\cdot 7z^6} - \cdots}.$$

He computed the zeros of these polynomials for $n = 2, 3, 4, 5, 6, 7$. Interestingly, Chebyshev was inspired to do this work by Hermite's 1873 Paris lectures[27] on the case $\phi(x) = \frac{1}{\sqrt{1-x^2}}$. We observe that, even before Hermite, Brice Bronwin gave the formula for this case in a paper of 1849 in the *Philosophical Magazine*.[28] In the chapter on numerical integration of their *Calculus of Observations*, Whittaker and Robinson discussed Chebyshev's method and noted that naval architects found it useful.[29]

## 9.2 Newton: Divided Difference Interpolation

Newton started his work on interpolation in the mid-1770s, but had made sufficient progress to make a brief mention of it in his letter for Leibniz, dated October 24, 1776. While discussing the problem of determining the area under a curve, especially when the expression for the curve led to difficult calculations of series, he wrote,[30]

> But I make little of this because, when simple series are not manageable enough, I have another method not yet communicated by which we have access to our solution at will. Its basis is a convenient, rapid and general solution of this problem, *To draw a geometrical curve which shall pass through any number of given points.* Euclid showed how to draw a circle through three given points. A conic section also can be described through five given points, and a curve of the third degree through eight given points; (so that I have it fully in my power to describe all the curves of that order, which can be determined by eight points only.) These things are done at once geometrically with no calculation intervening. But the above problem is of the second kind, and though at first it looks unmanageable, yet the matter turn out otherwise. For it ranks among the most beautiful of all that I could wish to solve.

[26] Chebyshev (1899–1907) pp. 165–180.
[27] Hermite (1873) pp. 452–454.
[28] Bronwin (1849).
[29] Whittaker and Robinson (1949) p. 158.
[30] Newton (1959–1960) vol. 2, p. 137.

Interestingly, Stirling quoted just this passage at the end of proposition 18 of his book. Clearly, Newton was pleased with the result of his researches on interpolation, so he did not neglect the chance to include at least one result in the *Principia*, as Lemma V, Book III. Newton gave his method of interpolation by divided differences in the *Principia* without proof; he provided details in his very short *Methodus Differentialis*.[31] The first proposition stated that if one started with a polynomial, then the divided differences would be polynomials of degree one less:[32]

> If the abscissa of a curve consist of a given quantity $A$ and an indeterminate quantity $x$, and if the ordinate consist of any number of quantities $b, c, d, e, \ldots$ multiplied respectively into a corresponding number of terms of the geometric progression $x, x^2, x^3, x^4, \ldots$ and if ordinates be erected at as many points of the abscissa; then the first differences of the ordinates are divisible by their intervals; and that the differences of the differences so divided are divisible by the intervals between alternate ordinates; and the differences of those differences so divided are divisible by the intervals between every third ordinate, and so on indefinitely.

We describe Newton's method in the Ampère–Cauchy notation: If $f(x)$ is a polynomial, then the first divided difference

$$f(x_1, x_2) = \frac{f(x_1) - f(x_2)}{x_1 - x_2} \tag{9.11}$$

is also a polynomial, as is the second divided difference

$$f(x_1, x_2, x_3) = \frac{f(x_1, x_2) - f(x_2, x_3)}{x_1 - x_3}, \tag{9.12}$$

and, in general, the so is the $n$th divided difference, defined inductively by

$$f(x_1, x_2, \ldots, x_n, x_{n+1}) = \frac{f(x_1, x_2, \ldots, x_n) - f(x_2, x_3, \ldots, x_{n+1})}{x_1 - x_{n+1}}. \tag{9.13}$$

Newton explicitly worked out all the divided differences for a fourth-degree polynomial. In the second proposition, he explained how the original polynomial or function could be constructed from the divided differences:[33]

> With the same suppositions, and taking the number of terms $b, c, d, e, \ldots$ to be finite, I assert that the last of the quotients will be equal to the last of the terms $b, c, d, e, \ldots$, and that the remaining terms $b, c, d, e, \ldots$ will be yielded by means of the remaining quotients; and that once these are determined there will be given the curve of parabolic kind which shall pass through the end-points of all the ordinates.

If this procedure is applied in general, we obtain Newton's divided difference formula (9.8). In the case of fourth differences we have

---

[31] Newton (1964–1967) vol. 2, pp. 165–173.
[32] ibid. p. 165.
[33] Newton (1967–1981) vol. 7, p. 247.

$$f(x, x_1) = \frac{f(x)}{x - x_1} - \frac{f(x_1)}{x - x_1},$$

$$f(x, x_1, x_2) = \frac{f(x, x_1)}{x - x_2} - \frac{f(x_1, x_2)}{x - x_2},$$

$$f(x, x_1, x_2, x_3) = \frac{f(x, x_1, x_2)}{x - x_3} - \frac{f(x_1, x_2, x_3)}{x - x_3},$$

$$f(x, x_1, x_2, x_3, x_4) = \frac{f(x, x_1, x_2, x_3)}{x - x_4} - \frac{f(x_1, x_2, x_3, x_4)}{x - x_4}. \tag{9.14}$$

Thus, in each step, the values from the previous equation are substituted for the terms on the right-hand side and the resulting equation is multiplied by $(x - x_1)(x - x_2)(x - x_3)(x - x_4)$, yielding

$$\begin{aligned}
f(x) = &\, f(x_1) + (x - x_1) f(x_1, x_2) + (x - x_1)(x - x_2) f(x_1, x_2, x_3) \\
&+ (x - x_1)(x - x_2)(x - x_3) f(x_1, x_2, x_3, x_4) \\
&+ (x - x_1)(x - x_2)(x - x_3)(x - x_4) f(x, x_1, x_2, x_3, x_4). \tag{9.15}
\end{aligned}$$

As we discuss in Section 9.4, the $n$th divided difference $f(x_0, x_1, \ldots, x_n)$ is symmetric in the variables $x_0, x_1, \ldots, x_n$. Note also that if the points $x_0 < x_1 < x_2 < \cdots < x_n$ are equidistant, with the distance between each being $h = x_i - x_{i-1}$ for $i = 1, 2, \ldots, n$, then we can prove by induction that

$$f(x_0, x_1, \ldots, x_n) = \frac{1}{n! \, h^n} \Delta^n f(x_0), \tag{9.16}$$

where

$$\Delta^i f(x_0) = \Delta\left(\Delta^{i-1} f(x_0)\right) = \Delta^{i-1} f(x_0 + h) - \Delta^{i-1} f(x_0).$$

For $n = 1$ the result holds true:

$$f(x_0, x_1) = \frac{f(x_0 + h) - f(x_0)}{h} = \frac{1}{h} \Delta f(x_0).$$

Next suppose that

$$f(x_0, x_1, \ldots, x_{n-1}) = \frac{1}{(n - 1)! \, h^{n-1}} \Delta^{n-1} f(x_0).$$

Then we have

$$\begin{aligned}
f(x_0, x_1, \ldots, x_n) &= \frac{f(x_1, x_2, \ldots, x_n) - f(x_0, x_1, \ldots, x_{n-1})}{nh} \\
&= \frac{1}{n! \, h^n}\left(\Delta^{n-1} f(x_0 + h) - \Delta^{n-1} f(x_0)\right) \\
&= \frac{1}{n! \, h^n} \Delta^n f(x_0),
\end{aligned}$$

from which the result is immediate.

In the third proposition, Newton derived his central difference formulas for the case where the points were equidistant. When the number of interpolating points was odd, he presented what is now known as the Newton–Stirling formula and, for the even case, the so-called Newton–Bessel formula. He did not write down details of the derivation, but it is most likely that he obtained it from his general divided difference formula, employed in modern textbooks.

The Newton–Stirling formula[34] was actually discovered by Newton, but it came to be widely known through Stirling, who stated it and worked out examples in his *Methodus Differentialis*. To state the formula: suppose that $f(t)$ is given for the values

$$\ldots, x_0 - 2h, x_0 - h, x_0, x_0 + h, x_0 + 2h, \ldots;$$

then

$$f(x_0 + xh) = f(x_0) + x\frac{\Delta f(x_0) + \Delta f(x_0 - h)}{2} + \frac{x^2}{2!}\Delta^2 f(x_0 - h)$$

$$+ \frac{x(x^2 - 1^2)}{3!}\frac{\Delta^3 f(x_0 - h) + \Delta^3 f(x_0 - 2h)}{2}$$

$$+ \frac{x^2(x^2 - 1^2)}{4!}\Delta^4 f(x_0 - 2h)$$

$$+ \frac{x(x^2 - 1^2)(x^2 - 2^2)}{5!}\frac{\Delta^5 f(x_0 - 2h) + \Delta^5 f(x_0 - 3h)}{2}$$

$$+ \frac{x^2(x^2 - 1^2)(x^2 - 2^2)}{6!}\Delta^6 f(x_0 - 3h) + \cdots. \qquad (9.17)$$

To prove (9.17), we apply (9.8) with $h > 0$ and

$$x = y_0 + yh, \ x_1 = y_0, \ x_2 = y_0 + h, \ x_3 = y_0 - h,$$
$$x_4 = y_0 + 2h, \ x_5 = y_0 - 2h, \ldots.$$

Then

$$x - x_1 = hy, \ x - x_2 = hy - h = h(y - 1),$$
$$x - x_3 = h(y + 1), \ x - x_4 = h(y - 2), \ldots,$$

and applying these values in (9.8) produces

$$f(y_0 + yh)$$
$$= f(y_0) + hy \, f(y_0, y_0 + h) + h^2 y(y - 1) \, f(y_0, y_0 + h, y_0 - h)$$
$$+ h^3 y(y - 1)(y + 1) \, f(y_0, y_0 + h, y_0 - h, y_0 + 2h)$$
$$+ h^4 y(y - 1)(y + 1)(y - 2) \, f(y_0, y_0 + h, y_0 - h, y_0 + 2h, y_0 - 2h) + \cdots.$$

$$(9.18)$$

---

[34] Newton (1964–1967) vol. 2, pp. 167–168. Stirling and Tweddle (2003) pp. 119–120. Whittaker and Robinson (1959) p. 38.

Now from the symmetry of the divided differences and (9.16), we obtain

$$f(y_0, y_0 + h)$$
$$= \frac{1}{h} \Delta f(y_0),$$
$$f(y_0, y_0 + h, y_0 - h)$$
$$= f(y_0 - h, y_0, y_0 + h)$$
$$= \frac{1}{2! h^2} \Delta^2 f(y_0 - h),$$
$$f(y_0, y_0 + h, y_0 - h, y_0 + 2h)$$
$$= f(y_0 - h, y_0, y_0 + h, y + 2h)$$
$$= \frac{1}{3! h^3} \Delta^3 (y_0 - h),$$
$$f(y_0, y_0 + h, y_0 - h, y_0 + 2h, y_0 - 2h)$$
$$= f(y_0 - 2h, y_0 - h, y_0, y_0 + h, y_0 + 2h)$$
$$= \frac{1}{4! h^4} \Delta^4 f(y_0 - 2h),$$
$$\cdots$$
$$(9.19)$$

When the values in (9.19) are substituted in (9.18), the result is the Newton–Gauss interpolation formula:

$$f(y_0 + yh) = f(y_0) + y \cdot \Delta f(y_0) + \frac{y(y-1)}{2!} \cdot \Delta f^2(y_0 - h)$$
$$+ \frac{(y+1)y(y-1)}{3!} \cdot \Delta^3 f(y_0 - h)$$
$$+ \frac{(y+1)y(y-1)(y-2)}{4!} \cdot \Delta^4 f(y_0 - 2h) + \cdots . \quad (9.20)$$

We can rearrange the terms of (9.20) as follows:

$$f(y_0 + yh) = f(y_0) + y \left( \Delta f(y_0) - \frac{1}{2} \Delta^2 f(y_0 - h) \right) + \frac{y^2}{2!} \cdot \Delta^2 f(y_0 - h)$$
$$+ \frac{y(y^2 - 1^2)}{3!} \left( \Delta^3 f(y_0 - h) - \frac{1}{2} \Delta^4 f(y_0 - 2h) \right)$$
$$+ \frac{y^2(y^2 - 1^2)}{4!} \cdot \Delta^4 f(y_0 - 2h) + \cdots . \quad (9.21)$$

Now observe that if the values

$$\Delta^2 f(y_0 - h) = \Delta f(y_0) - \Delta f(y_0 - h),$$
$$\Delta^4 f(y_0 - 2h) = \Delta^3 f(y_0 - h) - \Delta^3 f(y_0 - 2h), \cdots$$

are substituted in (9.21), the result is the Newton–Stirling formula (9.17).

The Newton–Bessel formula,[35] due to Newton but extensively applied and worked with by Bessel, may be given as

$$f\left(x_0 + \frac{1}{2}h + xh\right)$$

$$= \frac{1}{2}\left(f(x_0) + f(x_0 + h)\right) + x\Delta f(x_0)$$

$$+ \frac{x^2 - \frac{1}{4}}{2!} \cdot \frac{1}{2}\left(\Delta^2 f(x_0 - h) + \Delta^2 f(x_0)\right)$$

$$+ \frac{x(x^2 - \frac{1}{4})}{3!} \cdot \Delta^3 f(x_0 - h)$$

$$+ \frac{\left(x^2 - \frac{1}{4}\right)\left(x^2 - \frac{9}{4}\right)}{4!} \cdot \frac{1}{2}\left(\Delta^4 f(x_0 - 2h) + \Delta^4 f(x_0 - h)\right) + \cdots. \quad (9.22)$$

Stirling gave applications of both formulas (9.17) and (9.22) in proposition 20 of his *Methodus*. In the second example in proposition 21, he applied the Newton–Bessel formula to show that $\Gamma\left(\frac{3}{2}\right) = \frac{\sqrt{\pi}}{2}$. In fact, he obtained the approximation 0.8862269251 and then noted that this was $\frac{\sqrt{\pi}}{2}$. We remark that Stirling made this observation before Euler's paper on the gamma function had appeared.[36] These matters are further discussed in Section 17.2.

The reader might wish to read Chandrasekhar,[37] who shows in detail that the Newton–Stirling and Newton–Bessel interpolation formulas given in this chapter are identical with the original statements given by Newton in his *Methodus Differentialis* of 1711.

## 9.3　Gregory–Newton Interpolation Formula

The Gregory–Newton formula (9.7), or the Harriot–Briggs formula, is important not only in numerical analysis but also in the study of sequences whose $n$th differences, for some $n$, are constant. Apart from interpolation theory, these sequences are now studied as a part of combinatorial analysis, but in the seventeenth and early eighteenth centuries they arose in elementary number theory and in probability theory. It is therefore interesting to consider the methods by which mathematicians of that period proved this formula. Unfortunately, Gregory did not leave us a proof. It is possible that he had the simple inductive argument given by Stirling in proposition 19 of his *Methodus Differentialis*.[38] Stirling assumed that there existed some unknown coefficients, $A, B, C, D, \ldots$ such that

[35] Newton (1964–1967) vol. 2, p. 168. Stirling and Tweddle (2003) pp. 120–121. Whittaker and Robinson (1959) pp. 39–40.
[36] Stirling and Tweddle (2003) p. 127.
[37] Chandrasekhar (1995) pp. 495–498.
[38] Stirling and Tweddle (2003) pp. 112–114.

$$f(z) = A + Bz + C\frac{z(z-1)}{1 \cdot 2} + D\frac{z(z-1)(z-2)}{1 \cdot 2 \cdot 3} + \cdots .$$

Clearly $A = f(0)$. Moreover,

$$\Delta f(z) = f(z+1) - f(z)$$

$$= B\Delta z + C\Delta\frac{z(z-1)}{1 \cdot 2} + D\Delta\frac{z(z-1)(z-2)}{1 \cdot 2 \cdot 3} + \cdots .$$

Observing that for $n = 2, 3, 4, \ldots$

$$\Delta\frac{z(z-1)\cdots(z-n+1)}{1 \cdot 2 \cdots n} = \frac{z(z-1)\cdots(z-n+2)}{1 \cdot 2 \cdots (n-1)}, \tag{9.23}$$

and that $\Delta z = (z+1) - z = 1$, he obtained $B = \Delta f(0)$. Continuing this process, he got $C = \Delta^2 f(0), D = \Delta^3 f(0), \ldots$, completing the proof. Note that Gregory's version of the formula, given by (9.7), would be obtained by taking $A = 0$ and $z = \frac{a}{c}$.

Note that we can write the Gregory–Newton formula as

$$f(z) = f(0) + \sum_{n=1}^{\infty} \binom{z}{n} \Delta^n f(0), \tag{9.24}$$

where $\binom{z}{n} = \frac{z(z-1)\cdots(z-n+1)}{n!}$.

However, the seventeenth- and eighteenth-century mathematicians, including Newton and Stirling, did not write down the general expression for $\Delta^n f(0)$. They did not have a notation for the general expression, but they gave the values of the expression for small $n$:

$$\Delta f(0) = f(1) - f(0); \quad \Delta^2 f(0) = f(2) - 2f(1) + f(0);$$

$$\Delta^3 f(0) = f(3) - 3f(2) + 3f(1) - f(0), \ldots . \tag{9.25}$$

Observe that the absolute values of the coefficients on the right-hand sides of the expressions in (9.25) are $1, 1; 1, 2, 1; 1, 3, 3, 1; \ldots$. We recognize these as the binomial coefficients; we can thus write the general formula as

$$\Delta^n f(0) = (-1)^n \sum_{k=0}^{n} (-1)^k \binom{n}{k} f(k). \tag{9.26}$$

## 9.4 Waring, Lagrange: Interpolation Formula

Edward Waring and Joseph Lagrange independently but nearly simultaneously took up the interpolation problem of finding the polynomial of degree $n-1$, taking prescribed values at $n$ given points: $y_1, y_2, \ldots, y_n$ at $x_1, x_2, \ldots, x_n$. Their result is usually called Lagrange's interpolation formula, but it might be more accurate to call it the

Waring–Lagrange interpolation formula. Of course, this result may readily be derived by writing the Newton divided differences in symmetric form, but Lagrange and Waring gave the solution in a convenient and useful form. In fact, Waring remarked in his 1779 paper on the topic[39] that he could state and prove the result without any "recourse to finding the successive differences." We state Waring's theorem in modern notation: Let $y$ be a polynomial of degree $n-1$ and let the values of $y$ at $x_1, x_2, \ldots, x_n$ be given by $y_1, y_2, \ldots, y_n$. Then

$$y = \frac{(x-x_2)(x-x_3)\cdots(x-x_n)}{(x_1-x_2)(x_1-x_3)\cdots(x_1-x_n)}y_1 + \frac{(x-x_1)(x-x_3)\cdots(x-x_n)}{(x_2-x_1)(x_2-x_3)\cdots(x_2-x_n)}y_2$$

$$+\cdots+\frac{(x-x_1)(x-x_2)\cdots(x-x_{n-1})}{(x_n-x_1)(x_n-x_2)\cdots(x_n-x_{n-1})}y_n. \tag{9.27}$$

Waring's proof consisted in the observation that, when $x = x_1$, the first term on the right was $y_1$, while, because of the factor $x - x_1$, all the other terms were zero. Continuing this argument, taking successive values of $x$, Waring completed his proof. Lagrange published this result in 1795 in the last three pages of his *Leçons élémentaires sur les mathématiques donnés a l'École Normale*.[40] Observe that (9.27) can be written in a compact form: Let

$$f(x) \equiv (x-x_1)(x-x_2)\cdots(x-x_n)$$
$$\equiv (x-x_1)g(x). \tag{9.28}$$

Now taking the derivative of (9.28) gives us

$$f'(x) = g(x) + (x-x_1)g'(x).$$

Thus

$$f'(x_1) = g(x_1) = (x_1-x_2)(x_1-x_3)\cdots(x_1-x_n), \tag{9.29}$$

and so the denominator of the first term on the right-hand side of (9.27) is $f'(x_1)$. Similarly, the denominator of the second term is $f'(x_2)$, and so on. Thus, if $y(x)$ is a polynomial of degree $n - 1$ and $y(x_j) = y_j$, then (9.27) can be written as

$$y(x) = \sum_{j=1}^{n} \frac{f(x)}{f'(x_j)(x-x_j)}y_j. \tag{9.30}$$

In general, taking (9.27) as an interpolation formula, its right-hand side is an approximation for its left-hand side, provided $y(x)$ is a not polynomial of degree $\leq n - 1$. We note that $y(x)$ is useful even as an identity when it is a polynomial of degree $\leq n - 1$. We mention a corollary of (9.30) used by I. J. Good to find the

[39] Waring (1779).
[40] Lagrange (1867–1892) vol. 7, pp. 183–287, especially pp. 285–287.

constant term in a product conjectured by Dyson. Good took $y(x) \equiv 1$ in (9.30) and then set $x = 0$ to find[41]

$$\sum_{j=1}^{n} \prod_{\substack{1 \le k \le n \\ j \ne k}} \frac{1}{1 - \frac{x_j}{x_k}} = 1; \qquad (9.31)$$

we present Good's application of (9.31) in Section 17.13, and an application of (9.30) in Section 14.6.

To see how the Waring–Lagrange formula follows when Newton's $n$th divided difference (9.13) is written in symmetric form, first observe that the second difference can be expanded as

$$\begin{aligned}
y(x, x_1, x_2) &= \frac{y(x, x_1)}{x - x_2} - \frac{y(x_1, x_2)}{x - x_2} \\
&= \frac{y(x) - y(x_1)}{(x - x_1)(x - x_2)} - \frac{y(x_1) - y(x_2)}{(x_1 - x_2)(x - x_2)} \\
&= \frac{y(x)}{(x - x_1)(x - x_2)} - \frac{y(x_1)}{(x_1 - x_2)(x - x_1)} - \frac{y(x_2)}{(x_2 - x_1)(x - x_2)}. \quad (9.32)
\end{aligned}$$

We can then employ (9.32) to derive

$$\begin{aligned}
y(x, x_1, x_2, x_3) &= \frac{y(x)}{(x - x_1)(x - x_2)(x - x_3)} - \frac{y(x_1)}{(x - x_1)(x_1 - x_2)(x_1 - x_3)} \\
&\quad - \frac{y(x_2)}{(x - x_2)(x_2 - x_1)(x_2 - x_3)} - \frac{y(x_3)}{(x - x_3)(x_3 - x_2)(x_3 - x_1)}, \quad (9.33)
\end{aligned}$$

from which we see that $y(x, x_1, x_2, x_3)$ is symmetric in the four variables. We can, in fact, prove inductively that $y(x_0, x_1, \dots, x_n)$ is symmetric in the $n + 1$ variables. Next, using an inductive argument, we can show that, with $f(x)$ given by (9.28),

$$y(x, x_1, x_2, \dots, x_n) = \frac{y(x)}{f(x)} - \sum_{j=1}^{n} \frac{y(x_j)}{(x - x_j) f'(x_j)}, \qquad (9.34)$$

an equation that implies the Waring–Lagrange formula (9.30), because for $y(x)$, a polynomial of degree at most $n - 1$, the $n$th difference $y(x, x_1, x_2, \dots, x_n)$ is zero.

## 9.5 Euler on Interpolation

In a paper that he presented to the Petersburg Academy in 1772,[42] Euler obtained an interpolation formula similar to Newton's. Euler entitled his paper, "De eximio usu methodi interpolationum in serierum doctrina," or "The extraordinary use of the

---

[41] Good (1970).
[42] Eu. I-15 pp. 435–497. E 555.

method of interpolation in the doctrine of series." Euler considered an odd function $f(x)$ that at $x = a_1, a_2, a_3, a_4, \ldots$ takes the values $p_1, p_2, p_3, p_4, \ldots$ respectively. In section 2 of his paper, he offered the formula

$$f(x) = A_1 x + A_2 x(x^2 - a_1^2) + A_3 x(x^2 - a_1^2)(x^2 - a_2^2)$$
$$+ A_4 x(x^2 - a_1^2) \cdots (x^2 - a_3^2) + \cdots \tag{9.35}$$

where

$$A_1 = \frac{p_1}{a_1}$$

$$A_2 = \frac{p_1}{a_1(a_2^2 - a_1^2)} + \frac{p_2}{a_2(a_2^2 - a_1^2)}$$

$$A_3 = \frac{p_1}{a_1(a_1^2 - a_2^2)(a_1^2 - a_3^2)} + \frac{p_2}{a_2(a_2^2 - a_1^2)(a_2^2 - a_3^2)} + \frac{p_3}{a_3(a_3^2 - a_1^2)(a_3^2 - a_2^2)}, \text{ etc.}$$

At the end of section 6 of this paper, Euler added the remark that (9.35) was not the most general solution of the problem. Setting

$$Q = x \prod \frac{x^2 - a_n^2}{a_n^2},$$

he noted that a function of $Q$ when added to (9.35) would give the general solution.

## 9.6   Cauchy, Jacobi: Waring–Lagrange Interpolation Formula

The Waring–Lagrange interpolation formula is easy to prove, as Waring's demonstration shows. It is nevertheless interesting to consider other proofs such as those of Cauchy and Jacobi. Cauchy's argument, presented in his 1821 *Analyse algébrique* in the chapter on symmetric and alternating functions, was based on an interesting evaluation of the so-called Vandermonde determinant, without using modern notation for determinants. Lagrange had used this evaluation in a different context almost fifty years earlier. Cauchy was an expert on determinants, a term he borrowed from Gauss. He wrote an important 1812 paper on this topic, in which he also proved results on permutation groups and alternating functions. In his book, Cauchy considered the system of linear equations[43]

$$\alpha^j x + \alpha_1^j x_1 + \cdots + \alpha_{n-1}^j x_{n-1} = k_j, \tag{9.36}$$

where $j = 0, 1, \ldots, n - 1$. We have used subscripts more freely than Cauchy; he set

$$f(\alpha) = (\alpha - \alpha_1)(\alpha - \alpha_2) \cdots (\alpha - \alpha_{n-1}) = \alpha^{n-1} + A_{n-2}\alpha^{n-2} + \cdots + A_1\alpha + A_0,$$

---

[43] Cauchy (1989) Note V.

so that

$$\alpha_i^{n-1} + A_{n-2}\alpha_i^{n-2} + \cdots + A_1\alpha_i + A_0 = 0, \text{ for } i = 1, 2, \ldots, n-1.$$

Cauchy multiplied the first equation of the system (9.36) by $A_0$; the second, when $j = 1$, by $A_1$; ...; and the last, when $j = n-1$, by 1 and then added to get

$$(A_0 + A_1\alpha + \cdots + \alpha^{n-1})x = k_0 A_0 + k_1 A_1 + \cdots + k_{n-2}A_{n-2} + k_{n-1}$$

or

$$x = \frac{k_{n-1} + A_{n-2}k_{n-2} + \cdots + A_0 k_0}{f(\alpha)}. \tag{9.37}$$

He derived the values of $x_1, x_2, \ldots, x_{n-1}$ in a similar way. Cauchy applied this result to obtain the Lagrange interpolation polynomial. He supposed $u_0, u_1, \ldots, u_{n-1}$ to be values of some function at the numbers $x_0, x_1, \ldots, x_{n-1}$. It was required to find a polynomial of degree $n-1$

$$u = a_0 + a_1 x + a_2 x^2 + \cdots + a_{n-1}x^{n-1},$$

such that its values were $u_0, u_1, \ldots, u_{n-1}$ at $x_0, x_1, \ldots, x_{n-1}$, respectively. Then

$$u_j = a_0 + a_1 x_j + a_2 x_j^2 + \cdots + a_{n-1}x_j^{n-1}, \tag{9.38}$$

where $j = 0, 1, \ldots, n-1$. Note that the coefficient matrix of the system (9.38) is the transpose of the coefficient matrix of the system (9.36). So Cauchy multiplied these $n$ equation by unknowns $X_0, X_1, \ldots, X_{n-1}$ and subtracted their sum from the equation for $u$ to get

$$\begin{aligned}
u &- X_0 u_o - X_1 u_1 - X_2 u_2 - \cdots - X_{n-1}u_{n-1} \\
&= (1 - X_0 - X_1 - X_2 - \cdots - X_{n-1})a_0 \\
&\quad + (x - x_0 X_0 - x_1 X_1 - \cdots - x_{n-1}X_{n-1})a_1 \\
&\quad + (x^2 - x_0^2 X_0 - x_1^2 X_1 - \cdots - x_{n-1}^2 X_{n-1})a_2 + \cdots \\
&\quad + (x^{n-1} - x_0^{n-1}X_0 - x_1^{n-1}X_1 - \cdots - x_{n-1}^{n-1}X_{n-1})a_{n-1}. 
\end{aligned} \tag{9.39}$$

To determine $X_0, X_1, \ldots, X_{n-1}$ so that

$$u = X_0 u_0 + X_1 u_1 + \cdots + X_{n-1}u_{n-1},$$

he set equal to zero all the coefficients of $a_0, a_1, \ldots, a_{n-1}$ on the right-hand side of (9.39). Thus, he had the system of equations

$$x_0^j X_0 + x_1^j X_1 + \cdots + x_{n-1}^j X_{n-1} = x^j,$$

with $j = 0, 1, \ldots, n-1$. He could solve for $X_0, X_1, \ldots$, using the procedure for solving (9.36), to find

$$X_0 = \frac{f(x)}{f(x_0)} = \frac{(x - x_1)(x - x_2) \cdots (x - x_{n-1})}{(x_0 - x_1)(x_0 - x_2) \cdots (x_0 - x_{n-1})}, \qquad (9.40)$$

$$X_1 = \frac{(x - x_0)(x - x_2) \cdots (x - x_{n-1})}{(x_1 - x_0)(x_1 - x_2) \cdots (x_1 - x_{n-1})},$$

and so on, yielding him the Waring–Lagrange interpolation formula.

Jacobi's method employed partial fractions; he presented it in his doctoral dissertation on this topic as well as in his 1826 paper on Gauss quadrature.[44] He let

$$g(x) = (x - x_0)(x - x_1) \cdots (x - x_{n-1})$$

and $u(x)$ be the polynomials of degree $n - 1$ whose values were

$$u_0, \ldots, u_{n-1} \quad \text{at} \quad x_0, \ldots, x_{n-1},$$

respectively. Then by a partial fractions expansion he got

$$\frac{u(x)}{g(x)} = \frac{B_0}{x - x_0} + \frac{B_1}{x - x_1} + \cdots + \frac{B_{n-1}}{x - x_{n-1}}; \qquad (9.41)$$

by setting $x = x_j$, he obtained

$$B_j = \frac{u_j}{(x_j - x_0) \cdots (x_j - x_{j-1})(x_j - x_{j+1}) \cdots (x_j - x_{n-1})}. \qquad (9.42)$$

Jacobi arrived at Lagrange's formula by multiplying across by $g(x)$. We note that Jacobi's dissertation also discussed the case in which some of the $x_i$ were repeated.

## 9.7  Newton on Approximate Quadrature

The *Methodus Differentialis* stated the three-eighths rule for finding the approximate area under a curve when four values of the function were known; one proposition suggests that Newton most probably derived the formula by integrating the interpolating cubic for the four points. However, in October 1695, he wrote a very short manuscript,[45] though he left it incomplete; presenting his derivation of some rules for approximate quadrature. Surprisingly, he did not obtain these rules by integrating the interpolating polynomials but by means of heuristic and somewhat geometric reasoning. Since interpolation calculations tend to become very unwieldy, perhaps Newton sought a short cut, though it is not clear what stimulated him to write this short note. Whiteside conjectured that Newton may have been working on his contemporaneous amplified lunar theory where he used some of the results.

Newton wrote his results consecutively for two, three, four, ... ordinates. In Figure 9.1, he took equally spaced points $A, B, C, D, \ldots$ on the abscissa ($x$-axis) and points $K, L, M, N, \ldots$ on the curve ($y = f(x)$) such that $AK, BL, CM, DN, \ldots$ were

[44] Jacobi (1926).
[45] Newton (1967–1981) vol. 7, pp. 690–699.

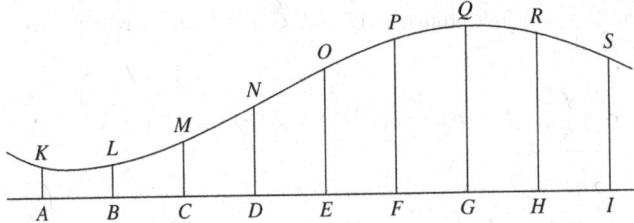

Figure 9.1 Newton's approximate quadrature.

the ordinates, or $y$ values of the corresponding points on the curve. For two points, he gave the trapezoidal rule labeled as Case 1.

If there be given two ordinates $AK$ and $BL$,
make the area $(AKLB) = \frac{1}{2}(AK + BL)AB$.

He next obtained Simpson's rule, published by Thomas Simpson in his *Mathematical Dissertations* of 1743;[46] Simpson gave an interesting geometric proof. We note that since Simpson's books were quite popular, his name got attached to the rule. In 1639, Cavalieri gave particular cases of this rule to determine the volume of a symmetrical wine cask. In 1668, in his *Exercitationes Geometricae*, Gregory too presented this rule to approximate $\int_0^h \tan x \, dx$.[47] Newton derived Simpson's and the three-eights rule as Cases 2 and 3, where the box notation denotes area:[48]

Case 2. If there be given three $AK, BL$ and $CM$, say that

$$\frac{1}{2}(AK + CM)AC = \square(AM), \quad \text{and again, by Case 1,}$$

$$\frac{1}{2}\left(\frac{1}{2}(AK + BL) + \frac{1}{2}(BL + CM)\right)AC$$

$$= \frac{1}{4}(AK + 2BL + CM)AC = \square(AM),$$

and that the error in the former solution is to the error in the latter as $AC^2$ to $AB^2$ or 4 to 1, and hence the difference $\frac{1}{4}(AK - 2BL + CM)AC$ of the solutions is to the error in the latter as 3 to 1, and the error in the latter will be

$$\frac{1}{12}(AK - 2BL + CM)AC.$$

Take away this error and the latter solution will come to be

$$\frac{1}{6}(AK + 4BL + CM)AC = \square(AM), \quad \text{the solution required.}$$

[46] Simpson (1743) pp. 109–110.
[47] Gregory (1668) pp. 25–27. For more information, see Newton (1967–1981) vol. 7, p. 692, footnote 8 and Whiteside (1961b) pp. 248–249.
[48] Newton (1967–1981) vol. 7, pp. 691–693.

Case 3. If there be given four ordinates $AK, BL, CM$ and $DN$, say that $\frac{1}{2}(AK + DN)AD = \square(AN)$; likewise, that

$$\frac{1}{3}\left(\frac{1}{2}(AK + BL) + \frac{1}{2}(BL + CM) + \frac{1}{2}(CM + DN)\right)AD,$$

that is, $\frac{1}{6}(AK + 2BL + 2CM + DN)AD = \square(AN)$. The errors in the solutions will be as $AD^2$ to $AB^2$ or 9 to 1, and hence the difference in the errors (which is the difference $\frac{1}{6}(2AK - 2BL - 2CM + 2DN)AD$ in the solutions) will be to the error in the latter as 8 to 1. Take away this error and the latter will remain as

$$\frac{1}{8}(AK + 3BL + 3CM + DN)AD = \square(AN).$$

We observe that in these three cases and others, Newton assumed without justification that when $n + 1$ equidistant ordinates were given, the corresponding ratio of the errors in using the trapezoidal rule would be $n^2 : 1$. Newton went on to consider cases with five, seven, and nine ordinates, but his results in the last two cases were not the same as the ones obtained by integrating the interpolating polynomials.

To describe Newton's proof of Simpson's formula in somewhat more analytic terms, let $[a, b]$ in Case 2 be the interval with $b = a + 2h$, and let $y = f(x)$ be the function on that interval. By the trapezoidal rule,

$$\int_a^b f(x)\,dx \approx \frac{1}{2}(f(a) + f(b))(2h) \equiv I_1.$$

If this rule is applied to each of the intervals $[a, a + h], [a + h, b]$, then

$$\int_a^b f(x)\,dx \approx \frac{1}{2}\left(\frac{1}{2}(f(a) + f(a + h)) + \frac{1}{2}(f(a + h) + f(b))\right)(2h)$$

$$= \frac{1}{4}(f(a) + 2f(a + h) + f(b))(2h) \equiv I_2.$$

Let the errors in the two formulas be $e_1$ and $e_2$, so that

$$\int_a^b f(x)\,dx = I_1 + e_1 = I_2 + e_2.$$

Newton assumed without proof that $\frac{e_1}{e_2} = 4$. Hence,

$$\frac{I_2 - I_1}{e_2} = \frac{e_1 - e_2}{e_2} = 3$$

so that

$$e_2 = \frac{1}{3}(I_2 - I_1) = -\frac{1}{12}(f(a) - 2f(a + h) + f(b))2h.$$

When this value of $e_2$ is added to $I_2$, we get Simpson's approximation

$$\frac{1}{6}(f(a) + 4f(a + h) + f(b)). \tag{9.43}$$

## 9.8 Hermite: Approximate Integration

The formulas of Newton, Cotes, and Stirling for numerical integration were used without change for a century. In the nineteenth century, mathematicians began to present new methods, starting with Gauss, whose work in this area is discussed in our treatment of orthogonal polynomials. Charles Hermite (1822–1901) was a professor at the École Polytechnique. He gave a series of analysis lectures in 1873; these and other such lectures were published and serve as a valuable resource even today. For example, Hermite discussed an original method[49] for the numerical evaluation of integrals of the form

$$\int_{-1}^{+1} \frac{\phi(x)\,dx}{\sqrt{1-x^2}}, \tag{9.44}$$

where $\phi(x)$ was an analytic function. He started with the $n$th-degree polynomial $F(x)$ defined by

$$F(x) = \cos n(\arccos x). \tag{9.45}$$

By taking the derivative, he obtained

$$F'(x) = n \sin n(\arccos x)\frac{1}{\sqrt{1-x^2}} = n\frac{\sqrt{1-F^2(x)}}{\sqrt{1-x^2}}.$$

Hence

$$\frac{1}{\sqrt{x^2-1}} = \frac{F'(x)}{nF(x)}\left(1 - \frac{1}{F^2(x)}\right)^{-\frac{1}{2}}$$

$$= \frac{F'(x)}{nF(x)}\left(1 + \frac{1}{2}\frac{1}{F^2(x)} + \frac{1\cdot 3}{2\cdot 4}\frac{1}{F^4(x)} + \cdots\right).$$

Hermite observed that the last expression without the first term could be written in decreasing powers of $x$ in the form

$$\frac{\lambda_0}{x^{2n+1}} + \frac{\lambda_1}{x^{2n+2}} + \frac{\lambda_2}{x^{2n+3}} + \cdots.$$

Consequently,

$$\frac{1}{\sqrt{x^2-1}} = \frac{F'(x)}{nF(x)} + \frac{\lambda_0}{x^{2n+1}} + \frac{\lambda_1}{x^{2n+2}} + \frac{\lambda_2}{x^{2n+3}} + \cdots.$$

At this point, Hermite invoked the integral formula

$$\int_{-1}^{+1} \frac{dz}{(x-z)\sqrt{1-z^2}} = \frac{\pi}{\sqrt{x^2-1}}, \tag{9.46}$$

[49] Hermite (1873) pp. 452–454.

to obtain

$$\frac{1}{\pi} \int_{-1}^{+1} \frac{dz}{(x-z)\sqrt{1-z^2}} = \frac{F'(x)}{nF(x)} + \frac{\lambda_0}{x^{2n+1}} + \frac{\lambda_1}{x^{2n+2}} + \cdots$$

$$= \frac{1}{n} \sum_{j=1}^{n} \frac{1}{x-a_j} + \frac{\lambda_0}{x^{2n+1}} + \frac{\lambda_1}{x^{2n+2}} + \cdots,$$

where $a_1, a_2, \ldots, a_n$ were the $n$ roots of $F(x) = 0$. An application of the geometric series $\frac{1}{x-z} = \sum \frac{z^n}{x^{n+1}}$ gave him

$$\frac{1}{\pi} \int_{-1}^{+1} \left( 1 + \frac{z}{x} + \frac{z^2}{x^2} + \cdots \right) \frac{dz}{\sqrt{1-z^2}}$$

$$= \frac{1}{n} \sum_{j=1}^{n} \left( 1 + \frac{a_j}{x} + \frac{a_j^2}{x^2} + \cdots \right) + \frac{\lambda_0}{x^{2n}} + \frac{\lambda_1}{x^{2n+1}} + \cdots.$$

Equating the coefficients of $\frac{1}{x^l}$ on both sides yielded

$$\frac{1}{\pi} \int_{-1}^{1} \frac{z^l}{\sqrt{1-z^2}} \, dz = \frac{1}{n} \sum_{j=1}^{n} a_j^l \qquad \text{when } l < 2n,$$

$$= \frac{1}{n} \sum_{j=1}^{n} a_j^{2n+s} + \lambda_s \qquad \text{when } l = 2n+s, \ s \geq 0.$$

So, Hermite wrote $\phi(z) = k_0 + k_1 z + k_2 z^2 + \cdots + k_n z^n + \cdots$ in order to obtain the formula

$$\frac{1}{\pi} \int_{-1}^{1} \frac{\phi(z)}{\sqrt{1-z^2}} \, dz = \frac{1}{n} \sum_{j=1}^{n} \phi(a_j) + R, \qquad (9.47)$$

where

$$R = \lambda_0 k_{2n} + \lambda_1 k_{2n+1} + \cdots.$$

Hermite also noted that since the roots $a_j$ of $F(x) = 0$ were given by

$$a_j = \cos \left( \frac{2j-1}{2n} \pi \right), \ j = 1, 2, \ldots, n,$$

he obtained

$$\int_0^\pi \phi(\cos \theta) \, d\theta = \frac{\pi}{n} \sum_{j=1}^{n} \phi \left( \cos \left( \frac{2j-1}{2n} \pi \right) \right) + \pi R. \qquad (9.48)$$

Observe that the expression for the error $R$ shows that it must be zero for any polynomial $\phi$ of degree less than $2n$. Hermite may have been unaware that in 1849, Brice Bronwin derived formula (9.48) by a different method, but without the error term.[50]

## 9.9 Chebyshev on Numerical Integration

A nice feature of the Bronwin-Hermite formula is that it allows us to find an approximate value of the integral by simply adding the values of the function $\phi(x)$ at the zeros of $F(x)$ and then multiplying by $\frac{\pi}{n}$. Chebyshev's interest in applications led him to seek similar formulas for other weight functions. Thus, the purpose of Chebyshev's 1874 paper[51] was to find a constant $k$, and numbers $x_1, x_2, \ldots, x_n$ such that $\int_{-1}^{1} F(x)\phi(x)\, dx$ could be approximated by $k(\phi(x_1) + \phi(x_2) + \cdots + \phi(x_n))$. Note that $\phi(x)$ was the function to be integrated with respect to the weight function $F(x)$. In general, Chebyshev required that the approximation be exact for polynomials of degree at most $n - 1$, so he looked for a formula of the form

$$\int_{-1}^{+1} \frac{F(x)}{z-x}\, dx$$

$$= k\left(\phi(x_1) + \phi(x_2) + \cdots + \phi(x_n)\right) + k_1 \phi^{(n+1)}(0) + k_2 \phi^{(n+2)}(0) + \cdots, \quad (9.49)$$

where $\phi^{(m)}$ denoted the $m$th derivative of $\phi$ and $k_1, k_2, \ldots$ were constants. Following Hermite, he considered the case $\phi(x) = \frac{1}{z-x}$ to obtain

$$\int_{-1}^{+1} \frac{F(x)}{z-x}\, dx = k\left(\frac{1}{z-x_1} + \cdots + \frac{1}{z-x_n}\right) + \frac{(n+1)!\, k_1}{z^{n+2}} + \frac{(n+2)!\, k_2}{z^{n+3}} + \cdots .$$

$$(9.50)$$

He set $f(z) = (z - x_1)(z - x_2) \cdots (z - x_n)$, so that after multiplying by $z$, the last relation became

$$z \int_{-1}^{+1} \frac{F(x)}{z-x}\, dx$$

$$= kz\frac{f'(z)}{f(z)} + \frac{1 \cdot 2 \cdot 3 \cdots (n+1)k_1}{z^{n+1}} + \frac{1 \cdot 2 \cdot 3 \cdots (n+2)k_2}{z^{n+2}} + \cdots . \quad (9.51)$$

He let $z \to \infty$ to get

$$\int_{-1}^{+1} F(x)\, dx = nk, \text{ or } k = \frac{1}{n} \int_{-1}^{+1} F(x)\, dx. \quad (9.52)$$

[50] Bronwin (1849).
[51] Chebyshev (1899–1907) vol. 2, pp. 165–180.

He thus had the value of $k$, and it remained for him to find the polynomial $f(z)$ whose zeros would be the numbers $x_1, x_2, \ldots, x_n$. For this purpose, he integrated equation (9.50) with respect to $z$ to obtain

$$\int_{-1}^{1} F(x) \ln(z - x)\, dx = k \ln \frac{f(z)}{C_1} - \frac{n! k_1}{z^{n+1}} - \frac{(n+1)! k_2}{z^{n+2}} - \cdots,$$

where $C$ was a constant. Hence, by exponentiation he could write

$$f(z)\, e^{\frac{-n! k_1}{k z^{n+1}} - \frac{(n+1)! k_2}{k z^{n+2}} - \cdots} = C \exp\left( \frac{1}{k} \int_{-1}^{+1} F(x) \ln(z - x)\, dx \right).$$

Chebyshev then noted that the exponential on the left differed from 1 by a series of powers of $z$ less than $z^{-n}$; hence, he noted that $f(z)$ was the polynomial part of the exponential on the right-hand side. He deduced Hermite's formula by taking $F(x) = \frac{1}{\sqrt{1-x^2}}$, so that

$$k = \frac{1}{n} \int_{-1}^{+1} \frac{dx}{\sqrt{1 - x^2}} = \frac{\pi}{n}$$

and

$$\int_{-1}^{+1} F(x) \ln(z - x)\, dx = \int_{-1}^{1} \frac{\ln(z - x)}{\sqrt{1 - x^2}}\, dx = \pi \ln \frac{z + \sqrt{z^2 - 1}}{2}. \tag{9.53}$$

Chebyshev could then conclude that the polynomial $f(z)$ in this case was in fact the polynomial part of

$$e^{n \ln \frac{z + \sqrt{z^2 - 1}}{2}} = \left( \frac{z + \sqrt{z^2 - 1}}{2} \right)^n,$$

and he wrote that it was equal to $\frac{1}{2^{n-1}} \cos(n \arccos z)$. He then considered the case where $F(x) = 1$, to obtain by (9.52): $k = \frac{2}{n}$ and

$$\int_{-1}^{+1} \ln(z - x)\, dx = \ln \frac{(z + 1)^{z+1}}{(z - 1)^{z-1}} - 2.$$

Thus, Chebyshev arrived at the result he wanted:

$$\int_{-1}^{1} \phi(x)\, dx = \frac{2}{n} (\phi(x_1) + \phi(x_2) + \cdots + \phi(x_n)), \tag{9.54}$$

where $x_1, x_2, \ldots, x_n$ were the zeros of the polynomial given by the polynomial part of the expression

$$(z + 1)^{\frac{n(z+1)}{2}} (z - 1)^{-\frac{n(z-1)}{2}} = z^n e^{-\frac{n}{2 \cdot 3 z^2} - \frac{n}{4 \cdot 5 z^4} - \frac{n}{6 \cdot 7 z^6} - \cdots}.$$

He also computed the cases in which $n = 2, 3, 4, 5, 6, 7$ to get the polynomials

$$z^2 - \frac{1}{3}, \quad z^3 - \frac{1}{2}z, \quad z^4 - \frac{2}{3}z^2 + \frac{1}{45}, \quad z^5 - \frac{5}{6}z^3 + \frac{7}{72}z,$$

$$z^6 - z^4 + \frac{1}{5}z^2 - \frac{1}{105}, \quad z^7 - \frac{7}{6}z^5 + \frac{119}{360}z^3 - \frac{149}{6480}z.$$

He calculated the zeros of these polynomials to six decimal places. At this juncture, Chebyshev pointed out that in (9.54) the sum of the squares of the coefficients had the smallest possible value, because they were all equal; thus, his formula might sometimes even be an improvement on Gauss's quadrature formula.

## 9.10   Exercises

(1) Let $A, B, C, E, \ldots$ be points on the $x$-axis and $K, L, M, N, \ldots$ corresponding points on the curve. Then the ordinates are given by $AK, BL, CM, DN, \ldots$. Newton described the following formulas for the approximate area under the curve in the case of seven and nine ordinates:

If there be seven ordinates there will come to be

$$\frac{1}{280}(17AK + 54BL + 51CM + 36DN + 51EO + 54FP + 17GQ)AG = \square AQ.$$

While if there be given nine there will come

$$\frac{(217AK + 1024BL + 352CM + 1024DN + 436EO + 1024FP + 352GQ + 1024HR + 217IS)AI}{5670} = \square AS.$$

Derive these formulas using Newton's ideas, as explained in the text and as presented by Newton in his *Of Quadrature by Ordinates*. Recall that these formulas, making use of Newton's assumption on the proportionality of the errors, differ from those obtained by integrating the interpolating polynomial. See Newton (1967–1981) vol. 7, p. 695, including Whiteside's footnotes.

(2) Suppose

$$u = \frac{a + bx + cx^2 + \cdots + hx^{n-1}}{\alpha + \beta x + \gamma x^2 + \cdots + \theta x^m},$$

and $u(x_k)$, $x = 0, 1, 2, \ldots, n + m - 1$. Determine the values of the coefficients $\frac{a}{\alpha}, \frac{b}{\alpha}, \ldots, \frac{h}{\alpha}, \frac{\beta}{\alpha}, \ldots \frac{\theta}{\alpha}$. In particular, show that when $m = 1$ and $n = 2$,

$$u = \frac{u_0 u_1 \frac{x - x_2}{(x_0 - x_2)(x_1 - x_2)} + u_0 u_2 \frac{x - x_1}{(x_0 - x_1)(x_2 - x_1)} + u_1 u_2 \frac{x - x_0}{(x_1 - x_0)(x_2 - x_0)}}{u_0 \frac{x_0 - x}{(x_0 - x_1)(x_0 - x_2)} + u_1 \frac{x_1 - x}{(x_1 - x_0)(x_1 - x_2)} + u_2 \frac{x_2 - x}{(x_2 - x_0)(x_2 - x_1)}}.$$

Cauchy discussed this interpolation by rational functions after he deduced the Waring–Lagrange formula in his lectures. See Cauchy (1989) pp. 527–529.

(3) Chebyshev computed the zeros of the polynomials, $z^5 - \frac{5}{6}z^3 + \frac{7}{72}z$ and $z^6 - z^4 + \frac{1}{5}z^2 - \frac{1}{105}$ for use in (9.54). His results were

$$\pm 0.832497, \ \pm 0.374541, 0 \quad \text{and} \quad \pm 0.866247. \ \pm 0.422519, \ \pm 0.266635.$$

Check Chebyshev's computations.

(4) Show that Chebyshev's result in (9.53) implies Hermite's formula (9.47).

(5) Prove the following formulas of Stieltjes:

$$\int_{-1}^{+1} \sqrt{1-x^2}\, f(x)\, dx = \frac{\pi}{n+1} \sum_{k=1}^{n} \sin^2 \frac{k\pi}{n+1} f\left(\cos \frac{k\pi}{n+1}\right) + \text{corr.,}$$

$$\int_{-1}^{+1} \sqrt{\frac{1-x}{1+x}}\, f(x)\, dx = \frac{4\pi}{2n+1} \sum_{k=1}^{n} \sin^2 \frac{k\pi}{2n+1} f\left(\cos \frac{2k\pi}{2n+1}\right) + \text{corr.}$$

The correction is zero when $f(x)$ is a polynomial of degree $\leq 2n-1$.

$$\int_0^1 \frac{f(x)}{\sqrt{x(1-x)}} = \frac{\pi}{n} \sum_{k=1}^{n} f\left(\cos^2 \frac{(2k-1)\pi}{4n}\right) + \text{corr.,}$$

$$\int_0^1 \sqrt{x(1-x)}\, f(x)\, dx = \frac{\pi}{4n+4} \sum_{k=1}^{n} \sin^2 \frac{k\pi}{n+1} f\left(\cos^2 \frac{k\pi}{2n+2}\right) + \text{corr.,}$$

$$\int_0^1 \sqrt{\frac{1-x}{x}}\, f(x)\, dx = \frac{2\pi}{2n+1} \sum_{k=1}^{n} \sin^2 \frac{k\pi}{2n+1} f\left(\cos^2 \frac{k\pi}{2n+1}\right) + \text{corr.}$$

See Stieltjes (1993) vol. 1, pp. 514–515.

## 9.11   Notes on the Literature

Chapters 10 and 11 of J. L. Chabert (1999) contain interesting observations on interpolation and quadrature with excerpts from original authors. Thomas Harriot's manuscript *De Numeris Triangularibus*, containing his derivations of symbolic interpolation formulae and their applications, has now been published in Beery and Stedall (2009), almost four centuries after it was written. Beery and Stedall provide a commentary to accompany the almost completely nonverbal presentation of Harriot. They also discuss the work on interpolation of several British mathematicians in the period 1610–1670.

# 10

## Series Transformation by Finite Differences

### 10.1 Preliminary Remarks

Around 1670, James Gregory found a large number of new infinite series, but his methods remain somewhat unclear. From circumstantial evidence and from the form of some of his series, it appears that he was the first mathematician to systematically make use of finite difference interpolation formulas in finding new infinite series. The work of Newton, Gregory, and Leibniz made the method of finite differences almost as important as calculus in the discovery of new infinite series. We observe that interpolation formulas usually deal with finite expressions because in practice the number of interpolating points is finite. By theoretically extending the number of points to infinity, Gregory found the binomial theorem, the Taylor series, and numerous interesting series involving trigonometric functions. Gregory most likely derived these theorems from the Gregory–Newton (or Harriot–Briggs) interpolation formula. Gregory's letter to Collins, of November 23, 1670,[1] explicitly mentions these results, and also contains some other series that were not direct consequences of the Harriot–Briggs result.[2] Instead, these other series seem to require the Newton–Gauss interpolation formula; one is compelled to conclude that Gregory must have obtained this interpolation formula, though it is not given anywhere in his surviving notes and letters. In a separate enclosure with his letter to Collins, Gregory wrote several formulas, including:[3]

> Given an arc whose sine is $d$, and sine of the double arc is $2d - e$, it is required to find another arc which bears to the arc whose sine is $d$ the ratio $a$ to $c$. The sine of the arc in question

$$= \frac{ad}{c} - \frac{be}{c} + \frac{ke^2}{cd} - \frac{le^3}{cd^2} + \cdots \qquad (10.1)$$

where $\quad \dfrac{b}{c} = \dfrac{a(a^2 - c^2)}{2 \cdot 3 \cdot c^3}, \quad \dfrac{k}{c} = \dfrac{a(a^2 - c^2)(a^2 - 4c^2)}{2 \cdot 3 \cdot 4 \cdot 5 \cdot c^5}$, etc.

---

[1] Turnbull (1938) pp. 118–132.

[2] ibid. pp. 129–130.

[3] ibid. p. 128.

213

In modern notation, $c = r\theta$, $d = r\sin\theta$, $2d - e = r\sin 2\theta$; hence, $e = 2r\sin\theta(1 - \cos\theta)$ and the series takes the form

$$\sin\frac{a\theta}{c} = \sin\theta\left(\frac{a}{c} - \frac{a(a^2 - c^2)}{3!\,c^3}2(1 - \cos\theta)\right.$$

$$\left. + \frac{a(a^2 - c^2)(a^2 - 4c^2)}{5!\,c^5}(2(1 - \cos\theta))^2 - \cdots\right). \qquad (10.2)$$

Gregory noted at the end of the enclosure that an infinite number of other ways of measuring circular arcs could be deduced from his calculations.

Gregory did not publish his work on series and his mathematical letters to Collins were not printed until later. Meanwhile, Newton developed his profound ideas on interpolation and finite differences starting in the mid-1670s. In the early 1680s, he applied the method of differences to infinite series and in June/July of 1684, he wrote two short treatises on the topic.[4] He was provoked into writing up his results upon receiving a work from David Gregory, the nephew of James, *Exercitatio Geometrica de Dimensione Figurarum*.[5] In this treatise, David Gregory discussed several aspects of infinite series, apparently without knowledge of Newton's work. Newton evidently wished to set the record straight; he first wrote *Matheseos Universalis Specimina*, in which he pointed out that James Gregory and Leibniz were indebted to him in their study of series. He did not finish this treatise, but instead started a new one, called *De Computo Serierum* in which he eliminated all references to Gregory and Leibniz. The first chapter of the second treatise dealt with infinite series in a manner similar to that of his early works of 1669 and 1671. However, the second chapter employed the entirely new idea of applying finite differences to derive an important transformation of infinite series,[6] often called Euler's transformation. In modern notation, this is given by

$$A_0 t + A_1 t^2 + A_2 t^3 + \cdots = A_0 z + \Delta A_0 z^2 + \Delta^2 A_0 z^3 + \cdots, \qquad (10.3)$$

where $z = \dfrac{t}{1 - t}$, $\Delta A_0 = A_1 - A_0$, $\Delta^2 A_0 = A_2 - 2A_1 + A_0$, etc.

Newton noted one remarkable special case of his transformation:

$$\tan^{-1} t = t - \frac{1}{3}t^3 + \frac{1}{5}t^5 - \frac{1}{7}t^7 + \cdots$$

$$= \frac{t}{1 + t^2}\left(1 + \frac{2}{1\cdot 3}\frac{t^2}{1 + t^2} + \frac{2\cdot 4}{1\cdot 3\cdot 5}\left(\frac{t^2}{1 + t^2}\right)^2\right.$$

$$\left. + \frac{2\cdot 4\cdot 6}{1\cdot 3\cdot 5\cdot 7}\left(\frac{t^2}{1 + t^2}\right)^3 + \cdots\right). \qquad (10.4)$$

---

[4] Newton (1967–1981) vol. 4, pp. 526–653.
[5] Gregory (1684).
[6] Newton (1967–1981) pp. 605–607.

Observe that when $t = 1$ we have $\frac{t^2}{1+t^2} = \frac{1}{2}$ so that while the first series converges very slowly for this value of $t$, the second series converges much more rapidly. Moreover when $t = \sqrt{3}$, the first series is divergent while the second is convergent. In fact, Newton wrote:[7]

> The chief use for these transformations is to turn divergent series into convergent ones, and convergent series into ones more convergent. Series in which all terms are of the same sign cannot diverge without simultaneously coming to be infinitely great and on that account false. These, consequently, have no need to be turned into convergent ones. Those, however, in which the terms alternate in sign and proceed regularly, are so moderated by the successive addition and subtraction of those terms as to remain true even in divergence. But in their divergent form their quantity cannot be computed and they must be turned into convergent ones by the rule introduced, while when they are sluggishly convergent the rule must be applied to make them converge more swiftly. Thus the series $y = x - \frac{1}{3}x^3 + \frac{1}{5}x^5 \cdots$ , when it converges or diverges slowly enough and has been turned into this one
>
> $$y = x - \frac{1}{3}x^3 + \frac{1}{5}x^5 - \frac{1}{7}x^7 + \frac{1}{9}x^9 - \frac{1}{11}x^{11} + \frac{1}{13}x^{13} - \frac{1}{15}x^{15}$$
> $$+ x^{15}\left(\frac{1}{17}z + \frac{2}{17 \cdot 19}z^2 + \frac{2 \cdot 4}{17 \cdot 19 \cdot 21}z^3 + \frac{2 \cdot 4 \cdot 6}{17 \cdot 19 \cdot 21 \cdot 23}z^4 + \cdots\right),$$
>
> will speedily enough be computed to many places of decimals. If the same series proves swiftly divergent it must be turned into the convergent
>
> $$xy = z + \frac{2}{1 \cdot 3}z^2 + \frac{2 \cdot 4}{1 \cdot 3 \cdot 5}z^3 \cdots \tag{10.5}$$
>
> and then by what is presented in the following chapter it can be computed. It is, however, frequently convenient to reduce the coefficients $A, B, C, \ldots$ to decimal fractions at the very start of the work.

We note that Newton's $A, B, C, \ldots$ are the $A_0, A_1, A_2 \ldots$ in (10.3) and $z = \frac{x^2}{1+x^2}$, as in (10.4). We do not know why Newton discontinued work on this treatise. Perhaps it was because Edmond Halley visited Cambridge in August 1684 and urged Newton to work on problems of planetary motion. As is well known, Newton started work on the *Principia* soon after this visit and for the next two years concentrated on this work with little respite.

Newton's transformation (10.4) for the arctangent series is obviously important, so it is not surprising that others rediscovered it, since Newton's paper did not appear in print until 1970. In August 1704, Jakob Bernoulli communicated[8] the $t = 1$ case of (10.4) to Leibniz as a recent discovery of Jean Christophe Fatio de Duillier. Jakob Hermann, a student of Bernoulli, found a proof for this and sent it to Leibniz in January 1705. This proof is identical with that of Newton's when specialized to $t = 1$. Johann Bernoulli, and probably others, succeeded in deriving the general form (10.4). Bernoulli, in fact, applied the general form in his 1742 notes on series[9] and thereby derived a remarkable series for $\pi^2$ found earlier by Takebe Katahiro

---

[7] ibid. p. 611.
[8] Newton (1967–1981) vol. IV, p. 608, footnote 44.
[9] Joh. Bernoulli (1742) vol. IV, p. 25.

by a different technique.[10] In 1717, the French mathematician Pierre Rémond de Montmort (1678–1719) rediscovered Newton's more general transformation (10.3) with a different motivation and method of proof.[11]

Montmort was born into a wealthy family of the French nobility and was self-taught in mathematics. An admirer of both Newton and Leibniz, he remained neutral but friendly with followers of both mathematicians during the calculus priority dispute in the early eighteenth century. He mainly worked in probability and combinatorics but also made contributions to the theory of series. His paper on series was inspired by Brook Taylor's 1715 work *Methodus Incrementorum*; consequently, Montmort's *De Seriebus Infinitis Tractatus* was published in the *Philosophical Transactions* with an appendix by Taylor, then Secretary of the Royal Society. Montmort's paper dealt with those finite as well as infinite series to which the method of differences could be applied. He first worked out the transformation of a finite power series and then obtained Newton's formula (10.3) as a corollary. He also quoted from a 1715 letter of Niklaus I Bernoulli,[12] showing that Niklaus had found a result similar to that of Newton.

In 1717, François Nicole (1683–1758), a pupil of Montmort, also published a paper[13] on finite differences. He too wrote that his ideas were suggested by Taylor's book of 1715. The title of his paper, *Traité du calcul des différences finies*, indicated that he viewed the calculus of finite differences as a new topic in mathematics, separate from geometry, calculus, or algebra. By means of examples, he showed that the shifted factorial expression

$$(x)_n = x(x + h)(x + 2h) \cdots (x + (n - 1)h) \tag{10.6}$$

behaved under differencing as $x^n$ under differentiation. Thus,

$$\Delta_h(x)_n = (x + h)_n - (x)_n = nh(x + h)_{n-1}. \tag{10.7}$$

Also, the difference relation

$$\frac{1}{(x)_n} - \frac{1}{(x + h)_n} = \frac{nh}{(x)_{n+1}} \tag{10.8}$$

showed that the analog of $x^{-n}$ was $\frac{1}{(x)_n}$. Moreover, the inverse of a difference would be a sum. And just as the derivative of a function indicated the integral of the derived function, so also one could use the difference to find the sum. He gave an example: From

$$f(x) = \frac{x(x + 1)(x + 2)}{3}, \; f(x + 1) - f(x) = (x + 1)(x + 2) \tag{10.9}$$

[10] Smith and Mikami (1914) pp. 148–150.
[11] Montmort (1717).
[12] ibid. pp. 674–675.
[13] Nicole (1717).

he got

$$1 \cdot 2 + 2 \cdot 3 + \cdots + x(x+1) = \frac{x(x+1)(x+2)}{3} + C. \tag{10.10}$$

By taking $x = 1$, he obtained the constant as zero.

In 1723, in his second memoir,[14] Nicole discussed the problem of computing the coefficients $a_0, a_1, a_2, \ldots$ in

$$f(x+m) - f(x) = a_0 + a_1(x+h) + a_2(x+h)(x+2h)$$
$$+ \cdots + a_{k-1}(x+h) \cdots (x+(k-1)h), \tag{10.11}$$

where $f(x) = x(x+h) \cdots (x+(k-1)h)$. His method employed a long inductive process, but simpler procedures have since been found. In his second memoir and in his third memoir of 1724,[15] Nicole solved a similar problem for the inverse factorial $\frac{1}{x(x+h) \cdots (x+(k-1)h)}$.

Both Montmort and Nicole mentioned Taylor as the source of their inspiration; we note that Taylor gave a systematic exposition of finite differences and derivatives with their inverses, sums, and integrals. Many of these ideas were already known but Taylor explicitly laid out some of the concepts, such as the method of summation by parts. In a letter of November 14, 1715,[16] Montmort also attributed to Taylor the summation formula

$$\frac{1}{b-a} = \frac{1}{b} + \frac{a}{b(b+d)} + \frac{a(a+d)}{b(b+d)(b+2d)} + \cdots. \tag{10.12}$$

There are several ways of proving (10.12), but it is likely that Taylor proved it by the Gregory–Newton interpolation formula, since he used this to prove his famous series.

The Scottish mathematician James Stirling (1692–1770) took Nicole's work much further. Stirling's book, *Methodus Differentialis*, is sometimes called the first book on the calculus of finite differences. Like all prominent British mathematicians of the early eighteenth century, he was a disciple of Newton. His first paper, *Lineae Tertii Ordinis Newtonianae*,[17] was an account with some extensions of Newton's theory of cubic curves. His second paper, *Methodus Differentialis Newtoniana Illustrata*,[18] developed some of Newton's ideas on interpolation. He later expanded this paper into the *Methodus*. Stirling received his early education in Scotland. In 1710 he traveled to Oxford and graduated from Balliol College the same year. He stayed on at Oxford on a scholarship, but he lost his support after the first Jacobite rebellion of 1715, as his family had strong Jacobite sympathies. He then spent several years in Italy and was unable to obtain a professorship there. Although details of his time in Italy are largely

---

[14] Nicole (1723).
[15] Nicole (1724).
[16] See Bateman (1907) p. 368.
[17] Stirling (1717).
[18] Stirling (1719).

unknown, his second paper was communicated from Venice. After returning to Britain in 1722, he was given assistance by Newton, making him one of Newton's devoted friends. After teaching in a London school, in 1735 Stirling began service as manager of the Leadhills Mines in Scotland where he was very successful, looking after the welfare of the miners as well as the interests of the shareholders. In the early 1750s, he also surveyed the River Clyde in preparation for a series of navigational locks.

Stirling started his book where Nicole ended. In the introduction, he defined the Stirling numbers of the first and second kinds. These numbers appeared as coefficients when $z^n$ was expanded in terms of $z(z-1)\cdots(z-k+1)$, and $\frac{1}{z^n}$ was expanded in terms of $\frac{1}{z(z+1)\cdots(z+k-1)}$. These expansions were required in order to apply the method of differences to functions or quantities normally expressed in terms of powers of $z$. Stirling constructed two small tables of these coefficients to make the transformation easy to use. In the first few propositions of his book, Stirling considered problems similar to those of Nicole, but he very quickly enlarged the scope of those methods. He applied his new method to the approximate summation of series such as $1 + \frac{1}{4} + \frac{1}{9} + \frac{1}{16} + \cdots$, whose approximate value had also been computed by Daniel Bernoulli, Goldbach and Euler in the late 1720s. It was a little later that Euler brilliantly found the exact value of the series to be $\frac{\pi^2}{6}$. Stirling also applied his method of differences to derive several new and interesting transformations of series. For example, observe propositions 7 and 8 of his *Methodus Differentialis* presented in modernized notation:

$$1 + \frac{z-n}{z(1-m)} + \frac{(z-n)(z-n+1)}{z(z+1)(1-m)^2} + \cdots = \frac{m-1}{m}\left(1 + \frac{n}{zm} + \frac{n(n+1)}{z(z+1)m^2} + \cdots\right)$$
$$(10.13)$$

and

$$1 + \frac{(z-m)(z-n)}{z(z-n+1)} + \frac{(z-m)(z-m+1)(z-n)(z-n+1)}{z(z+1)(z-n+1)(z-n+2)} + \cdots$$
$$= \frac{z-n}{m}\left(1 + \frac{nm}{z(m+1)} + \frac{n(n+1)m(m+1)}{z(z+1)(m+1)(m+2)} + \cdots\right).$$
$$(10.14)$$

Note that Newton's transformation (10.4) for arctan $t$ follows by taking $n = 1$, $z = \frac{3}{2}$, and $m = 1 + \frac{1}{t^2}$ in (10.13). As we discuss in Chapter 23, these formulas are particular cases of transformations of hypergeometric series. The hypergeometric generalization of (10.13) was discovered by Pfaff in 1797,[19] and the generalization of (10.14) was found by Kummer in 1836.[20] Thus, the methods of hypergeometric series provide the right context with the appropriate degree of generality to study the series (10.4), (10.13), and (10.14). Moreover, Gauss extended Stirling's finite difference method to the theory of hypergeometric series and derived his well-known and

[19] Pfaff (1797a).
[20] Kummer (1836).

important contiguous relations for hypergeometric series.[21] Even today, contiguous relations continue to provide unexplored avenues for research.

We also note that since expressions of the form $z(z - 1) \cdots (z - k + 1)$ appear in finite difference interpolation formulas, Stirling numbers of the first kind also appear in those formulas. For this reason, in the early 1600s, Harriot computed these numbers. Stirling numbers also cropped up in Lagrange's 1770s proof[22] of Wilson's theorem that $(p - 1)! + 1$ was divisible by $p$ if and only if $p$ was prime. In fact, Lagrange's proof gave the first number theoretic discussion of Stirling numbers of the first kind.

Like Gregory, Leibniz, Taylor, and Nicole, Euler saw the intimate connections between the calculus of finite differences and the calculus of differentiation and integration. His influential 1755 book on differential calculus, *Institutiones Calculi Differentialis*,[23] began with a chapter on finite differences. The second chapter on the use of differences in the summation of series discussed examples such as those in Nicole's work. In the second part of his book, Euler devoted the first chapter[24] to Newton's transformation (10.3). He gave a proof different from Newton's and from Montmort's; this in turn led him to a further generalization of the formula. Suppose

$$S = a_0 + a_1 x + a_2 x^2 + a_3 x^3 + a_4 x^4 + \cdots .$$

Euler then got the generalization

$$A_0 a_0 + A_1 a_1 x + A_2 a_2 x^2 + A_3 a_3 x^3 + \cdots$$

$$= A_0 S + \Delta A_0 \frac{x}{1!} \frac{dS}{dx} + \Delta^2 A_0 \frac{x^2}{2!} \frac{d^2 S}{dx^2} + \Delta^3 A_0 \frac{x^3}{3!} \frac{d^3 S}{dx^3} + \cdots . \qquad (10.15)$$

The Newton–Montmort formula followed by taking $a_0 = a_1 = a_2 = \cdots = 1$.

## 10.2 Newton's Transformation

In his 1684 *Matheseos*, Newton attempted to change slowly convergent series into more rapidly convergent ones. He considered the method of taking differences of the coefficients, but it was not until a little later that he arrived at the explicit and useful transformation (10.3) contained in the *De Computo*. He wrote the initial series as

$$v = At \pm Bt^2 + Ct^3 \pm Dt^4 + Et^5 \pm Ft^6 \, \&c. \qquad (10.16)$$

[21] Gauss (1813).
[22] Lagrange (1771).
[23] Eu. I-10. E 212.
[24] Eu. I-10$_2$, chapter 1. E 212.

Newton explained his transformation:[25]

Here $A, B, C, \ldots$ are to denote the coefficients of the terms whose ultimate ratio, if the series be extended infinitely, is one of equality, and 1 to $t$ is the remaining ratio of the terms; while the sign $\pm$ is ambiguous and the converse of the sign $\mp$. Collect the first differences $b, b_2, b_3, \ldots$ of the terms $A, B, C, \ldots$; then their second ones $c, c_2, c_3, \ldots$, third ones $d, d_2, d_3, \ldots$, and any number of following ones. Collect these, however, by always taking a latter term from the previous one: $B$ from $A$, $C$ from $B, \ldots$; $b_2$ from $b$, $b_3$ from $b_2, \ldots$; $d_2$ from $d$, $d_3$ from $d_2, \ldots$, and so on. Then make $\frac{t}{1 \mp t} = z$ and when the signs are appropriately observed there will be

$$v = Az \mp bz^2 + cz^3 \mp dz^4 + ez^5 \mp fz^6 + \cdots.$$

He took the differences in reverse order of the modern convention. He had $A - B$, $B - C, C - D, \ldots$ for the first differences instead of $B - A, C - B, D - C, \ldots$ and similarly for the higher-order differences. The revised version of the *De Computo* did not include a proof but notes of an earlier version suggest the following iterative argument:

$$
\begin{aligned}
v &= z(1 \mp t)(A \pm Bt + Ct^2 \pm Dt^3 + Et^4 \pm Ft^5 \cdots) \\
&= z(A \mp (A - B)t - (B - C)t^2 \mp (C - D)t^3 - (D - E)t^4 \mp \cdots) \\
&= Az \mp z((A - B)t \pm (B - C)t^2 + (C - D)t^3 \pm (D - E)t^4 + \cdots). \quad (10.17)
\end{aligned}
$$

Now in (10.17), the last series in parentheses is of the same form as the original series except that the coefficients are the differences of the coefficients of the original. So the procedure can be repeated to give

$$
\begin{aligned}
v &= Az \mp z(z(A - B) \mp z((A - 2B + C)t \\
&\quad \pm (B - 2C + D)t^2 + (C - 2D + E)t^3 \pm \cdots)) \\
&= Az \mp (A - B)z^2 + (A - 2B + C)z^3 \mp \cdots. \quad (10.18)
\end{aligned}
$$

The final formula results from an infinite number of applications of this procedure; Newton applied this formula to the logarithmic and arctangent series. In the case of the logarithm, the transformation amounted to the equation $\ln(1 + x) = -\ln\left(1 - \frac{x}{1+x}\right)$. Newton's main purpose was to use the transformation for numerical computation and this explains why he applied the transformation after the eighth term $\frac{1}{15}x^{15}$ in (10.5) rather than immediately at the outset. Note also that the first step in (10.17) was an example of the summation by parts discussed explicitly by Taylor and later used by Abel to study the convergence of series

## 10.3  Montmort's Transformation

In his 1717 paper, *De Seriebus Infinitis Tractatus*, Montmort started with elementary examples, but toward the end of the paper he posed the problem of summing or transforming the series

---

[25] Newton (1967–1981) vol. 4, pp. 605–607.

$$S = \frac{a_0}{h} + \frac{a_1}{h^2} + \frac{a_2}{h^3} + \cdots . \tag{10.19}$$

He also discussed partial sums of this series, written as

$$S_0 = \frac{A_0}{h}, \quad S_1 = \frac{A_1}{h^2}, \ldots, S_p = \frac{A_p}{h^{p+1}}, \ldots, \tag{10.20}$$

where

$$A_0 = a_0, \quad A_1 = a_0 h + a_1, \quad A_2 = a_0 h^2 + a_1 h + a_2, \ldots \tag{10.21}$$

He noted that a simple relation existed between the differences of the sequence $A_0, A_1, A_2, \ldots$ and the sequence $a_0, a_1, a_2, \ldots$. He wrote down just the first three cases: For $q = h - 1$,

$$\Delta A_0 = h a_0 + \Delta a_0,$$
$$\Delta^2 A_0 = q h a_0 + h \Delta a_0 + \Delta^2 a_0,$$
$$\Delta^3 A_0 = q^2 h a_0 + q h \Delta a_0 + h \Delta^2 a_0 + \Delta^3 a_0, \text{ etc.}$$

It is not difficult to write out the general relation from these examples. He then proved the result that if for $k \geq l$, $\Delta^k a_0 = 0$, then

$$\Delta^k A_0 = q^{k-l} \Delta^l A_0. \tag{10.22}$$

In fact, he verified this for $l = 1$ and 2, but noted that was sufficient to see that the result was true in general. Montmort then proceeded to evaluate the partial sums of (10.19) under the assumption that $\Delta^l a_0 = 0$ for some positive integer $l$. The basic result used here and in other examples was that for any sequence $b_0, b_1, b_2, \ldots$ and a positive integer $p$,

$$b_p = b_0 + \binom{p}{1} \Delta b_0 + \binom{p}{2} \Delta^2 b_0 + \cdots + \binom{p}{p} \Delta^p b_0. \tag{10.23}$$

For example, to evaluate $A_p$ for $p > l$ and $\Delta^l a_0 = 0$, he employed (10.22) to obtain

$$A_p = \left( A_0 + \binom{p}{1} \Delta A_0 + \cdots + \binom{p}{l-1} \Delta^{l-1} A_0 \right)$$
$$+ \binom{p}{l} \Delta^l A_0 + \binom{p}{l+1} \Delta^{l+1} A_0 + \cdots + \binom{p}{p} \Delta^p A_0$$
$$= A_0 + \binom{p}{1} \Delta A_0 + \cdots + \binom{p}{l-1} \Delta^{l-1} A_0$$
$$+ \frac{\Delta^l A_0}{q^l} \left( \binom{p}{l} q^l + \binom{p}{l+1} q^{l+1} + \cdots + \binom{p}{p} q^p \right)$$

$$= A_0 + \binom{p}{1} \Delta A_0 + \cdots + \binom{p}{l-1} \Delta^{l-1} A_0$$

$$+ \frac{\Delta^l A_0}{q^l} \left( h^p - 1 - \binom{p}{1} q - \cdots - \binom{p}{l-1} q^{l-1} \right). \qquad (10.24)$$

As an application of this formula, he gave the sum of the finite series

$$1 \cdot 3 + 3 \cdot 3^2 + 6 \cdot 3^3 + 10 \cdot 3^4 + 15 \cdot 3^5 + 21 \cdot 3^6.$$

Here $h = \frac{1}{3}$, $q = -\frac{2}{3}$, $p = 5$, and $\Delta^3 a_0 = 0$. Next, as a corollary, Montmort stated without proof the transformation formula

$$\sum_{k=0}^{\infty} \frac{a_k}{h^{k+1}} = \sum_{k=0}^{\infty} \frac{\Delta^k a_0}{(h-1)^{k+1}}. \qquad (10.25)$$

Of course, (10.25) is Newton's formula. Since equation (10.23) was Montmort's basic formula, we may assume that he applied it to give a formal calculation to justify (10.25). Indeed, this is easy to do:

$$\sum_{k=0}^{\infty} \frac{a_k}{h^{k+1}} = \sum_{k=0}^{\infty} \frac{1}{h^{k+1}} \sum_{j=0}^{k} \binom{k}{j} \Delta^j a_0$$

$$= \sum_{k=0}^{\infty} \Delta^k a_0 \sum_{j=0}^{\infty} \binom{k+j}{j} \frac{1}{h^{k+1+j}}$$

$$= \sum_{k=0}^{\infty} \frac{\Delta^k a_0}{(h-1)^{k+1}},$$

where the last step used the binomial theorem for negative integer exponents.

Unfortunately, he did not give any interesting examples of this formula. The three cases he explicitly mentioned follow from the binomial theorem just as easily. We mention that Zhu Shijie (c. 1260–1320), also known as Chu Shi-Chieh, knew (10.23) and used it to sum finite series in his *Siyuan Yujian* of 1303.[26]

## 10.4 Euler's Transformation Formula

As we mentioned earlier, Newton did not publish his transformation formula. It is not certain whether or not Euler saw Montmort's paper on this topic. In any case, Euler's approach differed from those of Newton and Montmort. Euler's proof of (10.3) applied the change of variables in the first step. We present the proof as Euler set it out.[27] He let

$$S = ax + bx^2 + cx^3 + dx^4 + ex^5 + \&c.$$

[26] Hoe (2007) p. 401.
[27] Eu. I-10$_2$ sections 2 and 3. E 212.

and let $x = \frac{y}{1+y}$ and replaced the powers of $x$ by the series

$$x = y - y^2 + y^3 - y^4 + y^5 - y^6 + \&c.,$$
$$x^2 = y^2 - 2y^3 + 3y^4 - 4y^5 + 5y^6 - 6y^7 + \&c.,$$
$$x^3 = y^3 - 3y^4 + 6y^5 - 10y^6 + 15y^7 - 21y^8 + \&c.,$$
$$x^4 = y^4 - 4y^5 + 10y^6 - 20y^7 + 35y^8 - 56y^9 + \&c.$$

Thus,

$$S = ay - ay^2 + ay^3 - ay^4 + ay^5 - \&c.$$
$$+b - 2b \quad +3b - 4b$$
$$+c \quad -3c + 6c$$
$$+d - 4d.$$

Note that the coefficients of the various powers of $y$ were presented in columns. Since $y = \frac{x}{1-x}$,

$$S = a\frac{x}{1-x} + (b-a)\frac{x^2}{(1-x)^2} + (c-2b+a)\frac{x^3}{(1-x)^3} + \&c.,$$

yielding the transformation formula.

Note that the series for $x$ is the geometric series, while the series for $x^2, x^3, \ldots$ can be obtained by the binomial theorem or by the differentiation of the series for $x$. In fact, Euler must have had differentiation in mind, since his proof of the second transformation formula (10.15) was obtained[28] by writing the right-hand side of (10.15) as

$$\alpha S + \beta\frac{x}{1!}\frac{dS}{dx} + \gamma\frac{x^2}{2!}\frac{d^2 S}{dx^2} + \delta\frac{x^3}{3!}\frac{d^3 S}{dx^3} + \cdots$$

and then substituting the series for $S$, $\frac{dS}{dx}$, $\frac{d^2 S}{dx^2}$, .... By equating the coefficients of the various powers of $x$, he found $\alpha = A_0$, $\beta = A_1 - A_0$, $\gamma = A_2 - 2A_1 + A_0$, and so on.

Euler gave several examples of these formulas in his differential calculus book. At times writing $xx$ and at other times using $x^2$, he considered the problem of summing the series

$$1x + 4xx + 9x^3 + 16x^4 + 25x^5 + \text{etc.} \tag{10.26}$$

The first and second differences of the coefficients $1, 4, 9, 16, 25, \ldots$ were $3, 5, 7, 9, \ldots$ and $2, 2, 2, \ldots$. Therefore, the third differences were zero, and by equation (10.3), the sum of the series (10.26) came out to be

---

[28] Eu. I-10₂, section 26. E 212.

$$\frac{x}{1-x} + \frac{3x^2}{(1-x)^2} + \frac{2x^3}{(1-x)^3} = \frac{x+xx}{(1-x)^3}. \tag{10.27}$$

To sum the finite series

$$1x + 4x^2 + 9x^3 + 16x^4 + \cdots + n^2 x^n, \tag{10.28}$$

Euler subtracted $(n+1)^2 x^{n+1} + (n+2)^2 x^{n+2} + \cdots$ from (10.26), the series in the previous example. He found the sum of this infinite series to be

$$\frac{x^{n+1}}{1-x}\left( (n+1)^2 + (2n+3)\frac{x}{1-x} + \frac{2x^2}{(1-x)^2} \right). \tag{10.29}$$

Euler observed the general rule, already established by Montmort, that if a power series had coefficients such that $\Delta^n A_0 = 0$ for $n \geq k$, then the series would sum to a rational function.

When Euler discovered a new method of summation or a new transformation of series, he applied it to divergent as well as convergent series. He believed that divergent series could be studied and used in a meaningful way. He explained that whenever he assigned a sum to a divergent series by a given method, he arrived at the same sum by alternative methods, leading him to conclude that divergent series could be legitimately summed. Applying the differences method, he found the sums of various divergent series, including $1 - 4 + 9 - 16 + 25 - \cdots$ ; $\ln 2 - \ln 3 + \ln 4 - \ln 5 + \cdots$ ; and $1 - 2 + 6 - 24 + 120 - 720 + \cdots$. Observe that the terms in the third example are $1!, 2!, 3!, 4!, \ldots$. This was one of Euler's favorite divergent series. By taking twelve terms of the transformed series and using (10.3), he found the sum to be 0.4036524077.[29] He must have later realized that this sum was very inaccurate, since he reconsidered it in a 1760 paper.[30] According to Jacobi's remarks, included in the Euler Archive in connection with this paper, this 1760 paper was read to the Berlin Academy in October, 1746. This information tallies with Euler's letter to Goldbach of August 7, 1746, in which he discussed the series $1 - 1 + 2 - 6 + 24 - 120 + \cdots$ ; to obtain the value of this series, he took the function

$$f(x) = 1 - 1!x + 2!x^2 - 3!x^3 + \cdots$$

and proceeded to express $f(x)$ as an integral and as a continued fraction:

$$f(x) = \frac{1}{x}\int_0^x \frac{1}{t}e^{-\frac{1}{t}}\,dt;$$

$$f(x) = \frac{1}{1+}\frac{x}{1+}\frac{x}{1+}\frac{2x}{1+}\frac{2x}{1+}\frac{3x}{1+}\frac{3x}{1+} \text{ etc.}$$

Euler then computed the value of the original series $f(1)$. To approximately evaluate the integral, he divided the interval $[0,1]$ into ten equal parts and used the

---

[29] ibid. section 10.
[30] Eu. I-14 pp. 585–617. E 247.

| 2 | | 5 | | 10 | | 17 | | 26 | | 37 |
|---|---|---|---|----|---|----|---|----|---|----|
| | 3 | | 5 | | 7 | | 9 | | 11 | |
| | | 2 | | 2 | | 2 | | 2 | | |

Figure 10.1 Difference table.

trapezoidal rule. By this method, $f(1) = 0.59637255$. The continued fraction, on the other hand, gave $f(1) = 0.59637362123$, correct to eight decimal places. Even earlier, in a letter to Goldbach of August 7, 1745,[31] Euler wrote that he had worked out the continued fraction for the divergent series $f(1)$ and found the value to be approximately 0.5963475922, adding the remark that in a small dispute with Niklaus I Bernoulli about the value of a divergent series, he himself had argued that all series such as $f(1)$ must have a definite value.[32]

In his differential calculus book, Euler gave a few applications of his more general formula (10.15), including the derivation of the exponential generating function of a sequence whose third difference was zero.[33] He began with the difference table in Figure 10.1.

From this, he derived

$$2 + 5x + \frac{10x^2}{2} + \frac{17x^3}{6} + \frac{26x^4}{24} + \text{etc.} = e^x(2 + 3x + xx) = e^x(1 + x)(2 + x).$$

More generally, he noted that when $S = e^x$, the result was

$$A_0 + A_1\frac{x}{1} + A_2\frac{x^2}{1\cdot 2} + A_3\frac{x^3}{1\cdot 2\cdot 3} + A_4\frac{x^4}{1\cdot 2\cdot 3\cdot 4} + \text{etc.}$$

$$= e^x\left(A_0 + \frac{x}{1}\Delta A_0 + \frac{x^2}{1\cdot 2}\Delta^2 A_0 + \frac{x^3}{1\cdot 2\cdot 3}\Delta^3 A_0 + \frac{x^4}{1\cdot 2\cdot 3\cdot 4}\Delta^4 A_0 + \text{etc.}\right).$$

In fact, this result is equivalent to $(1 + \Delta)^n A_0 = A_n$, that is (10.23), as is quickly verified by writing $e^x$ as a series and multiplying the two series. Jacobi gave a very interesting application of Euler's general transformation formula (10.15) to the derivation of the Pfaff–Gauss transformation for hypergeometric functions and we discuss this in Chapter 23.

## 10.5 Stirling's Transformation Formulas

Stirling's new generalization of Newton's transformation of the arctangent series (10.4) was a particular case of a hypergeometric transformation discovered by

---

[31] Fuss (1968) vol. I, pp. 323–328.
[32] For Euler's comments on summing divergent series, see his letters to Niklaus I Bernoulli: Eu. IVA-2 pp. 579–643, especially pp. 589–590 and pp. 604–606.
[33] Eu. I-10$_2$ section 27. E 212.

Pfaff in 1797.[34] Stirling stated his formula[35] as proposition 7 of his 1730 book and his proof made remarkable use of difference equations. Beginning with a series satisfying a certain difference equation, or recurrence relation, he then showed that the transformed series satisfied the same difference equation. Adhering closely to Stirling's exposition, we state the theorem: If the successive terms $T$ and $T'$ of a series $S$ satisfied $(z - n)T + (m - 1)zT' = 0$, then $S$ could be transformed to

$$S = \frac{m-1}{m}T + \frac{n}{z} \times \frac{A}{m} + \frac{n+1}{z+1} \times \frac{B}{m} + \frac{n+2}{z+2} \times \frac{C}{m} + \frac{n+3}{z+3} \times \frac{D}{m} + \text{etc.}$$

Stirling's notation made unusual use of the symbol $z$. In the equation $(z - n)T + (m - 1)zT' = 0$, $T$ and $T'$ represented any two successive terms of the series $S$. The value of $z$ changed by one when Stirling moved from one pair to the next. To see how this worked, take $S = T_0 + T_1 + T_2 + T_3 + \cdots$. The initial relation (between the first two terms) could then be expressed as $(z - n)T_0 + (m - 1)zT_1 = 0$ and, in general, the relation between two successive terms would be

$$T_{k+1} = -\frac{z - n + k}{(z + k)(m - 1)}T_k. \tag{10.30}$$

Thus, the relation between successive terms produced the series

$$S = T_0\left(1 + \frac{z - n}{z(1 - m)} + \frac{(z - n)(z - n + 1)}{z(z + 1)(1 - m)^2} + \cdots\right). \tag{10.31}$$

In Stirling's notation, $A, B, C, \ldots$ each represented the previous term so that

$$A = \frac{m-1}{m}T\left(= \frac{m-1}{m}T_0\right), B = \frac{n}{z} \times \frac{A}{m}, C = \frac{n+1}{z+1} \times \frac{B}{m}, \text{etc.}$$

Hence, the transformed series could be written as

$$T_0\left(1 + \frac{n}{z} \cdot \frac{1}{m} + \frac{n(n+1)}{z(z+1)} \cdot \frac{1}{m^2} + \frac{n(n+1)(n+2)}{z(z+1)(z+2)} \cdot \frac{1}{m^3} + \cdots\right). \tag{10.32}$$

Thus, Stirling's transformation formula is equivalent to the statement that the series in (10.31) equals the series in (10.32). We use modern notation to derive Stirling's difference equation. Let

$$S_k = \sum_{n=k}^{\infty} T_n = T_k y_k$$

so that

$$T_{k+1}y_{k+1} = S_{k+1} = S_k - T_k = T_k(y_k - 1). \tag{10.33}$$

[34] Pfaff (1797a).
[35] Stirling and Tweddle (2003) pp. 57–60.

By relation (10.30), (10.33) would become

$$(m-1)y_k + y_{k+1} - \frac{n}{z+k}y_{k+1} - m + 1 = 0. \tag{10.34}$$

Since Stirling wrote $y$ and $y'$ for any two successive $y_k$ and $y_{k+1}$, he could write $z$ instead of $z+k$ to get the recurrence relation

$$(m-1)y + y' - \frac{n}{z}y' - m + 1 = 0. \tag{10.35}$$

In proving his transformation formula, Stirling assumed that

$$y = a + \frac{b}{z} + \frac{c}{z(z+1)} + \frac{d}{z(z+1)(z+2)} + \cdots, \tag{10.36}$$

so he had

$$y' = a + \frac{b}{z+1} + \frac{c}{(z+1)(z+2)} + \frac{d}{(z+1)(z+2)(z+3)} + \cdots. \tag{10.37}$$

Before substituting these expressions for $y$ and $y'$ in (10.35), Stirling observed that

$$y' = a + \frac{b}{z} + \frac{c-b}{z(z+1)} + \frac{d-2c}{z(z+1)(z+2)} + \frac{e-3d}{z(z+1)(z+2)(z+3)} + \cdots, \tag{10.38}$$

an equation we can easily see to be true by taking the term-by-term difference of the series for $y$ and $y'$ in (10.36) and (10.37). Next, he substituted the expression (10.38) for $y'$ in (10.35) but used (10.37) for the term $-\frac{n}{z}y'$ in (10.35). After these substitutions, equation (10.35) became

$$ma - m + 1 + \frac{mb - na}{z} + \frac{mc - (n+1)b}{z(z+1)} + \frac{md - (n+2)c}{z(z+1)(z+2)} + \text{etc.} = 0.$$

On setting the like terms equal to zero, he got

$$a = \frac{m-1}{m}, \; b = \frac{n}{m}a, \; c = \frac{n+1}{m}b, \; d = \frac{n+2}{m}c, \ldots.$$

When these values were substituted back in $y$, Stirling got his result.
Stirling applied this transformation to the approximate summation of the series

$$\frac{1}{2}\left(1 + \frac{1}{3} + \frac{1 \cdot 2}{3 \cdot 5} + \frac{1 \cdot 2 \cdot 3}{3 \cdot 5 \cdot 7} + \frac{1 \cdot 2 \cdot 3 \cdot 4}{3 \cdot 5 \cdot 7 \cdot 9} + \cdots\right). \tag{10.39}$$

Now this is the series one gets upon taking $t = 1$ in Newton's formula (10.4), and its value is therefore $\frac{\pi}{4}$. It is interesting to note that, after posing the problem of approximately summing the series (10.39), Newton had abruptly ended the second chapter of his unpublished *Matheseos* of 1684 with the word "Inveniend" (to be found).[36] Thus, although the *Matheseos* went unpublished, Stirling took up the very problem left pending by Newton. Stirling began by adding the first twelve terms of the series and applied his transformation (10.13) to the remaining (infinite) part of the series, yielding

$$\frac{12!}{3 \cdot 5 \cdots 25} \left( 1 - \frac{1}{27} + \frac{1 \cdot 3}{27 \cdot 29} - \frac{1 \cdot 3 \cdot 5}{27 \cdot 29 \cdot 31} + \cdots \right). \qquad (10.40)$$

To approximate this sum, he took the first twelve terms of this series and found the value of $\frac{\pi}{4}$ as 0.78539816339. Since the terms are alternating and decreasing, Stirling could also have easily determined bounds on the error by using results of Leibniz dating from the 1680s. However, Leibniz's results, though communicated to Johann Bernoulli in 1713–14,[37] went unpublished for a long period.

Proposition 8[38] of Stirling's *Methodus Differentialis* was a transformation of what we now call a generalized hypergeometric series, noteworthy as an important particular case of a formula discovered by Kummer in 1836[39] in the course of his efforts to generalize Gauss's 1813 theory of hypergeometric series. After having shown the manner in which Stirling stated his propositions, we now state Stirling's eighth proposition in a form more immediately understandable to modern readers:

$$\frac{1}{z-n} + \frac{z-m}{z(z-n+1)} + \frac{(z-m)(z-m+1)}{z(z+1)(z-n+2)} + \frac{(z-m)(z-m+1)(z-m+2)}{z(z+1)(z+2)(z-n+3)} + \cdots$$

$$= \frac{1}{m} + \frac{n}{z(m+1)} + \frac{n(n+1)}{z(z+1)(m+2)} + \frac{n(n+1)(n+2)}{z(z+1)(z+2)(m+3)} + \cdots.$$

$$(10.41)$$

Let $S_k$ denote the sum of the series on the left after the first $k$ terms have been removed:

$$S_k = \frac{(z-m)\cdots(z-m+k-1)}{z(z+1)\cdots(z+k-1)}$$

$$\times \left( \frac{1}{z-n+k} + \frac{z-m+k}{(z+k)(z-n+k+1)} \right.$$

$$\left. + \frac{(z-m+k)(z-m+k+1)}{(z+k)(z+k+1)(z-n+k+2)} + \cdots \right).$$

Denote the sum in parentheses by $y_k$. It is simple to check that

$$y_k - y_{k+1} + \frac{m}{z+k} y_{k+1} - \frac{1}{z-n+k} = 0,$$

[36] Newton (1967–1981) vol. IV, pp. 552–553.
[37] Leibniz (1971) vol. 3, pp. 922–923.
[38] Stirling and Tweddle (2003) pp. 60–64.
[39] Kummer (1836).

a relation Stirling wrote as

$$y - y' + \frac{m}{z}y' = \frac{1}{z-n}. \tag{10.42}$$

He then assumed that for some $a, b, c, d, \ldots$

$$y = \frac{a}{m} + \frac{b}{(m+1)z} + \frac{c}{(m+2)z(z+1)} + \frac{d}{(m+3)z(z+1)(z+2)} + \cdots;$$

substituting this into the left side of (10.42), after a simple calculation, he obtained

$$y - y' + \frac{m}{z}y' = \frac{a}{z} + \frac{b}{z(z+1)} + \frac{c}{z(z+1)(z+2)} + \frac{d}{z(z+1)(z+2)(z+3)} + \cdots. \tag{10.43}$$

Stirling applied Taylor's formula (10.12) to the right side of (10.42) to get

$$\frac{1}{z-n} = \frac{1}{z} + \frac{n}{z(z+1)} + \frac{n(n+1)}{z(z+1)(z+2)} + \frac{n(n+1)(n+2)}{z(z+1)(z+2)(z+3)} + \cdots \tag{10.44}$$

and equated the coefficients in (10.43) and (10.44) to obtain

$$a = 1, \; b = n, \; c = n(n+1), \; d = n(n+1)(n+2), \ldots.$$

This proves the transformation formula (10.41). As we mentioned, a century later Kummer obtained a more general result but he did not seem to have been aware of Stirling's work.

Newton pointed out in his second letter for Leibniz,[40] of October 24, 1676, that $\sin^{-1}\frac{1}{2}$ was more convenient than $\sin^{-1}\frac{\sqrt{2}}{2}$ for computing $\pi$ because it converged rapidly. Clearly, $\sin^{-1} 1$ would not give a result even as good as $\sin^{-1}\frac{\sqrt{2}}{2}$. Nevertheless, Stirling's transformation was so powerful that, when applied to $\sin^{-1} 1$, it caused it to converge rapidly enough to be useful for computation. Stirling summed up the first twelve terms directly and then applied the transformation to the remainder, thereby achieving the value of $\pi$ to eight decimal places.

## 10.6  Nicole's Examples of Sums

The method of finite differences is also useful in the summation of series, as noted by Mengoli, Leibniz, Jakob Bernoulli, and Montmort. Nicole, student of Montmort, wished to establish a new subject devoted to the calculus of finite differences. Analogous to integration in the calculus, summation of series had to be developed within this new subject. Nicole attacked this problem by summing examples of certain kinds of series and Montmort gave similar examples. The basic idea was that, given a function $g(x)$ such that $g(x+h) - g(x) = f(x)$, the sum would be

[40] Newton (1959–1960) vol. II, p. 139.

$$f(a) + f(a + h) + \cdots + f(a + (n - 1)h) = g(a + nh) - g(a).$$

Nicole's examples took $f(x) = (x)_m$ or $f(x) = \frac{1}{(x)_m}$, where $(x)_m$ was given by (10.6). So, by (10.7) and (10.8), $g(x)$ could be taken to be, in the first case,

$$\frac{1}{h(m + 1)}(x - h)_{m+1}$$

and, in the second case,

$$\frac{-1}{h(m - 1)} \cdot \frac{1}{(x)_{m-1}}.$$

Thus, the sum $g(a + nh) - g(a)$ would be, in the first case,

$$\frac{1}{h(m + 1)}\left((a + (n - 1)h)_{m+1} - (a - h)_{m+1}\right)$$

and, in the second case,

$$\frac{1}{h(m - 1)}\left(\frac{1}{(a)_{m-1}} - \frac{1}{(a + nh)_{m-1}}\right).$$

Although he did not provide the general formulas, Nicole worked out special cases. When the sum was indefinite, Nicole wrote that the sum of the $f(x)$ was $g(x)$.[41] For example, the sum of the terms $(x + 2)(x + 4)(x + 6)$ was $\frac{x(x+2)(x+4)(x+6)}{8}$ because

$$\frac{(x + 2)(x + 4)(x + 6)(x + 8)}{8} - \frac{x(x + 2)(x + 4)(x + 6)}{8} = (x + 2)(x + 4)(x + 6).$$

In a similar manner, because

$$\frac{1}{x(x + 2)(x + 4)(x + 6)} = \frac{1}{6(x + 2)(x + 4)}\left(\frac{1}{x} - \frac{1}{x + 6}\right)$$

$$= \frac{1}{6x(x + 2)(x + 4)} - \frac{1}{6(x + 2)(x + 4)(x + 6)}, \quad (10.45)$$

he wrote that the sum of terms $\frac{1}{x(x+2)(x+4)(x+6)}$ would be $\frac{1}{6x(x+2)(x+4)}$.

As an application of the first type of sum, Nicole considered the series

$$1 \cdot 4 \cdot 7 \cdot 10 + 4 \cdot 7 \cdot 10 \cdot 13 + 7 \cdot 10 \cdot 13 \cdot 16 + 10 \cdot 13 \cdot 16 \cdot 19 + \text{ etc.}$$

He gave the general term as $(x + 3)(x + 6)(x + 9)(x + 12)$ and its integral (sum) as $\frac{x(x+3)(x+6)(x+9)(x+12)}{15}$. Now to find the constant of integration, note that the starting

[41] Nicole (1717).

value was $x = -2$ and the corresponding value of the integral was $\frac{(-2)\cdot 1\cdot 4\cdot 7\cdot 10}{15}$. Thus, Nicole got the value of the series as

$$\frac{x(x+3)(x+6)(x+9)(x+12)}{15} + \frac{2\cdot 1\cdot 4\cdot 7\cdot 10}{15}.$$

Similarly, he computed the infinite series

$$\frac{1}{1\cdot 3\cdot 5\cdot 7} + \frac{1}{3\cdot 5\cdot 7\cdot 9} + \frac{1}{5\cdot 7\cdot 9\cdot 11} + \frac{1}{7\cdot 9\cdot 11\cdot 13} +$$
$$\cdots + \frac{1}{x(x+2)(x+4)(x+6)} \text{ etc. } (10.46)$$

by using (10.45) and observing that

$$\frac{1}{x(x+2)(x+4)(x+6)} + \frac{1}{(x+2)(x+4)(x+6)(x+8)} + \text{ etc.}$$
$$= \frac{1}{6(x+2)(x+4)(x+6)}.$$

Since the original sum started at $x = 1$, Nicole gave its value as $\frac{1}{6\cdot 1\cdot 3\cdot 5} = \frac{1}{90}$. As another example, Nicole then considered a slightly more difficult sum:

$$\frac{4}{1\cdot 4\cdot 7\cdot 10\cdot 13\cdot 16} + \frac{49}{4\cdot 7\cdots 19} + \frac{225}{7\cdots 22} + \text{ etc.}$$

Note that the general term was $\frac{1}{36}\frac{(x+2)^2(x+3)^2}{x(x+3)\cdots(x+15)}$, where $x$ was replaced by $x+3$ as one moved from one term to the next. Nicole wrote the numerator as

$$\frac{1}{36}\left(A + Bx + Cx(x+3) + Dx(x+3)(x+6) + Ex(x+3)(x+6)(x+9)\right)$$

and found the coefficients $A$ to $E$ by taking $x = 0, -3, -6,$ and $-9$. Nicole then expressed the general term as

$$\frac{1}{36}\left(\frac{A}{x(x+3)\cdots(x+15)} + \cdots + \frac{E}{(x+12)(x+15)}\right)$$

and wrote the integral as

$$\frac{1}{36}\left(\frac{A}{15x(x+3)\cdots(x+12)} + \cdots + \frac{E}{3(x+12)}\right). \quad (10.47)$$

He found the sum by taking $x = 1$ in the sum, or integral, as he called it.

Series such as (10.46) are called inverse factorial series. Nicole treated these examples in a 1717 paper on finite differences.[42] He further developed inverse factorial

---

[42] Nicole (1717).

series in papers of 1723, 1724, and 1727. In his 1727 paper,[43] he gave a generalization of (10.12):

$$\frac{1}{b-a} = \frac{1}{b} + \frac{a}{b(b-a)} = \frac{1}{b} + \frac{a}{b(b+c)} + \frac{a(a+c)}{b(b+c)(b-a)}$$

$$= \frac{1}{b} + \frac{a}{b(b+c)} + \frac{a(a+c)}{b(b+c)(b+d)} + \frac{a(a+c)(a+d)}{b(b+c)(b+d)(b-a)}. \quad (10.48)$$

In modern terminology, with $c_0 = 0$, we can write (10.48) as

$$\frac{1}{b-a} = \frac{1}{b}\left(1 + \sum_{k=1}^{n} \frac{a(a+c_1)\cdots(a+c_{k-1})}{(b+c_1)(b+c_2)\cdots(b+c_k)} + \frac{a(a+c_1)\cdots(a+c_n)}{(b+c_1)\cdots(b+c_n)(b-a)}\right).$$

$$(10.49)$$

In a letter to Goldbach[44] dated July 26, 1746, Euler wrote, without reference to Nicole, that he had an infinite version of (10.49): Taking $c_i > 0$, $i = 1,2,3,\ldots$, and with $\frac{1}{c_1} + \frac{1}{c_2} + \frac{1}{c_3} + \cdots$ divergent,

$$\frac{1}{b-a} = \frac{1}{b} + \frac{a}{b(b+c_1)} + \frac{a(a+c_1)}{b(b+c_1)(b+c_2)} + \cdots.$$

As a special case, he took $c_1, c_2, c_3, \ldots$ to be primes starting with $1,2,3,5,\ldots$, $b = 4$, and $a = 3$ to obtain

$$1 = \frac{1}{4} + \frac{3}{4\cdot 5} + \frac{3\cdot 4}{4\cdot 5\cdot 7} + \frac{3\cdot 4\cdot 6}{4\cdot 5\cdot 7\cdot 9} + \cdots.$$

Euler gave no proof in his letter to Goldbach, but in his reply,[45] Goldbach reproduced, without reference, the argument of Nicole, but without Euler's condition that $\sum \frac{1}{c_n}$ must be divergent. Although Euler's letter did not refer to Nicole, he wrote that (10.49) was a generalization of Stirling's result

$$\frac{1}{b-a} = \frac{1}{b} + \frac{a}{b(b+1)} + \frac{a(a+1)}{b(b+1)(b+2)} + \frac{a(a+1)(a+2)}{b(b+1)(b+2)(b+3)} + \cdots,$$

given as Example 1 of Proposition 5[46] of Stirling's *Methodus Differentialis*. It is not clear why Euler specified that the series $\sum \frac{1}{c_i}$ must diverge, since clearly one could take, for example, $c_n = n^2$.

[43] Nicole (1727).
[44] Fuss (1968) vol. I, pp. 388–392, especially p. 392.
[45] ibid. pp. 394–396.
[46] Stirling and Tweddle (2003) p. 52.

## 10.7  Stirling Numbers

In the introduction to his *Methodus Differentialis*, Stirling explained that the series satisfying difference equations were best expressed by using terms of the factorial or inverse factorial form $z(z-1)\cdots(z-m+1)$, or $\frac{1}{z(z+1)\cdots(z+m-1)}$, instead of $z^m$ or $\frac{1}{z^m}$. He defined the Stirling numbers in order to facilitate the conversion of series expressed in powers of $z$ into series with terms in factorial form. He gave two tables of numbers, the second consisting of what we now call the signless Stirling numbers of the first kind:[47]

| | | | | | | | | |
|---|---|---|---|---|---|---|---|---|
| 1 | | | | | | | | |
| 1 | 1 | | | | | | | |
| 2 | 3 | 1 | | | | | | |
| 6 | 11 | 6 | 1 | | | | | |
| 24 | 50 | 35 | 10 | 1 | | | | |
| 120 | 274 | 225 | 85 | 15 | 1 | | | |
| 720 | 1764 | 1624 | 735 | 175 | 21 | 1 | | |
| 5040 | 13068 | 13132 | 6769 | 1960 | 322 | 28 | 1 | |
| 40320 | 109584 | 118124 | 67284 | 22449 | 4536 | 546 | 36 | 1 |

Stirling described how to construct the table: "Multiply the terms of this progression $n, 1+n, 2+n, 3+n$, etc. repeatedly by themselves, and let the results be arranged in the following table in order of the powers of the number $n$, only the coefficients having been retained." Thus, to get the fourth row take $n(n+1)(n+2)(n+3) = 6n + 11n^2 + 6n^3 + n^4$ and the coefficients will be the numbers 6, 11, 6, 1 in the fourth row.

Stirling applied these numbers to the expansion of $\frac{1}{z^{n+1}}$, $n = 1, 2, 3, \ldots$ as an inverse factorial series. The numbers in the first column then appeared as numerators in the expansion of $\frac{1}{z^2}$; the numbers in the second column appeared in that of $\frac{1}{z^3}$, and so on. He wrote down three expansions explicitly:

$$\frac{1}{z^2} = \frac{1}{z(z+1)}\left(1 + \frac{1}{z+2} + \frac{2}{(z+2)(z+3)} + \frac{6}{(z+2)(z+3)(z+4)} + \cdots\right),$$
(10.50)

$$\frac{1}{z^3} = \frac{1}{z(z+1)(z+2)}\left(1 + \frac{3}{z+3} + \frac{11}{(z+3)(z+4)} + \cdots\right),$$
(10.51)

$$\frac{1}{z^4} = \frac{1}{z(z+1)(z+2)(z+3)}$$
$$\times \left(1 + \frac{6}{z+4} + \frac{35}{(z+4)(z+5)} + \frac{225}{(z+4)(z+5)(z+6)} + \&c.\right).$$
(10.52)

---

[47] Stirling and Tweddle (2003) p. 29.

In his *Methodus*, Proposition 2, Example 6,[48] Stirling applied (10.50) to find an approximate value of $\sum_{k=1}^{\infty} \frac{1}{k^2}$; he observed that

$$\sum_{k=0}^{\infty} \frac{1}{(z+k)^2} = \sum_{k=0}^{\infty} \left( \frac{1}{(z+k)(z+k+1)} + \frac{1}{(z+k)(z+k+1)(z+k+2)} + \cdots \right)$$

$$= \sum_{k=0}^{\infty} \frac{1}{(z+k)(z+k+1)} + \sum_{k=0}^{\infty} \frac{1}{(z+k)(z+k+1)(z+k+2)} + \cdots .$$

$$(10.53)$$

Since the sums on the right-hand side of (10.53) can be evaluated by the Montmort-Nicole method mentioned earlier, the second of those sums would have the value

$$\sum_{k=0}^{\infty} \frac{1}{(z+k)(z+k+1)(z+k+2)}$$

$$= \sum_{k=0}^{\infty} \frac{1}{2} \left( \frac{1}{(z+k)(z+k+1)} - \frac{1}{(z+k+1)(z+k+2)} \right)$$

$$= \frac{1}{2z(z+1)}.$$

By treating the other infinite sums in a similar manner, Stirling could thus write that

$$\sum_{k=0}^{\infty} \frac{1}{(z+k)^2} = \frac{1}{z} + \frac{1}{2z(z+1)} + \frac{2}{3z(z+1)(z+2)} + \frac{6}{4z(z+1)(z+2)(z+3)} + \cdots$$

$$= \sum_{k=0}^{\infty} \frac{k!}{(k+1)z(z+1)\cdots(z+k)}. \qquad (10.54)$$

Stirling took $z = 13$ and then took thirteen terms of the right-hand side of (10.54) to get the approximation

$$\sum_{k=13}^{\infty} \frac{1}{k^2} \approx 0.079957427. \qquad (10.55)$$

He then added $\sum_{k=1}^{12} \frac{1}{k^2}$, a sum that approximately equals 1.564976638, to (10.55), to arrive at 1.644934065 as the approximate value for the series $1 + \frac{1}{4} + \frac{1}{9} + \frac{1}{16} +$ etc. Euler showed that this series was equivalent to $\frac{\pi^2}{6}$, implying that Stirling's value was correct to eight decimal places.

---

[48] ibid. p. 46.

To better understand Stirling's table, let us utilize Richard Stanley's notation[49] to denote by $c(m,k)$ the unsigned Stirling numbers defined by

$$z(z+1)(z+2)\cdots(z+m-1) = \sum_{k=1}^{m} c(m,k)z^k. \qquad (10.56)$$

When $m=4$, as Stirling observed, we get the numbers in the fourth row:

$$c(4,1)=6, \quad c(4,2)=11, \quad c(4,3)=6, \quad c(4,4)=1.$$

We set $c(m,0)=0$ and $c(m,k)=0$ for $k>m$. Thus, Stirling was basically stating that

$$\frac{1}{z^{n+1}} = \sum_{m=n}^{\infty} \frac{c(m,n)}{z(z+1)\cdots(z+m)}. \qquad (10.57)$$

Stirling did not write down a proof of (10.57), but it is obvious that he must have had one. Although we cannot know for sure, Stirling's argument may have gone something like this: First expand $\frac{1}{z^{n+1}}$ as a factorial series with unknown coefficients $b_{m,n}$:

$$\frac{1}{z^{n+1}} = \frac{b(n,n)}{z(z+1)\cdots(z+n)} + \frac{b(n+1,n)}{z(z+1)\cdots(z+n+1)} + \cdots + \frac{b(m,n)}{z\cdots(z+m)} + \cdots. \qquad (10.58)$$

Then multiply (10.58) by the left-hand side of (10.56) to obtain

$$\frac{z(z+1)\cdots(z+m-1)}{z^{n+1}} = b(n,n)(z+m)\cdots(z+n)$$

$$+\cdots+ b(m-1,n) + \frac{b(m,n)}{z+m} + \frac{b(m+1,n)}{(z+m)(z+m+1)} + \cdots. \qquad (10.59)$$

Substituting $\sum_{k=1}^{m} c(m,k)z^k$ for the numerator of the left-hand side of (10.59) produces

$$\frac{c(m,1)}{z^n} + \cdots + \frac{c(m,n)}{z} + \cdots + c(m,m)z^{m-n-1}.$$

When the right-hand side of (10.59) is expanded in powers of $z$ and $\frac{1}{z}$, then $\frac{1}{z}$ has $b(m,n)$ as its only coefficient; hence

$$b(m,n) = c(m,n).$$

[49] Stanley (2012) vol. I, pp. 26–27.

We note that the Stirling numbers of the first kind, $s(m,n)$, are defined by

$$s(m,n) = (-1)^{m-n} c(m,n). \qquad (10.60)$$

Stirling's first table[50] was used to construct what we now call Stirling numbers of the second kind:

$$
\begin{array}{ccccccccc}
1 & 1 & 1 & 1 & 1 & 1 & 1 & 1 & 1 \\
  & 1 & 3 & 7 & 15 & 31 & 63 & 127 & 255 \\
  &   & 1 & 6 & 25 & 90 & 301 & 966 & 3025 \\
  &   &   & 1 & 10 & 65 & 350 & 1701 & 7770 \\
  &   &   &   & 1 & 15 & 140 & 1050 & 6951 \\
  &   &   &   &   & 1 & 21 & 266 & 2646 \\
  &   &   &   &   &   & 1 & 28 & 462 \\
  &   &   &   &   &   &   & 1 & 36 \\
  &   &   &   &   &   &   &   & 1
\end{array}
$$

in which the first row represented the coefficients of the powers of $\frac{1}{z}$ in the expansion of $\frac{1}{z-1}$. The second row gave the coefficients of the powers of $\frac{1}{z}$ in the expansion of

$$\frac{1}{(z-1)(z-2)} = \frac{1}{z^2}\left(1 + \frac{1}{z} + \frac{1}{z^2} + \cdots\right)\left(1 + \frac{2}{z} + \frac{4}{z^2} + \cdots\right)$$
$$= \frac{1}{z^2} + \frac{3}{z^3} + \frac{7}{z^4} + \cdots.$$

Again using Stanley's notation,[51] the $m$th row was found by the expansion of

$$\frac{1}{(z-1)(z-2)\cdots(z-m)} = \sum_{n=m}^{\infty} \frac{S(n,m)}{z^n}. \qquad (10.61)$$

Stirling then gave an expansion of $z^m$ in terms of a factorial expansion using numbers of the second kind:

$$z^m = \sum_{k=1}^{m} S(m,k) z(z-1)\cdots(z-k+1) \qquad (10.62)$$

by giving examples for $m = 1, 2, \ldots, 5$. For example, the fourth column gave him the expansion for $z^4$:

$$z^4 = z + 7z(z-1) + 6z(z-1)(z-2) + z(z-1)(z-2)(z-3).$$

---

[50] Stirling and Tweddle (2003) p. 26.
[51] Stanley (2012) vol. I, p. 73.

Again, Stirling did not suggest a proof for formula (10.62), though it would appear that he must have had one in mind. Observe that (10.61) implies that for $n \geq m$,

$$\frac{z^n}{z(z-1)\cdots(z-m)} = S(m,m)z^{n-(m+1)}$$

$$+ S(m+1,m)z^{n-(m+2)} + \cdots + \frac{S(n,m)}{z} + \cdots. \quad (10.63)$$

Now we must prove (10.62); let us denote the coefficients on its right-hand side by $b(m,k)$. Then we can write the left-hand side of (10.63) as

$$\frac{\sum_{k=1}^{n} b(n,k)z(z-1)\cdots(z-m+1)}{z(z-1)\cdots(z-m)} = \frac{b(n,1)}{(z-1)\cdots(z-m)} + \cdots$$

$$+ \frac{b(n,m)}{z-m} + \cdots + b(n,n)\big(z - (m+1)\big)\cdots(z-n+1). \quad (10.64)$$

Comparing coefficients of $\frac{1}{z}$ on the right-hand sides of (10.63) and (10.64) produces $b(n,m) = S(n,m)$, verifying (10.62).

There is now a well-known formula for Stirling numbers of the second kind, of which Stirling may have been aware, namely:

$$S(m,n) = \frac{(-1)^n}{n!} \sum_{k=0}^{n} (-1)^k \binom{n}{k} k^m. \quad (10.65)$$

To validate this formula, take $f(z) = z^m$ in (9.24), which is the Gregory–Newton formula stated as Proposition 19 of the *Methodus Differentialis*, to obtain

$$z^m = \sum_{n=1}^{m} \frac{\Delta^n 0^m}{n!} z(z-1)\cdots(z-n+1), \quad (10.66)$$

because $\Delta^n 0^m = 0$ when $n > m$. Comparing (10.66) with (10.62), and using (9.26), the proof is complete:

$$S(m,n) = \frac{\Delta^n 0^m}{n!} = \frac{(-1)^n}{n!} \sum_{k=0}^{n} (-1)^k \binom{n}{k} k^m. \quad (10.67)$$

If we take the combinatorial interpretation of $S(m,n)$, rather than Stirling's definition given in (10.61), then we can say that (10.67) was derived by de Moivre in his "Mensura Sortis" of 1711/1712. I am indebted to James Pitman for pointing this out to me.

It is instructive to understand in more detail Stirling's construction of his tables. In general terms, for the signless numbers of the first kind $c(n,k)$, he wrote that one

*Series Transformation by Finite Differences*

should first obtain the first $(n-1)$ rows and then find the $n$th row by multiplying the polynomial for the $(n-1)$th row by $n-1+z$. We note that Stirling's description implies that

$$
\begin{aligned}
&\bigl(c(n-1,1)z + c(n-1,2)z^2 + \cdots + c(n-1,n-1)z^{n-1}\bigr)(n-1+z) \\
&= nc(n-1,1)z + \bigl(nc(n-1,2) + c(n-1,1)\bigr)z^2 + \cdots \\
&\quad + \bigl((n-1)c(n-1,k) + c(n-1,k-1)\bigr)z^k + \cdots + c(n-1,n-1)z^n \\
&= c(n,1)z + c(n,z)z^2 + \cdots + c(n,k)z^k + \cdots + c(n,n)z^n.
\end{aligned}
$$

Thus, if one takes $c(n,0)=0$, and $c(n,m)=0$ for $m>n$, then we arrive at the relations

$$
c(n,k) = c(n-1,k-1) + (n-1)\,c(n-1,k), \quad k=1,2,\ldots n. \tag{10.68}
$$

We thus perceive that Stirling must have known this relationship, though he did not explicitly mention it. Moreover, a similar recurrence relation holds for $S(n,m)$ with $n$ and $m$ positive integers and under the conditions $S(n,0)=0$ and $S(n,m)=0$ for $m>n$:

$$
S(n+1,m) = S(n,m-1) + mS(n,m). \tag{10.69}
$$

We follow Tweddle's argument[52] to verify (10.69), given in his notes on Stirling's work: Observe from (10.63) that $S(n+1,m)$ is the coefficient of $\frac{1}{z}$ in the expansion of

$$
\frac{z^{n+1}}{z(z-1)\cdots(z-m)}
$$

in powers of $z$ and $\frac{1}{z}$. This is the same coefficient as that of $\frac{1}{z^2}$ in the expansion of

$$
\frac{z^n}{z(z-1)\cdots(z-m)}.
$$

We see from (10.64), with $b(n,m)$ replaced by $S(n,m)$, that two terms, expanded in powers of $\frac{1}{z}$, will produce $\frac{1}{z^2}$:

$$
\begin{aligned}
&\frac{S(n,m-1)}{(z-m+1)(z-m)} + \frac{S(n,m)}{z-m} \\
&= \frac{S(n,m-1)}{z^2}\left(1-\frac{m-1}{z}\right)^{-1}\left(1-\frac{m}{z}\right)^{-1} + \frac{S(n,m)}{z}\left(1-\frac{m}{z}\right)^{-1} \\
&= \frac{S(n,m-1)}{z^2}\left(1+\frac{m-1}{z}+\cdots\right)\left(1+\frac{m}{z}+\cdots\right) + \frac{S(n,m)}{z}\left(1+\frac{m}{z}+\cdots\right).
\end{aligned}
$$

[52] Stirling and Tweddle (2003) p. 171.

The coefficient of $\frac{1}{z^2}$ is thus

$$S(n,m-1)+mS(n,m)$$

and (10.69) is proved. Tweddle noted[53] that "Apparently Stirling did not notice this fact which ... allows us to construct the table much more easily."

It is possible to extend the definition of Stirling numbers to negative integers, but when this is done for Stirling numbers of the first kind, the result is Stirling numbers of the second kind and vice versa. Interestingly, we will thus see that there is but one kind of Stirling numbers. Stirling was perhaps aware of this fact. The numbers $c(n,m)$ can be defined for positive as well as negative $m$ and $n$ by the initial conditions $c(0,0)=1$, $c(n,0)=c(0,m)=0$ when $n \neq 0$, $m \neq 0$ and the recurrence relation:

$$c(n,m)=c(n-1,m-1)+(n-1)c(n-1,m). \qquad (10.70)$$

Note that (10.70) and (10.68) actually represent the same recurrence relation, but with different initial conditions; in (10.70) $m$ and $n$ can take any integer values.

In the same manner, $S(n,m)$ can be defined for all integers $m$ and $n$ by the initial conditions $S(0,0)=1$, $S(n,0)=S(0,m)=0$, when $n \neq 0$, $m \neq 0$, and the recurrence relation

$$S(n,m)=S(n-1,m-1)+mS(n-1,m). \qquad (10.71)$$

Again, observe that (10.71) and (10.69) are actually identical, except that their initial conditions are different; (10.71) is defined for all integers $m$ and $n$.

We are now in a position to prove that the two kinds of Stirling numbers can be reduced to only one kind:

$$S(n,m)=c(-m,-n). \qquad (10.72)$$

Observe that by (10.70), $c(-m,-n)$ satisfies the recurrence relation

$$c(-m,-n)=c(-m-1,-n-1)+(-m-1)c(-m-1,-n). \qquad (10.73)$$

Let $a(n,m)=c(-m,-n)$ so that (10.73) can be written as

$$a(n+1,m+1)=a(n,m)+(m+1)a(n,m+1);$$

and by replacing $n+1$ by $n$ and $m+1$ by $m$ we get

$$a(n,m)=a(n-1,m-1)+m \cdot a(n-1,m). \qquad (10.74)$$

Therefore, $a(n,m)$ satisfies the same recurrence relation as $S(n,m)$ and has the same initial conditions; we can conclude that $a(n,m)=S(n,m)$ and (10.72) is proved.

[53] ibid. p. 171.

In his paper "Two notes on notation," Donald Knuth wrote, "... a rereading of Stirling's original treatment makes it clear that Stirling himself would not have found the duality law [(10.72)] at all surprising. From the very beginning, he thought of the numbers as two triangles hooked together in tandem." Knuth's paper[54] presents an interesting history of this duality law.

We also remark that Tweddle[55] has explained that Stirling, in an unpublished work, revealed that he was aware of a relation between Bernoulli numbers and Stirling numbers of the second kind. In his unpublished notes, Stirling referred to a formula of de Moivre:[56]

$$\sum_{k=0}^{\infty} \frac{1}{(z+k)^2} = \frac{1}{z} + \frac{1}{2z^2} + \sum_{k=1}^{\infty} \frac{B_{2k}}{z^{2m+1}}. \qquad (10.75)$$

Now de Moivre presented (10.75) in his *Supplement* to his 1730 *Miscellanea Analytica*. In the same *Supplement*, he discussed a method for calculating the Bernoulli numbers that was essentially the same as Jakob Bernoulli's own method. We discuss this formula in Chapter 18. But note that if Stirling's (10.61) is applied to his (10.54), the result is

$$\sum_{k=0}^{\infty} \frac{1}{(z+k)^2} = \sum_{k=0}^{\infty} \frac{k!}{k+1}(-1)^k \sum_{n=k}^{\infty} \frac{S(n,k)}{z^{n+1}}. \qquad (10.76)$$

Comparison of (10.75) and (10.76) reveals a relation between Bernoulli numbers and Stirling numbers of the second kind:

$$B_n = \sum_{k=1}^{n} \frac{(-1)^k k!}{k+1} S(n,k), \quad n \geq 2. \qquad (10.77)$$

Note that when $n$ is odd, $B_n = 0$, so that terms with even powers of $z$ in the denominators in the sum on the right-hand side of (10.75) do not appear. Now Stirling did not give formula (10.77), but rather remarked in connection with de Moivre's calculations that Bernoulli numbers could more easily be calculated by using Stirling numbers of the second kind.

The significance of the Stirling numbers was not fully realized until the twentieth century when they became very useful in combinatorics. In his 1939 book on the calculus of finite differences, the Hungarian mathematician Charles Jordan (1871–1959) wrote,[57] "Since Stirling's numbers are as important as Bernoulli's, or even more so, they should occupy a central position in the Calculus of Finite Differences. The demonstration of a great number of formulae is considerably shortened by using these numbers, and many new formulae are obtained by their aid; for instance, those which express differences by derivatives or vice versa."

---

[54] Knuth (1992).
[55] Tweddle (1988) pp. 15–16.
[56] de Moivre (1730b) p. 19.
[57] Jordan (1979) pp. vii–viii.

## 10.8   Lagrange's Proof of Wilson's Theorem

Lagrange was the first mathematician to investigate the arithmetical properties of Stirling numbers and he did so in the process of proving Wilson's theorem. This proposition, also found and published by al-Haytham in the tenth or eleventh century,[58] named after Edward Waring's best friend John Wilson (1741–1793), states that for $n > 1, (n-1)! +1$ is divisible by $n$ if and only if $n$ is prime. The statement of Wilson's theorem was also published in Waring's *Meditationes Algebraicae* of 1770.[59] Waring was certain of the truth of the theorem but was unable to prove it. Lagrange provided a proof,[60] using Stirling numbers, in 1771. For this purpose, he considered the product

$$(x+1)(x+2)(x+3)(x+4)\cdots(x+n-1)$$
$$= x^{n-1} + A'x^{n-2} + A''x^{n-3} + A'''x^{n-4} + \cdots + A^{(n-1)}.$$

We can see that the coefficients $A', A'', A''', \ldots A^{(n-1)}$ are in fact absolute values of Stirling numbers of the first kind, though Lagrange did not mention Stirling. Lagrange replaced $x$ by $x+1$ to get

$$(x+2)(x+3)(x+4)(x+5)\cdots(x+n)$$
$$= (x+1)^{n-1} + A'(x+1)^{n-2} + A''(x+1)^{n-3} + A'''(x+1)^{n-4} + \cdots + A^{(n-1)}.$$

It was then easy to see that

$$(x+n)(x^{n-1} + A'x^{n-2} + A''x^{n-3} + A'''x^{n-4} + \cdots + A^{(n-1)})$$
$$= (x+1)^n + A'(x+1)^{n-1} + A''(x+1)^{n-2} + A'''(x+1)^{n-3}$$
$$+ \cdots + A^{(n-1)}(x+1).$$

Expanding both sides of this equation in powers of $x$, Lagrange obtained

$$x^n + (n+A')x^{n-1} + (nA'+A'')x^{n-2} + (nA''+A''')x^{n-3} + \cdots$$
$$= x^n + (n+A')x^{n-1} + \left(\frac{n(n-1)}{2} + (n-1)A' + A''\right)x^{n-2}$$
$$+ \left(\frac{n(n-1)(n-2)}{2\cdot 3} + \frac{(n-1)(n-2)}{2}A' + (n-2)A'' + A'''\right)x^{n-3} + \cdots .$$

[58]  Rashed (1980).
[59]  Waring (1991).
[60]  Lagrange (1771).

He next equated the coefficients on both sides to get recurrence relations for the Stirling numbers of the first kind:

$$n + A' = n + A',$$

$$nA' + A'' = \frac{n(n-1)}{2} + (n-1)A' + A'',$$

$$nA'' + A''' = \frac{n(n-1)(n-2)}{2 \cdot 3} + \frac{(n-1)(n-2)}{2}A' + (n-2)A'' + A''', \cdots,$$

or

$$A' = \frac{n(n-1)}{2}, \tag{10.78}$$

$$2A'' = \frac{n(n-1)(n-2)}{2 \cdot 3} + \frac{(n-1)(n-2)}{2}A', \tag{10.79}$$

$$3A''' = \frac{n(n-1)(n-2)(n-3)}{2 \cdot 3 \cdot 4} + \frac{(n-1)(n-2)(n-3)}{2 \cdot 3}A'$$
$$+ \frac{(n-2)(n-3)}{2}A'', \cdots, \tag{10.80}$$

$$(n-1)A^{(n-1)} = 1 + A' + A'' + \cdots + A^{(n-2)}. \tag{10.81}$$

Lagrange noted that if $n$ were an odd prime $p$, then the equation (10.78) showed that $A'$ was divisible by $p$; (10.79) showed $A''$ divisible by $p$, and so on. The equation that would immediately precede (10.81) implied that $p$ divided $A^{(n-2)}$. Next, observing that $A^{(n-1)} = (n-1)! = (p-1)!$, Lagrange perceived that equation (10.81) implied Wilson's theorem that $A^{(n-1)} + 1 = (p-1)! + 1$ was divisible by $p$. As an application of this theorem, Lagrange determined the quadratic character of $-1$ modulo a prime $p$. That is, he deduced that if $p$ were a prime of the form $4n + 1$, then there had to exist an integer $x$ such that $x^2 + 1$ was divisible by $p$. Note that Euler had given a remarkable proof of this result using repeated differences of the sequence $1^n, 2^n, 3^n, 4^n, \ldots$.[61] See Exercise 13 at the end of this chapter.

## 10.9  Taylor's Summation by Parts

The method of summation by parts is usually attributed to Abel who used it in a rigorous discussion of convergence of series. However, a century earlier, in the 1717 *Philosophical Transactions*, Taylor explicitly worked out this idea as an analog of integration by parts. Moreover, one can see in the work of Newton, Leibniz, and others that they were implicitly aware of this method. Abel's summation by parts method consists in moving from (4.25) to (4.26) and this may be compared with the steps Newton took in (10.17).

[61] Fuss (1968) vol. I, pp. 493–497. See also Eu. I-2, section 5, pp. 328–337. E 242.

Taylor's result is actually an indefinite summation formula in which the constants of summation are not explicitly written. Because Taylor's presentation is obscure, we present the 1819 derivation below from Lacroix in which $\sum$ and $\Delta$ were taken as inverse operations. In Lacroix's notation, $P$ and $Q$ were functions of an integer variable $x$ and $P_1 = P + \Delta P$, $P_2 = P_1 + \Delta P_1$, etc. In this notation, Taylor's formula took the form

$$\sum PQ = Q \sum P - \sum \left( \Delta Q \sum P_1 \right). \qquad (10.82)$$

To prove this following Lacroix,[62] first suppose that

$$\sum PQ = Q \sum P + z.$$

Apply the difference operator $\Delta$ to both sides to get

$$\Delta \left( \sum PQ \right) = \Delta \left( Q \sum P + z \right),$$

or

$$PQ = (Q + \Delta Q) \sum (P + \Delta P) - Q \sum P + \Delta z$$
$$= Q \sum \Delta P + \Delta Q \sum (P + \Delta P) + \Delta z$$
$$= QP + \Delta Q \sum (P + \Delta P) + \Delta z.$$

Hence,

$$\Delta z = -\Delta Q \sum (P + \Delta P) \text{ or } z = - \sum \left( \Delta Q \sum (P + \Delta P) \right) = - \sum \left( \Delta Q \sum P_1 \right),$$

and Lacroix's proof of Taylor's formula (10.82) was complete. In his book, Lacroix attributed the result to Taylor.

Just as one can perform repeated integration by parts, one may also do repeated summation by parts if necessary. Thus, Lacroix gave this formula for repeated summation by parts:

$$\sum PQ = Q \sum P - \Delta \sum^2 P_1 + \Delta^2 Q \sum^3 P_2 - \Delta^3 Q \sum^4 P_3 + \text{ etc.,} \quad (10.83)$$

where $\sum^2$, $\sum^3$, ... denoted double, triple, ... summation. To derive this formula, he replaced $Q$ by $\Delta Q$ and $P$ by $\sum P_1$ in (10.82). He then had

$$\sum \left( \Delta Q \sum P_1 \right) = \Delta Q \sum^2 P_1 - \sum \left( \Delta^2 Q \sum^2 P_2 \right).$$

[62] Lacroix (1819) pp. 91–92 or Lacroix (1800) pp. 89–90.

Combining this with (10.82), he obtained

$$\sum PQ = Q \sum P - \Delta Q \sum{}^2 P_1 + \sum \left( \Delta^2 Q \sum{}^2 P_2 \right).$$

A continuation of this process would yield formula (10.83).

## 10.10   Exercises

(1) Show that for a finite sequence of positive decreasing numbers $a_0, a_1, \ldots, a_n$

$$a_0 = \sum_{i=0}^{n-1} (a_i - a_{i+1}) + a_n.$$

The sequence can be infinite; in that case "the last number" of the sequence $a_n$ should be replaced by the limit of the sequence. Then deduce the sum of the convergent infinite geometric series. For a reference to this 1644 result of Torricelli, see Weil (1989b).

(2) Use the inequality

$$\frac{1}{a-1} + \frac{1}{a} + \frac{1}{a+1} > \frac{3}{a}$$

to prove that

$$1 + \frac{1}{2} + \frac{1}{3} + \frac{1}{4} + \frac{1}{5} + \cdots$$

diverges. Apply Torricelli's formula in the previous problem to sum

$$\frac{1}{3} + \frac{1}{6} + \frac{1}{10} + \frac{1}{15} + \cdots .$$

See Weil (1989b) for the reference to these 1650 results of Mengoli.

(3) Show that

$$\frac{1}{3} + \frac{1}{15} + \frac{1}{35} + \frac{1}{63} + \frac{1}{99} + \cdots = \frac{1}{2}.$$

Leibniz also mentioned this result in his *Historia et Origo* of 1714, written in connection with the calculus controversy, where he explained that in a 1682 article in the *Acta Eruditorum*, he had extended the inverse relationship between differences and sums to differentials and integrals. See Leibniz (1971) vol. 5, p. 122.

(4) Find the sums of the reciprocals of the figurate numbers. For example, the sum of the reciprocals of the pyramidal numbers is given as

$$\frac{1}{1} + \frac{1}{4} + \frac{1}{10} + \frac{1}{20} + \frac{1}{35} + \cdots = \frac{3}{2}.$$

See Jakob Bernoulli (1993–99) vol. 4, p. 66.

(5) Show that if

$$x = \frac{1}{1^m} + \frac{1}{2^m} + \frac{1}{3^m} + \frac{1}{4^m} + \cdots \text{ and } y = \frac{1}{1^m} + \frac{1}{3^m} + \frac{1}{5^m} + \cdots,$$

then

$$x - y = \frac{x}{2^m}.$$

See Jakob Bernoulli (1993–99) vol. 4, p. 74.

(6) Let $m$, $n$, and $p$ be integers and $x = a + mn$. Show that

$$\sum_{k=0}^{m} (a + kn)(a + (k+1)n) \cdots (a + (k+p-1)n)$$

$$= \frac{x(x+n)\cdots(x+pn) - (a-n)a(a+n)\cdots(a+(p-1)n)}{(p+1)n}.$$

Deduce the values of the sums

$$1 + 2 + 3 + 4 + \cdots + x,$$

$$1 + 3 + 6 + 10 + \cdots \text{ to } x \text{ terms},$$

$$1 \cdot 3 \cdot 5 + 3 \cdot 5 \cdot 7 + 5 \cdot 7 \cdot 9 + \cdots \text{ to } x \text{ terms}.$$

This is the first result in Montmort's (1717) paper *De Seriebus Infinitis Tractatus*. He noted that this result was a generalization of a sum in Taylor's *Methodus Incrementorum*.

(7) Sum the series

$$\frac{5}{3 \cdot 5 \cdot 7 \cdot 9 \cdot 11 \cdot 13} + \frac{41}{5 \cdot 7 \cdot 9 \cdot 11 \cdot 13 \cdot 15}$$
$$+ \frac{131}{7 \cdot 9 \cdot 11 \cdot 13 \cdot 15 \cdot 17} + \frac{275}{9 \cdots 19} + \frac{473}{11 \cdots 21} + \cdots.$$

Montmort (1717) computed the sum to be $\frac{283}{80 \cdot 3 \cdot 5 \cdot 7 \cdot 9 \cdot 11}$.

(8) Prove Taylor's summation formula (10.12) by an application of the Harriot–Briggs, usually known as the Gregory–Newton, formula.

(9) Find the values of $A$, $B$, ..., $E$ to obtain the sum of Nicole's series (10.47).

(10) Derive Gregory's formula (10.1) or (10.2) from the Newton–Gauss interpolation.

(11) Use Wilson's theorem to prove that if $p = 4n + 1$ is a prime, there exists an integer $x$ such that $p$ divides $x^2 + 1$. See Lagrange (1867–1892) vol. 3, pp. 425–438.

(12) Prove Wilson's theorem by using Stirling numbers of the first kind and Fermat's little theorem that $z^{p-1} \equiv 1 \pmod{p}$ when $p$ is prime and $a$ is an integer not divisible by $p$:

(a) Let

$$(x - 1)(x - 2) \cdots (x - p + 1)$$
$$= x^{p-1} + A_1 x^{p-2} + A_x x^{p-3} + \cdots + A_{p-1},$$

and $A_0 = 1 + A_{p-1}$. Observe that $x^{p-1} + A_{p-1} \equiv A_0 \pmod{p}$ for $x = 1, 2, \ldots, p - 1$. Prove that

$$A_0 + k^{p-2} A_1 + k^{p-3} A_2 + \cdots + k A_{p-2} \equiv 0 \pmod{p}$$

for $k = 1, 2, \ldots, p - 1$.

(b) Show that the determinant of the system of $p - 1$ equations in $A_0, A_1, \ldots, A_{p-2}$ has a nonzero determinant modulo $p$.

(c) Deduce that $A_0 \equiv 0, A_1 \equiv 0, \ldots, A_{p-2} \equiv 0 \pmod{p}$. Sylvester published this result in 1854. See Sylvester (1973) vol. 2, p. 10.

(13) Show that, given a prime $p = 4n + 1$, the $2n$th difference of the sequence

$$1^{2n}, 2^{2n}, \ldots, (p - 1)^{2n}$$

is not divisible by $p$. Deduce that $a^{2n} - 1$ is not divisible by $p$ for all $1 \leq a \leq p - 1$. For this result of Euler, see his correspondence in Fuss (1968) p. 494 and Eu. I-2 section 5, pp. 328–337 (E 242). See also Weil (1984) p. 65, and Edwards (1977) p. 47.

## 10.11   Notes on the Literature

A brief summary of David Gregory's *Exercitatio* by Whiteside can be found in Newton (1967–1981) vol. IV, pp. 414–417. See Tweedie (1917–1918) for an evaluation of Nicole's researches in the calculus of finite differences. Whittaker may have been the first to notice the connection of proposition 7 of Stirling's *Methodus Differentialis* with the transformation of hypergeometric series. See p. 286 of Whittaker and Watson (1927). For an English translation by J. D. Blanton of the first part of Euler's 1755 book on differential calculus, see Euler (2000). An English translation of Fuss (1968) vol. 1, may be found in *Correspondence of Leonhard Euler with Christian Goldbach*, Part 2, published by Springer in 2015. Part 1 contains the untranslated letters.

# 11

## The Taylor Series

### 11.1 Preliminary Remarks

In 1715, Brook Taylor (1685–1731) published one of the most basic results in the theory of infinite series, now known as Taylor's formula. Taylor published his formula fifty years after Newton's seminal work on series and twenty-five years after Newton discovered, but did not publish, this same formula. In modern notation, Taylor's formula takes the form

$$f(x) = f(a) + (x - a)\frac{f'(a)}{1!} + (x - a)^2\frac{f''(a)}{2!} + \cdots . \qquad (11.1)$$

Taylor communicated this result to John Machin in a letter of July 26, 1712.[1] We note here that in 1706, Machin calculated $\pi$ to 100 digits by employing the formula

$$\frac{\pi}{4} = 4\arctan\frac{1}{5} - \arctan\frac{1}{239}.$$

In his letter to Machin, Taylor wrote

> I fell into a general method of applying Dr. Halley's Extraction of roots to all Problems, wherein the Abscissa is required, the Area being given which, for the service that it may be of calculations, (the only true use of all corrections) I cannot conceal. And it is comprehended in this Theorem. ... If $\alpha$ be any compound of the powers of $z$ and given quantities whether by a finite or infinite expression rational or surd. And $\beta$ be the like compound of $p$ and the same coefficients, and $z = p + x$, and $\dot{p} = 1 = \dot{z}$. Then will

$$\alpha - \beta = \frac{\dot{\beta}}{1}x + \frac{\ddot{\beta}}{1 \cdot 2}x^2 + \frac{\dddot{\beta}}{1 \cdot 2 \cdot 3}x^3 + \frac{\ddddot{\beta}}{1 \cdot 2 \cdot 3 \cdot 4}x^4 \quad \&c.$$

$$= \frac{\dot{\alpha}}{1}x - \frac{\ddot{\alpha}}{1 \cdot 2}x^2 + \frac{\dddot{\alpha}}{1 \cdot 2 \cdot 3}x^3 - \frac{\ddddot{\alpha}}{1 \cdot 2 \cdot 3 \cdot 4}x^4 \quad \&c.$$

---

[1] Bateman (1907).

Where $\dot{\alpha}$, $\ddot{\alpha}$, $\dddot{\alpha}$ &c. are formed in the same manner of $z$ and the given quantities, as $\dot{\beta}$, $\ddot{\beta}$, &c. are formed of $p$. &c. So that having given $\alpha$, $\beta$, and one of the abscissae $z$ or $p$, the other may be found by extracting $x$, their difference, out of this aequation. Or you may apply this to the invention of $\alpha$ or $\beta$, having given $z$, $p$ and $x$. But you will easily see the uses of this.

Newton discovered and extensively used infinite series in the period 1664–1670, but during that time it does not appear that he observed the connection between the derivatives of a function and the coefficients of its series expansion. This connection is the essence of the Taylor series, and it can be applied to obtain power series expansions of many functions. Newton discovered infinite series before he had investigated the concept of derivatives. Thus, he found expansions for several functions by using his binomial expansion, term-by-term integration, and reversion of series. He first indicated his awareness of the connection between derivatives and coefficients in his 1687 *Principia*, wherein he expanded $(e^2 - 2ao - o^2)^{\frac{1}{2}}$ in powers of $o$ and interpreted the coefficients as geometric quantities directly related to the derivatives of the function.[2] It is very possible that Newton was aware of the Taylor expansion at this time. In fact, forty years later, this *Principia* result inspired James Stirling to consider whether it could be generalized, leading him to an independent discovery of the Maclaurin series, published in 1717 in the *Philosophical Transactions*.[3]

In 1691–92, Newton gave an explicit statement of Taylor's formula as well as the particular case now named for Maclaurin. These appear in his *De Quadratura Curvarum*, composed in the winter of 1691–92 and never fully completed; in 1704 parts of this text were published under the title *Tractatus de Quadratura Curvarum*. Unfortunately, the published portions omitted the Taylor and Maclaurin theorems. As we shall see, Gregory used this result in 1670 to construct series for numerous functions, but Newton was the first to give its clear, though unpublished, statement. In his *De Quadratura*, finally published in 1976 by Whiteside, Newton discussed the problem of solving algebraic differential equations by means of infinite series. In this context, he stated the Taylor and Maclaurin expansions and then wrote the word "Example" and left a blank space. Apparently, he intended to give a solution of an algebraic differential equation but could not hit upon a satisfactory example. According to Whiteside, Newton's worksheets from this period show that he made several attempts to solve the equation $\sqrt{1 + \dot{y}^2} \times \ddot{y} = n\dot{y}$ without complete success.[4] This may explain why he did not include these results in the published work, although his corollaries three and four contain Newton's own formulation of the Taylor and Maclaurin series:[5]

*Corollary* 3. Hence, indeed, if the series proves to be of this form

$$y = az + bz^2 + cz^3 + dz^4 + ez^5 + \text{etc.}$$

(where any of the terms $a$, $b$, $c$, $d$ etc. can either be lacking or be negative), the fluxions of $y$, when $z$ vanishes, are had by setting $\dot{y}/\dot{z} = a$, $\ddot{y}/\dot{z}^2 = 2b$, $\dddot{y}/\dot{z}^3 = 6c$, $\ddddot{y}/\dot{z}^4 = 24d$, $\dddddot{y}/\dot{z}^5 = 120e$.

---

[2] See Proposition X of the second book of the *Principia*.
[3] Stirling (1717).
[4] Newton (1967–1981) vol. 7, p. 99, footnote 111.
[5] ibid. pp. 97–99.

*Corollary* 4. And hence if in the equation to be resolved there be written $w + x$ for $z$, as in Case 3, and by resolving the equation there should result the series $[y =]ex + fx^2 + gx^3 + hx^4 +$ etc. the fluxions of $y$ for any assumed magnitude of $z$ whatever will be obtained in finite equations by setting $x = 0$ and so $z = w$. For the equations of this sort gathered by the previous Corollary will be $\dot{y}/\dot{z} = e$, $\ddot{y}/\dot{z}^2 = 2f$, $\dddot{y}/\dot{z}^3 = 6g$, $\ddddot{y}/\dot{z}^4 = 24h$ etc.

Even before Newton, the Scottish mathematician James Gregory discovered and used Maclaurin's series to obtain power series expansions of some fairly complicated functions. In a letter to John Collins dated February 15, 1671,[6] Gregory gave the series expansions for $\arctan x$, $\tan x$, $\sec x$, $\ln \sec x$, $\ln \tan \left( \frac{\pi}{4} + \frac{x}{2} \right)$, $\operatorname{arcsec}(\sqrt{2}e^x)$, $2 \arctan \left( \tanh \left( \frac{x}{2} \right) \right)$. Of the seven series, the first two are inverses of each other, as are the fifth and the seventh; the fourth and sixth are inverses of each other, except for a constant factor. It does not seem that Gregory applied reversion of series, a technique used by Newton, to obtain the inverses. On the back of a letter from Shaw,[7] Gregory noted the first few derivatives of some of the seven functions. From this, we can see that Gregory derived the series for the second, third, sixth, and seventh functions by taking the derivatives; the series for the fourth and fifth using term-by-term integration of the series for the second and third functions; and the series for $\arctan x$ by integration of the series for $\frac{1}{1+x^2}$. As we shall see below, a key mistake in Gregory's calculations gives us evidence that he used the derivatives of a function to find its series.

Gregory knew that Newton had made remarkable advances in the theory of series, though he had seen only one example of Newton's work. He concluded that Newton must have known Taylor's expansion, since that could be used to find the power series expansion of any known function. Before giving his seven series expansions in his letter to Collins, Gregory wrote, "As for Mr. Newton's universal method, I imagine I have some knowledge of it, both as to geometrik and mechanick curves, however I thank you for the series [of Newton] ye sent me, and send you these following in requital."[8]

In 1694, Johann Bernoulli published a result in the *Acta Eruditorum* equivalent to Taylor's formula,[9] though it did not as easily produce the series expansion:

$$\int n \, dz = nz - \frac{z^2}{1 \cdot 2} \frac{dn}{dz} + \frac{z^3}{1 \cdot 2 \cdot 3} \frac{d \, dn}{dz^2} - \frac{z^4}{1 \cdot 2 \cdot 3 \cdot 4} \frac{d^3 n}{dz^3} + \text{etc.} \qquad (11.2)$$

He used this to solve first-order differential equations by infinite series and also applied it to find the series for $\sin x$ and $\ln(a + x)$ and to solve de Beaune's problem concerning the curve whose subtangent remained a constant. Unfortunately, Bernoulli could not use this formula to derive the series for $\sin x$; he obtained only a ratio of two series for $\frac{y}{\sqrt{a^2 - y^2}}$ where $y = a \sin x$. He commented that, though this method had this drawback, it was commendable for its universality. Bernoulli communicated his

---

[6] Turnbull (1939) pp. 168–176.
[7] Turnbull (1939) pp. 350–353. Thanks to the gracious help of the librarians at St. Andrews University, the author was able to view and obtain a copy of the original letter and Gregory's notes.
[8] ibid. p. 170.
[9] Bernoulli, Johann (1968) vol. 1, pp. 125–128.

series to Leibniz in a letter of September 2, 1694,[10] before the paper was printed. In reply, Leibniz observed that he had obtained similar results almost two decades earlier by using the method of differences of varying orders. He gave a detailed exposition of how that method would produce Bernoulli's series. We note that in 1704, Abraham de Moivre published an alternative proof of Bernoulli's series and four years later he communicated this to Bernoulli.[11]

It seems very likely that Gregory derived the Taylor expansion from the Gregory–Newton, or rather, the Harriot–Briggs, interpolation formula:

$$f(x) = f(a) + \frac{(x-a)}{h}\Delta f(a) + \frac{(x-a)(x-a-h)}{2!\,h^2}\Delta^2 f(a) + \cdots,$$

where

$$\Delta f(a) = f(a+h) - f(a), \quad \Delta^2 f(a) = f(a+2h) - 2f(a+h) + f(a), \quad \ldots.$$

As $h \to 0$, the number of interpolating points tends to infinity, and

$$\frac{\Delta f(a)}{h} \to f'(a), \quad \frac{\Delta^2 f(a)}{h^2} \to f''(a), \ldots.$$

The resulting series is Taylor's expansion. This proof is not rigorous, but the same argument was given by de Moivre in his letter to Bernoulli and then again by Taylor in his *Methodus Incrementorum Directa et Inversa* of 1715.[12] Leibniz too started with a formula involving finite differences to derive Bernoulli's series. On the other hand, the unpublished argument of Newton, also independently found by Stirling and Maclaurin, assumed that the function had a series expansion and then, by repeated differentiation of the equation, showed that the coefficients of the series were the derivatives of the function computed at specific values. This is called the method of undetermined coefficients.

We can see that there were three different methods by which the early researchers on the Taylor series discovered the expansion: (a) the method of taking the limit of an appropriate finite difference formula, by Gregory, Leibniz, de Moivre, and Taylor, (b) the method of undetermined coefficients, by Newton, Stirling, and Maclaurin, and (c) repeated integration by parts, or, equivalently, repeated use of the product rule, by Johann Bernoulli.

Infinite series, including power series, were used extensively in the eighteenth century for numerical calculations. On the basis of considerable experience, mathematicians usually had a good idea of the accuracy of their results even though they did not perform error analyses. It was only in the second half of the eighteenth century that a few mathematicians started considering an explicit error term. In the specific case of a binomial series, Jean d'Alembert (1717–1783) obtained bounds for the remainder

---

[10] Bernoulli and Leibniz (1745) vol. 1, pp. 13–16.
[11] See Feigenbaum (1981) chapter 4.
[12] Taylor (1715).

of the series after the first $n$ terms.[13] In 1754, as reported by Lacroix,[14] he also found a more general but not very useful method by which he expressed the remainder of a Taylor series using an iterative process, and when worked out, this would have resulted in an $n$-fold iterated integral. Surprisingly, in 1693 Newton proved a result that converted an iterated integral into a single integral. If d'Alembert had used this, he would have obtained the remainder given in many textbooks; Lagrange, de Prony, and Laplace discovered this remainder term using a different method.

In an undated manuscript determined by Whiteside to date from 1693,[15] apparently written while he was revising *De Quadratura*, Newton worked out the $n$th fluent (integral) of $\dot{y}$, the fluxion of $y$. The formula in modern notation takes the form

$$\frac{1}{n!}\left(z^n \int_0^z \dot{y}\,dt - nz^{n-1}\int_0^z t\,\dot{y}\,dt + \frac{n(n-1)}{2}z^{n-2}\int_0^z t^2\dot{y}\,dt - \cdots\right)$$

$$+ a_0\frac{z^n}{n!} + a_1\frac{z^{n-1}}{(n-1)!} + a_2\frac{z^{n-2}}{(n-2)!} + \cdots + a_n.$$

The expression without the polynomial part can be simplified by the binomial theorem so that we have

$$\frac{1}{n!}\int_0^z (z-t)^n\dot{y}\,dt.$$

If we instead take the $n$th iterated integral of $y$, then this expression takes the form

$$\frac{1}{(n-1)!}\int_0^z (z-t)^{n-1}y\,dt. \tag{11.3}$$

Newton's result is but one step away from Taylor's formula with the remainder as an integral. Compare Newton's result with Cauchy's work on the equation $\frac{d^n y}{dx^n} = f(x)$, given later in this chapter. It is not clear whether Newton was aware that the Taylor series followed easily from his result, but he certainly revised his monograph in that connection. Thus, it is possible that Newton was aware of the relation between his integral and Taylor series. Interestingly, Newton included an equivalent of this result in geometric garb, but without proof, in his 1704 *Tractatus*. In 1727, Benjamin Robins published a proof in the *Philosophical Transactions*.[16]

Newton's formula for the $n$th iterate of an integral appears to have escaped the notice of the continental mathematicians of the eighteenth century. Thus, it remained for Lagrange to discover an expression for the remainder term. This appeared in his *Théorie des fonctions analytiques* of 1797.[17] Owing to his algebraic conception of the calculus, Lagrange avoided the use of integrals in this work. So, to find bounds for the remainder, he wrote down an expression for its derivative. In a later work of 1801 entitled *Leçons sur les calcul des fonctions*, he generalized the mean value

[13] Grabiner (1981) pp. 60–64.
[14] Lacroix (1819) pp. 396–398. Note interesting footnote on p. 396.
[15] Newton (1967–1981) vol. 7, pp. 164–166.
[16] Robins (1727).
[17] Lagrange (1797) pp. 44–45.

theorem and consequently obtained the well-known expression of the remainder as an $n$th derivative, now called the Lagrange remainder.[18] He applied this to a discussion of the maximum or minimum of a function and also to his theory of the degree of contact between two curves. Without defining area, he also used the remainder to prove that the derivative of the area was the function itself.

Though the integral form of the remainder followed immediately from his work, Lagrange never explicitly stated it. In his lectures of 1823, Cauchy wrote that in 1805 Gaspard Riche de Prony (1755–1839) used integration by parts to obtain Taylor's theorem with the integral remainder.[19] De Prony was a noted mathematician of his time and taught at the École Polytechnique. He is now remembered as a leader in the construction of mathematical tables. To fill the need for the numerous human calculators required for this process, de Prony gave training in arithmetic to many hairdressers, left unemployed by the French Revolution. Pierre-Simon Laplace (1749–1827) included a derivation of Taylor's theorem with remainder, using integration by parts, in the second edition of his famous *Théorie analytique des probabilités*, published in 1814. The third volume of Lacroix's book on calculus, of 1819, referred to Laplace but not to de Prony;[20] Cauchy may have mentioned de Prony in order to set the record straight.

Lagrange's derivation of the remainder had significant gaps, though his outline was essentially correct. He regarded it as well known and therefore did not provide a proof that functions – by which he meant continuous functions, though he did not define continuity – had the intermediate value property as well as the maximum value property on a closed interval. It was not until about 1817 that Bolzano and Cauchy gave a precise definition of the continuity of a function and proved the intermediate value property. Bolzano's definition, similar to our modern definition, was that $f(x)$ was continuous if the difference $f(x + \omega) - f(x)$ could be made smaller than any given quantity, with $\omega$ chosen as small as desired.[21]

Bernard Bolzano studied philosophy and mathematics at Charles University in Prague from 1796 through 1800. Although he did not particularly enjoy his mathematics courses, he studied the work of Euler and Lagrange on his own. However, Bolzano was fully converted to mathematics by the study of Eudoxus in Eulcid's *Elements*. Bolzano served as professor of theology at Prague from 1807 to 1819, but published several mathematics papers during this period, including his 1817 work on the intermediate value theorem. He based this theorem on his lemma that if a property $M$ were true for all $x$ less than $u$, but not for all $x$, then, among all values of $u$ for which this was true, there existed a greatest, $U$. To prove this lemma, he applied a form of the Bolzano-Weierstrass theorem; he wrote that he had a proof of the latter theorem, though it has not been found among his papers.

A prolific author of scientific and other works, during the 1830s Bolzano wrote a work on the foundations of real analysis, *Functionenlehre*, finally published in 1930.[22]

---

[18] Lagrange (1867–1892) vol. 10, pp. 94–95.
[19] Cauchy (1823) p. 142.
[20] Lacroix (1819) pp. xvi and 399.
[21] For an English translation of Bolzano's 1817 paper, see Bolzano (1980).
[22] Bolzano (1930).

In this book, Bolzano proved the extreme value theorem, that states that a continuous function on a closed and bounded interval assumes its maximum/minimum value at some point in the interval. Note that eighteenth- and early nineteenth-century mathematicians did not feel a need to prove this theorem. In fact, as late as 1868, Serret's differential calculus assumed this fact without comment.[23] Bolzano proved the extreme value theorem in two steps. First, he proved that a continuous function on a closed and bounded interval has to be bounded; he then applied this result to verify that the function must assume the extreme values.

Bolzano also conceived of the important idea of uniform continuity[24] much before other mathematicians. Uniform continuity is needed, for one example, to prove that a continuous function on $[a, b]$ is Riemann integrable. Note that a function $f(x)$ is said to be uniformly continuous on an interval $I$ if for each $\epsilon > 0$ there exists a $\delta > 0$ such that $|f(x_1) - f(x_2)| < \epsilon$, when $|x_1 - x_2| < \delta$. In the 1830s, realizing that there was a need for a theory of real numbers, Bolzano made an unsuccessful attempt to develop it. However, Bolzano's insight into real analysis was deep; he was the first mathematician to construct an example of a continuous nowhere differentiable function.[25] About Bolzano, Abel wrote, amongst other doodles in his 1826 Paris notebook, "Bolzano is a clever fellow from what I have studied;"[26] note that Abel's comment was based only on Bolzano's early work.

Since Bolzano's work on the extreme value theorem was not published until 1930, Weierstrass tackled it, and proved the result in his Berlin lectures during the 1860s. Though Weierstrass did not publish these lectures, his students referred to them in their publications. Thus, we give our translation of a passage from Georg Cantor's 1870 paper on trigonometric series:[27]

> The proof [that Cantor gave in his paper] is essentially based on the frequently-occuring and proven theorem contained in the lectures of Mr. Weierstrass:
>
> "Given a real variable in an interval $(a, b)$ (including the limits), a continuous function $\phi(x)$ *reaches* the maximum value $g$ that it can assume, for at least one value of of a variable, $x_0$, so that $\phi(x_0) = g$."
>
> One similar proof based thereon for the fundamental theorem of differential calculus [the mean value theorem] was given by *Ossian Bonnet*; this can be found in "*Cours de calcul différentiel et intégral*, par *J. A. Serret*, Paris 1868" im ersten Bande, Seite 17–19.

Moreover, David Hilbert in his 1897 lecture in memory of Karl Weierstrass, a translated portion of which is given below, mentioned the extreme value theorem:[28]

> Of the highest importance, moreover, is the sharp distinction that Weierstrass made, as to whether at a point, a function reached a value or came only arbitrarily close, especially the distinction between the notion of a maximum or a minimum and the notion of the upper or lower limit of

[23]  Serret (1868) vol. I, p. 17.
[24]  Rusnock (2005).
[25]  For a discussion of Bolzano's example, see Strichartz (1995) pp. 403–406.
[26]  Stubhaug (2000) p. 505.
[27]  Cantor (1870b) p. 141, footnote.
[28]  Hilbert (1897) p. 63.

a function of a real variable. In his theorem, according to which a continuous function of a real variable always actually reaches its upper and lower limits, that it necessarily has a maximum and minimum, Weierstrass discovered a result that no mathematician engaged in research in higher analysis or arithmetic can do without.

In his lectures, Weierstrass also discussed the idea of uniform continuity. In fact, a proof of the theorem that a continuous function on a closed interval is uniformly continuous was published by E. Heine in 1872,[29] who wrote that this proof and other proofs in his paper had been orally communicated to him by Weierstrass and his students Schwarz and Cantor.

In modern calculus books, the remainder term for the Taylor series of a function $f(x)$ is used to determine the values of $x$ for which the series represents the function. This approach is due to Cauchy; in his courses at the École Polytechnique in the 1820s, he used this method to find series for the elementary functions. Cauchy's use of the remainder term for this purpose was consistent with his pursuit of rigor; we also note that in 1822 he discovered and published the fact that all the derivatives at zero of the function $f(x) = e^{-\frac{1}{x^2}}$ when $x \neq 0$ and $f(0) = 0$ were equal to zero. Thus, the Taylor series at $x = 0$ was identically zero; it therefore represented the function only at $x = 0$. This example would have come as a great surprise to Lagrange who believed that all functions could be represented as series and even attempted to prove it. He built the whole theory of calculus on this basis. He defined the derivative of $f(x)$, for example, as the coefficient of $h$ in the series expansion of $f(x + h)$. He was thereby attempting to eliminate vague concepts such as fluxions, infinitesimals, and limits in order to reduce all computations to the algebraic analysis of finite quantities. Cauchy, by contrast, rejected Lagrange's foundations for analysis but accepted with small changes some of Lagrange's proofs.

The proof of Taylor's theorem based on Rolle's theorem, now commonly given in textbooks, seems to have first been given in J.A. Serret's 1868 text on calculus;[30] he attributed the result to Pierre Ossian Bonnet (1819–1892). In fact, Rolle proved the theorem only for polynomials. Serret did not mention Rolle explicitly in the course of his proof, but did mention him in his algebra book.[31] Michel Rolle (1652–1719) was a paid member of the Academy of Sciences of Paris. In books published in 1690 and 1691,[32] Rolle established that the derivative of a polynomial $f(x)$ had a zero between two successive real zeros of $f(x)$. Since he did not initially accept the validity of calculus, Rolle worked out an algebraic procedure called the method of cascades, by which one could obtain the derivative of a polynomial. Euler, Lagrange, and Ruffini made mention of Rolle's result, but it did not occupy a central place in calculus at that time because it was seen as a theorem about polynomials. Once it was extended to all differentiable functions, its significance was greatly increased.

---

[29] Heine (1872).
[30] Serret (1868) pp. 17–19.
[31] Serret (1877) p. 271.
[32] Rolle (1690) and (1691). For an English translation of relevant passages, see Smith (1959) vol. 1, pp. 253–260.

The modern conditions for the validity of Rolle's theorem were given in substance by Bonnet, but they were more carefully and exactly stated by the Italian mathematician, Ulisse Dini (1845–1918) in his lectures at the University of Pisa in 1871–1872.[33] After this, mathematicians began investigating the consequences of relaxing the conditions. In the exercises, we state a 1909 result of W. H. Young and Grace C. Young, using left-hand and right-hand derivatives. Grace Chisholm (1868–1944) studied at Girton College, Cambridge and then went on to receive a doctorate in mathematics from the University of Göttingen in 1896. Her best work was done in real variables theory and she was among the very few women mathematicians of her generation with an international reputation. William Young (1863–1942) studied at Cambridge and became a mathematical coach there. He coached his future wife for the Tripos exam and took up mathematical research after their marriage in 1896. He published over 200 papers and was one of the most profound English mathematicians of the early twentieth century. It appears from a letter W. H. Young wrote to his wife that several papers published under his name alone were in fact joint efforts. In recognition of this, a volume of their selected papers was published in 2000 under both names.[34]

Returning to Bolzano and Cauchy's proofs of the intermediate value theorem, we note that they both had gaps. Bolzano assumed the existence of a least upper bound and Cauchy's argument produced a sequence of real numbers $a_n, n = 1, 2, 3, \ldots$ such that $a_{n+1} - a_n = \frac{1}{2}(a_n - a_{n-1})$; he assumed that such a sequence must have a limit. A theory of real numbers was required to shore up these proofs. Although it seems that by the 1830s, Bolzano had begun to understand the basic problem here,[35] it was not until the second half of the nineteenth century that mathematicians were able to construct the necessary framework. Richard Dedekind (1831–1916) was one of the first to develop it and he described his motivation in his famous paper on the theory of real numbers:[36]

As professor in the Polytechnique School in Zurich I found myself for the first time obliged to lecture upon the elements of the differential calculus and felt more keenly than ever before the lack of a really scientific foundation for arithmetic. In discussing the notion of the approach of a variable magnitude to a fixed limiting value, and especially in proving the theorem that every magnitude which grows continually, but not beyond all limits, must certainly approach a limiting value, I had recourse to geometric evidences. Even now such resort to geometric intuition in a first presentation of the differential calculus, I regard as exceedingly useful, from the didactic standpoint, and indeed indispensable, if one does not wish to lose too much time. But that this form of introduction into the differential calculus can make no claim to being scientific, no one will deny. For myself this feeling of dissatisfaction was so overpowering that I made the fixed resolve to keep meditating on the question till I should find a purely arithmetic and perfectly rigorous foundation for the principles of infinitesimal calculus.

[33] Dini (1878) p. 71.
[34] Young and Young (2000).
[35] Rootselaar (1964).
[36] Dedekind (1963) pp. 1–2.

Dedekind published his theory in 1872, though he had completed it by November 1858. Meanwhile, by 1872, the theories of Charles Méray,[37] Cantor,[38] and Eduard Heine,[39] equivalent to Dedekind's, were published. Though he had discovered it some years before, Weierstrass presented his own independently developed theory of real numbers as part of his lectures in Berlin during the 1860s.[40]

## 11.2 Gregory's Discovery of the Taylor Series

In 1671, Gregory gave power series expansions of the seven functions mentioned earlier. His notation was naturally different from the one we now use. For example, he described the series for $\tan x$ and $\ln \sec x$:

If radius $= r$, arcus $= a$, tangus $= t$, secans artificialis $= s$, then

$$t = a + \frac{a^3}{3r^2} + \frac{2a^5}{15r^4} + \frac{17a^7}{315r^6} + \frac{3233a^9}{181440r^8} + \quad \text{etc.,} \tag{11.4}$$

$$s = \frac{a^2}{2r} + \frac{a^4}{12r^3} + \frac{a^6}{45r^5} + \frac{17a^8}{2520r^7} + \frac{3233a^{10}}{1814400r^9} + \quad \text{etc.}$$

Gregory's descriptions of the $\ln \tan \left(\frac{\pi}{4} + \frac{x}{2}\right)$ and $\operatorname{arcsec}(\sqrt{2}e^x)$ functions were slightly more complicated. In his letter to Collins, he gave no indication of how he obtained his seven series, but H. W. Turnbull determined that, except for the series for $\arctan x$, Gregory obtained them by using their derivatives.[41] While examining Gregory's unpublished notes in the 1930s, Turnbull noticed that Gregory had written the successive derivatives of some trigonometric and logarithmic functions on the back of a letter, dated January 29, 1671, from Gideon Shaw, an Edinburgh stationer. For example, he gave the first seven derivatives of $r \tan \theta$ with respect to $\theta$ expressed as polynomials in $\tan \theta = \frac{q}{r}$. He denoted the function and its derivatives by $m$ so that he had

$$1^{st} \ m = q,$$

$$2^{nd} \ m = r + \frac{q^2}{r},$$

$$3^{rd} \ m = 2q + \frac{2q^3}{r^2},$$

$$4^{th} \ m = 2r + \frac{8q^2}{r} + \frac{6q^4}{r^3},$$

$$5^{th} \ m = 16q + \frac{40q^3}{r^2} + \frac{24q^5}{r^4},$$

[37] Méray (1869).
[38] Cantor (1872).
[39] Heine (1872).
[40] Dugac (1973) pp. 57–59. See also Snow (2003).
[41] Turnbull (1939) pp. 168–176.

$$6^{th} \quad m = 16r + \frac{136q^2}{r} + \frac{240q^4}{r^3} + \frac{120q^6}{r^5},$$

$$7^{th} \quad m = 272q + \frac{987q^3}{r^2} + 1680\frac{q^5}{r^4} + 720\frac{q^7}{r^6},$$

$$8^{th} \quad m = 272r + 3233\frac{q^2}{r} + 11361\frac{q^4}{r^3} + 13440\frac{q^6}{r^5} + 5040\frac{q^8}{r^7}.$$

Note that since the derivative of $\tan\theta$ is $\sec^2\theta = 1 + \frac{q^2}{r^2}$, one can move from one value of $m$ to the next by taking the derivative of the initial $m$ with regard to $q$ and then multiplying by $r + \frac{q^2}{r}$. This suggests that Gregory used a method equivalent to the chain rule; indeed, this conclusion is supported by his computational mistake in finding the seventh $m$ from the sixth:

$$\left(272\frac{q}{r} + 960\frac{q^3}{r^3} + 720\frac{q^5}{r^5}\right)\left(r + \frac{q^2}{r}\right).$$

The coefficient of $\frac{q^3}{r^2}$ is $272 + 960 = 1232$, whereas Gregory had 987. Evidently, he had miscopied 272 from the previous step as 27 to get $27 + 960 = 987$. This in turn produced an error in the coefficient of $\frac{q^2}{r}$ in the eighth $m$; this should be 3968 instead of Gregory's 3233. If this computation, with Gregory's mistake, is continued to the tenth $m$, that is, the ninth derivative of $r\tan\theta$, then the first term of the derivative would be $2 \times 3233 = 6466$, as given in his notes.

Note that the Maclaurin series for $\tan\theta$ is obtained by computing the derivatives at $\theta = 0$. According to Gregory's mistaken calculation, the coefficient of $\theta^9$ in the series for $\tan\theta$ in the letter to Collins would be $\frac{6466}{9!}$. This simplifies to $\frac{3233}{181440}$, just as Gregory noted in (11.4). As Turnbull pointed out, the appearance of this key error in Gregory's letter fortunately allows us to see that the calculations on the back of Shaw's letter were for the purpose of constructing the series. Thus, though no explicit statement of Maclaurin's formula has been found in Gregory's papers, we may conclude that Gregory was implicitly aware of it, since he made use of it in so many instances.

In 1713, Newton, then president of the Royal Society, insisted that the society publish relevant portions of Gregory's letters to Collins in the *Commercium Epistolicum* to prove his own absolute priority in the discovery of the calculus. Recall that Gregory's letters referred to the series of Newton communicated to him by Collins. But in the published accounts, Gregory's computational error was corrected.

Gregory found the series for $\mathrm{arcsec}(\sqrt{2}e^x)$ by taking the derivatives of $r\theta$ with respect to $\ln\sec\theta$. In his notes, he wrote down the first five derivatives employed to construct the series in the letter to Collins. If we write $y = r\theta$, $x = \ln\sec\theta$, $q = r\tan\theta$, then we have

$$\frac{dy}{dx} = \frac{1}{\tan\theta}\frac{dy}{d\theta} = \frac{r^2}{q} \quad \text{and} \quad \frac{dq}{dx} = \frac{r^2}{q} + q.$$

This implies that the successive derivatives can be found by taking the derivative with respect to $q$ and multiplying by $\frac{r^2}{q} + q$:

$$\frac{d^2 y}{dx^2} = \frac{d}{dq}\left(\frac{r^2}{q}\right) \cdot \frac{dq}{dx} = -\frac{r^2}{q^2}\left(\frac{r^2}{q} + q\right), \quad \frac{d^3 y}{dx^3} = \frac{r^2}{q} + \frac{4r^4}{q^3} + \frac{3r^6}{q^5}, \text{ etc.}$$

Except for the signs of the derivatives, Gregory wrote precisely these expressions in his notes. He also wrote, without signs, expressions for the next two derivatives (without signs):

$$\frac{r^2}{q} + 13\frac{r^4}{q^3} + 27\frac{r^6}{q^5} + 15\frac{r^8}{q^7}, \quad \frac{r^2}{q} + 40\frac{r^4}{q^3} + 174\frac{r^6}{q^5} + 240\frac{r^8}{q^7} + 105\frac{r^{10}}{q^9}.$$

Gregory then expanded $y = r\left(\theta - \frac{\pi}{4}\right)$ as a series in $x = \ln\frac{\sec\theta}{\sqrt{2}}$ about $x = 0$. Now when $x = 0$, then $\theta = \frac{\pi}{4}$ and $q = r$. Hence, he had the series, given in modern notation:

$$\theta = \frac{\pi}{4} + x - x^2 + \frac{4x^3}{3} - \frac{7x^4}{3} + \frac{14x^5}{3} - \frac{452x^6}{45} + \cdots.$$

To see how the constants in this series are produced, consider the coefficient of $x^3$ obtained from the third derivative with $q = r$. From the preceding expression for $\frac{d^3 y}{dx^3}$, this value can be given as $r + 4r + 3r = 8r$, and since this has to be divided by $3! = 6$, we arrive at $\frac{4r}{3}$ or simply $\frac{4}{3}$ after dividing by $r$.

## 11.3 Newton: An Iterated Integral as a Single Integral

Newton wrote up his evaluation of the $n$th iterated integral as a single integral sometime around 1693, but did not publish it. His main idea was to repeatedly use integration by parts to reduce a double integral to a single integral.[42] We reproduce his derivation, though we change his notation. He used the letters $A$, $B$, $C$, $\ldots$ to denote areas under $\dot{y}$, $z\dot{y}$, $z^2\dot{y}$, $\ldots$ but we shall use these letters to denote the areas under $y$, $zy$, $z^2 y$, $\ldots$ to obtain the result in standard form. We also employ the Fourier-Leibniz notation of a definite integral to denote area. Let $y = f(t)$ be a curve, and let $A = \int_0^z y\,dt$, $B = \int_0^z ty\,dt$, $C = \int_0^z t^2 y\,dt$, $D = \int_0^z t^3 y\,dt$, $\ldots$. Then for some constant $a$

$$\int_a^z y\,dt = A + g,$$

where $g$ is a constant. The second iterated integral of $y$ is

$$\int_a^z (A + g)\,dt = zA - \int_a^z t\frac{dA}{dt}\,dt + gz + h_1$$
$$= zA - B + gz + h,$$

42  Newton (1967–1981) vol. 7, pp. 164–166.

where $h$ is some constant. Integration of this expression gives

$$\frac{1}{2}z^2 A - zB - \int_a^z \left(\frac{1}{2}t^2 \frac{dA}{dt} - t\frac{dB}{dt}\right) dt + \frac{1}{2}gz^2 + hz + i_2$$

$$= \frac{1}{2}z^2 A - zB - \int_a^z \left(\frac{1}{2}\frac{dC}{dt} - \frac{dC}{dt}\right) dt + \frac{1}{2}gz^2 + hz + i_1$$

$$= \frac{1}{2}z^2 A - zB + \frac{1}{2}C + \frac{1}{2}gz^2 + hz + i.$$

This is the third iterated integral of $y$. The integral of its first three terms is

$$\frac{1}{2}\left(\frac{1}{3}z^3 A - z^2 B + zC\right) - \frac{1}{2}\int_a^z \left(\frac{1}{3}t^3 \frac{dA}{dt} - t^2 \frac{dB}{dt} + t\frac{dC}{dt}\right) dt + \text{constant}$$

$$= \frac{1}{6}\left(z^3 A - 3z^2 B + 3zC\right) - \frac{1}{2}\int_a^z \frac{1}{3}\frac{dD}{dt} dt + \text{constant}$$

$$= \frac{1}{6}\left(z^3 A - 3z^2 B + 3zC - D\right) + \text{constant}.$$

Hence the fourth iterated integral is

$$\frac{1}{6}\left(z^3 A - 3z^2 B + 3zC - D\right) + \frac{1}{6}gz^3 + \frac{1}{2}hz^2 + iz + k.$$

Newton worked out another iterate to obtain

$$\frac{1}{24}\left(z^4 A - 4z^3 B + 6z^2 C - 4zD + E\right) + \frac{1}{24}gz^4 + \frac{1}{6}hz^3 + \frac{1}{2}iz^2 + kz + l.$$

By induction, he wrote the general $n$th iterate, in our notation, as

$$\frac{1}{(n-1)!}\left(z^{n-1}\int_a^z y\,dt - (n-1)z^{n-2}\int_a^z ty\,dt + \frac{(n-1)(n-2)}{2}z^{n-3}\int_a^z t^2 y\,dt - \cdots\right)$$

$$+ \frac{1}{(n-1)!}gz^{n-1} + \frac{1}{(n-2)!}hz^{n-2} + \cdots. \tag{11.5}$$

Newton left the integral in this form, although it is clear that he could easily have applied the binomial theorem to obtain the integral in the form (11.3).

## 11.4 Bernoulli and Leibniz: A Form of the Taylor Series

Johann Bernoulli's 1794 result on series was stated in a paper[43] and in a letter to Leibniz[44] as

$$\text{Integr.} n\, dz = +nz - \frac{zz\, dn}{1 \cdot 2 \cdot dz} + \frac{z^2\, ddn}{1 \cdot 2 \cdot 3 \cdot dz^3} - \frac{z^3\, dddn}{1 \cdot 2 \cdot 3 \cdot 4 \cdot dz^4} \quad \text{etc.} \tag{11.6}$$

[43] Bernoulli, Johann (1968) vol. 1, pp. 125–128, especially p. 126.
[44] Leibniz and Bernoulli (1745) pp. 13–16.

Here Integr.$n\,dz$ stood for the integral of $n$, or $\int n\,dz$. In fact, the term "integral" was first used by the Bernoulli brothers, Jakob and Johann, who conceived of it as the antiderivative. In a letter to Johann, Leibniz once wrote that he preferred to think of the integral as a sum instead of as an antiderivative. Bernoulli's proof of this result was very simple:

$$n\,dz = +n\,dz + z\,dn - z\,dn - \frac{zz\,ddn}{1\cdot 2\cdot dz} + \frac{zz\,ddn}{1\cdot 2\cdot dz} + \frac{z^3\,dddn}{1\cdot 2\cdot 3\cdot dz^2} \quad \text{etc.}$$

He took the terms on the right in pairs to get

$$n\,dz = d(nz) - d\left(\frac{z^2}{1\cdot 2}\frac{dn}{dz}\right) + d\left(\frac{z^3}{1\cdot 2\cdot 3}\frac{ddn}{dz^2}\right) - \cdots.$$

The required result followed upon integration. This process amounts to repeated integration by parts applied to $\int n\,dz$. Bernoulli applied his formula to three questions: de Beaune's problem; determination of the series for $\ln(a+x)$; and the determination of the series for $\sin x$. He was not completely successful with the third problem and was only able to find $\frac{\sin x}{\cos x}$ as a ratio of two series.

In reply to Bernoulli's 1794 letter containing the result (11.6), Leibniz outlined his own derivation of the formula,[45] instructive as an illustration of Leibniz's conception of the analogy between finite and infinitesimal differences, leading to his characteristic approach to the calculus. We change Leibniz's notation slightly in the initial part of his derivation; he himself used neither subscripts nor the difference operator. Supposing the sequence $a_0, a_1, a_2, \ldots$ decreases to zero, Leibniz started with the equation

$$a_0 = -(\Delta a_0 + \Delta a_1 + \Delta a_2 + \cdots).$$

Since

$$a_n = (1+\Delta)^n a_0 = a_0 + \frac{n}{1}\Delta a_0 + \frac{n(n-1)}{1\cdot 2}\Delta^2 a_0 + \cdots,$$

Leibniz could rewrite the first equation as

$$\begin{aligned}
a_0 &= -\left(\Delta a_0 + \Delta a_0 + \Delta^2 a_0 + \Delta a_0 + 2\Delta^2 a_0 + \Delta^3 a_0 + \cdots\right)\\
&= -\left(\Delta a_0(1+1+1+\cdots) + \Delta^2 a_0(1+2+3+\cdots)\right)\\
&\quad - \left(\Delta^3 a_0(1+3+6+\cdots) + \cdots\right).
\end{aligned}$$

He then observed that a similar relation continued to hold when the differences were infinitely small and he replaced $a_0, -\Delta a_0, \Delta^2 a_0, -\Delta^3 a_0, \ldots$ by $y\,(=y(x))$, $dy, ddy, d^3y, \ldots$ respectively; moreover, by letting the infinitely small $dx$ become 1, he set

[45] ibid. pp. 18–24, especially pp. 21 and 22.

$$1 + 1 + 1 + \cdots = x,$$

$$1 + 2 + 3 + \cdots = \int x,$$

$$1 + 3 + 6 + 10 + \cdots = \int \int x, \text{etc.}$$

Since

$$\int x = \frac{1}{1 \cdot 2} xx, \quad \int \int x = \frac{1}{1 \cdot 2 \cdot 3} x^3, \ldots,$$

Leibniz obtained

$$y = \frac{1}{1} x \frac{dy}{dx} - \frac{1}{1 \cdot 2} xx \frac{ddy}{dx^2} + \frac{1}{1 \cdot 2 \cdot 3} x^3 \frac{d^3 y}{dx^3} - \frac{1}{1 \cdot 2 \cdot 3 \cdot 4} x^4 \frac{d^4 y}{dx^4} \text{ etc.}$$

He then noted that Bernoulli's formula followed upon replacing $y$, $dy$, $ddy$, etc. by $\int y$, $y$, $dy$, etc., respectively.

## 11.5  Taylor and Euler on the Taylor Series

In Taylor's book of 1715, he obtained his namesake series from the well-known interpolation formula by letting the distance between the equidistant points on the axis tend to zero. We shall follow Euler's exposition of 1736,[46] since Euler used a more convenient and easily understandable notation. Euler divided the interval from $x$ to $x + a$ into $m$ equal parts, each equal to $dx$. He let $y$ be a function of $x$ and then let $dy = y(x + dx) - y(x)$, $ddy = y(x + 2dx) - 2y(x + dx) + y(x), \ldots$ be the first, second, ... differences at $x$. He then had

$$y(x + 2dx) = y + 2dy + ddy, \quad y(x + 3dx) = y + 3dy + 3ddy + d^3 y,$$

$$\ldots \ldots$$

$$y(x + a) = y(x + mdx) = y + mdy + \frac{m(m-1)}{1 \cdot 2} ddy + \frac{m(m-1)(m-2)}{1 \cdot 2 \cdot 3} d^3 y + \text{ etc.}$$

Next, he let $m$ be an infinite number and $dx$ infinitely small so that $mdx$ was finite and equal to $a$. Then

$$y(x + a) = y + mdy + \frac{m^2}{1 \cdot 2} ddy + \frac{m^3}{1 \cdot 2 \cdot 3} d^3 y + \text{ etc.}$$

$$= y + a \frac{dy}{dx} + \frac{a^2}{1 \cdot 2} \frac{ddy}{dx^2} + \frac{a^3}{1 \cdot 2 \cdot 3} \frac{d^3 y}{dx^3} + \text{ etc.}$$

Gregory, de Moivre, and Taylor derived the Taylor series by means of essentially the same argument.

[46] Eu. I-14 pp. 108–123, especially pp. 108–110. E 47.

Euler then showed how Bernoulli's series could be derived from Taylor's formula. He set $y(0) = 0$ and $a = -x$ to get

$$0 = y - \frac{x}{1}\frac{dy}{dx} + \frac{x^2}{1\cdot 2}\frac{ddy}{dx^2} - \frac{x^3}{1\cdot 2\cdot 3}\frac{d^3 y}{dx^3} + \text{etc.}$$

This implied

$$y = \frac{x}{1}\frac{dy}{dx} - \frac{x^2}{1\cdot 2}\frac{ddy}{dx^2} + \frac{x^3}{1\cdot 2\cdot 3}\frac{d^3 y}{dx^3} - \text{etc.}$$

Euler then replaced $y$ by $\int y\,dx$, as Leibniz did, and obtained Bernoulli's formula. See the exercises for the converse.

## 11.6   Lacroix on D'Alembert's Derivation of the Remainder

In his 1754 book, *Recherches sur différents points importants du systme du monde*, Jean d'Alembert obtained the $n$-dimensional iterated integral for the remainder in the Taylor series. In his 1819 book, Sylvestre Lacroix (1765–1843) presented the essence of d'Alembert's proof in notation more familiar to us:[47]

Lacroix let $u' = u(x + h)$ and $u = u(x)$ and set $u' = u + P$. Then

$$\frac{du'}{dh} = \frac{dP}{dh},$$

and hence

$$P = \int \frac{du'}{dh}\,dh.$$

Note that the derivatives of $u'$ are partial derivatives; for now, we follow Lacroix's notation. Thus, he had

$$u' = u + \int \frac{du'}{dh}\,dh.$$

Next, he let

$$\frac{du'}{dh} = \frac{du}{dx} + Q,$$

so that

$$\frac{d^2 u'}{dh^2} = \frac{dQ}{dh}, \quad \text{or} \quad Q = \int \frac{d^2 u'}{dh^2}\,dh,$$

[47] Lacroix (1819) pp. 396–397.

$$\frac{du'}{dh} = \frac{du}{dx} + \int \frac{d^2u'}{dh^2}\,dh, \quad \int \frac{du'}{dh}\,dh = \frac{du}{dx}\frac{h}{1} + \int\int \frac{d^2u'}{dh^2}\,dh^2,$$

$$u' = u + \frac{du}{dx}\frac{h}{1} + \int\int \frac{d^2u'}{dh^2}\,dh^2.$$

Setting

$$\frac{d^2u'}{dh^2} = \frac{d^2u}{dx^2} + R,$$

he had

$$\frac{d^3u'}{dh^3} = \frac{dR}{dh}, \quad \text{or} \quad R = \int \frac{d^3u'}{dh^3}\,dh, \quad \frac{d^2u'}{dh^2} = \frac{d^2u}{dx^2} + \int \frac{d^3u'}{dh^3}\,dh,$$

$$u' = u + \frac{du}{dx}\frac{h}{1} + \frac{d^2u}{dx^2}\frac{h^2}{1\cdot 2} + \int\int\int \frac{d^3u'}{dh^3}\,dh^3.$$

Continuing in the same manner, he had in general

$$u' = u + \frac{du}{dx}\frac{h}{1} + \frac{d^2u}{dx^2}\frac{h^2}{1\cdot 2} + \cdots + \frac{d^{n-1}u}{dx^{n-1}}\frac{h^{n-1}}{1\cdot 2\cdots(n-1)} + \int^n \frac{d^nu'}{dh^n}\,dh^n,$$

where the $n$-fold multiple integral $\int^n$ was zero when $h = 0$. If Newton's formula (11.3) for $\int^n$ had been used here, then the remainder would have emerged in the form

$$R_n(h) = \frac{1}{(n-1)!}\int_0^h (h-t)^{n-1}\frac{d^n}{dx^n}u(x+t)\,dt. \tag{11.7}$$

In his 1823 lectures on calculus, Cauchy showed the equivalence of the two remainders by proving

$$\frac{d^n R_n}{dh^n} = \frac{d^n}{dx^n}u(x+t).$$

He applied the fundamental theorem of calculus and Leibniz's formula for the derivative of an integral to obtain

$$\frac{dR_n}{dh} = \frac{1}{(n-1)!}\int_0^h \frac{d}{dh}(h-t)^{n-1}\frac{d^n}{dx^n}u(x+t)\,dt$$

$$= \frac{1}{(n-2)!}\int_0^h (h-t)^{n-2}\frac{d^n}{dx^n}u(x+t)\,dt. \tag{11.8}$$

Cauchy derived the desired result, and in effect a proof of Newton's formula, by performing this process $n$ times. Lacroix did not provide a proof of (11.7), merely noting that

$$\int^n H\,dh^n = \frac{1}{1\cdot 2\cdots(n-1)}\left(h^{n-1}\int Hdh - \frac{n-1}{1}h^{n-2}\right.$$
$$\left.\int Hh\,dh + \frac{(n-1)(n-2)}{1\cdot 2}h^{n-3}\int Hh^2 dh - \cdots\right).$$

This result is equivalent to Newton's formula (11.5) and, as we have noted, Newton actually stated it in this form. Lacroix could have proved this inductively, using integration by parts. He did not mention Newton in this context; one may assume that he was not aware of Newton's work, since Lacroix was very meticulous in stating his sources.

## 11.7 Lagrange's Derivation of the Remainder Term

In his 1797 book, *Fonctions analytiques*, Joseph-Louis Lagrange (1736–1813) obtained the remainder term of the Taylor series as a single integral.[48] He started with

$$fx = f(x - xz) + xP,$$

where $P = 0$ at $z = 0$. The derivative with respect to $z$ of this equation gave $0 = -xf'(x - xz) + xP'$ or $P' = f'(x - xz)$. For the second-order remainder, Lagrange wrote

$$fx = f(x - xz) + xzf'(x - xz) + x^2 Q$$

and obtained

$$Q' = zf''(x - xz).$$

Similarly,

$$fx = f(x - xz) + xzf'(x - xz) + \frac{x^2 z^2}{2}f''(x - xz) + x^3 R$$

and, after taking the derivative and simplifying, he got $R' = \frac{z^2}{2}f'''(x - xz)$. Lagrange did not write the general expression for the remainder and gave only the recursive procedure. This method gives the remainder as an integral, though Lagrange did not write it in that form, since he avoided the use of integrals in this book. It is easy to see that

$$R(z) = \frac{1}{2!}\int_0^z t^2 f'''(x - xt)\,dt = \frac{1}{2!\,x^3}\int_{x-xz}^x (x - u)^2 f'''(u)\,du. \qquad (11.9)$$

[48] Lagrange (1797) pp. 43–45.

Had he stated it explicitly, the general form of Lagrange's formula would have been

$$fx = f(x - xz) + xzf'(x - xz) + \frac{x^2 z^2}{2} f''(x - xz) + \cdots$$
$$+ \frac{x^{n-1} z^{n-1}}{(n-1)!} f^{(n-1)}(x - xz) + x^n R_n,$$

where

$$x^n R_n = \frac{1}{(n-1)!} \int_{x-xz}^{x} (x - u)^{n-1} f^{(n)}(u)\, du.$$

If we replace $x - xz$ by $a$, Lagrange's formula becomes

$$f(x) = f(a) + (x - a)f'(a) + \frac{(x-a)^2}{2!} f''(a) + \cdots$$
$$+ \frac{(x-a)^{n-1}}{(n-1)!} f^{(n-1)}(a) + \frac{1}{(n-1)!} \int_{a}^{x} (x-t)^{n-1} f^{(n)}(t)\, dt.$$

Thus, here the multiple integral remainder of d'Alembert was replaced by a single integral. However, Lagrange himself gave only the derivative of the remainder. In his 1799 lectures on the calculus, published in 1801 as *Leons sur le calcul des fonctions*,[49] he presented the remainder as it appears in modern texts, as an $n$th derivative.

Lagrange proved a lemma for the purpose of determining bounds for $R_n$: If $f'x$ is positive for all values of $x$ between $x = a$ and $x = b$ with $b > a$, then $fb - fa > 0$. To prove this statement, Lagrange set $f(x + i) = fx - iP$, where $P$ was a function of $x$ and $i$, such that at $i = 0$, $P = f'(x) > 0$. So $P(x,i) > 0$ for $i$ sufficiently small, and it followed that $f(x + i) - f(x) > 0$ for small $i$. Next, he divided the interval $[a,b]$ into $n$ equal parts, each of length $(j = \frac{b-a}{n})$, with $n$ sufficiently large that in each subinterval

$$[a + kj, a + (k+1)j], \quad k = 0, 1, \ldots, n - 1,$$

he had

$$f(a + (k+1)j) - f(a + kj) > 0.$$

By adding up these inequalities, he got $fb - fa > 0$.

Lagrange's lemma was correct but his proof was obviously inadequate. For example, he assumed that the same $j$ would work in all parts of the interval. But he went on to use the result to derive a different form of the remainder. He supposed $f'(q)$ and $f'(p)$ to be the maximum and minimum values, respectively, of $f'(x)$ in an interval. Then $g'(i) = f'(x + i) - f'(p)$ and $h'(i) = f'(q) - f'(x + i)$ were both positive. Lagrange's lemma then gave

$$g(i) = f(x + i) - f(x) - if'(p) \geq 0, \quad h(i) = if'(q) - f(x + i) + f(x) \geq 0.$$

[49] Lagrange (1867–1892) vol. 10, pp. 91–95.

These inequalities implied bounds for $f(x + i)$:

$$f(x) + if'(p) \le f(x + i) \le f(x) + if'(q)$$

and thus

$$f(x + i) = f(x) + if'(x + i\theta), \quad 0 < \theta < 1,$$

by use of the intermediate value theorem, implicitly assumed by Lagrange; similarly, he assumed that $f'$ had a maximum/minimum in an interval. More generally, Lagrange showed that $f(x + i)$ lay between

$$f(x) + if'(x) + \frac{i^2}{2!} f''(x) + \cdots + \frac{i^u}{u!} f^{(u)}(p)$$

and

$$f(x) + if'(x) + \frac{i^2}{2!} f''(x) + \cdots + \frac{i^u}{u!} f^{(u)}(q),$$

where $p$ and $q$ were the values at which $f^{(u)}$ had a minimum and maximum, respectively, in the given interval. Once again, an application of the intermediate value theorem would yield Taylor's formula with the remainder as a derivative:

$$f(x + i) = f(x) + if'(x)\frac{i^2}{2!} f''(x) + \cdots + \frac{i^u}{u!} f^{(u)}(x + \theta i), \quad 0 < \theta < 1.$$

## 11.8  Laplace's Derivation of the Remainder Term

After being launched in his career by d'Alembert, Laplace used his tremendous command of analysis to make groundbreaking contributions in his areas of interest, celestial mechanics and probability. In the second edition of his his famous 1812 work, *Théorie analytique des probabilités*, published in 1814,[50] Laplace used repeated integration by parts in a direct way to obtain the remainder term. He started with the observation that

$$\int dz \, \phi'(x - z) = \phi(x) - \phi(x - z), \tag{11.10}$$

when the lower limit of integration was $z = 0$. This result, the fundamental theorem of calculus, would be written in modern notation:

$$\int_0^z \phi'(x - t) \, dt = \phi(x) - \phi(x - z).$$

[50] Laplace (1814), especially pp. 176–177.

Using Laplace's notation, integration by parts gave

$$\int dz\, \phi'(x-z) = z\phi'(x-z) + \int z\, dz\, \phi''(x-z),$$

$$\int z\, dz\, \phi''(x-z) = \frac{1}{2}z^2 \phi''(x-z) + \int \frac{1}{2}z^2\, dz\, \phi'''(x-z) \text{ etc.}$$

Hence, in general

$$\int dz\, \phi'(x-z) = z\phi'(x-z) + \frac{z^2}{1\cdot 2}\phi''(x-z) + \cdots + \frac{z^n}{1\cdot 2\cdot 3\cdots n}\phi^{(n)}(x-z)$$

$$+ \frac{1}{1\cdot 2\cdot 3\cdots n}\int z^n\, dz\, \phi^{(n+1)}(x-z).$$

$$(11.11)$$

Combined with (11.10), this equation provided Taylor's theorem with remainder. Laplace then converted this remainder to the Lagrange form. Since $\int z^n\, dz = \frac{z^{n+1}}{n+1}$, the integral in (11.11) lies between $\frac{mz^{n+1}}{n+1}$ and $\frac{Mz^{n+1}}{n+1}$ where $m$ and $M$ are the smallest and largest values of $\phi^{(n+1)}(x-z)$ in the interval of integration. Hence, the value of the integral in (11.11) lies in between these values and is given by

$$\frac{z^{n+1}}{n+1}\phi^{(n+1)}(x-u),$$

where $u$ is some value between 0 and $z$. Thus, the remainder term in (11.11) can be written as

$$\frac{z^{n+1}}{1\cdot 2\cdot 3\cdots (n+1)}\phi^{(n+1)}(x-u).$$

This completed Laplace's derivation of the two forms of the remainder in Taylor's theorem.

## 11.9 Cauchy on Taylor's Formula and l'Hôpital's rule

In his lectures published in 1823,[51] Cauchy took an interesting approach to Newton's $n$-fold integral. He started with the differential equation

$$\frac{d^n y}{dx^n} = f(x).$$

Repeated integration of this equation yielded

---

[51] Cauchy (1823) pp. 138–143.

$$\frac{d^{n-1}y}{dx^{n-1}} = \int_{x_0}^x f(z)\,dz + C, \quad \frac{d^{n-2}y}{dx^{n-2}} = \int_{x_0}^x (x-z)f(z)\,dz + C(x-x_0) + C_1,$$

. . . . . .

$$y = \int_{x_0}^x \frac{(x-z)^{n-1}}{1\cdot 2\cdots(n-1)}f(z)\,dz + \frac{C(x-x_0)^{n-1}}{1\cdot 2\cdots(n-1)} + \cdots + C_{n-2}(x-x_0) + C_{n-1},$$

where $C$, $C_1$, $\ldots$, $C_{n-1}$ were arbitrary constants. Here Cauchy used the result expressed in equation (11.8) to integrate in each step of the argument. The reader might compare this with Newton's formula (11.5). Cauchy then proceeded to obtain Taylor's theorem with remainder. He let $y = F(x)$ be a specific solution of $y^{(n)} = f(x)$ to obtain

$$C_{n-1} = F(x_0), \quad C_{n-2} = F'(x), \quad \ldots, \quad C = F^{(n-1)}(x_0),$$

and with these values, he had

$$F(x) = F(x_0) + \frac{F'(x_0)}{1!}(x-x_0) + \cdots + \frac{F^{(n-1)}(x_0)}{1\cdot 2\cdots(n-1)}(x-x_0)^{n-1}$$
$$+ \int_{x_0}^x \frac{(x-z)^{n-1}F^{(n)}(z)}{1\cdot 2\cdots(n-1)}\,dz.$$

Cauchy gave another proof,[52] that ran along the lines of Lagrange's second proof. He started with the lemma: Suppose $f(x)$ and $F(x)$ are continuously differentiable in $[x_0, x]$ with $f(x_0) = F(x_0) = 0$, and $F'(x_0) > 0$ in this interval. For $x$ in this interval, if

$$A \le \frac{f'(x)}{F'(x)} \le B,$$

then

$$A \le \frac{f(x)}{F(x)} \le B.$$

To prove this, Cauchy noted that since $F'(x) > 0$, he had

$$f'(x) - AF'(x) \ge 0 \quad \text{and} \quad f'(x) - BF'(x) \le 0.$$

He then applied Lagrange's lemma to the functions

$$f(x) - AF(x) \quad \text{and} \quad f(x) - BF(x)$$

to obtain the required result. Cauchy then took $x = x_0 + h$ and applied the intermediate value theorem to derive

---

[52] ibid. pp. 161–168.

$$\frac{f(x_0 + h)}{F(x_0 + h)} = \frac{f'(x_0 + \theta h)}{F'(x_0 + \theta h)}, \quad \text{where} \quad 0 < \theta < 1.$$

In the situation where $f(x_0)$ and $F(x_0)$ were nonzero, he replaced $f(x_0 + h)$ and $F(x_0 + h)$ by $f(x_0 + h) - f(x_0)$ and $F(x_0 + h) - F(x_0)$, respectively, to get the generalized mean value theorem:

$$\frac{f(x_0 + h) - f(x_0)}{F(x_0 + h) - F(x_0)} = \frac{f'(x_0 + \theta h)}{F'(x_0 + \theta h)}, \quad \text{where} \quad 0 < \theta < 1. \quad (11.12)$$

He next supposed $f'(x_0) = f''(x_0) = \cdots = f^{(n-1)}(x_0) = 0 = F'(x_0) = F''(x_0) = \cdots = F^{(n-1)}(x_0)$ and $F^{(n)} \neq 0$, and that all the derivatives were continuous. By an iteration of the process used to find the generalized mean value theorem, Cauchy deduced that

$$\frac{f(x_0 + h)}{F(x_0 + h)} = \frac{f^{(n)}(x_0 + \theta h)}{F^{(n)}(x_0 + \theta h)}, \quad \text{where} \quad 0 < \theta < 1. \quad (11.13)$$

He then let $h \to 0$ to deduce l'Hôpital's rule

$$\lim_{x \to x_0} \frac{f(x)}{F(x)} = \lim_{x \to x_0} \frac{f^{(n)}(x)}{F^{(n)}(x)}.$$

From this result, Cauchy derived Taylor's formula with Lagrange's remainder by taking $F(x) = (x - x_0)^n$ and replacing $f(x)$ by

$$g(x) = f(x) - f(x_0) - f'(x_0)(x - x_0) - \cdots - \frac{f^{(n-1)}(x_0)}{(n-1)!}(x - x_0)^{n-1};$$

this vanished at $x_0$, along with its first $n - 1$ derivatives. Then by (11.13),

$$g(x_0 + h) = \frac{h^n}{n!} g^{(n)}(x_0 + \theta h), \quad 0 < \theta < 1.$$

Since $g^{(n)}(x) = f^{(n)}(x)$, the required result followed.

Cauchy also obtained another form of the remainder[53] by defining a function $\phi(a)$ by the equation

$$f(x) = f(a) + \frac{x - a}{1} f'(a) + \frac{(x - a)^2}{1 \cdot 2} f''(a) + \cdots$$
$$+ \frac{(x - a)^{n-1}}{1 \cdot 2 \cdots (n - 1)} f^{(n-1)}(a) + \phi(a).$$

In (11.12), taking $F(x) = x, x_0 = a, h = x - a$, and $f = \phi$, he had

$$\phi(a) = \phi(x) + (a - x)\phi'(a + \theta(x - a)), \quad 0 < \theta < 1.$$

---

[53] ibid. pp. 147–148. Cauchy (1829) pp. 69–79, especially pp. 77 and 87.

Since

$$\phi(x) = 0 \quad \text{and} \quad \phi'(a) = -\frac{(x-a)^{n-1}}{1 \cdot 2 \cdots (n-1)} f^{(n)}(a),$$

he concluded

$$\phi(a) = \frac{(1-\theta)^{n-1}(x-a)^n}{1 \cdot 2 \cdots (n-1)} f^{(n)}(a + \theta(x-a)). \tag{11.14}$$

This remainder, called Cauchy's remainder, can also be obtained from the integral form of the remainder.

## 11.10   Cauchy: The Intermediate Value Theorem

Recall that the intermediate value theorem was regarded as intuitively or geometrically obvious by eighteenth-century mathematicians. For example, Lagrange and Laplace assumed it in their derivations of the remainder. Bolzano and Cauchy saw the need for a proof and each provided one. Cauchy stated and proved the theorem in his lectures,[54] published in 1821: Suppose $f(x)$ is a real function of $x$, continuous between $x_0$ and $X$. If $f(x_0)$ and $f(X)$ have opposite signs, then the equation $f(x) = 0$ is satisfied by at least one value between $x_0$ and $X$. In his proof, Cauchy first divided $[x_0, X]$ of length $h = X - x_0$ into $m$ parts to consider the sequence $f(x_0)$, $f(x_0 + \frac{h}{m})$, $f(x_0 + \frac{2h}{m})$, ..., $f(X - \frac{h}{m})$, $f(X)$. Since $f(x_0)$ and $f(X)$ had opposite signs, he had two consecutive terms, say, $f(x_1)$ and $f(X')$ with opposite signs. Clearly

$$x_0 \leq x_1 \leq X' \leq X \quad \text{and} \quad X' - x_1 = \frac{h}{m} = \frac{X - x_0}{m}.$$

We remark that Cauchy's notation was slightly different in that he used $<$ for $\leq$. He repeated the preceding process for the interval $[x_1, X']$ to get $x_1 \leq x_2 \leq X'' \leq X'$ with $X'' - x_2 = \frac{X - x_0}{m^2}$. Continuation of this procedure produced two sequences, $x_0 \leq x_1 \leq x_2 \leq \cdots$ and $X \geq X' \geq X'' \geq \cdots$, such that the differences between corresponding members of the two sequences became arbitrarily small. Thus, he had the two sequences converging to a common limit $a$. Now since $f$ was continuous between $x = x_0$ and $x = X$, the two sequences $f(x_0)$, $f(x_1)$, $f(x_2)$, ... and $f(X)$, $f(X')$, $f(X'')$, ... converged to $f(a)$. Since the signs of the numbers in the first sequence were opposite to the signs of the numbers in the second sequence, it followed by continuity that $f(a) = 0$. Since $x_0 \leq a \leq X$, Cauchy had the required result. Observe that Cauchy assumed that a sequence, now called a Cauchy sequence, must converge; this was later proved by Dedekind. Bolzano's slightly earlier proof of the intermediate value theorem had a similar deficiency, as he himself recognized in the 1830s.

---

[54] Cauchy (1989) Note V.

## 11.11 Exercises

(1) Following Johann Bernoulli, consider the differential equation $dy = \frac{\sqrt{a^2-y^2}}{a}\,dx$ for $y = \sin x$, and take $n = \frac{\sqrt{a^2-y^2}}{a}$, $dz = dx$ in Bernoulli's formula (11.6) to obtain

$$\frac{y}{\sqrt{a^2 - y^2}} = \frac{x - \frac{x^3}{2\cdot 3a^2} + \frac{x^5}{2\cdot 3\cdot 4\cdot 5a^4} - \cdots}{a - \frac{x^2}{2a} + \frac{x^4}{2\cdot 3\cdot 4a^3} - \cdots}.$$

Next consider the equation $dy = \frac{a\,dx}{a+x}$ and apply Bernoulli's method to obtain his series for $a\ln(\frac{a+x}{a})$:

$$y = \frac{ax}{a+x} + \frac{ax^2}{2(a+x)^2} + \frac{ax^3}{3(a+x)^3} + \frac{ax^4}{4(a+x)^4} + \cdots.$$

See Joh. Bernoulli (1968) vol. I, pp. 127–128. This paper was published in the *Acta Eruditorum* in 1694.

(2) Complete de Moivre's outline of a method to obtain Bernoulli's series for $\int y\,dz$. Note that this is similar to Newton's method of successive approximation. Let the fluent of $\dot{z}y$ be $zy - q$ so that $\dot{z}y = \dot{z}y + z\dot{y} - \dot{q}$ or $\dot{q} = z\dot{y}$. Now let $\dot{y} = \dot{z}v$ so that $\dot{q} = z\dot{z}v$. Take the fluent of each side to get $q = \frac{1}{2}zzv - r$ for some $r$. Then $z\dot{z}v = z\dot{z}v + \frac{1}{2}zz\dot{v} - \dot{r}$, so that $\dot{r} = \frac{1}{2}zz\dot{v}$. Set $\dot{v} = z\dot{s}$ and continue as before. De Moivre gave this argument in 1704. See Feigenbaum (1985) p. 93.

(3) Show that Bernoulli's series (11.6) is obtained by applying integration by parts to $\int n\,dz$ and then repeating the process infinitely often.

(4) In the Bernoulli series for $\int n\,dz$, set $n = f'(z)$ to obtain

$$f(z) - f(0) = zf'(z) - \frac{1}{2!}z^2 f''(z) + \frac{1}{3!}z^3 f'''(z) - \cdots. \tag{11.15}$$

Similarly, find the series for $f'(z) - f'(0)$, $f''(z) - f''(0)$, $f'''(z) - f'''(0), \ldots$ and use them to eliminate $f'(z)$, $f''(z)$, $f'''(z), \ldots$ from the right-hand side of (11.15). Show that the result is the Maclaurin series for $f(z)$. See Whiteside's footnote in Newton (1967–1981) vol. VII, p. 19.

(5) Show that all the derivatives of $f(x) = e^{-\frac{1}{x^2}}$ $(x \neq 0)$, $0(x = 0)$ are zero at $x = 0$. Cauchy remarked that the two functions $e^{-x^2}$ and $e^{-x^2} + e^{-\frac{1}{x^2}}$ had the same Maclaurin series. See Cauchy (1823) p. 230 and Cauchy (1829) p. 105.

(6) Show that the remainder in Taylor's theorem can be expressed in the form

$$\frac{h^n}{(n-1)!\,p}(1-\theta)^{n-p} f^{(n)}(x+\theta h), \quad 0 < \theta < 1,$$

due to Schlömilch and also published a decade later by E. Roche. See Prasad (1931) p. 90, Hobson (1957a) vol. 2, p. 200, and Schlömilch (1847) p. 179.

(7) Prove that if $f$, $\phi$, and $F$ are differentiable, then

$$\begin{vmatrix} f(x+h) & \phi(x+h) & F(x+h) \\ f(x) & \phi(x) & F(x) \\ f'(x+\theta h) & \phi'(x+\theta h) & F'(x+\theta h) \end{vmatrix} = 0,$$

for some $0 < \theta < 1$. This result and its generalization to $n+1$ functions is stated in Giuseppe Peano's *Calcolo differenzial* of 1884.

(8) Let $h > 0$. Set $m(x_1, x_2) = \frac{f(x_1)-f(x_2)}{x_1-x_2}$. Define the four derivatives of $f$:

$$f^+(x) = \overline{\lim_{h\to 0}}\, m(x+h, x),$$

$$f_+(x) = \underline{\lim_{h\to 0}}\, m(x+h, x),$$

$$f^-(x) = \overline{\lim_{h\to 0}}\, m(x-h, x),$$

$$f_-(x) = \underline{\lim_{h\to 0}}\, m(x-h, x).$$

Show that if $f(x)$ is continuous on $[a,b]$, then there is a point $x$ in $(a,b)$ such that either

$$f^+(x) \leq m(a,b) \leq f_-(x) \quad \text{or} \quad f^-(x) \geq m(a,b) \geq f_+(x).$$

The generalized mean value theorem is a corollary: If there is no distinction with respect to left and right with regard to the derivatives of $f(x)$, then there is a point $x$ in $(a,b)$ at which $f$ has a derivative and its value is equal to $m(a,b)$. See Young and Young (1909).

## 11.12   Notes on the Literature

Malet (1993) explains that Gregory could have obtained the Taylor rule without being in possession of a differential or equivalent technique. Feigenbaum (1985) presents a thorough discussion of Taylor's book, *Methodus Incrementorum*, as well as a treatment of the work of earlier mathematicians who contributed to the Taylor series. See also Feigenbaum (1981), containing an English translation of the *Methodus*. For later work on the Taylor series, especially the remainder term, see Pringsheim (1900).

Grabiner (1981) and (1990) are interesting sources for topics related to the work on series of Lagrange and Cauchy. Grabiner shows that, although Cauchy did not accept Lagrange's ideas on the foundations of calculus, Lagrange's use of algebraic inequalities nevertheless exerted a significant influence on Cauchy. She further points out that, a half century earlier, Maclaurin made brilliant use of inequalities to prove theorems in calculus. See her article, "Was Newton's Calculus a Dead End?" in Van Brummelen and Kinyon (2005). A look at Cauchy's 1820s lectures on calculus from a modern viewpoint is in Bressoud (2007).

# 12

# *Integration of Rational Functions*

## 12.1  Preliminary Remarks

The integrals of rational functions form the simplest class of integrals; they are included in a first course in calculus. Yet some problems associated with the integration of rational function have connections with the deeper aspects of algebra and of analysis. Examples are the factorization of polynomials and the evaluation of beta integrals. These problems have challenged mathematical minds as great as Newton, Johann Bernoulli, de Moivre, Euler, Gauss, and Hermite; indeed, they have their puzzles for us even today. For example, can a rational function be integrated without factorizing the denominator of the function?

Newton was the first mathematician to explicitly define and systematically attack the problem of integrating rational and algebraic functions. Of course, mathematicians before Newton had integrated some specific rational functions, necessary for their work. The Kerala mathematicians found the series for arctangent; N. Mercator[1] and Hudde worked out the series for the logarithm. In 1676, Leibniz met Hudde and one of Leibniz's short manuscripts contains notes, apparently made soon after that meeting, describing Hudde's mathematical work. The first few lines of these notes deal with the logarithm:[2]

> Hudde showed me that in the year 1662 he already had the quadrature of the hyperbola, which I found was the very same as Mercator also had discovered independently, and published. He showed me a letter written to a certain van Duck, of Leyden I think, on this subject.

Newton's work was made possible by his discovery, sometime in mid-1665, of the inverse relation between the derivative and the integral. At that time, he constructed tables, extending to some pages, of functions that could be integrated because they were derivatives of functions already explicitly or implicitly defined. He extended his tables by means of substitution or, equivalently, by use of the chain rule for derivatives. He further developed the tables by an application of the product rule for derivatives, or

---

[1] Mercator (1668).
[2] See Leibniz (1920) p. 123.

integration by parts, in his October 1666 tract on fluxions.[3] In this work, Newton viewed a curve dynamically: The variation of its coordinates $x$ and $y$ could be viewed as the motion of two bodies with velocities $p$ and $q$, respectively. He posed the problem of determining $y$ when $\frac{q}{p}$ was known and noted: "Could this ever bee done all problems whatever might bee resolved. But by $y^e$ following rules it may be very often done."[4]

After giving the already known rules for integrating $ax^{\frac{m}{n}}$ when $\frac{m}{n} \neq -1$ and when $\frac{m}{n} = -1$, Newton went on to consider examples, such as the integrals of $\frac{x^2}{ax+b}$, $\frac{x^3}{a^2-x^2}$, and $\frac{c}{a+bx^2}$. He did not take more complicated rational functions, perhaps because of a lack of an understanding of partial fractions. Instead, he evaluated integrals of some algebraic functions involving square roots. One result stated:[5]

If $\frac{cx^{3n}}{x\sqrt{ax^n+bx^{2n}}} = \frac{q}{p}$. Make $\sqrt{a+bx^n} = z$. $y^n$ [then] is

$$\frac{cx^n}{2nb}\sqrt{ax^n + bx^{2n}} - \square \frac{3ac}{2nbb}\sqrt{\frac{zz-a}{b}} = y. \tag{12.1}$$

The square symbol was the equivalent of our integral with respect to $z$, representing area.

Newton was led deeper into the integration of rational functions by a letter from Leibniz dated August 17, 1676, addressed to Oldenburg, but intended for all British mathematicians.[6] In this letter, Leibniz presented his series for $\pi$,

$$\frac{\pi}{4} = 1 - \frac{1}{3} + \frac{1}{5} - \frac{1}{7} + \cdots . \tag{12.2}$$

To obtain this series, Leibniz applied transmutation, a somewhat ad hoc method of finding the area of a figure by transforming it into another figure with the same area.[7] Grégoire St. Vincent, Pascal, Gregory, and others had employed this method before Leibniz.[8] In his reply to Leibniz, of October 1676,[9] Newton listed an infinitely infinite family of rational and algebraic functions, saying that he could integrate them. These included the four rational functions

$$\frac{dz^{\eta-1}}{e + fz^\eta + gz^{2\eta}}, \quad \frac{dz^{2\eta-1}}{e + fz^\eta + gz^{2\eta}} \quad \text{etc.,}$$

$$\frac{dz^{\frac{1}{2}\eta-1}}{e + fz^\eta + gz^{2\eta}}, \quad \frac{dz^{\frac{3}{2}\eta-1}}{e + fz^\eta + gz^{2\eta}} \quad \text{etc.,}$$

where $d$, $e$, $f$, and $g$ were constants and in the third and fourth expressions, in case $e$ and $g$ had the same sign, $4eg$ had to be $\leq f^2$. Newton went on to observe that the

[3] Newton (1967–1981) vol. 1, pp. 400–448.
[4] ibid. p. 403.
[5] ibid. p. 409.
[6] Newton (1959–1960) vol. 2, pp. 57–75.
[7] Leibniz (1920) pp. 42–47. See also Leibniz and Knobloch (1993) pp. 78–80.
[8] See Hofmann (1974) chapters 5–7.
[9] Newton (1959–1960) vol. 2, pp. 110–161.

expressions could become complicated,[10] "so that I hardly think they can be found by the transformation of the figures, which Gregory and others have used, without some further foundation. Indeed I myself could gain nothing at all general in this subject before I withdrew from the contemplation of figures and reduced the whole matter to the simple consideration of ordinates alone."

Newton then observed that Leibniz's series would be obtained by taking $\eta = 1$ and $f = 0$, and implicitly $e = g = 1$, in the first function. In fact,

$$\frac{\pi}{4} = \arctan 1 = \int_0^1 \frac{dz}{1 + z^2} = \int_0^1 (1 - z^2 + z^4 + \cdots)\, dz = 1 - \frac{1}{3} + \frac{1}{5} - \frac{1}{7} + \cdots .$$

As another series, he offered

$$\frac{\pi}{2\sqrt{2}} = 1 + \frac{1}{3} - \frac{1}{5} - \frac{1}{7} + \frac{1}{9} + \frac{1}{11} - \cdots \tag{12.3}$$

and explained that it could be obtained by means of a long calculation, setting $2eg = f^2$ and $\eta = 1$.[11] Again taking $e = g = 1$, this leads to the expression $\frac{1}{1 \pm \sqrt{2}x + x^2}$. He did not clarify any further and apparently Leibniz did not understand him, since even a quarter century later Leibniz had trouble with the integral arising in this situation. See Exercise 1 of this chapter.

Some of Newton's unpublished notes from this period suggest that he considered integrals of the form $\int \frac{1}{1 \pm x^m}$, since these integrals would lead to simple and interesting series. To express them in terms of standard integrals (that is, in terms of elementary functions such as the logarithm and arctangent), Newton had to consider the problem of factorizing $1 \pm x^m$ so that he could resolve the integrals into simpler ones by the use of partial fractions. We note that, in several examples in his 1670–71 treatise on fluxions and infinite series, he had broken up rational fractions into a sum of two fractions.[12] In the 1710s, Cotes and Johann Bernoulli and, to a lesser extent, Leibniz pursued the algebraic topic of partial fractions with more intensity than did Newton. It may be noted in this context that even in 1825 Jacobi was able to make an original contribution to partial fractions in his doctoral dissertation.[13]

Newton's method for finding the quadratic factors of $1 \pm x^m$ was to start with

$$(1 + nx + x^2)(1 - nx + px^2 - qx^3 + rx^4 - \cdots) = 1 \pm x^m \tag{12.4}$$

and then determine the pattern of the algebraic equations satisfied by $n$ for different values of $m$. In this way he factored $1 - x^3$, $1 + x^4$, $1 - x^5$, $1 + x^6$, $1 \pm x^8$, $1 \pm x^{12}$, though he apparently was unable to resolve the equation for $n$ when $m = 10$. As an example, Newton found the equation for $1 \pm x^4$ to be $n^3 - 2n = 0$, or $n = \pm\sqrt{2}$ and $n = 0$.[14] This would yield

[10] ibid. p. 138.
[11] ibid.
[12] Newton (1967–1981) vol. 3, p. 246.
[13] Jacobi (1969) vol. 3, pp. 1–44.
[14] Newton (1967–1981) vol. IV, p. 207.

$$x^4 + 1 = (x^2 + \sqrt{2}x + 1)(x^2 - \sqrt{2}x + 1), \tag{12.5}$$

and, of course

$$x^4 - 1 = (x^2 - 1)(x^2 + 1),$$

though Newton did not bother to write this last explicitly. Note that this factorization of $x^4 + 1$ was just what he needed to derive his series for $\frac{\pi}{2\sqrt{2}}$ in (12.3). He also recognized that values of $n$ were related to cosines of appropriate angles.[15] He was just a step away from Cotes's factorization of $x^n \pm a^n$.

Newton also considered the binomial $1 \pm x^7$ and found the equation for $n$ to be $n^6 - 5n^4 + 6nn - 1 = 0$. Note that the solution involved cube roots; Newton did not write the values of $n^2$, apparently because he wanted to consider only those values expressible, at worst, by quadratic surds. One wonders whether it occurred to him to ask which values of $m$ would lead to equations in $n$ solvable by quadratic radicals. In 1796, Gauss resolved this problem in his theory of constructible regular polygons.[16]

In 1702, since Newton's work remained unpublished, Johann Bernoulli and Leibniz in separate publications discussed the problem of factorizing polynomials, in connection with the integration of rational functions. In general, Leibniz and Bernoulli were of the opinion that integration of rational functions could be carried out by partial fractions, but the devil lay in the details. In his paper, "Specimen novum analyseos pro scientia infiniti circa summas et quadraturas," Leibniz factored[17]

$$x^4 + a^4 = (x + a\sqrt{\sqrt{-1}})(x - a\sqrt{\sqrt{-1}})(x + a\sqrt{-\sqrt{-1}})(x - a\sqrt{-\sqrt{-1}}). \tag{12.6}$$

He was puzzled by this factorization and wondered whether the integrals $\int \frac{dx}{x^4 + a^4}$ and $\int \frac{dx}{x^8 + a^8}$ could be expressed in terms of logarithms and inverse trigonometric functions. Bernoulli's paper, "Solution d'un problème concernant le calcul intégral"[18] also observed that the arctangent was related to the logarithm of imaginary values because

$$\int \frac{dx}{a^2 + x^2} = \frac{1}{2a} \left( \int \frac{dx}{a + ix} + \int \frac{dx}{a - ix} \right). \tag{12.7}$$

Cotes made the connection between the logarithm and the trigonometric functions even more explicit with his discovery of the formula

$$\log(\cos\theta + i\sin\theta) = i\theta. \tag{12.8}$$

Roger Cotes (1679–1716) is known for his factorization theorem, his work on approximate quadrature, and for editing the 1713 edition of the *Principia*. He studied

[15] ibid. p. 208.
[16] Gauss (1966) pp. 407–460, especially pp. 457–460.
[17] Leibniz (1971) vol. 5, pp. 350–361, especially p. 360.
[18] Bernoulli (1968) vol. 1, pp. 393–400.

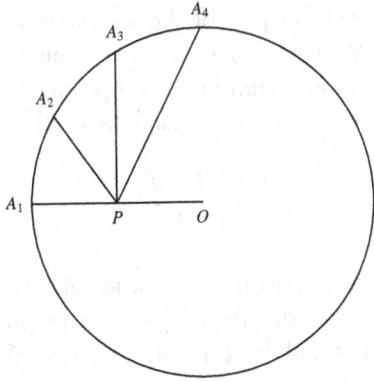

Figure 12.1 Cotes's factorization of $x^n - a^n$ as a property of the circle.

at Cambridge and became Fellow of Trinity College in 1704 and Plumian Professor of Astronomy and Experimental Philosophy in 1705. Unfortunately, he published only one paper in his lifetime, on topics related to the logarithmic function. Robert Smith published Cotes's mathematical writings in a 1722 work titled *Harmonia Mensurarum*. Formula (12.8) was stated geometrically:[19]

> For if some arc of a quadrant of a circle described with radius $CE$ has sine $CX$, and sine of the complement of the quadrant $XE$, taking radius $CE$ as modulus, the arc will be the measure of the ratio between $EX + XC\sqrt{-1}$ and $CE$, the measure having been multiplied by $\sqrt{-1}$.

Observe that this statement translates to $iR \ln(\cos\theta + i\sin\theta) = R\theta$, which is not quite correct, because the $i$ should be on the other side of the equation.

Cotes's factorization theorem was stated as a property of the circle. In Figure 12.1, the circumference of a circle of radius $a$ with center $O$ is divided into $n$ equal parts. A point $P$ lying on the line $OA_1$ and inside the circle is joined to each division point $A_1, A_2, A_3, \ldots, A_n$. Then, with $x = OP$,

$$PA_1 \cdot PA_2 \cdots PA_n = a^n - x^n. \tag{12.9}$$

Cotes noted that if $P$ lay outside the circle, product equaled $x^n - a^n$, and he had a similar result for the factorization of $x^n + a^n$.

Cotes wrote to William Jones on May 5, 1716,[20] that he had resolved by a general method the questions raised by Leibniz in his 1702 paper on the integration of rational functions. Unfortunately, Cotes died two months later, but Smith searched among his papers and unearthed the new method. Smith's note in his copy of the *Harmonia* stated:[21] "Sir Isaac Newton, speaking of Mr. Cotes said 'if he had lived we might have known something.'"

It is very likely that the source of Cotes's inspiration was Bernoulli's paper on the integration of rational functions, pointing out the connection between the logarithm

---

[19] Gowing (1983) p. 50.
[20] Rigaud (1841) vol. I, p. 271.
[21] Gowing (1983) p. 141.

and the arctangent. In the second part of the *Logometria* on integration published posthumously in 1722, Cotes wrote that the close connection between the measure of angles and measure of ratios (logarithms) had persuaded him to propose a single notation to designate the two measures. He used the symbol

$$R \left| \frac{R+T}{S} \right.$$

(12.10)

to stand for $R \ln \frac{R+T}{S}$ when $R^2$ was positive; when $R^2$ was negative, it represented $|R|\theta$, where $\theta$ was an angle such that the radius, tangent, and secant were in the ratio $R : T : S$. We should keep in mind that for Cotes, tangent and secant stood for $R \tan \theta$ and $R \sec \theta$. In his tables, he gave the single value

$$\frac{2}{e} R \left| \frac{R+T}{S} \right.$$

for the fluent of $\frac{\dot{x}}{e+fx^2}$; that is, for $\int \frac{dx}{e+fx^2}$, when $R = \sqrt{-\frac{e}{f}}$, $T = x$ and $S = \sqrt{\left(x^2 + \frac{e}{f}\right)}$. Recall that when $f < 0$, the integral is a logarithm and when $f > 0$, the integral is an arctangent. Cotes's notation distinguished the two cases by the interpretation of the symbols depending on $R$. This notation implies that when $R$ is replaced by $iR$ in the logarithm, we get the angular measure provided that $S$ and $T$ are replaced by $R \sec \theta$ and $R \tan \theta$. Thus, we have, where $C$ is the constant of integration,

$$i R \ln \frac{iR + R \tan \theta}{R \sec \theta} = R\theta + C.$$

When we take $C = -R\frac{\pi}{2}$ and $\theta - \frac{\pi}{2} = \phi$, this yields

$$\ln(\cos \phi + i \sin \phi) = i\phi.$$

It may be of interest to note that, as de Moivre and Euler showed, this result connecting logarithms with angles also served as the basis for Cotes's factorization formula. Surprisingly, Johann Bernoulli did not make any use of his discovery of the connection between the logarithm and the arctangent (12.7). British mathematicians such as Newton and Cotes were ahead of the Continental European mathematicians in the matter of integration of rational functions, but by 1720 the Continental mathematicians had caught up. In 1718, Brook Taylor challenged them to integrate rational functions of the form

$$\frac{x^{m-1}}{e + fx^m + gx^{2m}}.$$

Johann Bernoulli and Jakob Hermann, a former student of Jakob Bernoulli, responded with solutions.[22] In particular, they explained how the denominator could

---

[22] Bernoulli (1968) vol. 2, pp. 402–419 and Herman (1719).

be factored into two trinomials of the form $a + bx^{\frac{m}{2}} + cx^m$. And in 1720, Niklaus I Bernoulli showed how to deal with the integral of $\frac{x^{kq-1}}{(e+fx^q)^t}$.[23]

We can describe the Newton–Cotes–Bernoulli–Leibniz algorithm for integrating a rational function $f(x)$ with real coefficients by writing

$$f(x) = P(x) + \frac{N(x)}{D(x)},$$

where $P, N, D$ are polynomials with degree $N <$ degree $D$ and where $N$ and $D$ have 1 as their greatest common divisor. Factorize $D(x)$ into linear and quadratic factors so that their coefficients are real:

$$D(x) = c \prod_{i=1}^{n} (x - a_i)^{e_i} \prod_{j=1}^{m} (x^2 + b_j x + c_j)^{f_j}. \tag{12.11}$$

Then there are real numbers $A_{ik}$, $B_{jk}$, and $C_{jk}$ such that

$$f(x) = P + \sum_{i=1}^{n} \sum_{k=1}^{e_i} \frac{A_{ik}}{(x - a_i)^k} + \sum_{j=1}^{m} \sum_{k=1}^{f_j} \frac{B_{jk}x + C_{jk}}{(x^2 + b_j x + c_j)^k}. \tag{12.12}$$

From this it is evident that the result of the integration of $f(x)$ contains an algebraic part, consisting of a rational function; and a transcendental part, consisting of arctangents and logarithms.

Though Leibniz and Bernoulli had in principle solved the problem of the integration of rational functions, the practical problem of computing the constants $a, b, c$ and $A, B, C$ was formidable. In 1744, Euler tackled this problem in two long papers, running to 150 pages of the Petersburg Academy Journal (or 125 pages of vol. 17 of Euler's *Opera Omnia*). In these papers he explained in detail how to compute $A, B, C$ in (12.12) when the roots of the denominator were known. He also worked out a large number of special integrals of the form

$$\int \frac{x^m}{1 \pm x^n} \, dx, \tag{12.13}$$

where $m$ and $n$ were integers. By evaluating these integrals, Euler gained insight into several important topics. In fact, they provided him with new proofs of partial fractions expansions of trigonometric functions; evaluations of zeta and L-series values; the reflection formula for the gamma function; and infinite products for trigonometric functions. It is not surprising that Euler published several hundred pages on the integration of rational functions!

The problem of factoring a polynomial is a difficult one, so the partial fractions method has its drawbacks. A question raised in the nineteenth century was whether a part or all of the integral of a rational function could be obtained without factorizing the denominator. The Russian mathematician Mikhail Vasilyevich Ostrogradsky

---

[23] Bernoulli (1968) vol. 2 pp. 419–422.

published an algorithm in 1845 by which the rational part after integration could be obtained without factorization.[24] In 1873, Charles Hermite published a different algorithm and taught it in his courses at the École Polytechnique.[25] Ostrogradsky's algorithm was essentially rediscovered by E. Horowitz in his University of Wisconsin doctoral thesis.[26]

With the development of general computer algebra systems, the problem of mechanizing integration, including the integration of rational functions, has received new attention. The methods of Ostrogradsky and Hermite, along with others, have been important in the development of symbolic integration. The question of obtaining the logarithmic or arctangent portion of the integral of a rational function, without factorization of the denominator, has been resolved by a host of researchers. In these symbolic integration methods, the problem of factorization is replaced by the much more accessible problems of obtaining the greatest common divisors and/or resultants of polynomials. These last procedures in turn require polynomial division and the elimination of variables. Contributors to symbolic integration are many, including M. Bronstein, R. Risch, and M. F. Singer.[27]

## 12.2 Newton's 1666 Basic Integrals

In the beginning sections of his October 1666 tract on calculus, Newton tackled the problems of finding the areas under the curves $y = \frac{1}{c+x}$ and $y = \frac{c}{a+bx^2}$, equivalent to evaluating integrals of those functions. Recall, however, that seventeenth-century mathematicians thought in terms of curves, even those defined by equations, rather than functions. The variables in an equation were regarded as quantities or magnitudes on the same footing, rather than dependent and independent variables. Newton's two integrals were the building blocks for the more general integrals of rational functions. It is interesting to read what he said about these integrals. He first noted the rule that if

$$\frac{q}{p} = ax^{\frac{m}{n}}, \quad \text{then} \quad y = \frac{na}{m+n}x^{\frac{m+n}{n}}.$$

Note that in Newton's 1690s notation $\frac{q}{p}$ was written as $\frac{\dot{y}}{\dot{x}}$, whereas Leibniz wrote $\frac{dy}{dx}$. Newton next observed:[28]

Soe [so] if $\frac{q}{x} = \frac{q}{p}$. Then is $\frac{a}{0}x^0 = y$. soe $y^t$ [that] y is infinite. But note $y^t$ in this case $x$ & $y$ increase in $y^e$ [the] same proportion $y^t$ numbers & their logarithmes doe [do], $y$ being like a logarithme added to an infinite number $\frac{a}{0}$. [That is, $\int_0^x \frac{a}{t} dt = a \ln x - a \ln 0 = a \ln x + \frac{a}{0}$.] But if $x$ bee diminished by $c$, as if $\frac{a}{c+x} = \frac{q}{p}$, $y$ is also diminished by $y^e$ infinite number $\frac{a}{0}c^0$ & becomes finite like a logarithme of $y^e$ number $x$. & so $x$ being given, $y$ may bee mechanically found by a Table of logarithmes, as shall be hereafter showne.

[24] Ostrogradsky (1845).
[25] Hermite (1905–1917) vol. 3, pp. 40–44.
[26] Horowitz (1969).
[27] See Bronstein (1997).
[28] Newton (1967–1981) vol. 1, p. 403.

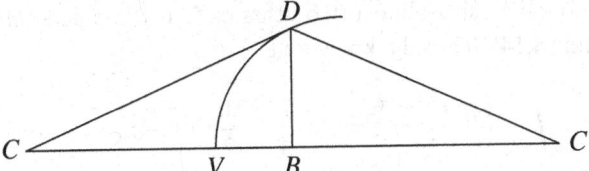

Figure 12.2 Newton's integration of a rational function.

Here Newton was explaining how the logarithm could be obtained by an application of the power rule, since he regarded this as the fundamental rule for integration. Newton clearly saw the difficulty of division by zero when the rule was applied to $\frac{a}{x}$. His method of dealing with this stumbling block can now be seen as an attempt to define the logarithm as a limit. We may say that Newton was describing the calculation:

$$y = \int_0^x \frac{a}{c+t} \, dt = \lim_{\epsilon \to 0} \int_0^x \frac{a}{(c+t)^{1-\epsilon}} \, dt$$

$$= \lim_{\epsilon \to 0} \left( \frac{a}{\epsilon} \left( (c+x)^\epsilon - c^\epsilon \right) \right) = a \ln \frac{c+x}{c}. \tag{12.14}$$

As for the integral of $\frac{1}{1+x^2}$, we would now give the result as arctan $x$. In Newton's time, the trigonometric quantities or functions were conceived of as line segments and their ratios constructed in relation to arcs of circles. It was therefore natural for Newton to connect the area under $y = \frac{1}{1+x^2}$ to the area of simpler or more well-known geometric objects such as conic sections. For this reason, he reduced the integral to the area of a sector of an ellipse.[29]

Consider the diagram in Newton's tract (Figure 12.2). Set $BD = v(x)$ and $CB = z(x)$, where $C$ is the point on the right-hand side. Let $z(t) = \frac{1}{\sqrt{1+t^2}}$ so that $tz(t) = \sqrt{1 - z^2(t)} = \frac{v(t)}{2}$. Thus, the curve $VD$ is an arc of the ellipse $\left(\frac{v}{2}\right)^2 + z^2 = 1$. Note that $CV = z(0) = 1$. In a one-line argument, Newton showed that $\int_0^x \frac{dt}{1+t^2}$ was equal to the area of sector $CVD$. To see this, observe that

$$\int_0^x \frac{1}{1+t^2} \, dt = \int_0^x z^2 \, dt = xz^2(x) - \int_1^{z(x)} 2zt \, dz. \tag{12.15}$$

Since $2zt = v$, the rightmost integral represents the area under the ellipse from $V$ to $B$; when the negative sign is included with the integral, the area under the ellipse from $B$ to $V$ is obtained. Moreover, $xz^2(x) = \frac{v(x)z(x)}{2} = BD \cdot \frac{CB}{2}$, and hence $xz^2(x)$ represents the area of the triangle $DBC$, completing the proof.

At this point, Newton may also have known that he could relate this area to an arc of the circle. Since $2zt = 2\sqrt{1 - z^2}$, the integral on the right of (12.15) is twice the area under the circle $y = \sqrt{1 - z^2}$ from $z(x)$ to 1. Recall that Newton had already

[29] ibid. p. 405.

related this area to the arcsine almost two years earlier, when generalizing a result of Wallis,[30] to obtain (8.14). Thus, he knew that

$$\int_{z(x)}^{1} 2zt\, dz = 2\int_{z(x)}^{1} \sqrt{1-z^2}\, dz = 2\left(\frac{\pi}{4} - \frac{1}{2}z\sqrt{1-z^2} - \frac{1}{2}\arcsin z\right). \quad (12.16)$$

When (12.15) is combined with (12.16), we get the integral in terms of the arctangent:

$$\int_{0}^{x} \frac{1}{1+t^2}\, dt = \frac{\pi}{2} - \arcsin z(x) = \arccos\frac{1}{\sqrt{1+x^2}} = \arctan x. \quad (12.17)$$

## 12.3  Newton's Factorization of $x^n \pm 1$

Most probably around 1676, Newton wrote his very sketchy notes[31] on this factorization, of which Whiteside has given a very helpful clarification. From these sources we learn that Newton's method of factoring $1 \pm x^m$ was to write

$$\left(1 + nx + x^2\right)\left(1 - nx + px^2 - qx^3 + rx^4 - \cdots \pm x^{m-2}\right) \equiv 1 \pm x^m \quad (12.18)$$

and equate the coefficients of $x^i$ for $2 \le i \le m-1$ to 0 in order to obtain equations satisfied by $n, p, q, r, \ldots$. By eliminating $p, q, r, \ldots$, he obtained the algebraic equation satisfied by $n$. For example, when $m = 4$, Newton's equations were $p + 1 - n^2 = 0$ and $pn - n = 0$. Note that the first equation multiplied by $n$ gives $pn + n - n^3 = 0$, and hence, by the second equation, $n^3 - 2n = 0$. One can then write $n = 0, \pm\sqrt{2}$. The first case gives the factorization

$$1 - x^4 = (1 - x^2)(1 + x^2)$$

and the second gives

$$1 + x^4 = \left(1 - \sqrt{2}x + x^2\right)\left(1 + \sqrt{2}x + x^2\right). \quad (12.19)$$

Recall that Newton applied this factorization to derive his series for $\frac{\pi}{2\sqrt{2}}$. His cryptic remark on his method was apparently insufficient for Leibniz to decipher, so in 1702 Leibniz could obtain only (12.6), leading him to wonder if $\int \frac{dx}{a^4+x^4}$ could be expressed in terms of logarithms and arctangents. One may get a sense of the ill will existing at that time between the supporters of Newton and those of Leibniz from a remark in a 1716 letter from Roger Cotes to William Jones:[32]

[30] Newton (1967–1981) vol. I, p. 108
[31] Newton (1967–1981) vol. 4, pp. 205–213.
[32] Rigaud (1848) vol. I, p. 271.

M. Leibnitz, in the Leipsic Acts of 1702 p. 218 and 219, has very rashly undertaken to demonstrate that the fluent of $\frac{\dot{x}}{x^4+a^4}$ cannot be expressed by measures of ratios and angles; and he swaggers upon the occasion (according to his usual vanity), as having by this demonstration determined a question of the greatest moment.

Using the same method as before, $m = 5$ gave Newton the equation $n^4 - 3n^2 + 1 = 0$, or $n^2 = \frac{3}{2} \pm \sqrt{\frac{5}{4}}$, or $n = \frac{1 \pm \sqrt{5}}{2}$. Thus, he got the factorization

$$1 - x^5 = \left(1 + \frac{1+\sqrt{5}}{2}x + x^2\right)\left(1 - \frac{1+\sqrt{5}}{2}x + \frac{1+\sqrt{5}}{2}x^2 - x^3\right).$$

Of course, the second factor could be further factorized as $(1 - x)(1 + \frac{1-\sqrt{5}}{2}x + x^2)$, though Newton did not write this out explicitly. Newton explicitly gave the factorization of $x^6 + 1$. Here the equations satisfied by the coefficients $n$, $p$, $q$, $r$ are $q = nr$, $p = qn - r$, $n = pn - q$, and $n^2 - p = 1$. This implies that

$$p = n^2 - 1, \quad q = n^3 - 2n, \quad r = n^4 - 3n^2 + 1, \qquad (12.20)$$

and hence the equation satisfied by $n$ is

$$n^5 - 4n^3 + 3n = 0.$$

This indicates the values of $n$ to be $0, \pm 1, \pm\sqrt{3}$. When $n = 0$, we have $r = 1$, $q = 0$, $p = -1$, and the factorization

$$1 + x^6 = (1 + x^2)(1 - x^2 + x^4). \qquad (12.21)$$

When $n = \sqrt{3}$, we have $r = 1$, $q = \sqrt{3}$, $p = 2$, and the factorization given by Newton was

$$1 + x^6 = \left(1 + \sqrt{3}x + x^2\right)\left(1 - \sqrt{3}x + 2x^2 - \sqrt{3}x^3 + x^4\right). \qquad (12.22)$$

It follows that the second factor in (12.21) is $1 - x^2 + x^4 = (1 + \sqrt{3}x + x^2)(1 - \sqrt{3}x + x^2)$, and the second factor in (12.22) can be written as $(1 + x^2)(1 - \sqrt{3}x + x^2)$.

Newton wrote down the polynomials satisfied by $n$ for $m = 3$ to $m = 12$ and solved the polynomials for those cases when $n$ could be expressed in terms of quadratic surds. In the case $m = 7$, he had the equation $n^6 - 5n^4 + 6n^2 - 1 = 0$. He wrote "$n^2 =$" next to the equation and filled in no value when he realized it would involve cube roots. For $m = 11$, he did not appear to expect the solutions to be in terms of quadratic surds and wrote nothing after the equation for $n$. He appears to have missed the factorization, noted by Whiteside,

$$n^9 - 8n^7 + 21n^5 - 20n^3 + 5n = n(n^4 - 5n^2 + 5)(n^4 - 3n^2 + 1),$$

yielding the quadratic surds for $n$ when $m = 10$.

Newton seems to have grasped the connection between the values of $n$ and the cosines of $\frac{2k\pi}{m}$ or $\frac{(2k-1)\pi}{m}$. He drew a diagram of a right triangle with one angle as $22\frac{1}{2}^\circ = \frac{\pi}{8}$ and noted that $2\cos\frac{\pi}{8}$ gave a value of $n$ when $m = 8$. At this point, he was just one step away from Cotes's factorization of $a^n \pm x^n$. Moreover, the number $\cos\frac{2\pi}{m}$ is related to the length of a side of a regular polygon of $m$ sides. Such a polygon is constructible when $\cos\frac{2\pi}{m}$ can be expressed in terms of quadratic surds. It is unlikely that Newton considered constructible polygons, but he may have wondered about conditions for $n$ to be expressed in quadratic surds. Thus, we have an interesting connection between Newton and Gauss, although Gauss could not have been aware of it because Newton did not publish his work, since it was incomplete.

## 12.4   Cotes and de Moivre's Factorizations

De Moivre presented his method of factorizing the more general trinomial

$$x^{2n} - 2\cos n\theta\, x^n + 1$$

in his *Miscellanea Analytica* of 1730.[33] His method depended on a formula he stated without proof: Let $l$ and $x$ be cosines of arcs $A$ and $B$, respectively, of the unit circle where $A$ is to $B$ as the integer $n$ to one. Then

$$x = \frac{1}{2}\sqrt[n]{l + \sqrt{l^2 - 1}} + \frac{1}{2}\frac{1}{\sqrt[n]{l + \sqrt{l^2 - 1}}}. \tag{12.23}$$

Note that this is equivalent to the formula named after de Moivre:

$$\cos\theta = \frac{1}{2}\left((\cos n\theta + i\sin n\theta)^{\frac{1}{n}} + (\cos n\theta - i\sin n\theta)^{\frac{1}{n}}\right). \tag{12.24}$$

De Moivre had published a similar result without proof for sine in a *Philosophical Transactions* paper of 1707.[34] In Chapter 13, we present a proof by Daniel Bernoulli in which he solved a difference equation obtained from the addition formula for cosine. Here we present Euler's simple proof given in his *Introductio in Analysin Infinitorum* of 1748.[35] Observe that by the addition formulas for sine and cosine

$$(\cos y \pm i\sin y)(\cos z \pm i\sin z) = \cos(y + z) \pm i\sin(y + z). \tag{12.25}$$

By taking $y = z$, Euler had

$$(\cos y \pm i\sin y)^2 = (\cos 2y \pm i\sin 2y).$$

When both sides were multiplied by $\cos y \pm i\sin y$, he got

$$(\cos y \pm i\sin y)^3 = \cos 3y \pm i\sin 3y.$$

[33]  de Moivre (1730a) pp. 1–2.
[34]  de Moivre (1707).
[35]  Euler (1988) p. 106.

Finally, it followed by induction that for a positive integer $n$,

$$(\cos y \pm i \sin y)^n = \cos ny \pm i \sin ny, \qquad (12.26)$$

completing the proof.

To obtain the factorization, de Moivre set $z = \sqrt[n]{l + \sqrt{l^2 - 1}}$ so that

$$z^n - l = \sqrt{l^2 - 1} \quad \text{or} \quad z^{2n} - 2lz^n + 1 = 0, \quad \text{where} \quad l = \cos n\theta.$$

By de Moivre's formula (12.23), $x = \frac{z + \frac{1}{z}}{2}$, or $z^2 - 2zx + 1 = 0$, where $x = \cos\theta$. De Moivre's theorem was therefore equivalent to the statement that $z^{2n} - 2\cos n\theta z^n + 1 = 0$, when $z^2 - 2\cos\theta z + 1 = 0$. Thus, $z^2 - 2xz + 1$ was a factor of $z^{2n} - 2lz^n + 1$. To obtain the other $n - 1$ factors, de Moivre observed that

$$(\cos A \pm i \sin A)^{\frac{1}{n}} = \cos\left(\frac{2k\pi \pm A}{n}\right) + i \sin\left(\frac{2k\pi \pm A}{n}\right), \quad k = 0, 1, 2, \dots.$$

The factorization thus obtained after taking $\theta = \frac{A}{n}$ may be written in modern notation as

$$z^{2n} - (2\cos A)z^n + 1 = \prod_{k=0}^{n-1}\left(z^2 - 2\cos\left(\frac{2k\pi + A}{n}\right)z + 1\right). \qquad (12.27)$$

We note that de Moivre used the symbol $C$ for $2\pi$. Cotes's factorization theorems are actually corollaries of de Moivre's (12.27). For example, let $C$ be a circle of radius $a$ and center $O$ with $B$ a point on the circumference and $P$ a point on $OB$ such that $OP = x$. Also let $A_1, A_2, \dots, A_n$ be points on the circumference such that, for $k = 1, 2, \dots, n$, the angle $BOA_k = \frac{(2k-1)\pi}{n}$. Then the product $A_1P \cdot A_2P \cdots A_nP$ is equal to $x^n + a^n$. This result of Cotes can be derived by taking $A = \pi$ in (12.27). But by taking $A = 0$, we obtain Cotes's result (12.9).

Cotes's factorization is most useful when viewed analytically; we will use it in the next section to integrate a specific rational function and later when discussing Euler's factorization of trigonometric functions. Set $A = 0$ in (12.27) so that the left-hand side is $(z^n - 1)^2$. Observe that the factor corresponding to $k = 0$ on the right-hand side is $(z-1)^2$ and the factor corresponding to $k \geq 1$ is identical with the factor corresponding to $n - k$, because

$$\cos\frac{2\pi k}{n} = \frac{2\pi(n-k)}{n}.$$

Thus

$$z^n - 1 = (z - 1) \prod_{k=1}^{\lfloor \frac{n-1}{2} \rfloor}\left(z^2 - 2\cos\frac{2\pi k}{n}z + 1\right).$$

Now, putting $z = \frac{x}{y}$ and multiplying by $y^n$, we obtain

$$x^n - y^n = (x - y) \prod_{k=1}^{\lfloor \frac{n-1}{2} \rfloor} \left( x^2 - 2\cos\frac{2\pi k}{n} xy + y^2 \right);$$

when $n$ is odd, we can write

$$x^n \pm y^n = (x \pm y) \prod_{k=1}^{\lfloor \frac{n-1}{2} \rfloor} \left( x^2 \pm 2\cos\frac{2\pi k}{n} xy + y^2 \right).$$

## 12.5   Euler: Integration of Rational Functions

In a letter to Niklaus I Bernoulli dated January 16, 1742,[36] Euler wrote of his efforts to evaluate integrals of the form

$$\int_0^\infty \frac{x^{m-1}dx}{1 \pm x^n},$$

where $m$ and $n$ were integers, and how such integrals could be applied to obtain, for the first time, the partial fractions expansions of trigonometric functions. Since Euler was in the habit of communicating his important ideas to his interested colleagues almost immediately, this letter indicates that Euler probably began work on integration of rational functions sometime in 1741. One difficulty in this task is the factorization of the denominator. Now Euler was aware from the work of de Moivre that a polynomial of degree $2n$ could be factorized if it took the form $x^{2n} + 2\cos\theta\, x^n + 1$. Thus, Euler considered integrals of rational functions,

$$\frac{x^{m-1}}{x^{2n} + 2\cos\theta\, x^n + 1}, \tag{12.28}$$

and in so doing, found a gold mine. He discovered results in this area that were intimately connected with many of his most significant and famous results in mathematical analysis: the gamma function, the product and partial fractions representations of trigonometric functions, the evaluation of zeta values, $\sum_{n=1}^\infty \frac{1}{n^{2k}}$, and the values of some $L$-series such as $\sum_{n=0}^\infty \frac{(-1)^n}{(2n+1)^{2k+1}}$. It is no wonder that Euler wrote many papers on the integration of rational functions and its connections with these topics, as he returned to them again and again, working out alternative proofs and extensions of his results. In 1742, he presented two papers to the Berlin Academy,[37] published in 1743. In these papers, Euler evaluated the integral of the expression (12.28) and, in particular, he showed that

---

[36] Eu. IV A-2 pp. 483–490.
[37] Eu. I-17 pp. 1–34, 35–69. E 59, E 60.

$$\int_0^\infty \frac{x^{m-1}}{1+x^{2n}}dx = \frac{\pi}{2n}\csc\frac{m\pi}{2n} = \int_0^1 \frac{x^{m-1}+x^{2n-m-1}}{1+x^{2n}}dx$$

$$= \frac{1}{m} + \frac{1}{2n-m} - \frac{1}{2n+m} - \frac{1}{4n-m} + \frac{1}{4n+m} + \cdots, \quad (12.29)$$

as well as a similar result for the principal value of $\int_0^\infty \frac{x^{m-1}}{1-x^{2n}}dx$. In his paper *De inventione integralium*,[38] he evaluated the integral by factorizing $1+x^{2n}$ using Cotes's formula and then obtained the indefinite integral as a sum of logarithms and arctangents. In his *Theoremata circa reductionem formularum integralium ad quadraturam circuli*,[39] Euler followed a different route to the evaluation of the integral in (12.29). He first showed that

$$\frac{\pi}{2n}\csc\frac{m\pi}{2n} = \frac{1}{m} + \frac{1}{2n-m} - \frac{1}{2n+m} - \frac{1}{4n-m} + \frac{1}{4n+m} - \cdots \quad (12.30)$$

by a method to be discussed in our Chapters 15 and 16. The series on the right-hand side of (12.30) is called the partial fractions expansion of $\frac{\pi}{2n}\csc\frac{m\pi}{2n}$. Next, Euler showed how to convert the series of fractions on the right-hand side of (12.30) into an integral. For that purpose, he considered the series

$$S(x) = \frac{x^m}{m} + \frac{x^{2n-m}}{2n-m} - \frac{x^{2n+m}}{2n+m} - \frac{x^{4n-m}}{4n-m} + \frac{x^{4n+m}}{4n+m} + \cdots$$

whose derivative would be the sum of two geometric series:

$$\frac{dS}{dx} = x^{m-1} + x^{2n-m-1} - x^{2n+m-1} - x^{4n-m-1} + x^{4n+m-1} + \cdots$$

$$= x^{m-1} - x^{2n+m-1} + x^{4n+m-1} - \cdots + x^{2n-m-1} - x^{4n-m-1} + \cdots$$

$$= x^{m-1}(1 - x^{2n} + x^{4n} - \cdots) + x^{2n-m-1}(1 - x^{2n} + x^{4n} - \cdots)$$

$$= \frac{x^{m-1}}{1+x^{2n}} + \frac{x^{2n-m-1}}{1+x^{2n}}. \quad (12.31)$$

Upon integrating (12.31), he obtained

$$S(1) - S(0) = \frac{1}{m} + \frac{1}{2n-m} - \frac{1}{2n+m} - \cdots$$

$$= \int_0^1 \frac{x^{m-1}}{1+x^{2n}}dx + \int_0^1 \frac{x^{2n-m-1}}{1+x^{2n}}dx$$

$$= \int_0^1 \frac{x^{m-1}}{1+x^{2n}}dx + \int_1^\infty \frac{y^{m-1}}{1+y^{2n}}dy \quad \left(\text{where } y = \frac{1}{x}\right)$$

$$= \int_0^\infty \frac{x^{m-1}}{1+x^{2n}}dx.$$

[38] E 60.
[39] E 59.

As we discuss in Chapter 16, Euler's first derivation of (12.30) met with objections based on questions of convergence and zeros of the sine function. He therefore sought and found an alternative proof using integration; for this proof, see Section 12.6. First note that in his papers during the 1740s, Euler habitually assumed that the reader knew the value of the integral

$$\int \frac{(Ax + B)dx}{b^2x^2 - 2ab\, x \cos \zeta + a^2}$$

in terms of logarithms and arctangents. However, in chapter 1 of the first volume of his book on integral calculus, presented to the Petersburg Academy in 1766 and published in 1768,[40] he gave the details of this calculation. He first noted that

$$\frac{d}{dx}(b^2x^2 - 2ab\, x \cos \zeta + a^2) = 2b^2x - 2ab\, \cos \zeta.$$

Thus

$$\int \frac{(Ax + B)dx}{b^2x^2 - 2ab\, x \cos \zeta + a^2}$$

$$= \frac{A}{2b^2} \int \frac{(2b^2x - 2ab \cos \zeta)dx}{b^2x^2 - 2ab\, x \cos \zeta + a^2} + \int \frac{(B + \frac{Aa}{b} \cos \zeta)dx}{(bx - a \cos \zeta)^2 + a^2 \sin^2 \zeta}. \quad (12.32)$$

Euler observed that the first integral on the right-hand side of (12.32) was

$$\frac{A}{2b^2} \ln(b^2x^2 - 2abx \cos \zeta + a^2); \quad (12.33)$$

for the second integral, he set $bx - a \cos \zeta = (a \sin \zeta)v$ so that $dx = \frac{a \sin \zeta}{b}dv$ and the integral became

$$\frac{Bb + Aa \cos \zeta}{ab^2 \sin \zeta} \int \frac{dv}{v^2 + 1} = \frac{Bb + Aa \cos \zeta}{ab^2 \sin \zeta} \arctan \frac{bx - a \cos \zeta}{a \sin \zeta}. \quad (12.34)$$

Euler then gave the arctangent in an alternate form, noting that since $\arctan \left( \frac{\cos \zeta}{\sin \zeta} \right)$ was a constant, it could be added to the arctangent in (12.34) without affecting the value of the indefinite integral. He had

$$\arctan \frac{bx - a \cos \zeta}{a \sin \zeta} + \arctan \left( \frac{\cos \zeta}{\sin \zeta} \right) = \arctan \frac{bx \sin \zeta}{a - bx \cos \zeta}; \quad (12.35)$$

by employing the formula

$$\arctan A + \arctan B = \arctan \frac{A + B}{1 - AB}.$$

---

[40] Eu. I-11. E 342.

Euler devoted two long papers, coming to a total of 150 pages, to the evaluation of indefinite integrals of rational functions.[41] Both were presented to the Petersburg Academy on the same day in 1748 and published consecutively in 1751 in the Academy journal. Both papers tackle nearly the same problems, though the second offers simplified methods in some cases. For example, in the first paper, Euler wrote the polynomial in the denominator of the integrand in the form

$$N(x) = (1 + px)(1 + qx)(1 + rx) \cdots ,$$

where some factors could be repeated. Suppose $M(x)$ is a polynomial whose degree is less than that of $N(x)$. Now allow that $1 + px$ is repeated exactly $n$ times and that

$$N(x) = (1 + px)^n A(x),$$

$$\frac{M(x)}{N(x)} = \frac{C_1}{1 + px} + \frac{C_2}{(1 + px)^2} + \cdots + \frac{C_n}{(1 + px)^n} + \frac{D(x)}{A(x)}, \qquad (12.36)$$

and

$$V(p) = p^{n-1} \frac{M\left(-\frac{1}{p}\right)}{A\left(-\frac{1}{p}\right)}. \qquad (12.37)$$

Euler showed after some work that

$$C_k = \frac{p}{(n-k)!} \frac{d^{n-k}}{dp^{n-k}} \left(\frac{V}{p^k}\right), \quad k = 1, 2, \ldots, n. \qquad (12.38)$$

In the second paper, Euler presented the formula for the case where the factors were of the form $(p + qx)^n$. He let $N(x) = (p + qx)^n S$ and

$$\frac{M(x)}{N(x)} = \frac{b_0}{(p + qx)^n} + \frac{b_1}{(p + qx)^{n-1}} + \cdots + \frac{b_{n-1}}{p + qx} + \frac{D(x)}{S(x)}. \qquad (12.39)$$

To find $b_j$, he then multiplied (12.39) by $(p + qx)^n$ to get

$$\frac{M(x)}{S(x)} = b_0 + b_1(p + qx) + \cdots + b_j(p + qx)^j + \cdots$$

$$+ b_{n-1}(p + qx)^{n-1} + \frac{D(x)}{S(x)}(p + qx)^n$$

and thus

$$b_j = \frac{1}{j! \, q^j} \left(\frac{d^j}{dx^j}\left(\frac{M}{S}\right)\right)_{x = -\frac{p}{q}}, \quad j = 0, 1, 2, \ldots, n - 1. \qquad (12.40)$$

Returning to our discussion of Euler's paper on definite integrals, "De inventione integralium," note that he started off with a consideration of sums of sines and cosines

[41] Eu. I-17 pp. 70–148, 149–194. E 162, 163.

that he required to evaluate definite integrals. For example, in Problem 5 of his paper he summed

$$S = a \sin \alpha + (a + b) \sin(\alpha + u) + (a + 2b) \sin(\alpha + 2u) + \cdots$$
$$+ (a + (p - 1)b) \sin(\alpha + (p - 1)u)$$

to obtain

$$\frac{(a - b) \sin \alpha - a \sin(\alpha - u) + (a + pb) \sin(\alpha + (p - 1)u) - (a + (p - 1)b) \sin(\alpha + pu)}{2 - 2 \cos u}.$$
$$(12.41)$$

To prove this, observe that

$$(2 \cos u)S = \sum_{k=0}^{p-1} (a + kb) \, 2 \sin(\alpha + ku) \cos u$$

$$= \sum_{k=0}^{p-1} (a + kb)\big( \sin(\alpha + (k - 1)u) + \sin(\alpha + (k + 1)u) \big)$$

$$= a \, \sin(\alpha - u) - (a - b) \sin \alpha + 2 \sum_{k=0}^{p-1} (a + kb) \sin(\alpha + ku)$$

$$- (a + pb) \sin(\alpha + (p - 1)u) + (a + (p - 1)b) \sin(\alpha + pu).$$

Rearranging the terms yields (12.41).

Later in the paper, Euler showed that for $0 < m < 2n$

$$\int_0^\infty \frac{x^{m-1}dx}{1 - x^{2n}} = \frac{\pi}{2n} \cot \frac{\pi}{2n};$$
$$(12.42)$$

in which we note that the integral is improper at $x = 1$, being undefined at that point. The first definition of an improper integral was later given by Cauchy in the 1820s.[42] In fact, the integral in (12.42) does not exist; Euler had in fact found its principal value, also defined by Cauchy[43] as

$$\lim_{\substack{\epsilon \to 0^+ \\ b \to \infty}} \left( \int_0^{1-\epsilon} + \int_{1+\epsilon}^b \right) \frac{x^{m-1}}{1 - x^{2n}} \, dx.$$

Euler noted that, by Cotes's factorization,

$$1 - x^{2n} = (1 - x^2) \prod_{k=1}^{n-1} \left( 1 + 2x \cos \frac{k\pi}{n} + x^2 \right).$$
$$(12.43)$$

---

[42] Cauchy (1823) p. 93.
[43] ibid. p. 96.

He first determined the value of the indefinite integral; note that we will give the details of this evaluation in the next section, when we consider a more general integral. For the indefinite integral, he had

$$\int \frac{x^{m-1}}{1-x^{2n}}\,dx = \frac{(-1)^{m-1}}{2n}\big(\ln(1+x) - \ln(1-x)\big)$$

$$+ \frac{(-1)^{m-1}}{2n}\sum_{k=1}^{n-1}\cos\frac{km\,\pi}{n}\ln\left(1+2x\cos\frac{(2k-1)\pi}{n}+x^2\right)$$

$$+ \frac{(-1)^{m-1}}{2n}\sum_{k=1}^{n-1}\sin\frac{km\,\pi}{n}\arctan\frac{x\sin\frac{k\pi}{n}}{1+x\cos\frac{k\pi}{n}}+C. \quad (12.44)$$

He observed that if $x$ were replaced by $\frac{1}{x}$, then

$$\int \frac{x^{m-1}}{1-x^{2n}}\,dx = -\int \frac{x^{2n-m-1}}{1-x^{2n}}\,dx,$$

so that

$$\int_0^\infty \frac{x^{m-1}}{1-x^{2n}}\,dx = -\int_0^1 + \int_1^\infty \frac{x^{m-1}}{1-x^{2n}}\,dx$$

$$= \int_0^1 \frac{x^{m-1}-x^{2n-m-1}}{1-x^{2n}}\,dx.$$

He next expanded $\frac{1}{1-x^{2n}}$ as a geometric series, $1+x^{2n}+x^{4n}+\cdots$, to obtain

$$\int_0^\infty \frac{x^{m-1}}{1-x^{2n}}\,dx = \int_0^1 (x^{m-1}-x^{2n-m-1})(1+x^{2n}+x^{4n}+\cdots)dx$$

$$= \frac{1}{m} - \frac{1}{2n-m} + \frac{1}{2n+m} - \frac{1}{4n-m} + \frac{1}{4n+m} - \frac{1}{6n-m} + \cdots. \quad (12.45)$$

Since $(-1)^{2n-m} = (-1)^m$ and $\cos k\frac{(2n-m)\pi}{n} = \cos\frac{km\pi}{n}$, by applying (12.44), Euler had

$$\int \frac{x^{2n-m-1}}{1-x^{2n}}\,dx = \frac{(-1)^{m-1}}{2n}\big(\ln(1+x) - \ln(1-x)\big)$$

$$+ \frac{(-1)^{m-1}}{2n}\sum_{k=1}^{n-1}\cos\frac{km\,\pi}{n}\ln\left(1+2x\cos\frac{(2k-1)x}{n}+x^2\right)$$

$$- \frac{(-1)^{m-1}}{n}\sum_{k=1}^{n-1}\sin\frac{km\,\pi}{n}\arctan\frac{x\sin\frac{k\pi}{n}}{1+x\cos\frac{k\pi}{n}}+C. \quad (12.46)$$

When (12.46) was subtracted from (12.42), all logarithmic terms cancelled; hence

$$\int \frac{x^{m-1} - x^{2n-m-1}}{1 - x^{2n}} dx = 2\frac{(-1)^{m-1}}{n} \sum_{k=1}^{n-1} \sin \frac{km\pi}{n} \arctan \frac{x \sin \frac{k\pi}{n}}{1 + x \cos \frac{k\pi}{n}}$$

and thus

$$\int_0^1 \frac{x^{m-1} - x^{2n-m-1}}{1 - x^{2n}} dx = 2\frac{(-1)^{m-1}}{n} \sum_{k=1}^{n-1} \sin \frac{km\pi}{n} \arctan \frac{\sin \frac{k\pi}{n}}{1 + \cos \frac{k\pi}{n}}.$$

Next, since

$$\frac{\sin \frac{k\pi}{n}}{1 + \cos \frac{k\pi}{n}} = \frac{2 \sin \frac{k\pi}{2n} \cos \frac{k\pi}{2n}}{2 \cos^2 \frac{k\pi}{2n}} = \tan \frac{k\pi}{2n},$$

$$\arctan \frac{\sin \frac{k\pi}{n}}{1 + \cos \frac{k\pi}{n}} = \arctan \left( \tan \frac{k\pi}{2n} \right) = \frac{k\pi}{2n}.$$

Thus, Euler got

$$\int_0^\infty \frac{x^{m-1}}{1 - x^{2n}} dx = \int_0^1 \frac{x^{m-1} - x^{2n-m-1}}{1 - x^{2n}} dx$$

$$= \frac{(-1)^{m-1}\pi}{n^2} \sum_{k=1}^{n-1} k \sin \frac{km\pi}{n}. \qquad (12.47)$$

Now Euler evaluated the sum contained in (12.47) by taking $p = n$, $\alpha = 0$, $a = 0$, $b = 1$, and $u = \frac{m\pi}{n}$ in (12.41):

$$\sum_{k=1}^{n-1} k \sin \frac{km\pi}{n} = \frac{n \sin \left( (n-1)\frac{m\pi}{n} \right) - (n-1) \sin(m\pi)}{2 - 2 \cos \frac{m\pi}{n}}$$

$$= \frac{n \sin \left( m\pi - \frac{m\pi}{n} \right)}{4 \sin^2 \frac{m\pi}{2n}}$$

$$= \frac{n(-1)^{m-1} \sin \frac{m\pi}{n}}{4 \sin^2 \frac{m\pi}{2n}}$$

$$= (-1)^{m-1} \frac{n \cos \frac{m\pi}{2n}}{2 \sin \frac{m\pi}{2n}}.$$

Hence

$$\int_0^\infty \frac{x^{m-1}}{1 - x^{2n}} dx = \frac{\pi}{2n} \cot \frac{m\pi}{2n}. \qquad (12.48)$$

Euler combined (12.45) with (12.48) to discover his famous partial fractions expansion of the cotangent function:

$$\frac{\pi}{2n}\cot\frac{m\pi}{2n} = \frac{1}{m} - \frac{1}{2n-m} + \frac{1}{2n+m} - \frac{1}{4n-m} + \frac{1}{4n+m} - \cdots . \quad (12.49)$$

We remark that this identity led Euler to one of his half-dozen or more evaluations of $\sum_{n=1}^{\infty}\frac{1}{n^{2k}}$, where $k$ is a positive integer. We work out some of these evaluations in Chapters 15 and 16. In his "De inventione integralium," Euler considered the integral of (12.28), but he did not include a result for this integral similar to (12.45). Whether or not he was aware of such a result in 1748, he did not publish it until thirty years later and we discuss it in the next section.

## 12.6 Euler's "Investigatio Valoris Integralis"

Euler's paper "Investigatio valoris integralis $\int \frac{x^{m-1}dx}{1-2x^k\cos\theta+x^{2k}}$ a termino $x = 0$ ad $x = \infty$ extensi" was presented to the Petersburg Academy in 1775 but was published ten years later by the Academy in the second volume of his *Opuscula analytica*.[44] To evaluate the integral in the title of the paper, he used methods very similar to those he employed in his 1743 "De inventione integralium." However, he found some new corollaries such as the Fourier series for $\cos\lambda x$ and $\sin\lambda x$; these appeared as the partial fractions expansion for the integral. Note that the denominator of the integrand can be factorized by de Moivre's formula (12.27) and the integrand may be expressed as a sum of partial fractions when $m \leq 2k$:

$$\frac{x^{m-1}}{1 - 2x^k\cos\theta + x^{2k}} = \sum_{s=0}^{k-1} \frac{A_s + B_s x}{1 - 2x\cos\left(\frac{2s\pi+\theta}{k}\right) + x^2}. \quad (12.50)$$

To calculate the $A$s and $B$s, Euler observed that

$$1 - 2x\cos\omega_s + x^2 = (x - e^{i\omega_s})(x - e^{-i\omega_s}).$$

Here note that Euler did not employ subscripts and except on rare occasions he wrote $\cos\omega_s + \sqrt{-1}\sin\omega_s$ instead of $e^{i\omega_s}$. In any case, he let $\omega_s = \frac{2s\pi+\theta}{k}$ and set

$$\frac{A_s + B_s x}{1 - 2x\cos\omega_s + x^2} = \frac{f_s}{x - e^{i\omega_s}} + \frac{g_s}{x - e^{-i\omega_s}}. \quad (12.51)$$

Now by multiplying equation (12.51) by the denominator of its left-hand side, he had

$$A_s + B_s x = f_s(x - e^{i\omega_s}) + g_s(x - e^{i\omega_s})$$

[44] Eu. I-18 pp. 190–208. E 589.

and equating coefficients of the powers of $x$, Euler arrived at

$$B_s = f_s + g_s \quad \text{and} \quad A_s = -(f_s e^{i\,\omega_s} + g_s e^{i\,\omega_s}). \tag{12.52}$$

To find $f_s$, Euler noted that (12.50) when multiplied by $x - e^{i\,\omega_s}$ yielded

$$\frac{x^{m-1}(x - e^{i\,\omega_s})}{1 - 2x^k \cos\theta + x^{2k}} = f_s + R_s(x - e^{i\,\omega_s}), \tag{12.53}$$

where the $R_s$ consisted of the remaining partial fractions. He let $x \to e^{i\,\omega_s}$. Now denoting $f(x) = x^{2k} - 2x^k \cos\theta + 1$ and $a = e^{i\,\omega_s}$, one gets $f(a) = 0$. Also, since $\omega_s = \frac{2s\pi + \theta}{k}$, and using l'Hôpital's rule:

$$f_s = \lim_{x \to a} \frac{x^{m-1}(x - a)}{f(x) - f(a)} = \frac{a^{m-1}}{f'(a)} = \frac{a^m}{a f'(a)} \tag{12.54}$$

$$= \frac{e^{im\omega_s}}{2k e^{2ik\omega_s} - 2k e^{ik\omega_s}\cos\theta}$$

$$= \frac{e^{im\omega_s}}{2k e^{2i\theta} - 2k e^{i\theta}\cos\theta}, \tag{12.55}$$

simplifying to

$$= \frac{e^{im\omega_s}}{2ki\,\sin\theta e^{i\theta}}. \tag{12.56}$$

As Euler remarked, $g_s$ could be obtained from $f_s$ by changing every $\sqrt{-1}$ to $-\sqrt{-1}$ or, in other words, $i$ to $-i$, since $g_s$ is the complex conjugate of $f_s$. So

$$g_s = \frac{e^{-im\,\omega_s}}{-2ki\,\sin\theta\,e^{-i\theta}}$$

and then by (12.52)

$$B_s = \frac{e^{i(m\,\omega_s - \theta)} - e^{-i(m\,\omega_s - \theta)}}{2i(k\,\sin\theta)} = \frac{\sin(m\,\omega_s - \theta)}{k\,\sin\theta} \tag{12.57}$$

$$A_s = -\left(\frac{e^{i(m-1)\omega_s - i\theta} - e^{-i(m-1)+i\theta}}{2i(k\,\sin\theta)}\right) = -\frac{\sin\big((m-1)\omega_s - \theta\big)}{k\,\sin\theta}. \tag{12.58}$$

Now

$$\int \frac{(A_s + B_s x)\,dx}{1 - 2x\cos\omega_s + x^2} = \int \left(\frac{B_s(x - \cos\omega_s)}{1 - 2x\cos\omega_s + x^2} + \frac{(A_s + B_s\cos\omega_s)}{1 - 2x\cos\omega_s + x^2}\right) dx$$

$$= \frac{1}{2}B_s \ln(1 - 2x\cos\omega_s + x^2) + \frac{A_s + B_s\cos\omega_s}{\sin\omega_s} \arctan\frac{x\sin\omega_s}{1 - x\cos\omega_s}. \tag{12.59}$$

Thus, by (12.50),

$$\int_0^\infty \frac{x^{m-1}dx}{1 - 2x^k \cos\theta + x^{2k}} \tag{12.60}$$

was shown to be a sum of the integrals in (12.59) as $s$ ranged from 0 to $k - 1$. Now note that the integrals are on the interval $(0, \infty)$. At $x = 0$, the value of the logarithm is zero, as is the value of the arctangent. Euler observed that for large $x$

$$\ln(1 - 2x \cos\omega_s + x^2) = \ln x^2 + \ln\left(\frac{1}{x^2} - \frac{2\cos\omega_s}{x} + 1\right) \approx 2\ln x.$$

It is now clear that, by (12.60) and because $\omega_s = \frac{2s\pi+\theta}{k}$, the logarithmic part of the integral in that formula must behave as two times the sum

$$\ln x \sum_{s=0}^{k-1} B_s = \frac{\ln x}{k \sin\theta} \sum_{s=0}^{k-1} \sin\left((m - k)\frac{\theta}{k} + 2s\frac{m\pi}{k}\right)$$

$$= \frac{\ln x}{k \sin\theta} \sum_{s=0}^{k-1} \sin(2s\alpha + \zeta), \tag{12.61}$$

where $\alpha = \frac{m\pi}{k}$ and $\zeta = \frac{(m-k)\theta}{k}$. Euler noted that

$$2\sin\alpha \sin(2s\alpha + \zeta) = \cos\left((2s - 1)\alpha - \zeta\right) - \cos\left((2s + 1)\alpha + \zeta\right), \tag{12.62}$$

meaning that when (12.62) was summed over the range $s = 0$ to $s = k-1$, cancellation would leave only the first and last terms remaining. Thus

$$2\sin\alpha \sum_{s=0}^{k-1} \sin(2s\alpha + \zeta) = \cos(\alpha - \zeta) - \cos\left((2k - 1)\alpha + \zeta\right). \tag{12.63}$$

Now the sum of the angles $\alpha - \zeta$ and $(2k - 1)\alpha + \zeta$ is $2k\alpha = 2m\pi$; hence, the cosines of these two angles are the same and the sums in (12.63) and (12.61) are zero, or

$$\ln x \sum_{s=0}^{k-1} B_s = 0.$$

Euler thus demonstrated that (12.60) had a vanishing logarithmic part. He proceeded to calculate the arctangent part in (12.59); though he did not employ the language of limits, we use them for convenience:

$$\lim_{x\to\infty} \arctan\left(\frac{x \sin\omega_s}{1 - x \cos\omega_s}\right) = \arctan(-\tan\omega_s)$$

$$= \arctan(\tan(\pi - \omega_s))$$

$$= \pi - \omega_s.$$

In addition, by (12.57) and (12.58), using the addition formula for sine, and noting that $(m-1)\omega_s - \theta = (m\,\omega_s - \theta) - \omega_s$, Euler did a short calculation to obtain

$$\frac{A_s + B_s \cos \omega_s}{\sin \omega_s} = \frac{\cos(m\omega_s - \theta)}{k \sin \theta}.$$

Therefore, the sum of the arctangents at $x = \infty$ is

$$\frac{1}{k \sin \theta} \sum_{s=0}^{k-1} (\pi - \omega_s) \cos(m\omega_s - \theta). \tag{12.64}$$

Set $\frac{\pi}{k} = \beta$, $\pi - \frac{\theta}{k} = \gamma$ and let $\alpha, \zeta$ be as before. Euler denoted the sum in (12.64), omitting the factor $\frac{1}{k \sin \theta}$, by $S$ and noted that, since

$$\pi - \omega_s = \gamma - 2s\beta \quad \text{and} \quad m\,\omega_s - \theta = 2s\alpha + \zeta,$$

the addition formula for the sine function implied

$$2S \sin \alpha = \sum_{s=0}^{k-1} 2(\gamma - 2s\beta) \cos(2s\alpha + \zeta) \sin \alpha$$

$$= \sum_{s=0}^{k-1} (\gamma - 2s\beta)\big(\sin((2s+1)\alpha + \zeta) - \sin((2s-1)\alpha + \zeta)\big)$$

$$= -\gamma \sin(-\alpha + \zeta) + \big(\gamma - 2(k-1)\beta\big) \sin\big((2k-1)\alpha + \zeta\big)$$

$$+ \sum_{s=1}^{k-1} \big(\gamma - 2s\beta - (\gamma - 2(s+1)\beta)\big) \sin\big((2s-1)\alpha + \zeta\big)$$

$$= \gamma \sin(\alpha - \zeta) + (\gamma - 2(k-1)\beta) \sin((2k-1)\alpha + \zeta) + \beta T \tag{12.65}$$

where $\quad T = 2(\sin(\alpha + \zeta) + \sin(3\alpha + \zeta) + \cdots + \sin((2k-3)\alpha + \zeta)$.

Euler summed $T$ by using the addition formula for cosine, arriving at

$$T \sin \alpha = \cos \zeta - \cos(2\alpha + \zeta) + \cos(2\alpha + \zeta) - \cos(4\alpha + \zeta) + \cdots$$

$$+ \cos\big((2k-4)\alpha + \zeta\big) - \cos\big((2k-2) + \zeta\big)$$

$$= \cos \zeta - \cos\big((2k-2)\alpha + \zeta\big)$$

$$= \cos \frac{\theta(k-m)}{k} - \cos \frac{2m\pi + \theta(k-m)}{k}$$

$$= 2 \sin \frac{m\pi + \theta(k-m)}{k} \sin \frac{m\pi}{k}$$

$$= 2 \sin(\alpha - \zeta) \sin \alpha.$$

Thus, $T = 2 \sin(\alpha - \zeta)$.

Euler substituted this value of $T$ in (12.65) and applied the identity $\sin A + \sin B = 2 \sin \frac{A+B}{2} \cos \frac{A-B}{2}$ so that

$$
\begin{aligned}
2S \sin \alpha &= (\gamma + 2\beta) \sin(\alpha - \zeta) + (\gamma - 2(k-1)\beta) \sin((2k-1)\alpha + \zeta) \\
&= (\gamma + 2\beta)(\sin(\alpha - \zeta) + \sin((2k-1)\alpha + \zeta)) - 2k\,\beta \sin((2k-1)\alpha + \zeta) \\
&= (2\gamma + 4\beta) \sin \alpha k \cos((k-1)\alpha + \zeta) - 2k\beta \sin((2k-1)\alpha + \zeta).
\end{aligned}
$$
$$(12.66)$$

Now in (12.66), $\sin \alpha k = \sin m\pi = 0$. So

$$
S = -k\beta \frac{\sin((2k-1)\alpha + \zeta)}{\sin \alpha} = \pi \frac{\sin \frac{m\pi + \theta(k-m)}{k}}{\sin \frac{m\pi}{k}}.
$$

Thus, Euler had the final result

$$
\int_0^\infty \frac{x^{m-1}}{1 - 2x^k \cos\theta + x^{2k}}\,dx = \frac{\pi \sin \frac{m(\pi-\theta)+k\theta}{k}}{k \sin\theta \sin \frac{m\pi}{k}}.
$$
$$(12.67)$$

The special case $\theta = \frac{\pi}{2}$ gave Euler the value of the beta integral

$$
\int_0^\infty \frac{x^{m-1}}{1 + x^{2k}}\,dx = \frac{\pi}{2k \sin \frac{m\pi}{2k}},
$$
$$(12.68)$$

and using integration by parts, or from $\theta = \pi$ in (12.67), he obtained

$$
\int_0^\infty \frac{x^{m-1}\,dx}{(1 + x^k)^2} = \frac{(1 - \frac{m}{k})\pi}{k \sin \frac{m\pi}{k}}.
$$
$$(12.69)$$

In fact, Euler himself mentioned his use of integration by parts in this case in section 34 of his 1743 paper on integration, published by the Berlin Academy.[45]

At this point, Euler set $\pi - \theta = \eta$, so that $\sin\theta = \sin\eta$ and $\cos\theta = -\cos\eta$ in (12.67), and obtained

$$
\int_0^\infty \frac{x^{m-1}\,dx}{1 + 2x^k \cos\eta + x^{2k}} = \frac{\pi \sin(1 - \frac{m}{k})\eta}{k \sin\eta \sin \frac{m\pi}{k}}.
$$
$$(12.70)$$

He next gave an interesting method for expanding the integrand of (12.70) as a series in powers of x. He wrote

$$
\frac{\sin\eta}{1 + 2x^k \cos\eta + x^{2k}} = \sin\eta + Ax^k + Bx^{2k} + Cx^{3k} + Dx^{4k} + \cdots,
$$
$$(12.71)$$

multiplied this by the denominator of its left-hand side $(1 + 2x^k \cos\eta + x^{2k})$, and arrived at

---

[45] Eu. I-17 pp. 35–69. E 60.

$$\sin \eta = (1 + 2x^k \cos \eta + x^{2k})(\sin \eta + Ax^k + Bx^{2k} + Cx^{3k} + \cdots)$$
$$= \sin \eta + (A + 2 \sin \eta \cos \eta)x^k + (B + 2A \cos \eta + \sin \eta)x^{2k} + \cdots.$$

Equating coefficients of the powers of $x$, Euler could write the relations:

$$2A + 2 \sin \eta \cos \eta = 0 \quad \text{or} \quad A = - \sin 2\eta$$
$$B + 2A \cos \eta + \sin \eta = 0 \quad \text{or} \quad B = 2 \sin 2\eta \cos \eta - \sin \eta$$
$$= \sin(2\eta + \eta) + \sin(2\eta - \eta) + \sin \eta$$
$$= \sin 3\eta$$
$$C + 2B \cos \eta + A = 0 \quad \text{or} \quad C = -(2 \sin 3\eta \cos \eta - \sin 2\eta)$$
$$= -\big( \sin(3\eta + \eta) + \sin(3\eta - \eta) - \sin 2\eta \big)$$
$$= - \sin 4\eta$$
$$D + 2 \cos \eta + B = 0 \quad \text{or} \quad D = \sin 5\eta,$$

and so on. Thus, after substituting these values in (12.71):

$$\frac{\sin \eta}{1 + 2x^k \cos \eta + x^{2k}} = \sin \eta - \sin 2\eta \, x^k + \sin 3\eta \, x^{2k} - \sin 4\eta \, x^{3k} + \cdots . \quad (12.72)$$

Euler noted that

$$\sin \eta \int_0^\infty \frac{x^{m-1}dx}{1 + 2x^k \cos \eta + x^{2k}} = \sin \eta \left( \int_0^1 + \int_1^\infty \right) \frac{x^{m-1}dx}{1 + 2x^k \cos \eta + x^{2k}}$$
$$= \sin \eta \int_0^\infty \frac{x^{m-1} + x^{2k-m-1}}{1 + 2x^k \cos \eta + x^{2k}} dx, \quad (12.73)$$

where the term $x^{2k-m-1}$ was obtained by changing $x$ to $\frac{1}{x}$ in the integral $\int_1^\infty$. On term by term integration and using (12.72) in (12.73), and applying (12.70), he had

$$\frac{\pi \sin \left( 1 - \frac{m}{k} \right) \eta}{k \sin \frac{m\pi}{k}} = \int_0^1 \left( x^{m-1} + x^{2k-m-1} \right) \sum_{s=1}^\infty (-1)^{s-1} x^{(s-1)k} \sin s\eta$$
$$= \sum_{s=1}^\infty (-1)^{s-1} \left( \frac{1}{m + (s-1)k} + \frac{1}{(s+1)k - m} \right) \sin s\eta. \quad (12.74)$$

Euler next assumed that $\eta$ was an infinitesimal, so that

$$\sin \left( 1 - \frac{m}{k} \right) \eta = \left( 1 - \frac{m}{k} \right) \eta, \quad \sin \eta = \eta, \quad \sin 2\eta = 2\eta, \quad \text{and so on;}$$

after division by $\eta$, he obtained

$$\frac{\left( 1 - \frac{m}{k} \right) \pi}{k \sin \frac{m\pi}{k}} = \sum_{s=1}^\infty (-1)^{s-1} \frac{2s^2 k}{\big(m + (s-1)k\big)\big((s+1)k - m\big)}. \quad (12.75)$$

In modern terms, Euler obtained (12.75) by dividing (12.74) by $\eta$, allowing $\eta \to 0$, and applying $\lim_{\eta \to 0} \frac{\sin s\eta}{\eta} = s$. He set $m = k-n$ to put (12.75) into more symmetrical form:

$$\frac{\pi \sin \frac{n\eta}{k}}{2k^2 \sin \frac{n\pi}{k}} = \sum_{s=1}^{\infty} \frac{(-1)^{s-1}s \, \sin s\eta}{s^2 k^2 - n^2} \qquad (12.76)$$

and then differentiated to derive

$$\frac{\pi n \cos \frac{n\eta}{k}}{2k^3 \sin \frac{n\pi}{k}} = \sum_{s=1}^{\infty} \frac{(-1)^{s-1}s^2 \cos s\eta}{s^2 k^2 - n^2}.$$

He also integrated (12.76) to obtain

$$\frac{\pi \cos \frac{n\eta}{k}}{2nk \sin \frac{n\pi}{k}} = \frac{1}{2n^2} + \sum_{s=1}^{\infty} \frac{(-1)^{s-1}s \cos s\eta}{s^2 k^2 - n^2}.$$

However, he also required (12.49) to determine the value of the constant of integration $\frac{1}{2n^2}$.

Now in 1775, at the time he wrote the paper we are discussing, Euler was well aware of the factorization

$$1 + 2x^k \cos \eta + x^{2k} = (1 + e^{i\eta}x^k)(1 + e^{-i\eta}x^k).$$

See, for example, his paper "Summatio progressionum,"[46] presented to the Petersburg Academy in 1773, published in 1774. He gives this factorization explicitly in exemplum 1, near the end of the paper, where he also gives the formula, that may be compared with (12.72):

$$\frac{x \sin \phi}{1 - 2x \cos \phi + x^2} = x \sin \phi + x^2 \sin 2\phi + x^3 \sin 3\phi + \cdots,$$

that involves only the geometric series, equivalent to (12.72):

$$\frac{2i \sin \phi}{(1 - e^{i\phi}x)(1 - e^{-i\phi}x)} = \frac{1}{1 - e^{-i\phi}x} - \frac{1}{1 - e^{i\phi}x}$$
$$= 1 + (e^{i\phi} - e^{-i\phi})x + (e^{2i\phi} - e^{-2i\phi})x^2 + \cdots,$$

that would have yielded a simpler derivation. But Euler's purpose here was rather to give multiple derivations of the same result.

## 12.7 Hermite's Rational Part Algorithm

As a professor at the École Polytechnique, Hermite lectured on analysis. This gave him the opportunity to rethink several elementary topics. He often came up with new

[46] Eu. I-15 pp. 168–184. E 447.

proofs and presentations of old material. In his lectures, published in 1873, Hermite gave a method for finding the rational part of the integral of a rational function, by employing the Euclidean algorithm.[47] He first found the square-free factorization of the denominator $Q(x)$ of the rational function $\frac{P(x)}{Q(x)}$:

$$Q = Q_1 Q_2^2 \cdots Q_n^n, \qquad (12.77)$$

where $Q_1, Q_2, \ldots, Q_n$ were the relatively prime polynomials with simple roots. This decomposition could be accomplished by the Euclidean algorithm, but Hermite did not give details in his published lectures. Note that there existed polynomials $P_1, P_2, \ldots, P_n$ such that

$$\frac{P}{Q} = \frac{P_1}{Q_1} + \frac{P_2}{Q_2^2} + \cdots + \frac{P_n}{Q_n^n}. \qquad (12.78)$$

As a first step in the derivation of this relation, Hermite observed that $U = Q_1$, and that $V = Q_2^2 \cdots Q_n^n$ were relatively prime and hence by the Euclidean algorithm, there existed polynomials $P_1$ and $\tilde{P}_1$ such that

$$P = P_1 V + \tilde{P}_1 U$$

or

$$\frac{P}{Q} = \frac{P_1}{Q_1} + \frac{\tilde{P}_1}{Q_2^2 \cdots Q_n^n}. \qquad (12.79)$$

Hermite obtained the required result by a repeated application of this procedure. Since

$$\int \frac{P}{Q} = \int \frac{P_1}{Q_1} + \int \frac{P_2}{Q_2^2} + \cdots + \int \frac{P_n}{Q_n^n}, \qquad (12.80)$$

he needed a method to reduce $\int \frac{P_k}{Q_k^k}$ to $\int \frac{E}{Q_k^{k-1}}$, for some polynomial $E$, when $k > 1$. Since $Q_k$ had simple roots, $Q_k$ and its derivative $Q_k'$ were relatively prime. Thus, there existed polynomials $C$ and $D$ such that

$$P_k = C Q_k + D Q_k'$$

and

$$\frac{P_k}{Q_k^k} = \frac{C}{Q_k^{k-1}} + \frac{D Q_k'}{Q_k^k} = \frac{C}{Q_k^{k-1}} - \frac{D}{k-1} \frac{d}{dx} \left( \frac{1}{Q_k^{k-1}} \right).$$

[47] Hermite (1905–1917) vol. 3, pp. 40–44. See also Hardy (1905) pp. 13–16.

After integration by parts, he obtained the necessary reduction

$$\int \frac{P_k}{Q_k^k} = -\frac{D}{(k-1)Q_k^{k-1}} + \int \frac{C + \frac{D'}{k-1}}{Q_k^{k-1}}. \tag{12.81}$$

Again, by a repeated application of this algorithm, Hermite had

$$\int \frac{P}{Q} = R + \int \frac{P_1}{Q_1} + \int \frac{\tilde{P}_2}{Q_2} + \int \frac{\tilde{P}_3}{Q_3} + \cdots + \int \frac{\tilde{P}_n}{Q_n}, \tag{12.82}$$

where $R$ was a rational function.

Since $Q_1, Q_2, \cdots, Q_n$ were pairwise relatively prime and had simple roots, the integrals on the right-hand side formed the transcendental part of the original integral.

## 12.8 Johann Bernoulli: Integration of $\sqrt{ax^2 + bx + c}$

We know that Isaac Barrow geometrically evaluated the integrals $\int \sqrt{x^2 + a^2}\, dx$ and $\int \frac{dx}{\sqrt{x^2+a^2}}$ and that his results could be immediately converted to analytic form.[48] Roger Cotes included in his tables of integrals those of the form

$$\int R(x,t)\, dx \quad \text{with} \quad t = \sqrt{ax^2 + bx + c},$$

where the integrand was a rational function of $x$ and $t$.[49] Clearly, seventeenth- and eighteenth-century mathematicians knew how to handle such integrals. But Johann Bernoulli pointed out in his very first lecture on integration, contained in vol. 3 of his *Opera Omnia*, that there was another method, related to Diophantine problems. By a substitution used in the study of Diophantine equations, the integral $\int R(x,t)\, dx$ could be rationalized. At the end of his lecture, Bernoulli illustrated this idea by means of an example:[50] His problem was to integrate $a^3\, dx : x\sqrt{ax - x^2}$; his method was to rewrite the quantity within the root as a square containing $x$ and a newly introduced variable. In this case, he had $ax - x^2 = a^2x^2 : m^2$. Thus, $x = am^2 : (m^2 + a^2)$, $dx = 2a^3m\, dm : (m^2 + a^2)^2$ and

$$\int \frac{a^3\, dx}{x\sqrt{ax - x^2}} = \int \frac{2a^3\, dm}{m^2} = -\frac{2a^3}{m}.$$

We note that a general substitution of the form $ax^2 + bx + c = (u + \sqrt{a}x)^2$ could be used to rationalize integrals involving $\sqrt{ax^2 + bx + c}$.

[48] Barrow (1916) pp. 160–161 and p. 185.
[49] Gowing (1983) pp. 46–48.
[50] Joh. Bernoulli (1968) vol. 3, p. 393.

## 12.9  Exercises

(1) Prove Newton's formula (12.3) by showing

(i) $\displaystyle\int_0^x \frac{1+t^2}{1+t^4}\,dt = \frac{1}{\sqrt{2}}\arctan\frac{x\sqrt{2}}{1-x^2}$,

(ii) $\displaystyle\int_0^x \frac{1+t^2}{1+t^4}\,dt = \frac{x}{1}+\frac{x^3}{3}-\frac{x^5}{5}-\frac{x^7}{7}+\frac{x^9}{9}+\frac{x^{11}}{11}-\cdots$, for $0 \le x \le 1$.

Use Newton's factorization (12.19).

(2) In his letter to Oldenburg dated October 24, 1676, Newton remarked that Leibniz's series and his own variant of it were unsuitable for the approximate evaluation of $\pi$: "For if one wished by the simple calculation of this series $1+\frac{1}{3}-\frac{1}{5}-\frac{1}{7}+\frac{1}{9}+$ etc. to find the length of the quadrant to twenty decimal places, it would need about 5000000000 terms of the series, for the calculation of which 1000 years would be required." He recommended his series for arcsin for this purpose. He suggested another formula to evaluate $\pi$:

$$\frac{\pi}{4} = \frac{a}{1}-\frac{a^3}{3}+\frac{a^5}{5}-\frac{a^7}{7}+\text{ etc.}$$
$$+\frac{a^2}{1}+\frac{a^5}{3}-\frac{a^8}{5}-\frac{a^{11}}{7}+\frac{a^{14}}{9}+\frac{a^{17}}{11}-\text{ etc.}$$
$$+\frac{a^4}{1}-\frac{a^{10}}{3}+\frac{a^{16}}{5}-\frac{a^{22}}{7}+\frac{a^{28}}{9}-\text{ etc.},$$

where $a = \frac{1}{2}$. Prove this formula and show that it is equivalent to

$$\frac{\pi}{4} = \arctan\frac{1}{2}+\frac{1}{2}\arctan\frac{4}{7}+\frac{1}{2}\arctan\frac{1}{8}.$$

Also prove Newton's formula:

$$1-\frac{1}{7}+\frac{1}{9}-\frac{1}{15}+\frac{1}{17}-\frac{1}{23}+\frac{1}{25}-\frac{1}{31}+\frac{1}{33}+\text{ etc. }= \frac{\pi}{4}(1+\sqrt{2}).$$

(3) Derive Newton's equations for $n$ defined by (12.18) when $m = 3, 4, 5, 6, 7, 8, 9, 10, 11$, and $12$. For these values of $m$, Newton had, respectively,

$$nn - 1 = 0, \quad n^3 - 2n = 0, \quad n^4 - 3nn + 1 = 0, \quad n^5 - 4n^3 + 3n = 0,$$
$$n^6 - 5n^4 + 6nn - 1 = 0, \quad n^7 - 6n^5 + 10n^3 - 4n = 0,$$
$$n^8 - 7n^6 + 15n^4 - 10nn + 1 = 0,$$
$$n^9 - 8n^7 + 21n^5 - 20n^3 + 5n = 0,$$
$$n^{10} - 9n^8 + 28n^6 - 35n^4 + 15nn - 1 = 0,$$
$$n^{11} - 10n^9 + 36n^7 - 56n^5 + 35n^3 - 6n = 0.$$

Use Newton's equation for $m = 7$ to show that the cubic equation satisfied by $2\cos\frac{2\pi}{7}$ is $x^3 + x^2 - 2x - 1 = 0$.

(4) Prove Newton's integration formula (12.1).

(5) Show that for $e > 0$ and $f < 0$

$$\int \frac{dx}{d + fx^2} = \frac{1}{e} R \ln \frac{R+T}{S} + \frac{1}{e} R \ln \frac{1}{i},$$

where $R = \sqrt{-\frac{e}{f}}$, $T = x$, $S = \sqrt{x^2 + \frac{e}{f}}$. Then show that for $e > 0$ and $f > 0$

$$\int \frac{dx}{e + fx^2} = \frac{1}{e} R \arctan \frac{x}{R},$$

where $R = \sqrt{\frac{e}{f}}$. Compare these results with comments on Cotes's notation for integrals of rational functions in the preliminary remarks for this chapter.

(6) Prove that

$$\int \frac{dx}{e + fx^2 + gx^4} =$$

(i) (when $4eg < f^2$ and $b^2 = e$) $\int \frac{\alpha\, dx}{b + mx^2} + \int \frac{\beta\, dx}{b + nx^2}$, where $\alpha, \beta, m, n$

can be determined in terms of $e, f, g$,

(ii) (when $4eg \geq f^2$ and $b^2 = e$) $\int \frac{(\alpha + \gamma x)\, dx}{b + nx + mx^2} + \int \frac{(\beta + \epsilon x)\, dx}{b - nx + nx^2}$,

where $\alpha, \beta, \gamma, \epsilon, m, n$ can be determined in terms of $e, f$ and $g$.

Bernoulli published an entertaining paper containing this result in the *Acta Eruditorum* in response to a challenge from Brook Taylor. See Joh. Bernoulli (1968) vol. II, p. 409.

(7) Use Hermite's reduction formula (12.81) to show that the integral

$$\int \frac{4x^9 + 21x^6 + 2x^3 - 3x^2 - 3}{(x^7 - x + 1)^2}\, dx$$

has only the rational part $-\frac{x^3+3}{x^7-x+1}$ and no transcendental part. See G. H. Hardy (1905) pp. 14–15.

(8) Prove that

$$a^{2n-2} \int \frac{dx}{(x^2 - a^2)^n} = (-1)^{n-1} \frac{1 \cdot 3 \cdots (2n-3)}{2 \cdot 4 \cdots (2n-2)} \left( \int \frac{dx}{x^2 - a^2} + f_{n-1}(x) \right),$$

where

$$f_n(x) = \frac{x}{x^2 - a^2}\left(1 + \sum_{k=1}^{n-1}(-1)^k \frac{2 \cdot 4 \cdots (2k)}{3 \cdot 5 \cdots (2k+1)} \frac{a^{2k}}{(x^2 - a^2)^k}\right).$$

See Hermite (1905–1917) vol. 3, p. 50.

(9) Show that

$$\int \frac{(1-x)\,dx}{x^4(2x-1)^3(3x-2)^2(4x-3)}$$

$$= C - \frac{1}{36x^3} - \frac{7}{18xx} + \frac{1879}{432x} + \frac{24499}{1296}\ln x + \frac{8}{(2x-1)^2} + \frac{48}{2x-1}$$

$$-272\ln(2x-1) + \frac{729}{16(3x-2)} + \frac{3645}{16}\ln(3x-2) + \frac{2048}{81}\ln(4x-3).$$

See Eu. I-17 p. 165.

(10) Show that

$$\int \frac{dx}{1+x^5} = \frac{1}{5}\ln(1+x) + \frac{1}{5}\cos\frac{2\pi}{5}\ln\left(1 + 2x\cos\frac{2\pi}{5} + x^2\right)$$

$$+ \frac{2}{5}\sin\frac{2\pi}{5}\arctan\frac{x\sin\left(\frac{2\pi}{5}\right)}{1 + x\cos\left(\frac{2\pi}{5}\right)}$$

$$- \frac{1}{5}\cos\frac{\pi}{5}\ln\left(1 - 2x\cos\frac{\pi}{5} + x^2\right)$$

$$+ \frac{2}{5}\sin\frac{\pi}{5}\arctan\frac{x\sin\left(\frac{\pi}{5}\right)}{1 - x\cos\left(\frac{\pi}{5}\right)},$$

$$\cos\frac{\pi}{5} = \frac{1+\sqrt{5}}{4}, \quad \sin\frac{\pi}{5} = \frac{\sqrt{(10-2\sqrt{5})}}{4},$$

$$\cos\frac{2\pi}{5} = \frac{-1+\sqrt{5}}{4}, \quad \sin\frac{2\pi}{5} = \frac{\sqrt{(10+2\sqrt{5})}}{4}.$$

Also show that

$$\int \frac{dx}{1+x^6} = \frac{\sqrt{3}}{12}\ln\frac{1+x\sqrt{3}+x^2}{1-x\sqrt{3}+x^2} + \frac{1}{3}\arctan x$$

$$+ \frac{1}{6}\arctan\frac{x}{2+x\sqrt{3}} + \frac{1}{6}\arctan\frac{x}{2-x\sqrt{3}},$$

and note that this implies

$$1 - \frac{1}{7} + \frac{1}{13} - \frac{1}{19} + \frac{1}{25} - \frac{1}{31} + \cdots = \frac{\pi}{6} + \frac{\sqrt{3}}{12}\ln\frac{2+\sqrt{3}}{2-\sqrt{3}}.$$

See Eu. I-17 pp. 131 and 120, respectively, for the two formulas.

(11) Show that for $0 < m < 2k$,

$$\int_0^\infty \frac{x^{m-1}\,dx}{1 + 2x^k \cos\theta + x^{2k}} = \frac{\pi \sin\left(1 - \frac{m}{k}\right)\theta}{k \sin\theta \sin\frac{m\pi}{k}}.$$

See Eu. I-18 p. 202.

(12) Show that for $0 < m < k$

$$\int_0^\infty \frac{x^{m-1}\,dx}{(1 + x^k)^n} = \frac{\pi}{k \sin\frac{m\pi}{k}} \prod_{s=1}^{n-1}\left(1 - \frac{m}{sk}\right).$$

See Eu. I-18 p. 188.

(13) Show that

$$\int_0^\infty \frac{dx}{(x^4 + 2ax^2 + 1)^{m+1}} = \frac{\pi}{2^{m+\frac{3}{2}}(a+1)^{m+\frac{1}{2}}} P_m(a),$$

where

$$P_m(a) = 2^{-2m} \sum_{k=0}^m 2^k \binom{2m - 2k}{m - k}\binom{m + k}{m}(a + 1)^k.$$

See Moll (2002) or Boros and Moll (2004) p. 154 for this example and for some intriguing open questions related to the integration of rational functions.

## 12.10 Notes on the Literature

The *Acta Eruditorum* was founded in 1682 by Otto Mencke (1644–1707), a Leipzig professor of moral and political philosophy. Leibniz and the two elder Bernoullis published many of their papers in this journal. Newton's two letters were first published in full in *Commercium Epistolicum D. Johannis Collins, et aliorum de analysi promota* of 1712. Newton had these letters published in order to document his priority in the calculus, partially in response to Leibniz's 1705 review of Newton's "Tractatus de Quadratura Curvarum" in which Leibniz claimed priority for the calculus. In his review, Leibniz charged Newton with appropriating his results by changing Leibniz's differentials to fluxions. Smith (1959), on pp. 440–454 of vol. II, gives an English translation of parts of de Moivre's works on factorization and related topics.

# 13

---

## *Difference Equations*

### 13.1 Preliminary Remarks

Difference equations occur in discrete problems, such as are encountered in probability theory, where recursion is an oft-used method. In the mid-seventeenth century, probability was developing as a new discipline; Pascal and Huygens used recursion, or first-order difference equations, in working out some elementary probability problems. Later, in the early eighteenth century, Niklaus I Bernoulli, Montmort, and de Moivre made use of more general difference equations. By the 1710s, it was clear that a general method for solving linear difference equations would be of great significance in probability and in analysis. Bernoulli and Montmort corresponded on this topic, discussing their methods for solving second-order difference equations with constant coefficients. In particular, they found the general term of the Fibonacci sequence. In 1712, Bernoulli also solved a special homogeneous linear equation of general degree with constant coefficients. He accomplished this in the course of tackling the well-known Waldegrave problem, involving the probability of winning a game, given players of equal skill.[1] Then in 1715, Montmort rediscovered and communicated to de Moivre Newton's transformation, (10.3). This revealed the connection between difference equations and the summation of infinite series.[2] It was an easy consequence of the Newton–Montmort transformation formula that the difference equation

$$\Delta^n A_k = A_{n+k} - \binom{n}{1} A_{n+k-1} + \binom{n}{2} A_{n+k-2} - \cdots + (-1)^n A_k = 0, \qquad (13.1)$$

$k = 0, 1, 2, \ldots$, implied that the series $\sum_{k=0}^{\infty} A_k x^k$ was a rational function with $(1-x)^n$ as denominator. More generally, de Moivre called a series recurrent if its coefficients satisfied the recurrence relation

$$a_0 A_{n+k} + a_1 A_{n+k-1} + \cdots + a_n A_k = 0, \qquad (13.2)$$

---

[1] See Hald (1990) pp. 375–393.
[2] Montmort (1717).

where $a_0, a_1, \ldots, a_n$ were constants and $k = 0, 1, 2, \ldots$. De Moivre was the first to present a general theory of recurrent series. He proved that such a series could be represented by a rational function and showed how to find this function. He then applied partial fractions to obtain the general expression for $A_n$ in terms of the roots of the denominator of the rational function. De Moivre was therefore the first mathematician to solve a general linear difference equation with constant coefficients by generating functions. He expounded this theory without proofs in the first edition of his *Doctrine of Chances*, published in 1717,[3] but he provided proofs in his 1730 *Miscellanea Analytica*.[4]

In the 1720s, several mathematicians turned their attention toward recurrent series. Daniel Bernoulli (1700–1782) made some very early investigations into this topic without making much headway, being unaware of the results of de Moivre, Niklaus I Bernoulli, and Montmort. In his *Exercitationes* of 1724, he stated that there was no formula for the general term of the sequence $1, 3, 4, 7, 11, 18, \ldots$. Niklaus informed his cousin Daniel that this was false and that the general term should be $\left(\frac{1+\sqrt{5}}{2}\right)^n + \left(\frac{1-\sqrt{5}}{2}\right)^n$. Apparently, Daniel Bernoulli subsequently became familiar with the work of de Moivre and others. At the end of 1728, he wrote a paper explaining the method, still contained in our textbooks, for giving special solutions for homogeneous linear difference equations with constant coefficients, in which the form of the solution is assumed and then substituted into the equation. The values of the parameter in the assumed solution can then be determined by means of an algebraic equation. He obtained the general solution by taking an arbitrary linear combination of the special solutions.[5]

Though the connection between differentials and finite differences had become clear by 1720, simultaneous advances in the two topics did not occur; one area seemed to make progress in alternation with the other. For example, D. Bernoulli's 1728 method of solution was not matched by a similar advance in the area of differential equations with constant coefficients until 1740. Euler, having defined the number $e$ and the corresponding exponential function, then gave the general solution of a differential equation as a combination of special exponential functions. He used the exponential function as the form of the solution in giving a method for solving a differential equation with constant coefficients.[6]

Then, from the 1730s through the 1750s, the theory of differential equations made great strides, partially due to the application of this subject to physics problems, such as hanging chains and vibration of strings. In fact, d'Alembert, Euler, and Clairaut initiated the study of partial differential equations in this context. However, no corresponding progress took place in the area of difference equations until 1759 when Lagrange had the inspiration of applying the progress made in differential equations to difference equations.[7] He found a technique, analogous to d'Alembert's

---

[3] For a reprint of the third edition, see de Moivre (1967).
[4] de Moivre (1730a).
[5] Bernoulli (1982–1996) vol. 2, pp. 49–64.
[6] Eu. I-22 pp. 108–149. E 62.
[7] Lagrange (1867–1892) vol. 1, pp. 23–36.

method for differential equations, for solving a nonhomogeneous difference equation by reducing its degree by one. A repeated application of this technique reduced a general $n$th degree difference equation to a first-degree equation, already treated by Taylor in 1715.[8] Similarly, Lagrange adapted his method of variation of parameters for solving differential equations to the case of difference equations. In the 1770s, Laplace published several papers, extending Lagrange's method and using other techniques to solve linear difference equations with variable coefficients. In a paper written in 1780 and published in 1782,[9] he introduced the term "generating functions." He developed this theory by the symbolic methods introduced in 1772 by Lagrange, courtesy of Leibniz. Laplace also applied generating functions of two variables to solve partial difference equations.[10]

De Moivre's work on recurrent series also contained interesting, albeit implicit, results on infinite series, a topic of his earliest research and a life-long interest. In the 1737 second edition of his *Doctrine of Chances*, he solved the problem of summing the series $\sum_{n=0}^{\infty} a_{np+k}x^n$, where $p$ and $k$ were integers and $0 \leq k < p$, when the sum of $\sum_{n=0}^{\infty} a_n x^n$ was known. In his solution, de Moivre dealt only with recurrent series, but in 1758, Simpson published a paper tackling the general problem. Even a year earlier, Waring also gave a general solution for summing the series $\sum_{n=0}^{\infty} a_{np+k}x^n$. An expert in the area of symmetric functions, he obtained his solution by taking specific symmetric functions of roots of unity. He did not publish the paper, but communicated it to the Royal Society. Waring later wrote that he believed Simpson's proof was based on this result, since Simpson was an active member of the society.

Another of de Moivre's results on series implied that if the recurrent series $\sum a_n x^n$ and $\sum b_n x^n$ had singularities at $\alpha$ and $\beta$, respectively, then $\sum a_n b_n x^n$ had a singularity at $\alpha\beta$. In 1899 Hadamard extended this result to arbitrary power series,[11] though it does not seem that he was motivated by de Moivre's theorem.

## 13.2   De Moivre on Recurrent Series

In his *Doctrine of Chances*, de Moivre wrote that the summation of series was required for the solution of several problems relating to chance, that is, to probabilistic problems.[12] He then presented a list of nine propositions connected with recurrent series, series whose coefficients satisfied a linear recurrence relation. Thus, a series $\sum_{n=0}^{\infty} a_n x^n$ was a recurrent series if there were constants $\alpha_1, \alpha_2, \ldots, \alpha_k$ such that $a_n$ satisfied the difference equation

$$a_n + \alpha_1 a_{n-1} + \alpha_2 a_{n-2} + \cdots + \alpha_k a_{n-k} = 0, \tag{13.3}$$

[8]  Taylor (1715).
[9]  Laplace (1782) p. 5.
[10] For the early history of generating functions, see Seal (1949).
[11] Hadamard (1899).
[12] de Moivre (1967) pp. 220–229.

for $n = k, k+1, k+2, \ldots$. De Moivre started with the example of the series

$$1 + 2x + 3xx + 10x^3 + 34x^4 + 97x^5 + \cdots, \qquad (13.4)$$

whose coefficients satisfied the equation

$$a_n - 3a_{n-1} + 2a_{n-1} - 5a_{n-3} = 0, \qquad (13.5)$$

for $n = 3, 4, 5, \ldots$. His terminology is no longer used. For example, instead of the recurrence relation (13.5), he called $3x - 2xx + 5x^3$, or simply $3 - 2 + 5$, the scale of relation of the series. This scale of relation was used to sum the series. Thus, if we let $S$ denote the series (13.4), then

$$-3xS = -3x - 6x^2 - 9x^3 - 30x^4 - 102x^5 - \cdots,$$
$$+2x^2 S = \qquad 2x^2 + 4x^3 + 6x^4 + 20x^5 + \cdots,$$
$$-5x^3 S = \qquad \qquad - 5x^3 - 10x^4 - 15x^5 - \cdots.$$

Now add these three series to the original series (13.4) for $S$. Because of the recurrence relation (13.5) satisfied by the coefficients, or because of the scale of relation of the series, we get

$$(1 - 3x + 2x^2 - 5x^3)S = 1 - x - x^2.$$

All other terms on the right-hand side cancel and we have the sum of $S$:

$$S = \frac{1 - x - x^2}{1 - 3x + 2x^2 - 5x^3}. \qquad (13.6)$$

De Moivre called the expression in the denominator the differential scale, since it was obtained by subtracting the scale of relation from unity.

De Moivre's purpose in summing $S$ was to find the numerical value of the coefficient $a_n$ of the series, or, in modern terms, to solve the difference equation (13.5). Once he had $S$, he could factorize the denominator and obtain the partial fractions decomposition of the rational function $S$. Actually, he did not discuss this algebraic process in his book. He merely noted the form of $a_n$ when the denominator was a polynomial of degree $m$ with roots $\alpha_1, \alpha_2, \ldots, \alpha_m$ in the cases $m = 2, 3, 4$. Moreover, he wrote the solutions for only those cases in which the roots were distinct. One can be sure that he knew how to handle the case of repeated roots, because only a knowledge of the binomial theorem for negative integral powers was required. Thus, all series $\sum_{n=0}^{\infty} a_n x^n$, whose coefficients $a_n$ satisfy a linear difference equation with constant coefficients as in (13.3), must be rational functions. Conversely, the power series expansions of rational functions whose numerators are of degree less than the corresponding denominators are recurrent series. Euler devoted a chapter of his *Introductio in Analysin Infinitorum* to recurrent series.[13] Since de Moivre gave

---

[13] Euler (1988) chapter 17.

very few examples, we consider two examples from Euler's exposition, illustrating the method of using generating functions to solve linear difference equations such as (13.3). In the first example, the recurrence relation was the same as the one satisfied by the Fibonacci sequence, though the initial values were different. The coefficients of the series

$$\sum_{n=0}^{\infty} a_n x^n = 1 + 3x + 4x^2 + 7x^3 + 11x^4 + 18x^5 + 29x^6 + 47x^7 + \cdots$$

satisfied the recurrence relation $a_n = a_{n-1} + a_{n-2}$ for $n \geq 2$. Note from our discussion of de Moivre's work that the above series would sum to a rational function whose denominator would be $1 - x - x^2$. In fact, the sum was

$$\frac{1 + 2x}{1 - x - x^2} = \frac{\frac{1+\sqrt{5}}{2}}{1 - \left(\frac{1+\sqrt{5}}{2}\right)x} + \frac{\frac{1-\sqrt{5}}{2}}{1 - \left(\frac{1-\sqrt{5}}{2}\right)x}.$$

Hence Euler had the solution of the difference equation:

$$a_n = \left(\frac{1 + \sqrt{5}}{2}\right)^{n+1} + \left(\frac{1 - \sqrt{5}}{2}\right)^{n+1}.$$

In the second example, there were repeated roots as well as complex roots. Euler explained earlier in his book precisely how to obtain the partial fractions in this situation. The difference equation would be

$$a_n - a_{n-1} - a_{n-2} + a_{n-4} + a_{n-5} - a_{n-6} = 0, \tag{13.7}$$

and the initial conditions would yield the sum of the series as

$$\frac{1}{1 - x - x^2 + x^4 + x^5 - x^6} = \frac{1}{(1-x)^3(1+x)(1+x+x^2)}$$

$$= \frac{1}{6(1-x)^3} + \frac{1}{4(1-x)^2} + \frac{17}{72(1-x)} + \frac{1}{8(1+x)} + \frac{2+x}{9(1+x+x^2)}. \tag{13.8}$$

Euler obtained the general term $a_n$ by expanding the partial fractions using the binomial theorem. Thus, he had

$$a_n = \frac{n^2}{12} + \frac{n}{2} + \frac{47}{72} \pm \frac{1}{8} \pm \frac{4\sin\frac{(n+1)\pi}{3} - 2\sin\frac{n\pi}{3}}{9\sqrt{3}}, \tag{13.9}$$

where a positive sign was used for $n$ even and negative sign for $n$ odd.

   In his *Doctrine of Chances*, de Moivre stated a few specific examples but did not work out details for obtaining the general term. Of his nine theorems, the first six dealt in general terms with the ideas in the above two examples from Euler. The last three propositions applied to more general series, though de Moivre worked wholly in terms of recurrent series. In the seventh proposition, on the even and odd parts of a rational function, de Moivre supposed that $\sum_{n=0}^{\infty} a_n x^n$ was a recurrent series and hence representable as a rational function. He then gave a method for representing the even series $\sum_{n=0}^{\infty} a_{2n} x^{2n}$ and the odd series $\sum_{n=0}^{\infty} a_{2n+1} x^{2n+1}$ as rational functions. In this connection, he explained that if $A(x)$ was the denominator, or differential scale, for $\sum_{n=0}^{\infty} a_n x^n$, then the common differential scale for the two series with the even and odd powers was the polynomial obtained by eliminating $x$ from the equations $A(x) = 0$ and $x^2 = z$. More generally, de Moivre wrote that if

$$a_0 + a_1 x + a_2 x^2 + \cdots = \frac{B(x)}{A(x)}, \qquad (13.10)$$

then the $m$ series

$$a_j x^j + a_{m+j} x^{m+j} + a_{2m+j} x^{2m+j} + \cdots, \quad j = 0, 1, \ldots, m-1, \qquad (13.11)$$

had the common differential scale obtained by eliminating $x$ from $A(x) = 0$ and $x^m = z$. We may state a more general problem: Given $f(x) = \sum_{n=0}^{\infty} a_n x^n$, express $g(x) = \sum_{k=0}^{\infty} a_{km+j} x^{km+j}$ in terms of values of $f(\alpha x)$, $\alpha$ a root of unity. This was the problem solved by Simpson and Waring in the late 1750s. The essence of their method was to use appropriate $m$th roots of unity and those roots were implicit in de Moivre's use of the equation $x^m = z$.

The eighth proposition of de Moivre explained how to find the differential scale for $\sum (a_n + b_n) x^n$ when the differential scales of $\sum a_n x^n$ and $\sum b_n x^n$ were known. This was straightforward. In the very interesting ninth proposition, de Moivre worked out the differential scale for $\sum a_n b_n x^n$, but only for the case where the scales for $\sum a_n x^n$ and $\sum b_n x^n$ were quadratic polynomials. His result is stated as an exercise at the end of this chapter. An immediate consequence of this result is that $\sum a_n b_n x^n$ has a singularity at $\alpha\beta$ if $\sum a_n x^n$ and $\sum b_n x^n$ have singularities at $\alpha$ and $\beta$ respectively. In 1899, Hadamard, probably unaware of this result of de Moivre, stated and proved a beautiful generalization, usually called Hadamard's multiplication of singularities theorem:[14] If $\sum_{n=0}^{\infty} a_n z^n$ has singularities at $\alpha_1, \alpha_2, \ldots,$ and $\sum_{n=0}^{\infty} b_n z^n$ at $\beta_1, \beta_2, \ldots,$ then the singularities of $\sum_{n=0}^{\infty} a_n b_n z^n$ are among the points $\alpha_i \beta_j$.

## 13.3 Simpson and Waring on Partitioning Series

In 1758, Thomas Simpson gave a solution of the general problem of determining the values of the $m$ series in (13.11) when (13.10) was replaced by a general series

$$f(x) = a_0 + a_1 x + a_2 x^2 + a_3 x^3 + \cdots. \qquad (13.12)$$

Simpson's paper,[15] "The invention of a general method for determining the sum of every 2d, 3d, 4th, or 5th, &c, term of a series, taken in order; the sum of the whole series being known," was published in the *Philosophical Transactions* of

[14] Hadamard (1899).
[15] Simpson (1759).

the Royal Society. To accomplish the task of his paper, Simpson employed roots of unity as well as some theory of symmetric functions. In fact, one can avoid the use of symmetric functions here, but the origins of this topic are relevant, to the extent that Simpson employed it in his arguments.

In his 1629 book, *Invention nouvelle en l'algébra*, Albert Girard (1595–1632) defined the elementary symmetric functions:[16]

> When several numbers are proposed, the entire sum may be called the first <u>faction</u>; the sum of all the products taken two by two may be called the second <u>faction</u>; and always thus to the end, but the product of all numbers is the last <u>faction</u>. Now there are as many <u>factions</u> as proposed numbers.

Thus, if $\alpha_1, \alpha_2, \ldots, \alpha_n$ are $n$ quantities, the elementary symmetric functions of these $n$ quantities would be defined by

$$\sigma_1 = \sum_{i=1}^{n} \alpha_i, \quad \sigma_2 = \sum_{1 \le i < j \le n} \alpha_i \alpha_j, \quad \sigma_3 = \sum_{1 \le i < j < k \le n} \alpha_i \alpha_j \alpha_k, \quad \cdots, \sigma_n = \alpha_1 \alpha_2 \cdots \alpha_n.$$

Girard referred to $\sigma_1$ as the first faction, $\sigma_2$ as the second faction, and $\sigma_n$ as the last faction. This concept has connections with algebraic equations:

$$(x - \alpha_1)(x - \alpha_2) \cdots (x - \alpha_n) = x^n - \sigma_1^{n-1} + \sigma_2 x^{n-2} - \sigma_3 x^{n-3} + \cdots + (-1)^n \sigma_n,$$
$$(13.13)$$

a result verifiable by induction.

Girard explained the connection between the elementary symmetric functions of the roots of a polynomial and the coefficients of the polynomial:

> Every algebraic equation except the incomplete ones admits of as many solutions as the denomination of the highest quantity indicates. And the first faction of the solutions is equal to the number of the first mixed quantity, the second faction of them is equal to the number of the second mixed quantity, the third to the third, and so on, so that the last faction is equal to the closing quantity–all this according to the signs that can be noted in the alternating order.

Now the symmetric functions result invoked by Simpson maintained that the sum of a given power of roots

$$s_m = \alpha_1^m + \alpha_2^m + \cdots + \alpha_n^m, \quad m \text{ a positive integer,} \qquad (13.14)$$

must be a polynomial in $\sigma_1, \sigma_2, \ldots, \sigma_n$. Girard stated this result as:

> It might seem to some that the *factions* would be also explicable otherwise than above. That instead of saying the sum, the products two by two, the products three by three, etc., one could say more simply, the sum, the sum of the squares, the sum of the cubes, etc., which however is not so, for when there are several solutions, the sum will be for the first mixed quantity, the sum of the products two by two for the second, etc., as has been sufficiently explicated. But it is not the case for any *factions* of the powers that someone might offer.

---

[16] For the Girard material, see Girard (1884), a later printing in which no page numbers seem to be given; also see translation by Robert Smith, in de Beaune et al. (1986).

Example: Let:

$A$ be the first quantity,

$B$ the second,

$C$ the third,

$D$ the fourth,

etc.

Then, in every type of equation,

$A$

$A$ sq $- B2$

$A$ cub $- AB3 + C3$

$A$ sq -sq $- A$ sq$B4 + AC4 + B$ sq$2 - D4$

will be the sum, respectively, of the

solutions

squares

cubes

square-squares.

In modern notation, using (13.14), Girard had

$$s_1 = \sigma_1, \quad s_2 = \sigma_1^2 - 2\sigma_2, \quad s_3 = \sigma_1^3 - 3\sigma_1\sigma_2 + 3\sigma_3,$$
$$s_4 = \sigma_1^4 - 4\sigma_1^2\sigma_2 + 4\sigma_1\sigma_3 + 2\sigma_2^2 - 4\sigma_4, \tag{13.15}$$

a result rediscovered by Newton in 1665–1666.[17] Newton had learned algebra by reading Oughtred, Viète, and Descartes; he read Descartes in van Schooten's Latin translation. Thus, it appears that Newton's discovery was independent.[18] Moreover, Newton also provided a recurrence rule for determining $s_k$ when $s_1, s_2, \ldots, s_{k-1}$ were known. In modern notation, this rule is

$$s_k - \sigma_1 s_{k-1} + \sigma_2 s_{k-2} - \cdots + (-1)^{k-1}\sigma_{k-1}s_1 + (-1)^k k\, \sigma_k = 0. \tag{13.16}$$

Newton stated his rule within his earliest researches in algebra.[19] He included these and other algebraic results from 1665–1666 in his lectures on algebra, given in the 1670s and early 1680s, published in 1707 as *Arithmetica Universalis*.[20] A proof of the rule was not published by Newton, but was given by Maclaurin in his posthumous *Treatise of Algebra*.[21] Euler gave two different proofs in 1750.[22]

In his *Meditationes Algebraicae*, Edward Waring (c. 1736–1798) accused Simpson of stealing his idea for partitioning series into $n$ parts. The English translation by

[17] Newton (1967–81) vol. 1 pp. 517–520.
[18] ibid. p. 518, footnote 12.
[19] ibid. p. 519. See also footnote 15.
[20] For a reprint of this book, see Newton (1964–67), vol. 2.
[21] Maclaurin (1748).
[22] Eu. I-6, pp. 20–30. E 153. See also pp. 263–286. E 406.

Dennis Weeks of the 1782 edition of Waring's *Meditationes* presents Waring's point of view:[23]

De Moivre gives a method of generating the sum of a series of terms that are equal, or alternating, or cyclic over an interval of distance two, three, or more, $a + bx + cx^2 + \cdots$ through division of unity by a rational multinomial expression $p + qx + rx^2 + \cdots$. In 1757 I sent the first version of this work to the Royal Society of London, which Simpson read, then in 1758 he inserted in the *Philosophical Transactions* a short piece containing a rule that was in the work I submitted, *viz* let $S$ be a given function of the quantity $x$, which is expanded into a series proceeding according to the dimensions of $x$, say $a + bx + cx^2 + dx^3 + \cdots$; in $S$ now substitute for $x$ respectively $\alpha x, \beta x, \gamma x, \delta x, \ldots$ where $\alpha, \beta, \gamma, \delta, \ldots$ are roots of the equation $x^n - 1 = 0$, resulting in a total of $n$ quantities $A, B, C, D$, etc., then $\frac{A+B+C+D+\cdots}{n}$ will be the sum of the first term and of those whose position is respectively $n, 2n, 3n$, etc. beyond the first term. Nevertheless at the end of his paper he says of the series $a + bx + cx^2 + \cdots$ that he has given a solution of an example of this problem by a method which differs a little from another one where a general method is indicated. But I say that no one, before my submission to the Royal Society in 1757, had ever claimed to have devised a general method, and it was my notes that Simpson had read, in which the above method was contained.

It is possible that Waring could have been mistaken. Though not formally trained, Simpson was an able mathematician, capable of conceiving of the idea of partitioning series. Also, it is not at all clear that Waring's communication to the Royal Society was read by Simpson, or that it was indeed available to him. However, Waring provided an important result, allowing one to calculate $s_m$ directly, stated and proved as the first result in his *Meditationes*:

$$s_m = m \sum (-1)^{m+k_1+k_2+\cdots+k_m} \cdot \frac{(k_1 + k_2 + \cdots + k_m - 1)!}{k_1! \, k_2! \cdots k_m!} \sigma_1^{k_1} \sigma_2^{k_2} \cdots \sigma_m^{k_m},$$

$$(13.17)$$

where the sum was to be taken over all $k_1, \ldots, k_m$ such that $k_1 + 2k_2 + \cdots + mk_m = m$. The number of terms in the sum yielded the number of partitions of $m$, denoted by $p(m)$. A definition of partitions is provided in our Chapter 26. To get an idea of the usefulness of Waring's formula (13.17), we here apply it to efficiently derive $s_6$; note that it took Newton much more effort to do this in 1665–1666.

We first find all eleven solutions of

$$k_1 + 2k_2 + 3k_3 + 4k_4 + 5k_5 + 6k_6 = 6:$$

$$k_1 = 6; \quad k_1 = 4, k_2 = 1; \quad k_1 = 3, k_3 = 1;$$

$$k_1 = 2, k_2 = 2; \quad k_1 = 2, k_4 = 1; \quad k_1 = 1, k_5 = 1;$$

$$k_1 = 1, k_2 = 1, k_3 = 1; \quad k_2 = 3;$$

$$k_2 = 1, k_4 = 1; \quad k_3 = 2; \quad k_8 = 1.$$

[23] Waring and Weeks (1991) pp. xlii–xliii.

Note that these solutions indicate that $p(6) = 11$. Now observe that the solution $k_1 = 3$, $k_3 = 1$, for example, leads to the term

$$6(-1)^{6+3+1} \frac{3!}{3! \ 1!} \sigma_1^3 \sigma_3 = 6\sigma_1^3 \sigma_3$$

and the complete formula for $s_6$ is given by

$$s_6 = \sigma_1^6 - 6\sigma_1^4 \sigma_2 + 6\sigma_1^3 \sigma_3 + 9\sigma_1^2 \sigma_2^2 - 6\sigma_1^2 \sigma_4$$
$$- 12\sigma_1 \sigma_2 \sigma_3 + 6\sigma_1 \sigma_5 - 2\sigma_2^3 + 6\sigma_2 \sigma_4 + 3\sigma_3^2 - 6\sigma_6.$$

Returning to Simpson's method, he gave an illustration of his approach by partitioning the series (13.12) into three parts; for numbers $p$, $q$, and $r$, he obtained

$$\frac{1}{3} f(px) = \frac{1}{3}(a_0 + a_1 px + a_2 p^2 x^2 + a_3 p^3 x^3 + \cdots),$$
$$\frac{1}{3} f(qx) = \frac{1}{3}(a_0 + a_1 qx + a_2 q^2 x^2 + a_3 q^3 x^3 + \cdots),$$
$$\frac{1}{3} f(rx) = \frac{1}{3}(a_0 + a_1 rx + a_2 r^2 x^2 + a_3 r^3 x^3 + \cdots).$$

Simpson wrote that

$$\frac{1}{3}(f(px) + f(qx) + f(rx)) = a_0 + a_3 x^3 + a_6 x^6 + \cdots$$

if

$$p+q+r = 0, \quad p^2 + q^2 + r^2 = 0, \quad p^3 + q^3 + r^3 = 3, \quad p^4 + q^4 + r^4 = 0, \quad \text{etc.}$$

Thus he required that

$$p^k + q^k + r^k = 0 \quad \text{when} \quad 3 \nmid k$$

and

$$p^k + q^k + r^k = 3 \quad \text{when} \quad 3 \mid k. \qquad (13.18)$$

He noted that the relations in (13.18) would hold if $p, q, r$ were the roots of the equation

$$x^3 - 1 = 0.$$

He observed that the methods of algebra showed that $p + q + r = 0$ and $p^2 + q^2 + r^2 = 0$, presumably since the elementary symmetric functions $p + q + r$ and $pq + pr + rq$ of the roots $p, q, r$ were zero. He also noted that $p^3 + q^3 + r^3 = 3$, $p^4 + q^4 + r^4 = p + q + r$, $p^5 + q^5 + r^5 = p^2 + q^2 + r^2$, and so on. In this way, he verified his result for the case where the series was trisected.

On the general case, Simpson wrote:[24]

And, by the very same reasoning, and the process above laid down, it is evident, that, if every $n$th term (instead of every third term) of the given series be taken, the values $p, q, r, s$, &c. will be the roots of the equation $z^n - 1 = 0$; and that, the sum of all the terms so taken, will be truly obtained by substituting $px, qx, rx, sx$, &c. successively for $x$, in the given value of $S$ [the sum of the series], and then dividing the sum of all the quantities thence arising from the given number $n$.

In order to verify Simpson's contention for the general case, one must show that, given $p_1, p_2, \ldots, p_m$ as the $m$ roots of

$$x^m - 1 = 0, \tag{13.19}$$

$$s_k = p_1^k + p_2^k + \cdots + p_m^k = 0 \text{ when } k < m$$
$$= m \text{ when } k = m.$$

Note that the coefficients of $\sigma_1, \sigma_2, \ldots, \sigma_{m-1}$ are all zero, while $\sigma_m = 1$. It follows from Newton's relation (13.16) that for $k < m$

$$s_k = \sigma_1 s_{k-1} - \sigma_2 s_{k-2} + \cdots + (-1)^k \sigma_{k-1} s_1 + (-1)^{k+1} k \sigma_k = 0.$$

Recalling that $s_m = m$, it follows from (13.19) that

$$s_n = p_1^n + p_2^n + \cdots + p_m^n = p_1^k + p_2^k + \cdots + p_m^k,$$

completing the proof.

In the middle of his paper, while he was considering some examples, Simpson suddenly mentioned that the known values of $p_1, p_2, p_3, \ldots$ were of the form $\alpha + \sqrt{\alpha^2 - 1}$, where $\alpha = \cos \frac{2\pi k}{m}$. Thus, if

$$p_1 = \cos \frac{2\pi}{m} + i \sin \frac{2\pi}{m}$$

then we can take

$$p_k = \cos \frac{2\pi k}{m} + i \sin \frac{2\pi k}{m},$$

because, by de Moivre's theorem,

$$\left( \cos \frac{2\pi}{m} + i \sin \frac{2\pi}{m} \right)^k = \cos \frac{2\pi k}{m} + i \sin \frac{2\pi k}{m}.$$

This indicates that all the $m$th roots of unity are given by $1, \omega, \omega^2, \ldots, \omega^{m-1}$, where $\omega = \cos \frac{2\pi}{m} + i \sin \frac{2\pi}{m}$.

24 Simpson (1759) pp. 759–760.

Also, it is clear that

$$1 + \omega + \omega^2 + \cdots + \omega^{m-1} = \frac{1 - \omega^m}{1 - \omega} = 0. \qquad (13.20)$$

This confirms the fact that the theory of symmetric functions can be avoided, as we saw in the case $m = 3$ in Section 2.11, where (13.20) is used in the particular case given by (2.65).

## 13.4 Stirling's Method of Ultimate Relations

Stirling extended de Moivre's recurrent series method to sequences satisfying difference equations with nonconstant coefficients. In the preface to his 1730 *Methodus*, he wrote,[25]

> For I was not unaware that De Moivre had introduced this property of the terms into algebra with the greatest success, as the basis for solving very difficult problems concerning recurrent series: And so I decided to find out whether it could also be extended to others, which of course I doubted since there is so great a difference between recurrent and other series. But, the practical test having been made, the matter has succeeded beyond hope, for I have found out that this discovery of De Moivre contains very general and also very simple principles not only for recurrent series but also for any others in which the relation of the terms varies according to some regular law.

In the statement of the proposition 14, Stirling explained the term ultimate relation:[26] Let $T$ be the $z$th term of a series and $T'$ the next term; let $r, s, a, b, c, d$, be constants. Suppose that the relation

$$r(z^2 + az + b)T + s(z^2 + cz + d)T' = 0 \qquad (13.21)$$

held between the successive terms. Then the ultimate relation of the terms was defined as

$$rT + sT' = 0. \qquad (13.22)$$

Stirling used the term ultimate because he understood that $z$ was a very large integer, so that $az + b$ and $cz + d$ could be neglected in comparison to $z^2$. This made it clear that (13.22) followed from (13.21). Similarly, if the equation were

$$r(z + a)T + s(z + b)T' + t(z + c)T'' = 0, \qquad (13.23)$$

then the ultimate relation would be

$$rT + sT' + tT'' = 0. \qquad (13.24)$$

---

[25] Stirling and Tweddle (2003) p. 18.
[26] ibid. p. 88.

In modern notation, if $\sum A_n$ is the series, then (13.23) takes the form

$$r(k+a)A_k + s(k+b)A_{k+1} + t(k+c)A_{k+2} = 0. \qquad (13.25)$$

Stirling stated his theorem as proposition 14:

Every series $A + B + C + D + E + \&c.$ in which the ultimate relation of the terms is $rT + sT' + tT'' = 0$ splits into the following

$$(s+t) \times \left( \frac{A}{n} + \frac{A_2}{n^2} + \frac{A_3}{n^3} + \frac{A_4}{n^4} + \frac{A_5}{n^5} + \&c. \right)$$

$$+ t \times \left( \frac{B}{n} + \frac{B_2}{n^2} + \frac{B_3}{n^3} + \frac{B_4}{n^4} + \frac{B_5}{n^5} + \&c. \right)$$

where $n = r + s + t$ and

$$A_2 = rA + sB + tC, \qquad A_3 = rA_2 + sB_2 + tC_2, \quad A_4 = rA_3 + sB_3 + tC_3, \&c.$$
$$B_2 = rB + sC + tD, \qquad B_3 = rB_2 + sC_2 + tD_2, \quad B_4 = rB_3 + sC_3 + tD_3, \&c.$$
$$C_2 = rC + sD + tE, \qquad C_3 = rC_2 + sD_2 + tE_2^*. \quad C_4 = rC_3 + sD_3 + tE_3, \&c.$$
$$D_2 = rD + sE + tF, \qquad D_3 = rD_2 + sE_2 + tF_2, \quad D_r = rD_3 + sE_3 + tF_3, \&c.$$
$$E_2 = rE + sF + tG, \qquad E_3 = rE_2 + sF_2 + tG_2, \quad E_4 = rE_3 + sF_3 + tG_3, \&c.$$
$$\&c.$$

This result generated some interest in its time. A reviewer of the *Methodus* wrote in 1732 that the result was very powerful and complicated. In a letter to Stirling dated June 8, 1736, Euler wrote[27]

But before I wrote to you, I searched all over with great eagerness for your excellent book on the method of differences, a review of which I had seen a short time before in the *Actae Lipslienses*, until I achieved my desire. Now that I have read through it diligently, I am truly astonished at the great abundance of excellent methods contained in such a small volume, by means of which you show how to sum slowly converging series with ease and how to interpolate progressions which are very difficult to deal with. But especially pleasing to me was prop. XIV of Part I in which you give a method by which series, whose law of progression is not even established, may be summed with great ease using only the relation of the last terms; certainly this method extends very widely and is of the greatest use. In fact the proof of this proposition, which you seem to have deliberately withheld, caused me enormous difficulty, until at last I succeeded with very great pleasure in deriving it from the preceding results, which is the reason why I have not yet been able to examine in detail all the subsequent propositions.

Stirling gave three examples of this theorem. The first example, similar to the second, was the summation of the series

$$1 + 4x + 9x^2 + 16x^3 + 25x^4 + 36x^5 + \text{ etc.} \qquad (13.26)$$

Recall that Euler summed this series in his *Institutiones Calculi Differentialis* of 1755 by applying Newton or Montmort's transformation. Stirling was aware that the

[27] Tweddle (1988) p. 141.

series could be summed by that method and mentioned Montmort explicitly. Stirling observed that the difference equation for the terms of the series was

$$(z^2 + 2z + 1)xT - z^2T' = 0. \tag{13.27}$$

For example, for the third term, $z = 3$, $T = 9x^2$, and $T' = 16x^3$. The ultimate relation was $xT - T' = 0$, so that $r = x$, $s = -1$, $t = 0$, and $n = x - 1$. It followed that $A = 1$, $A_2 = -3x$, $A_3 = 2x^2$, $A_4 = 0$, and the series transformed to

$$-1\left(\frac{1}{x-1} - \frac{3x}{(x-1)^2} + \frac{2x^2}{(x-1)^3}\right) = \frac{1+x}{(1-x)^3}.$$

In the third example, he considered the series

$$1 - 6x + 27x^2 - 104x^3 + 366x^4 - 1212x^5 + 3842x^6 - 11784x^7 + \text{etc.}$$

defined by the difference equation

$$x^2(z+4)T - 2x(z+2)T' - zT'' = 0; \tag{13.28}$$

the ultimate relation was

$$x^2T - 2xT' - T'' = 0. \tag{13.29}$$

Hence $r = x^2$, $s = -2x$, $t = -1$ and $n = x^2 - 2x - 1$. Stirling computed the values of the $A$ and $B$ as

$$A = 1, \ A_2 = -14x^2, \ A_3 = 29x^4, \ A_4 = 0,$$

$$B = -6x, \ B_2 = 44x^3, \ B_3 = -70x^5, \ B_4 = 0.$$

Thus, the sum of the series was

$$-(2x+1)\left(\frac{1}{x^2 - 2x - 1} - \frac{14x^2}{(x^2 - 2x - 1)^2} + \frac{29x^4}{(x^2 - 2x - 1)^3}\right)$$

$$-\left(\frac{-6x}{x^2 - 2x - 1} + \frac{44x^3}{(x^2 - 2x - 1)^2} - \frac{70x^5}{(x^2 - x - 1)^3}\right).$$

Note that in (13.28) $z$ takes the values $2, 3, 4, \ldots$ while in (13.27) $z$ starts at 1. Thus, in the second series when $T = 1$, $T' = -6x$, and $T'' = 27x^2$, we take $z = 2$. Stirling's normal practice was to start at $z = 1$.

## 13.5  Daniel Bernoulli on Difference Equations

In 1728, while at the St. Petersburg Academy, Bernoulli presented to the academy a method for solving a difference equation in which the form of the solution was assumed; this particular approach is often given in elementary textbooks.[28] Unlike Bernoulli, we use subscripts to write the equation

$$a_n = \alpha_1 a_{n-1} + \alpha_2 a_{n-2} + \cdots + \alpha_k a_{n-k}, \tag{13.30}$$

with $\alpha_1, \alpha_2, \ldots, \alpha_k$ constants. Bernoulli assumed $a_x = \lambda^x$, substituted in the equation and divided by $\lambda^{n-k}$ to arrive at

$$\lambda^k = \alpha_1 \lambda^{k-1} + \alpha_2 \lambda^{k-2} + \cdots + \alpha_k. \tag{13.31}$$

Bernoulli stated that if $\lambda_1, \lambda_2, \ldots, \lambda_k$ were the $k$ distinct solutions of the algebraic equation (13.31), then the general solution of (13.30) would be an arbitrary linear combination of the particular solutions $\lambda_i^x$, that is

$$a_x = A_1 \lambda_1^x + A_2 \lambda_2^x + \cdots + A_k \lambda_k^x. \tag{13.32}$$

However, if $\lambda_1 = \lambda_2$, then the first two terms of (13.32) would be replaced by $(A_1 + A_2 x)\lambda_1^x$. More generally, if a root $\lambda_j$ was repeated $m$ times, then that part of the solution (13.32) corresponding to $\lambda_j$ would be replaced by

$$(A_j + A_{j+1}x + \cdots + A_{m+j-1})\lambda_j^x.$$

Daniel Bernoulli considered examples of distinct roots and of repeated roots. He first took the Fibonacci sequence $0, 1, 1, 2, 3, 5, 8, 13, \ldots$, leading to the difference equation $a_n = a_{n-1} + a_{n-2}$ and the algebraic equation $\lambda^2 = \lambda + 1$. The solutions were $\lambda_1 = \frac{1+\sqrt{5}}{2}$, $\lambda_2 = \frac{1-\sqrt{5}}{2}$ so that

$$a_x = A_1 \lambda_1^x + A_2 \lambda_2^x.$$

To find $A_1$ and $A_2$, Bernoulli took $x = 0$ and $x = 1$ to get

$$A_1 + A_2 = 0 \text{ and } A_1 \left(\frac{1+\sqrt{5}}{2}\right) + A_2 \left(\frac{1-\sqrt{5}}{2}\right) = 1.$$

Solving these equations, Bernoulli found $A_1 = \frac{1}{\sqrt{5}}$ and $A_2 = -\frac{1}{\sqrt{5}}$. Recall that Montmort and Niklaus I Bernoulli in their correspondence of 1718–1719 had already solved the problem of the general term in the Fibonacci sequence.

As an example of a difference equation leading to repeated roots, Daniel Bernoulli considered the sequence $0, 0, 0, 0, 1, 0, 15, -10, 165, -228$, etc., generated by the difference equation

$$a_n = 0a_{n-1} + 15a_{n-2} - 10a_{n-3} - 60a_{n-4} + 72a_{n-5}.$$

[28] Bernoulli (1982–1996) vol. 2, pp. 49–64.

He found the roots of the corresponding algebraic equation to be $2, 2, 2, -3, -3$; the general term of the sequence was then

$$\frac{(1026 - 1035x + 225xx) \cdot 2^x + (224 - 80x) \cdot (-3)^x}{9000}.$$

As a final example, Bernoulli set $a_n = \sin nx$ and applied the addition formula for sine to get, in modern notation,

$$a_{n+1} + a_{n-1} = \sin(n+1)x + \sin(n-1)x = 2\cos x \sin nx = 2\cos x a_n.$$

This produced the algebraic equation $\lambda^2 - 2\cos x\lambda + 1 = 0$ whose roots were given by

$$\lambda_1 = \cos x + \sqrt{\cos^2 x - 1} = \cos x + \sqrt{-1}\sin x,$$
$$\lambda_2 = \cos x - \sqrt{\cos^2 x - 1} = \cos x - \sqrt{-1}\sin x.$$

This gave him the formula for sine:

$$a_n = \sin nx = \frac{(\cos x + \sqrt{-1}\sin x)^n - (\cos x - \sqrt{-1}\sin x)^n}{2\sqrt{-1}}.$$

Note that this gives a new proof of de Moivre's formula. Bernoulli also made an interesting observation about the root largest in absolute value of the algebraic equation (13.31), noting that such a root could be obtained from the sequence satisfying the corresponding difference equation. Taking $\lambda$ to be the root largest in absolute value, and writing the sequence as $a_1, a_2, a_3, \ldots, a_m, \ldots$, he observed that as $m$ went to infinity, $\frac{a_{m+1}}{a_m}$ would approach the value $\lambda$. Also, the root smallest in absolute value could be found by setting $\lambda = \frac{1}{\mu}$ in (13.31). Bernoulli was quite proud of this result; he wrote a letter to Goldbach on February 20, 1728,[29] that even if it were not useful, it was among the most beautiful theorems on the topic. Euler must have agreed with Bernoulli, since he devoted a whole chapter of his *Introductio* of 1748 to finding roots of algebraic equations by solving difference equations.[30]

Illustrating that his beautiful theorem was in fact useful, Bernoulli showed how to find the approximate solution of $xx = 26$. He began by setting $x = y + 5$ to get $1 = 10y + yy$. Of the two roots of this last equation, he needed the smaller in absolute value, so he set $y = \frac{1}{z}$ to obtain $z^2 = 10z + 1$. The corresponding difference equation was $a_n = 10a_{n-1} + a_{n-2}$, and Bernoulli took the two initial values of the sequence to be 0 and 1. The difference equation then gave him the sequence 0, 1, 10, 101, 1020, 10301, 104030, .... To obtain an approximate value of $y$, Bernoulli took the ratio of the seventh and sixth terms of the sequence, obtaining $x = \sqrt{26} = 5 + \frac{10301}{104030} = 5.09901951360$. He then computed $\sqrt{26}$ by the usual method and got 5.0990151359. Bernoulli employed this idea to find the smallest roots of Laguerre polynomials of low

[29] Fuss (1968) vol. 2, pp. 250–253.
[30] Euler (1988) chapter 17.

degree. In his work with hanging chains, the roots of these polynomials yielded the frequencies of the oscillations. See Exercises 3 and 4 of Chapter 14.

## 13.6  Lagrange: Nonhomogeneous Equations

In 1759, Lagrange published a method for solving a nonhomogeneous linear difference equation with constant coefficients and this method can be seen as the analog of d'Alembert's method for the corresponding differential equation.[31] Lagrange started with a third-order equation to illustrate the technique. In brief, let the equation be $y + A\Delta y + B\Delta^2 y + C\Delta^3 y = X$; set $\Delta y = p$ and $\Delta p = q$ so that the equation can be written as $y + Ap + Bq + C\Delta q = X$. For the arbitrary constants $a$ and $b$ we have

$$y + (A + a)p + (B + b)q - a\Delta y - b\Delta p + C\Delta q = X. \tag{13.33}$$

Choose $a$ and $b$ such that

$$\Delta y + (A + a)\Delta p + (B + b)\Delta q = \Delta y + \frac{b}{a}\Delta p - \frac{C}{a}\Delta q. \tag{13.34}$$

Then $A + a = \frac{b}{a}$, $B + b = -\frac{C}{a}$, implying that $a$ satisfies the cubic $a^3 + Aa^2 + Ba + C = 0$. Moreover, by (13.34), equation (13.33) is reduced to the first-order equation

$$z - a\Delta z = X, \tag{13.35}$$

where

$$z = y + (A + a)p + (B + b)q. \tag{13.36}$$

The problem is now reduced to solving (13.35). Suppose it has been solved for each of the three values $a_1, a_2, a_3$ of $a$, obtained from the cubic. Let $z_1, z_2$ and $z_3$ be the corresponding values of $z$ from (13.35). Now we have three linear equations

$$y + (A + a_1)p + (B + b_1)q = z_1, \quad y + (A + a_2)p + (B + b_2)q = z_2,$$

$$y + (A + a_3)p + (B + b_3)q = z_3,$$

and these can be solved to obtain $y = Fz_1 + Gz_2 + Hz_3$ for some constants $F, G$, and $H$. Finally, to solve the first-order equation (13.35), Lagrange considered the more general equation

$$\Delta y + My = N, \tag{13.37}$$

[31] Lagrange (1867–1892) vol. 1, pp. 23–36.

where $M$ and $N$ were functions of an integer variable $x$. He set $y = uz$ to get

$$u\Delta z + z\Delta u + Mzu = N. \tag{13.38}$$

He let $u$ be such that $(\Delta u + Mu)z = 0$, or $u$ was a solution of the homogeneous part of (13.37). Thus,

$$u(x) - u(x-1) = -M(x-1)u(x-1) \quad \text{or} \quad u(x) = (1 - M(x-1))u(x-1).$$

By iteration,

$$u(x) = (1 - M(x-1)(1 - M(x-2)) \cdots (1 - M(1))).$$

For this $u$, equation (13.38) simplified to

$$z(x) - z(x-1) = \frac{N(x-1)}{u(x-1)}.$$

Therefore

$$z(x) = \frac{N(x-1)}{u(x-1)} + z(x-1) = \frac{N(x-1)}{u(x-1)} + \frac{N(x-2)}{u(x-2)} + \cdots + \frac{N(1)}{u(1)} + z(1).$$

Laplace later observed that (13.37) could be solved directly by iteration:

$$y(x) = y(x-1) + N(x-1) - M(x-1)y(x-1)$$
$$= (1 - M(x-1))y(x-1) + N(x-1)$$
$$= N(x-1) + (1 - M(x-1))N(x-2) + (1 - M(x-2))y(x-2) \text{ etc.}$$

As Lagrange himself pointed out, this method could obviously be generalized to a nonhomogeneous equation of any order. Of course, the question of solving the corresponding algebraic equation of arbitrary degree would be a separate problem.

Lagrange found another method of solving difference equations, using the device of the variation of parameters.[32] Again presenting Lagrange's work in brief, suppose we have a third-order difference equation

$$y_{x+3} + P_x y_{x+2} + Q_x y_{x+1} + R_x y_x = V_x. \tag{13.39}$$

Let $z_x$, $z'_x$, $z''_x$ be three independent solutions of the corresponding homogeneous equation

$$y_{x+3} + P_x y_{x+2} + Q_x y_{x+1} + R_x y_x = 0. \tag{13.40}$$

The general solution of this equation is $Cz_x + C'z'_x + C''z''_x$ where $C$, $C'$, and $C''$ are constants. Now suppose $C_x$, $C'_x$ and $C''_x$ are functions of $x$, determined by the condition that

---

[32] ibid. vol. 4, pp. 151–160.

$$y_x = C_x z_x + C'_x z'_x + C''_x z''_x \tag{13.41}$$

is a solution of the nonhomogeneous equation. Changing $x$ to $x + 1$, we have

$$
\begin{aligned}
y_{x+1} &= C_{x+1} z_{x+1} + C'_{x+1} z'_{x+1} + C''_{x+1} z''_{x+1} \\
&= C_x z_{x+1} + C'_x z'_{x+1} + C''_x z''_{x+1} + \Delta C_x z_{x+1} + \Delta C'_x z'_{x+1} + \Delta C''_x z''_{x+1}.
\end{aligned}
$$

Now suppose that $C_x$, $C'_x$, $C''_x$ are such that

$$z_{x+1} \Delta C_x + z'_{x+1} \Delta C'_x + z''_{x+1} \Delta C''_x = 0.$$

Then

$$y_{x+1} = C_x z_{x+1} + C'_x z'_{x+1} + C''_x z''_{x+1}. \tag{13.42}$$

If in the equation for $y_{x+2}$, we again change $x$ to $x + 1$, the result is

$$y_{x+3} = C_x z_{x+3} + C'_x z'_{x+3} + C''_x z''_{x+3} + \Delta C_x z_{x+3} + \Delta C'_x z'_{x+3} + \Delta C''_x z''_{x+3}.$$

Thus,

$$y_{x+3} = C_x z_{x+3} + C'_x z'_{x+3} + C''_x z''_{x+3}. \tag{13.43}$$

We also have an equation for $y_{x+1}$ resembling the equation for $y_x$. If we make a similar $x \to x + 1$ change in the equation for $y_{x+1}$, we can require that

$$z_{x+2} \Delta C_x + z'_{x+2} \Delta C'_x + z''_{x+2} \Delta C''_x = 0. \tag{13.44}$$

Multiply equation (13.41) by $R_x$; multiply equation (13.42) by $Q_x$; multiply (13.43) by $P_x$. Now add the results to (13.44). From (13.39) and the fact that $z_x$, $z'_x$, $z''_x$ satisfy (13.40), it follows that

$$z_{x+3} \Delta C_x + z'_{x+3} \Delta C'_x + z''_{x+3} \Delta C''_x = V_x. \tag{13.45}$$

Consider (13.45) together with the two equations, that we required be satisfied by $\Delta C$:

$$z_{x+2} \Delta C_x + z'_{x+2} \Delta C'_x + z''_{x+2} \Delta C''_x = 0,$$

$$z_{x+1} \Delta C_x + z'_{x+1} \Delta C'_x + z''_{x+1} \Delta C''_x = 0.$$

Thus, we have three linear equations yielding $\Delta C_x$, $\Delta C'_x$, and $\Delta C''_x$. Suppose we obtain $\Delta C_x = H_x$, $\Delta C'_x = H'_x$, and $\Delta C''_x = H''_x$. These first-order equations can be solved for $C_x$, $C'_x$, and $C''_x$, and hence we have $y_x$ from (13.41).

As an example of the method of variation of parameters, Lagrange considered a nonhomogeneous equation with constant coefficients. In this case, briefly, $z_x$ will be of the form $m^x$ for some constant $m$. Suppose we have a second-order equation for which $z_x = m^x$ and $z'_x = m_1^x$. Then the equations for $\Delta C_x$ and $\Delta C'_x$ are

$$m^{x+1} \Delta C_x + m_1^{x+1} \Delta C'_x = 0,$$

$$m^{x+2} \Delta C_x + m_1^{x+2} \Delta C'_x = V_x.$$

Solving for $\Delta C_x$ and $\Delta C'_x$, we have

$$\Delta C_x = \frac{V_x}{m^{x+1}(m - m_1)},$$

$$\Delta C'_x = \frac{V_x}{m_1^{x+1}(m_1 - m)};$$

therefore

$$C_x = \frac{V_x(m^x - 1)}{(m - 1)(m - m_1)m^x} + C_0 \text{ and } C'_x = \frac{V_x(m_1^x - 1)}{(m_1 - 1)(m_1 - m)m_1^x} + C'_0.$$

## 13.7 Laplace: Nonhomogeneous Equations

The method given by Laplace for solving a nonhomogeneous equation differed from the variation of parameters of Lagrange, but was analogous to Lagrange's method for equations with constant coefficients (13.34). Suppose the equation to be

$$y_{x+n} + P_x y_{x+n-1} + Q_x y_{x+n-2} + \cdots + T_x y_{x+1} + U_x y_x = V_x.$$

Laplace assumed that there existed functions $p_x$ and $q_x$ such that $y_{x+1} = p_x y_x + q_x$. This implied

$$y_{x+2} = p_{x+1} y_{x+1} + q_{x+1} \cdots,$$

$$y_{x+n} = p_{x+n-1} y_{x+n-1} + q_{x+n-1}.$$

Laplace introduced functions $\alpha_1, \alpha_2, \ldots, \alpha_{n-1}$ to obtain

$$\begin{aligned}
y_{x+n} &= p_{x+n-1} y_{x+n-1} + q_{x+n-1} \\
&= (p_{x+n-1} - \alpha_{n-1}) y_{x+n-1} + (\alpha_{n-1} p_{x+n-2} - \alpha_{n-2}) y_{x+n-2} \\
&\quad + (\alpha_{n-2} p_{x+n-3} - \alpha_{n-3}) y_{x+n-3} + \cdots + \alpha_1 p_x y_x \\
&\quad + q_{x+n-1} + \alpha_{n-1} q_{x+n-2} + \cdots + \alpha_1 q_x.
\end{aligned}$$

He then chose $\alpha_1, \alpha_2, \ldots, \alpha_{n-1}$ such that

$$P_x = p_{x+n-1} - \alpha_{n-1},$$
$$Q_x = \alpha_{n-1} p_{x+n-2} - \alpha_{n-2},$$
$$R_x = \alpha_{n-2} p_{x+n-3} - \alpha_{n-3},$$
$$\cdots$$
$$T_x = \alpha_2 p_{x+1} - \alpha_1,$$
$$U_x = \alpha_1 p_x.$$

Therefore, $p_x$ satisfied

$$\prod_{i=0}^{n-1} p_{x+i} = P_x \prod_{i=0}^{n-2} p_{x+i} + Q_x \prod_{i=0}^{n-3} p_{x+i} + \cdots + p_x T_x + U_x;$$

$q_x$ satisfied an equation of order $m - 1$:

$$V_x = q_{x+n-1} + \alpha_{n-1} q_{x+n-2} + \cdots + \alpha_1 q_x.$$

By successive reduction, $q_x$ could be determined, although the equation satisfied by $p_x$ was more difficult to handle.

## 13.8   Exercises

(1) In Proposition VII for recurrent series of the *Doctrine of Chances*, de Moivre showed that if $a_0 + a_1 x + a_2 x^2 + \cdots = \frac{B(x)}{1-fx+gx^2}$, with $B(x)$ a linear function, then the two series $\sum_{n=0}^{\infty} a_{2n} x^{2n}$ and $\sum_{n=0}^{\infty} a_{2n+1} x^{2n+1}$ sum to a rational function with denominator $1 - (f^2 - 2g)x^2 + g^2 x^4$. If the denominator of the original series was $1 - fx + gx^2 - hx^3$, then the denominator of the two series would be $1 - (f^2 - 2g)x^2 - (2fh - g^2)x^4 - h^2 x^6$. Work out the details by following de Moivre's method described in the text. Extend the results to the case where the original series is divided into three parts. See de Moivre (1967).

(2) Simpson showed that if $p$, $q$, and $r$ were the three cube roots of unity and $f(x) = \sum_{n=0}^{\infty} a_n x^n$, then $p + q + r = p^2 + q^2 + r^2 = 0$, $p^3 + q^3 + r^3 = 3$, and

$$\frac{f(px) + f(qx) + f(rx)}{3} = \sum_{n=0}^{\infty} a_{3n} x^{3n}.$$

He also explained how to generalize this to sum

$$\sum_{n=0}^{\infty} a_{mn+j} x^{mn+j}, \quad j = 0, 1, \ldots, m - 1,$$

by using $m$th roots of unity. Prove Simpson's result and obtain the generalization. Compare Simpson's results with de Moivre's in Exercise 1. See Simpson (1759). Thomas Simpson (1710–1761) was a self-taught mathematician who contributed to the popularization of mathematics and other intellectual pursuits during that period in England. He was an editor of the *Ladies Diary* and was one of the earliest mathematics professors at the Royal Military Academy at Woolwich. See the excellent account by Clarke (1929).

(3) In Proposition IX for recurrent series of his *Doctrine*, de Moivre stated that if $\sum a_n x^n$ and $\sum b_n x^n$ have the differential scales $1 - fx + gx^2$ and

$1 - mx + px^2$, respectively, then the differential scale of $\sum a_n b_n x^n$ is $1 - fmx + (f^2 p + m^2 g - 2gp)x^2 - fgmpx^3 + g^2 p^2 x^4$. Prove this result. Compare with Hadamard's theorem on the multiplication of singularities in Hadamard (1899). See de Moivre (1967).

(4) Solve the recurrence relation $a_{n+5} = a_{n+4} + a_{n+1} - a_n$, with $a_0 = 1, a_1 = 2$, $a_2 = 3, a_3 = 3, a_4 = 4$ by recurrent series (generating function) as well as by letting $a_x = \lambda^x$. See Euler (1988) p. 195.

(5) Use recurrent series to find the largest root of the equation $y^3 - 3y + 1 = 0$. See Euler (1988) p. 288.

(6) Find the smallest root of $y^3 - 6y^2 + 9y - 1 = 0$. This value is $2(1 - \sin 70°)$. See Euler (1988) p. 290.

## 13.9   Notes on the Literature

In D. Bernoulli (1982–1996) vol. 1, pp. 133–189, U. Bottazzini has discussed Daniel Bernoulli's early mathematical work, including difference equations, and has put it into historical perspective. Hald (1990) includes an interesting chapter on the use of difference equations to solve problems in probability theory by de Moivre, Lagrange, and Laplace. De Beaune, Girard, and Viète (1986) contains English translations by Robert Schmidt of algebra texts by de Beaune, Girard, and Viète, giving easy access to the methods and notation of early seventeenth-century French algebraists.

# 14

## Differential Equations

### 14.1 Preliminary Remarks

In the seventeenth century, before the development of calculus, problems reducible to differential equations began to appear in the study of general curves and in navigation. Interestingly, these differential equations were often related to the logarithm or exponential curve. For example, Harriot obtained the logarithmic spiral by projecting the loxodrome onto the equatorial plane.[1] And in 1638, I. F. de Beaune (1601–1652) posed to Descartes the problem of finding a curve such that the subtangent at each point was a constant.[2] Note that the problem actually leads to the simple differential equation $\frac{dy}{dx} = \frac{y}{a}$. Descartes replied with a solution involving the logarithmic function, though he did not explicitly recognize it.[3] In a paper of 1684, Leibniz gave the first published solution by explicitly stating the problem as a differential equation.[4]

Newton understood the significance of the differential equation as soon as he started developing calculus. In his October 1666 tract on calculus, written a year after he graduated from Cambridge, he wrote[5]

> If two Bodys $A$ & $B$, by their velocitys $p$ & $q$ describe $y^e$ [the] lines $x$ and $y$. & an Equation bee given expressing $y^e$ relation twixt one of $y^e$ lines $x$, & $y^e$ ratio $\frac{q}{p}$ of their motions $q$ & $p$; To find the other line $y$. Could this ever bee done all problems whatever might be resolved.

So Newton's problem to solve all problems could be stated as follows: Given $f\left(x, \frac{dy}{dx}\right) = 0$, find $y$. In a treatise prepared five years later, Newton gave a classification of first-order differential equations $\frac{dy}{dx} = f(x, y)$.

In the 1660s, Isaac Barrow and James Gregory too dealt with differential equations, arising from geometric problems. Gregory considered the question of determining a

---

[1] See Pepper (1968).
[2] Descartes (1897–1913) vol. IV, pp. 229–230.
[3] ibid. vol. II, pp. 514–517.
[4] Leibniz (1684); also see Scriba (1961).
[5] Newton (1967–1981) vol. 1, p. 403.

curve whose area of surface of revolution produced a given function.[6] This translates to the differential equation

$$y^2 \left(1 + \left(\frac{dy}{dx}\right)^2\right) = f(x),$$

where $f(x)$ is the given function. In connection with this, Barrow gave a geometric solution for the differential equation

$$y \left(1 + \frac{dy}{dx}\right) = a,$$

by expressing the solution in terms of areas under hyperbolas.[7] Geometrically, the problem would be to find a curve $y = f(x)$ such that the sum of the ordinate $y$ and the subnormal $yy'$ is a constant.

By the 1690s, it was clear to Newton, Leibniz, the two Bernoullis, and the other mathematicians with whom they corresponded that differential equations were intimately connected with curves and their properties. If they knew some property of a geometric object, such as the subtangent or curvature, then the problem of finding the curve itself usually led them to a differential equation. They had begun to recognize or get a glimpse of some general methods of solving these equations, such as separation of variables and multiplication of the equation by an integrating factor.

Newton encountered differential equations in the geometrical and astronomical problems of the *Principia*; in his *De Quadratura Curvarum* of 1691–1692, Newton once again emphasized the importance of differential equations, or fluxional equations, discussing a number of special methods for solving them, as well as the general separation of variables method. He wrote, "Should the equation involve both fluent quantities, but can be arranged so that one side of the equation involves but a single one together with its fluxion and the other the second alone with its fluxions."[8] The term separation of variables was first used by Johann Bernoulli in his letter to Leibniz of May 9, 1694, and then in a related paper published in November 1694.[9] Bernoulli also noted that there were important equations unable to be solved by this method, such as $aady = xxdx + yydx$. Observe that this is an equation between the differentials $dx$ and $dy$; it is hence given the name "differential equation." We would now write it as $a^2 \frac{dy}{dx} = x^2 + y^2$, a particular case of Riccati's equation, to which we will return later.

The first person to discover the integrating factor technique seems to be the Swiss mathematician Nicolas Fatio de Duillier (1664–1753). In 1687, he mastered the elements of differential and integral calculus by his own unaided efforts.[10] Since so little on this subject had been published, Fatio's achievement was remarkable.

---

[6] Turnbull (1939) pp. 167 and 174.
[7] ibid. p. 167.
[8] Newton (1967–1981) vol. 7, p. 73.
[9] Bernoulli and Leibniz (1745) pp. 5–9, especially p. 7 and Bernoulli (1742) vol. 1, pp. 123–128.
[10] Newton (1967–1981) vol. 7, pp. 78–79, footnote 68.

He exchanged several letters on calculus with Huygens, to whom in February 1691 Fatio first communicated his method of multiplying an equation by $x^\mu y^\nu$ to possibly put it into integrable form. Huygens in turn wrote Leibniz concerning Fatio's method for solving the differential equations

$$-2xy\,dx + 4x^2\,dy - y^2\,dy = 0 \quad \text{and} \quad -3a^2y\,dx + 2xy^2\,dx - 2x^2y\,dy + a^2x\,dy = 0.$$

Observe that, after multiplying across by $y^{-5}$, the first equation becomes the differential of $-x^2y^{-4} + \frac{1}{2}y^{-2} = c$. And the second equation, when multiplied by $x^{-4}$, can be integrated to yield $a^2x^{-3}y - x^{-2}y^2 = c$. Fatio later told Newton about his method, and Newton included it in his *De Quadratura*, giving credit to Fatio. The technique, as Newton explained and generalized it, was to multiply $f_1(x,y)\dot{x} + f_2(x,y)\dot{y} = 0$ by $x^\mu y^\nu$ to get $M(x,y)\dot{x} + N(x,y)\dot{y} = 0$, where $M$ and $N$ were polynomials or even algebraic functions of $x$ and $y$. The basic idea was to compute $\frac{\partial}{\partial y} \int M(x,y)\,dx$ and choose $\mu$ and $\nu$ so that this quantity became equal to $N(x,y)$. If this was possible, then $\int M(x,y)\,dx = c$ was the solution of the differential equation. Newton even extended this method to second, third- and higher-order differential equations. Regrettably, Newton did not include these results on differential equations in the published version of *De Quadratura*; they were rediscovered by Leibniz and the two elder Bernoullis. It should be noted that Newton's acknowledgment[11] of Fatio's contribution was unusual; it showed the depth of their friendship at that time. Unfortunately, it appears that in 1693, this friendship was abruptly and emotionally terminated.[12]

In the tenth of his 1691–92 lectures on integral calculus to l'Hôpital,[13] Johann Bernoulli gave an ingenious application of integrating factors to solve the separable equation $axdy - ydx = 0$. He multiplied the equation by $\frac{y^{a-1}}{x^2}$ to obtain $\frac{ay^{a-1}}{x}dy - \frac{y^a}{x^2}dx = 0$. Since the left-hand side was the differential of $y^{\frac{a}{x}}$, integration yielded $y^{\frac{a}{x}} = c$. Note that, after separation of variables in the original differential equation, one gets a logarithm on each side. But it seems that at that time Bernoulli found some difficulty in working with logarithms in an analytic setting and found the solution by a method avoiding the logarithm. It was only after an exchange of letters with Leibniz that Bernoulli understood logarithms; in fact, in 1697 he published a paper on exponentials and logarithms.[14]

In the 1690s, Leibniz and the Bernoullis also learned to handle first-order linear differential equations. In a 1695 paper, Jakob Bernoulli raised the question of how to solve the nonlinear equation $ady = yp\,dx + by^n q\,dx$ where $p$ and $q$ were functions of $x$ and $a$, $b$ were constants.[15] In response, Leibniz as well as Johann Bernoulli observed[16] that the equation could be linearized by the substitution $v = y^{1-n}$.

[11] ibid. p. 79.
[12] Westfall (1980) pp. 538–539.
[13] Bernoulli (1742) vol. 3, pp. 385–558.
[14] Bernoulli (1742) vol. 1, pp. 179–187.
[15] Bernoulli (1744) vol. 1, p. 663.
[16] Bernoulli and Leibniz (1745) p. 199.

Bernoulli found an interesting method, applicable to linear equations as well, for solving the equation by setting $y = mz$. This technique showed that in the case of linear equations, $m$ could be chosen to be $e^{-\int p\,dx}$, the reciprocal of the integrating factor. Three decades later, in 1728, Euler wrote a paper, published in 1732, in which he solved the linear equation by making use of an integrating factor, making due reference to his teacher Bernoulli.[17]

The theory of linear differential equations with constant coefficients was developed much more slowly than one might expect. Recall that in 1728 Daniel Bernoulli solved linear difference equations with constant coefficients[18] by substituting $x^n$ in the difference equation to obtain an algebraic equation in $x$, whose solutions $x_1, x_2, \ldots$ determined the possible values of $x$. The general solution was then a linear combination $c_1 x_1^n + c_2 x_2^n + \cdots$ of the special solutions. Yet it took Euler nearly a decade to perceive that he could solve the corresponding differential equation in a similar way. For more discussion on the alternating development of difference and differential equations, see Chapter 13.

The search for a general solution for a linear differential equation with constant coefficients seems to have started with Daniel Bernoulli's letter to Euler[19] of May 4, 1735, describing his work on the transverse vibration of a hanging elastic band fixed at one end to a wall. Bernoulli wrote that he found the equation for the curve of vibration to be $nd^4 y = y\,dx^4$, $n$ a constant. He requested Euler's help in solving the equation, noting that if $p$ divided $m$, then the solutions of $\alpha d^p y = y\,dx^p$ were contained in those of $nd^m y = y\,dx^m$. It followed, he observed, that the logarithm satisfied both his equation and $n^{\frac{1}{2}}ddy = y\,dx^2$, but that it was not general enough for his purpose. Euler too was unable to solve the equation except as an infinite series. Commenting on this, C. Truesdell remarked, "These are two great mathematicians who have just shown themselves not fully familiar with the exponential function; we must recall that this is 1735!"[20]

In Proposition XXIV, Theorem XIX of Book II of his *Principia*, Newton gave a full description of his geometric treatment of simple harmonic motion. In a 1728 paper on simple harmonic motion, Johann Bernoulli gave a more analytic treatment, solving the second-order differential equation $\frac{d^2 y}{dx^2} = -y$ by reducing it to a first-order equation.[21] Hermann did similar work in a 1716 paper in the *Acta Eruditorum*.[22] These seem to be the earliest treatments of simple harmonic motion by the integration of the differential equation describing the motion.

In a letter to Johann Bernoulli of May 5, 1739, Euler wrote that he had succeeded in solving the third-order equation $a^3 dy^3 = y\,dx^3$, where $dx$ was assumed constant.[23] Note that this meant that $x$ was the independent variable. Euler gave the solution as

[17] Eu. I-22 pp. 1–14. E 10 § 15.
[18] Bernoulli (1982–1996) vol. 2, pp. 49–64.
[19] Fuss (1968) vol. 2, pp. 419–423, especially p. 422.
[20] Truesdell (1960) p. 167.
[21] Bernoulli (1742) vol. III, p. 210.
[22] Hermann (1716).
[23] Eu. IVA-2, pp. 287–305, especially p. 302.

$$y = be^{\frac{x}{a}} + ce^{-\frac{x}{2a}}. \text{ Sinum Arcus } \frac{(f+x)\sqrt{3}}{2a}.$$

He gave no indication of how he found this, but he probably did not use a general method, because in his letter to Bernoulli of September 15, 1739,[24] he wrote that he had recently found a general method for solving in finite terms the equation

$$y + a\frac{dy}{dx} + b\frac{ddy}{dx^2} + c\frac{d^3y}{dx^3} + d\frac{d^4y}{dx^4} + e\frac{d^5y}{dx^5} + \text{ etc. } = 0.$$

He noted that the solution depended on the roots of the algebraic equation

$$1 - ap + bp^2 - cp^3 + dp^4 - ep^5 + \text{ etc. } = 0.$$

As an example, he explained that the solution of Daniel Bernoulli's equation $d^4y = k^4 y dx^4$ was determined by the algebraic equation $1 - k^4 p^4 = 0$. Thus, the solution of the differential equation emerged as[25]

$$y = Ce^{-\frac{x}{k}} + De^{\frac{x}{k}} + E\,\sin\left(\frac{x}{k}\right) + F\cos\left(\frac{x}{k}\right).$$

In a letter of January 19, 1740,[26] Euler mentioned that he could also solve

$$0 = y + ax\frac{dy}{dx} + bx^2\frac{d^2y}{dx^2} + cx^3\frac{d^3y}{dx^3} + \cdots. \tag{14.1}$$

Johann Bernoulli replied to Euler in a letter of April 16, 1740,[27] that he too had solved (14.1). He reduced its order by multiplying it by $x^p$ and choosing $p$ appropriately. He remarked that he had actually done this before 1700 and also wrote that he had found a special solution similar to that found by Euler for the equation with constant coefficients. However, he was puzzled as to how the imaginary roots could lead to sines and cosines. For about a year, he discussed this with Euler. Finally, Euler pointed out that the equation $ddy + y dx^2 = 0$ had the obvious solution $y = 2\cos x$, also taking the form $y = e^{x\sqrt{-1}} + e^{-x\sqrt{-1}}$. Bernoulli ended his letter by asking whether it was possible to reduce the equation

$$yxx\,dx^2 + addy = 0 \left(\text{i.e., } a\frac{d^2y}{dx^2} + x^2 y = 0\right),$$

to a first-order equation. Euler answered[28] that by the substitution $y = e^{\int z\,dx}$, the equation reduced to

$$xx\,dx + a\,dz + azz\,dx = 0 \tag{14.2}$$

[24] ibid. p. 314.
[25] ibid. p. 315.
[26] ibid. p. 369.
[27] Fuss (1968) vol. 2, pp. 33–41.
[28] Eneström (1905).

and noted that this was a particular case of the Riccati equation $dy = yydx + ax^m dx$ on which he had already written papers.

Interestingly, in a paper on curves and their differential equations, published 46 years earlier, Bernoulli stated an equation almost identical to (14.2) and wrote that he had not solved it; he noted that, of course, separation of variables would not work.[29] His older brother Jakob made persistent efforts to solve the equation and finally succeeded in 1702. In a letter to Leibniz dated November 15, 1702,[30] some of which was devoted to the relation of the sum $\sum \frac{1}{n^2}$ with integrals of the form $\int x^l \ln(1+x)\, dx$, he mentioned in passing that he could solve $dy = yy\, dx + xx\, dx$ by reducing it to $ddy = x^2 y dx^2$ and then applying separation of variables. When Leibniz asked for details, Jakob provided them in a letter of October 3, 1703.[31] Here given in modern notation, he defined a new function $z$ by the equation $y = -\frac{1}{z}\frac{dz}{dx}$ to reduce

$$\frac{dy}{dx} = x^2 + y^2$$

to the form

$$\frac{d^2 z}{dx^2} + x^2 z = 0.$$

He solved this second-order linear equation by an infinite series for $z$ from which he obtained $y$ as a quotient of two infinite series. After performing the division of one series by the other, his result was

$$y = \frac{x^3}{3} + \frac{x^7}{3 \cdot 3 \cdot 7} + \frac{2x^{11}}{3 \cdot 3 \cdot 3 \cdot 7 \cdot 11} + \frac{13x^{15}}{3 \cdot 3 \cdot 3 \cdot 3 \cdot 5 \cdot 7 \cdot 7 \cdot 11} + \cdots . \quad (14.3)$$

Now Newton would have been satisfied with this infinite series solution, whereas Leibniz and the Bernoullis had a different general outlook. They strove to find solutions in finite form, using the known elementary functions. Perhaps this may explain why mathematicians in the Leibniz-Bernoulli school took scant note of Jakob Bernoulli's new method for dealing with the Riccati equation. However, Euler himself later rediscovered this method and generalized it.

Jacopo Riccati (1676–1754) studied law at the University of Padua, but was encouraged to pursue mathematics by Stephano degli Angeli, who had earlier taught James Gregory. Riccati became interested in the equation named after him upon studying Gabriele Manfredi's treatise *De Constructione Aequationum Differentialium* in which he considered the equation

$$nxxdx - nyydx + xxdy = xydx,$$

[29] Bernoulli (1742) vol. 1, pp. 123–125.
[30] Leibniz (1971) vol. 3/1, pp. 62–66.
[31] ibid. pp. 72–79.

a special case of what is now known as the generalized Riccati equation

$$\frac{dy}{dx} = P + Qy + Ry^2,$$

where $P$, $Q$, and $R$ are functions of $x$. Around 1720, Riccati and others worked on the special case

$$ax^m dx + yy dx = bdy, \tag{14.4}$$

and Riccati attempted a solution using separation of variables. Amusingly, Riccati corresponded on this topic with all the then living Bernoulli mathematicians: Niklaus I, Johann, and his two sons Niklaus II and Daniel. The latter two determined by different methods a sequence of values of $m$ for which the equation could be solved in finite terms. Riccati published his work in 1724[32] with a note by D. Bernoulli, who gave the announcement of the solution of the Riccati equation (14.4) as an anagram. About a year later, without reference to his anagram, D. Bernoulli published details of his solution. Briefly, his result was that the equation could be solved in terms of the logarithmic, exponential and algebraic functions when $n = \frac{-4m}{2m\pm1}$, that is, when $n$ was a number in the sequence

$$0; \ -\frac{4}{1}, \ -\frac{4}{3}; \ -\frac{8}{3}, \ -\frac{8}{5}; \ -\frac{12}{5}, \ -\frac{12}{7}; \ -\frac{16}{7}, \ -\frac{16}{9}, \ \dots .$$

His method was to show that the substitutions $\frac{x^{n+1}}{n+1} = u$, $y = -\frac{1}{v}$ in the equation $\frac{dy}{dx} = ax^n + by^2$ produced another equation of the same form, but with $n$ changed to $-\frac{n}{n+1}$. Then again, the substitutions $x = \frac{1}{u}$, $y = -\frac{u}{b} - vu^2$ also produced another equation of the same form, but with $n$ changed to $-n - 4$. Now when $n = 0$, the equation was the integrable $\frac{dy}{dx} = a + by^2$. It followed that when $n = -4$, the equation would still be integrable, and the same was true when $n = -\frac{-4}{-4+1} = -\frac{4}{3}$. The result was the sequence given by Daniel Bernoulli. Note that when $m \to \infty$, $n = \frac{-4m}{2m\pm1} \to -2$; it turns out that when $n = -2$, the equation is still integrable. In that case, $\frac{dy}{dx} = \frac{a}{x^2} + by^2$ and the substitution $y = \frac{v}{x}$ produces the separable equation

$$x\frac{dv}{dx} = a + v + bv^2.$$

Euler published many papers on the Riccati equation. In the 1730s, he elucidated its relation to continued fractions,[33] and solved it as a ratio of two infinite series[34] in the manner of Jakob Bernoulli. Euler's method here also demonstrated that these series would be finite for those values of $m$ defined by Daniel Bernoulli and his brother. In the 1760s, Euler published a demonstration that the generalized Riccati equation could be transformed to a linear second-order equation, and conversely.[35] The paper was read

---

[32] Riccati (1724); an English translation available through the online Euler archive.
[33] Eu. I-14, pp. 187–216. E 71, § 28.
[34] Eu. I-22, pp. 19–35. E 31.
[35] Eu. I-22, pp. 403–418. E 284.

to the Berlin Academy in 1742. He also showed that if one particular solution, $y_0$, of the generalized Riccati equation were known, then, by the substitution $y = y_0 + \frac{1}{v}$, Riccati's equation could be reduced to the linear equation

$$\frac{dv}{dx} + (Q + 2Ry_0)v + R = 0. \tag{14.5}$$

On the other hand, if two solutions, $y_0$ and $y$, were known, then $w = \frac{y-y_0}{y-y_1}$ satisfied the simpler equation

$$\frac{1}{w}\frac{dw}{dx} = R(y_0 - y_1). \tag{14.6}$$

Interestingly, in 1841, Joseph Liouville proved the converse of D. Bernoulli's theorem on the Riccati equation.[36] Liouville was very interested in the general problem of integration in finite terms, using the elementary and algebraic functions, a topic now seeing renewed interest in the area of symbolic integration. Liouville proved that if $\frac{dy}{dx} = ax^n + by^2$ could be solved in finite terms, then $n$ had to be one of the numbers determined by Bernoulli.

In a 1753 paper, Euler solved nonhomogeneous linear equations by the technique of multiplying the equation by an appropriate function to reduce its order.[37] Later, in 1762,[38] Lagrange found another method for reducing the order of such an equation, leading him to the concept of an adjoint, a label apparently first used in this context by Lazarus Fuchs about a century later. Briefly, Lagrange took the differential equation to be

$$Ly + M\frac{dy}{dt} + N\frac{d^2y}{dt^2} + \cdots = T, \tag{14.7}$$

where $L, M, N, \ldots, T$ were functions of $t$. He then multiplied the equation by some function $z(t)$ and integrated by parts. Since

$$\int Mz\frac{dy}{dt}\, dt = Mzy - \int \frac{d}{dt}(Mz)y\, dt,$$

$$\int Nz\frac{d^2y}{dt^2}\, dt = Nz\frac{dy}{dt} - \frac{d}{dt}(Nz)y + \int \frac{d^2}{dt^2}(Nz)y\, dt,$$

and so on, the original equation was transformed to

$$y\left(Mz - \frac{d}{dt}(Nz)\right) + \frac{dy}{dt}Nz + \cdots$$

$$+ \int \left(Lz - \frac{d}{dt}(Mz) + \frac{d^2}{dt^2}(Nz) + \cdots\right)y\, dt = \int Tz\, dt. \tag{14.8}$$

[36] Liouville (1841).
[37] Eu. I-22, pp. 181–213. E 188.
[38] Lagrange (1867–1892) vol. I, pp. 471–478.

Lagrange then took $z$ to be the function satisfying

$$Lz - \frac{d}{dt}(Mz) + \frac{d^2}{dt^2}(Nz) - \cdots = 0. \tag{14.9}$$

This was the adjoint equation, and if $z$ satisfied it, then the expression within parentheses in the integral on the left-hand side of (14.8) would vanish. The remaining equation would then be of order $n - 1$. In this way, the order of the equation (14.7) was reduced by one, and the process could be continued. When Lagrange applied this procedure to the adjoint equation (14.9) to reduce its order, he obtained the homogeneous part of the equation (14.7). Thus, he saw that the adjoint of a homogeneous equation was, in fact, that equation itself. Lagrange also discovered the general method of variation of parameters in order to obtain the solution of a nonhomogeneous equation, once the solution of the corresponding homogeneous equation was known. Lagrange did this work around 1775,[39] but in 1739, Euler applied this same method to the special equation $\frac{d^2y}{dx^2} + ky = X$.

We have seen that in his 1743 paper, Euler introduced the concepts of general and particular solutions of linear equations.[40] By choosing appropriate constants in the general solution, any particular solution could be obtained. Taylor in 1715 and Clairaut in 1734 found solutions for some special nonlinear equations, solutions not producible by choosing constants in the general solutions.[41] Of one solution, Taylor remarked that it was singular, and so they were named. Euler also studied singular solutions; he found it paradoxical that they could not be obtained from the general solutions. He first encountered such a situation in the course of his study of mechanics in the 1730s. In his paper of 1754,[42] he posed a number of geometric problems leading to singular solutions, commenting that the paradox of singular solutions was not a mere aberration of mechanics.

It appears that the French mathematician Alexis Claude Clairaut (1713–1765) was the first to give a geometric interpretation of a singular solution; this appeared in his 1734 paper on differential equations. He considered the equation

$$y = (x+1)\frac{dy}{dx} - \left(\frac{dy}{dx}\right)^2 \equiv (x+1)p - p^2.$$

Note here that the more general solution $y = xy' + f(y')$ is now called Clairaut's equation. Briefly describing his solution, we differentiate with respect to $p$ and simplify the equation to get

$$dp\,(x+1-2p) = 0.$$

The first factor gives $p = c$, a constant, so that $y = c(x+1) - c^2$. The second factor gives $p = \frac{x+1}{2}$ so that $y = \frac{(x+1)^2}{4}$ is a solution. Now the envelope (or, in a plane, the

[39] ibid. vol. 4, pp. 5–108.
[40] Eu. I-22, pp. 108–149, E 62.
[41] Taylor (1715) and Clairaut (1734).
[42] Eu. I-22, pp. 214–236. E 236.

curve that is tangent to each one of a family of curves) of the family of straight lines $y = c(x + 1) - c^2$ is found by first eliminating $c$ from this equation and its derivative with respect to $c$, given by $0 = x + 1 - 2c$. Thus, the envelope is given by $y = \frac{(x+1)^2}{4}$ and in this case, the singular solution $y = \frac{(x+1)^2}{4}$ is the envelope of the family of integral curves.

D'Alembert, Euler, and Laplace also studied singular solutions. These efforts culminated in the general theory developed in the 1770s by Lagrange. Considering equations of the form

$$f\left(x, y, \frac{dy}{dx}\right) = a_n(x, y)\left(\frac{dy}{dx}\right)^n + \cdots + a_1(x, y)\frac{dy}{dx} + a_0(x, y) = 0,$$

he gave two approaches to the study of singular solutions. In the second of these methods, he differentiated the equation with respect to $x$ to obtain the expression for the second derivative

$$\frac{d^2 y}{dx^2} = -\frac{\frac{\partial f}{\partial x} + \left(\frac{\partial f}{\partial y}\right)\frac{dy}{dx}}{\frac{\partial f}{\partial y'}}.$$

Lagrange asserted that at the points of the singular solutions, the numerator and the denominator both vanish. Hence, both the equations

$$\frac{\partial f}{\partial y'} = 0 \quad \text{and} \quad \frac{\partial f}{\partial x} + \frac{\partial f}{\partial y}\frac{dy}{dx} = 0$$

had to be satisfied along the singular solutions. Note that these relations are true for Clairaut's singular solution. In fact, the second equation is identically true in that case.

In 1835, Cauchy was the first to treat the question of the existence of a solution of a differential equation.[43] Earlier mathematicians had presented methods for solving the equation on the assumption that solutions existed. Cauchy worked with a system of first-order differential equations; given in its simplest form with only one equation, his result can be stated: Suppose $f(x, y)$ is analytic in the variables $x$ and $y$ in the neighborhood of a point $(x_0, y_0)$; then the equation $\frac{dy}{dx} = f(x, y)$ has a unique analytic solution $y(x)$ in a neighborhood of $x_0$ such that $y(x_0) = y_0$. Cauchy's first proof of this theorem had gaps and in the 1840s, he published papers giving a more detailed exposition of the result. The French mathematicians C. Briot and J. Bouquet worked out a clearer and more complete presentation of this method, called the method of majorant.[44] For a brief description of this method, take $x_0 = y_0 = 0$ and suppose

$$f(x, y) = \sum a_{kl} x^l y^k \quad \text{for } |x| \le A, \ |y| \le B.$$

Also, let $M$ be the maximum of $|f(x, y)|$ in this region. Then the function

[43] Cauchy (1840–1841) vol. 1, pp. 327–384.
[44] Briot and Bouquet (1856).

$$F(x, y) = \frac{M}{\left(1 - \frac{x}{A}\right)\left(1 - \frac{y}{B}\right)}$$

is such that if

$$F(x, y) = \sum A_{kl} x^l y^k$$

then

$$|a_{kl}| \leq |A_{kl}|.$$

The differential equation $\frac{dy}{dx} = F(x, y)$ can now be solved explicitly. First, it can be shown that the coefficients of the formal power series solution of $\frac{dy}{dx} = f(x, y)$ are bounded by the coefficients of the explicit solution of $\frac{dy}{dx} = F(x, y)$, and this fact may then be used to show that the formal solution is an actual solution and is unique. In the 1870s, Kovalevskaya showed that this method could be extended to a certain system of partial differential equations.[45] The paper containing this result formed a part of her doctoral dissertation, supervised by Weierstrass.

## 14.2  Leibniz: Equations and Series

Leibniz's approach, as contrasted with Newton's, called for only occasional use of infinite series. Nevertheless, Leibniz discussed the connection between series and calculus; he derived series for elementary functions in several ways. In a paper of 1693, referring to Mercator and Newton, he derived the logarithmic and exponential series. As early as 1674, while working his way toward his final conception of the calculus, Leibniz discovered[46] the series for $\exp(x)$. He started with the series

$$y = x + \frac{x^2}{2!} + \frac{x^3}{3!} + \cdots,$$

and integrated to get

$$\int_0^x y\,dx = \frac{x^2}{2!} + \frac{x^3}{3!} + \frac{x^4}{4!} + \cdots = y - x. \tag{14.10}$$

Note that at the time Leibniz did this calculation, he was still using the "omn." notation for the integral. By taking the differential of both sides, he obtained

$$y\,dx = dy - dx \quad \text{or} \quad dx = \frac{dy}{y+1}.$$

Now Leibniz knew from the work of N. Mercator that this last equation implied $x = \ln(y + 1)$ so that

---

[45]  von Kowalevsky (1875).
[46]  See Scriba (1964).

$$\exp(x) = 1 + y = 1 + x + \frac{x^2}{2!} + \frac{x^3}{3!} + \cdots .$$

Although Leibniz did not explicitly write the last equation, he clearly understood the relation of the series to the logarithm.

Leibniz wrote a letter to Huygens,[47] dated September 4/14, 1694, explaining his new calculus techniques. Here note that Leibniz gave two dates because of the ten-day difference between the Julian and Gregorian calendars during that period. Huygens, at the close of an outstanding scientific career, was still eager to learn of the new advances in mathematics. He had already glimpsed the power of calculus from the work of Leibniz and the two Bernoullis on the catenary problem. Leibniz's first example for Huygens was the derivation of the infinite series for $\cos x$ from its differential equation. He started by observing that if $y$ was the arc of a circle of radius $a$, and $x$ denoted $\cos y$, then

$$y = a \int \frac{dx}{\sqrt{a^2 - x^2}}, \tag{14.11}$$

and

$$dy = \frac{a\,dx}{\sqrt{a^2 - x^2}}, \tag{14.12}$$

or

$$\sqrt{a^2 - x^2}\,dy = a\,dx. \tag{14.13}$$

Note here that the integral (14.11) was Leibniz's definition of arcsine or arccosine, as given in his 1686 paper.[48] We follow Leibniz almost word for word. He set

$$v = \sqrt{a^2 - x^2} \tag{14.14}$$

so that

$$v\,dy = a\,dx. \tag{14.15}$$

By differentiating this equation he found

$$v\,ddy + dvdy = a\,ddx. \tag{14.16}$$

Leibniz then assumed that the arcs $y$ increased uniformly, that is, $dy$ was a constant or $ddy$ was zero. Recall that in our terms this meant that he was taking $y$ as the independent variable and $x$ and $v$ as functions of $y$. Thus, he had $ddy = 0$, and equation (14.16) reduced to

$$dv\,dy = a\,ddx. \tag{14.17}$$

[47] Leibniz (1971) vol. 2, pp. 195–196.
[48] ibid. vol. 3, pp. 226–235.

To eliminate $v$, he observed that from (14.14), $v^2 = a^2 - x^2$, and therefore $v\,dv = -x\,dx$ or

$$dv = -x\frac{dx}{v}. \tag{14.18}$$

By (14.15) and (14.18),

$$dv = -\frac{x\,dy}{a}; \tag{14.19}$$

then, by (14.17) and (14.19), he arrived at the required differential equation,

$$-x\,dy\,dy = a^2\,ddx. \tag{14.20}$$

In order to derive the series for $a\cos\left(\frac{y}{a}\right)$ from this equation, Leibniz set

$$x = a + by^2 + cy^4 + ey^6 + \cdots.$$

He substituted this in (14.20) and equated coefficients. After a detailed calculation, he arrived at

$$x = \frac{1}{1}a - \frac{1}{1\cdot2\cdot a}y^2 + \frac{1}{1\cdot2\cdot3\cdot4\cdot a^3}y^4 - \frac{1}{1\cdot2\cdot3\cdot4\cdot5\cdot6\cdot a^5}y^6 + \text{etc.}$$

## 14.3 Newton on Separation of Variables

To get an idea of Newton's thinking and notation, we consider one simple example on separation of variables from *De Quadratura*.[49] He began with the equation $-ax\dot{x}y^2 = a^4\dot{y} + a^3x\dot{y}$. He separated the variables

$$\left(\frac{aa}{a+x} - a\right)\dot{x} = \frac{a^3}{y^2}\dot{y}$$

and then integrated to get

$$\boxed{\frac{aa}{a+x}} - ax = -\frac{a^3}{y}.$$

In Newton's notation, the square box denoted an integral. Sometimes, he replaced the box by the letter $Q$, denoting the Latin expression for area. He had no special notation for the logarithm and merely referred to it as the area of the hyperbola with ordinate $\frac{aa}{a+x}$ and abscissa $x$. He then rewrote the last equation, omitting the constant term, as an infinite series

$$\frac{a^2}{y} = \frac{1}{2}xx - \frac{x^3}{3a} + \frac{x^4}{4aa} - \cdots.$$

---

[49] Newton (1967–1981) vol. 7, p. 73.

"Or again, if $c^2$ is any given quantity, then is

$$\frac{a^2}{y} = c^2 + \frac{1}{2}xx - \frac{1}{3}\frac{x^3}{a} + \frac{1}{4}\frac{x^4}{aa} \cdots$$

the equation to be found."

We note in passing that in a 1691 letter to Huygens,[50] Leibniz discussed an example of a separable equation Apparently, Johann Bernoulli first used the expression separation of variables in a May 1694 letter to Leibniz. He wrote,[51] "Ut in aequationibus differentialibus indeterminatae $x$ cum suis differentialibus $dx$ separentur ab indeterminatis $y$ and $dy$."

Newton illustrated Fatio's method of integrating factors,[52] starting with

$$9x\dot{x}y - 18\dot{x}y^2 - 18xy\dot{y} + 5x^2\dot{y} = 0. \tag{14.21}$$

Applying Fatio's technique, he multiplied the equation by $x^\mu y^\nu$ to get

$$9\dot{x}x^{\mu+1}y^{\nu+1} - 18\dot{x}x^\mu y^{\nu+2} - 18x^{\mu+1}y^{\nu+1}\dot{y} + 5x^{\mu+2}y^\nu\dot{y} = 0.$$

He then integrated with respect to $x$ the terms forming the coefficient of $\dot{x}$ to obtain

$$\frac{9}{\mu+2}x^{\mu+2}y^{\nu+1} - \frac{18}{\mu+1}x^{\mu+1}y^{\nu+2} + g.$$

The fluxion of this expression would then reproduce the terms containing $\dot{x}$, but the terms with $\dot{y}$ would be given by

$$\frac{9(\nu+1)}{\mu+2}x^{\mu+2}y^\nu\dot{y} - \frac{18(\nu+2)}{\mu+1}x^{\mu+1}y^{\nu+1}\dot{y}.$$

For this to agree with (14.21), Newton required that $\frac{9(\nu+1)}{\mu+2} = 5$ and $\frac{\nu+2}{\mu+1} = 1$; or that $\mu = \frac{5}{2}$ and $\nu = \frac{3}{2}$. Hence, with $g$ a constant, the solution of the fluxional equation was

$$2x^{\frac{9}{2}}y^{\frac{5}{2}} - \frac{36}{7}x^{\frac{7}{2}}y^{\frac{7}{2}} + g = 0.$$

## 14.4 Johann Bernoulli's Solution of a First-Order Equation

We have seen that quite early in the study of differential equations, mathematicians noticed that separation of variables and integrating factors were applicable in many special situations. At the same time, they observed that there were simple first-order equations to which these methods could not be directly applied. Jakob Bernoulli, in

[50] Gerhardt (1898) p. 680.
[51] Bernoulli an Leibniz (1745) pp. 5–9, especially p. 7.
[52] Newton (1967–1981) vol. 7, pp. 78–81. See also Whiteside's footnote concerning Fatio on pp. 78–79.

the November 1695 issue of the *Acta Eruditorum*, posed the problem[53] of solving the differential equation

$$a\,dy = yp\,dx + by^n q\,dx,\tag{14.22}$$

where $p$ and $q$ were functions of $x$ and $a$ was a constant. In 1696, Leibniz noted that this problem could be reduced to a linear equation, though he did not give details. A year later, Johann Bernoulli wrote[54] $y = v^{\frac{1}{1-n}}$ would reduce the equation to the linear equation

$$\frac{1}{1-n}a\,dv = vp\,dx + bq\,dx.\tag{14.23}$$

However, his alternative method was to take $y$ to be a product of two new variables so that the extra variable could be appropriately chosen. Thus, he set $y = mz$ so that $dy = mdz + zdm$ and equation (14.22) would take the form

$$az\,dm + am\,dz = mzp\,dx + bm^n z^n q\,dx.\tag{14.24}$$

He then set $am\,dz = mzp\,dx$ or

$$adz : z = p\,dx.\tag{14.25}$$

Hence, $z$ could be found in terms of $x$. Bernoulli denoted this function by $\xi$. In the notation developed by Euler in the late 1720s, one would write $\xi = c^{\frac{1}{a}\int p\,dx}$ or $e^{\frac{1}{a}\int p\,dx}$. Note that Euler changed the $c$ to $e$ in the 1730s. Equation (14.24) was then reduced to

$$az\,dm = bm^n z^n q\,dx \text{ or } a\xi\,dm = bm^n \xi^n q\,dx$$

or

$$am^{-n}\,dm = b\xi^{n-1}q\,dx.\tag{14.26}$$

After integration, he had

$$\frac{a}{-n+1}m^{-n+1} = b\int \xi^{n-1}q\,dx.\tag{14.27}$$

In 1728, Euler made use of the integrating factor suggested by Bernoulli's method to solve nonhomogeneous linear equations of the first order.[55] The specific equation he faced was

$$dz + \frac{2z\,dt}{t-1} + \frac{dt}{tt-t} = 0.$$

[53] Bernoulli (1744) vol. 1, p. 663.
[54] ibid. pp. 174–179. Also see his letter to Leibniz in which he gave exactly the same proofs, except that he wrote $\xi$ as $X$.
[55] Eu. I-22, pp. 1–14, especially pp. 10–12. E 10.

He found the integrating factor by taking the exponential of the integral of the coefficient of $z$. So he multiplied the equation by $c^{2\int \frac{dt}{t-1}} = (t-1)^2$ and then integrated to obtain the solution

$$(t-1)^2 z + \int \frac{t-1}{t} \, dt = a.$$

In the general situation, given by Euler in a presentation of 1750,[56] this would require the solution of $\frac{dy}{dx} + py = q$. Multiplying by $e^{\int p \, dx}$, and using $e$ instead of $c$, Euler wrote

$$\frac{d}{dx}\left(y e^{\int p \, dx}\right) = q e^{\int p \, dx}$$

and hence

$$y = e^{-\int p \, dx} \int q e^{\int p \, dx} \, dx.$$

## 14.5 Euler on General Linear Equations with Constant Coefficients

Euler seems to have been the first mathematician to apply linear superposition of special solutions to obtain the general solution of a linear differential equation. One may contrast this with the method employed by Johann Bernoulli[57] to solve the equation for simple harmonic motion,

$$n^2 a^2 \frac{d^2 x}{dy^2} = a - x.$$

Note that we have slightly modernized Bernoulli's notation; he wrote the equation as

$$nnaaddx : dy^2 = a - x.$$

Note also that $dy^2$ stands for $(dy)^2$. Bernoulli multiplied the equation by $dx$ $\left(\text{or } \frac{dx}{dy}\right)$ and integrated to get

$$\frac{n^2 a^2}{2} \left(\frac{dx}{dy}\right)^2 = ax - \frac{1}{2}x^2.$$

Here observe that Bernoulli used $\int dx \, ddx = \frac{1}{2}(dx)^2$; next, he had

$$n \int \frac{a \, dx}{\sqrt{2ax - x^2}} = \int dy = y, \quad \text{or} \quad y = na \arcsin \frac{x-a}{a} + C.$$

56   Eu. I-22 pp. 181–213. E 188, § 7.
57   Bernoulli (1742) vol. 3, p. 210.

Bernoulli presented his solution in this form with no mention of linear superposition of particular sine and cosine solutions. In a paper presented to the Berlin Academy in 1742 and published in 1743,[58] Euler considered an equation of the form

$$0 = Ay + B\frac{dy}{dx} + C\frac{d^2y}{dx^2} + D\frac{d^3y}{dx^3} + \cdots + N\frac{d^ny}{dx^n} \tag{14.28}$$

and observed that if $y = u$ was a solution, then so was $y = \alpha u$, with $\alpha$ a constant. Moreover, if $n$ particular solutions $y = u, y = v, \ldots$ could be found, then the general solution would be $y = \alpha u + \beta v + \cdots$. To obtain these special solutions, he took $y = e^{\int p\,dx}$; Euler then wrote out the derivatives of $y$:

$$y = e^{\int p\,dx}$$

$$\frac{dy}{dx} = e^{\int p\,dx}\,p$$

$$\frac{d^2y}{dx^2} = e^{\int p\,dx}\left(pp + \frac{dp}{dx}\right)$$

$$\frac{d^3y}{dx^3} = e^{\int p\,dx}\left(p^3 + 3p\frac{dp}{dx} + \frac{d^2p}{dx^2}\right)$$

$$\frac{d^4y}{dx^4} = e^{\int p\,dx}\left(p^4 + 6pp\frac{dp}{dx} + 4p\frac{d^2p}{dx^2} + \frac{d^3p}{dx^3}\right).$$

When these values were substituted in the differential equation, the expression would be simplest when $p$ was a constant, for in that case the derivatives of $p$ would vanish; $p$ would then satisfy the algebraic equation

$$0 = A + Bz + Czz + Dz^3 + \cdots + Nz^n.$$

If $z = \frac{s}{t}$ was a root of this equation, then $s - tz = 0$ satisfied the $n$th-degree algebraic equation, and the solution $y = \alpha e^{\frac{sx}{t}}$ of the differential equation

$$sy - t\frac{dy}{dx} = 0$$

also satisfied the $n$th-order differential equation. When there was a repeated root, so that $(s - tz)^2 = ss - 2stz + ttzz = 0$ was a factor of the $n$th-degree polynomial, then there would be a corresponding factor

$$ssy - 2st\frac{dy}{dx} + tt\frac{ddy}{dx^2} = 0$$

of (14.28). To solve this second-order equation, Euler set $y = e^{\frac{sx}{t}}u$ to find that $u$ satisfied

$$\frac{ddu}{dx^2} = 0.$$

[58] Eu. I-22, pp. 108–149. E 62.

Thus, $u = \alpha x + \beta$ and $y = e^{\frac{sx}{r}}(\alpha x + \beta)$. Similarly, when there were three repeated roots, then $y = e^{\frac{sx}{r}}(\alpha x^2 + \beta x + \gamma)$. In the general case where a root was repeated $k$ times, the solution would turn out to be $e^{\frac{sx}{r}}$ times a general polynomial of degree $k-1$. Euler then considered complex roots, making use of a result he had published three years earlier, that the general solution of the equation

$$\frac{d^2y}{dx^2} + ky = 0$$

was

$$A \cos \sqrt{k}x + B \sin \sqrt{k}x.$$

So he supposed that $a - bz + cz^2 = 0$ was a quadratic factor of the polynomial; this yielded $z = \frac{-b \pm \sqrt{b^2 - 4ac}}{2c}$. Then he assumed $\frac{b}{2\sqrt{ac}} < 1$ and set $\cos \phi = \frac{b}{2\sqrt{ac}}$. Euler wrote the quadratic factor as $a - 2z\sqrt{ac} \cos \phi + cz^2$ and the corresponding differential equation as

$$0 = ay - 2\sqrt{ac} \cos \phi \frac{dy}{dx} + c \frac{d^2y}{dx^2}.$$

The substitution $y = e^{\sqrt{ac}x \cos \phi} u$ reduced the equation to

$$c \frac{d^2u}{dx^2} + (ac^2 \cos^2 \phi - 2ac \cos^2 \phi + a)u = 0;$$

note that this was of the required form $\frac{d^2u}{dx^2} + ku = 0$. Euler then proceeded to the case where there were repeated complex roots.

## 14.6 Euler: Nonhomogeneous Equations

In his 1750 paper, published in 1753,[59] Euler described two methods for solving the nonhomogeneous equation

$$A_0 y + A_1 \frac{dy}{dx} + A_2 \frac{d^2y}{dx^2} + \cdots + A_n \frac{d^ny}{dx^n}. \tag{14.29}$$

In section 6 of this paper, he assumed $X$ to be a polynomial and found a particular solution by taking $y$ to be a polynomial of the same degree as $X$, then substituting in the differential equation to find the coefficients. He then arrived at the general solution by adding the general solution of the homogeneous equation to the particular solution.

[59] Eu. I-22 pp. 181–213. E 188.

In sections 7 through 22, Euler described a second method of solution; in section 23, he presented four examples. Now in section 7, he considered the first-order equation

$$X = A_0 y + A_1 \frac{dy}{dx}, \tag{14.30}$$

to solve which he first multiplied by $e^{\alpha x} \, dx$ to obtain

$$e^{\alpha x} X \, dx = A_0 e^{\alpha x} y \, dx + A_1 e^{\alpha x} \, dy. \tag{14.31}$$

To determine $\alpha$ he assumed

$$\int e^{\alpha x} X \, dx = A_1 e^{\alpha x} y; \tag{14.32}$$

he then took the differential of equation (14.32) to obtain

$$e^{\alpha x} X \, dx = A_1 \alpha e^{\alpha x} y \, dx + A_1 e^{\alpha x} \, dy. \tag{14.33}$$

By equating (14.31) and (14.33), he got

$$\alpha = \frac{A_0}{A_1} \quad \text{or} \quad A_1 \alpha - A_0 = 0. \tag{14.34}$$

and using (14.32) Euler had the solution of (14.30):

$$y = \frac{e^{-\alpha x}}{A_1} \int e^{\alpha x} X \, dx. \tag{14.35}$$

In section 8, Euler considered the second-order equation

$$X = A_0 y + A_1 \frac{dy}{dx} + A_2 \frac{d^2 y}{dx^2}, \tag{14.36}$$

that he multiplied by $e^{\alpha x} dx$ to obtain

$$e^{\alpha x} X \, dx = A_0 e^{\alpha x} y \, dx + A_1 e^{\alpha x} \, dy + A_2 e^{\alpha x} \frac{d^2 y}{dx}. \tag{14.37}$$

Euler next assumed

$$\int e^{\alpha x} X \, dx = e^{\alpha x} \left( B_0 y + B_1 \frac{dy}{dx} \right) \tag{14.38}$$

and took its differential to get

$$e^{\alpha x} X \, dx = e^{\alpha x} \left( \alpha B_0 y \, dx + B_0 \, dy + B_1 \frac{d^2 y}{dx} + \alpha B_1 \, dy \right). \tag{14.39}$$

Equating (14.37) and (14.39) yielded

$$B_1 = A_2, \quad B_0 = A_1 - \alpha A_2, \quad A_0 = \alpha A_1 - \alpha^2 A_2. \tag{14.40}$$

Thus, $\alpha$ was a solution of the second-degree equation

$$A_2\alpha^2 - A_1\alpha + A_0 = 0, \tag{14.41}$$

so that Euler had next to solve the equation of order one arising from (14.38),

$$e^{-\alpha x}\int e^{\alpha x} X\, dx = B_0 y + B_1 \frac{dy}{dx}. \tag{14.42}$$

Using (14.35), the solution of (14.42) would be given by

$$y = \frac{e^{-\beta x}}{B_1}\int e^{(\beta-\alpha)x}\left(\int e^{\alpha x} X\, dx\right) dx, \tag{14.43}$$

where, from (14.34) and (14.40),

$$\beta = \frac{B_0}{B_1} = \frac{A_1 - \alpha A_2}{A_2} = \frac{A_1}{A_2} - \alpha \quad \text{or} \quad \alpha + \beta = \frac{A_1}{A_2}.$$

Clearly then, $\beta$ must be a root of the quadratic equation (14.41). Here Euler assumed that the two roots $\alpha$ and $\beta$ were distinct. He next applied integration by parts to equation (14.43) to obtain

$$y = \frac{e^{-\beta x}}{A_2}\left(\frac{e^{(\beta-\alpha)x}}{\beta-\alpha}\int e^{\alpha x} X\, dx - \frac{1}{\beta-\alpha}\int e^{(\beta-\alpha)x} e^{\alpha x} X\, dx\right)$$

or

$$y = \frac{e^{-\alpha x}}{A_2(\beta-\alpha)}\int e^{\alpha x} X\, dx + \frac{e^{-\beta x}}{A_2(\alpha-\beta)}\int e^{\beta x} X\, dx \tag{14.44}$$

as the solution of the second-order equation (14.36). Note that if $P(x) = A_2 x^2 + A_1 x + A_0$, then by (14.40)

$$\frac{P(x)}{x+\alpha} = A_2 x + A_1 - A_2\alpha = B_1 x + B_0. \tag{14.45}$$

Moreover, taking $P'(x)$ as the derivative of $P(x)$, (14.44) may be written:

$$y = \frac{e^{-\alpha x}}{P'(-\alpha)}\int e^{\alpha x} X\, dx + \frac{e^{-\beta x}}{P'(-\beta)}\int e^{\beta x} X\, dx.$$

Euler next considered the $n$th-order equation

$$X = A_0 y + A_1\frac{dy}{dx} + \cdots + A_n\frac{d^n y}{dx^n}, \tag{14.46}$$

first working with the case in which the $n$th-degree polynomial

$$P(x) = A_0 + A_1 x + \cdots + A_n x^n \tag{14.47}$$

had $n$ distinct roots $-\alpha_1, -\alpha_2, \ldots, -\alpha_n$. He gave the solution of (14.46) as

$$y = \frac{e^{-\alpha_1 x}}{P'(-\alpha_1)} \int e^{\alpha_1 x} X \, dx + \frac{e^{-\alpha_2 x}}{P'(-\alpha_2)} \int e^{\alpha_2 x} X \, dx + \cdots + \frac{e^{-\alpha_n x}}{P'(-\alpha_n)} \int e^{\alpha_n x} X \, dx,$$

$$(14.48)$$

although he did not prove it in general. Rather, he essentially worked out the case for an equation of order 3, noticed the pattern in the solutions for $n = 1, 2, 3$ and then stated the general case. We here present a general proof, whose interesting feature is that it requires the application of Lagrange's interpolation formula. This is a proof that Euler could have given, and he had himself discovered a formula from which Lagrange's interpolation formula followed. See his paper, "De eximio usu methodi interpolationum in serierum doctrina."[60]

Following Euler, we reduce the order of equation (14.46) by taking

$$e^{-\alpha_1 x} \int e^{\alpha_1 x} X \, dx = B_0 y + B_1 \frac{dy}{dx} + \cdots + B_{n-1} \frac{d^{n-1} y}{dx^{n-1}}, \qquad (14.49)$$

where $-\alpha_1$ is a root of the polynomial (14.47). By taking the derivative of

$$e^{\alpha_1 x} \left( B_0 y + B_1 \frac{dy}{dx} + \cdots + B_{n-1} \frac{d^{n-1} y}{dx^{n-1}} \right),$$

and, following the procedure by which he obtained (14.45) from (14.38), Euler showed that

$$\frac{P(x)}{x + \alpha_1} = B_0 + B_1 x + \cdots + B_{n-1} x^{n-1} \equiv Q(x). \qquad (14.50)$$

Assuming (14.48) to be true up to $n - 1$ and with

$$Y = e^{-\alpha_1 x} \int e^{\alpha_1 x} X \, dx, \qquad (14.51)$$

we have the general solution of (14.49) as

$$y = \frac{e^{-\alpha_2 x}}{Q'(-\alpha_2)} \int e^{\alpha_2 x} Y \, dx + \cdots + \frac{e^{-\alpha_n x}}{Q'(-\alpha_n)} \int e^{\alpha_n x} Y \, dx. \qquad (14.52)$$

Observe that

$$P(x) = (x + \alpha_1) Q(x)$$

and hence

$$P'(x) = Q(x) + (x + \alpha_1) Q'(x).$$

---

[60] Eu. I-15 pp. 435–497. E 555.

Next note that since $-\alpha_j$, $j = 2, \ldots n$, are roots of $Q(x)$ we may conclude that

$$P'(-\alpha_j) = (\alpha_1 - \alpha_j)\, Q'(-\alpha_j). \tag{14.53}$$

With $Y$ as given by (14.51), consider the general term on the right-hand side of (14.52):

$$\frac{e^{-\alpha_j x}}{Q'(-\alpha_j)} \int e^{\alpha_j x} Y \, dx. \tag{14.54}$$

Integration by parts then shows that (14.54) is equal to

$$\frac{e^{-\alpha_j x}}{Q'(-\alpha_j)} \left( \frac{e^{(\alpha_j - \alpha_1)x}}{\alpha_j - \alpha_1} \int e^{\alpha_1 x} X \, dx - \frac{1}{\alpha_j - \alpha_1} \int e^{\alpha_j x} X \, dx \right). \tag{14.55}$$

By using (14.53), we may rewrite (14.55) as

$$-\frac{e^{-\alpha_1 x}}{P'(-\alpha_j)} \int e^{\alpha_1 x} X \, dx + \frac{e^{-\alpha_j x}}{P'(-\alpha_j)} \int e^{\alpha_j x} X \, dx. \tag{14.56}$$

Substituting (14.56) for the right-hand side of (14.52), we can arrive at (14.48), providing that we can verify that

$$-\sum_{j=2}^{n} \frac{1}{P'(-\alpha_j)} = \frac{1}{P'(\alpha_1)}. \tag{14.57}$$

We can prove (14.57) by applying the Waring–Lagrange interpolation formula (9.27), with $T$ a polynomial of degree $\leq n - 1$:

$$T(x) = \sum_{j=1}^{n} \frac{P(x)}{P'(-\alpha_j)(x + \alpha_j)} T(-\alpha_j).$$

Taking $T(x) \equiv 1$, we obtain

$$1 = \sum_{j=1}^{n} \frac{P(x)}{P'(-\alpha_j)(x + \alpha_j)},$$

in which we equate the coefficient of $x^{n-1}$ on both sides to arrive at

$$0 = \sum_{j=1}^{n} \frac{1}{P'(-\alpha_j)},$$

concluding our proof of (14.57).

In this same paper, Euler also considered the case for which roots, real or complex, were repeated. He first supposed a real root $\alpha_1$ occurred exactly twice, with $\alpha_1 = \alpha_2$. He set

$$P(x) = (x + \alpha_1)^2 \, Q(x)$$

and took the derivative twice to find that

$$\frac{1}{2} P''(-\alpha_1) = Q(-\alpha_1) = (\alpha_3 - \alpha_1)(\alpha_4 - \alpha_1) \cdots (\alpha_n - \alpha_1). \qquad (14.58)$$

He then let $\alpha_2 = \alpha_1 + \omega$, where $\omega$ was an infinitesimal. Then

$$P'(-\alpha_1) = \omega \, Q(-\alpha_1) \quad \text{and} \quad P'(-\alpha_2) = -\omega \, Q(-\alpha_1),$$

and the first two terms on the right-hand side of (14.48) would take the form

$$\frac{e^{-\alpha_1 x}}{\omega \, Q(-\alpha_1)} \int e^{\alpha_1 x} X \, dx - \frac{e^{-(\alpha_1 + \omega)x}}{\omega \, Q(-\alpha_1)} \int e^{(\alpha_1 + \omega)x} X \, dx. \qquad (14.59)$$

Euler observed that

$$e^{-(\alpha_1 + \omega)x} = e^{-\alpha_1 x}(1 - \omega x) \quad \text{since} \quad e^{-\omega x} = (1 - \omega x), \qquad (14.60)$$

where he neglected the second and higher-order terms in $\omega$. Similarly,

$$e^{(\alpha_1 + \omega)x} = e^{\alpha_1 x}(1 + \omega x). \qquad (14.61)$$

Substituting (14.60) and (14.61) into (14.59), simplifying the expression, and neglecting the term involving $\omega$, Euler arrived at

$$\frac{e^{-\alpha_1 x}}{Q(-\alpha_1)} \left( x \int e^{\alpha_1 x} X \, dx - \int e^{\alpha_1 x} X x \, dx \right) = \frac{e^{-\alpha_1 x}}{Q(-\alpha_1)} \int \left( \int e^{\alpha_1 x} X \, dx \right) dx. \qquad (14.62)$$

Using a similar approach, Euler then considered the case in which there were three or more repeated roots, for the real as well as the complex case.

## 14.7   Lagrange's Use of the Adjoint

In 1761–62, Lagrange illustrated how to use the knowledge of the solutions of a homogeneous equation of second-order to solve the corresponding nonhomogeneous equation.[61] He supposed that $y_1$ and $y_2$ were (independent) solutions of the homogeneous part of the second-order equation

$$Ly + M\frac{dy}{dt} + N\frac{d^2 y}{dt^2} = T. \qquad (14.63)$$

---

[61]  Lagrange (1867–1892) vol. 1, pp. 471–478.

By multiplying by $z$ and applying integration by parts, the adjoint equation was obtained:

$$y\left(Mz - \frac{dNz}{dt}\right) + \frac{dy}{dt}Nz + \int\left(Lz - \frac{d(Mz)}{dt} + \frac{d^2(Nz)}{dt^2}\right)dt = \int Tz\,dt.$$

$$(14.64)$$

He then set

$$Lz - \frac{d(Mz)}{dt} + \frac{d^2(Nz)}{dt^2} = 0,$$

so that

$$y\left(Mz - \frac{d(Nz)}{dt}\right) + \frac{dy}{dt}Nz = \int Tz\,dt. \qquad (14.65)$$

Multiplying this adjoint equation by a function $y$ and applying the same integration by parts procedure, Lagrange arrived at

$$y\left(Mz - \frac{d(Nz)}{dt}\right) + \frac{dy}{dt}Nz - \int\left(Ly + M\frac{dy}{dt} + N\frac{d^2y}{dt^2}\right)z\,dt = \text{constant}.$$

When $y = y_1$ or $y = y_2$, the integral vanished, and he got the two relations

$$z\left[\left(M - \frac{dN}{dt}\right)y_1 + N\frac{dy_1}{dt}\right] - \frac{dz}{dt}Ny_1 = c_1, \qquad (14.66)$$

$$z\left[\left(M - \frac{dN}{dt}\right)y_2 + N\frac{dy_2}{dt}\right] - \frac{dz}{dt}Ny_2 = c_2, \qquad (14.67)$$

where $c_1$ and $c_2$ were constants. Lagrange solved for $z$ to obtain

$$z = \frac{c_1 y_2 - c_2 y_1}{N\left(y_2\frac{dy_1}{dt} - y_1\frac{dy_2}{dt}\right)}. \qquad (14.68)$$

It may be helpful to note that the denominator here is $N$ times the Wronskian of $y_1$ and $y_2$. Lagrange then took $c_1 = 0$ and then $c_2 = 0$, denoting the corresponding values of $z$ by $z_1$ and $z_2$, respectively. When these were substituted in (14.65), he obtained

$$y\left(Mz_1 - \frac{d(Nz_1)}{dt}\right) + \frac{dy}{dt}Nz_1 = \int Tz_1\,dt,$$

$$y\left(Mz_2 - \frac{d(Nz_2)}{dt}\right) + \frac{dy}{dt}Nz_2 = \int Tz_2\,dt.$$

He solved for $y$ and thereby arrived at a solution for (14.63):

$$y = \frac{z_2 \int T z_1 \, dt - z_1 \int T z_2 \, dt}{N \left( z_1 \frac{dz_2}{dt} - z_2 \frac{dz_1}{dt} \right)}.$$

Lagrange then considered the case where just one solution $y_1$ was known so that he had only equation (14.66). Now (14.66) was a linear first-order equation in $z$ solvable by Euler's integrating factor method, so that

$$z = \frac{y_1}{N} e^{\int \frac{M}{N} dt} \left( c - c_1 \int \frac{e^{-\int \frac{M}{N} dt}}{y_1^2} dt \right).$$

Again, by taking $c_1 = 0$ and then $c = 0$, he obtained two values, $z_1$ and $z_2$, of $z$. Thus, Lagrange taught us that when two solutions of the homogeneous equation are known, the solution of the nonhomogeneous equation may be obtained by solving two sets of linear equations. But when only one solution is known, one must solve a first-order equation and a pair of linear equations. Lagrange pointed out that when $L, M$, and $N$ were constants, then the homogeneous equation could easily be solved by Euler's method, and hence the nonhomogeneous equation could also be solved in general for this case. In the case for which $L, M$, and $N$ were constants and $k_1$ and $k_2$ were solutions of $L + Mk + Nk^2 = 0$, $(k_1 \neq k_2)$, Lagrange gave the solution of (14.63) by the formula

$$y = \frac{e^{k_2 t} \int T e^{-k_2 t} dt - e^{k_1 t} \int T e^{-k_2 t} dt}{N(k_2 - k_1)}.$$

## 14.8  Jakob Bernoulli and Riccati's Equation

In his 1703 letter to Leibniz,[62] Bernoulli gave the derivation for the solution of Riccati's equation, $dy = yy dx + xx dx$. Part of the problem here was to reduce the equation to a separable one. Bernoulli used an interesting substitution to accomplish this, ending up with a second-order instead of a first-order equation. In order to solve the second order equation, he had to use infinite series. He began by setting $y = -dz : z dx \left( y = -\frac{1}{z} \frac{dz}{dx} \right)$, so that by the quotient rule for differentials, the differential equation took the form

$$dx dz^2 - z dx ddz : zz dx^2 = dy = yy dx + xx dx = dz^2 : zz dx + xx dx;$$

this simplified to

$$-z dx ddz = xxzz dx^3,$$

[62] Leibniz (1971) vol. 3, part 2, pp. 74–78.

a separable equation expressible as

$$-ddz : z = xxdx^2.  \tag{14.69}$$

Now this was the second-order equation $-\frac{1}{z}\frac{d^2z}{dx^2} = x^2$. Bernoulli observed that if you had an equation $-z^e ddz = x^v dx^2$ and you sought a solution $z = ax^m$, then by substituting this in the equation you would find that $m = (v+2) : (e+1)$. However, in (14.69), $e = -1$ and therefore no solution of the form $ax^m$ would be possible. He then drew an analogy with the first-order equation $dz : z = x^v dx$, for which no algebraic solution was possible when $v = -1$. So Bernoulli concluded that no algebraic solution was possible for (14.69) and that he must take recourse in infinite series. He obtained the series solution as

$$z = 1 - \frac{x^4}{3\cdot4} + \frac{x^8}{3\cdot4\cdot7\cdot8} - \frac{x^{12}}{3\cdot4\cdot7\cdot8\cdot11\cdot12} + \frac{x^{16}}{3\cdot4\cdot7\cdot8\cdot11\cdot12\cdot15\cdot16} - \cdots.$$

Since $y = -\frac{1}{z}\frac{dz}{dx}$, he could write his solution as a ratio of two infinite series. By dividing the series for $-\frac{dz}{dx}$ by the series for $z$, he obtained the first few terms of the series for $y$ given in (14.3).

## 14.9  Riccati's Equation

We have seen that in the 1730s, Euler wrote on Riccati's equation, and returned to the topic sometime around 1760, then composing an important paper on first-order differential equations, published in 1763.[63] In that paper, he explained how to obtain the general solution of the Riccati equation if one particular solution were known. His method was to use the known solution to reduce the Riccati equation to a first-order linear differential equation. He supposed $v$ to be a solution of the equation

$$dy + Py\,dx + Qyy\,dx + R\,dx = 0,  \tag{14.70}$$

and observed that $y = v + \frac{1}{z}$ reduced this equation to

$$-\frac{dz}{zz} + \frac{P\,dx}{z} + \frac{2Qv\,dx}{z} + \frac{Q\,dx}{zz} = 0$$

or

$$dz - (P + 2Qv)z\,dx - Q\,dx = 0.  \tag{14.71}$$

He then noted that $S = e^{-\int (P+2Qv)\,dx}$ was an integrating factor. Hence his solution of (14.71) was $Sz - \int QS\,dx = $ Constant. Later in the paper, Euler considered the particular case of (14.70), discussed by Riccati and the Bernoullis:

$$dy + yy\,dx = ax^m\,dx.  \tag{14.72}$$

---

[63] Eu. I-22 pp. 334–394. E 269.

Euler could find a special solution of this equation, by use of which he could determine the general solution by the already described method. He set $a = cc$, $m = -4n$ and

$$y = cx^{-2n} + \frac{1}{z}\frac{dz}{dx} \qquad (14.73)$$

so that (14.72) was converted to the linear second-order equation

$$\frac{ddz}{dx^2} + \frac{2c}{x^{2n}}\frac{dz}{dx} - \frac{2nc}{x^{2n+1}}z = 0. \qquad (14.74)$$

He then solved this equation as a series:

$$z = Ax^n + Bx^{3n-1} + Cx^{5n-2} + Dx^{7n-3} + Ex^{9n-4} + \text{etc.}$$

After substituting this in (14.74), he found

$$B = \frac{-n(n-1)A}{2(2n-1)c}, \quad C = \frac{-(3n-1)(3n-2)B}{4(2n-1)c}, \quad D = \frac{-(5n-2)(5n-3)C}{6(2n-1)c}, \text{ etc.}$$

Euler did not write the general case but if we let $A_k$ denote the coefficient of $x^{(2k+1)n-k}$, starting at $k = 0$, then the recurrence relation would be

$$A_k = \frac{-((2k-1)n-k+1)((2k-1)n-k)}{2k(2n-1)c}A_{k-1}, \quad k = 1,2,3,\dots. \qquad (14.75)$$

Note that if for some $k$, $A_k = 0$, then $A_n = 0$ for $n = k+1, k+2, \dots$. In this case, the series reduced to a polynomial and from (14.75) one could determine the general condition to be $n = \frac{k-1}{2k-1}$, or $n = \frac{k}{2k-1}$. For these values, the solution could be written in finite form.

### 14.10  Singular Solutions

In his 1715 book, *Methodus Incrementorum*, Brook Taylor presented some techniques for solving differential equations.[64] In proposition VIII, he explained that solutions in finite form might be found if the equation could be suitably transformed. In describing one method of doing this, he considered the differential equation

$$4x^3 - 4x^2 = (1+z^2)^2\dot{x}^2 \qquad (14.76)$$

and found a singular solution. He set $x = v^\theta y^\gamma$, where $\theta$ and $\gamma$ were parameters to be chosen appropriately later on, so that

$$\dot{x} = (\theta\dot{v}y + \gamma\dot{y}v)v^{\theta-1}y^{\gamma-1}. \qquad (14.77)$$

[64] Taylor (1715) pp. 26–27. For an English translation, see Fiegenbaum (1981).

He substituted these values of $x$ and $\dot{x}$ in (14.76) to obtain

$$4v^{3\theta}y^{3\gamma} - 4v^{2\theta}y^{2\gamma} = (1+z^2)^2(\theta \dot{v}y + \gamma \dot{y}v)^2 v^{2\theta-2}y^{2\gamma-2}. \tag{14.78}$$

Taylor then chose $v = 1 + z^2$ and assumed that $z$ was flowing uniformly, that is $\dot{z} = 1$ so that $\dot{v} = 2z$. Substituting these values in (14.78), he arrived at

$$4v^{\theta}y^{\gamma+2} - 4y^2 = (2\theta zy + \gamma \dot{y}v)^2. \tag{14.79}$$

Taylor then took $\gamma = -2$ to eliminate $y$ in the first term on the left-hand side to obtain

$$v^{\theta} - y^2 = (\theta zy - \dot{y}v)^2$$

or

$$v^{\theta} = (\theta^2 z^2 + 1)y^2 - 2\theta zvy\dot{y} + v^2\dot{y}^2. \tag{14.80}$$

At this point, he set $\theta = 1$ so that $\theta^2 z^2 + 1 = z^2 + 1 = v$; then dividing by $v$, equation (14.80) reduced to $1 = y^2 - 2zy\dot{y} + v\dot{y}\dot{y}$. Taking fluxions, he found

$$0 = 2y\dot{y} - 2y\dot{y} - 2zy\ddot{y} - 2z\dot{y}\dot{y} + \dot{v}\dot{y}\dot{y} + 2v\dot{y}\ddot{y}.$$

Since

$$\dot{v} = 2z,$$

he had

$$-2zy\ddot{y} + 2\dot{v}\ddot{y} = 0. \tag{14.81}$$

Thus, (14.81) implied that either $\ddot{y} = 0$ or $-2zy + 2v\dot{y} = 0$. Then the latter gave him

$$-\dot{v}y + 2v\dot{y} = 0 \quad \text{or} \quad y^2 = v. \tag{14.82}$$

Now since he had taken $\theta = 1$ and $\gamma = -2$, he got a solution,

$$x = v^{\theta}y^{\gamma} = vy^{-2} = 1,$$

where the last relation followed from (14.82). At this point, he remarked that $x = 1$ was "a certain singular solution of the problem." For the equation $\ddot{y} = 0$, he picked the initial values in such a way as to obtain as solution

$$y = a + \sqrt{1 - a^2}z,$$

and hence

$$x = vy^{-2} = \frac{1 + z^2}{a + \sqrt{1 - a^2}z}. \tag{14.83}$$

Observe that the solution $x = 1$ is not obtained from the general (14.83) for any value of $a$.

Euler discussed singular solutions in a 1754 paper on paradoxes in integral calculus.[65] He there noted the paradoxical fact was that there were differential equations easier to solve by differentiating than by integrating. He wrote that he had encountered such equations in his work on mechanics but that his purpose was to explain that there were easily stated geometric problems from which similar types of equations could arise. Euler started by presenting the problem of finding a curve such that the length of the perpendicular from a given point to any tangent to the curve was a constant. By using similar triangles, he found that the differential equation would be

$$y\,dx - x\,dy = a\sqrt{dx^2 + dy^2},\qquad(14.84)$$

where $a$ denoted a constant. After squaring and solving for $dy$ he obtained

$$(a^2 - x^2)\,dy + xy\,dx = a\,dx\sqrt{x^2 + y^2 - a^2}.\qquad(14.85)$$

He substituted $y = u\sqrt{a^2 - x^2}$ to transform the equation into the separable equation

$$\frac{du}{\sqrt{u^2 - 1}} = \frac{a\,dx}{a^2 - x^2}.\qquad(14.86)$$

After integration he obtained

$$\ln(u + \sqrt{u^2 - 1}) = \frac{1}{2}\ln\frac{n^2(a + x)}{a - x},$$

where $n$ was a constant. He simplified this to

$$u = \frac{n}{2}\sqrt{\frac{a + x}{a - x}} + \frac{1}{2n}\sqrt{\frac{a - x}{a + x}}$$

or

$$y = u\sqrt{a^2 - x^2} = \frac{n}{2}(a + x) + \frac{1}{2n}(a - x).\qquad(14.87)$$

Note that equation (14.86), when written as

$$du = \frac{a\sqrt{u^2 - 1}\,dx}{a^2 - x^2},$$

shows that $u \equiv 1$ is also a solution because both sides vanish. When Euler used this solution in the first equation in (14.87), he got $y = \sqrt{a^2 - x^2}$, or

$$x^2 + y^2 = a^2. \tag{14.88}$$

Thus, the solution of (14.85) turned out to be a family of straight lines (14.87) as well as the circle (14.88).

In the same paper, Euler next set out to show that he could derive these solutions by differentiation. Now note that this paper was written before his book on differential calculus in which he explained how higher differentials could be completely replaced by higher differential coefficients. So he explained that he would assume that $dy = p\,dx$, with $p$ a differential coefficient, to remove difficulties associated with further differentiation. His equation (14.84) then became

$$y = px + a\sqrt{1 + p^2}. \tag{14.89}$$

Euler then differentiated (instead of integrated) this equation to get

$$dy = p\,dx + x\,dp + \frac{ap\,dp}{\sqrt{1 + p^2}}$$

or

$$0 = x\,dp + \frac{ap\,dp}{\sqrt{1 + p^2}} \tag{14.90}$$

or

$$x = -\frac{ap}{\sqrt{1 + p^2}}.$$

Hence by (14.89), $y = \frac{a}{\sqrt{1+p^2}}$. By eliminating $p$, he obtained the solution $x^2 + y^2 = a^2$ and noted that he could also find the family of straight lines by this method. For that purpose, he observed that (14.90) also had the solution $dp = 0$. This implied $p = \text{constant} = n$, and so by (14.89) he obtained $y = nx + a\sqrt{1 + n^2}$, the required system of straight lines.

Euler then remarked that equation (14.84) could be modified in such a way that the new equation could be solved more easily by the second method than the first. He considered the equation

$$y\,dx - x\,dy = a(dx^3 + dy^3)^{\frac{1}{3}}, \tag{14.91}$$

or $y = px + a(1 + p^3)^{\frac{1}{3}}$. As solutions, he found a sixth-order curve

$$y^6 + 2x^3y^3 + x^6 - 2a^3y^3 + 2a^3x^3 + a^6 = 0 \tag{14.92}$$

and the family of straight lines

$$y = nx + a(1 + n^3)^{\frac{1}{3}}. \tag{14.93}$$

Euler gave three other geometric examples leading to differential equations with singular solutions. One of them yielded

$$ydx - (x-b)dy = \sqrt{(a^2-x^2)dx^2 + a^2dy^2},\qquad(14.94)$$

an equation difficult to integrate. He solved it by differentiation to obtain the solutions: the ellipse

$$\frac{(x-b)^2}{a^2} + \frac{y^2}{a^2-b^2} = 1,$$

and the family of straight lines

$$y = -n(b-x) + \sqrt{a^2(1+n^2) - b^2}.$$

Later in the paper, Euler remarked that he found it strange that integration, which introduced arbitrary constants, did not produce the general solution, while differentiation did.

## 14.11 Mukhopadhyay on Monge's Equation

In his letter to Huygens of September 4/14, 1694,[66] Leibniz noted that the differential equation for the circle $x^2 + y^2 = a^2$ could be expressed as $\frac{dy}{dx} = -\frac{x}{y}$. Now start with the equation for a general circle,

$$(x-a)^2 + (y-b)^2 = c^2,$$

or

$$x^2 + y^2 = 2ax + 2by + c^2 - a^2 - b^2.\qquad(14.95)$$

To obtain the differential equation, we temporarily let $p,q,r$ denote the first three derivatives of $y$ with respect to $x$; differentiation of the last equation gives $x + yp = a+bp$ and hence $1+p^2+yq = bq$. Therefore, $1+p^2 = (b-y)q$ and $x-a = (b-y)p$; then, after a short calculation,

$$c = \frac{(1+p^2)^{\frac{3}{2}}}{q}.\qquad(14.96)$$

In his book on differential equations, Boole remarked[67] that since the right hand side of (14.96) was the expression for the radius of curvature, this equation was the differential equation for a circle of radius $c$. To obtain the equation for a circle of arbitrary radius, take the derivative of (14.96) to get

[66] Leibniz (1971) vol. 2, pp. 195–196.
[67] Boole (1877) p. 20.

$$3pq^2 - r(1 + p^2) = 0. \tag{14.97}$$

George Salmon (1819–1904) offered an interesting geometric interpretation of this equation[68] in his book *Higher Plane Curves*, first published in 1852, also in later editions. Salmon defined "aberrancy of a curve" where the curve was $y = f(x)$. Let $P$ be a point on the curve and $V$ the midpoint of a chord $AB$ drawn parallel to the tangent at $P$. Let $\delta$ denote the limit of the angles made by the normal at $P$ with the line $PV$ as $A$ and $B$ tend to $P$. Salmon called $\delta$ the aberrancy because $\delta = 0$ for a circle. He noted that

$$\tan \delta = p - \frac{(1 + p^2)r}{3q^2}. \tag{14.98}$$

Thus, the geometric meaning of (14.97) was that the aberrancy vanished at any point of any circle.

Boole also stated the differential equation of a general conic

$$ax^2 + 2nxy + by^2 + 2gx + 2fy + c = 0:$$

$$9\left(\frac{d^2y}{dx^2}\right)^2 \frac{d^5y}{dx^5} - 45\frac{d^2y}{dx^2}\frac{d^3y}{dx^3}\frac{d^4y}{dx^4} + 40\left(\frac{d^3y}{dx^3}\right)^3 = 0. \tag{14.99}$$

This differential equation was published in 1810 by the French geometer Gaspard Monge (1746–1818). Boole remarked on this equation, "But here our powers of geometrical interpretation fail, and results such as this can scarcely be otherwise useful than as a registry of integrable forms."[69]

The Indian mathematician Asutosh Mukhopadhyay (also Mookerjee) (1864–1924) published an 1889 paper[70] in the *Journal of the Asiatic Society of Bengal*, showing an interesting geometric interpretation of Monge's equation (14.99). Concerning this result, much appreciated by some British mathematicians, Edwards wrote in the 1892 second edition of his treatise on differential calculus:[71]

A remarkable interpretation which calls for notice has, however, been recently offered by Mr. A. Mukhopadhyay, who has observed that the expression for the radius of curvature of the locus of the centre of the conic of five pointic [*sic*] contact with any curve (called the centre of aberrancy) contains as a factor the left-hand member of Monge's equation, and this differential equation therefore expresses that the "*radius of curvature of the 'curve of aberrancy' vanishes* for any point of any curve."

Mukhopadhyay received an M.A. in mathematics from Calcutta University in 1886. He studied much on his own, as is shown by entries in his diary:[72] "Rose at 6.15 a.m. Read Statesman [Newspaper], and Boole's Diff. Equations in the morning. Read Fourier's Heat at noon." And "At noon read from Messenger of Math. Vol. 2,

---

[68] Salmon (1879) p. 369.
[69] Boole (1877) p. 20.
[70] Mukhopadhyay (1889a).
[71] Edwards (1954a) p. 436.
[72] Mukhopadhyay (1998).

Prof. Cayley's Memoir on Singular Solutions–to my mind, the simplest but the most philosophical account of the subject yet given; read from Forsyth on the same subject." Mukhopadhyay published papers on topics in differential geometry, elliptic functions and hydrodynamics. The following abstract of his 1889 paper, *On a Curve of Aberrancy*, may give a sense of his mathematical work:[73]

> The object of this note is to prove that the aberrancy curve (which is the locus of the centre of the conic of closest contact) of a plane cubic of Newton's fourth class is another plane cubic of the same class, the invariants of which are proportional to the invariants of the original cubic; it is also proved that the two cubics have only one common point of intersection, which is the point of inflection for both.

Mukhopadhyay thus gave evidence of a fine mathematical mind but his dream of spending his life in mathematical research could not be realized because there was no support for such endeavors in nineteenth-century Indian universities. As one biographer wrote, "Sir Asutosh's contributions to mathematical knowledge were due to his unaided efforts while he was only a college student." Interestingly, after serving as a judge, in 1906 Mukhopadhyay became Vice-Chancellor of Calcutta University and his first order of business was to have the University "combine the functions of teaching and original investigation." He appointed Syamadas Mukhopadhyay (1866–1937) professor of mathematics and encouraged him to pursue research. S. Mukhopadhyay subsequently produced several interesting results, including the well-known four vertex theorem,[74] published in the *Bulletin of the Calcutta Mathematical Society*, founded by Asutosh. Syamadas stated the theorem: "The minimum number of cyclic points on an oval is four." Asutosh created a physics department at the university with the appointment of the experimental physicist C. V. Raman, whom he persuaded to leave his post as an officer in the Indian Accounts Department. Raman went on to win a Nobel Prize in physics. In applied mathematics, he appointed S. N. Bose, known for his statistical derivation of Planck's law leading to the Bose-Einstein statistics and M. N. Saha who discovered the Saha ionization law in astrophysics. These discoveries by Saha and Bose were made several years after their appointments, validating Asutosh's confidence in them.

## 14.12   Exercises

(1) (a) Solve the equation $\frac{a^4 d^4 t}{dx^4} - y = 0$. Euler gave

$$y = \alpha e^{\frac{x}{a}} + \beta e^{-\frac{x}{a}} + \gamma \sin\left(\frac{x}{a} + A\right).$$

(b) Solve the equation $\frac{a^4 d^4 t}{dx^4} + y = 0$.

---

[73] Mukhopadhyay (1889b).
[74] Mukhopadhyay (1909).

Euler gave

$$y = \alpha e^{-\frac{x}{\sqrt{2}a}} \sin\left(\frac{x}{\sqrt{2}a} + A\right) + \beta e^{\frac{x}{\sqrt{2}a}} \sin\left(\frac{x}{\sqrt{2}a} + B\right).$$

See sections 33 and 34 of Euler's paper E 62.

(2) (a) Solve the equation $X = y - \frac{d^2y}{dx^2}$. Euler gave the solution

$$y = \frac{1}{2}e^{-x}\int e^x X \, dx - \frac{1}{2}e^x \int e^{-x} X \, dx.$$

(b) Solve the equation $X = y + \frac{a^3 d^3 y}{dx^3}$. Euler gave

$$y = \frac{1}{3a} e^{-\frac{x}{a}} \int e^{\frac{x}{a}} X \, dx - \frac{2}{3a} e^{\frac{x}{2a}} \cos\left(\frac{\sqrt{3}x}{2a} + \frac{\pi}{3}\right) \int e^{-\frac{x}{2a}} \cos\frac{\sqrt{3}x}{2a} X \, dx$$

$$- \frac{2}{3a} e^{\frac{x}{2a}} \sin\left(\frac{\sqrt{3}x}{2a} + \frac{\pi}{3}\right) \int 3^{-\frac{x}{2a}} \sin\frac{\sqrt{3}x}{2a} X \, dx.$$

These are examples given in section 23 of Euler's paper E 188.

(3) Show that the polynomial

$$y_k = L_k(x) = \sum_{j=0}^{k} \frac{k(k-1)\cdots(k-j+1)}{j! \, j!}(-x)^j,$$

the $k$th Laguerre polynomial, is a solution of the recurrence relation

$$(k+1)y_{k+1} - (2k+1-x)y_k + ky_{k-1} = 0, \quad k = 0,1,2,\ldots.$$

D. Bernoulli and Euler encountered this equation in their works on the discrete analog of the problem of the small oscillations of a hanging chain. They discussed the discreet and the continuous forms of the problem while they were colleagues at the Petersburg Academy in the late 1720s and early 1730s. Bernoulli submitted his results to the academy before his departure in 1733 and a year later presented his proofs. Upon seeing Bernoulli's work, Euler, who had obtained similar results, submitted his work. They took $x = \frac{a}{\alpha}$, where $a$ was the distance between the weights and $\alpha$ was related to the angular frequency $\omega$ by $\omega^2 = \frac{g}{\alpha}$; $g$ was the acceleration due to gravity. The $y_k$ was the simultaneous displacement of the $k$th weight. The chain was assumed to hang from the $n$th weight, so $y_n = 0$. This gave $L_n(\frac{a}{\alpha}) = 0$ as the equation determining the frequencies. Euler discovered the polynomial solution of the difference equation; in that sense, he and D. Bernoulli were the first mathematicians to use Laguerre polynomials. Bernoulli found the smallest roots of these polynomials for some values of $k$, using his method of sequences; see Chapter 13 for Daniel Bernoulli's work on difference equations.

(4) The differential equation satisfied by the displacement $y$ at a distance $x$ from the point of suspension of a heavy chain was determined by D. Bernoulli to be

$$\alpha x \frac{d^2 y}{dx^2} + \alpha \frac{dy}{dx} + y = 0.$$

Show that

$$y = A J_0 \left( 2\sqrt{\frac{x}{\alpha}} \right) = A \sum_{j=0}^{\infty} \frac{1}{j!\, j!} \left( \frac{-x}{\alpha} \right)^j$$

is a solution of this equation. Note that Bernoulli did not use the $J_0$ notation but gave the corresponding series. The value of $\alpha$ is determined from the equation $J_0 \left( 2\sqrt{\frac{l}{\alpha}} \right) = 0$. Bernoulli stated that this equation had an infinite number of roots and gave the first value $\frac{\alpha}{l} = 0.691$, a good approximation. About 50 years later, Euler gave the first three roots. Bernoulli and Euler may have conjectured the existence of infinitely many roots because the solution of the difference equation in the previous exercise, with a slight change of variables, approximates the solution of the differential equation given in this exercise. We remark that this is not surprising, as the first is a discrete analog of the second. In fact, $L_k(\frac{x}{k}) \to J_0(2\sqrt{x})$. And since $L_k(x) = 0$ has $k$ zeros, $J_0$ must have infinitely many roots; in fact, the $r$th root of $L_k$ must tend to the $r$th root of $J_0$. This can be proved rigorously by the theory of analytic functions of a complex variable, though Bernoulli and Euler obviously had no knowledge of this theory.

(5) In the works mentioned in Exercises 1 and 2, Euler also treated the equation

$$\frac{x}{n+1} \frac{d^2 y}{dx^2} + \frac{dy}{dx} + \frac{y}{\alpha} = 0.$$

Show, with Euler, that the equation has the series solution

$$y = A q^{-\frac{n}{2}} I_n(2\sqrt{q}), \quad \text{where} \quad q = -\frac{(n+1)x}{\alpha},$$

and

$$I_n(x) = \left( \frac{x}{2} \right)^n \sum_{k=0}^{\infty} \frac{\left( \frac{x}{2} \right)^{2k}}{k!\, \Gamma(k+n+1)}.$$

This appears to be the first occurrence of the Bessel function of arbitrary real index $n$. Euler also proved that for $n = -\frac{1}{2}$, the solution was $y = A \cos \sqrt{\frac{2x}{\alpha}}$.

Prove this and show that this is equivalent to the result $I_{-\frac{1}{2}}(x) = \sqrt{\frac{2}{\pi x}} \cos x$.

(6) Euler also obtained the solution of the differential equation in the previous exercise as the definite integral

$$\frac{y}{A} = \frac{\int_0^1 (1 - t^2)^{\frac{2n-1}{2}} \cosh\left(2t\sqrt{\frac{(n+1)x}{\alpha}}\right) dt}{\int_0^1 (1 - t^2)^{\frac{2n-1}{2}} dt}.$$

Prove Euler's result. It is equivalent to the Poisson integral representation

$$J_n(x) = \frac{2}{\sqrt{\pi}} \frac{\left(\frac{x}{2}\right)^n}{\Gamma\left(n + \frac{1}{2}\right)} \int_0^{\frac{\pi}{2}} \cos(x \sin \phi) \cos^{2n} \phi \, d\phi.$$

Prove this. According to Truesdell, this may be the earliest example of solution of a second-order differential equation by a definite integral. For this and the previous three exercises, see Truesdell (1960) pp. 154–165 and Cannon and Dostrovsky (1981) pp. 53–64. The references to the original papers of Euler and Bernoulli may be found there.

(7) Solve the equation

$$(ydx - xdy)(ydx - xdy + 2bdy) = c^2(dx^2 + dy^2)$$

by Euler's method for finding singular solutions.

(8) Let $f(x, y)$ be bounded and continuous on a domain $G$. Show that then at least one integral curve of the differential equation $\frac{dy}{dx} = f(x, y)$ passes through each interior point $(x_0, y_0)$ of $G$. This result is due to the Italian mathematician Giuseppe Peano (1838–1932) who graduated from the University of Turin (Torino), where he heard lectures by Angelo Genocchi and Faà di Bruno. Peano developed some aspects of mathematical logic in order to bring a higher degree of clarity to proofs of theorems in analysis. This led him to produce several counterexamples to intuitive notions in mathematics; his most famous example is that of a space-filling curve, dating from 1890. Bertrand Russell wrote that Peano's ideas on logic had a profound impact on him. See Peano (1973) pp. 51–57 for Peano's 1885 formulation and not completely rigorous proof of the theorem stated in this exercise. A more stringent proof may be found in Petrovski (1966) pp. 29–33. In 1890, Peano generalized this theorem to systems of differential equations. That paper also contained the first explicit formulation of the axiom of choice; interestingly, Peano rejected it as a possible component of the logic of mathematics. He wrote, "But as one cannot apply infinitely many times an *arbitrary* rule by which one assigns to a class $A$ an individual of this class, a *determinate* rule is stated here." See Moore (1982) p. 4. Nevertheless, a logical equivalent of the axiom of choice, in the form of Zorn's lemma, has turned out to be of fundamental importance in algebra. For the origins of Zorn's lemma, see Paul Campbell (1978).

## 14.13 Notes on the Literature

Eneström (1897) has a good history of differential equations with constant coefficients. For Leibniz's early work on series, see Scriba (1964). An English translation of Taylor's *Methodus Incrementorum* was a part of Feigenbaum's (1981) Yale doctoral dissertation. See Edwards (1954a) p. 436, for his remarks on Mukhopadhyay's equation. For some discussion of Mukhopadhyay's role in the development of mathematics in India, see Narasimhan (1991); for biographical information, see Sen Sen Gupta (2000). Katz (1998) and (1987) contain a discussion of how, in May 1739, Euler may have solved $a^3 d^3 y - y dx^3 = 0$. A lively history of the Riccati equation is available in Bottazzini's article, "The Mathematical Writings from Daniel Bernoulli's Youth," contained in D. Bernoulli (1982–1996) vol. I, pp. 142–166. The reader may also wish to see Burn (2001) for the development of the concept of the logarithm in the second half of the seventeenth century, starting with the 1649 work of Alphonse de Sarasa. This topic is also discussed in Hofmann's article on differential equations in the seventeenth century; see Hofmann (1990) vol. 2, pp. 277–316.

# 15

---

# Series and Products for Elementary Functions

## 15.1 Preliminary Remarks

Euler was the first mathematician to give a systematic and coherent account of the elementary functions, although earlier mathematicians had certainly paved the way. These functions are comprised of the circular or trigonometric, the logarithmic, and the exponential functions. Euler's approach was a departure from the prevalent, more geometric, point of view. On the geometric perspective, the elementary functions were defined as areas under curves, lengths of chords, or other geometric conceptions. Euler's 1748 *Introductio in Analysin Infinitorum* defined the elementary functions arithmetically and algebraically, as functions.

At that time, infinite series were regarded as a part of algebra, though they had been obtained through the use of calculus. The general binomial theorem was considered an algebraic theorem. So in his *Introductio*, Euler used the binomial theorem to produce new derivations of the series for the elementary functions. Interestingly, in this book, where he avoided using calculus, Euler gave no proof of the binomial theorem itself; perhaps he had not yet found any arguments without the use of calculus. In a paper written in the 1730s, Euler derived the binomial theorem from the Taylor series,[1] as Stirling had done in 1719.[2] It was only much later, in the 1770s, that Euler found an argument for the binomial theorem depending simply on the multiplication of series. We discuss this argument in Chapter 4.

We recall from Chapter 8 that before Euler, between 1664 and 1666, Newton found the series for all the elementary functions, using a combination of geometric arguments, integration, and reversion of series. In the course of his work, he also discovered the general binomial theorem. Later on, Gregory, Leibniz, Johann Bernoulli, and others used methods of calculus to obtain infinite series for elementary functions. Even before Newton, unknown to European mathematicians, the Kerala school had derived infinite series for some trigonometric functions, also using a form of integration.

---

[1] Eu. I-14 pp. 108–123. E 47 § 6.
[2] Stirling (1719).

Although the series for $e^x$ and $e$ were already known, one of Euler's major inno-vations was to explicitly define the exponential function. To understand this peculiar fact, recall that Newton and N. Mercator discovered the series for $y = \ln(1 + x)$ in the mid-1660s.[3] Soon afterward, Newton applied reversion of series to obtain $x$ as a series in $y$.[4] Also note that for the eighteenth-century mathematician who took the geometric point of view, the basic object of study was not the function, but the curve. From this perspective, there was hardly any need to distinguish between the function and its inverse, since both curves would take same form, although with differing orientations.

In a 1714 paper published in the *Philosophical Transactions*,[5] Roger Cotes, in the spirit of Halley's earlier work of 1695,[6] took the step of setting up an analytic definition of the logarithm. Cotes used this definition to derive the logarithmic series and then, by inversion, the series for the exponential. He proceeded to use the series for $e$ to compute its value to thirteen decimal places. Incidentally, he also gave continued fractions for $e$ and $\frac{1}{e}$ to obtain rational approximations of $e$. However, Cotes focused on the logarithm, rather than its inverse. To understand how the lack of a clear conception of the exponential handicapped mathematics, consider that in the early 1730s, as discussed in Chapter 14, Daniel Bernoulli was unable to fully solve the differential equation $\frac{K^4 d^4 y}{dx^4} = y$. He observed that the logarithm, meaning the inverse or exponential, satisfied this equation as well as the equation $\frac{K^2 d^2 y}{dx^2} = y$, but that no such logarithm was sufficiently general. Euler was also stumped by this problem, until he gave an explicit definition of the exponential and developed its properties in the mid-1730s.

To derive the series for elementary functions, Euler made considerable use of infinitely large and infinitely small numbers. This method can be made rigorous by an appropriate use of limits, as accomplished by Cauchy in the 1820s.[7] Following Euler's style, Cauchy divided analysis into two parts, algebraic analysis and calculus. The former dealt with infinite series and products without using calculus, yet employed the ideas of limits and convergence. It is interesting to note here that in his *Fonctions analytiques* of 1797, Lagrange had attempted to make differential calculus a part of algebraic analysis by defining the derivative of $f(x)$ as the coefficient of $h$ in the series expansion of $f(x + h)$. Gauss, Cauchy, and their followers rejected this idea as invalid. Besides providing greater rigor, Cauchy's lectures presented original and insightful derivations of some of Euler's results.

In addition to defining elementary functions, Euler also showed that functions could be represented by infinite products and partial fractions. The latter could be obtained from products by applying logarithmic differentiation, a process he worked out in his correspondence with Niklaus I Bernoulli.[8] In his *Introductio*, Euler presented

[3] Mercator (1668).

[4] Newton (1967–1981) vol. 1, pp. 112–115.

[5] Cotes (1714).

[6] Halley (1695).

[7] Cauchy (1989).

[8] Eu. IV A-2 pp. 483–550.

fascinating ways of avoiding the methods of calculus involving differentiation and integration. He also gave an exposition on the connection, discovered earlier by Cotes, between the trigonometric and the exponential functions. As discussed in Chapter 12, Cotes had found the relation $\log(\cos x + i \sin x) = ix$, although this equation was more useful when Euler wrote it as $\cos x + i \sin x = e^{ix}$. Of course, Cotes was unable to take this last step because he did not explicitly define the exponential $e^x$. Euler, on the other hand, made use of this relationship to derive important results such as the infinite products for the trigonometric functions.

At the very beginning of his career, Euler discovered the simple and useful dilogarithm function. The dilogarithm is defined by

$$\text{Li}_2(x) = -\int_0^x \frac{\ln(1-t)}{t}\, dt = \frac{x}{1^2} + \frac{x^2}{2^2} + \frac{x^3}{3^2} + \cdots, \tag{15.1}$$

where the series converges for $|x| \leq 1$. Euler initially investigated this function in 1729-30[9] to evaluate the series $\sum_{n=1}^{\infty} \frac{1}{n^2}$. He succeeded at that time only in determining its approximate value, but in the 1730s he found the exact value by the factorization of $\sin x$.

In the 1740s, the English mathematician and surveyor John Landen began publishing his mathematical problems in the *Ladies Diary*. In the late 1750s, he discovered that the dilogarithm could be used to exactly evaluate $\sum_{n=1}^{\infty} \frac{1}{n^2}$,[10] provided that logarithms of negative numbers were employed. Euler had already developed his theory of logarithms of complex numbers at that time but his work had not appeared in print. So Landen's determination of $\ln(-1) = \pm\sqrt{-1}\pi$ in 1758 was an important and independent discovery. And he went further, by repeated integration, to define the more general polylogarithm,

$$\text{Li}_k(x) = \frac{x}{1^k} + \frac{x^2}{2^k} + \frac{x^3}{3^k} + \cdots, \tag{15.2}$$

for $k = 1, 2, 3, \ldots$. He could then evaluate the series $\sum_{n=1}^{\infty} \frac{1}{n^{2k}}$ for $k = 1, 2, 3, \ldots$. We will discuss Euler's and Landen's evaluations of $\sum \frac{1}{n^{2k}}$ in Chapter 16.

The polylogarithm was further studied by the Scottish mathematician William Spence (1777–1815) who published his book, *Essay on Logarithmic Transcendents*, in 1809.[11] He derived several interesting results on dilogarithms, including the theorem for which he is known today. As a student in the 1820s, Abel rediscovered this formula, having been inspired to study the dilogarithm by reading Legendre's three volumes on the integral calculus. This work discussed numerous results of Euler. Spence was apparently self-taught but, unlike many other British mathematicians of his time, he read Bernoulli, Euler, Lagrange, and other continental mathematicians. In the preface to his work, Spence commented[12] on the disadvantage of British insularity:

[9] Eu. I-14 pp. 25–41, especially pp. 38–41. E 20 § 22–23.
[10] Landen (1760).
[11] Spence (1809).
[12] Spence (1809) p. xi.

> Our pupils are taught the science by means of its applications; and when their minds should be occupied with the contemplation of general methods and operations, they are usually employed on particular processes and results, in which no traces of the operations remain. On the Continent, Analysis is studied as an independent science. Its general principles are first inculcated; and then the pupil is led to the applications; and the effects have been, that while we have remained nearly stationary during the greater part of the last century, the most valuable improvements have been added to the science in almost every other part of Europe. The truth of this needs no illustration. Let any person who has studied Mathematics only in British authors look into works of the higher analysts of the Continent, and he will soon perceive that he has still much to learn.

Interestingly, other British mathematicians were independently arriving at this conclusion. In 1813, a few students at Cambridge University formed the Analytical Society in order to promote broadening mathematical studies to include the works of non-British mathematicians. Among the members of this new Society was John Herschel, who collected, published, and annotated the works of Spence,[13] an example of the progress facilitated by broader mathematical horizons. In one of his extensive notes on Spence's work, Herschel presented, without fanfare, his own discovery of the Schwarzian derivative. Interestingly, in 1781 Lagrange also found this derivative, but in the context of cartography.[14] We also note that Kummer, before he became a committed number theorist, wrote a very long 1840 paper on the dilogarithm;[15] this paper contained a wealth of results, including the rediscovery of Spence's formula.

## 15.2   Euler: Series for Elementary Functions

Euler defined the exponential functions in about 1748[16] by explaining the meaning of $a^z$, first for $z$ as an integer and then for a $z$ as a rational number. He remarked that for irrational $z$ the concept was more difficult to understand but that $a^{\sqrt{7}}$, for example, had a value between $a^2$ and $a^3$ when $a > 1$. He noted that the study of $a^z$ for $0 < a < 1$ could be reduced to the case where $a > 1$. He then defined the logarithm: If $a^z = y$, then $z$ is called the logarithm of $y$ to the base $a$. Euler did not have a notation for the base of a logarithm and always expressed the base in words. According to Cajori,[17] in 1821, A. L. Crelle, founder of the famous journal and a friend of Abel, introduced the notation for the base of the logarithm, writing the base $a$ on the upper left-hand side of the log. However, we employ the modern notation: $\log_a$.

To obtain the series for $a^z$, Euler observed that since $a^0 = 1$, he could write $a^w = 1 + k\omega$, where $\omega$ was an infinitely small number. Thus, by the binomial theorem, called the universal theorem by Euler,

$$a^{j\omega} = (1 + k\omega)^j = 1 + \frac{j}{1}k\omega + \frac{j(j-1)}{1\cdot 2}k^2\omega^2 + \frac{j(j-1)(j-2)}{1\cdot 2\cdot 3}k^3\omega^3 + \cdots .$$

$$(15.3)$$

[13] Spence (1819).
[14] Lagrange (1781).
[15] Kummer (1840).
[16] Euler (1988) chapters 6 and 7.
[17] Cajori (1993) vol. 2, p. 107.

He took $j$ to be infinitely large so that $j\omega = z$, a finite number, and equation (15.3) was transformed to

$$a^z = \left(1 + \frac{kz}{j}\right)^j = 1 + \frac{1}{1}kz + \frac{1(j-1)}{1\cdot 2j}k^2z^2 + \frac{1(j-1)(j-2)}{.1\cdot 2j\cdot 3j}k^3z^3 + \cdots .$$

$$(15.4)$$

For infinitely large $j$, he concluded that $\frac{j-1}{2j} = \frac{1}{2}$, $\frac{(j-1)(j-2)}{2j\cdot 3j} = \frac{1}{2\cdot 3}$ etc. and hence he had the series

$$a^z = 1 + \frac{kz}{1} + \frac{k^2z^2}{1\cdot 2} + \frac{k^3z^3}{1\cdot 2\cdot 3} + \cdots .$$

$$(15.5)$$

He then set $z = 1$ to obtain the equation for $k$:

$$a = 1 + \frac{k}{1} + \frac{k^2}{1\cdot 2} + \frac{k^3}{1\cdot 2\cdot 3} + \cdots .$$

$$(15.6)$$

He denoted by $e$ the value of $a$ when $k = 1$ and computed it to 23 decimal places. From (15.5 ) and (15.6), with $a = e$, Euler obtained these famous equations:

$$e^z = 1 + z + \frac{z^2}{1\cdot 2} + \frac{z^3}{1\cdot 2\cdot 3} + \cdots ,$$

$$(15.7)$$

$$e = 1 + 1 + \frac{1}{1\cdot 2} + \frac{1}{1\cdot 2\cdot 3} + \cdots .$$

$$(15.8)$$

It also followed from (15.6) and (15.7) that $k = \ln a$ where ln stands for the natural logarithm. To find the series for $\log_a(1 + x)$, Euler set $a^{j\omega} = (1 + k\omega)^j = 1 + x$, so that

$$j\omega = \log_a(1 + x) = \frac{j}{k}\left((1 + x)^{\frac{1}{j}} - 1\right)$$

$$(15.9)$$

and he then expanded the expression in parentheses by the binomial theorem to get

$$\log_a(1 + x) = \frac{j}{k}\left(\frac{x}{j} - \frac{1(j-1)}{j\cdot 2j}x^2 + \frac{1(j-1)(2j-1)}{j\cdot 2j\cdot 3j}x^3 - \cdots\right)$$

$$= \frac{1}{k}\left(x - \frac{x^2}{2} + \frac{x^3}{3} - \cdots\right).$$

$$(15.10)$$

The second equation in (15.10) followed from the condition that $j$ was an infinitely large number. When $k = 1$,

$$\ln(1 + x) = \log_e(1 + x) = x - \frac{x^2}{2} + \frac{x^3}{3} - \cdots .$$

$$(15.11)$$

To obtain the series for $\sin x$ and $\cos x$, Euler started with de Moivre's formulas

$$\cos nz = \frac{(\cos z + \sqrt{-1}\sin z)^n + (\cos z - \sqrt{-1}\sin z)^n}{2},$$

$$(15.12)$$

$$\sin nz = \frac{(\cos z + \sqrt{-1}\sin z)^n - (\cos z - \sqrt{-1}\sin z)^n}{2\sqrt{-1}}. \tag{15.13}$$

By the binomial theorem, equation (15.12) could be written as

$$\cos nz = (\cos z)^n - \frac{n(n-1)}{1\cdot 2}(\cos z)^{n-2}(\sin z)^2$$
$$+ \frac{n(n-1)(n-2)(n-3)}{1\cdot 2\cdot 3\cdot 4}(\cos z)^{n-4}(\sin z)^4 - \cdots.$$

Euler took $n$ infinitely large and $z$ infinitely small, such that that $nz = x$ was finite. He then concluded that $\sin z = z = \frac{x}{n}$ and $\cos z = 1$, and hence

$$\cos x = 1 - \frac{x^2}{1\cdot 2} + \frac{x^4}{1\cdot 2\cdot 3\cdot 4} - \frac{x^6}{1\cdot 2\cdot 3\cdot 4\cdot 5\cdot 6} + \cdots. \tag{15.14}$$

Similarly, Euler found the series for $\sin x$ from (15.13). Since we have discussed at some length the series for sine and cosine in connection with the works of Madhava, Newton, and Leibniz, we will not reproduce Euler's derivation of the series for sine, very similar to his derivation of (15.14).

## 15.3 Euler: Products for Trigonometric Functions

Euler derived the infinite products for the sine and cosine functions[18] from the Cotes factorization of $x^n \pm y^n$. Note that Cotes's formula for $n$ odd, given in Section 12.4, may be expressed as

$$x^n \pm y^n = (x \pm y)\prod_{k=1}^{\frac{(n-1)}{2}}\left(x^2 \pm 2xy\cos\frac{2k\pi}{n} + y^2\right). \tag{15.15}$$

Euler observed that the series for the exponential, cosine and sine functions, with $j$ infinite, yielded the relations:

$$\cos x = \frac{e^{xi} + e^{-xi}}{2} = \frac{\left(1 + \frac{xi}{j}\right)^j + \left(1 - \frac{xi}{j}\right)^j}{2}, \tag{15.16}$$

$$\sin x = \frac{e^{xi} - e^{-xi}}{2i} = \frac{\left(1 + \frac{xi}{j}\right)^j - \left(1 - \frac{xi}{j}\right)^j}{2i}. \tag{15.17}$$

He first determined the factors of $e^x - 1 = \left(1 + \frac{x}{j}\right)^j - 1$ by Cotes's formula. He noted that one factor was $\left(1 + \frac{x}{j}\right) - 1 = \frac{x}{j}$ and the quadratic factors were of the form $\left(1 + \frac{x}{j}\right)^2 - 2\left(1 + \frac{x}{j}\right)\cos\left(2k\frac{\pi}{j}\right) + 1$. He also noted that every factor could be obtained by taking all positive even integers $2k$. Euler then set $\cos\left(2k\frac{\pi}{j}\right) = 1 - \frac{2k^2\pi^2}{j^2}$ by taking

---

[18] ibid. chapter 9.

the first two nonzero terms of the series expansion for cosine and then simplified the quadratic factor to

$$\frac{x^2}{j^2} + \frac{4k^2}{j^2}\pi^2 + \frac{4k^2}{j^3}\pi^2 x = \frac{4k^2\pi^2}{j^2}\left(1 + \frac{x}{j} + \frac{x^2}{4k^2\pi^2}\right).$$

He observed at this point that though $\frac{x}{j}$ was infinitesimal, it could not be neglected because there were $\frac{j}{2}$ factors, producing a nonzero term $\frac{x}{2}$. This remark shows us that Euler had some pretty clear ideas about the convergence of infinite products. To eliminate this difficulty, Euler then considered the factors of $e^x - e^{-x} = \left(1 + \frac{x}{j}\right)^j - \left(1 - \frac{x}{j}\right)^j$. In this case, he simplified the general quadratic factor to

$$1 + \frac{x^2}{k^2}\pi^2 - \frac{x^2}{j^2}.$$

The contribution of the term $\frac{x^2}{j^2}$ after multiplication of $\frac{j}{2}$ factors was $\frac{x^2}{j}$ and he could now neglect this. So Euler determined the quadratic factors to be $1 + \frac{x^2}{k^2\pi^2}$ and wrote the formula

$$\frac{e^x - e^{-x}}{2} = x\left(1 + \frac{x^2}{\pi^2}\right)\left(1 + \frac{x^2}{4\pi^2}\right)\left(1 + \frac{x^2}{9\pi^2}\right)\left(1 + \frac{x^2}{16\pi^2}\right)\cdots. \qquad (15.18)$$

Similarly, he got

$$\frac{e^x + e^{-x}}{2} = \left(1 + \frac{4x^2}{\pi^2}\right)\left(1 + \frac{4x^2}{9\pi^2}\right)\left(1 + \frac{4x^2}{25\pi^2}\right)\left(1 + \frac{4x^2}{49\pi^2}\right)\cdots. \qquad (15.19)$$

To obtain the products for $\sin x$ and $\cos x$, he changed $x$ to $ix$ to find

$$\sin x = x\left(1 - \frac{x^2}{\pi^2}\right)\left(1 - \frac{x^2}{4\pi^2}\right)\left(1 - \frac{x^2}{9\pi^2}\right)\left(1 - \frac{x^2}{16\pi^2}\right)\cdots \qquad (15.20)$$

and

$$\cos x = \left(1 - \frac{4x^2}{\pi^2}\right)\left(1 - \frac{4x^2}{9\pi^2}\right)\left(1 - \frac{4x^2}{25\pi^2}\right)\left(1 - \frac{4x^2}{49\pi^2}\right)\cdots. \qquad (15.21)$$

From (15.20) it appears that the constant factors somehow cancelled. To see this more clearly, take $n = 2m + 1$, $X = 1 + \frac{x\sqrt{-1}}{2m+1}$, $Y = 1 - \frac{x\sqrt{-1}}{2m+1}$ in (15.15) to arrive at

$$\frac{X^{2m+1} - Y^{2m+1}}{2\sqrt{-1}}$$

$$= \frac{x}{2m+1}\prod_{k=1}^{m} 2\left(1 - \frac{x^2}{(2m+1)^2} - \cos\frac{2k\pi}{2m+1}\left(1 + \frac{x^2}{(2m+1)^2}\right)\right)$$

$$= \frac{2^m \prod_{k=1}^{m}\left(1 - \cos\frac{2k\pi}{2m+1}\right)}{2m+1} x \prod_{k=1}^{m}\left(1 - \frac{1 + \cos\frac{2k\pi}{2m+1}}{1 - \cos\frac{2k\pi}{2m+1}} \cdot \frac{x^2}{(2m+1)^2}\right).$$

To simplify, observe that

$$\lim_{x \to 1} \frac{x^{2m+1} - 1}{x - 1} = \lim_{x \to 1} \prod_{k=1}^{m} \left( x^2 + 1 - \cos \frac{2k\pi}{2m+1} x \right)$$

or

$$2m + 1 = 2^m \prod_{k=1}^{m} \left( 1 - \cos \frac{2k\pi}{2m+1} \right).$$

Thus

$$\frac{X^{2m+1} - Y^{2m+1}}{2\sqrt{-1}} = x \prod_{k=1}^{m} \left( 1 - \frac{1 + \cos \frac{2k\pi}{2m+1}}{1 - \cos \frac{2k\pi}{2m+1}} \cdot \frac{x^2}{(2m+1)^2} \right).$$

Now for each $k$ and with $m$ infinitely large, we see that

$$1 - \frac{1 + \cos \frac{2k\pi}{2m+1}}{1 - \cos \frac{2k\pi}{2m+1}} \cdot \frac{x^2}{(2m+1)^2} = 1 - \frac{2}{\frac{2k^2\pi^2}{(2m+1)^2}} \cdot \frac{x^2}{(2m+1)^2}$$

$$= 1 - \frac{x^2}{k^2\pi^2}$$

and therefore

$$\sin x = x \prod_{k=1}^{m} \left( 1 - \frac{x^2}{k^2\pi^2} \right).$$

## 15.4   Euler's Finite Product for $\sin nx$

In his 1748 *Introductio in Analysin Infinitorum*, Euler also gave an elegant proof of

$$\sin(2m+1)x = \pm 2^{2m} \sin x \, \sin \left( x + \frac{2\pi}{2m+1} \right)$$

$$\times \sin \left( x + \frac{4\pi}{2m+1} \right) \cdots \sin \left( x + \frac{4m\pi}{2m+1} \right),$$

$$(15.22)$$

with $n = 2m + 1$, an odd integer, and where the positive sign would be employed when $m$ was even and the negative sign when $m$ was odd. Though there are various methods of proof, Euler's was not only very nice, but also could be applied to derive other identities. Euler's proof relied upon a result well-known at that time: Suppose $a_1, a_2, \ldots, a_{2m+1}$ are the roots of

$$1 - c_1 y + c_2 y^2 - \cdots + (-1)^m c_{2m+1} y^{2m+1} = 0.$$

Then

$$1 - c_1 y + c_2 y^2 - \cdots + (-1)^{2m+1} c_{2m+1} y^{2m+1}$$
$$= \left(1 - \frac{y}{a_1}\right)\left(1 - \frac{y}{a_2}\right) \cdots \left(1 - \frac{y}{a_{2m+1}}\right). \tag{15.23}$$

Moreover, multiplying out the factors and then equating the coefficients, one obtains Girard's relations:

$$\frac{1}{a_1} + \frac{1}{a_2} + \cdots + \frac{1}{a_{2m+1}} = c_1, \tag{15.24}$$

$$\cdots \cdots \cdots$$

$$\frac{1}{a_1} \cdot \frac{1}{a_2} \cdots \cdots \frac{1}{a_{2m+1}} = c_{2m+1}; \tag{15.25}$$

in this connection, see Section 13.3.

Euler began his proof[19] by noting the result of Viète, later stated by Newton, in more general terms: With $n = 2m + 1$,

$$\sin nx = n \sin x - \frac{n(n^2 - 1^2)}{1 \cdot 2 \cdot 3} \sin^3 x + \frac{n(n^2 - 1^2)(n^2 - 3^2)}{1 \cdot 2 \cdot 3 \cdot 4 \cdot 5} \sin^5 x - \cdots$$
$$+ (-1)^m \frac{n(n^2 - 1^2) \cdots (n^2 - (2m - 1)^2)}{1 \cdot 2 \cdot 3 \cdots (2m + 1)} \sin^n x. \tag{15.26}$$

For a discussion of (15.26), see Sections 8.5–8.7. Euler observed that when $x$ is assigned the values $\frac{z}{n}, \frac{z+2\pi}{n}, \cdots \frac{z+2(n-1)\pi}{n}$, $\sin nx$ keeps the same value. Observe that if $n$ and $x$ are fixed, then $\sin nx$, the left-hand side of (15.26), is a constant and (15.26) is a polynomial of degree $n = 2m + 1$ in $y = \sin x$, with roots

$$a_1 = \sin \frac{z}{n}, \quad a_2 = \sin \frac{z + 2\pi}{n}, \quad \cdots, \quad a_n = \sin \frac{z + 2(n - 1)\pi}{n}.$$

Thus, $a_1, a_2, \ldots, a_n$ are the roots of the equation obtained after dividing (15.26) by $\sin nx$. Euler wrote this equation as

$$1 - \frac{n}{\sin nx} y + \frac{n(n^2 - 1^2)}{1 \cdot 2 \cdot 3 \cdot \sin nx} y^3 - \cdots \pm \frac{2^{n-1}}{\sin nx} y^n = 0, \tag{15.27}$$

since the coefficient of $y^n$ in (15.26) has the absolute value

$$\frac{n(n^2 - 1^2) \cdots (n - (2m - 1)^2)}{n!} = \frac{2m(2m + 2)(2m - 2)(2m + 4) \cdots 2(4m)}{(2m)!}$$

$$= 2^{2m} \frac{(2m)!}{(2m)!} = 2^{n-1}.$$

[19] Euler (1988) pp. 205–206.

So with $a_1 = \sin \frac{z}{n}, \cdots, a_n = \sin \frac{z+2(n-1)\pi}{n}$, equations (15.25) and (15.27) imply (15.22). On the other hand, as Euler noted, (15.24) and (15.27) produce

$$\frac{n}{\sin nx} = \frac{1}{\sin x} + \frac{1}{\sin\left(x + \frac{2\pi}{n}\right)} + \frac{1}{\sin\left(x + \frac{4\pi}{n}\right)} + \cdots + \frac{1}{\sin\left(x + \frac{2(n-1)\pi}{n}\right)}.$$

$$(15.28)$$

Now observe that, except for the constant 1, only odd powers of $y$ appear in equation (15.27); thus, the coefficient of $y^{n-1} = y^{2m}$ is zero. This gave Euler the relation

$$0 = \sin x + \sin\left(x + \frac{2\pi}{n}\right) + \sin\left(x + \frac{4\pi}{n}\right) + \cdots + \sin\left(x + \frac{2(n-1)\pi}{n}\right).$$

$$(15.29)$$

Euler made use of all these formulas to establish many more trigonometric identities, as given in chapter 14 of his *Introductio*, and Cauchy employed (15.22) and (15.28) to determine the infinite product for the sine function by a method different from Euler's, without using Cotes's factorization.

## 15.5  Cauchy's Derivation of the Product Formulas

In his lectures at the École Polytechnique, published in 1821 under the title *Analyse algébrique*,[20] Cauchy gave a treatment of infinite series and products more rigorous than Euler's. Cauchy then applied these ideas to a discussion of elementary functions; his discourse on infinite products was presented in note IX as the last topic in the work. Suppose $u_0, u_1, u_2, \ldots$ to be real numbers with $u_n \geq -1$. Cauchy started his discussion of infinite products with the observation that if the series

$$\ln(1 + u_0) + \ln(1 + u_1) + \ln(1 + u_2) + \cdots \qquad (15.30)$$

converged to $s$, then the sequence

$$p_n = (1 + u_0)(1 + u_1) \cdots (1 + u_{n-1}), \quad n = 1, 2, 3 \ldots,$$

converged to a finite limit different from zero, equal to $e^s$. Thus, according to Cauchy, the infinite product

$$(1 + u_0)(1 + u_1)(1 + u_2) \cdots \qquad (15.31)$$

was said to converge if $\lim\limits_{n \to \infty} \left((1 + u_0) \cdots (1 + u_n)\right)$ existed and was different from zero. He then stated the theorem: Suppose the series (15.30) converges. If the series

---

[20] Cauchy (1989).

$$u_0 + u_1 + u_2 + \cdots \tag{15.32}$$

and

$$u_0^2 + u_1^2 + u_2^2 + \cdots \tag{15.33}$$

are convergent, then the infinite product (15.31) converges. However, if (15.32) is convergent and (15.33) is divergent, then the infinite product diverges to zero.

In his proof, Cauchy first noted that the convergence of (15.30) implied that $\ln(1 + u_n) \to 0$ as $n \to \infty$, and this in turn implied that $u_n \to 0$ as $n \to \infty$. He next observed that for large enough $n$,

$$\ln(1 + u_n) = u_n - \frac{1}{2}u_n^2 + \frac{1}{3}u_n^3 - \cdots = u_n - \frac{1}{2}u_n^2(1 \pm \epsilon_n),$$

with $\epsilon_n$ infinitesimally small. He concluded that

$$\ln(1 + u_n) + \ln(1 + u_{n+1}) + \cdots + \ln(1 + u_{n+m-1})$$
$$= u_n + u_{n+1} + \cdots + u_{n+m-1} - \frac{1}{2}(u_n^2 + u_{n+1}^2 + \cdots + u_{n+m-1}^2)(1 \pm \epsilon),$$

$$\tag{15.34}$$

when all the $u$ had absolute value less than one and $1 \pm \epsilon$ was the average of $1 \pm \epsilon_n$, $1 \pm \epsilon_{n+1}, \ldots$. Formula (15.34) completed the proof of the theorem, because the infinite product converged if and only if the series $\ln(1 + u_0) + \ln(1 + u_1) + \ln(1 + u_2) + \cdots$ converged. As examples of this theorem, Cauchy noted that the product

$$(1 + x^2)\left(1 + \frac{x^2}{2^2}\right)\left(1 + \frac{x^2}{3^2}\right)\cdots$$

converged for all $x$, while the product

$$(1 + 1)\left(1 - \frac{1}{\sqrt{2}}\right)\left(1 + \frac{1}{\sqrt{3}}\right)\left(1 - \frac{1}{\sqrt{4}}\right)\cdots$$

diverged to zero.

Although earlier mathematicians did not explicitly state such a theorem on infinite products, it can hardly be doubted that Euler, with his enormous experience manipulating and calculating with infinite products and series, intuitively understood this result. However, Cauchy's presentation – with clearer, more precise and explicit definitions of fundamental concepts such as limits, continuity, and convergence – paved the way for future generations of mathematicians. His work led to higher standards of clarity and rigor in definitions, statements of theorems, and proofs.

While Cauchy was a pioneer in the establishment of rigor in mathematical arguments, we comment that his lectures nevertheless contain many arguments in the looser style of that time. This was perhaps a concession to his students, who apparently chafed at the rigorous approach. Thus, Cauchy's proof of the infinite product formula for $\sin x$, located between equations (11) and (16) inclusive of note 9 in his book, contained less-than-rigorous statements about the averages of various quantities. In his main argument, that includes equations (12) and (13), for example, he dealt with $2n$ quantities, $A_1 \approx B_1, A_2 \approx B_2, \ldots A_n \approx B_n$. He wrote that if $1 + \alpha$ denoted the average of the quantities

$$\frac{A_1}{B_1}, \frac{A_2}{B_2}, \ldots, \frac{A_n}{B_n},$$

then

$$A_1 + A_2 + \cdots + A_n = (B_1 + B_2 + \cdots + B_n)(1 + \alpha).$$

However, because the two $\alpha$'s are not necessarily identical, it would require a good amount of work to verify this. Cauchy makes a similar jump in reasoning, that would have required a very complicated argument to justify, in a result involving a quantity he labels $1 + \beta$. Nevertheless, Cauchy actually had all the results necessary for a complete proof of the infinite product formula for $\sin x$ and our presentation here is based exclusively upon Cauchy's own results and methods, as presented in his *Analyse Algébrique*, especially note 8.

For our proof, first note that

$$\sin\left(x + \frac{2\pi}{n}\right) \sin\left(x + \frac{2(n-1)\pi}{n}\right) = \sin^2 x - \sin^2 \frac{2\pi}{n},$$

$$\sin\left(x + \frac{4\pi}{n}\right) \sin\left(x + \frac{2(n-2)\pi}{n}\right) = \sin^2 x - \sin^2 \frac{4\pi}{n}, \quad \text{etc.}$$

Hence, by Euler's formula (15.22) for odd $n$,

$\sin nx$

$$= 2^{n-1} \sin x \left(\sin^2 \frac{2\pi}{n} - \sin^2 x\right)\left(\sin^2 \frac{4\pi}{n} - \sin^2 x\right) \cdots \left(\sin^2 \frac{(n-1)\pi}{n} - \sin^2 x\right).$$

$$(15.35)$$

It is easy to see that the set of $\frac{n-1}{2}$ numbers $\sin^2 \frac{2\pi}{n}, \sin^2 \frac{4\pi}{n}, \ldots, \sin^2 \frac{(n-1)\pi}{n}$ is identical to the set of $\frac{n-1}{2}$ numbers $\sin^2 \frac{\pi}{n}, \sin^2 \frac{2\pi}{n}, \ldots, \sin^2 \frac{(n-1)\pi}{2n}$. Thus,

$$\sin nx = 2^{n-1} \sin x \left(\sin^2 \frac{\pi}{n} - \sin^2 x\right)\left(\sin^2 \frac{2\pi}{n} - \sin^2 x\right) \cdots \left(\sin^2 \frac{(n-1)\pi}{2n} - \sin^2 x\right),$$

in we which let $x \to 0$ to get

$$n = 2^{n-1} \sin^2 \frac{\pi}{n} \sin^2 \frac{2\pi}{n} \cdots \sin^2 \frac{(n-1)\pi}{2n}.$$

After replacing $nx$ by $x$,

$$\sin x = n \sin \frac{x}{n} \left(1 - \frac{\sin^2 \frac{x}{n}}{\sin^2 \frac{\pi}{n}}\right)\left(1 - \frac{\sin^2 \frac{x}{n}}{\sin^2 \frac{2\pi}{n}}\right) \cdots \left(1 - \frac{\sin^2 \frac{x}{n}}{\sin^2 \frac{(n-1)\pi}{2n}}\right),$$

given as equation (18) in note 8 of Cauchy's *Analyse Algébrique*.

When $n \to \infty$, we get

$$1 - \frac{\sin^2 \frac{x}{n}}{\sin^2 \frac{k\pi}{n}} \longrightarrow 1 - \frac{x^2}{k^2\pi^2}.$$

Take $x > 0$ and let $m$ be any fixed number less than $\frac{n-1}{2}$, chosen so that $(m+1)\pi^2 > 4x^2$. Then write

$$\sin x = n \sin \frac{x}{n} \left(1 - \frac{\sin^2 \frac{x}{n}}{\sin^2 \frac{\pi}{n}}\right) \cdots \left(1 - \frac{\sin^2 \frac{x}{n}}{\sin^2 \frac{m\pi}{n}}\right)$$

$$\times \left(1 - \frac{\sin^2 \frac{x}{n}}{\sin^2 \frac{(m+1)\pi}{n}}\right) \cdots \left(1 - \frac{\sin^2 \frac{x}{n}}{\sin^2 \frac{(n-1)\pi}{2n}}\right). \qquad (15.36)$$

Now because $\prod_1^n (1 - \alpha_i) > 1 - \sum_1^n \alpha_i$, we have

$$\left(1 - \frac{\sin^2 \frac{x}{n}}{\sin^2 \frac{(m+1)\pi}{n}}\right) \cdots \left(1 - \frac{\sin^2 \frac{x}{n}}{\sin^2 \frac{(n-1)\pi}{2n}}\right)$$

$$> 1 - \sin^2 \frac{x}{n}\left(\csc^2 (m+1)\frac{\pi}{n} + \cdots + \csc^2 \frac{(n-1)\pi}{2n}\right). \qquad (15.37)$$

Our calculations are now similar to Cauchy's derivation of equation (30) from (29) in his note 8. Since for $0 < x < \frac{\pi}{2}$, we have $x < 2 \sin x$; it follows that

$$\csc^2 \frac{k\pi}{n} < \frac{4n^2}{k^2\pi^2}, \text{ for } k = m+1, \ldots, \frac{n-1}{2}.$$

Now because $\sin x < x$ for $x > 0$, we get

$$\sin^2 \frac{x}{n} \csc^2 \frac{k\pi}{n} < \frac{4x^2}{\pi^2 k^2}. \qquad (15.38)$$

Thus, the product in (15.37) is greater than

$$1 - \frac{4x^2}{\pi^2}\left(\frac{1}{(m+1)^2} + \frac{1}{(m+2)^2} + \cdots + \frac{4}{(n-1)^2}\right) > 1 - \frac{4x^2}{\pi^2(m+1)} > 0,$$

and less than one. Therefore, it is equal to $1 - \frac{4x^2}{\pi^2(m+1)}\theta_n$, where $\theta_n$ lies between 0 and 1. Thus, (15.36) can be written as

$$\sin x = n \sin \frac{x}{n} \left(1 - \frac{\sin^2 \frac{x}{n}}{\sin^2 \frac{\pi}{n}}\right) \cdots \left(1 - \frac{\sin^2 \frac{x}{n}}{\sin^2 \frac{m\pi}{n}}\right) \left(1 - \frac{4x^2 \theta_n}{\pi^2(m+1)}\right). \quad (15.39)$$

Also observe that

$$\frac{\sin^2 \frac{x}{n}}{\sin^2 \frac{k\pi}{n}} \rightarrow \frac{x^2}{k^2\pi^2} \quad \text{as } n \rightarrow \infty$$

and

$$\theta_n \rightarrow \theta \quad \text{(say) as } n \rightarrow \infty.$$

Finally, letting $n \rightarrow \infty$ in (15.37), we arrive at

$$\sin x = x \left(1 - \frac{x^2}{\pi^2}\right) \cdots \left(1 - \frac{x^2}{m^2\pi^2}\right) \left(1 - \frac{4x^2}{\pi^2(m+1)} \theta\right).$$

The result follows by letting $m \rightarrow \infty$. The restriction $x > 0$ can now be removed.

## 15.6   Euler and Niklaus I Bernoulli: Partial Fraction Expansions

Recall that Euler obtained the partial fractions expansions of $\csc x$ and $\cot x$ by the use of integration methods and, later on, by other methods. The former approach depended upon the evaluation of the integrals $\int_0^\infty \frac{x^{p-1} dx}{1 \pm x^q}$ in two different ways. First, Euler used Cotes's factorization of $1 \pm x^q$ to express $\frac{1}{1 \pm x^q}$ as partial fractions; he then integrated to find

$$\int_0^\infty \frac{x^{p-1}}{1 + x^q} \, dx = \frac{\pi}{q \sin \frac{p\pi}{q}} \quad (15.40)$$

and

$$\int_0^\infty \frac{x^{p-1}}{1 - x^q} \, dx = \frac{\pi}{q \tan \frac{p\pi}{q}}, \quad (15.41)$$

where $0 < p < q$ and $p, q$ were integers. For Euler's derivation of (15.40) and (15.41), see Section 12.6. The integral in (15.41) is a principal value, although this concept was not explicitly defined until Cauchy introduced it in the 1820s. Dedekind gave a number of proofs of (15.40), including a streamlined form of Euler's proof, and we present that in Section 17.8.

For Euler's derivation of the partial fractions expansion of $\csc x$, note that the change of variable $y = \frac{1}{x}$ shows that

$$\int_1^\infty \frac{x^{p-1}}{1 + x^q} \, dx = \int_0^1 \frac{y^{q-p-1}}{1 + y^q} \, dy.$$

So Euler could rewrite the integral (15.40) as

$$\int_0^1 \frac{x^{p-1} + x^{q-p-1}}{1 + x^q}\, dx = \int_0^1 (x^{p-1} + x^{q-p-1})(1 - x^q + x^{2q} - x^{3q} + \cdots)\, dx$$

$$= \int_0^1 (x^{p-1} + x^{q-p-1} - x^{q+p-1} - x^{2q-p-1} + x^{2q+p-1} + x^{3q-p-1} - \cdots)\, dx$$

$$= \frac{1}{p} + \frac{1}{q-p} - \frac{1}{q+p} - \frac{1}{2q-p} + \frac{1}{2q+p} + \frac{1}{3q-p} - \cdots$$

$$= \frac{\pi}{q \sin \frac{p\pi}{q}}.$$

Thus, he had the partial fractions expansion for $\csc x$

$$\frac{\pi}{\sin \pi x} = \frac{1}{x} + \frac{1}{1-x} - \frac{1}{1+x} - \frac{1}{2-x} + \frac{1}{2+x} + \frac{1}{3-x} - \text{etc.}$$

$$= \frac{1}{x} - \frac{2x}{x^2 - 1^2} + \frac{2x}{x^2 - 2^2} - \frac{2x}{x^2 - 3^2} + \text{etc.}$$

(15.42)

In a similar way, he obtained the partial fractions expansion of $\cot x$:

$$\frac{\pi}{\tan \pi x} = \frac{1}{x} - \frac{1}{1-x} + \frac{1}{1+x} - \frac{1}{2-x} + \frac{1}{2+x} - \frac{1}{3-x} + \frac{1}{3+x} - \text{etc.}$$

$$= \frac{1}{x} + \frac{2x}{x^2 - 1^2} + \frac{2x}{x^2 - 2^2} + \frac{2x}{x^2 - 3^2} + \text{etc.}$$

(15.43)

Note that by integrating (15.43), Euler had another way to obtain the product for $\sin x$.

In a letter of January 16, 1742,[21] Euler communicated these results to Niklaus I Bernoulli, also observing that partial fractions expansions of other functions could be found by repeated differentiation of (15.42) and (15.43). In particular, he had

$$\frac{\pi^2 \cos \pi x}{(\sin \pi x)^2} = \frac{1}{x^2} - \frac{1}{(1-x)^2} - \frac{1}{(1+x)^2} + \frac{1}{(2-x)^2} + \frac{1}{(2+x)^2} - \frac{1}{(3-x)^2} - \text{etc.}$$

(15.44)

and

$$\frac{\pi^2}{(\sin \pi x)^2} = \frac{1}{x^2} + \frac{1}{(1-x)^2} + \frac{1}{(1+x)^2} + \frac{1}{(2-x)^2} + \frac{1}{(2+x)^2} + \text{etc.}$$ (15.45)

In his reply of July 13, 1742,[22] Bernoulli noted that the logarithmic differentiation of Euler's product for $\sin x$ would immediately produce (15.43). To see this, we note that

[21] Eu. IV A-2 pp. 483–490.
[22] Eu. IV A-2 pp. 491–500.

$$\ln \sin \pi x = \ln(\pi x) + \sum_{n=1}^{\infty} \ln\left(1 - \frac{x^2}{n^2}\right);$$

differentiation yields the required formula. In his subsequent letter of October 24, 1742,[23] Bernoulli noted that (15.42) could also be obtained by logarithmic differentiation. In his letter, Bernoulli wrote the differential $d \ln x$ as differ.$\ln x$. His notation for the natural logarithm was log. We maintain our practice of writing $\ln x$, and present Bernoulli's argument as he wrote it:

$$
\begin{aligned}
\text{differ. } \ln \frac{\sin \pi x}{\cos \pi x} &= \frac{\text{differ. } \sin \pi x}{\sin \pi x} - \frac{\text{differ. } \cos \pi x}{\cos \pi x} \\
&= \frac{\pi \, dx \cos \pi x}{\sin \pi x} + \frac{\pi \, dx \sin \pi x}{\cos \pi x} = \frac{\pi \, dx}{\sin \pi x \cos \pi x} \\
&= \frac{2\pi \, dx}{\sin 2\pi x} = \text{differ. } \ln \frac{\pi x(1 - xx)\left(1 - \frac{1}{4}xx\right)\left(1 - \frac{1}{9}xx\right) \text{ etc.}}{(1 - 4xx)\left(1 - \frac{4}{9}xx\right)\left(1 - \frac{4}{25}xx\right) \text{ etc.}}
\end{aligned}
$$

Replace $2x$ by $x$ to get

$$
\begin{aligned}
\frac{\pi \, dx}{\sin \pi x} &= \text{differ. } \ln \frac{\frac{1}{2}\pi x\left(1 - \frac{1}{4}xx\right)\left(1 - \frac{1}{16}xx\right)\left(1 - \frac{1}{36}xx\right) \text{ etc.}}{(1 - xx)\left(1 - \frac{1}{9}xx\right)\left(1 - \frac{1}{25}xx\right) \text{ etc.}} \\
&= dx\left(\frac{1}{x} + \frac{1}{1 - x} - \frac{1}{1 + x} - \frac{1}{2 - x} + \frac{1}{2 + x} + \text{etc.}\right),
\end{aligned}
$$

which yields (15.42) after division by $dx$.

In his *Introductio*,[24] Euler gave an alternate derivation of the partial fractions expansions of the trigonometric functions, avoiding the use of integration and differentiation. He first showed by Cotes's formula that the quadratic factors of

$$e^y + e^{c-y} = \left(1 + \frac{y}{j}\right)^j + \left(1 + \frac{c - y}{j}\right)^j$$

were of the form

$$1 - \frac{4cy - 4y^2}{m^2\pi^2 + c^2},$$

with odd $m$, and hence

$$\frac{e^y + e^c \cdot e^{-y}}{1 + e^c} = \left(1 - \frac{4cy - y^2}{\pi^2 + c^2}\right)\left(1 - \frac{4cy - y^2}{9\pi^2 + c^2}\right)\left(1 - \frac{4cy - y^2}{25\pi^2 + c^2}\right) \quad \text{etc.,}$$

[23] Eu. IV A-2 pp. 532–550.
[24] Euler (1988) pp. 130, 134, 135.

where the denominator on the left-hand side was chosen so that the value of both sides of the equation was 1 for $y = 0$. Euler then took $c = i\pi x$ and $y = i\pi \frac{v}{2}$, so that the left-hand side reduced to $\cos(\pi \frac{v}{2}) + \tan(\pi \frac{x}{2}) \sin(\pi \frac{v}{2})$. This gave him the formula

$$\cos \frac{\pi v}{2} + \tan \frac{\pi x}{2} \sin \frac{\pi v}{2} = \left(1 + \frac{v}{1-x}\right)\left(1 - \frac{v}{1+x}\right)\left(1 + \frac{v}{3-x}\right)\left(1 - \frac{v}{3+x}\right) \text{ etc.}$$

In a similar way hē derived

$$\cos \frac{\pi v}{2} + \cot \frac{\pi x}{2} \sin \frac{\pi v}{2}$$
$$= \left(1 + \frac{v}{x}\right)\left(1 - \frac{v}{2-x}\right)\left(1 + \frac{v}{2+x}\right)\left(1 - \frac{v}{4-x}\right)\left(1 + \frac{v}{4+x}\right) \text{ etc.}$$

By equating the coefficients of $v$ in the two equations, he had respectively

$$\frac{\pi}{2} \tan \frac{\pi x}{2} = \frac{1}{1-x} - \frac{1}{1+x} + \frac{1}{3-x} - \frac{1}{3+x} + \frac{1}{5-x} - \frac{1}{5+x} + \text{ etc.}$$

and

$$\frac{\pi}{2} \cot \frac{\pi x}{2} = \frac{1}{x} - \frac{1}{2-x} + \frac{1}{2+x} - \frac{1}{4-x} + \frac{1}{4+x} \text{ etc.}$$

Since

$$\frac{\pi}{2}\left(\tan \frac{\pi x}{2} + \cot \frac{\pi x}{2}\right) = \frac{\pi}{\sin \pi x},$$

Euler was also able to obtain the partial fractions expansion of $\frac{\pi}{\sin \pi x}$. Note that in in section 225 of the second part of his 1755 differential calculus book, Euler replaced $x$ by $\frac{1}{2} - x$ in (15.42), and if we do that we get

$$\pi \sec \pi x = \frac{4}{1 - 4x^2} - \frac{4 \cdot 3}{3^2 - 4x^2} + \frac{4 \cdot 5}{5^2 - 4x^2} - \frac{4 \cdot 7}{7^2 - 4x^2} + \cdots . \quad (15.46)$$

## 15.7 Euler: Logarithm

Euler defined the logarithm of a number as a multivalued function in his 1751 paper, "De la controverse entre Mrs. Leibniz et Bernoulli sur les logarithmes des nombres negatifs et imaginaires."[25] Euler started his paper by discussing the positions of Johann Bernoulli and Leibniz on the logarithms of negative numbers; he proceeded to show that each point of view led to a contradiction. He then stated the theorem that resolved the difficulties: If $y$ denotes the logarithm of any number $x$, then $y$ must contain an infinity of different values.

[25] Eu. I-17 pp. 195–232. E 168.

To prove this theorem, Euler first observed that if $\omega$ were an infinitely small number, then $\log(1 + \omega) = \omega$, so that

$$\log(1 + \omega)^n = n\,\omega.$$

He reasoned that $n$ must be infinitely large because $(1 + \omega)^n = x$, where $x$ denoted any given number. With $n$ infinitely large,

$$x = (1 + \omega)^n \quad \text{and} \quad \log x = n\,\omega = y.$$

To express $y$ in terms of $x$, he observed that $1 + \omega = x^{\frac{1}{n}}$. Hence, $\omega = x^{\frac{1}{n}} - 1$ and $y = n\,\omega = nx^{\frac{1}{n}} - n = \log x$. Thus, as the value of $n$ in $nx^{\frac{1}{n}} - n$ became larger and larger, $nx^{\frac{1}{n}} - n$ would come closer and closer to its true value, $\log x$. Euler argued that $x^{\frac{1}{2}}$ had two values, $x^{\frac{1}{3}}$ has three values, and so on; therefore, $x^{\frac{1}{n}}$ would take an infinity of different values with $n$ infinitely large. Thus, the theorem was that $\log x$ must have infinitely many values.

After proving this theorem, the first problem posed by Euler was to find all the values of $\log a$, with $a$ a positive number. He noted that since $a$ was positive, one of the values of $\log a$ had to be a real number $A$. Since $\log a = \log 1 \cdot a = \log 1 + A$, he needed to find all the values of $\log 1 = n1^{\frac{1}{n}} - n = y$. He thus had to determine all values of $y$ that were solutions of the equation

$$\left(1 + \frac{y}{n}\right)^n = 1 \quad \text{or} \quad \left(1 + \frac{y}{n}\right)^n - 1 = 0. \tag{15.47}$$

To factorize $\left(1 + \frac{y}{n}\right)^n$, he noted that a typical factor of $p^n - q^n$ would be $p^2 - 2pq\,\cos\frac{2\lambda\pi}{n} + q^2$, where $\lambda = 0, \pm 1, \pm 2, \ldots$. He observed that the solutions of equation (15.47) with $p = 1 + \frac{y}{n}$ and $q = 1$ were the solutions of

$$\left(1 + \frac{y}{n}\right)^2 - 2\left(1 + \frac{y}{n}\right)\cos\frac{2\lambda\pi}{n} + 1 = 0, \quad \lambda = 0, \pm 1, \pm 2, \pm 3, \ldots \tag{15.48}$$

and these were:

$$1 + \frac{y}{n} = \cos\frac{2\lambda\pi}{n} \pm \sqrt{-1}\,\sin\frac{2\lambda\pi}{n}.$$

Because $\frac{2\lambda\pi}{n}$ was an infinitesimal, $\cos\frac{2\lambda\pi}{n} = 1$ and $\sin\frac{2\lambda\pi}{n} = \frac{2\lambda\pi}{n}$; thus,

$$1 + \frac{y}{n} = 1 \pm \sqrt{-1}\,\frac{2\lambda\pi}{n} \quad \text{or} \quad y = \pm 2\lambda\pi\sqrt{-1}, \quad \lambda = 0, 1, 2, \ldots.$$

Euler had thereby found all the values of $\log a$ as: $A \pm 2\lambda\pi\sqrt{-1}$, $\lambda = 0, 1, 2, \ldots$.

The second problem posed by Euler was: Find all values of $\log(-a)$, where $a > 0$. This amounted to finding all values of $\log(-1)$. In that case, all values $y$ of $\log(-1)$ would be solutions of

$$\left(1 + \frac{y}{n}\right)^n + 1 = 0. \tag{15.49}$$

Euler noted that all solutions of $p^n + q^n = 0$ could be found by solving the quadratic equations

$$p^2 - 2pq \cos \frac{(2\lambda - 1)\pi}{n} + q^2 = 0, \quad \lambda = 0, \pm 1, \pm 2, \ldots.$$

Thus, (15.49) had solutions

$$1 + \frac{y}{n} = \cos \frac{(2\lambda - 1)\pi}{n} \pm \sqrt{-1} \sin \frac{(2\lambda - 1)\pi}{n}, \quad \lambda = 0, 1, 2, \ldots$$

and, since $n$ was an infinite number,

$$y = \frac{(2\lambda - 1)\pi}{n} \sqrt{-1}, \quad \lambda = 0, \pm 1, \pm 2, \ldots.$$

As a third problem, Euler proposed to find all values for the logarithm of a complex number $a + b\sqrt{-1}$, beginning by noting that

$$a + b\sqrt{-1} = c(\cos \phi + \sqrt{-1} \sin \phi), \quad \text{where } c = \sqrt{a^2 + b^2} \text{ and } \tan \phi = \frac{b}{a}.$$

He reasoned that since $c$ was positive, one of the values of $\log c$ must be real. Call this value $C$ so that

$$\log(a + b\sqrt{-1}) = C + \log(\cos \phi + \sqrt{-1} \sin \phi).$$

Now since

$$\cos \phi + \sqrt{-1} \sin \phi = e^{\sqrt{-1}\phi} = \left(1 + \frac{\sqrt{-1}\,\phi}{n}\right)^n,$$

the values of $\log(\cos \phi + \sqrt{-1} \sin \phi)$ would be given by the solutions of

$$p^n - q^n = 0, \quad \text{where } p = 1 + \frac{y}{n} \text{ and } q = 1 + \frac{\sqrt{-1}\,\phi}{n}.$$

Moreover, because

$$p = q \left(\frac{\cos 2\lambda\pi}{n} \pm \sqrt{-1} \sin \frac{2\lambda\pi}{n}\right),$$

Euler arrived at

$$1 + \frac{y}{n} = \left(1 + \frac{\sqrt{-1}\,\phi}{n}\right)\left(\cos \frac{2\lambda\pi}{n} \pm \sqrt{-1} \sin \frac{2\lambda\pi}{n}\right)$$

$$= \left(1 + \frac{\sqrt{-1}\,\phi}{n}\right)\left(1 \pm \sqrt{-1} \frac{2\lambda\pi}{n}\right).$$

or

$$y = (\phi \pm 2\lambda\pi)\sqrt{-1}, \quad \lambda = 0, \pm 1, \pm 2, \ldots.$$

With $C = \frac{1}{2}\log(a^2 + b^2)$ and $\tan\phi = \frac{b}{a}$, he obtained his result:

$$\log(a + b\sqrt{-1}) = C + \phi\sqrt{-1} + 2\lambda\pi\sqrt{-1}, \quad \lambda = 0, \pm 1, \pm 2, \ldots.$$

In modern notation and terminology, if we set $z = a + ib$, then $|z| = (a^2 + b^2)^{\frac{1}{2}}$ and $\arg z = \phi + 2\lambda\pi$, where $\arg z$ denotes the argument of $z$, a multivalued function. We now designate the value of $\arg z$ where $-\pi < \arg z \le \pi$ as the principal value of $\log z$:

$$\log z = \frac{1}{2}\log|z| + i\arg z, \quad -\pi < \arg z \le \pi.$$

## 15.8 Euler: Dilogarithm

The dilogarithm function $\text{Li}_2(x)$ can be defined for $-1 \le x \le 1$ by the series

$$\text{Li}_2(x) = x + \frac{x^2}{2^2} + \frac{x^3}{3^2} + \cdots. \tag{15.50}$$

This series is obtained when the series for $\frac{-1}{t}\log(1 - t)$ is integrated term by term. Thus, we have

$$\text{Li}_2(x) = -\int_0^x \frac{\ln(1 - t)}{t}\,dt = \int_0^x \left(1 + \frac{t}{2} + \frac{t^2}{3} + \cdots\right)dt. \tag{15.51}$$

Note that the first integral in (15.51) does not require that $x$ be restricted to the interval $-1 \le x \le 1$. In fact, if we use the integral definition of $\text{Li}_2(x)$, as given by the first integral in (15.51), then $x$ can be given any complex value and, because the logarithm is many-valued, $\text{Li}_2(x)$ must be a many-valued function. Now observe that since

$$\log(1 - t) = \log|1 - t| + i\arg(1 - t),$$

the principal value of $\text{Li}_2(x)$ has the condition $-\pi < \arg(1 - x) \le \pi$. Observe that for $x > 1$, the principal value of $\log(1 - x)$ is

$$\log(-1) + \log(x - 1) = i\pi + \log(x - 1).$$

Thus $\text{Li}_2(x)$ contains

$$-i\pi\int \frac{dx}{x} = -i\pi\log x$$

in its imaginary part.

The dilogarithm series (15.50) arose in a 1731 paper of Euler,[26] the purpose of which was to evaluate, when $s = 2$,

$$\zeta(s) = \sum_{n=1}^{\infty} \frac{1}{n^s}, \quad s > 1.$$

Though Euler succeeded in finding only an approximate value for $\zeta(2)$, he obtained an interesting and useful formula for the dilogarithm:

$$\text{Li}_2(x) + \text{Li}_2(1 - x) = \sum_{n=1}^{\infty} \frac{1}{n^2} - \ln x \, \ln(1 - x). \tag{15.52}$$

Euler wished to find an approximation for the sum of the right-hand side of (15.52). Note that this series converges very slowly; one thousand terms of the series would yield an approximation to three decimal places. Euler took $x = \frac{1}{2}$ and got

$$\sum_{n=1}^{\infty} \frac{1}{n^2} = \sum_{n=1}^{\infty} \frac{1}{2^{n-1} n^2} + (\log 2)^2. \tag{15.53}$$

The series on the right-hand side in (15.53) converges much more rapidly and Euler found its approximate value to be 1.164481, though he did not mention how many terms of the series $\sum_{n=1}^{\infty} \frac{1}{2^{n-1} n^2}$ were required to obtain this approximation. He approximated $(\log 2)^2$ as 0.480453 and obtained

$$\sum_{n=1}^{\infty} \frac{1}{n^2} \approx 1.644934.$$

Euler's proof of (15.52) was slightly complicated, partly because he first proved a more general result and then derived his result as a particular case. We therefore reproduce the brief argument given by Abel:[27]

$$
\begin{aligned}
\text{Li}_2(x) + \text{Li}_2(1 - x) &= -\int_0^x \frac{\log(1 - t)}{t} \, dt - \int_0^{1-x} \frac{\log(1 - t)}{t} \, dt \\
&= -\int_0^1 \frac{\log(1 - t)}{t} \, dt - \int_0^x \frac{\log(1 - t)}{t} \, dt + \int_{1-x}^1 \frac{\log(1 - t)}{t} \, dt \\
&= \sum_{n=1}^{\infty} \frac{1}{n^2} - \int_0^x \left( \frac{\log(1 - t)}{t} - \frac{\log t}{1 - t} \right) dt \\
&= \sum_{n=1}^{\infty} \frac{1}{n^2} - \log x \, \log(1 - x),
\end{aligned}
$$

---

[26] Eu. I-14 pp. 25–41, especially section 22. E 20.
[27] Abel (1965) vol. 2, pp. 189–193.

where the final step followed from the observation that the integrand in the last integral was the derivative of $\log t \, \log(1-t)$. This same argument was given by the British mathematician William Spence[28] in his 1809 work on what he then called logarithmic transcendents. At that time, Spence was unaware of John Landen's 1760 paper containing the identical reasoning. In fact, Spence learned of Landen's work just as the book was going into publication, so he mentioned Landen in the preface.

## 15.9   Spence: Two-Variable Dilogarithm Formula

In his essay on logarithmic transcendents,[29] William Spence remarked that Euler and Bernoulli had only one variable in their formulas for the dilogarithm, whereas he himself used more unknowns. Spence worked with the function defined by

$$\overset{2}{L}(1+x) = \int_0^x \frac{dt}{t} \ln(1+t). \tag{15.54}$$

Note that he wrote $\ln(1+t)$ as $\overset{1}{L}(1+t)$. Spence observed that when $|x| < 1$:

$$\frac{\ln(1+t)}{t} = 1 - \frac{1}{2}t + \frac{1}{3}t^2 - \frac{1}{4}t^3 + \cdots \quad -1 < t < 1,$$

and so

$$\overset{2}{L}(1+x) = \frac{x}{1^2} - \frac{x^2}{2^2} + \frac{x^3}{3^2} - \frac{x^4}{4^2} + \cdots \quad -1 \leq x \leq 1.$$

He then gave a simple proof of the formula

$$\overset{2}{L}((1+mx)(1+nx)) = \overset{2}{L}(1+mx) + \overset{2}{L}(1+nx) - \overset{2}{L}\left(\frac{m+n+mnx}{m}\right)$$
$$- \overset{2}{L}\left(\frac{m+n+mnx}{n}\right) + \ln\left(\frac{m+n+mnx}{m}\right) \cdot \ln\left(\frac{n(1+mx)}{m}\right)$$
$$+ \ln\left(\frac{m+n+mnx}{n}\right) \cdot \ln\left(\frac{m(1+nx)}{n}\right) - \frac{1}{1\cdot 2}\left(\ln\frac{m}{n}\right)^2 + 2\overset{2}{L}(2). \tag{15.55}$$

He expressed the formula (15.54) as

$$\overset{2}{L}(1+x) = \int \frac{dx}{x} \ln(1+x), \tag{15.56}$$

and worked with the integral as if it were an indefinite integral where the constant of integration was computed in the final step. With this in mind, he replaced $x$ by $(m+n)x + mnx^2$ to get

[28] Spence (1809).
[29] Spence (1809) sections 9 and 10.

$$\overset{2}{L}\left((1+mx)(1+nx)\right) = \int\left(\frac{dx}{x} + \frac{mndx}{m+n+mnx}\right)\ln\left((1+mx)(1+nx)\right)$$

$$= \int\frac{dx}{x}\ln(1+mx) + \int\frac{dx}{x}\ln(1+nx) + \int\frac{dx}{m+n+mnx}\ln(1+mx)$$

$$+ \int\frac{dx}{m+n+mnx}\ln(1+nx).$$

$$(15.57)$$

By definition, the first two integrals were $\overset{2}{L}(1+mx)$ and $\overset{2}{L}(1+nx)$, respectively. Letting $z$ denote the sum of the last two integrals and setting $v = m+n+mnx$, he obtained

$$z = \int\frac{dv}{v}\ln\left(\frac{v-m}{n}\right) + \int\frac{dv}{v}\ln\left(\frac{v-n}{m}\right)$$

$$= \int\frac{dv}{v}\ln\left(\frac{v}{m}-1\right) + \int\frac{dv}{v}\ln\left(\frac{v}{n}-1\right)$$

$$= \ln\left(\frac{v}{m}\right)\ln\left(\frac{v}{m}-1\right) - \overset{2}{L}\left(\frac{v}{m}\right) + \ln\left(\frac{v}{n}\right)\ln\left(\frac{v}{n}-1\right) - \overset{2}{L}\left(\frac{v}{n}\right) + C,$$

where the last step involved integration by parts and the value of the constant $C$ was found by setting $x = 0$. This completed Spence's proof.

Abel rediscovered Spence's formula, with a different proof; it first appeared in his collected papers in 1839.[30] Abel stated his formula as

$$\mathrm{Li}_2\left(\frac{x}{1-x}\cdot\frac{y}{1-y}\right) = \mathrm{Li}_2\left(\frac{y}{1-x}\right) + \mathrm{Li}_2\left(\frac{x}{1-y}\right)$$

$$- \mathrm{Li}_2(y) - \mathrm{Li}_2(x) - \ln(1-y)\ln(1-x). \qquad (15.58)$$

Abel gave a simple and elegant proof of this formula: He let $a$ denote a constant. Then it was easy to check that

$$\mathrm{Li}_2\left(\frac{a}{1-a}\cdot\frac{y}{1-y}\right) = -\int\left(\frac{dy}{y} + \frac{dy}{1-y}\right)\ln\frac{1-a-y}{(1-a)(1-y)} \qquad (15.59)$$

$$= -\int\frac{dy}{y}\ln\left(1-\frac{y}{1-a}\right) + \int\frac{dy}{y}\ln(1-y) - \int\frac{dy}{1-y}\ln\left(1-\frac{a}{1-y}\right)$$

$$+ \int\frac{dy}{1-y}\ln(1-a)$$

$$= \mathrm{Li}_2\left(\frac{y}{1-a}\right) - \mathrm{Li}_2(y) - \ln(1-a)\ln(1-y) - \int\frac{dy}{1-y}\left(1-\frac{a}{1-y}\right);$$

$$(15.60)$$

30 Abel (1965) vol. 2, pp. 189–193.

and (15.59) could be verified by taking the derivative of both sides. To evaluate the last integral, in (15.60), Abel set

$$z = \frac{a}{1-y} \quad \text{or} \quad 1 - y = \frac{a}{z} \quad \text{and} \quad dy = \frac{a\,dz}{z^2}$$

so that

$$-\int \frac{dy}{1-y} \ln\left(1 - \frac{a}{1-y}\right) = -\int \frac{dz}{z} \ln(1-z) = \text{Li}_2(z) + C = \text{Li}_2\left(\frac{a}{1-y}\right) + C.$$

Thus, he had

$$\text{Li}_2\left(\frac{a}{1-a} \cdot \frac{y}{1-y}\right) = \text{Li}_2\left(\frac{y}{1-a}\right) + \text{Li}_2\left(\frac{a}{1-y}\right)$$

$$- \text{Li}_2(y) - \ln(1-a)\,\ln(1-y) + C.$$

To find $C$, Abel let $y = 0$ to get $C = -\text{Li}_2(a)$. After $a$ was replaced by the variable $x$, this proved the formula.

## 15.10   Schellbach: Products to Series

In an 1854 paper[31] published in *Crelle's Journal*, Karl Schellbach gave a method for converting the infinite products for sine and cosine into series of partial fractions. His method was based on the formal identity

$$\frac{\alpha}{\alpha - a} \cdot \frac{\beta}{\beta - b} \cdot \frac{\gamma}{\gamma - c} \cdot \frac{\delta}{\delta - d} \cdots$$

$$= 1 + \frac{a}{\alpha - a} + \frac{\alpha b}{(\alpha - a)(\beta - b)} + \frac{\alpha\beta c}{(\alpha - a)(\beta - b)(\gamma - c)} + \cdots. \quad (15.61)$$

To prove (15.61), Schellbach first observed that

$$\frac{1}{1-a} = 1 + \frac{a}{1-a}, \quad \frac{1}{(1-a)(1-b)} = 1 + \frac{a}{1-a} + \frac{b}{(1-a)(1-b)},$$

$$\frac{1}{(1-a)(1-b)(1-c)} = 1 + \frac{a}{1-a} + \frac{b}{(1-a)(1-b)} + \frac{c}{(1-a)(1-b)(1-c)},$$

and so on; to obtain (15.61), replace $a$ by $\frac{a}{\alpha}$, $b$ by $\frac{b}{\beta}$, etc. Schellbach went on to note that the infinite product for $\sin x$ in (15.20) yielded

$$\sin \frac{\pi x}{2} = \frac{\pi x}{2} \prod_{n=1}^{\infty} \left(1 - \frac{x^2}{(2n)^2}\right), \quad (15.62)$$

[31]  Schellbach (1854).

and taking $x = 1$ produced Wallis's formula

$$\frac{2}{\pi} = \prod_{n=1}^{\infty} \left(1 - \frac{1}{(2n)^2}\right). \tag{15.63}$$

He next multiplied the reciprocal of (15.62) by (15.63) to arrive at

$$x \csc \frac{\pi x}{2} = \frac{3}{3 - (x^2 - 1)} \cdot \frac{15}{15 - (x^2 - 1)} \cdot \frac{35}{35 - (x^2 - 1)} \cdots. \tag{15.64}$$

Then, applying the partial fractions formula (15.61) to the infinite product on the right-hand side of (15.64), obtain

$$x \csc \frac{\pi x}{2} = 1 + \frac{x^2 - 1}{2^2 - x^2} + \frac{(2^2 - 1^2)(x^2 - 1^2)}{(2^2 - x^2)(4^2 - x^2)} + \frac{(2^2 - 1^2)(4^2 - 1^2)(x^2 - 1^2)}{(2^2 - x^2)(4^2 - x^2)(6^2 - x^2)} + \cdots$$

$$= 1 + \frac{x^2 - 1}{4 - x^2} \left(1 + \sum_{n-1}^{\infty} \frac{(2^2 - 1^2)(4^2 - 1^2) \cdots ((2n)^2 - 1^2)}{(4^2 - x^2)(6^2 - x^2) \cdots ((2n+2)^2 - x^2)}\right). \tag{15.65}$$

Schellbach let $x \to 0$ in (15.65) to derive the interesting formula

$$\frac{2}{\pi} = 1 - \left(\frac{1}{2}\right)^2 - \frac{1}{3}\left(\frac{1 \cdot 3}{2 \cdot 4}\right)^2 - \frac{1}{5}\left(\frac{1 \cdot 3 \cdot 5}{2 \cdot 4 \cdot 6}\right)^2 - \frac{1}{7}\left(\frac{1 \cdot 3 \cdot 5 \cdot 7}{2 \cdot 4 \cdot 6 \cdot 8}\right)^2 - \cdots. \tag{15.66}$$

In fact, (15.66) is a particular case of Gauss's formula (17.65) for the sum of a hypergeometric series, but Schellbach's derivation was elementary.

In a manner similar to that used to derive (15.65), Schellbach obtained

$$\sec \frac{\pi x}{2} = 1 + \frac{x^2}{1 - x^2} + \frac{x^2}{1 - x^2} \sum_{n=1}^{\infty} \frac{\left(1 \cdot 3 \cdots (2n-1)\right)^2}{(3^2 - x^2) \cdots \left((2n+1)^2 - x^2\right)}. \tag{15.67}$$

Dividing equation (15.67) by $x^2$ and letting $x \to 0$, he obtained the famous formula of Euler:

$$\frac{\pi^2}{8} = 1 + \frac{1}{3^2} + \frac{1}{5^2} + \frac{1}{7^2} + \cdots, \tag{15.68}$$

a formula for which we give several alternative proofs in Chapter 16.

Schellbach discussed another method for converting products into series. This method is due to Euler, though Schellbach does not mention this fact; it is not clear whether or not he was aware of Euler's work. Euler applied this method to convert the product $\prod_{n=1}^{\infty} (1 - x^n)$ into a series. In Chapter 25, we discuss this method and example; the approach depends, as Euler himself explained, on the algebraic identity

$$(1 - \alpha)(1 - \beta)(1 - \gamma)(1 - \delta) \cdots$$
$$= 1 - \alpha - \beta(1 - \alpha) - \gamma(1 - \alpha)(1 - \beta) - \delta(1 - \alpha)(1 - \beta)(1 - \gamma) - \cdots .$$

$$(15.69)$$

Schellbach observed that since

$$\cos\frac{\pi x}{2} = (1 - x^2)\left(1 - \frac{x^2}{3^2}\right)\left(1 - \frac{x^2}{5^2}\right) \cdots ,$$

it followed from (15.69) that

$$\cos\frac{\pi x}{2} = 1 - x^2 - (1 - x^2)\frac{x^2}{3^2} - (1 - x^2)\left(1 - \frac{x^2}{3^2}\right)\frac{x^2}{5^2} - \cdots .$$

$$(15.70)$$

Similarly, since by (15.64)

$$\sin\frac{\pi x}{2} = x\left(1 - \frac{x^2 - 1}{3}\right)\left(1 - \frac{x^2 - 1}{15}\right)\left(1 - \frac{x^2 - 1}{35}\right) \cdots ,$$

he had

$$\sin\frac{\pi x}{2} = x + \frac{x(1 - x^2)}{1 \cdot 3} + \frac{x(1 - x^2)(2^2 - x^2)}{(1 \cdot 3)^2 \cdot 5} + \frac{x(1 - x^2)(2^2 - x^2)(4^2 - x^2)}{(1 \cdot 3 \cdot 5)^2 \cdot 7}$$
$$+ \frac{x(1 - x^2)(2^2 - x^2)(4^2 - x^2)(6^2 - x^2)}{(1 \cdot 3 \cdot 5 \cdot 7)^2 \cdot 9} + \cdots . \quad (15.71)$$

In the last section of his paper, Schellbach gave yet another method for obtaining series for trigonometric functions, a tentative method that he nevertheless endorsed on the grounds that it gave good approximations for specific values of trigonometric functions. However, he found two very nice formulas by this method, involving $\cos\frac{\pi x}{3}$ and $\sin\frac{\pi x}{3}$.

Regard the function $\cos\frac{\pi x}{3}$: At the points $x = 0, \pm 1, \pm 2, \pm 3, \pm 4, \pm 5, \ldots$, it takes the values $1, \frac{1}{2}, -\frac{1}{2}, -1, -\frac{1}{2}, \frac{1}{2}, \ldots$, respectively; thus, the values repeat themselves. So Schellbach considered the series

$$\cos\frac{\pi x}{3} = A_1 + A_2 x^2 + A_3 x^2(x^2 - 1^2) + A_4 x^2(x^2 - 1^2)(x^2 - 2^2)$$
$$+ A_5 x^2(x^2 - 1^2)(x^2 - 2^2)(x^2 - 3^2) + \cdots$$

and set $x = 0, 1, 2, 3, 4, \ldots$ successively to find that

$$A_1 = 1, \quad A_2 = -\frac{1}{2!}, \quad A_3 = \frac{1}{4!}, \quad A_4 = -\frac{1}{6!}, \quad A_5 = \frac{1}{8!}, \quad \ldots .$$

He accordingly obtained the formula

$$\cos\frac{\pi x}{3} = 1 - \frac{x^2}{2!} + \frac{x^2(x^2 - 1^2)}{4!} - \frac{x^2(x^2 - 1^2)(x^2 - 2^2)}{6!}$$
$$+ \frac{x^2(x^2 - 1^2)(x^2 - 2^2)(x^2 - 3^2)}{8!} - \cdots \qquad (15.72)$$

and following a similar argument he had

$$\sin\frac{\pi x}{3} = \frac{\sqrt{3}}{2}\left( x - \frac{x(x^2 - 1^2)}{3!} + \frac{x(x^2 - 1^2)(x^2 - 2^2)}{5!} \right.$$
$$\left. - \frac{x(x^2 - 1^2)(x^2 - 2^2)(x^2 - 3^2)}{7!} + \cdots \right). \qquad (15.73)$$

Now to prove (15.72) by Schellbach's method amounts to showing that, when $n$ is an integer,

$$\cos\frac{n\pi}{3} = 1 - \frac{n^2}{2!} + \frac{n^2(n^2 - 1^2)}{4!} - \cdots + (-1)^n \frac{n^2(n^2 - 1^2)\cdots(n^2 - (n-1)^2)}{(2n)!};$$
$$\qquad (15.74)$$

similarly, proving (15.73) requires us to prove that, when $n$ is an integer,

$$\frac{2}{\sqrt{3}}\sin\frac{n\pi}{3} = n - \frac{n(n^2 - 1^2)}{3!} + \frac{n(n^2 - 1^2)(n^2 - 2^2)}{5!} -$$
$$\cdots + (-1)^{n-1}\frac{n(n^2 - 1^2)\cdots(n^2 - (n-1)^2)}{(2n - 1)!}. \qquad (15.75)$$

Formulas (15.74) and (15.75) can, of course, be verified, though Schellbach did not give the proofs in his paper.

Note that Schellbach's formula (15.72) is a particular case of (8.17), a formula most likely known to Newton. To see this, replace $n$ by $2x$ and $\theta$ by $\frac{\pi}{6}$. Also observe that (15.73) may be obtained by taking the derivative of (8.17) with respect to $\theta$ and then replacing $n$ by $2x$ and $\theta$ by $\frac{\pi}{6}$. Note also that if we take $\theta = \frac{\pi}{2}$ in (8.16) and (8.17), we arrive at the series

$$\sin\frac{\pi x}{2} = x - \frac{x(x^2 - 1^2)}{3!} + \frac{x(x^2 - 1^2)(x^2 - 3^2)}{5!}$$
$$- \frac{x(x^2 - 1^2)(x^2 - 3^2)(x^2 - 5^2)}{7!} + \cdots \qquad (15.76)$$

and

$$\cos\frac{\pi x}{3} = 1 - \frac{x^2}{2!} + \frac{x^2(x^2 - 2^2)}{4!} - \frac{x^2(x^2 - 2^2)(x^2 - 4^2)}{6!} + \cdots. \qquad (15.77)$$

Glaisher employed formulas (15.72), (15.73), (15.76), and (15.77) to obtain nice formulas for $\pi^3$ and $\pi^4$:[32]

First, by equating coefficients of $x^3$ in (15.76), he had

$$\frac{\pi^3}{48} = \sum_{n=1}^{\infty} \frac{1 \cdot 3 \cdots (2n-1)}{2 \cdot 4 \cdots 2n} \frac{1}{2n+1} \left(1 + \frac{1}{3^2} + \cdots + \frac{1}{(2n-1)^2}\right);$$

equating coefficients of $x^3$ in (15.73) gave him

$$\frac{\pi^3}{3^4\sqrt{3}} = \sum_{n=1}^{\infty} \frac{n!}{(n+1)\cdots(2n+1)} \left(1 + \frac{1}{2^2} + \cdots + \frac{1}{n^2}\right);$$

next, by equating coefficients of $x^4$ in (15.77), he got

$$\frac{\pi^4}{48} = \sum_{n=1}^{\infty} \frac{2 \cdot 4 \cdots (2n)}{3 \cdot 5 \cdots (2n+1)} \frac{1}{n+1} \left(1 + \frac{1}{2^2} + \cdots + \frac{1}{n^2}\right);$$

finally, he equated coefficients of $x^4$ in (15.72) to obtain

$$\frac{\pi^4}{1944} = \sum_{n=1}^{\infty} \frac{n!}{(n+1)\cdots(2n+2)} \left(1 + \frac{1}{2^2} + \cdots + \frac{1}{n^2}\right).$$

## 15.11 Exercises

(1) Show that the series for $\cos x$ can be obtained by a repeated integration of the equation $\cos x = 1 - \int_0^x \int_0^t \cos u \, du \, dt$. This method of deriving the series for cosine is due to Leibniz. See Newton (1959–1960) vol. 2, p. 74. See also Chapter 1, where Madhava's similar, but earlier, method is discussed.

(2) Let $n = 2m + 1$. Show that

$$\sin x - \sin\left(x + \frac{\pi}{n}\right) - \sin\left(x - \frac{\pi}{n}\right) + \sin\left(x - \frac{2\pi}{n}\right)$$
$$+ \sin\left(x + \frac{2\pi}{n}\right) - \sin\left(x - \frac{3\pi}{n}\right) - \sin\left(x + \frac{3\pi}{n}\right) + \cdots$$
$$\pm \sin\left(x + \frac{m\pi}{n}\right) \pm \sin\left(x - \frac{m\pi}{n}\right) = 0,$$

where the plus sign is used when $m$ is even and the minus sign otherwise. See Euler (1988) p. 208.

(3) With $n$ as in the previous problem, prove Euler's formula

$$n \csc nx = \csc x - \csc\left(x + \frac{\pi}{n}\right) - \csc\left(x - \frac{\pi}{n}\right) + \csc\left(x + \frac{2\pi}{n}\right)$$

$$+ \csc\left(x - \frac{2\pi}{n}\right) - \cdots \pm \csc\left(x + \frac{m\pi}{n}\right) \pm \csc\left(x - \frac{m\pi}{n}\right).$$

See Euler (1988) p. 209.

(4) Prove the following formulas:

$$\cos nx = 2^{n-1} \cos\left(x + \frac{n-1}{n}\pi\right) \cos\left(x - \frac{n-1}{n}\pi\right)$$

$$\cdot \cos\left(x + \frac{n-3}{n}\pi\right) \cos\left(x - \frac{n-3}{n}\pi\right) \cdots,$$

where there are $n$ factors;

$$n \cot nx = \cot x + \cot\left(x + \frac{\pi}{n}\right) + \cot\left(x + \frac{2\pi}{n}\right)$$

$$+ \cdots + \cot\left(x + \frac{n-1}{n}\pi\right).$$

Also show that the sum of the squares of the cotangents is $\frac{n^2}{(\sin x)^2} - n$. See Euler (1988) pp. 214 and 218.

(5) Set Clausen's function as $Cl_2(\theta) = \sum_{k=1}^{\infty} \frac{\sin(k\theta)}{k^2}$. First show that

$$Cl_2(\theta) == -\int_0^\theta \log |2\sin\frac{t}{2}| \, dt.$$

Then show that Landen's formula (16.91) can be correctly and comprehensively stated by Kummer's 1840 result:

$$Li_2(re^{i\theta}) = Li_2(r,\theta) + \frac{i}{2}[2\omega \log r + Cl_2(2\omega) + Cl_2(2\theta) - Cl_2(2\omega + 2\theta)],$$

where $\tan\omega = \frac{r\sin\theta}{1 - r\cos\theta}$, and

$$Li_2(r,\theta) = -\frac{1}{2}\int_0^r \frac{\log(1 - 2r\cos\theta + r^2)}{r} \, dr.$$

See Kummer (1840) pp. 74–90 and Lewin (1981) pp. 120–121.

(6) Prove Kummer's formula

$$Li_2\left(\frac{x(1-y)^2}{y(1-x)^2}\right) = Li_2\left(\frac{x - xy}{x - 1}\right) + Li_2\left(\frac{1-y}{xy - y}\right) + Li_2\left(\frac{x - xy}{y - xy}\right)$$

$$+ Li_2\left(\frac{1-y}{1-x}\right) + \frac{1}{2}(\ln y)^2.$$

See Clausen (1832) and Kummer (1975) vol. 2, p. 238.

(7) Prove that Spence's formula (15.55) and Abel's formula (15.58) are equivalent.

(8) Show that

$$\frac{\sin x}{\sin y} = \frac{x}{y}\left(\frac{\pi - x}{\pi - y}\right)\left(\frac{\pi + x}{\pi + y}\right)\left(\frac{2\pi - x}{2\pi - y}\right)\left(\frac{2\pi + x}{2\pi + y}\right)\left(\frac{3\pi - x}{3\pi - y}\right)\left(\frac{3\pi + x}{3\pi + y}\right)\cdots.$$

Derive the product for $\cos x$ by replacing $x$ by $\frac{\pi}{2} - x$, $y$ by $\frac{\pi}{2}$, and applying Wallis's formula. See Cauchy's *Analyse algébrique*, note IV.

(9) Prove that

$$\text{Li}_2(x) + \text{Li}_2(y) - \text{Li}_2(xy) = \text{Li}_2\left(\frac{x(1-y)}{1-xy}\right) + \text{Li}_2\left(\frac{y(1-x)}{1-xy}\right)$$
$$+ \ln\left(\frac{1-x}{1-xy}\right)\ln\left(\frac{1-y}{1-xy}\right).$$

See L. J. Rogers (1907).

(10) Prove the two inequalities for $0 < x \leq \frac{\pi}{2}$ used in Cauchy's proof of (15.35):

$$\sin x < x \quad \text{and} \quad \frac{x}{\sin x} < 2.$$

See note IX of Cauchy's *Analyse algébrique*.

(11) Prove that for even $m$

$$\cos mx = \prod_{k=1}^{\frac{m}{2}}\left(1 - \frac{\sin^2 x}{\sin^2 \frac{(2k-1)\pi}{2m}}\right),$$

$$\sin mx = m \sin x \cos x \prod_{k=1}^{\frac{m-2}{2}}\left(1 - \frac{\sin^2 x}{\sin^2 \frac{2k\pi}{2m}}\right).$$

State and prove a similar formula for odd $m$. See Cauchy, *Analyse algébrique*, note IX.

(12) Use the formulas in Exercise 11 and (15.35) to show that for $m$ even

$$\cos mx = 2^{\frac{m}{2}-1} \prod_{k=1}^{\frac{m}{2}}\left(\cos 2x - \cos\frac{(2k-1)\pi}{m}\right),$$

$$\sin mx = 2^{\frac{m}{2}-1} \sin 2x \prod_{k=2}^{\frac{m}{2}}\left(\cos 2x - \cos\frac{(2k-1)\pi}{m}\right).$$

State and prove similar results for $m$ odd. See Cauchy, *Analyse algébrique*, note IX.

(13) Suppose $\phi_0(x) = \phi_0\left(\frac{1}{x}\right)$ and

$$\phi_n(x) = \int \frac{dx}{x} \phi_{n-1}(x), \quad n = 1, 2, 3, \ldots.$$

Prove that

(a) $\phi_{2n}(x) - \phi_{2n}\left(\frac{1}{x}\right) = 2 \sum_{k=0}^{n-1} \phi_{2n-2k-1}(1)(\ln x)^{2k+1}$,

(b) $\phi_{2n+1}(x) - \phi_{2n+1}\left(\frac{1}{x}\right) = 2 \sum_{k=0}^{n} \phi_{2n-2k+1}(1)(\ln x)^{2k}$,

(c) $\phi_{2n+1}(1) - \phi_{2n+1}(-1) = \sum_{k=1}^{n}(-1)^{k-1}\frac{\pi^{2k}}{(2k)!}\phi_{2n-2k+1}(1)$.

Find $\phi_1(1)$, when $\phi_0(x) = \frac{x^p + x^{-p}}{x^m + a + x^{-m}}$. See Spence (1819) pp. 139–143.

## 15.12  Notes on the Literature

Spence (1819), edited by John Herschel, contains Spence's 1809 book on logarithmic transcendents. Lewin (1981) gives a modern treatment of the dilogarithm and polylogarithm as functions of complex variables.

# 16

# Zeta Values

## 16.1  Preliminary Remarks

One of the outstanding and most difficult mathematical problems of the early eighteenth century was the summation of the series $\sum_{n=0}^{\infty} \frac{1}{n^2}$. In his 1650 book *Novae Quadraturae Arithmeticae Seu De Additione Fractionum*, Pietro Mengoli (1626–1686) considered the sum of the reciprocals of figurate numbers: natural numbers, triangular numbers, square numbers, and so on.[1] For the natural numbers, Mengoli showed that the sum of their reciprocals diverged, or that the harmonic series was divergent. For triangular numbers, he showed how the reciprocal of each triangular number could be written as a difference of two fractions, thus summing the series. In the next step, square numbers, Mengoli posed the problem of summing their reciprocals, but could not solve it. He expressed surprise that the series of triangular reciprocals could be summed more easily than the series of square reciprocals, saying that a "richer intellect" would be required to solve this problem.

Leibniz, Jakob and Johann Bernoulli, and James Stirling all later attempted to sum this series.[2] In fact, this question became known as the Basel problem because it frustrated the very best efforts of Jakob Bernoulli of Basel, who wrote that he would be greatly indebted to anyone who would send him a solution. Unfortunately, the solution was not found until thirty years after Bernoulli's death in 1705; in a famous paper of 1735, Euler became the first to sum this series.[3] With characteristic brilliance, Euler made use of the formulas relating the roots to the coefficients of algebraic equations and he boldly applied them to equations of infinite degree. When Euler communicated his results without proof to Stirling in 1736,[4] the latter wrote in response that Euler

---

[1] Mengoli (1650).

[2] Leibniz (1971) vol. 2, part 1, pp. 118–122. Bernoulli (1744) vol. 1, pp. 375–402, 517–542. Bernoulli (1742) vol. 4, pp. 19–25. Stirling (1730) pp. 28–29 or for an English translation of this letter, see Stirling and Tweddle (2003) p. 46. Johann Bernoulli (1742) presented this material in 1742 in a somewhat misleading manner, without mention of Euler, from whom he learned the key ideas. Indeed, for this reason, Spence had the impression that Johann Bernoulli was the original discoverer of the formula for the sums of even powers of the reciprocals of integers.

[3] Eu. I-14 pp. 73–86. E 41.

[4] For an English translation, see Tweddle (1988) pp. 141–144, especially p. 143.

must have tapped a new source, since the old methods were insufficient. Euler had indeed found a new source, beginning with the equation

$$0 = 1 - \frac{1}{y}\sin x = 1 - \frac{1}{y}\left(x - \frac{x^3}{3!} + \frac{x^5}{5!} - \cdots\right) \tag{16.1}$$

where $y$ was a constant between $-1$ and $1$. He argued that $x = 2n\pi + A$ and $x = (2n+1)\pi - A$ gave a complete list of the roots of (16.1), provided that $\sin A = y$ and that $n$ assumed all possible integer values. Hence, he factored the right-hand side of (16.1) as

$$\left(1 - \frac{x}{A_1}\right)\left(1 - \frac{x}{A_2}\right)\left(1 - \frac{x}{A_3}\right)\cdots, \tag{16.2}$$

where $A_1, A_2, A_3, \ldots$ comprised all the roots. He equated the coefficients of $x$ to obtain the sum of the reciprocals of the roots:

$$\frac{1}{A_1} + \frac{1}{A_2} + \frac{1}{A_3} + \cdots = \frac{1}{y}. \tag{16.3}$$

In a similar manner, Euler argued, the sum of all the products of the reciprocals of the roots, taken two at a time, was equal to the coefficient of $x^2$ and was thus zero; the sum of the products of the reciprocals taken three at a time was equal to the negative of the coefficient of $x^3$ and was thus $-\frac{1}{y\cdot 3!}$; and so on.

To obtain the sums of the squares, cubes, fourth powers, etc., of the reciprocals of the roots, he applied the Girard–Newton formulas. For a discussion of these formulas, see Section 13.3, particularly the material contained between equations (13.14) and (13.17). Thus, when $y = 1$ and $A = \frac{\pi}{2}$, Euler had the Madhava–Leibniz formula

$$\frac{4}{\pi} - \frac{4}{3\pi} + \frac{4}{5\pi} - \cdots = 1 \tag{16.4}$$

or

$$1 - \frac{1}{3} + \frac{1}{5} - \cdots = \frac{\pi}{4}. \tag{16.5}$$

Note here that the roots of order 2 of the equation $\sin x = 1$ are $x = \frac{\pi}{2}, -\frac{3\pi}{2}, \frac{5\pi}{2}, \cdots$. Thus, each root occurs twice: $A_1 = \frac{\pi}{2}, A_2 = \frac{\pi}{2}, A_3 = -\frac{3\pi}{2}, A_4 = -\frac{3\pi}{2}$, and so on. This explains why there is a factor 4 in the numerators of the series in (16.4).

From the Girard–Newton formulas, Euler obtained

$$1 + \frac{1}{3^2} + \frac{1}{5^2} + \cdots = \frac{\pi^2}{8}, \tag{16.6}$$

$$1 - \frac{1}{3^3} + \frac{1}{5^3} - \cdots = \frac{\pi^3}{32}, \tag{16.7}$$

$$1 + \frac{1}{3^4} + \frac{1}{5^4} + \cdots = \frac{\pi^4}{96}. \tag{16.8}$$

······ .

He then observed, as Jakob Bernoulli had also seen,[5] that

$$1 + \frac{1}{2^2} + \frac{1}{3^2} + \frac{1}{4^2} + \cdots = 1 + \frac{1}{3^2} + \frac{1}{5^2} + \cdots + \frac{1}{4}\left(1 + \frac{1}{2^2} + \frac{1}{3^2} + \cdots\right)$$

and hence by (16.6)

$$\sum_{n=1}^{\infty} \frac{1}{n^2} = \frac{\pi^2}{6}. \tag{16.9}$$

Similarly, Euler found

$$\sum_{n=1}^{\infty} \frac{1}{n^4} = \frac{\pi^4}{90}, \tag{16.10}$$

and so on. To derive other series, he took $y$ as constants other than one. For example, for $y = \frac{1}{\sqrt{2}}$ and $y = \frac{\sqrt{3}}{2}$ he had, respectively,

$$1 + \frac{1}{3} - \frac{1}{5} - \frac{1}{7} + \frac{1}{9} + \frac{1}{11} - \frac{1}{13} - \frac{1}{15} + \cdots = \frac{\pi}{2\sqrt{2}} \tag{16.11}$$

and

$$1 + \frac{1}{2} - \frac{1}{4} - \frac{1}{5} + \frac{1}{7} + \frac{1}{8} - \frac{1}{10} - \frac{1}{11} + \cdots = \frac{2\pi}{3\sqrt{3}}. \tag{16.12}$$

Observe that the roots of the equation $\sin x = \frac{1}{\sqrt{2}}$ are $\frac{\pi}{4}, \frac{3\pi}{4}, -\frac{5\pi}{4}, -\frac{7\pi}{4}, \cdots$. Similarly, the roots $\frac{\pi}{3}, \frac{2\pi}{3}, -\frac{4\pi}{3}, -\frac{5\pi}{3}, \cdots$ of the equation $\sin x = \frac{\sqrt{3}}{2}$ produce (16.12).

Recall that Newton had earlier proved (16.11) by the integration of the rational function $\frac{1+x^2}{1+x^4}$ over $[0,1]$. See Section 12.1 and equation (12.3). Indeed, Euler credited Newton with this result. Of course, Madhava and Leibniz found (16.5) by integrating the rational function $\frac{1}{1+x^2}$. In the same manner, (16.12) can by obtained by the integration of $\frac{1+x}{1+x^3}$. Thus, integration of rational functions is a powerful method for evaluating many series that sum to multiples of $\pi$ or to logarithms of numbers. However, this method is not as effective for series such as (16.6) through (16.10). Here Euler provided a new insight, so that one could efficiently sum up many series.

When Euler communicated his method to some of his mathematical correspondents, there were objections to his procedure. How did he know, for example, that $n\pi$ were the only roots of $\sin x = 0$? It was possible that some roots were complex! How could he employ the Girard–Newton formulas, applicable to polynomials, for equations of infinite degree? In addition, there were also convergence questions concerning some of Euler's series. But a year after writing his famous paper summing

[5] Bernoulli (1744) vol. 1, pp. 529–533. Stirling also discovered this independently. See his reply to Euler in Tweddle (1988) pp. 144–145.

the series $\sum_{n=0}^{\infty} \frac{1}{n^2}$, Euler proved his product formula for the sine function,[6] justifying his contention that with $n$ an integer, $n\pi$ gave all the roots of the sine function, of which there were thus no complex roots.

Euler was well aware of these inevitable objections, but he believed in the correctness of his formulas. His methods had succeeded in rederiving known formulas and, moreover, numerical methods such as the Euler–Maclaurin formula showed him that his results were correct to many decimal places. So Euler made great efforts to resolve any doubts about his method, as well as to prove his formulas using alternative procedures. For example, by proving the product formulas for $\sin x$ and $\cos x$, (15.20) and (15.21), he showed that these functions had only the well-known zeros and no others. And in 1737 Euler gave an ingenious method for deriving (16.6) and (16.9) by computing $\int_0^1 \frac{\arcsin x}{\sqrt{1-x^2}}\, dx$ in two different ways. Unfortunately, this method did not extend to formulas such as (16.7), (16.8), and (16.10) where the powers of the denominators were greater than two. Meanwhile, in 1738, Niklaus I Bernoulli gave an extremely clever proof of (16.6) by squaring the series (16.5).[7] Euler soon simplified the argument and in a letter of July 30, 1738, communicated the simplification to Johann Bernoulli.[8] A few years later, Euler generalized the essential idea of the simplification and communicated it to his friend Goldbach as a theorem.[9] In his next letter of October 27, 1742, Euler explained with details how he had used that theorem to obtain (16.6) from (16.5). Goldbach responded to the theorem in two letters of February 1743, suggesting the problem of the summation of some new series, leading Euler to the study of series now known as double-zeta values.[10]

It also turned out that series methods could be used to evaluate $\sum_{n=1}^{\infty} \frac{1}{n^{2k}}$ for $k > 1$. In particular, one could call upon the series of the polylogarithmic function defined by $\sum_{n=1}^{\infty} \frac{x^n}{n^{2k}}$, for complex $x$ with $|x| \leq 1$. John Landen was the first to use a polylogarithmic function with complex values to sum this series, though his work shows that he did not fully grasp the difficulties connected with multivalued functions. Euler had a clearer idea of how to deal with such difficulties, though he did not work out the details of this method for the summation of $\sum_{n=1}^{\infty} \frac{1}{n^{2k}}$, a derivation first accomplished by William Spence.[11]

In a paper of 1743 published in Berlin, Euler used the partial fractions expansions of $\cot \pi x$ and $\csc \pi x$ and their derivatives to find the sum of $\sum \frac{1}{n^{2k}}$ and related series.[12] This is essentially the method often used in modern textbooks, although Euler, Daniel Bernoulli, and Landen found other significant proofs. In 1752, Euler discovered the "Fourier" expansion[13]

---

[6] Eu. I-14 pp. 3–6. E 61 § 7–10.
[7] Bernoulli (1738).
[8] Eu. IV A-2 pp. 230–247, especially p. 243.
[9] Fuss (1968) vol. I, pp. 144–153, especially p. 153.
[10] ibid. pp. 193–208.
[11] Spence (1809) § 31.
[12] Eu. I-17 pp. 1–34. E 59.
[13] Eu. I-14 pp. 542–584. E 246.

$$\sin\phi - \frac{1}{2}\sin 2\phi + \frac{1}{3}\sin 3\phi - \frac{1}{4}\sin 4\phi + \frac{1}{5}\sin 5\phi - \text{etc.} = \frac{\phi}{2}, \qquad (16.13)$$

yielding (16.5) when $\phi = \frac{\pi}{2}$. Integration of (16.13) gave him

$$\cos\phi - \frac{1}{2^2}\cos 2\phi + \frac{1}{3^2}\cos 3\phi - \frac{1}{4^2}\cos 4\phi + \text{etc.} = C - \frac{\phi^2}{4}. \qquad (16.14)$$

As we discuss in Section 20.6, in 1773 Daniel Bernoulli showed that the values of $\sum \left(\frac{1}{n}\right)^{2k}$ and $\sum \frac{(-1)^n}{(2n+1)^{2k+1}}$ could be obtained from (16.14).[14] Euler further improved on these results, and their joint work produced the Fourier expansions for Bernoulli polynomials, related to the polylogarithmic function

$$\sum_{n=1}^{\infty} \frac{e^{in\theta}}{n^k}.$$

Note that in 1758 Landen evaluated the sums $\sum \frac{1}{n^{2k}}$ and $\sum \frac{(-1)^n}{(2n+1)^{2k+1}}$ by means of the polylogarithmic functions.[15] To do this, he employed complex numbers through the use of $\log(-1)$. In the 1770s, Euler made further strides in this area, though his work had a few errors, especially where complex numbers were used. Thus, by the 1770s, Euler had worked out many ways of evaluating $\sum \frac{1}{n^2}$. In his papers written during that period, he described his methods and even added a new one, employing integration and differentiation under the integral sign.

From the start of his work on what we would now call zeta values, Euler observed that the numbers appearing in the values of $\sum \frac{1}{n^{2k}}$, $k = 1, 2, 3, \dots$ also presented themselves as coefficients in the Euler–Maclaurin summation formula. He was intrigued by this puzzle and wrote in 1738 in his second letter to Stirling that an explanation of this would be a significant advancement.[16] In a 1739 paper, finally published in 1750, Euler began to understand this, as he used differential equations to obtain the Taylor series expansions of $\cot \pi x$ and $\frac{xe^x}{e^x-1}$. The coefficients of the series for $\cot \pi x$ involved the sums $\sum \frac{1}{n^{2k}}$; on the other hand, the coefficients of $\frac{xe^x}{e^x-1}$ involved the Bernoulli numbers and were the coefficients in the Euler–Maclaurin series. Thus, Euler made his own "significant advancement." It was around 1740 that Euler more precisely understood the relation between the trigonometric functions and the exponential function, noting that

$$\cot x = i\left(1 + \frac{2}{e^{2ix}-1}\right), \text{ and } \frac{xe^x}{e^x-1} = x\left(1 + \frac{2}{e^x-1}\right) \qquad (16.15)$$

gave a simpler reason for the appearance of Bernoulli numbers in the summation of $\sum \frac{1}{n^{2k}}$. He gave details in his differential calculus book of 1755.[17]

---

[14] Bernoulli (1982–1996) vol. 2, pp. 119–134, especially pp. 120–121.
[15] Landen (1758).
[16] For a translation of this letter, see Tweddle (1988) pp. 145–154.
[17] Eu. I-10$_2$. E 212.

Euler was also mystified by the fact that, even though he could sum $\sum \frac{1}{n^{2k}}$ or $\sum \frac{(-1)^{n+1}}{n^{2k}}$ in various ways, he was unable to find the sum of the series with odd powers, $\sum \frac{1}{n^{2k+1}}$. Of course, this is a problem that has baffled mathematicians to the present day. To shed some light on this question, Euler considered the divergent series $\sum(-1)^{n+1}n^k$. In a 1739 paper,[18] he noted that

$$1 - 1 + 1 - 1 + 1 - \text{etc.} = \frac{1}{2}, \tag{16.16}$$

$$1 - 2^{2k} + 3^{2k} - 4^{2k} + \text{etc.} = 0, \tag{16.17}$$

$$1 - 2^{2k-1} + 3^{2k-1} - 4^{2k-1} + \text{etc.}$$
$$= (-1)^{k-1}2 \cdot \frac{1 \cdot 2 \cdots (2k-1)}{\pi^{2k}}\left(1 + \frac{1}{3^{2k}} + \frac{1}{5^{2k}} + \cdots\right), \tag{16.18}$$

$k = 1, 2, 3, \dots$. Note that the series in parentheses in (16.18) can also be written as

$$\frac{2^{2k} - 1}{2^{2k} - 2}\left(1 - \frac{1}{2^{2k}} + \frac{1}{3^{2k}} - \frac{1}{4^{2k}} + \text{etc.}\right). \tag{16.19}$$

Thus, the series with even/odd powers in the denominator were related to the series with odd/even powers in the numerator; however, the series with even powers in the numerator summed to zero (except for $k = 0$), and hence gave no information about the odd series $\sum \frac{(-1)^{n+1}}{n^{2k+1}}$, $k = 1, 2, 3, \dots$. In the exceptional case, for $k = 0$, one gets $\sum \frac{(-1)^{n+1}}{n^{2k+1}} = \sum \frac{(-1)^{n-1}}{n} = \ln 2$. We observe that Euler was well aware that the series on the left-hand side were divergent, but he plunged right in anyway, since this work was yielding him insight into a very challenging problem. Indeed, he was justified in his audacity, since this approach led him to the functional equation for the zeta function.

It appears that at some point in the 1740s, Euler started thinking of the series $\sum \frac{1}{n^{2k}}$ and other related series as particular values of the functions defined by $\zeta(s) = \sum \frac{1}{n^s}$ etc. Here note, however, that the label $\zeta(s)$ was later given by Riemann. In a paper presented to the Berlin Academy in 1749 but published in 1768, Euler drew a relation between $\zeta(s)$ and $\zeta(1 - s)$, writing the equation,[19] with $m$ a real number:

$$\frac{1 - 2^{m-1} + 3^{m-1} - 4^{m-1} + 5^{m-1} - 6^{m-1} + \text{etc.}}{1 - 2^{-m} + 3^{-m} - 4^{-m} + 5^{-m} - 6^{-m} + \text{etc.}}$$
$$= -\frac{1 \cdot 2 \cdot 3 \cdots (m-1)(2^m - 1)}{(2^{m-1} - 1)\pi^m}\cos\frac{m\pi}{2}. \tag{16.20}$$

In this way, Euler found a generalization of (16.18).

[18] E 130 § 30, Eu. I-14, pp. 407–462. See also E 212 § 185–187, Eu. 1-10$_2$, pp. 384–386.
[19] E 352, § 13. Eu. I-15 pp. 70–90, especially p. 83.

In 1826, Abel wrote his friend Holmboe that equation (16.17) was a laughable equation to write.[20] Abel's early training had been in the formal mathematical tradition of which Euler was considered the model. After he studied Cauchy's lectures on analysis, Abel changed his view of mathematics; he believed it illegitimate to use divergent series at all and therefore wished to abolish formulas such as (16.17), (16.18), and (16.20). However, Euler had a very clear idea that the definition of the divergent series in these formulas amounted to a limit:

$$1 - 2^n + 3^n - 4^n + \cdots = \lim_{x \to 1^-} (1 - 2^n x + 3^n x^2 - 4^n x^3 + \cdots). \qquad (16.21)$$

He found the values of these limits, for nonnegative integer values of $n$, by repeated multiplication by $x$ followed by differentiation of the geometric series formula

$$1 - x + x^2 - x^3 + x^4 - \cdots = \frac{1}{1+x}. \qquad (16.22)$$

Euler's technique of summing (16.21) became an important summability method in the theory of divergent series developed after 1890. Ironically, it is called the Abel sum.

Euler verified (16.20) for all integer values of $m$ and for two fractional values, $m = \frac{1}{2}$ and $m = \frac{3}{2}$. In general, he meant $1 \cdot 2 \cdot 3 \cdots (m-1)$ to stand for the gamma function $\Gamma(m)$. For positive integer values of $m$, he made illegitimate but successful use of the Euler–Maclaurin summation formula to sum the divergent series

$$x^m - (x+1)^m + (x+2)^m - (x+3)^m + \cdots. \qquad (16.23)$$

By doing this, he could bring into play the Bernoulli numbers that appear in the Euler–Maclaurin summation, using them to evaluate the sums on the left-hand side of (16.21). In fact, he could have done this without using Euler–Maclaurin, had he first applied the change of variables $x = e^{-y}$ in (16.22). Euler believed that a divergent series, especially an alternating series, had a definite value, obtainable by varying methods.

In order to verify (16.20) for $m = 0$, Euler used (16.16) and the series for $\ln 2$. For negative integers, he noted that under the transformation $m \to 1 - m$, both sides were converted into their reciprocals. For the right-hand side, he required the reflection formula for the gamma function $\Gamma(m)\Gamma(1-m) = \frac{\pi}{\sin \pi m}$, and it appears that Euler explicitly stated this formula for the first time in this paper. For $m = \frac{1}{2}$, Euler used the value $\Gamma(\frac{1}{2}) = \sqrt{\pi}$; he had known this value since 1729, but it also followed immediately from his reflection formula. Finally, for $m = \frac{3}{2}$, Euler computed both sides to several decimal places, checking that the results were identical. To do this, he applied the Euler–Maclaurin formula to sum the divergent series $1 - \sqrt{2} + \sqrt{3} - \sqrt{4} + \cdots = 0.380129$, as well as the convergent series

---

[20]  See Stubhaug (2000) pp. 343–344 or Ore (1974) p. 97.

$$1 - \frac{1}{2\sqrt{2}} + \frac{1}{3\sqrt{3}} - \frac{1}{4\sqrt{4}} + \cdots = 0.765158.$$

G. Faber, the editor of vol. 15 of Euler's *Opera Omnia* in which this paper was reprinted, noted that the values could be expressed more exactly as 0.380105 and 0.765147.

At the end of his 1749 paper,[21] Euler mentioned without proof the functional equation for a special *L*-function:

$$\frac{1 - 3^{n-1} + 5^{n-1} - 7^{n-1} + \text{etc.}}{1 - 3^{-n} + 5^{-n} - 7^{-n} + \text{etc.}} = \frac{1 \cdot 2 \cdot 3 \cdots (n-1) 2^n}{\pi^n} \sin \frac{n\pi}{2}. \qquad (16.24)$$

These results on the functional relations for the zeta and *L*-functions apparently went unnoticed. In the 1840s, the functional equation (16.20) was given a complete proof for $0 < m < 1$ by the Swedish mathematician Carl Johan Malmsten (1814–1886),[22] who mentioned seeing (16.20) somewhere in Euler. Oscar Schlömilch independently found a proof and stated it as a problem in a journal in 1849; he gave a solution in 1858.[23] A solution was published by Thomas Clausen (1801–1885) in 1858[24] and another was noted by Eisenstein on the last blank page in his copy of the *Disquisitiones* in the French translation of 1807. André Weil conjectured that Eisenstein discussed this topic with Riemann,[25] providing the impetus for Riemann's well-known paper of 1859 on the zeta function. Indeed, Riemann and Eisenstein had been close friends in Berlin, and Eisenstein's note is dated April 1849, just before Riemann left Berlin for Göttingen.

## 16.2 Euler's First Evaluation of $\sum \frac{1}{n^{2k}}$

Euler's evaluation was based on the factorization given by (16.1) and (16.2):

$$1 - \frac{1}{y}\left(\frac{x}{1} - \frac{x^3}{1 \cdot 2 \cdot 3} + \frac{x^5}{1 \cdot 2 \cdot 3 \cdot 5} - \cdots\right) = \left(1 - \frac{x}{A_1}\right)\left(1 - \frac{x}{A_2}\right)\left(1 - \frac{x}{A_3}\right)\cdots. \qquad (16.25)$$

Note that Euler wrote $A, B, C, D, \ldots$ instead of $A_1, A_2, A_3, A_4, \ldots$. He observed that the coefficients of the powers of $x$ were the elementary symmetric functions of the infinitely many variables $\frac{1}{A_1}, \frac{1}{A_2}, \frac{1}{A_3}, \ldots$. Therefore, by equating coefficients, he had

$$\alpha \equiv \sum \frac{1}{A_i} = \frac{1}{y}, \quad \beta \equiv \sum_{i<j} \frac{1}{A_i A_j} = 0, \quad \gamma \equiv \sum_{i<j<k} \frac{1}{A_i A_j A_k} = \frac{1}{6y}, \text{ etc.,} \qquad (16.26)$$

---

[21] ibid. § 20.
[22] Malmsten (1849).
[23] Schlömilch (1849) and (1858).
[24] Clausen (1858).
[25] Weil (1989a).

where $\alpha, \beta, \gamma, \delta$, etc. denoted the elementary symmetric functions. Euler then applied the Girard–Newton formulas (13.15) connecting the sums of the squares, cubes, fourth powers, etc., of $\frac{1}{A_1}, \frac{1}{A_2}, \frac{1}{A_3}, \ldots$ with the symmetric functions $\alpha, \beta, \gamma$, etc. Thus, he got

$$\sum \frac{1}{A_i^2} = \alpha^2 - 2\beta, \quad \sum \frac{1}{A_i^3} = \alpha^3 - 3\alpha\beta + 3\gamma,$$

$$\sum \frac{1}{A_i^4} = \alpha^4 - 4\alpha^2\beta + 4\alpha\gamma + 2\beta^2 - 4\delta, \text{ etc.}$$

For $y = 1$, the roots $A_1, A_2, A_3, \ldots$ of the equation $\sin x = 1$ were

$$\frac{\pi}{2}, \frac{\pi}{2}, -\frac{3\pi}{2}, -\frac{3\pi}{2}, \frac{5\pi}{2}, \frac{5\pi}{2}, \ldots,$$

since $\sin x - 1 = 0$ had double roots. Thus, Euler obtained equations (16.6), (16.7), and (16.8). Clearly, he could continue the calculations to arbitrarily large powers of the reciprocals of the roots. Euler explicitly wrote the values of $\sum \frac{1}{n^{2k}}$ for $k = 1, 2, \ldots, 6$, and the last of these turned out to be

$$1 + \frac{1}{2^{12}} + \frac{1}{3^{12}} + \frac{1}{4^{12}} + \cdots = \frac{691\pi^{12}}{6825 \cdot 93555}. \tag{16.27}$$

The appearance of the fairly large prime 691 may have alerted Euler to the connection of zeta values with Bernoulli numbers. Recall that this prime had already appeared in the Euler–Maclaurin series he had found only two or three years earlier, and at the time he discovered (16.27), he was still intensely studying the Euler–Maclaurin summation.

## 16.3   Euler: Bernoulli Numbers and $\sum \left(\frac{1}{n}\right)^{2k}$

In a paper presented to the Petersburg Academy in October 1739, but published in 1750,[26] Euler explained the connection between the Bernoulli numbers appearing in the Euler–Maclaurin formula and the sums $\sum \frac{1}{n^{2k}}$. A year earlier, he had found the partial fractions expansion of $\cot x$ and he made use of this in his explanation. Euler started with a generating function for the sums $\sum \frac{1}{n^{2k}}$ and changed the order of summation to obtain a partial fractions expansion that he could recognize as a cotangent function. Denoting the generating function by $S$, he had by a rearrangement of terms

$$S = \left(\sum_{n=1}^{\infty} \frac{1}{n^2}\right) x^2 + \left(\sum_{n=1}^{\infty} \frac{1}{n^4}\right) x^4 + \left(\sum_{n=1}^{\infty} \frac{1}{n^6}\right) x^6 + \cdots$$

$$= x^2 + x^4 + x^6 + \cdots + \frac{x^2}{2^2} + \frac{x^4}{2^4} + \frac{x^6}{2^6} + \cdots + \frac{x^2}{3^2} + \frac{x^4}{3^4} + \frac{x^6}{3^6} + \cdots$$

$$= \frac{x^2}{1 - x^2} + \frac{x^2}{2^2 - x^2} + \frac{x^2}{3^2 - x^2} + \cdots = \frac{1}{2} - \frac{\pi x}{2 \tan \pi x}. \tag{16.28}$$

[26] Eu. I-14 pp. 407–462. E 130 §§ 22–24.

At this point, Euler could have used equation (16.15) to derive the value of $\sum\frac{1}{n^{2k}}$ in terms of Bernoulli numbers. By expressing $\tan\pi x$ in terms of the exponential function, he would have obtained

$$1 - \frac{\pi x}{\tan\pi x} = 1 + \pi i x - \frac{2\pi i x e^{2\pi i x}}{e^{2\pi i x} - 1}. \tag{16.29}$$

He next could have used his generating function for the Bernoulli numbers appearing in the Euler–Maclaurin formula to express the right-hand side as

$$\frac{1}{2}\sum_{n=1}^{\infty}(-1)^{n-1}\frac{B_{2n}}{(2n)!}(2\pi x)^{2n}.$$

In 1740 Euler was just beginning to delve into the connection between the circular and exponential functions; he was not yet ready to make full use of it. For example, in a letter to Johann Bernoulli written during this period,[27] he explained the equality of $2\cos x$ and $e^{ix} + e^{-ix}$ by means of differential equations. Similarly, in his 1750 paper, he proved (16.29) through the use of differential equations. Thus, Euler continued his argument, proceeding to define $A, B, C, \ldots$ by $A = \frac{1}{\pi^2}\sum\frac{1}{n^2}$, $B = \frac{1}{\pi^4}\sum\frac{1}{n^4}$, etc. Since

$$S = \frac{1}{2}\left(1 - \frac{\pi x}{\tan\pi x}\right), \quad u = \arctan\frac{u}{1 - 2S},$$

where $u = \pi x$, a simple calculation showed Euler that $S$ satisfied

$$2u\frac{dS}{du} + 2S = u^2 + 4S^2.$$

He substituted the series $S = Au^2 + Bu^4 + Cu^6 + Du^8 + \cdots$ into the differential equation and determined that

$$A = \frac{1}{6}, \quad B = \frac{2A^2}{5}, \quad C = \frac{4AB}{7},$$

$$D = \frac{AC + 2B^2}{9}, \quad E = \frac{4AD + BC}{11}, \text{ etc.} \tag{16.30}$$

He then observed[28] that the coefficients in the Euler–Maclaurin series were generated by

$$s = \frac{xe^x}{e^x - 1} \equiv 1 + \alpha x + \beta x^2 + \gamma x^3 + \delta x^4 + \text{ etc.}$$

and saw that $s$ satisfied the differential equation

$$x\frac{ds}{dx} - s - sx + s^2 = 0.$$

[27] Eu. IV-2 p. 392.
[28] Eu. I-14 pp. 407–462, especially §§ 27–28. E 130.

By substituting the series for $s$, he obtained relations for the coefficients $\alpha$, $\beta$, $\gamma$, $\delta$, etc. He noted that except for $\alpha = \frac{1}{2}$, the coefficients of the odd powers were zero. To see this more easily, the reader may consider the fact that $-\frac{x}{2} + \frac{xe^x}{e^x-1}$ is an even function. Euler next set $\beta = \frac{A}{2}$, $\delta = -\frac{B}{2^3}$, $\zeta = \frac{C}{2^5}$, $\theta = -\frac{D}{2^7}$, $\chi = \frac{E}{2^9}$, etc., where $\zeta, \theta, \chi$ denoted the coefficients of $x^6, x^8, x^{10}$, respectively. He then showed that these $A, B, C, \ldots$ also satisfied the relations (16.30). Thus, Euler had the formula we now write as

$$\zeta(2n) = 1 + \frac{1}{2^{2n}} + \frac{1}{3^{2n}} + \frac{1}{4^{2n}} + \cdots = (-1)^{n-1} \frac{2^{2n-1}\pi^{2n}}{(2n)!} B_{2n}. \qquad (16.31)$$

## 16.4  Euler's Evaluation of Some L-Series Values by Partial Fractions

Euler's essential idea in the derivation of his famous zeta value formula, proved in the last section, was the partial fractions expansion of $\frac{\pi}{\tan \pi x}$. After his move to Berlin in 1741, Euler followed this up with a paper in the Berlin Academy journal of 1743.[29] There he showed how the same partial fractions could also be applied to the derivation of several $L$-series values. See Chapter 28 for a definition of $L$-series. Euler started with

$$\frac{\pi}{\sin s\pi} = \frac{1}{s} + \frac{1}{1-s} - \frac{1}{1+s} - \frac{1}{2-s} + \frac{1}{2+s} + \frac{1}{3-s} - \frac{1}{3+s} - \text{etc.,} \quad (16.32)$$

$$\frac{\pi}{\tan s\pi} = \frac{1}{s} - \frac{1}{1-s} + \frac{1}{1+s} - \frac{1}{2-s} + \frac{1}{2+s} - \frac{1}{3-s} + \frac{1}{3+s} - \text{etc.,} \quad (16.33)$$

where he took $s$ to be a rational number $s = \frac{p}{q}$. He assigned specific integer values to $p$ and $q$ and evaluated several series, including (16.5), (16.11), and (16.12).

To get the series for the squares of the partial fractions, Euler took the derivatives of (16.32) and (16.33) to obtain

$$\frac{\pi^2 \cos \pi s}{(\sin \pi s)^2} = \frac{1}{s^2} - \frac{1}{(1-s)^2} - \frac{1}{(1+s)^2} + \frac{1}{(2-s)^2} + \frac{1}{(2+s)^2} - \frac{1}{(3-s)^2} - \text{etc.}$$
$$(16.34)$$

$$\frac{\pi^2}{(\sin \pi x)^2} = \frac{1}{s^2} + \frac{1}{(1-s)^2} + \frac{1}{(1+s)^2} + \frac{1}{(2-s)^2} + \frac{1}{(2+s)^2} + \frac{1}{(3-s)^2} + \text{etc.}$$
$$(16.35)$$

Among examples of these relations, for $s = \frac{1}{3}$ Euler gave

$$\frac{2\pi^2}{27} = 1 - \frac{1}{2^2} - \frac{1}{4^2} + \frac{1}{5^2} + \frac{1}{7^2} - \frac{1}{8^2} - \frac{1}{10^2} + \text{etc.} \qquad (16.36)$$

$$\frac{4\pi^2}{27} = 1 + \frac{1}{2^2} + \frac{1}{4^2} + \frac{1}{5^2} + \frac{1}{7^2} + \frac{1}{8^2} + \frac{1}{10^2} + \text{etc.} \qquad (16.37)$$

---

[29] Eu. I-14 pp. 138–155, E 61 §§ 18–20.

and for $s = \frac{1}{4}$ in (16.34) he obtained

$$\frac{\pi^2}{8\sqrt{2}} = 1 - \frac{1}{3^2} - \frac{1}{5^2} + \frac{1}{7^2} + \frac{1}{9^2} - \frac{1}{11^2} - \frac{1}{13^2} + \text{etc.} \qquad (16.38)$$

He knew that he could obtain (16.37) directly from (16.9). Near the end of the paper he noted that if $P$ and $Q$ denoted the left-hand sides of (16.32) and (16.33), respectively, then

$$\frac{(-1)^{n-1}}{(n-1)!} \frac{d^{n-1} P}{ds^{n-1}} = \frac{1}{s^n} + (-1)^{n-1} \frac{1}{(1-s)^n} - \frac{1}{(1+s)^n} + (-1)^n \frac{1}{(2-s)^n} \qquad (16.39)$$
$$+ \frac{1}{(2+s)^n} + (-1)^{n-1} \frac{1}{(3-s)^n} - \text{etc.,}$$

$$\frac{(-1)^{n-1}}{(n-1)!} \frac{d^{n-1} Q}{ds^{n-1}} = \frac{1}{s^n} + (-1)^{n-1} \frac{1}{(1-s)^n} + \frac{1}{(1+s)^n} + (-1)^{n-1} \frac{1}{(2-s)^n}$$
$$+ \frac{1}{(2+s)^n} + (-1)^{n-1} \frac{1}{(3-s)^n} + \text{etc.}$$
$$(16.40)$$

## 16.5   Euler's Evaluation of $\sum \frac{1}{n^2}$ by Integration

Because some mathematicians raised objections to his first evaluation of $\sum \frac{1}{n^2}$, Euler looked for other methods. His evaluations by partial fractions were immune to these objections, but even before finding the partial fractions method, Euler discovered an ingenious technique using integration. In 1737, Euler worked out and communicated to Johann Bernoulli his integration method.[30] But the paper containing this method appeared in 1743 in the *Journal littéraire d'Allemagne*.[31] It was unusual for Euler to use this journal; consequently, the paper received very scant notice. It was finally reprinted in the 1907-1908 *Bibliotheca Mathematica* as a forgotten work of Euler.[32] However, Euler's evaluation of $\sum \frac{1}{(2n+1)^2}$ was presented in a 1743 calculus book by Simpson, although Simpson did not mention Euler.[33]

Briefly and in modern notation, Euler started with Newton's series

$$\arcsin x = x + \frac{1}{2} \frac{x^3}{3} + \frac{1 \cdot 3}{2 \cdot 4} \frac{x^5}{5} + \frac{1 \cdot 3 \cdot 5}{2 \cdot 4 \cdot 6} \frac{x^7}{7} + \cdots$$

to get

$$\frac{1}{2}(\arcsin x)^2 = \int_0^x \frac{\arcsin t}{\sqrt{1-t^2}} dt = \int_0^x \left( t + \frac{1}{2} \frac{t^3}{3} + \cdots \right) \frac{dt}{\sqrt{1-t^2}}. \qquad (16.41)$$

[30] Eu. IV A-2 pp. 161–175, especially pp. 170–171.
[31] Eu. I-14 pp. 177–186. E 63.
[32] Stäckel (1908).
[33] Simpson (1743) pp. 140–141.

Assuming $n$ was odd, since only odd powers appeared in the series, integration by parts gave him

$$\int_0^1 \frac{x^{n+2}}{\sqrt{1-x^2}}dx = \frac{n+1}{n+2}\int_0^1 \frac{x^n}{\sqrt{1-x^2}}dx$$
$$= \frac{(n+1)(n-1)(n-3)\cdots 2}{(n+2)(n)(n-2)\cdots 3}.$$

(16.42)

Euler applied this to (16.41) with $x = 1$ to obtain

$$\frac{\pi^2}{8} = 1 + \frac{1}{3^2} + \frac{1}{5^2} + \frac{1}{7^2} + \cdots.$$

Unfortunately, this method could not be extended to $\sum \frac{1}{n^{2k}}$ for $k = 2,3,\ldots$, though Euler attempted it. For example, he considered the series for $(\arcsin x)^2$ divided by $\sqrt{1-x^2}$ and integrated over (0,1). After some similar calculations, he obtained

$$\frac{\pi^3}{48} = \frac{1}{2^2}\cdot\frac{\pi}{2} + \frac{1}{4^2}\cdot\frac{\pi}{2} + \frac{1}{6^2}\cdot\frac{\pi}{2} + \frac{1}{8^2}\cdot\frac{\pi}{2} + \text{etc.,}$$

a result equivalent to $\frac{\pi^2}{6} = \sum \frac{1}{n^2}$.

In a letter dated August 27, 1737,[34] of the Russian or Julian calendar, Euler communicated to his former teacher, Johann Bernoulli, the evaluation of $\sum \frac{1}{n^2}$ by means of (16.41). In his reply of November 6 (Gregorian), Bernoulli expressed his admiration for this method and observed that it had led him to find a new series for $\frac{\pi^2}{8}$.[35]

$$\frac{1}{1\cdot 2} + \frac{2}{1\cdot 3\cdot 4} + \frac{2\cdot 4}{1\cdot 3\cdot 5\cdot 6} + \frac{2\cdot 4\cdot 6}{1\cdot 3\cdot 5\cdot 7\cdot 8} + \frac{2\cdot 4\cdot 6\cdot 8}{1\cdot 3\cdot 5\cdot 7\cdot 9\cdot 10} + \cdots = \frac{\pi^2}{8}.$$

(16.43)

We note that Bernoulli wrote $C$ instead of $\pi$. Since the Russian calendar at that time was about ten days behind the Gregorian calendar, keeping track of correspondence can be challenging.

Bernoulli gave no further details in his letter, but in 1742 he offered an explanation in the fourth volume of his *Opera Omnia*.[36] His method was to divide Newton's transformed series for $\arctan t$, (10.4), written up in the *De Computo*, by $1 + t^2$ and then integrate. Recall that Newton had not published his work, so the alternative series for $\arctan t$ in powers of $\frac{t}{1+t^2}$ was Bernoulli's rediscovery. Thus, Bernoulli had

[34] Eu. IV A-2, pp. 170–171.
[35] ibid. pp. 185–187.
[36] Bernoulli (1743) vol. 4, p. 25.

$$\frac{(\arctan t)^2}{2} = \int \frac{\arctan t}{1+t^2} \, dt$$

$$= \int \left( \frac{t}{(1+t^2)^2} + \frac{2t^3}{1\cdot 3\cdot(1+t^2)^3} + \frac{2\cdot 4t^5}{1\cdot 3\cdot 5(1+t^2)^4} + \cdots \right) dt. \tag{16.44}$$

He obtained (16.43) by integrating this formula over $(0,\infty)$. Concerning (16.43), Euler noted in a letter dated December 10, 1737 (Julian), that he had found the more general series:[37]

$$\frac{1}{2}(\arcsin x)^2 = \frac{x^2}{1\cdot 2} + \frac{2\cdot x^4}{1\cdot 3\cdot 4} + \frac{2\cdot 4\cdot x^6}{1\cdot 3\cdot 5\cdot 6} + \frac{2\cdot 4\cdot 6\cdot x^8}{1\cdot 3\cdot 5\cdot 7\cdot 8} + \cdots. \tag{16.45}$$

Euler went on to remark that Bernoulli's formula followed from this by taking $x = 1$. Moreover, he observed that other interesting series would result by taking $x = \frac{1}{2}, \frac{1}{\sqrt{2}},$ or $\frac{\sqrt{3}}{2}$.

In his 1743 paper, Euler gave a derivation of (16.45) by observing that $(\arcsin x)^2$ satisfied the second-order linear differential equation

$$(1-x^2)\frac{d^2y}{dx^2} - x\frac{dy}{dx} - 2 = 0. \tag{16.46}$$

He then solved this equation by infinite series to prove (16.45). Observe, however, that Bernoulli could have obtained Euler's formula (16.45) from his own (16.44): Rewrite (16.44) as

$$\frac{1}{2}(\arctan z)^2 = \int_0^z \frac{\arctan t}{1+t^2} \, dt$$

$$= \sum_{n=0}^{\infty} \frac{2^{2n}(n!)^2}{(2n+1)!} \int_0^z \left(\frac{t^2}{1+t^2}\right)^n \frac{t\,dt}{(1+t^2)^2}$$

$$= \sum_{n=0}^{\infty} \frac{2^{2n}(n!)^2}{2(2n+1)!} \int_0^{\frac{z^2}{1+z^2}} u^n \, du$$

$$= \sum_{n=0}^{\infty} \frac{2^{2n}(n!)^2}{(2n+2)!} \left(\frac{z^2}{1+z^2}\right)^{n+1}. \tag{16.47}$$

Now set $x^2 = \frac{z^2}{1+z^2}$ so that $z = \tan(\arcsin x)$ and (16.45) follows.

We here note the remarkable fact that the Japanese mathematician Takebe Katahiro (1664–1739) published both Bernoulli's (16.43) and Euler's (16.45) series in his 1722 treatise, *Tetsujutsu Sankei*.[38] Takebe's approach was very different, since he apparently made his discoveries of these series on the basis of a considerable amount of numerical work. He related the length of an arc of a circle determined by a chord to the height

[37] Eu. IV A-2 pp. 191–198, especially p. 196.
[38] Smith and Mikami (1914) pp. 148–153.

of the chord. The latter would be the distance between the midpoint of the arc and the midpoint of the chord. After finding the series, Takebe sought an analytic justification for it. The *Tetsujutsu* exerted great influence on the development of mathematics in eighteenth-century Japan,[39] spurring Japanese mathematicians to discover other series 'for $\pi$.

## 16.6   N. Bernoulli's Evaluation of $\sum \frac{1}{(2n+1)^2}$

Euler was eager to find many different evaluations of $\sum \frac{1}{n^2}$ and in this he was assisted by Niklaus I Bernoulli who in 1738 published a very interesting method[40] by squaring the Madhava–Leibniz series for $\frac{\pi}{4}$, given by (16.5). Bernoulli's derivation involved many transformations of series but in a July 1738 letter to Johann Bernoulli, Euler gave a shorter proof by greatly simplifying the second portion.[41] We present this simplified proof, whose fundamental idea remained Bernoulli's.

Bernoulli first observed that by squaring equation (16.5) he had

$$
\frac{\pi^2}{16} = \left(1 - \frac{1}{3} + \frac{1}{5} - \frac{1}{7} + \cdots\right)^2 = \sum_{n=0}^{\infty} \frac{1}{(2n+1)^2} - 2\sum_{n=0}^{\infty} \frac{1}{2n+1} \cdot \frac{1}{2n+3}
$$

$$
+ 2\sum_{n=0}^{\infty} \frac{1}{2n+1} \cdot \frac{1}{2n+5} - 2\sum_{n=0}^{\infty} \frac{1}{2n+1} \cdot \frac{1}{2n+7} + \cdots .
$$

$$(16.48)$$

The first series on the right was the sum of the squares of $1, \frac{1}{3}, \frac{1}{5}, \ldots$. The other series were the sums of the mixed terms obtained by squaring. He then noted that

$$
2\sum_{n=0}^{\infty} \frac{1}{2n+1} \cdot \frac{1}{2n+3} = \sum_{n=0}^{\infty} \left(\frac{1}{2n+1} - \frac{1}{2n+3}\right) = 1,
$$

$$
2\sum_{n=0}^{\infty} \frac{1}{2n+1} \cdot \frac{1}{2n+5} = \frac{1}{2}\sum_{n=0}^{\infty} \left(\frac{1}{2n+1} - \frac{1}{2n+5}\right) = \frac{1}{2}\left(1 + \frac{1}{3}\right),
$$

$$
2\sum_{n=0}^{\infty} \frac{1}{2n+1} \cdot \frac{1}{2n+7} = \frac{1}{3}\sum_{n=0}^{\infty} \left(\frac{1}{2n+1} - \frac{1}{2n+7}\right) = \frac{1}{3}\left(1 + \frac{1}{3} + \frac{1}{5}\right),
$$

and so on. Hence,

$$
\frac{\pi^2}{16} = \sum_{n=0}^{\infty} \frac{1}{(2n+1)^2} - \left(1 - \frac{1}{2}\left(1 + \frac{1}{3}\right) + \frac{1}{3}\left(1 + \frac{1}{3} + \frac{1}{5}\right) - \cdots\right). \quad (16.49)
$$

---

[39] Ogawa and Morimoto (2018).
[40] Bernoulli (1738).
[41] Eu. IV A-2 pp. 230–247, especially pp. 242–243.

Euler's simplification took effect at this point. To sum the series within the parentheses in (16.49), he observed that

$$\frac{\arctan x}{1+x^2} = (1 - x^2 + x^4 - x^6 + \cdots)\left(x - \frac{x^3}{3} + \frac{x^5}{5} - \frac{x^7}{7} + \cdots\right)$$

$$= x - x^3\left(1 + \frac{1}{3}\right) + x^5\left(1 + \frac{1}{3} + \frac{1}{5}\right) - x^7\left(1 + \frac{1}{3} + \frac{1}{5} + \frac{1}{7}\right) + \cdots.$$

$$(16.50)$$

Euler then integrated over (0,1) to obtain

$$\frac{1}{2}(\arctan 1)^2 = \frac{1}{2}\left(\frac{\pi}{4}\right)^2 = \frac{1}{2} - \frac{1}{4}\left(1 + \frac{1}{3}\right) + \frac{1}{6}\left(1 + \frac{1}{3} + \frac{1}{5}\right) - \cdots.$$

Thus, he showed that the series within parentheses in (16.49) was equal to $\frac{\pi^2}{16}$; hence, $\sum \frac{1}{(2n+1)^2} = \frac{\pi^2}{8}$, as was required. Not only did Euler simplify N. Bernoulli's proof but he also obtained a more general result. This result and the inspiration and assistance of his friend Christian Goldbach eventually lead him to a fruitful study of double zeta values.

## 16.7 Euler and Goldbach: Double Zeta Values

The route toward the consideration of double zeta values started with a theorem communicated by Euler to Goldbach on August 28, 1742.[42] Note that this is a generalization of the last equation in N. Bernoulli's evaluation, (16.49). Euler's theorem states that if

$$s = 1 + \frac{a}{n+1} + \frac{aa}{2n+1} + \frac{a^3}{3n+1} + \frac{a^4}{4n+1} + \cdots, \qquad (16.51)$$

then

$$\frac{ss}{2} = \frac{1}{2} + \frac{a}{n+2}\left(1 + \frac{1}{n+1}\right) + \frac{aa}{2n+2}\left(1 + \frac{1}{n+1} + \frac{1}{2n+1}\right)$$

$$+ \frac{a^3}{3n+2}\left(1 + \frac{1}{n+1} + \frac{1}{2n+1} + \frac{1}{3n+1}\right)$$

$$+ \frac{a^4}{4n+2}\left(1 + \frac{1}{n+1} + \frac{1}{2n+1} + \frac{1}{3n+1} + \frac{1}{4n+1}\right) + \text{etc.} \qquad (16.52)$$

Goldbach was intrigued by this result and raised some questions about it in a letter dated October 1, 1742.[43] Euler responded by explaining how Niklaus I Bernoulli's result, discussed in the previous section, could be obtained from the theorem. This led

---

[42] Fuss (1968) vol. 1, pp. 144–153, especially p. 153.
[43] ibid. pp. 154–159, especially p. 156.

Goldbach to consider the series now known as double zeta values. In a letter to Euler of December 24, 1742, Goldbach wrote[44] that the type of series in Euler's theorem suggested the study of the series

$$1 + \frac{1}{2^n}\left(1 + \frac{1}{2^m}\right) + \frac{1}{3^n}\left(1 + \frac{1}{2^m} + \frac{1}{3^m}\right) + \frac{1}{4^n}\left(1 + \frac{1}{2^m} + \frac{1}{3^m} + \frac{1}{4^m}\right) + \cdots .$$

(16.53)

In fact, Goldbach wrote only examples of such series for $n = 1, m = 1; n = 2, m = 1; n = 1, m = 2$. He also considered alternating series.

Let us denote this series by $\zeta_G(n,m)$. In modern terminology, this is almost the double zeta value defined by

$$\zeta(n,m) = 1 + \frac{1}{2^n} + \frac{1}{3^n}\left(1 + \frac{1}{2^m}\right) + \frac{1}{4^n}\left(1 + \frac{1}{2^m} + \frac{1}{3^m}\right) + \cdots$$
$$= \zeta_G(n,m) - \zeta(m+n).$$

(16.54)

Goldbach further wrote that he had found

$$\zeta_G(3,1) = \frac{\pi^4}{72}, \quad \text{and} \quad 2\zeta_G(5,1) + \zeta_G(4,2) = \frac{19\pi^6}{2 \cdot 5 \cdot 7 \cdot 3^4}.$$

(16.55)

He also mentioned that, while he could not evaluate $\zeta_G(n,m)$, he could handle $\zeta_G(n,m) + \zeta_G(m,n)$. Euler must have been greatly fascinated by these series, for on January 5, 1743, he responded with details of the proof of his theorem and then two weeks later gave a number of evaluations of particular cases of Goldbach's series.[45] Thus, he had

$$\zeta_G(3,1) = \frac{1}{2}(\zeta(2))^2; \; \zeta_G(5,1) = \zeta(2)\zeta(4) - \frac{1}{2}(\zeta(3))^2;$$

$$\zeta_G(7,1) = \zeta(2)\zeta(6) - \zeta(3)\zeta(5) + \frac{1}{2}(\zeta(4))^2;$$

$$\zeta_G(9,1) = \zeta(2)\zeta(8) - \zeta(3)\zeta(7) + \zeta(4)\zeta(6) - \frac{1}{2}(\zeta(5))^2;$$

$$\zeta_G(2,2) = \frac{1}{2}(\zeta(2))^2 + \frac{1}{2}\zeta(4); \; \zeta_G(4,2) = (\zeta(3))^2 - \frac{1}{3}\zeta(6);$$

$$\zeta_G(6,2) = 2\zeta(3)\zeta(5) - \frac{3}{2}(\zeta(4))^2 + \frac{1}{4}\zeta(8);$$

$$\zeta_G(8,2) = 2\zeta(3)\zeta(7) - 3\zeta(4)\zeta(6) + \frac{4}{2}(\zeta(5))^2 - \frac{1}{5}\zeta(10);$$

$$\zeta_G(3,3) = \frac{1}{2}(\zeta(3))^2 + \frac{1}{2}\zeta(6);$$

$$\zeta_G(5,3) = \frac{3}{2}(\zeta(4))^2 - \frac{5}{8}\zeta(8) = \frac{\pi^8}{16 \cdot 3 \cdot 25 \cdot 7}.$$

(16.56)

[44] ibid. pp. 172–175, especially p. 175.
[45] ibid. pp. 188–192, especially pp. 189–190.

Euler showed how Goldbach's results (16.55) could be derived from these values and, concerning Goldbach's remark on $\zeta_G(m,n,) + \zeta_G(n,m)$, he observed that

$$\zeta(m)\zeta(n) + \zeta(m+n) = \zeta_G(m,n) + \zeta_G(n,m). \tag{16.57}$$

Presumably, Goldbach had also found this elementary but basic formula now written as

$$\zeta(m)\zeta(n) - \zeta(m+n) = \zeta(m,n) + \zeta(n,m). \tag{16.58}$$

In a letter of February 26, 1743,[46] Euler explained that his results depended on the partial fractions identity:

$$\frac{1}{x^m(x+a)^n} = \frac{A_0}{x^m} + \frac{A_1}{x^{m-1}} + \cdots + \frac{A_{m-1}}{x} + \frac{B_0}{(x+a)^n} + \frac{B_1}{(x+a)^{n-1}} + \cdots + \frac{B_{n-1}}{x+a} \tag{16.59}$$

where

$$A_k = \frac{(-1)^k n(n+1)\cdots(n+k-1)}{k!\,a^{n+k}}; \quad B_k = \frac{(-1)^m m(m+1)\cdots(m+k-1)}{k!\,a^{m+k}}. \tag{16.60}$$

The identity (16.59) was needed when he considered the series

$$\zeta(m)\zeta(n) - \zeta(m+n) = \sum_{x=1}^{\infty}\frac{1}{x^m}\sum_{a=1}^{\infty}\frac{1}{(x+a)^n} + \sum_{x=1}^{\infty}\frac{1}{x^n}\sum_{a=1}^{\infty}\frac{1}{(x+a)^m}. \tag{16.61}$$

Needless to say, Euler did not use this notation in his letter or in the paper he wrote up more than three decades later, in 1775. He wrote several terms of the expressions in (16.59) and (16.61) to make it clear how the series progressed. In his 1775 paper,[47] Euler noted that (16.59) could be obtained by the method he had presented in his *Introductio* of 1748. Here note that Euler's notation for $\zeta(s)$ and $\zeta_G(m,n)$ used the integral sign for summation:

$$\zeta(s) = \int \frac{1}{x^s}, \quad \zeta_G(m,n) = \int \frac{1}{z^m}\left(\frac{1}{y^n}\right). \tag{16.62}$$

To evaluate (16.61), Euler started with the simple observation that for a positive integer $a$

$$\sum_{x=1}^{\infty}\frac{1}{(x+a)^t} = \sum_{x=1}^{\infty}\frac{1}{x^t} - \sum_{k=1}^{a}\frac{1}{k^t}. \tag{16.63}$$

[46] ibid. pp. 200–208, especially p. 204.
[47] Eu. I-15 pp. 217–267. E 477.

*Zeta Values*

He then applied (16.59) and (16.60) to transform the series $s$:

$$s = \sum_{x=1}^{\infty} \frac{1}{x^m(x+a)^n}$$

$$= \frac{1}{a^n}\sum_{z=1}^{\infty}\frac{1}{z^m} - \frac{n}{a^{n+1}}\sum_{z=1}^{\infty}\frac{1}{z^{m-1}} + \frac{n(n+1)}{2!\,a^{n+2}}\sum_{z=1}^{\infty}\frac{1}{z^{m-2}} - \cdots$$

$$+ (-1)^m \left( \frac{1}{a^m}\sum_{z=1}^{\infty}\frac{1}{z^n} + \frac{m}{a^{m+1}}\sum_{z=1}^{\infty}\frac{1}{z^{n-1}} + \frac{m(m+1)}{2!\,a^{m+2}}\sum_{z=1}^{\infty}\frac{1}{z^{m-2}} + \cdots \right)$$

$$+ (-1)^{m+1} \left( \frac{1}{a^m}\sum_{k=1}^{a}\frac{1}{k^n} + \frac{m}{a^{m+1}}\sum_{k=1}^{a}\frac{1}{k^{n-1}} + \frac{m(m+1)}{2!\,a^{m+2}}\sum_{k=1}^{a}\frac{1}{k^{m-2}} + \cdots \right).$$

$$(16.64)$$

By summing over $a$, he then obtained an expression for the first series on the right-hand side of (16.61):

$$\sum_{x=1}^{\infty}\frac{1}{x^m}\sum_{a=1}^{\infty}\frac{1}{(x+a)^n}$$

$$= \zeta(n)\zeta(m) - n\zeta(n+1)\zeta(m-1) + \frac{n(n+1)}{2!}\zeta(n+2)\zeta(m-2) - \cdots$$

$$+ (-1)^m\left( \zeta(m)\zeta(n) + m\zeta(m+1)\zeta(n-1) + \frac{m(m+1)}{2!}\zeta(m+2)\zeta(n-2) + \cdots \right)$$

$$+ (-1)^{m-1}\left( \zeta_G(m,n) + m\zeta_G(m+1,n-1) + \frac{m(m+1)}{2!}\zeta_G(m+2,n-2) + \cdots \right).$$

$$(16.65)$$

By interchanging $m$ and $n$, Euler immediately got the formula for $\sum\frac{1}{x^n}\sum\frac{1}{(x+a)^m}$. Thus, he had established all the formulas necessary to evaluate the results in (16.56).

Euler evaluated several specific examples in his paper. He took $m+n=3$ with $m=2, n=1$ and by applying (16.65) and using an analogous formula with $m$ and $n$ interchanged, he got

$$\zeta(2)\zeta(1) - \zeta(3)$$
$$= 2\zeta(2)\zeta(1) - \zeta_G(2,1) - 2\zeta(2)\zeta(1) + \zeta_G(1,2) + \zeta_G(2,1)$$
$$= \zeta_G(1,2).$$

$$(16.66)$$

Then by (16.57), he obtained

$$\zeta(2)\zeta(1) + \zeta(3) = \zeta_G(2,1) + \zeta_G(1,2).$$

$$(16.67)$$

Again, by subtracting (16.66) from (16.67), Euler could conclude that

$$\zeta_G(2,1) = 2\zeta(3),\qquad(16.68)$$

or in modern notation for the double-zeta function:

$$\zeta(2,1) = \zeta(3).\qquad(16.69)$$

Observe that series $\zeta(1)$ and $\zeta(1,2)$ are both logarithmically divergent, though it is possible to suitably modify Euler's argument to avoid the use of divergent series. To compute $\zeta_G(4,1)$, Euler took $m = 4, n = 1$ to get, after simplification,

$$\zeta_G(2,3) + \zeta_G(3,2) - \zeta_G(4,1) = 2\zeta(2)\zeta(3) - 2\zeta(5).$$

He then took $m = 2, n = 3$ in (16.57) to obtain

$$\zeta(2)\zeta(3) + \zeta(5) = \zeta_G(2,3) + \zeta_G(3,2).\qquad(16.70)$$

Combined with the previous equation, this gave $\zeta_G(4,1) = 3\zeta(5) - \zeta(2)\zeta(3)$. Euler noted that he was unable to obtain $\zeta_G(2,3)$ and $\zeta_G(3,2)$ by this method. When he set $m = 3, n = 2$ in (16.65), the result was once again (16.70). Euler effectively remedied this drawback by developing a new algebraic method in the later part of the paper.

The Euler-Goldbach double-zeta values can be generalized to multizeta values, defined as

$$\zeta(s_1, s_2, \ldots, s_k) = \sum_{n_1 > n_2 > \cdots n_k > 0} \frac{1}{n_1^{s_1} n_2^{s_2} \cdots n_k^{s_k}},\qquad(16.71)$$

where $s_1, s_2 \ldots, s_k$ are natural numbers. These values have been found to have connections with basic objects in number theory, algebraic geometry, and topology. They have been studied by the use of methods from combinatorics, real and complex analysis, algebra, and number theory, creating great current interest in this topic.

Christian Goldbach, who was the first to see the potential for double-zeta series, was a Prussian but moved to Russia in the 1720s and remained there until his death. In 1725 he became secretary of conferences at the Petersburg Academy. He was a man of diverse talents and his chief hobbies were languages, number theory, differential calculus, and infinite series. He was one of the educators of the young Tsar Peter (1715–1730). As we have seen, Euler frequently communicated important results to his friend Goldbach, who proposed problems of possible interest to Euler and made helpful comments. Goldbach informed Euler of Fermat's unproved theorems and proposed the well-known Goldbach conjecture. Thus, he succeeded in directing Euler's extraordinary talents toward the field of number theory, in which Euler's other colleagues had minimal interest before Lagrange entered the scene. The first volume of P. Fuss's *Correspondance mathématique et physique de quelques célèbres géomètres du XVIIIème siècle* contains 176 letters written by Euler and Goldbach to each other over more than thirty-four years. Euler apparently regarded Goldbach as his close friend, writing him an urgent letter in 1738 when his eyesight was threatened. Goldbach then made unsuccessful attempts to relieve his friend of his burdensome responsibilities in geographical studies. We note that it was in 1766, when Euler returned to St. Petersburg after a twenty-five year stay in Berlin, that he became blind for all practical purposes.

### 16.8 Secant and Tangent Numbers and $\zeta(2m)$

The evaluation of the series

$$\zeta(2m) = \sum_{n=1}^{\infty} \frac{1}{n^{2m}} \tag{16.72}$$

was a major problem faced by early eighteenth-century mathematicians. First Mengoli[48] and then Jakob Bernoulli[49] attempted to sum the series for $m = 1$, but they were not successful. In 1735, Euler became the first to evaluate the sum (16.72).[50] Eventually, he gave more than six methods of summing the series exactly, discussed in this chapter and elsewhere. One of his methods was to apply the geometric series to the partial fractions expansions of the trigonometric functions given in, or example, (15.42) through (15.45).[51] Recall that the geometric series formula can be written

$$\frac{1}{a-x} = \frac{1}{a} \cdot \frac{1}{1-\frac{x}{a}} = \frac{1}{a}\left(1 + \frac{x}{a} + \frac{x^2}{a^2} + \cdots\right)$$

$$= \frac{1}{a} + \frac{x}{a^2} + \frac{x^2}{a^3} + \cdots, \quad \left|\frac{x}{a}\right| < 1. \tag{16.73}$$

Observe that (15.44) requires the series for $\frac{1}{(a-x)^2}$ and this can be found by taking the derivative with respect to $x$ of the equation (16.73). Thus we have

$$\frac{d}{dx}(a-x)^{-1} = (a-x)^{-2} = \frac{d}{dx}\left(\frac{1}{a} + \frac{x}{a^2} + \frac{x^2}{a^3} + \cdots\right)$$

$$= \frac{1}{a^2} + \frac{2}{a^3}x + \frac{3}{a^4}x^2 + \cdots, \quad \left|\frac{x}{a}\right| < 1. \tag{16.74}$$

[48] Mengoli (1650).
[49] Jakob Bernoulli (1744) vol. I, pp. 377–399, epecially p. 398.
[50] Eu. I-14 pp. 73–86. E 41.
[51] See, for example, Eu. I-10$_2$ § 124, in part 2 of his 1755 book on differential calculus.

Euler could then write $x$ times the right-hand side of (15.43) as

$$\left(1 - 2\frac{x^2}{1^2}\left(1 - \frac{x^2}{1^2}\right)^{-1} - 2\frac{x^2}{2^2}\left(1 - \frac{x^2}{2^2}\right)^{-1} - \cdots\right)$$

$$= \left(1 - 2\sum_{k=1}^{\infty}\frac{x^{2k}}{1^{2k}} - 2\sum_{k=1}^{\infty}\frac{x^{2k}}{2^{2k}} - 2\sum_{k=1}^{\infty}\frac{x^{2k}}{3^{2k}} - \cdots - 2\sum_{k=1}^{\infty}\frac{x^{2k}}{n^{2k}} - \cdots\right)$$

and, by changing the order of summation,

$$= \left(1 - 2\sum_{n=1}^{\infty}\frac{1}{n^2}x^2 - 2\sum_{n=1}^{\infty}\frac{1}{n^4}x^4 - \cdots\right). \tag{16.75}$$

Now if the left-hand side of (16.75) can be independently expressed as a power series, then we can find $\sum_{n=1}^{\infty}\frac{1}{n^{2k}}$ by equating the coefficient of $x^{2k}$ on each side. Euler achieved this step by using his generating function for Bernoulli numbers (2.36):

$$\frac{t}{e^t - 1} = B_0 + \frac{B_1}{1!}t + \frac{B_2}{2!}t^2 + \frac{B_3}{3!}t^3 + \cdots$$

$$= 1 - \frac{1}{2}t + \frac{B_2}{2!}t^2 + \frac{B_4}{4!}t^4 + \frac{B_6}{6!}t^6 + \cdots, \quad |t| < 2\pi, \tag{16.76}$$

since $B_3 = B_5 = B_7 = \cdots = 0$. Recall that the odd Bernoulli numbers are all zero because

$$\frac{t}{e^t - 1} + \frac{t}{2}$$

is an even function. Now $x$ times the left-hand side of (15.43) produces $\pi x \cot \pi x$; (16.76) and (15.17) then produce

$$\pi x \cot \pi x = \pi x i \frac{e^{2\pi x i} + 1}{e^{2\pi x i} - 1} = \pi x i + \frac{2\pi x i}{e^{2\pi x i} - 1}$$

$$= \sum_{k=0}^{\infty}\frac{B_{2k}}{(2k)!}(-1)^k(2\pi x)^{2k}, \quad |x| < 1. \tag{16.77}$$

We point out that Euler treated series without too much concern for convergence conditions and thus did not state the condition $|x| < 1$. However, his reasoning shows that he could easily have proved that the series converged for $|x| < 1$. The explicit treatment of convergence and divergence was at that time still developing.

Next, equate the coefficient of $x^{2k}$ in (16.75) and (16.77) to arrive at

$$\sum_{n=1}^{\infty}\frac{1}{n^{2k}} = (-1)^{k-1}\frac{B_{2k}2^{2k-1}\pi^{2k}}{(2k)!}, \tag{16.78}$$

thus completing one of Euler's proofs of (16.78).

Euler used $\sqrt{-1}$ in an exponent a few times in his *Introductio*, but most of the time he avoided using $\sqrt{-1}$ in an exponent. Instead, he usually wrote $\cos x + \sqrt{-1} \sin x$. We summarize his derivation of (16.77), given in his work on differential calculus.[52] He first used (2.36) to show that

$$\frac{x\left(e^{\frac{x}{2}} + e^{-\frac{x}{2}}\right)}{e^{\frac{x}{2}} - e^{-\frac{x}{2}}} = 1 + B_2 \frac{x^2}{2!} + B_4 \frac{x^4}{4!} + B_6 \frac{x^6}{6!} + \cdots, \qquad (16.79)$$

noting that when the exponentials were expanded in a series, then the left-hand side could be written as

$$\frac{1 + \frac{x^2}{2! 2^2} + \frac{x^4}{4! 2^4} + \cdots}{1 + \frac{x^2}{3! 2^3} + \frac{x^4}{5! 2^5} + \cdots}. \qquad (16.80)$$

When Euler changed $x^2$ to $-x^2$ in (16.80), the series in the numerator became $\cos \frac{x}{2}$ and the series in the denominator became $\frac{\sin \frac{x}{2}}{x}$. Moreover, the series in (16.79) changed to

$$1 - B_2 \frac{x^2}{2!} + B_4 \frac{x^4}{4!} - \cdots.$$

Thus, after replacing $x$ by $\pi x$, he had (16.77).

Now in order to express $\csc x$, appearing on the left-hand side of (15.42), as a Taylor series, Euler observed that[53]

$$\csc x = \cot \frac{x}{2} - \cot x$$

and then used (16.77) to obtain

$$\pi x \csc \pi x = 1 + \sum_{k=1}^{\infty} (-1)^{k-1} \frac{B_{2k}(2^{2k} - 2) \pi^{2k}}{(2k)!} x^{2k}. \qquad (16.81)$$

Next, expanding the right-hand side of (15.42) as a geometric series and then multiplying by $x$ gave him

$$1 + 2 \sum_{k=1}^{\infty} \left( \sum_{n=1}^{\infty} \frac{(-1)^{n-1}}{n^{2k}} \right) x^{2k}. \qquad (16.82)$$

Equating the coefficients of $x^{2k}$ in (16.81) and (16.82), Euler arrived at

$$\sum_{n=1}^{\infty} \frac{(-1)^{n-1}}{n^{2k}} = \frac{(-1)^{k-1} B_{2k}(2^{2k-1} - 1)\pi^{2k}}{(2k)!}. \qquad (16.83)$$

---

[52] Eu. I-10$_2$ § 113–127. E 212.
[53] Eu. I-10$_2$ § 223.

Although Euler was well aware that (16.83) could be obtained directly from (16.78), he here wished to show that the alternating series could be obtained from the partial fractions expansion of $\csc \pi x$.

He was able to obtain the Taylor expansion of $\tan x$ by using the fact that $\tan x = \cot x - \cot 2x$:

$$\tan x = \sum_{k=1}^{\infty} \frac{(-1)^{k-1} B_{2k} 2^{2k} (2^{2k} - 1)}{2k} \cdot \frac{x^{2k-1}}{(2k-1)!}. \tag{16.84}$$

The coefficients $\frac{x^{2k-1}}{(2k-1)!}$ in (16.84) are now called tangent numbers, denoted by $T_k$. Thus

$$T_k = (-1)^{k-1} \frac{B_{2k} 2^{2k} (2^{2k} - 1)}{2k}. \tag{16.85}$$

Euler did not appear to have noted the fact that tangent numbers were integers. But in section 222 of the second part of his differential calculus book,[54] he explicitly noted that

$$\tan x = \frac{2Ax}{1 \cdot 2} + \frac{2^3 Bx^3}{1 \cdot 2 \cdot 3 \cdot 4} + \frac{2^5 Cx^5}{1 \cdot 2 \cdots 6} + \frac{2^7 Dx^7}{1 \cdot 2 \cdots 8} + \text{etc.,}$$

where $A, B, C, D$ were integers, as he had shown in section 182. Thus, he actually had

$$2(2^{2k} - 1)B_{2k} = \text{integer.} \tag{16.86}$$

Based on the manner in which he treated the Taylor series for $\sec x$, it would certainly appear that Euler could have proved that $T_k$, as given by (16.85), was an integer.

Euler also found a series expression for $\sec x$, expressed here with subscripts, though that was not his notation:

$$\sec x = \frac{1}{\cos x} = E_0 + \frac{E_2}{2!}x^2 + \frac{E_4}{4!}x^4 + \cdots$$

so he could then obtain

$$\left(E_0 + \frac{E_2}{2!}x^2 + \frac{E_4}{4!}x^4 + \cdots\right)\left(1 - \frac{x^2}{2!} + \frac{x^4}{4!} - \cdots\right) = 1. \tag{16.87}$$

To determine $E_0, E_2, E_4, \ldots$, he equated coefficients of powers of $x$ to obtain equations for $E_{2k}$, $k = 0, 1, 2, \ldots$, to find that the numbers were integers. These numbers are now called secant numbers; in section 224,[55] Euler accurately calculated them to be

[54] Eu. I-10$_2$ § 222. E 212.
[55] ibid § 224.

$$E_0 = 1, \ E_2 = 1, \ E_4 = 5, \ E_6 = 61, \ E_8 = 1385, \ E_{10} = 50521, \ E_{12} = 2702765,$$
$$E_{14} = 199360981, \ E_{16} = 19391512145, \ E_{18} = 2404879661671, \ \ldots .$$

We may imagine that such complex calculations must have afforded the mathematicians of the past not just skill, but very deep insight into the inner workings of their formulas.

That $E_{2k}$ is a positive integer can be proved in general by induction. Note that $E_0 = 1$ and equate the coefficients of $x^{2n}$ in (16.87) to obtain

$$E_{2n} = \binom{2n}{2n-2} E_{2n-2} - \binom{2n}{2n-4} E_{2n-4} + \cdots + (-1)^{n-1} E_0. \qquad (16.88)$$

If all the numbers up to $E_{2n-2}$ are integers, then by (16.88), $E_{2n}$ is an integer. Based on the fact that $\tan x = \frac{\sin x}{\cos x} = \sin x \sec x$, one can show that the series for $\sin x$ multiplied by the series for $\sec x$ produces a series in which the coefficient of $\frac{x^{2k-1}}{(2k-1)!}$ is a positive integer. Thus, the tangent numbers are integers.

The Euler numbers are given by $E_k$ in the formula

$$\sec x + \tan x = \sum_{k=0}^{\infty} E_k \frac{x^k}{k!}.$$

With even subscripts, Euler numbers are called secant numbers; with odd subscripts, tangent numbers. Euler determined[56] that

$$1 - \frac{1}{3^{2k+1}} + \frac{1}{5^{2k+1}} - \frac{1}{7^{2k+1}} + \cdots = \frac{\pi^{2k+1}}{2^{2k+2}(2k)!} E_{2k}. \qquad (16.89)$$

To prove (16.89), Euler expanded the partial fractions on the right-hand side of (15.46) as geometric series. Note that this can be done for $|2x| < 1$ or $|x| < \frac{1}{2}$. Euler completed the proof by a change in the order of summation.

## 16.9   Landen and Spence: Evaluation of $\zeta(2k)$

Landen used the logarithm of $-1$ to evaluate $\sum_{n=1}^{\infty} \frac{1}{n^2}$ and, more generally, $\sum_{n=1}^{\infty} \frac{1}{n^{2k}}$. He showed that the dilogarithm could be used for this purpose but complex numbers were required. Thus, Landen here succeeded where Euler had failed. Landen started his paper of 1760,[57] "A New Method of Computing the Sums of Certain Series," with the determination of the values of $\log(-1)$. He observed that if $x = \sin z$, then

$$\dot{z} = \frac{\dot{x}}{\sqrt{1 - x^2}}$$

---

[56] Eu. I-10$_2$ § 224. E 212.
[57] Landen (1760).

or

$$\frac{\dot{z}}{\sqrt{-1}} = \frac{\dot{x}}{\sqrt{x^2 - 1}},$$

where the dot notation indicates the derivative.

He integrated, taking $z = 0$ where $x = 0$, to get

$$\frac{z}{\sqrt{-1}} = \log \frac{x + \sqrt{x^2 - 1}}{\sqrt{-1}}.$$

For $z = \frac{\pi}{2}$ and $x = 1$, he had

$$\log \frac{1}{\sqrt{-1}} = \frac{\pi}{2\sqrt{-1}}$$

and consequently

$$\log \sqrt{-1} = -\frac{\pi}{2\sqrt{-1}}.$$

Presumably because the square root of a number must have two values, Landen concluded that

$$\log(-1) = 2 \log \sqrt{-1} = \pm \frac{\pi}{\sqrt{-1}}. \qquad (16.90)$$

Landen's fundamental relation for the dilogarithm was

$$\mathrm{Li}_2(x) = \frac{\pi^2}{3} + \frac{\pi}{\sqrt{-1}} \log x - \frac{1}{2} (\log x)^2 - \mathrm{Li}_2 \left( \frac{1}{x} \right). \qquad (16.91)$$

His proof was straightforward; note that he decided to use the minus sign in (16.90):

$$x^{-1} + \frac{x^{-2}}{2} + \frac{x^{-3}}{3} + \cdots = \log \frac{1}{1 - \frac{1}{x}}$$

$$= \log x + \log \frac{1}{1 - x} + \log(-1)$$

$$= -\frac{\pi}{\sqrt{-1}} + \log x + \log \frac{1}{1 - x}.$$

Landen divided the equation by $x$ and integrated to get

$$-\frac{x^{-1}}{1} - \frac{x^{-2}}{2^2} - \frac{x^{-3}}{3^2} - \cdots = -\frac{\pi}{\sqrt{-1}} \log x + \frac{1}{2} (\log x)^2 + \mathrm{Li}_2(x) + C. \qquad (16.92)$$

He set $x = 1$ to find that

$$C = -2 \sum_{n=1}^{\infty} \frac{1}{n^2},$$

the value of which he derived by setting $x = -1$ in (16.92), thus obtaining

$$-C = 2\sum_{n=1}^{\infty} \frac{1}{n^2} = 2\text{Li}_2(-1) - \frac{\pi}{\sqrt{-1}}\log(-1) + \frac{1}{2}\left(\log(-1)\right)^2. \qquad (16.93)$$

Since

$$2\text{Li}_2(-1) = -2\left(1 - \frac{1}{2^2} + \frac{1}{3^2} - \cdots\right) = -2\left(1 - \frac{2}{2^2}\right)\left(1 + \frac{1}{2^2} + \frac{1}{3^2} + \cdots\right)$$

$$= -\sum_{n=1}^{\infty} \frac{1}{n^2} = \frac{C}{2},$$

equation (16.93) became

$$-C = \frac{C}{2} - \frac{\pi}{\sqrt{-1}} \cdot \frac{\pi}{\sqrt{-1}} + \frac{1}{2}\left(\frac{\pi}{\sqrt{-1}}\right)^2. \qquad (16.94)$$

Landen took $\log(-1) = -\frac{\pi}{\sqrt{-1}}$ to derive (16.92), but he took $\log(-1) = \frac{\pi}{\sqrt{-1}}$ in equation (16.94). Although he did not explain this, he clearly wished to obtain a positive value for the series $\sum_{n=1}^{\infty} \frac{1}{n^2}$. So (16.94) simplified to

$$-\frac{1}{2}C = \sum_{n=1}^{\infty} \frac{1}{n^2} = \frac{\pi^2}{6}, \qquad (16.95)$$

completing Landen's proof of (16.91).

Landen also considered the polylogarithmic functions, for $-1 \le x \le 1$ with $n$ a positive integer, defined by the series

$$\text{Li}_n(x) = \frac{x}{1^n} + \frac{x^2}{2^n} + \frac{x^3}{3^n} + \frac{x^4}{4^n} + \cdots. \qquad (16.96)$$

The $n = 2$ case would denote the dilogarithm. More generally, for all complex values of $x$

$$\text{Li}_n(x) = \int_0^x \frac{\text{Li}_{n-1}(t)}{t}\, dt. \qquad (16.97)$$

To obtain a formula for $\text{Li}_3(x)$ corresponding to (16.91), Landen divided (16.91) by $x$ and integrated. Observe that if we set $u = \frac{1}{x}$, we have

$$\int \frac{\text{Li}_2\left(\frac{1}{x}\right)}{x}\, dx = -\int \frac{\text{Li}_2(u)}{u}\, du = -\text{Li}_3(u) + C = -\text{Li}_3\left(\frac{1}{x}\right) + C.$$

Note that because Landen wrote $\text{Li}_2\left(\frac{1}{x}\right)$ as a series, he had

$$\text{Li}_3\left(\frac{1}{x}\right) = \frac{x^{-1}}{1} + \frac{x^{-2}}{2^3} + \frac{x^{-3}}{3^3} + \cdots.$$

After integration he arrived at

$$\mathrm{Li}_3(x) = \frac{\pi^2}{3}\log x + \frac{\pi}{2\sqrt{-1}}(\log x)^2 - \frac{1}{2\cdot 3}(\log x)^3 + \mathrm{Li}_3\left(\frac{1}{x}\right) + C,$$

where the constant of integration $C$ can be seen to be zero upon setting $x = 1$; thus Landen found

$$\mathrm{Li}_3(x) = \frac{\pi^2}{3}\log x + \frac{\pi}{2\sqrt{-1}}(\log x)^2 - \frac{1}{2\cdot 3}(\log x)^3 + \mathrm{Li}_3\left(\frac{1}{x}\right). \tag{16.98}$$

Similarly, by successive integration he discovered

$$\mathrm{Li}_4(x) = 2\sum_{n=1}^{\infty}\frac{1}{n^4} + \frac{\pi^2}{6}(\log x)^2 + \frac{\pi}{2\sqrt{-1}}\frac{1}{3}(\log x)^3 - \frac{1}{4!}(\log x)^4 - \mathrm{Li}_4\left(\frac{1}{x}\right) \tag{16.99}$$

and

$$\mathrm{Li}_5(x) = \left(2\sum_{n=1}^{\infty}\frac{1}{n^4}\right)\log x + \frac{\pi^2}{18}(\log x)^3$$
$$+ \frac{\pi}{2\sqrt{-1}}\frac{1}{3\cdot 4}(\log x)^4 - \frac{1}{5!}(\log x)^5 + \mathrm{Li}_5\left(\frac{1}{x}\right). \tag{16.100}$$

Landen put $x = \frac{1}{\sqrt{-1}}$ in (16.98) and saw that

$$\mathrm{Li}_3\left(\frac{1}{\sqrt{-1}}\right) - \mathrm{Li}_3(\sqrt{-1}) = -2\sqrt{-1}\left(1 - \frac{1}{3^3} + \frac{1}{5^3} - \cdots\right);$$

using (16.90) he could write

$$\frac{\pi^3}{3}\log\frac{1}{\sqrt{-1}} + \frac{\pi}{2\sqrt{-1}}\left(\log\frac{1}{\sqrt{-1}}\right)^2 - \frac{1}{2\cdot 3}\left(\log\frac{1}{\sqrt{-1}}\right)^3 = -\sqrt{-1}\frac{\pi^3}{16}.$$

Thus, he arrived at the formula

$$1 - \frac{1}{3^3} + \frac{1}{5^3} - \frac{1}{7^3} + \cdots = \frac{\pi^3}{32}. \tag{16.101}$$

On the other hand, $x = -1$ in (16.99) and $\log(-1) = -\pi\sqrt{-1}$ gave him

$$1 + \frac{1}{2^4} + \frac{1}{3^4} + \frac{1}{4^4} + \cdots = \frac{\pi^4}{90}. \tag{16.102}$$

Landen pointed out that these formulas could be continued indefinitely, but he did not indicate a connection with Bernoulli numbers.

To further understand Landen's work, write equation (16.91) in the form

$$\text{Li}_2(x) + \text{Li}_2\left(\frac{1}{x}\right) = \frac{\pi^2}{3} + \frac{\pi}{\sqrt{-1}}\log x - \frac{1}{2}(\log x)^2. \tag{16.103}$$

Since he was working with the series for $\text{Li}_2(x)$ and $\text{Li}_2(\frac{1}{x})$, Landen noted that the values of $x$ were those for which both series converged. Such values of $x$ are given by $x = e^{i\theta}$, though he did not write that specifically. As we have seen, he used only $x = \pm 1$, $\pm i$ and the corresponding principal values of $\theta$ are, respectively, $\theta = 0, \pi, \frac{\pi}{2}, -\frac{\pi}{2}$, since the principal values fall in the range $-\pi < \theta \leq \pi$. However, we have also seen that Landen took $e^{i\theta} = -1$, sometimes with $\theta = \pi$ and at other times with $\theta = -\pi$. The reason for this can be understood when one takes $x = e^{i\theta}$ in (16.103) and arrives at a mistake:

$$\text{Li}_2(e^{i\theta}) + \text{Li}_2(e^{-i\theta}) = 2\left(\cos\theta + \frac{\cos 2\theta}{2^2} + \frac{\cos 3\theta}{3^2} + \cdots\right). \tag{16.104}$$

Substituting

$$2\cos k\theta = e^{ik\theta} + e^{-ik\theta}$$

in (16.103) yields

$$\sum_{k=1}^{\infty}\frac{\cos k\theta}{k^2} = \frac{\pi^2}{6} + \frac{\pi}{2}\theta + \frac{1}{4}\theta^2, \tag{16.105}$$

where the right-hand side should actually take the form, as discussed in Section 20.6, of the second Bernoulli polynomial, multiplied by a constant; in other words, the term $+\frac{\pi}{2}\theta$ on the right-hand side should actually be $-\frac{\pi}{2}\theta$.

Now in this connection, we point out that in section 53 of his paper, presented to the Petersburg Academy in 1753 and published in 1760, "Subsidium calculi sinuum,"[58] Euler derived the formula

$$\sum_{k=1}^{\infty}(-1)^{k-1}\frac{\cos k\theta}{k^2} = \frac{\pi^2}{12} - \frac{\phi^2}{4}, \quad 0 \leq \phi < 2\pi. \tag{16.106}$$

Now set $\phi = \pi - \theta$ to obtain a correct version of Landen's (16.105):

$$\sum_{k=1}^{\infty}\frac{\cos k\theta}{k^2} = -\frac{1}{4}(\pi - \theta)^2 - \frac{\pi^2}{12} = \frac{1}{4}\theta^2 - \frac{\pi}{2}\theta + \frac{\pi^2}{6}. \tag{16.107}$$

Initial appearances to the contrary, Landen's formula (16.103) holds for $1 \leq x < \infty$, but not for $0 < x \leq 1$; the left-hand side is not invariant under the transformation $x \to \frac{1}{x}$. When $x$ is changed to $\frac{1}{x}$, the result is a different formula, valid for $0 < x \leq 1$.

---

[58] Eu. I-14 pp. 542–584. E 246.

Now in a paper presented to the Petersburg Academy in 1778 but published in 1811,[59] Euler presented a dilogarithm formula that was invariant under $x \to \frac{1}{x}$:

$$\text{Li}_2(-x) + \text{Li}_2\left(-\frac{1}{x}\right) = -\frac{\pi^2}{6} - \frac{1}{2}(\log x)^2. \qquad (16.108)$$

In section 27 of his paper, he gave this result in the form:

If $Y = \dfrac{x}{1} - \dfrac{x^2}{4} + \dfrac{x^3}{9} - \dfrac{x^4}{16} + \cdots$ and $Y = \dfrac{1}{x} - \dfrac{1}{4x^2} + \dfrac{1}{9x^3} - \dfrac{1}{16x^4} + \cdots$,

$$\text{then } X + Y = \frac{\pi^2}{6} + \frac{1}{2}(\log x)^2$$

and proved it thus:

$$X = \int \frac{\log(1+x)}{x}\,dx$$

$$Y = \int \frac{1}{x}\log\left(\frac{x}{1+x}\right)dx$$

$$= \int \frac{\log(1+x)}{x}\,dx + \int \frac{\log x}{x}\,dx.$$

Hence

$$X + Y = \frac{1}{2}(\log x)^2 + C.$$

To find $C$, Euler set $x = 1$ and used the fact, that he had discovered and had been using for more than forty years, that

$$\sum_{n=1}^{\infty} \frac{(-1)^{n-1}}{n^2} = \frac{\pi^2}{12}. \qquad (16.109)$$

Thus, $C = \frac{\pi^2}{6}$ and (16.108) was proved. Observe that (16.109) follows from

$$\sum_{n=1}^{\infty} \frac{1}{n^2} = \frac{\pi^2}{6}; \qquad (16.110)$$

Euler could have found this by taking $x = -1$ and the principal value of $\log(-1) = \pi\sqrt{-1}$:

$$X + Y = -2\sum_{n=1}^{\infty} \frac{1}{n^2} = \frac{\pi^2}{6} + \frac{1}{2}\left(\log(-1)\right)^2 = \frac{\pi^2}{6} - \frac{\pi^2}{2} = -\frac{\pi^2}{3}.$$

[59] Eu. I-16 pp. 117–138. E 736.

In this way, Euler could have given another proof of (16.110). Here note that Euler was not resistant to the use of complex numbers in this context. In fact, he remarked[60] that $\mathrm{Li}_2(x)$, when $x > 1$, was complex-valued, as can also be perceived from Landen's formula (16.103). Euler had already given a half dozen proofs of the result (16.110), so perhaps he did not see a need for another.

Interestingly, Spence published (16.108) in his 1809 book[61] two years before Euler's paper was published. Spence also proved corresponding results for polylogarithms in a slightly different notation, though in this section we adhere to a uniform notation. William Spence was a Scottish mathematician who does not appear to have been affiliated with any university. As we mentioned in Section 15.1, unlike most British mathematicians at that time, Spence was familiar with the works of continental mathematicians, including the Bernoullis, Euler, and Lagrange. Spence was not familiar with Landen's 1760 paper in 1809 when he wrote his work on logarithmic transcendents, or polylogarithms, though he makes reference to Landen in his preface,[62] having discovered his work at the last minute. Now recall there was a mistake in Landen's work, since he used a formula outside its domain of applicability. This led to incorrect results on the polylogarithms, though he managed to deduce the correct values of

$$\zeta(2k) = \sum_{n=1}^{\infty} \frac{1}{n^{2k}}$$

by judiciously choosing the principal value $\log(-1) = \pi\sqrt{-1}$ or a non–principal value $\log(-1) = -\pi\sqrt{-1}$, so as to yield a positive value for the sum. Spence, on the other hand, avoided this pitfall by working with $\mathrm{Li}_2(-x)$ instead of $\mathrm{Li}_2(x)$.

In § 29 of his book, Spence stated the formulas

$$\mathrm{Li}_2(-x) = -\mathrm{Li}_2\left(-\frac{1}{x}\right) - \frac{1}{2}(\log x)^2 - 2\,\mathrm{Li}_2(-1), \qquad (16.111)$$

$$\mathrm{Li}_3(-x) = \mathrm{Li}_3\left(-\frac{1}{x}\right) - \frac{1}{2\cdot 3}(\log x)^3 - 2\,\mathrm{Li}_2(-1)\log x. \qquad (16.112)$$

He verified (16.111) in an earlier section in exactly the same way that Euler had proved it. He proved (16.112) by dividing (16.111) by $x$ and then integrating. To derive the general formulas for $\mathrm{Li}_{2n}(-x)$ and $\mathrm{Li}_{2n-1}(-x)$, he first assumed

$$\mathrm{Li}_{2n}(-x) = -\mathrm{Li}_{2n}\left(-\frac{1}{x}\right) + A_0^{(n)} + A_1^{(n)}\log x + \cdots + A_{2n}^{(n)}(\log x)^{2n}. \quad (16.113)$$

He took the derivative of (16.113) and multiplied by $x$ to find

$$\mathrm{Li}_{2n-1}(-x) = \mathrm{Li}_{2n-1}\left(-\frac{1}{x}\right) + A_1^{(n)} + 2A_2^{(n)}\log x + \cdots + 2nA_{2n}^{(n)}(\log x)^{2n-1},$$

$$(16.114)$$

[60] Eu. I-16 section 21, pp. 117–138. E 736.
[61] Spence (1809).
[62] ibid. p. xii.

then took the derivative and multiplied by $x$ to arrive at

$$\mathrm{Li}_{2n-2}(-x) = -\mathrm{Li}_{2n-2}\left(-\frac{1}{x}\right) + 2A_2^{(n)} - 2\cdot 3A_3^{(n)}\log x$$

$$+ \cdots + (2n-1)2nA_{2n}^{(n)}(\log x)^{2n-2}. \tag{16.115}$$

Spence next changed $n$ to $n-1$ in (16.113) to obtain

$$\mathrm{Li}_{2n-2}(-x) = -\mathrm{Li}_{2n-2}\left(-\frac{1}{x}\right) + A_0^{(n-1)} + A_1^{(n-1)}\log x + \cdots + A_{2n-2}^{(n-1)}(\log x)^{2n-2}. \tag{16.116}$$

Equating the coefficients of the powers of $\log x$ in (16.115) and (16.116), he could write

$$A_2^{(n)} = \frac{A_0^{(n-1)}}{1\cdot 2}, \quad A_3^{(n)} = \frac{A_1^{(n-1)}}{2\cdot 3}, \quad A_4^{(n)} = \frac{A_2^{(n-1)}}{3\cdot 4}, \quad \cdots, \quad A_{2n}^{(n)} = \frac{A_{2n-2}^{(n-1)}}{(2n-1)2n} \tag{16.117}$$

and these equations produce

$$A_4^{(n)} = \frac{A_0^{(n-2)}}{4!}, \quad A_5^{(n)} = \frac{A_1^{(n-2)}}{5!}, \quad A_6^{(n)} = \frac{A_0^{(n-3)}}{6!}, \quad A_7^{(n)} = \frac{A_1^{(n-3)}}{7!}, \quad \cdots .$$

Spence also observed that the last equation in (16.117), when iterated, produced

$$A_{2n}^{(n)} = \frac{A_2^{(1)}}{3\cdot 4\cdots 2n}.$$

From (16.111), $A_2^{(1)} = -\frac{1}{2}$ and hence $A_{2n}^{(n)} = -\frac{1}{(2n)!}$. Combining his results, Spence obtained the relations

$$\mathrm{Li}_{2n}(-x) + \mathrm{Li}_{2n}\left(-\frac{1}{x}\right) = A_0^{(n)} + A_1^{(n)}\log x + \frac{A_0^{(n-1)}}{2!}(\log x)^2 + \frac{A_1^{(n-1)}}{3!}(\log x)^3$$

$$+ \frac{A_0^{(n-2)}}{4!}(\log x)^4 + \frac{A_1^{(n-2)}}{5!}(\log x)^5 + \cdots - \frac{1}{(2n)!}(\log x)^{2n} \tag{16.118}$$

and

$$\mathrm{Li}_{2n-1}(-x) - \mathrm{Li}_{2n-1}\left(-\frac{1}{x}\right) = A_1^{(n)} + \frac{A_0^{(n-1)}}{1!}(\log x) + \frac{A_1^{(n-1)}}{2!}(\log x)^2$$

$$+ \frac{A_0^{(n-2)}}{3!}(\log x)^3 + \cdots - \frac{1}{(2n-1)!}(\log x)^{2n-1}. \tag{16.119}$$

To completely determine the right-hand sides of equations (16.118) and (16.119), Spence needed only the values of $A_0^{(n)}$, and $A_1^{(n)}$. To find $A_0^{(n)}$ and $A_1^{(n)}$, he set $x = 1$ in (16.118) and (16.119) to obtain

$$A_0^{(n)} = 2\operatorname{Li}_{2n}(-1) \quad \text{and} \quad A_1^{(n)} = 0.$$

Thus, he had the final formulas of section 29 of his book:

$$\operatorname{Li}_{2n}(-x) + \operatorname{Li}_{2n}\left(-\frac{1}{x}\right) = 2\operatorname{Li}_{2n}(-1) + 2\operatorname{Li}_{2n-2}(-1)\frac{(\log x)^2}{2!}$$

$$+ 2\operatorname{Li}_{2n-4}(-1)\frac{(\log x)^4}{4!} + \cdots - \frac{(\log x)^{2n}}{(2n)!}, \quad (16.120)$$

$$\operatorname{Li}_{2n-1}(-x) - \operatorname{Li}_{2n-1}\left(-\frac{1}{x}\right) = 2\operatorname{Li}_{2n-2}(-1)\log x + 2\operatorname{Li}_{2n-4}(-1)\frac{(\log x)^3}{3!}$$

$$+ 2\operatorname{Li}_{2n-6}(-1)\frac{(\log x)^5}{5!} + \cdots - \frac{(\log x)^{2n-1}}{(2n-1)!}.$$

$$(16.121)$$

Formulas (16.120) and (16.121) are true for all integers $n > 1$. For $n = 1$, one must define $\operatorname{Li}_1(-x)$ and $\operatorname{Li}_0(-x)$. Observe that $\operatorname{Li}_{n-1}(-x)$ can be obtained from $\operatorname{Li}_n(-x)$ for $n \geq 3$ by differentiating $\operatorname{Li}_n(-x)$ and multiplying the result by $-x$. Spence therefore defined

$$\operatorname{Li}_1(-x) = -\log(1+x) \quad \text{and} \quad \operatorname{Li}_0(-x) = \frac{-x}{1+x}. \quad (16.122)$$

In his section 31, Spence considered the problem of finding $C = \sum_{k=1}^{\infty}\frac{1}{k^{2n}}$. For this purpose, he observed that

$$-\operatorname{Li}_{2n}(-1) = \sum_{k=1}^{\infty}\frac{(-1)^{k-1}}{k^{2n}} = \sum_{k=1}^{\infty}\frac{1}{k^{2n}} - 2\sum_{k=1}^{\infty}\frac{1}{(2k)^{2n}}$$

$$= C - \frac{2}{2^{2n}}C = \left(1 - \frac{1}{2^{2n-1}}\right)C. \quad (16.123)$$

To determine the value of $C$ when $n = 1$, Spence set $n = 1$ and $x = -1$ in (16.120) to obtain

$$2\operatorname{Li}_2(1) = 2\operatorname{Li}_2(-1) + 2\operatorname{Li}_0(-1)\frac{(\log(-1))^2}{2!}.$$

Taking the principal value of $\log(-1)$ as $\pi\sqrt{-1}$, he found

$$\sum_{k=1}^{\infty}\frac{1}{k^2} = \frac{\pi^2}{6};$$

and with $n = 2$ he had

$$2\operatorname{Li}_4(1) = 2\operatorname{Li}_4(-1) + 2\operatorname{Li}_2(-1)\frac{(\pi\sqrt{-1})^2}{2!} + 2\operatorname{Li}_0(-1)\frac{(\pi\sqrt{-1})^4}{4!}$$

$$= -2\left(1 - \frac{1}{8}\right)\operatorname{Li}_4(1) + \left(1 - \frac{1}{2}\right)\operatorname{Li}_2(1)\frac{\pi^2}{2} - \frac{\pi^4}{24}.$$

Since he had already found that $\mathrm{Li}_2(1) = \frac{\pi^2}{6}$, Spence could conclude

$$\frac{15}{4} \mathrm{Li}_4(1) = \frac{\pi^4}{12} - \frac{\pi^4}{24} = \frac{\pi^4}{24} \quad \text{or} \quad \mathrm{Li}_4(1) = \frac{\pi^4}{90}.$$

Using similar reasoning, Spence successively determined the values of $\mathrm{Li}_6(1)$, $\mathrm{Li}_8(1)$, and $\mathrm{Li}_{10}(1)$. However, Spence was very interested in the numerical values of these quantities and he focused on methods that would yield good approximations. In any case, Spence obtained the correct formula for $\mathrm{Li}_{2n}(-x) + \mathrm{Li}_{2n}(-\frac{1}{x})$, the formula he used for finding $\mathrm{Li}_{2n}(1)$ for some values of $n$.

Recall from our Chapter 2 the formula (2.35) of Euler, or (16.78):

$$\sum_{n=1}^{\infty} \frac{1}{n^{2k}} = \frac{(-1)^{k-1} 2^{2k-1} \pi^{2k} B_{2k}}{(2k)!}.$$

One might raise the question: Could Spence's formulas be employed to produce Euler's general result? A hint of this possibility appears when we rewrite (16.107) as

$$2! \, 2 \sum_{n=1}^{\infty} \frac{\cos 2\pi n\phi}{(2\pi n)^2} = \phi^2 - \phi + \frac{1}{6}. \tag{16.124}$$

Since the first three Bernoulli numbers are $B_0 = 1$, $B_1 = -\frac{1}{2}$, $B_2 = \frac{1}{6}$, and using the definition (2.48), we see that the right-hand side of (16.124) can be rewritten as

$$\binom{2}{0} B_0 \, \phi^2 + \binom{2}{1} B_1 \, \phi + \binom{2}{2} B_2 = B_2 (\phi),$$

the Bernoulli polynomial of degree 2. This connection between Bernoulli polynomials shown in equation (16.124) was explicitly illustrated in 1834, when Jacobi ascertained the remainder term for the Euler–Maclaurin series in terms of Bernoulli polynomials; this was clearly the same as the expression for this remainder in terms of Fourier series, found by Poisson in 1826. For a discussion of the remainder term found by Poisson and by Jacobi, see Sections 20.4 and 20.5. Based on the results of Jacobi and Poisson, we have the formula

$$B_{2k}(\phi) = 2(-1)^{k-1}(2k)! \sum_{n=1}^{\infty} \frac{\cos 2\pi \, n\phi}{(2\pi n)^{2k}}, \quad 0 \le \phi \le 1, \ k \ge 1, \tag{16.125}$$

and then by differentiation,

$$B_{2k-1}(\phi) = 2(-1)^k (2k-1)! \sum_{n=1}^{\infty} \frac{\sin 2\pi \, n\phi}{(2\pi n)^{2k-1}}, \quad 0 \le \phi \le 1, \ k > 1, \tag{16.126}$$

where the range of $\phi$ in (16.126) is $0 < \phi < 1$ when $k = 1$. A further discussion of this is contained in Chapter 20.

In his paper, presented to the Petersburg Academy in 1774 and published in 1775, Euler repeatedly integrated (16.107) to obtain formulas (16.125) and (16.126) for $1 \leq k \leq 4$. He worked with the variable $\theta = 2\pi\phi$ so that he got the polynomials $B_n\left(\frac{\theta}{2\pi}\right)$; he apparently did not perceive that these polynomials were related to the polynomials obtained by Jakob Bernoulli when he expressed $1^n + 2^n + \cdots + (m-1)^n$ as polynomials in $m$.

To verify this relationship inductively, suppose (16.125) true up to some value $k \geq 1$. Apply the operation $(2k+1)\int_0^x$ to both sides of this equation for $0 \leq x \leq 1$ to obtain

$$x^{2k+1} + \binom{2k+1}{1} B_1 x^{2k} + \cdots + \binom{2k+1}{2k} B_{2k} x + C$$

$$= 2(-1)^k (2k+1)! \sum_{n=1}^{\infty} \frac{\sin 2\pi n\phi}{(2\pi n)^{2k+1}}, \quad (16.127)$$

where $C$ is the constant of integration, equal to zero when $x = 0$. Thus, the left-hand side of (16.127) is $B_{2k+1}(x)$ and applying $(2k+2)\int_0^x$ to both sides, we get

$$x^{2k+2} + \binom{2k+2}{1} B_1 x^{2k+1} + \cdots + \binom{2k+2}{2k+1} B_{2k} x^2 + C$$

$$= 2(-1)^k (2k+2)! \sum_{n=1}^{\infty} \frac{\cos 2\pi n\phi}{(2\pi n)^{2k+2}}, \quad (16.128)$$

where $C$ is again the constant of integration. Setting $x = 0$, we see that

$$C = \frac{2(-1)^k (2k+2)!}{(2\pi)^{2k+2}} \sum_{n=1}^{\infty} \frac{1}{n^{2k+2}}.$$

One more integration produces a formula similar to (16.127), in which we set $x = 1$ to conclude that $C$ also satisfies

$$1 + \binom{2k+3}{1} B_1 + \cdots + \binom{2k+3}{3} B_{2k} + \binom{2k+3}{1} C = 0,$$

thus showing that $C = B_{2k+2}$ and the polynomial on the left-hand side of (16.128) is $B_{2k+2}(x)$. We have thus proved (16.125) and (16.126) and have provided another proof of (2.35).

Recall that Raabe defined Bernoulli polynomials, that he called Jacob Bernoulli functions, as the right-hand sides of equations (16.125) and (16.126). Moreover, he defined the Bernoulli numbers by equation (2.35). These definitions allowed him to easily prove some interesting properties of Bernoulli polynomials. For example, from (16.125) and (16.126) it was immediately clear that

$$B_{2k}(x) = B_{2k}(1-x), \quad B_{2k-1}(x) = -B_{2k-1}(1-x). \tag{16.129}$$

As we saw in Chapter 2, (16.129) was proved by Jacobi; see (2.54). It can also be verified by using Jacobi's method of generating functions for Bernoulli polynomials, a method defined by

$$\frac{t\,e^{xt}}{e^t - 1} = \sum_{k=0}^{\infty} B_k(x)\,\frac{t^k}{k!}.$$

A special case of Raabe's (2.56) is given by:

$$B_k(2x) = 2^{k-1}\left( B_k(x) + B_k\left(x + \frac{1}{2}\right) \right). \tag{16.130}$$

Next observe that (16.129) and (16.130) can be employed to show that (16.125) follows from Spence's formula (16.120), taking (2.35) as the definition of Bernoulli numbers. Though Spence did not prove this, it is interesting to work it out: Set $x = e^{2\pi i(\frac{1}{2} - \phi)}$ in the left-hand side of (16.120), yielding the series

$$2\sum_{k=1}^{\infty} \frac{\cos 2k\pi\phi}{k^{2n}}$$

so that the right-hand side of (16.120) becomes

$$\frac{(-1)^n (2\pi)^{2n}}{(2n)!}\left[ \left(1 - \frac{1}{2^{2n-1}}\right) B_{2n} + \left(1 - \frac{1}{2^{2n-3}}\right)\binom{2n}{2}\left(\phi - \frac{1}{2}\right)^2 B_{2n-2} \right.$$
$$+ \left(1 - \frac{1}{2^{2n-5}}\right)\binom{2n}{4}\left(\phi - \frac{1}{2}\right)^4 B_{2n-4} + \cdots$$
$$\left. + \left(1 - \frac{1}{2}\right)\binom{2n}{2}\left(\phi - \frac{1}{2}\right)^{2n-2} B_2 - \left(\phi - \frac{1}{2}\right)^{2n} \right]. \tag{16.131}$$

Now note that

$$\frac{1}{2}\left(B_{2n}(x) + B_{2n}(-x)\right) = x^{2n} + \binom{2n}{2} B_2\, x^{2n-2} + \cdots + B_{2n}$$

so that

$$\left(\phi - \frac{1}{2}\right)^{2n} - \binom{2n}{2}\left(\phi - \frac{1}{2}\right)^{2n-2} B_2 - \cdots - B_{2n}$$
$$= 2\left(\phi - \frac{1}{2}\right)^{2n} - \frac{1}{2}\left(B_{2n}\left(\phi - \frac{1}{2}\right) + B_{2n}\left(\frac{1}{2} - \phi\right)\right) \tag{16.132}$$

and

$$-2\left(\frac{B_{2n}}{2^{2n}} + \binom{2n}{2}\left(\phi - \frac{1}{2}\right)^2 \frac{B_{2n-2}}{2^{2n-2}} + \cdots\right)$$

$$= -\frac{2}{2^{2n}}\left(B_{2n} + \binom{2n}{2}(2\phi - 1)^2 B_{2n-2} + \cdots\right)$$

$$= -\frac{1}{2^{2n-1}}\left((2\phi - 1)^{2n} - \frac{1}{2}\left(B_{2n}(2\phi - 1) + B_{2n}(2 - 2\phi)\right)\right).  \quad (16.133)$$

Upon adding (16.132) and (16.133), we see that the expression in (16.131) is equal to

$$\frac{1}{2^{2n}}\left(B_{2n}(2\phi - 1) + B_{2n}(1 - 2\phi)\right) - \frac{1}{2}\left(B_{2n}\left(\phi - \frac{1}{2}\right) + B_{2n}\left(\frac{1}{2} - \phi\right)\right).$$

$$(16.134)$$

Using (16.130) we have

$$\frac{1}{2^{2n}}\left(B_{2n}(2\phi - 1) + B_{2n}(1 - 2\phi)\right)$$

$$= \frac{1}{2}\left(B_{2n}\left(\phi - \frac{1}{2}\right) + B_{2n}(\phi) + B_{2n}\left(\frac{1}{2} - \phi\right) + B_{2n}(1 - \phi)\right);$$

then, by means of equation (16.129), (16.134) simplifies to

$$\frac{1}{2}\left(B_{2n}(\phi) + B_{2n}(1 - \phi)\right) = B_{2n}(\phi).$$

This proves that Spence's formula (16.120) implies the Euler–Raabe formula (16.125) and hence also (16.126).

## 16.10   Exercises

(1) Show that

$$1 - \frac{1}{3^5} + \frac{1}{5^5} - \frac{1}{7^5} + \frac{1}{9^5} - \text{etc.} = \frac{5\pi^5}{1536},$$

$$1 + \frac{1}{3^6} + \frac{1}{5^6} + \frac{1}{7^6} + \frac{1}{9^6} + \text{etc.} = \frac{\pi^6}{960},$$

$$1 - \frac{1}{3^7} + \frac{1}{5^7} - \frac{1}{7^7} + \frac{1}{9^7} - \text{etc.} = \frac{61\pi^7}{184320},$$

$$1 + \frac{1}{3^8} + \frac{1}{5^8} + \frac{1}{7^8} + \frac{1}{9^8} + \text{etc.} = \frac{17\pi^8}{161280}.$$

See Eu. I-14 p. 81.

(2) Express Newton's series (16.11) as a Dirichlet $L$-series. Do the same with Euler's series (16.36) and (16.38).

(3) Divide Takebe's series

$$\frac{1}{2}(\arcsin x)^2 = \frac{x^2}{2} + \frac{2}{3}\cdot\frac{x^4}{4} + \frac{2\cdot 4}{3\cdot 5}\frac{x^6}{6} + \frac{2\cdot 4\cdot 6}{3\cdot 5\cdot 7}\frac{x^8}{8} + \cdots$$

by $\sqrt{1-x^2}$ and integrate term by term over $(0,1)$ to obtain

$$\frac{\pi^2}{6} = 1 + \frac{1}{2^2} + \frac{1}{3^2} + \frac{1}{4^2} + \cdots.$$

See Eu. I-14 p. 184.

(4) Show that

$$1 + \frac{1}{2^{24}} + \frac{1}{3^{24}} + \frac{1}{4^{24}} + \frac{1}{5^{24}} + \text{etc.} = \frac{2^{23}}{1\cdot 2\cdot 3\cdots 25}\cdot\frac{1181820455}{546}\pi^{24},$$

$$1 + \frac{1}{2^{26}} + \frac{1}{3^{26}} + \frac{1}{4^{26}} + \frac{1}{5^{26}} + \text{etc.} = \frac{2^{25}}{1\cdot 2\cdot 3\cdots 27}\cdot\frac{76977927}{2}\pi^{26}.$$

See Eu. I-14 p. 185.

(5) Prove Eisenstein's formula:

$$e^{\beta\pi i}\sum_{\sigma=0}^{\infty}\frac{(-1)^\sigma}{(\sigma+\beta)^{1-q}} = \frac{\Gamma(q)}{(2\pi)^q}\left(e^{\frac{q\pi i}{2}} + e^{-\frac{2\beta\pi i - q\pi i}{2}}\right)\sum_{\sigma=0}^{\infty}\frac{e^{-\sigma\beta\cdot 2\pi i}}{(\sigma+\frac{1}{2})^q}.$$

See Weil (1989a).

(6) Prove that for $|a| < \pi$ and $0 < s < 1$,

$$\sum_{k=0}^{\infty}(-1)^k\left(\frac{1}{((2k+1)\pi+a)^s} - \frac{1}{((2k+1)\pi-a)^s}\right)$$

$$= \frac{1}{\Gamma(s)\sin\frac{s\pi}{2}}\sum_{k=1}^{\infty}\frac{(-1)^{k-1}\sin ka}{k^{1-s}}.$$

Deduce the functional equation for $L(s) = \sum_{k=0}^{\infty}(-1)^k(2k+1)^{-s}$.

See Malmsten (1849). Carl Malmsten became professor of mathematics in Uppsala in 1841; during his career, he made significant contributions to the development of the Swedish mathematical tradition. See Gårding (1994).

(7) Prove Goldbach's formula

$$1 + \frac{1}{2^3}\left(1 + \frac{1}{2}\right) + \frac{1}{3^3}\left(1 + \frac{1}{2} + \frac{1}{3}\right) + \cdots = \frac{\pi^4}{72}.$$

See Fuss (1968) p. 197.

434

*Zeta Values*

(8) Prove Euler's formula

$$1 + \frac{1}{2^5}\left(1 + \frac{1}{2^3}\right) + \frac{1}{3^5}\left(1 + \frac{1}{2^3} + \frac{1}{3^3}\right) + \cdots = \frac{\pi^8}{16\cdot 3\cdot 25\cdot 7}.$$

See Fuss (1968) p. 190.

(9) Prove Euler's formula (16.52). See Fuss (1968) pp. 181–182.

(10) Show that for $t = s - \frac{1}{2}$ and $0 < s < 1$,

$$\int_0^1 \frac{x^{s-1} - x^{-s}}{1 - x}\, dx = -\pi \tan \pi t.$$

Euler took successive derivatives of both sides with respect to $s$ and set $s = \frac{1}{2}$ or $t = 0$. Verify that after taking the first derivative and setting $s = \frac{1}{2}$, the result is

$$\int_0^1 \frac{2\ln x}{1 - x}\frac{dx}{x^{\frac{1}{2}}} = -\pi^2,$$

or

$$\int_0^1 \frac{\ln y}{1 - y^2}\, dy = -\frac{\pi^2}{8}.$$

More generally, show that

$$\int_0^1 \frac{(\ln y)^{2k-1}}{1 - y^2}\, dy = (-1)^k(2^{2k} - 1)\frac{B_{2k}}{4k}\pi^{2k}. \tag{16.135}$$

Euler wrote down the formulas for $k = 1, 2$ and $3$. See Eu. I-17 p. 406.

(11) Show that for $t = s - \frac{1}{2}$ and $0 < s < 1$

$$\int_0^1 \frac{x^{s-1} + x^{-s}}{1 + x}\, dx = \pi \sec \pi t.$$

Use the method of the previous problem and the series for $\sec u$ given in this chapter to prove the formula for Euler numbers

$$\int_0^1 \frac{(\ln y)^{2k}}{1 + y^2}\, dy = \frac{E_{2k}\pi^{2k+1}}{2^{2k+2}}.$$

Euler computed the Euler numbers $E_{2k}$ for $k = 0, 1, 2, 3, 4$ to obtain $1, 1, 5, 61, 1385$, respectively. See Eu. I-17 pp. 401, 405.

(12) Let $P = \int_0^1 \frac{y(\ln y)^{2k-1}}{1-y^2}\, dy$, and let $Q$ denote the integral in (16.135). Observe that

$$Q \pm P = \int_0^1 \frac{(\ln y)^{2k-1}}{1 \mp y}\, dy,$$

$$P = \frac{1}{2^{2k}} \int_0^1 \frac{(\ln y)^{2k-1}}{1-y} \, dy.$$

Deduce that

$$\int_0^1 \frac{(\ln y)^{2k-1}}{1-y} \, dy = (-1)^k \frac{B_{2k}}{4k} (2\pi)^{2k},$$

$$\int_0^1 \frac{(\ln y)^{2k-1}}{1+y} \, dy = (-1)^k (2^{2k-1} - 1) \frac{B_{2k}}{2k} \pi^{2k}.$$

Euler gave this argument in Eu. I-17 pp. 406–407.

(13) Show that

$$\int_0^1 y^m (\ln y)^{2k-1} \, dy = \frac{\Gamma(2k)}{m^{2k}}.$$

From this, compute the $\zeta$ and $L$-series values

$$\sum_{n=1}^\infty \frac{1}{n^{2k}}, \quad \sum_{n=1}^\infty \frac{(-1)^n}{n^{2k}}, \quad \sum_{n=1}^\infty \frac{(-1)^{n-1}}{(2n-1)^{2k-1}}.$$

In 1737 Euler used integration to exactly evaluate $\sum_{n=1}^\infty \frac{1}{n^2}$ but he regretted that the method did not extend to $k \geq 2$. In 1774, he finally found what he was looking for. See Eu. I-17 pp. 428–451.

# 17

# The Gamma Function

## 17.1 Preliminary Remarks

In the 1720s, several mathematicians attempted to interpolate the sequence

$$1 = 0!, 1 = 1!, 2 = 2!, 6 = 3!, \ldots, n!, (n+1)!, \ldots.$$

Clearly, the interpolating function would have to satisfy $n! = n\big((n-1)!\big)$ and $0! = 1$. To avoid confusion, we give the definition of the interpolating function $f(x)$, satisfying modern requirements:

$$f(x+1) = xf(x), \quad f(1) = 1, \tag{17.1}$$

implying that when $n$ is a positive integer, $f(n) = (n-1)!$. Now in his 1811 *Exercices de calcul intégral*,[1] Legendre denoted the function $f$ by $\Gamma$; in a paper on the hypergeometric function,[2] Gauss represented $f(x+1)$ by $\Pi(x)$. Since some results on definite integrals could be more conveniently stated using Legendre's definition, even influential German mathematicians such as Jacobi and Dirichlet began to employ $\Gamma$-function notation. Thus, the interpolating function for the factorial became known as the gamma function.

Euler and Stirling made significant contributions to this problem starting in the late 1720s. They worked independently, Euler in Russia and Stirling in Scotland; their approaches and aims were also distinct. In the mid-1730s, they came to know of each other's work and had a brief correspondence. Always the algorist, Euler was interested in obtaining analytic expressions for the interpolating function $f(x)$. His first paper on the subject, written in 1730, gives two different representations of $f(x)$, one as an infinite product and the other as a definite integral. On the other hand, Stirling was a numerical analyst interested in finding efficient methods for computing $f(x)$. He was undoubtedly extremely experienced in computation and demonstrated that he knew the values of many mathematical constants to several decimal places. Without

---

[1] Legendre (1811–1817) vol. 1, p. 277.
[2] Gauss (1813). See also Kikuchi (1891) for an English translation.

giving an explicit analytic formula for $f(x)$, but making use of Newton's method of interpolation, he calculated the value of $\frac{1}{2}! = f\left(\frac{3}{2}\right)$ as 0.8862269251. He recognized this as $\frac{\sqrt{\pi}}{2}$ and, indeed, this is the correct value of $\Gamma\left(\frac{3}{2}\right)$.

There was a common feature in the thinking of Euler and Stirling: They both believed that there was only one reasonable or logical interpolating function $f(x)$. Thus, Euler did not prove in his first paper that the integral and infinite product representations of $f(x)$ were equal for all $x > 0$, but merely that they were equal for positive integral values of $x$. Similarly, Stirling thought that his numerical methods gave the value of the unique interpolating function $f(x)$. The later work of H. Bohr and J. Mollerup from around 1920 showed that to obtain uniqueness of the interpolating function one must assume the convexity of $\ln f(x)$, in addition to the two above-mentioned properties $f(x + 1) = x f(x)$ and $f(1) = 1$.

Leonhard Euler (1707–1783) was born in Basel, Switzerland, and studied at the University of Basel from 1720–1724. After this he studied independently, concentrating on mathematics, physics, and astronomy, under the guidance of Johann Bernoulli, with whom he met once a week. Adhering to this regime, Euler quickly became an excellent mathematician and by 1725 he began seeking a position. Failing to find one in Switzerland, he moved to Russia in 1727 to join his friends Daniel and Niklaus (also Nicolaus) II Bernoulli at the newly founded Petersburg Academy. The Bernoulli brothers received their appointments when their father Johann declined a position and persuaded the Academy to employ his sons. Euler was originally appointed to a position in medicine, prompting him to brush up on his anatomy, but he ended up getting a situation in mathematics when Niklaus II died unexpectedly before Euler arrived in St. Petersburg. Euler enjoyed a very stimulating scientific collaboration with Daniel until the latter returned to Basel in 1834. Euler also developed a friendship with Christian Goldbach (1690–1764) from Prussia, whom Clifford Truesdell described as "an energetic and intelligent Prussian for whom mathematics was a hobby, the entire realm of letters an occupation, and espionage a livelihood."[3] Euler and Goldbach corresponded extensively with each other, and Goldbach sometimes suggested problems, stimulating Euler to important mathematical discoveries. Euler spent 1741–66 in Berlin and then returned to St. Petersburg where he died, mathematically active until the end.

Euler became interested in the interpolation problem when it appeared in a 1728 paper presented by Goldbach to the St. Petersburg Academy. Goldbach also mentioned the problem in his letters to Daniel Bernoulli who may have discussed the matter with Euler. Bernoulli outlined a solution in a postscript to a letter to Goldbach dated October 6, 1729.[4] He let $A$ stand for an infinite number. Then the general $x$th term of the factorial sequence was given by

$$\left(A + \frac{x}{2}\right)^{x-1} \left(\frac{2}{1+x} \cdot \frac{3}{2+x} \cdot \frac{4}{3+x} \cdots \frac{A}{A-1+x}\right).$$

[3] Truesdell (1984) p. 345.
[4] Fuss (1968) vol. II, pp. 324–325.

He noted that when $x = \frac{3}{2}$ and $A = 8$, the value of the preceding expression was approximately 1.3005. He had made a computational error here, and the value should have been 1.329, as he observed in a letter two weeks later. This value of $\left(\frac{3}{2}\right)!$ is correct to three decimal places. Even at this early stage of his career, Daniel Bernoulli did not pursue this problem any further, and it was left to Euler to initiate and develop the theory of the gamma function. In fact, Daniel was primarily a mathematical physicist and after middle age, his interest in pure mathematical questions waned.

Euler's letter to Goldbach, dated October 15, 1729,[5] gave the value of the interpolating function $\Gamma(m+1)$, using Legendre's notation, as an infinite product

$$\frac{1 \cdot 2^m}{1+m} \cdot \frac{2^{1-m} \cdot 3^m}{2+m} \cdot \frac{3^{1-m} \cdot 4^m}{3+m} \cdot \frac{4^{1-m} \cdot 5^m}{4+m} \cdots . \tag{17.2}$$

He observed that the infinite product reduced to $m!$ when $m$ was a positive integer, though he verified this only for $m = 2$ and $m = 3$. He also noted in the letter that the infinite product (17.2), when terminated after $n$ terms, and after cancellation of terms $k^m, k^{-m}, k = 2, 3, \ldots, n$, could be written as

$$\frac{1 \cdot 2 \cdot 3 \cdots n \cdot (n+1)^m}{(1+m)(2+m) \cdots (n+m)}. \tag{17.3}$$

This implied that the product (17.2) was equal to

$$\lim_{n \to \infty} \frac{1 \cdot 2 \cdots n \cdot (n+1)^m}{(1+m)(2+m) \cdots (n+m)}. \tag{17.4}$$

In his 1812 paper on hypergeometric functions,[6] a paper published in 1813, Gauss denoted the function defined by the limit (17.4) as $\Pi(m)$. Thus

$$\lim_{n \to \infty} \frac{1 \cdot 2 \cdots n \cdot (n+1)^m}{(1+m)(2+m) \cdots (n+m)} = \Pi(m) = \Gamma(m+1). \tag{17.5}$$

Euler recognized importance of this interpolating function but he did not have a notation for it. Euler simply wrote $1 \cdot 2 \cdot 3 \cdots x$, for some real positive number $x$. On some occasions he used the symbol $[x]$ for $\Gamma(x+1)$, but this was a temporary situational device; he used square brackets for other functions as well.

Soon after he wrote his letter to Goldbach, Euler presented to the Petersburg Academy a long paper on the subject, although the Academy did not publish it until 1738.[7] In this paper, Euler wrote that after he found the product (17.2), he set $m = \frac{1}{2}$ to obtain an expression that, as he recalled from Wallis, had the value $\frac{\sqrt{\pi}}{2}$.

Since Wallis had obtained this result while investigating the area under $y = \sqrt{1 - x^2}$ on the interval $[0, 1]$, Euler was led to consider the integral $\int_0^1 x^e (1-x)^n dx$ and eventually arrived at the formula

[5] Fuss (1968) vol. I, pp. 3–7.
[6] Gauss (1813) or Gauss (1863–1927) vol. 3, pp. 123–162, especially pp. 144–152.
[7] Eu. I-14 pp. 1–24. E 19.

$$\Gamma(m+1) = \int_0^1 (-\ln x)^m dx. \tag{17.6}$$

Euler proved only that the integral (17.6) was equal to the product (17.2) when $m$ was a positive integer. From this, he concluded that the integral would equal the product for all real $m > 0$. He then took $m = \frac{1}{2}$ in (17.6) so that, based on Wallis's value, he got

$$\int_0^1 (-\ln x)^{\frac{1}{2}} dx = \frac{\sqrt{\pi}}{2}. \tag{17.7}$$

Note that a change of variables $x = e^{-t^2}$, followed by integration by parts, gives the probability integral

$$\int_0^\infty e^{-t^2} dt = \frac{\sqrt{\pi}}{2}, \tag{17.8}$$

so that Euler had in fact found the value of the probability integral in the form (17.7).

Recall from Sections 3.1 and 3.2 that, in modern notation, Wallis had guessed the value of the integral

$$\int_0^1 \left(1 - x^{\frac{1}{p}}\right)^q dx = \frac{p!}{(q+1)\cdots(q+p)} = \frac{q!}{(p+1)\cdots(p+q)}. \tag{17.9}$$

When $p$ was a positive integer, Wallis could use the first equation in (17.9) to compute the integral; when $q$ was a positive integer, he could use the second equation. In a paper presented to the Petersburg Academy in 1739,[8] Euler gave a proof of (17.9), extended to the case where both $p$ and $q$ were not necessarily integers. Interestingly, his proof was modeled on Wallis's technique of defining two sequences and then combining them. In effect, Euler proved that

$$\int_0^1 t^{p-1}(1-t)^{q-1} dt = \frac{\Gamma(p)\Gamma(q)}{\Gamma(p+q)}, \tag{17.10}$$

though he stated the result in this form only in 1766.[9] In his 1739 paper, Euler did not clearly specify when he was taking $p$ or $q$ to be a positive integer and when he was not. Interestingly, in 1772,[10] Euler was very clear about this, proving the results for $p$ and $q$ integers. But then he invoked a form of "the principle of permanence of equivalent forms" and wrote that the results would hold for $p$ and $q$ real values. We note that in section 52 of this paper, Euler gave a result equivalent to Gauss's multiplication formula (17.15).

---

[8] Eu. 1-14, pp. 260–290. E 112.
[9] Eu. I-17 pp. 268–288. E 321.
[10] Eu. I-17 pp. 316–357. E 421.

Sometime in the 1740s or perhaps earlier, Euler discovered a connection between the gamma function and the trigonometric functions, in the form of his reflection formula:

$$\Gamma(x)\,\Gamma(1-x) = \frac{\pi}{\sin \pi x}. \qquad (17.11)$$

Note that in the integral (17.10), $p$ and $q$ are positive. Therefore, if we take $p = x$ and $q = 1 - x$, so that $0 < x < 1$, then (17.10) changes to

$$\int_0^1 \frac{t^{x-1}}{(1-t)^x}\,dt = \Gamma(x)\,\Gamma(1-x), \quad 0 < x < 1. \qquad (17.12)$$

Now, following Euler,[11] we set $t = \frac{y}{1+y}$; then (17.12) is transformed to

$$\int_0^\infty \frac{y^{x-1}}{1+y}\,dy = \Gamma(x)\,\Gamma(1-x), \quad 0 < x < 1. \qquad (17.13)$$

Recall that Euler had proved that the integral in (17.13) was equal to $\frac{\pi}{\sin \pi x}$; see our equation (15.40).

Euler made a curious observation in his 1729 letter to Goldbach. He wrote that the value of (17.2) when $m = \frac{1}{2}$ was $\frac{1}{2}\sqrt{\sqrt{-1}\log(-1)}$, equal to the square root of the area of a circle with diameter 1. This amounted to $\sqrt{-1}\ln(-1) = \pi$. At that time, mathematicians did not have a clear idea about how the logarithm of a negative number should be defined. Leibniz and Johann Bernoulli had some correspondence on this point in the 1710s, but these discussions brought forth nothing of real value. Eventually, Euler produced a complete definition of the logarithm of a complex number, including its property of being multivalued. The question is, did Euler have a good understanding of this definition in 1729? Perhaps he did not. Roger Cotes's posthumous work, *Harmonia Mensurarum*, published in 1722, had the formula

$$\sqrt{-1}\log(\cos \theta + i \sin \theta) = \theta$$

and Euler's formula covered the particular case when $\theta = \pi$. Moreover, Cotes's result had an error in sign and this error reappeared in Euler, if we take the principal value of $\log(-1)$ to be $i\pi$. It seems reasonable to draw the conclusion that Euler got his result from Cotes. However, the formula of Cotes set Euler on the right track toward his own more conclusive results, finally written up in the 1740s.

Euler did not deal with the question of the convergence of the infinite product (17.2). It was not the practice among mathematicians of the eighteenth century to go into the details of convergence problems. However, the manner of Euler's expression in some cases leads us to believe that he had clear ideas about what was meant by convergence. For example, in (17.2) Euler did not cancel the factors, showing us that

---

[11] Eu. I-15 p. 558; E 575, §38.

here he was not unaware of convergence issues. One may easily check that the $n$th term of the product is

$$\frac{n^{1-m} \cdot (n+1)^m}{n+m} = \left(1+\frac{1}{n}\right)^m \left(1+\frac{m}{n}\right)^{-1} = 1 + \frac{m(m-1)}{2n^2} + O\left(\frac{1}{n^3}\right),$$

and thus that the infinite product converges.

The eighteenth-century mathematicians produced an enormous body of analytical results without a substantial discussion of convergence. The first mathematician to seriously think about convergence issues was Carl Friedrich Gauss (1777–1855). Like Euler, he had an extremely broad range of interests, covering almost every area of pure and applied mathematics. His paper on the gamma function was a part of a larger work on hypergeometric series published in 1813. He founded his study of convergence on the theory of limits of sequences. In an unpublished early work, he discussed concepts such as the upper and lower limits of sequences. It is difficult to determine the influences informing Gauss's work. Of course, he was extremely well read and was very familiar with the works of his great predecessors. But he appeared to prefer to work in isolation. So it is not clear what motivated him to study convergence of infinite series and products, besides a desire for greater mathematical rigor. Thus, in the 1813 paper mentioned earlier, Gauss showed that the limit in (17.4) existed. He also gave a new method of deriving Euler's results (17.7), (17.10), and (17.11). At the heart of Gauss's new method was the summation formula

$$1 + \frac{a \cdot b}{1 \cdot c} + \frac{a(a+1) \cdot b(b+1)}{1 \cdot 2 \cdot c(c+1)} + \frac{a(a+1)(a+2) \cdot b(b+1)(b+2)}{1 \cdot 2 \cdot 3 \cdot c(c+1)(c+2)} + \cdots$$

$$= \frac{\Gamma(c)\Gamma(c-a-b)}{\Gamma(c-a)\Gamma(c-b)}, \tag{17.14}$$

where $a$, $b$, $c$ were complex numbers with $\operatorname{Re}(c-a-b) > 0$. He gave a completely satisfactory proof of this formula, given in Chapter 23.

Gauss also found the multiplication formula for the gamma function:

$$n^{nz-\frac{1}{2}}\Gamma(z)\Gamma\left(z+\frac{1}{n}\right)\Gamma\left(z+\frac{2}{n}\right)\cdots\Gamma\left(z+\frac{n-1}{n}\right)$$

$$= (2\pi)^{\frac{n-1}{2}}\Gamma(nz), \tag{17.15}$$

where $n$ was a positive integer. The reflection formula (17.11) suggests that the inspiration for this must have been the similar formula for $\sin n\pi z$ discovered and published by Euler in his 1748 *Introductio in Analysin Infinitorum*. In slightly modified form, Euler's formula was

$$\sin n\pi z = 2^{n-1}\sin \pi z \sin \pi\left(z+\frac{1}{n}\right)\sin \pi\left(z+\frac{2}{n}\right)\cdots\sin \pi\left(z+\frac{n-1}{n}\right). \tag{17.16}$$

Euler also gave a special case of (17.15):[12]

$$\sqrt{n}\,\Gamma\left(\frac{1}{n}\right)\Gamma\left(\frac{2}{n}\right)\cdots\Gamma\left(\frac{n-1}{n}\right) = (2\pi)^{\frac{(n-1)}{2}}. \qquad (17.17)$$

Slightly before Gauss's paper was published, Legendre discovered the duplication formula,[13] the $n = 2$ case of Gauss's formula (17.15). Recall that Euler had a formula equivalent to (17.15) in a paper of 1772.[14] Legendre's proof employed the integral representation of the gamma function, and this in turn suggested the problem of deriving the properties of the gamma function using definite integrals. At that time, definite integrals were appearing in many areas of mathematics and its applications. For Euler, this topic was a life-long interest; he had already evaluated several definite integrals by means of a variety of techniques. S. P. Laplace and Legendre also pursued the study of definite integrals, of great usefulness in solving problems in probability theory and mechanics. The method of Fourier transforms, originated by Fourier in his work on heat conduction and its applications to wave phenomena, also produced numerous definite integrals.

By 1810, several French mathematicians had published papers whose aim was to evaluate classes of definite integrals. In 1814, Cauchy wrote a long memoir on definite integrals, published in 1827, the first of his many contributions to what would become complex function theory. A decade later, Cauchy gave a precise definition of a definite integral in his lectures at the École Polytechnique; he then proceeded to define improper integrals and their convergence.

Dirichlet, though a Prussian, studied in Paris in the mid-1820s. He mastered Cauchy's ideas on rigor and applied them to the series introduced into mathematics and mathematical physics by his friend Fourier. Even in his first paper on Fourier series,[15] Dirichlet recognized the importance of extending the definite integral to include highly discontinuous functions. He even made use of improper integrals in his number theoretic work. He employed the integrals $\int_0^\infty \cos x^2\, dx$ and $\int_0^\infty \sin x^2\, dx$, closely related to the gamma function, to obtain a remarkable evaluation of the quadratic Gauss sum. We discuss this in Section 19.6. In his famous work on primes in arithmetic progressions, Dirichlet used Euler's integral formula for the gamma function in the form

$$\int_0^1 x^{n-1}\left(\ln\left(\frac{1}{x}\right)\right)^{s-1} dx = \frac{\Gamma(s)}{n^s} \qquad (17.18)$$

to represent certain Dirichlet series as integrals. We discuss this in detail in Chapter 28. Dirichlet's number theoretic work motivated him to further investigate the gamma function within the theory of definite integrals. He wrote several papers on the topic,

---

[12]  Eu. I-19 pp. 439–490, especially § 46. E 816.
[13]  Legendre (1811–1817) vol. I, p. 284.
[14]  Eu. I-17 pp. 316–357. E 421 section 52.
[15]  Dirichlet (1829b).

including one dealing with a multidimensional generalization of Euler's beta integral (17.10).[16] In a paper of 1829,[17] he also studied the gamma integral with a complex parameter.

Dirichlet's interest in definite integrals, expressed through his publications and his lectures at Berlin University, created an interest in this topic among German mathematicians. Thus, in 1852, Richard Dedekind, who did great work in number theory, wrote his Göttingen doctoral thesis on Eulerian integrals. Riemann, greatly influenced by Dirichlet, made brilliant use of definite integrals, and the gamma integral in particular, in his great 1859 paper on the distribution of primes. In this paper, he expressed the zeta function as a contour integral from which he derived the functional equation for the zeta function. This work of Riemann inspired his student Hermann Hankel (1839–1873) to find, in 1864, a contour integral representation for $\Gamma(z)$, valid for all complex $z$ except the negative integers.[18]

The gamma function also played a significant role in the development of the theory of infinite products. In an 1848 paper,[19] the English mathematician F. W. Newman explained how an exponential factor $e^{-\frac{x}{n}}$ in $(1 + \frac{x}{n})e^{-\frac{x}{n}}$ ensured the convergence of the product $\prod_{n=1}^{\infty}(1 + \frac{x}{n})e^{-\frac{x}{n}}$. Using this product, he obtained a new representation for the gamma function. Oscar Schlömilch (1823–1901), a student of Dirichlet, published this result in 1843,[20] taking an integral of Dirichlet as a starting point of the proof. Schlömilch's work was based on the evaluation of definite integrals. In 1856, Weierstrass gave a foundation to the theory of the gamma function by defining it in terms of an infinite product.[21] In fact, the ideas of this paper inspired him to construct entire functions with a prescribed sequence of zeros.

The gamma function is one of the basic special functions, cropping up again and again. Consequently, mathematicians have tried to derive its properties from several different points of view. In 1930,[22] Emil Artin observed that the concept of logarithmic convexity, used by Bohr and Mollerup to prove the equivalence of the product and integral representations of the gamma function, could be employed to characterize and develop the properties of this function. While Artin worked with real variables, in 1939 Helmut Wielandt gave a complex analytic characterization. The defining property other than the obvious $f(z + 1) = zf(z)$ was that $f(z)$ was bounded in the vertical strip $1 \leq \mathrm{Re}\, z < 2$. Wielandt, a group theorist, did not publish his theorem. Instead, he showed it to Konrad Knopp, who included it in the fifth edition of his *Funktionentheorie II*.[23]

---

[16] Dirichlet (1839a) and (1839b).
[17] Dirichlet (1829a).
[18] Hankel (1864).
[19] Newman (1848).
[20] Schlömilch (1843).
[21] Weierstrass (1856).
[22] Artin (1964).
[23] Knopp (1941) pp. 47–49.

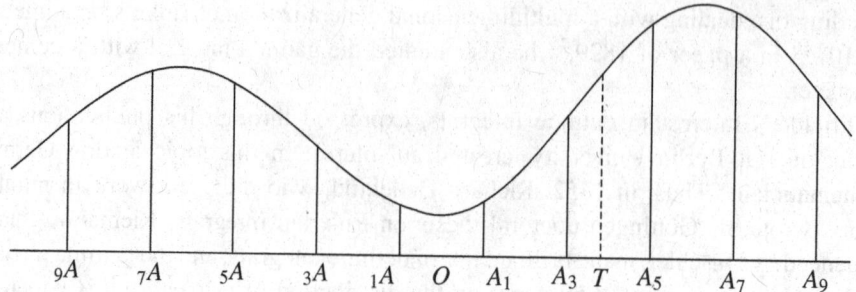

Figure 17.1 Stirling: gamma values by interpolation.

## 17.2 Stirling: $\Gamma\left(\frac{1}{2}\right)$ by Newton–Bessel Interpolation

James Stirling gave a remarkable numerical evaluation of $\Gamma\left(\frac{1}{2}\right)$. He tabulated the base 10 logarithms of the twelve numbers $5!, 6!, \ldots, 16!$ and then applied the Newton–Bessel interpolation formula to obtain the middle value $\left(\frac{1}{2}\right)!$. Then by successive division, he computed $\Gamma\left(\frac{1}{2}\right) = \left(-\frac{1}{2}\right)!$ to ten decimal places and recognized it to be $\sqrt{\pi}$. See Section 9.2 for a discussion of interpolation formulas. In the *Methodus Differentialis*,[24] proposition 20, Stirling described the interpolation formula:

He first supposed an even number of equidistant ordinates and, using a diagram similar to Figure 17.1, he denoted them $_9A, _7A, _5A, \ldots, A_5, A_7, A_9$. Note that these values refer to heights of the line segments. He called $_1A$ and $A_1$ the middle values and set $A$ to be their sum. Thus, he had nine differences of these ten numbers, such as $_7A - {}_9A, {}_5A - {}_7A, \ldots, A_9 - A_7$. He called the middle difference $a$. Taking the eight differences of these nine differences, he denoted the sum of the middle two terms as $B$. Next, Stirling called the middle term of the seven differences of the eight second differences $b$, and so on. He took $O$ to be the midpoint of $_1A$ and $A_1$ and let $T$ be an arbitrary ordinate; he let $\frac{z}{2}$ be the ratio of the distance between $O$ and that point whose ordinate was $T$, and the distance between $_1A$ and $A_1$. Note that this last distance is also the distance between any two successive ordinates. Stirling wrote the formula as

$$T = \frac{A + az}{2} + \frac{3B + bz}{2} \times \frac{z^2 - 1}{4 \cdot 6} + \frac{5C + cz}{2} \times \frac{z^2 - 1}{4 \cdot 6} \times \frac{z^2 - 9}{8 \cdot 10}$$
$$+ \frac{7D + dz}{2} \times \frac{z^2 - 1}{4 \cdot 6} \times \frac{z^2 - 9}{8 \cdot 10} \times \frac{z^2 - 25}{12 \cdot 14}$$
$$+ \frac{9E + ez}{2} \times \frac{z^2 - 1}{4 \cdot 6} \times \frac{z^2 - 9}{8 \cdot 10} \times \frac{z^2 - 25}{12 \cdot 14} \times \frac{z^2 - 49}{16 \cdot 18} + \cdots. \qquad (17.19)$$

Stirling also noted that $z$ was positive when $T$ was on the right side of the middle point $O$, as in Figure 17.1, and negative when it lay on the left-hand side. To write this in modern form, set

---

[24] Stirling and Tweddle (2003).

$$_1A = f\left(s - \frac{h}{2}\right),\ A_1 = f\left(s + \frac{h}{2}\right),\ _2A = f\left(s - \frac{3}{2}h\right),\ A_2 = f\left(s + \frac{3}{2}h\right),\ \text{etc.}$$

Then the Newton–Bessel formula, discussed in Section 9.2, is given by

$$f\left(s + \left(x - \frac{1}{2}\right)h\right) = \frac{1}{2}\left(f\left(s - \frac{h}{2}\right) + f\left(s + \frac{h}{2}\right)\right) + \left(x - \frac{1}{2}\right)\Delta f\left(s - \frac{h}{2}\right)$$

$$+ \frac{x(x-1)}{2!}\frac{1}{2}\left(\Delta^2 f\left(s - \frac{3h}{2}\right) + \Delta^2 f\left(s - \frac{h}{2}\right)\right)$$

$$+ \frac{x(x-1)(x-\frac{1}{2})}{3!}\Delta^3 f\left(s - \frac{3h}{2}\right) + \cdots . \tag{17.20}$$

Stirling's expression is obtained by setting $z = 2x - 1$ and combining pairs of terms in (17.20).

In Example 2 of Proposition 21, he explained his approach to the evaluation of $\Gamma(x)$ for $x > 0$:

Let the series [sequence] to be interpolated be $1, 1, 2, 6, 24, 120, 720$, etc. whose terms are generated by repeated multiplication of the numbers $1, 2, 3, 4, 5$, etc. Since these terms increase very rapidly, their differences will form a divergent progression, as a result of which the ordinate of the parabola does not approach the true value. Therefore, in this and similar cases I interpolate the logarithms of the terms, whose differences can in fact form a rapidly convergent series, even if the terms themselves increase very rapidly as in the present example.

Stirling interpolated the sequence

$$\ln 0!,\ \ln 1!,\ \ln 2!,\ldots,\ \ln n!,\ \ln(n+1)!,\ldots.$$

The first difference would be

$$\ln(n+1)! - \ln n! = \ln(n+1);$$

the second difference would be written as

$$\ln(n+1) - \ln n = \ln\left(1 + \frac{1}{n}\right) \approx \frac{1}{n};$$

and the third difference would be written as

$$\ln(n+1) - 2\ln(n) + \ln(n-1) = \ln\left(1 - \frac{1}{n^2}\right) \approx -\frac{1}{n^2}.$$

Thus, we can see that the successive differences get small rapidly when $n$ is large enough, though $\ln n!$ increases. This means that if one desires to find the value of $\Gamma(n+1)$ for a small value, $n = \frac{1}{2}$, then one must first compute $\ln \Gamma(n+1)$ for a larger value of $n$ by the method of differences and then apply the functional equation $\Gamma(n+1) = n\Gamma(n)$. Stirling thought that the interpolating function must satisfy the same functional relation as the successive terms of the original sequence. In fact, in Proposition 16 of his book, he attempted to explain why this must be so. Stirling then continued:

> Now I propose to find the term which stands in the middle between the first two 1 and 1. And since the logarithms of the initial terms have slowly convergent differences, I first seek the term standing in the middle between two terms which are sufficiently far removed from the beginning, for example, that between the eleventh term 3628800 and the twelfth term 39916800: and when this is given, I may go back to the term sought by means of Proposition 16. And since there are some terms located on both sides of the intermediate term which is to be determined first, I set up the operation by means of the second case of Proposition 20.

Actually, Stirling worked with $\log_{10}$ since the logarithmic tables were often in base 10. He therefore took the twelve known ordinates to be $\log_{10} \Gamma(\frac{1}{2}(z + 23))$ for $z = \pm 1, \pm 3, \pm 5, \pm 7, \pm 9,$ and $\pm 11$ and used Newton–Bessel to find the value at $z = 0$. From this he found $\Gamma(11.5)$ and after successive division by $10.5, 9.5, \ldots,$ down to 1.5, he computed $\Gamma(\frac{3}{2})$ to ten decimal places. After this amazing calculation Stirling commented, "From this is established that the term between 1 and 1 [referring to 0! and 1!] is .8862269251, whose square is .7853... etc., namely, the area of a circle whose diameter is one. And twice that, 1.7724538502, namely the term which stands before the first principal term by half the common interval, is equal to the square root of the number 3.1415926... etc., which denotes the circumference of a circle whose diameter is one."

Here Stirling gave the value of $\Gamma(\frac{1}{2})$, obtained from $\Gamma(\frac{3}{2}) = \frac{1}{2}\Gamma(\frac{1}{2})$. His value for $\Gamma(\frac{3}{2})$ is incorrect only in the tenth decimal place; when rounded, the tenth place should be 5 instead of 1.

## 17.3  Euler's Integral for the Gamma Function

In his 1730 paper "De Progressionibus Transcendentibus," Euler wrote[25] that he found the infinite product

$$\frac{1 \cdot 2^m}{1 + m} \cdot \frac{2^{1-m} \cdot 3^m}{2 + m} \cdot \frac{3^{1-m} \cdot 4^m}{3 + m} \cdot \frac{4^{1-m} \cdot 5^m}{4 + m} \cdots \qquad (17.21)$$

as an expression that reduced to $m!$ when $m$ was a positive integer. Note that the product of a finite number of factors of the numerators, $2^m \cdot 2^{1-m}, 3^m \cdot 3^{1-m}, \ldots,$ $n^m \cdot n^{1-m}$, is $n!$. If $n > m$ then

$$n! = m!\,(1 + m)(2 + m) \cdots (n - m + m);$$

so that the product (17.21) reduces to $m!$ because of the cancellation of the denominator terms with the terms in the numerator following $m!$.

---

[25] Eu. I-14 pp. 1–24. E 19.

Euler then observed that for $m = \frac{1}{2}$, (17.21) reduced to

$$\sqrt{\frac{2\cdot 4}{3\cdot 3}\cdot\frac{4\cdot 6}{5\cdot 5}\cdot\frac{6\cdot 8}{7\cdot 7}\cdots}. \qquad (17.22)$$

Euler remarked that when he obtained this result, he recalled having seen in Wallis's work that its value was $\frac{\sqrt{\pi}}{2}$. Wallis was evaluating, in modern notation, $\int_0^1 (1-x^2)^{\frac{1}{2}}dx$. Euler was thus inspired to consider integrals of the form

$$\int_0^1 x^e(1-x)^n dx.$$

Next, by the binomial theorem, from

$$x^e(1-x)^n = x^e\left(1 - nx + \frac{n(n-1)}{1\cdot 2}x^2 - \frac{n(n-1)(n-2)}{1\cdot 2\cdot 3}x^3 + \cdots\right)$$

$$= x^e - nx^{e+1} + \frac{n(n-1)}{1\cdot 2}x^{e+2} - \frac{n(n-1)(n-2)}{1\cdot 2\cdot 3}x^{e+3} + \cdots$$

term-by-term integration gave him

$$\int_0^1 x^e(1-x)^n dx = \frac{1}{e+1} - \frac{n}{1\cdot(e+2)} + \frac{n(n-1)}{1\cdot 2(e+3)} - \frac{n(n-1)(n-2)}{1\cdot 2\cdot 3(e+4)} + \cdots. \qquad (17.23)$$

Euler evaluated the series in (17.23) for $n = 0, 1, 2, 3$ to obtain

$$\frac{1}{e+1}, \frac{1}{(e+1)(e+2)}, \frac{1\cdot 2}{(e+1)(e+2)(e+3)}, \frac{1\cdot 2\cdot 3}{(e+1)(e+2)(e+3)(e+4)}.$$

He concluded that

$$(e+n+1)\int_0^1 x^e(1-x)^n dx = \frac{1\cdot 2\cdot 3\cdots n}{(e+1)(e+2)\cdots(e+n)}. \qquad (17.24)$$

For a modern mathematician, (17.24) would require an inductive proof, but during the eighteenth century, mathematicians found this reasoning sufficient.

Wishing to make the denominator of (17.24) equal to 1, Euler set $e = \frac{f}{g}$ and rewrote (17.24) as

$$\frac{f+(n+1)g}{g^{n+1}}\int_0^1 x^{\frac{f}{g}}(1-x)^n dx = \frac{1\cdot 2\cdot 3\cdots\cdot n}{(f+g)(f+2g)\cdots(f+ng)}. \qquad (17.25)$$

He next effected a change of variables, replacing $x$ by $x^{\frac{g}{f+g}}$, to find that

$$\frac{f+(n+1)g}{f+g}\int_0^1 \frac{\left(1-x^{\frac{g}{f+g}}\right)^n}{g^n}dx = \frac{1\cdot 2\cdot 3\cdots\cdot n}{(f+g)(f+2g)\cdots(f+ng)}. \qquad (17.26)$$

He could then obtain $n!$ as an integral by taking $f = 1$ and $g = 0$ in the integral in (17.26); in that case, he observed, the integrand in (17.26) became

$$\left(\frac{1 - x^0}{0}\right)^n.$$

In modern terms, Euler meant that the integrand was

$$\lim_{g \to 0} \left(\frac{1 - x^g}{g}\right)^n.$$

Euler stated that the limit could be evaluated by a well-known rule; he was referring to l'Hôpital's rule. Thus, in our modern notation, since

$$-\ln x = \lim_{g \to 0} -\frac{x^g \ln x}{1} = \lim_{g \to 0} \frac{1 - x^g}{g},$$

he got

$$\int_0^1 (-\ln x)^n dx = 1 \cdot 2 \cdot 3 \cdots n. \tag{17.27}$$

Euler assumed that because the product (17.21) and the integral in (17.27) both interpolated the sequence of factorials, they must be equal. With this assumption, he could write (17.7):

$$\int_0^1 (-\ln x)^{\frac{1}{2}} dx = \frac{\sqrt{\pi}}{2}.$$

At the end of the paper, he explained how the gamma function could be used to define fractional derivatives. This was a problem already raised by Leibniz, as Euler may have been aware. He observed that when $n$ was a nonnegative integer,

$$\frac{d^n z^e}{dz^n} = e(e - 1) \cdots (e - n + 1) z^{e-n};$$

and so one could define, for any positive real number $n$,

$$\frac{d^n z^e}{dz^n} = \frac{\int_0^1 (-\ln x)^e dx}{\int_0^1 (\ln x)^{e-n} dx} z^{e-n}. \tag{17.28}$$

By taking $e = 1$ and $n = \frac{1}{2}$, he had

$$\frac{d^{\frac{1}{2}} z}{dz^{\frac{1}{2}}} = 2\sqrt{\frac{z}{\pi}}. \tag{17.29}$$

Euler did not do any more with this concept, later rediscovered by Abel who in 1823 applied it to more succinctly express the solution of an integral equation.[26]

[26] See Smith (1959) vol. 2, pp. 656–662 for an English translation of Abel's paper.

In 1832, Liouville showed[27] that the integral equation considered by Abel could be easily solved by means of the fractional derivative.

Several proofs of (17.11) follow immediately from formulas appearing in Euler's papers. He did not explicitly state this result very often, perhaps because he had not developed a convenient notation for $\Gamma(x)$. In his simple proof published in 1772,[28] Euler used the products for $\Gamma(x)$ and $\sin \pi x$: He let

$$[m] = \frac{1 \cdot 2^m}{1+m} \cdot \frac{2^{1-m} \cdot 3^m}{2+m} \cdot \frac{3^{1-m} \cdot 4^m}{3+m} \cdot \text{etc.}$$

Then

$$[-m] = \frac{1 \cdot 2^{-m}}{1-m} \cdot \frac{2^{1+m} \cdot 3^{-m}}{2-m} \cdot \frac{3^{1+m} \cdot 4^{-m}}{3-m} \cdot \text{etc.},$$

$$[m][-m] = \frac{1}{1-m^2} \cdot \frac{2^2}{2^2-m^2} \cdot \frac{3^2}{3^2-m^2} \cdot \text{etc.} = \frac{\pi m}{\sin \pi m}. \tag{17.30}$$

Note that (17.30) followed from the infinite product for $\sin x$. Here Euler used the symbol $[m]$ for $\Gamma(m+1)$, but its use was merely provisional. Observe that equation (17.30) is identical with

$$\Gamma(1+m)\Gamma(1-m) = \frac{\pi}{\sin \pi m},$$

an equation that we can see is equivalent to (17.11). The result (17.30) was first stated by Euler without proof in a 1749 presentation to the Berlin Academy of a paper eventually published in 1768.[29] Euler finally published a proof in his 1772 paper. This result offers an example of the complexity involved in tracing and dating mathematical results, especially those to which Euler contributed.

## 17.4 Euler's Evaluation of the Beta Integral

Euler did not prove (17.10) in his first paper on the gamma function, but the second paper, "De Productis ex Infinitis Factoribus Ortis," presented to the Petersburg Academy in 1739,[30] contained an argument upon which a proof can be worked out. He started with the observation that

$$\int_0^1 x^{m-1}(1-x^{nq})^{\frac{p}{q}}dx = \frac{m+(p+q)n}{m} \int_0^1 x^{m+nq-1}(1-x^{nq})^{\frac{p}{q}}dx. \tag{17.31}$$

Wallis had stated a similar functional relation without proof, but with the development of calculus, this relation could easily be proved by using integration by parts. In his 1739 paper, Euler wrote that the result (17.31) was easy to prove, and when he returned to the subject over three decades later in his integral calculus book,

27 Liouville (1832).
28 Eu. I-17 pp. 316–357, especially § 43. E 421.
29 Eu. I-15 pp. 70–90, especially § 13. E 352.
30 Eu. I-14 pp. 260–290, especially pp. 282–284. E 122 § 38–45.

he provided the details,[31] revealing his explanation of the technique of integration by parts: He supposed

$$\int x^{f-1}(1-x^g)^m dx = A \int x^{f-1}(1-x^g)^{m-1} dx + B x^f (1-x^g)^m. \quad (17.32)$$

Euler then took the derivative to get

$$x^{f-1}(1-x^g)^m = A x^{f-1}(1-x^g)^{m-1} - Bmg x^{f+g-1}(1-x^g)^{m-1} \\ + Bf x^{f-1}(1-x^g)^m,$$

or

$$1 - x^g = A - Bmg x^g + Bf(1-x^g) = A - Bmg + B(f+gm)(1-x^g).$$

Thus,

$$A - Bmg = 0 \quad \text{and} \quad B(f+mg) = 1$$

or

$$A = \frac{mg}{f+gm} \quad \text{and} \quad B = \frac{1}{f+mg}. \quad (17.33)$$

Next, equation (17.31) followed from (17.32) by choosing the parameters appropriately. Euler applied the functional relation (17.32) infinitely often to arrive at

$$\int_0^1 x^{m-1}(1-x^{nq})^{\frac{p}{q}} dx$$
$$= \frac{(m+(p+q)n)(m+(p+2q)n) \cdots (m+(p+\infty q)n)}{m(m+nq)(m+2nq) \cdots (m+\infty nq)}$$
$$\times \int_0^1 x^{m+\infty nq-1}(1-x^{nq})^{\frac{p}{q}} dx; \quad (17.34)$$

we note that the infinite product diverges and the integral on the right-hand side vanishes. We can, however, define the right-hand side as a limit. Let us continue to go along with Euler, who followed Wallis again by taking another integral similar to the one on the left-hand side of (17.34) but which could be exactly evaluated. He took $m = nq$ to obtain

$$\frac{1}{(p+q)n} = \int_0^1 x^{nq-1}(1-x^{nq})^{\frac{p}{q}} dx$$
$$= \frac{(nq+(p+q)n)(nq+(p+2q)n) \cdots (nq+(p+\infty q)n)}{nq(2nq)(3nq) \cdots (\infty nq)}$$
$$\times \int_0^1 x^{nq+\infty nq-1}(1-x^{nq})^{\frac{p}{q}} dx. \quad (17.35)$$

---

[31] Eu. I-11 chapter 2. E 342.

Euler then observed that if $k$ was an infinite number and $\alpha$ finite, then

$$\int_0^1 x^k(1-x^{nq})^{\frac{p}{q}}dx = \int_0^1 x^{k+\alpha}(1-x^{nq})^{\frac{p}{q}}dx. \tag{17.36}$$

So, dividing equation (17.34) by equation (17.35), the integrals on the right-hand side cancelled, and the result was

$$\int_0^1 x^{m-1}(1-x^{nq})^{\frac{p}{q}}dx$$
$$= \frac{1}{(p+q)n} \cdot \frac{nq(m+(p+q)n)\cdot 2nq(m+(p+2q)n)\cdot 3nq(m+(p+3q)n)\cdots}{m(p+2q)n(m+nq)(p+3q)n(m+2nq)\cdots}.$$

Replacing $n$ by $\frac{n}{q}$, the form of the relation became

$$\int_0^1 x^{m-1}(1-x^n)^{\frac{p}{q}}dx$$
$$= \frac{q}{(p+q)n} \cdot \frac{1(mq+(p+q)n)}{m(p+2q)} \cdot \frac{2(mq+(p+2q)n)}{(m+n)(p+3q)} \cdot \frac{3(mq+(p+3q)n)}{(m+2n)(p+4q)}\cdots. \tag{17.37}$$

This was Euler's final result, from which we can derive (17.10), although we have the benefit of hindsight. Observe that if we set $n=q=1$ and then replace $p$ by $n-1$, we get

$$\int_0^1 x^{m-1}(1-x)^{n-1}dx = \frac{1\cdot(m+n)\cdot 2\cdot(m+n+1)\cdot 3\cdot(m+n+2)\cdots}{n\cdot m\cdot(n+1)\cdot(m+1)\cdot(n+2)\cdot(m+2)\cdots}. \tag{17.38}$$

Thus, in his 1739 paper, Euler did not prove (17.10), but merely showed that the beta integral could be written as a quotient of infinite products. In a paper of 1766,[32] he derived some properties of the beta integral. He started with the formula obtained from (17.37) and replaced $m$ by $p$, $p$ by $q-n$, and $q$ by $n$:

$$\int_0^1 x^{p-1}(1-x^n)^{\frac{q}{n}-1}dx$$
$$= \frac{p+q}{pq} \cdot \frac{n(p+q+n)}{(p+n)(q+n)} \cdot \frac{2n(p+q+2n)}{(p+2n)(q+2n)} \cdot \frac{3n(p+q+3n)}{(p+3n)(q+3n)}\cdots. \tag{17.39}$$

Euler denoted the integral in (17.39) by $\left(\frac{p}{q}\right)$; although its value clearly depended also on $n$, Euler's notation did not take account of this. Note that when $n=1$, this integral is the beta integral $B(p,q)$, in Legendre's notation. Euler's first observation was that the right-hand side of (17.39) showed that the integral was symmetric in $p$ and $q$, that is

[32] Eu. I-17 pp. 268–288. E 321.

$$\left(\frac{p}{q}\right) = \left(\frac{q}{p}\right).$$

A little later in the paper, Euler set $q = n - p$ in (17.39) to arrive at

$$\left(\frac{p}{n-p}\right) = \left(\frac{n-p}{p}\right)$$

$$= \frac{n}{p(n-p)} \cdot \frac{n \cdot 2n}{(n+p)(2n-p)} \cdot \frac{2n \cdot 3n}{(2n+p)(3n-p)} \cdot \frac{3n \cdot 4n}{(3n+p)(4n-p)} \cdot \text{etc.}$$

$$= \frac{1}{p} \cdot \frac{nn}{nn-pp} \cdot \frac{4nn}{4nn-pp} \cdot \frac{9nn}{9nn-pp} \cdots;$$

comparing with the product formula for $\sin x$, (15.20), we obtain with Euler

$$\left(\frac{p}{n-p}\right) = \left(\frac{n-p}{p}\right) = \frac{\pi}{n \sin \frac{p\pi}{n}},$$

a result Euler also discussed in an earlier paper on the beta integral.[33]

After this, Euler devoted the remaining portions of his paper to working out some particular cases, when $n = 1, 2, \ldots, 9$.

Euler did not explicitly evaluate the beta integral in terms of the gamma integral until 1772, when he wrote a paper[34] dealing with the integral

$$\int_0^1 x^{f-1}(\ln x)^{\frac{m}{n}} dx.$$

An unusual feature of this paper was that Euler explicitly specified when he was using a variable that took integer values and when he was using a variable taking arbitrary real values; he did not do this in his earlier papers on the gamma function. He began the paper with the theorem that if $n$ denoted a positive integer, then

$$\int_0^1 x^{f-1}(1-x^g)dx = \frac{g^n}{f} \cdot \frac{1 \cdot 2 \cdots n}{(f+g)(f+2g)\cdots(f+ng)}. \tag{17.40}$$

To prove this, he first performed the integration by parts calculations given in (17.32) through (17.33), and then wrote that this implied that

$$\int_0^1 x^{f-1}(1-x^g)^n dx = \frac{ng}{f+ng} \int_0^1 x^{f-1}(1-x^g)^{n-1} dx. \tag{17.41}$$

To obtain (17.40), he iterated (17.41) $n$ times and used the fact that

$$\int_0^1 x^{f-1}(1-x^g)^0 dx = \frac{1}{f}.$$

[33] See Eu. I-17 pp. 233–267. E 264, § 45.
[34] Eu. I-17 pp. 316–357, E 421.

In section 22 of this paper, Euler stated and proved what he called a general theorem: If $n$ is a positive integer and $f$ and $g$ denote any positive numbers, then

$$
\frac{1 \cdot 2 \cdot 3 \cdots n}{(\lambda n + 1)(\lambda n + 2)\cdots(\lambda n + n)}
$$
$$
= \frac{\lambda}{n+1} ng \int_0^1 x^{f+\lambda ng - 1}(1 - x^g)^{n-1} dx \cdot \frac{\int_0^1 x^{f-1}(1 - x^g)^{\lambda n - 1} dx}{\int_0^1 x^{f-1}(1 - x^g)^{(\lambda+1)n - 1} dx}. \quad (17.42)
$$

Euler proved this theorem by using (17.40) to find the values of integrals on the right-hand side of (17.42). In corollary 1 of this theorem, he substituted the integral

$$
\frac{\lambda n}{\lambda + 1} \int_0^1 x^{\lambda n - 1}(1 - x)^{n-1} dx
$$

for the left-hand side of (17.42).

Next, in section 24, Euler made a remarkable claim: This theorem holds, even if $n$ is not an integer. Since $\lambda$ is arbitrary, $\lambda n$ can be replaced by $m$, to obtain

$$
\int_0^1 x^{m-1}(1 - x)^{n-1} dx = g \int_0^1 x^{f+mg-1}(1 - x^g)^{n-1} dx \cdot \frac{\int_0^1 x^{f-1}(1 - x^g)^{m-1} dx}{\int_0^1 x^{f-1}(1 - x^g)^{m+n-1} dx}.
$$
$$(17.43)$$

Euler's reason for this conclusion appears to be that, though the theorem had been proved for the case in which $n$ was an integer, the integrals made sense in the case where $n$ was any positive real number, and hence the result must be true for any positive real number. Legendre followed Euler very closely in his exposition of the Eulerian integrals; in the 1811 first volume of his *Exercices de calcul intégral*, Legendre gave an additional comment on this point.[35] On page 279, he wrote that both sides of (17.43) equal the same function of $n$ and $m = \lambda n$:

$$
\frac{1}{\lambda n} - \frac{n-1}{1} \cdot \frac{1}{\lambda n + 1} + \frac{(n-1)(n-2)}{1 \cdot 2} \cdot \frac{1}{\lambda n + 2} - \cdots . \quad (17.44)
$$

Legendre's point was that in (17.44), $n$ and $\lambda n$ need not be integers and could be replaced by any positive real numbers. Therefore, the extension of the formula from positive integers to positive real numbers made sense to him. To continue Euler's argument, he next observed that $1 - x^g = g \ln \frac{1}{x}$ when $g = 0$, so that, setting $f = 1$, (17.43) gave him

$$
\int_0^1 x^{m-1}(1 - x)^{n-1} dx = \frac{\int_0^1 \left(\ln \frac{1}{x}\right)^{n-1} dx \int_0^1 \left(\ln \frac{1}{x}\right)^{m-1} dx}{\int_0^1 \left(\ln \frac{1}{x}\right)^{m+n-1} dx}. \quad (17.45)
$$

[35] Legendre (1811–1817) vol. I, p. 279.

Note that from (17.6),

$$\Gamma(s) = \int_0^1 \left( \ln \frac{1}{x} \right)^{s-1} dx,$$

(17.45) can be written as

$$\int_0^1 x^{m-1}(1-x)^{n-1} dx = \frac{\Gamma(n)\,\Gamma(m)}{\Gamma(m+n)}.$$

Concerning the extension of formulas from integer values to real or even complex values, we recall that Euler had made a similar argument in proving the formula (4.18):

$$\binom{n}{0}\binom{m}{k} + \binom{n}{1}\binom{m}{k-1} + \cdots + \binom{n}{k}\binom{m}{0} = \binom{m+n}{k}.$$

This argument of Euler and Legendre is an early example of George Peacock's flawed principle of permanence of equivalent forms. George Peacock stated his principle in his book on symbolic algebra, the second volume of his algebra book:[36] "Whatever algebraical forms are equivalent, when the symbols are general in form but specific in value, will be equivalent likewise when the symbols are general in value as well as in form." According to such a principle, since the two sides of (17.43) are identical with algebraic form (17.44) when $n$ and $\lambda n$ are integers, they remain equivalent when $m$ and $\lambda n$ are general, that is, real or complex values. It is surprising to see Peacock make a claim of this kind after Gauss and Cauchy had already clarified the matter. See, for example, our Chapter 4 for Cauchy's proof of (4.18), and Chapter 23 for Gauss's remarks on analytic continuation.

In fact, Fritz Carlson's uniqueness or interpolation theorem is required to generalize from integers to all real or complex values in the half-plane Re $z > 0$. Carlson's theorem[37] states that if: $f(z)$ is analytic in Re $z > \delta > 0$; $f(z) = O(e^{k|z|})$, $k < \pi$; and $f(z) = 0$ for $z = 1, 2, 3, \ldots$, then: $f(z)$ is identically zero.

For the case here at hand, it would be sufficient to assume that $f(z)$ was bounded. Atle Selberg gave a simple proof for this case using only Cauchy's residue theorem.[38] To prove Carlson's theorem in general, the machinery of the Phragmén–Lindelöf theorems is required. To apply Carlson's theorem here, first observe that the formula

$$\int_0^1 x^{m-1}(1-x)^{n-1} dx = \frac{\Gamma(m)\,\Gamma(n)}{\Gamma(m+n)}$$

holds when $n$ is a positive integer; the integral is bounded and analytic in Re $n \geq \delta > 0$, as is the right-hand side. Thus, by Carlson's theorem, the formula holds when $n$ is complex and Re $n \geq \delta > 0$.[39]

---

[36] Peacock (1842–1845) vol. 2, p. 59.

[37] Carlson (1914) or see Titchmarsh (1962) sections 5.6–5.8.

[38] Selberg (1944).

[39] Andrews et al. (1999) pp. 110–111.

## 17.5    Newman and the Product for $\Gamma(x)$

Francis W. Newman (1805–1897) earned a double first class from Oxford University in 1826 and was elected a fellow of Balliol college the same year. He was a professor of Latin at University College, London, when he published his insightful 1848 paper on the gamma function.[40] Reflecting the breadth and depth of his scholarship, that same year Newman published *A History of the Hebrew Monarchy* and, a year later, *The Soul, Her Sorrows and Aspirations*. His interests included political economy, grammar, and languages. For example, in 1872, he published a two-volume dictionary of modern Arabic.

In the course of reworking Euler's formula (17.38), Newman arrived at the result: For any real or complex number $m$

$$\Gamma(m) = \frac{e^{-\gamma m}}{m} \prod_{k=1}^{\infty} \left(1 + \frac{m}{k}\right)^{-1} e^{\frac{m}{k}}, \qquad (17.46)$$

where $\gamma$ was Euler's constant defined by

$$\gamma = \lim_{k \to \infty} \left(1 + \frac{1}{2} + \frac{1}{3} + \cdots + \frac{1}{k} - \ln k\right). \qquad (17.47)$$

The proof that this limit exists is given in our Chapter 20. This representation of the gamma function is sometimes attributed to Weierstrass, who defined the function by a similar infinite product in 1856. In fact, the German mathematician Schlömilch discovered (17.46) even earlier, in 1843;[41] his proof appears in the exercises. Newman wrote the product in (17.38) as

$$\frac{m+n}{mn} \cdot \frac{1+(m+n)}{1+m \cdot 1+n} \cdot \frac{1+\frac{1}{2}(m+n)}{1+\frac{1}{2}m \cdot 1+\frac{1}{2}n} \cdot \frac{1+\frac{1}{3}(m+n)}{1+\frac{1}{3}m \cdot 1+\frac{1}{3}n} \cdots$$

and then observed that the product

$$\psi(m) = m(1+m)\left(1 + \frac{1}{2}m\right)\left(1 + \frac{1}{3}m\right)\cdots$$

was divergent because its logarithm was

$$\ln m + \left(1 + \frac{1}{2} + \frac{1}{3} + \cdots\right)m - \frac{1}{2}\left(1 + \frac{1}{2^2} + \frac{1}{3^2} + \cdots\right)m^2$$
$$+ \frac{1}{3}\left(1 + \frac{1}{2^3} + \cdots\right)m^3 - \cdots$$

and the coefficient of $m$ was a divergent series. To remove this defect, he defined a new product,

---

[40] Newman (1848).
[41] Schlömilch (1843).

$$m \cdot \frac{1+m}{e^m} \cdot \frac{1+\frac{1}{2}m}{e^{\frac{1}{2}m}} \cdot \frac{1+\frac{1}{3}m}{e^{\frac{1}{3}m}} \cdots , \tag{17.48}$$

whose logarithm did not have this divergent part, so that the product converged.

Newman denoted the left-hand side of (17.38) by $B(m,n)$, although Newman actually used $F$ instead of $B$, so that he could write

$$B(m,n) = \frac{\psi(m+n)}{\psi(m)\psi(n)}, \tag{17.49}$$

where the $\psi$s now represented convergent products, that is, $\psi(m)$ was defined by the product in (17.49). He then changed $m$ to $m+k$ in (17.49) to obtain

$$B(m+k,n) = \psi(m+k+n) \div \big(\psi(m+k)\psi(n)\big),$$

whence

$$\frac{B(m,n)}{B(m+k,n)} = \frac{\psi(m+n)\psi(m+k)}{\psi(m)\psi(m+n+k)}. \tag{17.50}$$

Since interchanging $n$ and $k$ did not change the right-hand side of (17.50), Newman found out that

$$\frac{B(m,n)}{B(m+k,n)} = \frac{B(m,k)}{B(m+n,k)}. \tag{17.51}$$

Now a change of variables $t$ to $1-t$ showed that

$$B(m,n) = \int_0^1 t^{m-1}(1-t)^{n-1}dt = \int_0^1 t^{n-1}(1-t)^{m-1}dt = B(n,m);$$

hence by (17.51)

$$B(m,n) = \frac{B(k,m)B(k+m,n)}{B(k,m+n)}. \tag{17.52}$$

Newman let $k \to \infty$ and used (17.36) to obtain $B(k+m,n) = B(k,n)$ so that he got

$$B(m,n) = \frac{B(k,m)B(k,n)}{B(k,m+n)}, \quad k = \infty. \tag{17.53}$$

At this point, Newman remarked that (17.53) "had a close analogy" with (17.49) and this led him to further investigate $B(k,m)$ when $k = \infty$. He set $t = \frac{y}{k}$ to find

$$B(k+1,m) = B(m,k+1) = \int_0^1 t^{m-1}(1-t)^k dt = \int_0^\infty \left(1 - \frac{y}{k}\right)^k y^{m-1}k^{-m}dy$$

$$= k^{-m}\int_0^\infty e^{-y}y^{m-1}dy. \tag{17.54}$$

Following Legendre, Newman denoted the integral by $\Gamma(m)$. He then applied (17.54) in (17.53) to obtain

$$B(m,n) = \frac{\Gamma(m)\Gamma(n)}{\Gamma(m+n)}. \tag{17.55}$$

Comparing (17.49) with (17.55), Newman concluded that $\Gamma(m) = \chi(m)(\psi(m))^{-1}$, where $\chi$ represented an unknown function. Substituting this in (17.55) and comparing with (17.49), he saw that

$$\chi(m+n) = \chi(m)\chi(n),$$

an equation solvable by differentiation or by even more elementary methods to get

$$\chi(m) = e^{\gamma m},$$

with $\gamma$ an unknown constant. Thus, he could write

$$\Gamma(m) = \frac{e^{\gamma m}}{m} \prod_{k=1}^{\infty} \left(1 + \frac{m}{k}\right)^{-1} e^{\frac{m}{k}}. \tag{17.56}$$

To determine the value of $\gamma$, Newman took the logarithm of equation (17.56) to find

$$\log\Gamma(m) = -\log m - \gamma m + (m - \log(1+m)) + \left(\frac{m}{2} - \log\left(1 + \frac{m}{2}\right)\right)\cdots$$

and then set $m = 1$ to discover that

$$\gamma = \lim_{k\to\infty} (1 - \log 2) + \left(\frac{1}{2} - \log\frac{3}{2}\right) + \cdots + \left(\frac{1}{k} - \log\left(1 + \frac{1}{k}\right)\right)$$

$$= \lim_{k\to\infty} \left(1 + \frac{1}{2} + \cdots + \frac{1}{k} - \log(k+1)\right).$$

Hence, $\gamma$ was Euler's constant.

In this way, Newman started from the beta integral $B(m,n) = \int_0^1 t^{m-1}(1-t)^{n-1}dt$, $m, n > 0$, and obtained the basic formulas for the gamma function, based on the functional relation that Euler had found by integration by parts:

$$B(m,n) = \frac{m+n}{m} B(m+n,n). \tag{17.57}$$

Both Euler and Newman iterated the functional relation (17.57) to end up with a divergent product multiplied by a vanishing integral, as in (17.34). For this reason Euler needed another sequence of beta integrals, (17.35), that also tended to zero, multiplied by a divergent product, so that he could take the ratio and thereby find a finite nonzero value. See, for example, (17.38). However, Newman's work, particularly equation (17.54), shows that it might be possible to avoid this difficulty.

To see this modification of the Euler–Newman argument, found by Richard Askey,[42] first note that from (17.57) it is easy to obtain the functional relation

$$B(m,n) = \frac{m+n}{n} B(m, n+1). \tag{17.58}$$

Iterate (17.58) $k$ times and change variables as was done in (17.54) to arrive at

$$B(m,n) = \frac{(m+n)(m+n+1)\cdots(m+n+k-1)}{n(n+1)\cdots(n+k-1)}$$
$$\times \int_0^k \left(\frac{t}{k}\right)^{m-1} \left(1-\frac{t}{k}\right)^{n+k-1} \frac{dt}{k}$$

$$= \frac{(m+n)\cdots(m+n+k-1)}{k!\, k^{m+n-1}} \cdot \frac{k!\, k^{n-1}}{n\cdots(n+k-1)}$$
$$\times \int_0^k t^{m-1}\left(1-\frac{t}{k}\right)^{k+n-1} dt. \tag{17.59}$$

Taking (17.4) as the definition of the gamma function, we can then take the limit in (17.59) as $k \to \infty$ to see that

$$B(m,n) = \frac{\Gamma(n)}{\Gamma(m+n)} \int_0^\infty e^{-t} t^{m-1} dt. \tag{17.60}$$

Taking $n = 1$ in (17.60) and observing that $\Gamma(m+1) = m\Gamma(m)$ and that $B(m,1) = \frac{1}{m}$, we get

$$\Gamma(m) = \int_0^\infty e^{-t} t^{m-1} dt. \tag{17.61}$$

Newman's product for the gamma function can then be obtained directly:

$$\Gamma(m) = \lim_{k\to\infty} \frac{k!\, k^{m-1}}{m(m+1)\cdots(m+k-1)}$$

$$= \lim_{k\to\infty} \frac{k^m}{m\left(1+\frac{m}{1}\right)\left(1+\frac{m}{2}\right)\cdots\left(1+\frac{m}{k-1}\right)}$$

$$= \lim_{k\to\infty} \frac{e^{m\log k}}{m} \prod_{s=1}^{k-1}\left(1+\frac{m}{s}\right)^{-1}$$

$$= \lim_{k\to\infty} \frac{1}{m} e^{-m(1+\frac{1}{2}+\cdots+\frac{1}{k-1}-\log k)} \prod_{s=1}^{k-1}\left(1+\frac{m}{s}\right)^{-1} e^{\frac{m}{s}}$$

$$= \frac{e^{-\gamma m}}{m} \prod_{s=1}^{\infty}\left(1+\frac{m}{s}\right)^{-1} e^{\frac{m}{s}}.$$

[42] Andrews et al. (1999) pp. 5–6.

Euler could have obtained (17.55) by using (17.4) in (17.38). Thus, the left-hand side of (17.38) is clearly $B(m,n)$, leading us to

$$
\begin{aligned}
B(m,n) &= \lim_{s\to\infty} \frac{s!\,(m+n)\ldots(m+n+s)}{n(n+1)\cdots(n+s)\,m(m+1)\cdots(m+s)} \\
&= \lim_{s\to\infty} \left( \frac{s!\,s^n}{n(n+1)\cdots(n+s)} \cdot \frac{s!\,s^m}{m(m+1)\cdots(m+s)} \right. \\
&\qquad\qquad \left. \cdot \frac{(m+n)\cdots(m+n+s)}{s!\,s^{m+n}} \right) \\
&= \frac{\Gamma(n)\Gamma(m)}{\Gamma(m+n)}.
\end{aligned}
$$

## 17.6 Gauss's Theory of the Gamma Function

Gauss's work on the gamma function[43] was marked by his systematic approach and a greater sense of rigor than was earlier practiced. He discussed the convergence of series and products but did not justify changing the order of limits, or term-by-term integration of infinite series. He started with the finite product

$$
\Pi(k,z) = \frac{1\cdot 2\cdot 3\cdots k}{(z+1)(z+2)\cdots(z+k)}k^z, \tag{17.62}
$$

where $k$ was a positive integer and $z$ a complex number not equal to a negative integer. He first proved that the limit as $k \to \infty$ existed. For this purpose, he noted that

$$
\Pi(k+n,z) = \Pi(k,z)\frac{\left(1-\frac{1}{k+1}\right)^{-z}}{1+\frac{z}{k+1}}\cdots\frac{\left(1-\frac{1}{k+n}\right)^{-z}}{1+\frac{z}{k+n}}, \tag{17.63}
$$

and that the logarithm of the product written after $\Pi(k,z)$ remained finite as $n \to \infty$. This proved that the limit existed.

Wallis and Euler had perceived the significance of the gamma function for the evaluation of certain definite integrals. Gauss's important contribution here was to use the gamma function to sum series; some early hints of this also appeared in the works of Stirling, Euler, and Pfaff. Gauss's insight opened up the subject of summation of series of the hypergeometric type. Moreover, Gauss used (17.14) to establish the basic results on the gamma function. Interestingly, in this connection Gauss made use of two series discovered by Newton while he was a student at Cambridge:

$$
\arcsin x = x + \frac{\frac{1}{2}}{1!}\frac{x^3}{3} + \frac{\frac{1}{2}\cdot\frac{3}{2}}{2!}\frac{x^5}{5} + \frac{\frac{1}{2}\cdot\frac{3}{2}\cdot\frac{5}{2}}{3!}\frac{x^7}{7} + \cdots
$$

[43] Gauss (1813).

and

$$\sin n\theta = n\sin\theta - \frac{n(n^2 - 1^2)}{3!}\sin^3\theta + \frac{n(n^2 - 1^2)(n^2 - 3^2)}{5!}\sin^5\theta - \cdots,$$

where $n$ was not necessarily an integer. Gauss's analysis of the convergence of the hypergeometric series showed that the first formula was true for $|x| \leq 1$ and the second for $|\theta| \leq \frac{\pi}{2}$. By contrast, Newton took a much more cavalier approach toward convergence questions. He is known to have discussed the convergence of the geometric series on one occasion, but his remarks contained no new insights. To write (17.14) in a compact form, we employ the modern notation for a shifted factorial:

$$\begin{aligned}
(a)_n &= a(a+1)(a+2)\cdots(a+n-1), & &\text{for } n > 0, \\
&= 1, & &\text{for } n = 0. \qquad (17.64)
\end{aligned}$$

We can now write Gauss's formula in the form

$$\sum_{n=0}^{\infty} \frac{(a)_n(b)_n}{n!\,(c)_n} = \frac{\Gamma(c)\Gamma(c-a-b)}{\Gamma(c-a)\Gamma(c-b)}, \qquad (17.65)$$

when $\mathrm{Re}(c - a - b) > 0$. To obtain the value of $\Gamma(\frac{1}{2})$, Gauss took $x = 1$ in Newton's series for $\arcsin x$ and used (17.65) to get

$$\frac{\pi}{2} = \sum_{n=0}^{\infty} \frac{(\frac{1}{2})_n(\frac{1}{2})_n}{n!\,(\frac{3}{2})_n} = \frac{\Gamma(\frac{3}{2})\Gamma(\frac{1}{2})}{\Gamma(1)\Gamma(1)} = \frac{1}{2}\left(\Gamma\left(\frac{1}{2}\right)\right)^2 \qquad (17.66)$$

or

$$\Gamma\left(\frac{1}{2}\right) = \sqrt{\pi}. \qquad (17.67)$$

He then derived Euler's reflection formula (17.11) from (17.65) by taking $\theta = \frac{\pi}{2}$ in Newton's series for $\sin n\theta$ where $n$ was any real number. In that case,

$$\begin{aligned}
\sin\frac{n\pi}{2} &= n - \frac{n(n-1)(n+1)}{2\cdot 3} + \frac{n(n-1)(n+1)(n-3)(n+3)}{2\cdot 3\cdot 4\cdot 5}\cdots \\
&= n\sum_{k=0}^{\infty} \frac{\left(\frac{-n+1}{2}\right)_k \left(\frac{n+1}{2}\right)_k}{k!\,(\frac{3}{2})_k} \\
&= n\frac{\Gamma(\frac{3}{2})\Gamma(\frac{1}{2})}{\Gamma(\frac{3}{2}+\frac{n}{2}-\frac{1}{2})\Gamma(\frac{3}{2}-\frac{n}{2}-\frac{1}{2})} \\
&= \frac{n\pi}{2\Gamma(1+\frac{n}{2})\Gamma(1-\frac{n}{2})}. \qquad (17.68)
\end{aligned}$$

Note that in the last two steps, (17.67) and (17.65) were employed. Gauss next set $x = \frac{n}{2}$ in (17.68) to obtain

$$\Gamma(1+x)\Gamma(1-x) = \frac{\pi x}{\sin \pi x} \qquad (17.69)$$

or

$$\Gamma(x)\Gamma(1-x) = \frac{\pi}{\sin \pi x}. \qquad (17.70)$$

Finally, to derive (17.10) from (17.65), he wrote

$$\int_0^1 x^{m-1}(1-x)^{n-1}\,dx$$

$$= \int_0^1 x^{m-1}\left(1-(n-1)x + \frac{(n-1)(n-2)}{2!}x^2 - \cdots\right)dx$$

$$= \frac{1}{m} - \frac{n-1}{m+1} + \frac{(n-1)(n-2)}{2!\,(m+2)} - \frac{(n-1)(n-2)(n-3)}{3!\,(m+3)} + \cdots$$

$$= \frac{1}{m}\left[1 + \frac{(-n+1)m}{m+1} + \frac{(-n+1)(-n+2)m(m+1)}{2!\,(m+1)(m+2)} + \cdots\right]$$

$$= \frac{1}{m}\cdot\frac{\Gamma(m+1+n-1-m)\Gamma(m+1)}{\Gamma(m+1+n-1)\Gamma(m+1-m)} = \frac{\Gamma(m)\Gamma(n)}{\Gamma(m+n)}. \qquad (17.71)$$

Gauss derived the integral representation for the gamma function by setting $y = nx$ in (17.71), where $n$ was an integer, to obtain

$$\int_0^n y^{m-1}\left(1 - \frac{y}{n}\right)^n dy = \frac{n!\,n^{m-1}}{m(m+1)\cdots(m+n-1)}.$$

The limit as $n \to \infty$ gave

$$\int_0^\infty y^{m-1}e^{-y}\,dy = \Gamma(m),$$

a result also due to Euler. Gauss did not justify that $\lim_{n\to\infty}\int = \int \lim_{n\to\infty}$ in this situation.

Gauss also defined a new function $\Psi(z)$, given by

$$\Psi(z) = \frac{d}{dz}\ln\Gamma(z+1) = \frac{\Gamma'(z+1)}{\Gamma(z+1)}. \qquad (17.72)$$

He observed that $\Psi(z)$, the digamma function, was almost as remarkable a function as $\Gamma(z)$ and noted some of its more important properties. See the exercises for some of these properties.

In section 26 of his 1813 paper, Gauss gave an elegant proof of the multiplication formula (17.15). First set

$$\Gamma(k,z) = \frac{k!\, k^{z-1}}{z(z+1)\cdots(z+k-1)}.$$

Gauss started by observing that for a positive integer $n$, the expression

$$\frac{n^{nz}\Gamma(k,z+1)\,\Gamma\left(k,z+1-\frac{1}{n}\right)\Gamma\left(k,z+1-\frac{2}{n}\right)\cdots\Gamma\left(k,z+1-\frac{n-1}{n}\right)}{\Gamma(nk,nz+1)} \tag{17.73}$$

could be reduced to

$$\frac{(k!)^n n^{nk}}{(nk)!\, k^{\frac{n-1}{2}}},$$

and hence was independent of $z$. Therefore, the value of the expression (17.73) was equal to its value at $z=0$, that is, since $\Gamma(k,1)=\Gamma(nk,1)=1$,

$$\Gamma(k,1)\,\Gamma\left(k,1-\frac{1}{n}\right)\Gamma\left(k,1-\frac{2}{n}\right)\cdots\Gamma\left(k,1-\frac{n-1}{n}\right)$$
$$= \Gamma\left(k,1-\frac{1}{n}\right)\Gamma\left(k,1-\frac{2}{n}\right)\cdots\Gamma\left(k,\frac{1}{n}\right). \tag{17.74}$$

He then let $k\to\infty$ in (17.73) and (17.74) to find that

$$\frac{n^{nz}\Gamma(z+1)\,\Gamma\left(z+1-\frac{1}{n}\right)\cdots\Gamma\left(z+\frac{1}{n}\right)}{\Gamma(nz+1)}$$
$$= \Gamma\left(\frac{1}{n}\right)\Gamma\left(\frac{2}{n}\right)\cdots\Gamma\left(1-\frac{2}{n}\right)\Gamma\left(1-\frac{1}{n}\right). \tag{17.75}$$

To determine the value of the right-hand side of (17.75), Gauss multiplied it by itself in reverse order, and used the reflection formula (17.11) to get

$$\Gamma\left(\frac{1}{n}\right)\Gamma\left(1-\frac{1}{n}\right)\Gamma\left(\frac{2}{n}\right)\Gamma\left(1-\frac{2}{n}\right)\cdots\Gamma\left(1-\frac{1}{n}\right)\Gamma\left(\frac{1}{n}\right)$$
$$= \frac{\pi}{\sin\frac{\pi}{n}}\cdot\frac{\pi}{\sin\frac{2\pi}{n}}\cdots\frac{\pi}{\sin\frac{(n-1)\pi}{n}} = \frac{(2\pi)^{n-1}}{n}. \tag{17.76}$$

Next, following Gauss, write $\Gamma(z+1)=z\Gamma(z)$ and $\Gamma(nz+1)=nz\Gamma(nz)$ in (17.75). Then, use (17.76) in the form

$$\Gamma\left(\frac{1}{n}\right)\Gamma\left(\frac{2}{n}\right)\cdots\Gamma\left(\frac{n-1}{n}\right) = \frac{(2\pi)^{\frac{n-1}{2}}}{\sqrt{n}}$$

to obtain the multiplication formula.

## 17.7 Euler: Series to Product

In a very interesting paper,[44] "De mirabilibus propriatibus unciarum quae in evolutione binomii occurrent," Euler stated two theorems, given here in modern notation.

Theorem 1

$$1 + \binom{n}{1}^2 + \binom{n}{2}^2 + \binom{n}{3}^2 + \cdots = \frac{2}{\pi} \cdot 2^{2n} \int_0^1 \frac{x^{2n} \, dx}{\sqrt{1 - x^2}}. \tag{17.77}$$

Theorem 2

$$1 + \binom{n}{1}\binom{n'}{1} + \binom{n}{2}\binom{n'}{2} + \binom{n}{3}\binom{n'}{3} + \cdots = \frac{1}{n \int_0^1 x^{n'}(1 - x)^{n-1} \, dx}. \tag{17.78}$$

In this paper, Euler wrote $\begin{bmatrix} n \\ k \end{bmatrix}$ for $\binom{n}{k}$ so that his notation here is very similar to our modern notation. Of course, the first theorem is a special case of the second, though this may not be immediately obvious. But to see this, observe that an application of (17.10) shows that the right-hand side of (17.78) is equal to

$$\frac{\Gamma(n + n' + 1)}{\Gamma(n + 1) \Gamma(n' + 1)}. \tag{17.79}$$

By making the substitution $y = x^2$ in the integral in (17.77) and using (17.10) and (17.7), we perceive that the right-hand side of (17.77) can be rewritten as

$$\frac{2^{2n} \Gamma\left(n + \frac{1}{2}\right)}{\sqrt{\pi} \, \Gamma(n + 1)}. \tag{17.80}$$

Finally, it follows from Legendre's duplication formula, i.e. the formula for the case $n = 2$ in (17.15), that when $n = n'$, the expression in (17.79) is identical with that in (17.80).

In fact, Euler was unable to prove his two formulas (17.77) and (17.78). After stating theorem 1, he wrote that he found it remarkable that there was no direct way to prove it in general. Instead, in his paper he verified a large number of particular cases and did the same thing for theorem 2. In particular, he checked the case for theorem 2 in which $n' = -n, 0 < n < 1$. His verification here is highly interesting because he showed how a particular sum can be converted into a product. First note that for $n' = -n$, (17.11) shows that (17.78) is equal to

$$\frac{1}{\Gamma(1 + n) \Gamma(1 - n)} = \frac{\sin n\pi}{n\pi}.$$

---

[44] Eu. I-15 pp. 528–568. E 575.

But Euler evaluated the integral in (17.78) for $n' = -n$ in a different manner. He transformed the integral into

$$\int_0^\infty \frac{y^{n'-1}\,dy}{1+y},$$

the value of which he already knew from his work on the integration of rational functions: $\frac{\pi}{\sin n'\pi}$. In this connection, see (17.13). Thus, in section 44 of his paper, Euler found that he needed to verify the formula

$$1 - \frac{n^2}{1^2} + \frac{n^2(n^2-1)}{1^2\cdot 2^2} - \frac{n^2(n^2-1)(n^2-2^2)}{1^2\cdot 2^2\cdot 3^2} + \cdots = \frac{\sin n\pi}{n\pi}, \qquad (17.81)$$

in which the series is obtained after taking $n' = -n$ on the left-hand side of (17.78). To verify the sum in (17.81), he denoted the expression on its left-hand side by $S$ and observed that $1-n^2$ was a common factor of the sum. Dividing by this common factor, he arrived at

$$\frac{S}{1-n^2} = 1 - \frac{n^2}{2^2} + \frac{n^2(n^2-2^2)}{1^2\cdot 2^2\cdot 3^2} - \frac{n^2(n^2-2^2)(n^2-3^2)}{1^2\cdot 2^2\cdot 3^2\cdot 4^2} + \cdots. \qquad (17.82)$$

One can now see that $1 - \frac{n^2}{2^2}$ is a common factor in the sum in (17.82). Thus, rewrite the sum contained in (17.82) as

$$\frac{S}{(1-n^2)\left(1 - \frac{n^2}{2^2}\right)} = 1 - \frac{n^2}{3^2} + \frac{n^2(n^2-3^2)}{1^2\cdot 3^2\cdot 4^2} - \cdots, \qquad (17.83)$$

where $1 - \frac{n^2}{3^2}$ is clearly the common factor. Repeating this process infinitely often, Euler found that

$$\frac{S}{(1-n^2)\left(1 - \frac{n^2}{2^2}\right)\left(1 - \frac{n^2}{3^2}\right)\cdots} = 1; \qquad (17.84)$$

since the product was $\frac{\sin n\pi}{n\pi}$, Euler had succeeded in verifying the case in his theorem 2 for $n' = -n$. Observe that in moving from the series in (17.81) to the product in (17.84), Euler employed (15.69) in reverse, that is

$$1 - \alpha - \beta(1-\alpha) - \gamma(1-\alpha)(1-\beta) - \delta(1-\alpha)(1-\beta)(1-\gamma) - \cdots$$
$$= (1-\alpha)(1-\beta)(1-\gamma)(1-\delta)\cdots .$$

Based on this work of Euler, we point out that if (17.81) could be proved by a different method, we would have a new and remarkable proof of the product formula for $\sin \pi x$.

It turns out that (17.78) is a particular case of Gauss's summation formula (17.65). Observe that the series on the left-hand side of (17.78) may be written as

$$1 + \frac{(-n)(-n')}{1!\,1!} + \frac{(-n)(-n+1)(-n!)(-n'+1)}{2!\,2!} + \cdots;$$

use the notation defined in (17.64) to rewrite as

$$\sum_{k=0}^{\infty} \frac{(-n)_k(-n')_k}{k!\,(1)_k} = \frac{\Gamma(1)\,\Gamma(n+n'+1)}{\Gamma(n+1)\,\Gamma(n'+1)}.$$

This completes our proof of (17.78) and therefore also of (17.81).

## 17.8  Euler: Products to Continued Fractions

In his 1739 paper "De fractionibus continuis, observationes,"[45] Euler investigated the relation of certain infinite products with continued fractions. For example, he considered the infinite product

$$\frac{p(p+2q+r)(p+2r)(p+2q+3r)\cdots}{(p+2q)(p+r)(p+2q+2r)(p+3r)\cdots}. \tag{17.85}$$

This and other similar products had earlier appeared prominently in Euler's work on the gamma and beta functions. In this paper, Euler noted that the product (17.85) was a ratio of beta integrals:

$$\frac{\int_0^1 y^{p+2q-1}(1-y^{2r})^{-\frac{1}{2}}\,dy}{\int_0^1 y^{p-1}(1-y^{2r})^{-\frac{1}{2}}\,dy}. \tag{17.86}$$

In an analysis quite similar to one carried out by Wallis, he associated a sequence $A_0, A_1, A_2, \ldots$ (he wrote $A, B, C, \ldots$) with the product. The sequence was defined by the relations

$$A_0 A_1 = \frac{p}{p+2q}, \quad A_2 A_3 = \frac{p+2r}{p+2q+2r}, \quad A_4 A_5 = \frac{p+4r}{p+2q+4r}, \ldots \tag{17.87}$$

He added the requirement that

$$A_1 A_2 = \frac{p+r}{p+2q+r}, \quad A_3 A_4 = \frac{p+3r}{p+2q+3r}, \quad A_5 A_6 = \frac{p+5r}{p+2q+5r}, \ldots \tag{17.88}$$

Observe that the relations (17.87) and (17.88) gave Euler

$$
\begin{aligned}
A_0 &= \frac{p}{p+2q} \cdot \frac{1}{A_1} = \frac{p}{p+2q} \cdot \frac{p+2q+r}{p+r} \cdot A_2 \\
&= \frac{p}{p+2q} \cdot \frac{p+2q+r}{p+r} \cdot \frac{p+2r}{p+2q+2r} \cdot \frac{1}{A_3} \\
&= \cdots\cdots.
\end{aligned}
$$

45  E. I-14 pp. 219–349. E123, § 21–24.

Euler desired a continued fraction representation for the infinite product $A_0$, given by (17.85). To eliminate the denominators, Euler defined another sequence $a_0, a_1, a_2, \ldots$ by the relations

$$A_0 = \frac{a_0}{p + 2q - r}, \quad A_1 = \frac{a_1}{p + 2q}, \quad A_2 = \frac{a_2}{p + 2q + r}, \quad A_3 = \frac{a_3}{p + 2q + 2r}, \cdots,$$

$$(17.89)$$

so that

$$a_0 a_1 = (p + 2q - r)p, \quad a_1 a_2 = (p + 2q)(p + r), \tag{17.90}$$

$$a_2 a_3 = (p + 2q + r)(p + 2r), \cdots. \tag{17.91}$$

He then set

$$a_0 = m - r + \frac{1}{\alpha_1}, \quad a_1 = m + \frac{1}{\alpha_2}, \tag{17.92}$$

$$a_2 = m + r + \frac{1}{\alpha_3}, \quad a_4 = m + 2r + \frac{1}{\alpha_4}, \cdots \tag{17.93}$$

to obtain a continued fraction for $a_0$ dependent on $m$. Thus, Euler potentially had an infinite number of continued fractions depending on the parameter $m$ and each of these continued fractions thus had the value $a_0$. He then chose special values of $m$ such as $p - r$, $p + q$, $p + 2q$ to obtain several interesting continued fractions for specific $a_0$.

To simplify the equations satisfied by $\alpha_1, \alpha_2, \alpha_3, \ldots$, Euler set

$$P = p(p + 2q - r) - m(m - r) \quad \text{and} \quad Q = 2r(p + q - m).$$

Then he could write

$$P\alpha_1 \alpha_2 - (m - r)\alpha_1 = m\alpha_2 + 1, \tag{17.94}$$

$$(P + Q)\alpha_2 \alpha_3 - m\alpha_2 = (m + r)\alpha_3 + 1, \tag{17.95}$$

$$(P + 2Q)\alpha_3 \alpha_4 - (m + r)\alpha_3 = (m + 2r)\alpha_4 + 1, \tag{17.96}$$

$$(P + 3Q)\alpha_4 \alpha_5 - (m + 2r)\alpha_4 = (m + 3r)\alpha_5 + 1, \tag{17.97}$$

$$\cdots \cdots, \tag{17.98}$$

relations that are easy to obtain. Euler did not give the details; for the convenience of the reader, we indicate how they may be derived. First, multiplying the equations in (17.92) produces

$$a_0 a_1 = \left(m - r + \frac{1}{\alpha_1}\right)\left(m + \frac{1}{\alpha_2}\right).$$

Then, using the first equation in (17.90), we have

$$p(p + 2q - r) = \left(m - r + \frac{1}{\alpha_1}\right)\left(m + \frac{1}{\alpha_2}\right)$$

or

$$p(p + 2q - r)\alpha_1\alpha_2 = m(m - r)\alpha_1\alpha_2 + (m - r)\alpha_1 + m\alpha_2 + 1$$

or

$$P\alpha_1\alpha_2 - (m - r)\alpha_1 = m\alpha_2 + 1,$$

the first of Euler's relations, (17.94). The second, (17.95), can be verified in a similar manner by multiplying the second equation in (17.92) and the first equation in (17.93) to find

$$a_1a_2 = \left(m + \frac{1}{\alpha_2}\right)\left(m + r + \frac{1}{\alpha_3}\right); \tag{17.99}$$

combining (17.99) with the second equation in (17.90) then produces

$$(p + 2q)(p + r)\alpha_2\alpha_3 = m(m + r)\alpha_2\alpha_3 + m\alpha_2 + (m + r)\alpha_3 + 1,$$

leading us to

$$(P + Q)\alpha_2\alpha_3 - m\alpha_2 = (m + r)\alpha_3 + 1,$$

or (17.95). Thus we see how to derive Euler's relations (17.94) through (17.97) and so on, from which he deduced

$$\alpha_1 = \frac{m\alpha_2 + 1}{P\alpha_2 - (m - r)} = \frac{m}{P} + \frac{p(p + 2q - r) : P^2}{-(m - r) : P + \alpha_2}, \quad \text{where} \quad x : y = \frac{x}{y},$$

$$\alpha_2 = \frac{(m + r)\alpha_3 + 1}{(P + Q)\alpha_3 - m} = \frac{m + r}{P + Q} + \frac{(p + r)(p + 2q) : (P + Q)^2}{-m : (P + Q) + \alpha_3},$$

$$\alpha_3 = \frac{(m + 2r)\alpha_4 + 1}{(P + 2Q)\alpha_4 - (m + r)} = \frac{m + 2r}{P + 2Q} + \frac{(p + 2r)(p + 2q + r) : (P + 2Q)^2}{-(m + r) : (P + 2Q) + \alpha_4}, \quad \text{etc.}$$

To write the resulting continued fraction for $\alpha_1$ in simpler form, Euler set

$$R = p^2 + 2pq - mp - mq + qr \quad \text{and} \quad S = pr + qr - mr$$

and obtained the continued fraction for $\alpha_1$ as

$$\alpha_1 = \frac{m}{P} + \frac{p(p + 2q - r) : P^2}{2rR : P(P + Q)+} \frac{(p + r)(p + 2q) : (P + Q)^2}{2r(R + S) : (P + Q)(P + 2Q)+}$$
$$\frac{(p + 2r)(p + 2q + r) : (P + 2Q)^2}{2r(R + 2S) : (P + 2Q)(P + 3Q)+} \cdots . \tag{17.100}$$

Euler's notation may be easier to understand:

$$\alpha_1 = \frac{m}{P} + \cfrac{\frac{p(p+2q-r)}{P^2}}{\frac{2rR}{P(P+Q)} + \cfrac{\frac{(p+r)(p+2q)}{(P+Q)^2}}{\frac{2r(R+S)}{(P+Q)(P+2Q)} + \cdots}}.$$

He could therefore write down the continued fraction for $a_0$ after transforming the denominators in the fractions of $\alpha_1$:

$$a_0 = m - r + \cfrac{1}{\alpha_1}$$

$$= m - r + \cfrac{P}{m + \cfrac{P(p+2q-r)(P+Q)}{2rR + \cfrac{(p+r)(p+2q)P(P+2Q)}{2r(R+S)+\cdots}}}$$

or in modern notation

$$a_0 = m - r + \frac{P}{m+} \; \frac{p(p+2q-r)(P+Q)}{2rR+}$$
$$\frac{(p+r)(p+2q)P(P+2Q)}{2r(R+S)+} \; \frac{(p+2r)(p+2q+r)(P+Q)(P+3Q)}{2r(R+2S)+} \cdots .$$

$$(17.101)$$

To derive the continued fractions related to Wallis's product, Euler took $p = 2q = r = 1$. In this case the ratio of the beta integrals was

$$A_0 = a_0 = \frac{\int_0^1 \frac{y\,dy}{\sqrt{1-y^2}}}{\int_0^1 \frac{dy}{\sqrt{1-y^2}}} = \frac{2}{\pi};$$

the values of $P, Q, R, S$ were, respectively, $1 + m - m^2$, $3 - 2m$, $\frac{5-3m}{2}$, $\frac{3-2m}{2}$. Thus,

$$a_0 = m - 1 + \frac{1+m-m^2}{m+} \; \frac{1^2(4-m-m^2)}{5-3m+}$$
$$\frac{2^2(1+m-m^2)(7-3m-m^2)}{8-5m+} \; \frac{3^2(4-m-m^2)(10-5m-m^2)}{11-7m+} \cdots . \quad (17.102)$$

To get the continued fraction for $\frac{\pi}{2}$ in a nice form, Euler took $m = 1$ in (17.102); he skipped a step that we fill in:

$$a_0 = \frac{2}{\pi} = \cfrac{1}{1 + \cfrac{1^2 \cdot 2}{2 + \cfrac{2^2 \cdot 1 \cdot 3}{3 + \cfrac{3^2 \cdot 2 \cdot 4}{4+}}}}$$

$$= \cfrac{1}{1 + \cfrac{1}{1 + \cfrac{1 \cdot 2}{1 + \frac{2 \cdot 3}{1+}}}}$$

Thus, Euler had $\frac{\pi}{2}$ as the reciprocal of $a_0$:

$$\frac{\pi}{2} = 1 + \frac{1}{1+} \ \frac{1 \times 2}{1+} \ \frac{2 \times 3}{1+} \ \frac{3 \times 4}{1+} \cdots . \tag{17.103}$$

Thus, the continued fraction produced by Euler's method for the product for $\pi$ was different from the continued fraction found by Brouncker.

## 17.9 Sylvester: A Difference Equation and Euler's Continued Fraction

In 1869, J. J. Sylvester rediscovered Euler's formula (17.103) while investigating a difference equation arising out of the successive involutes to a circle.[46] The successive convergents of the continued fraction in Euler's formula produced the partial products of Wallis's infinite product. This infinite product was not useful for deriving approximations of $\pi$, but Sylvester showed that its continued fraction could be modified to yield good approximations. The continued fraction representation often provides better approximations than other representations, as was noted by Euler in his first paper on the topic. In his work on successive involutes, Sylvester was led to study the difference equation

$$v_{n+1} - v_{n-1} = \frac{1}{n} v_n. \tag{17.104}$$

He found two sequences as particular solutions of this equation:

$$\beta_1 = 1, \ \beta_{2n} = \beta_{2n+1} = \frac{2 \cdot 4 \cdot 6 \cdots 2n}{1 \cdot 3 \cdot 5 \cdots 2n - 1}, \ n = 1, 2, 3, \ldots$$

$$\alpha_{2n-1} = \alpha_{2n} = \frac{3 \cdot 5 \cdot 7 \cdots 2n - 1}{2 \cdot 4 \cdot 6 \cdots 2n - 2}, \quad n = 1, 2, 3, \ldots .$$

From Wallis's formula for $\pi$, Sylvester concluded that

$$\frac{\pi}{2} = \lim_{n \to \infty} \frac{\beta_n}{\alpha_n}.$$

From equations (3.27), (3.28), and (17.104), we see that $\frac{\beta_n}{\alpha_n}$ is the $n$th convergent of a continued fraction

$$b_0 + \frac{a_1}{b_1+} \ \frac{a_2}{b_2+} \ \frac{a_3}{b_3+} \cdots ,$$

where $a_n = 1$ and $b_n = \frac{1}{n}$ for $n \geq 1$. Thus, Sylvester could write that

$$\frac{\pi}{2} = 1 + \frac{1}{1+} \ \frac{1}{2^{-1}+} \ \frac{1}{3^{-1}+} \cdots = 1 + \frac{1}{1+} \ \frac{1 \cdot 2}{1+} \ \frac{2 \cdot 3}{1+} \ \frac{3 \cdot 4}{1+} \ \frac{4 \cdot 5}{1+} \cdots . \tag{17.105}$$

---

[46] Sylvester (1869).

To give a sense of Sylvester's inimitable style, we quote the sentence immediately following the last continued fraction, from Sylvester's paper: "This is obviously the simplest form of continued fraction for $\pi$ that can be given, and yet, strange to say, has not, I believe, before been observed. Truly wonders never cease!" Though Sylvester's result was of course not new, his method was original, and he also explained that this continued fraction could be used to improve the approximation obtained from Wallis's product or, equivalently, from the Madhava–Leibniz series.

Note that if $u_n$ is the remainder after $n$ terms of the fraction, then

$$u_n = \frac{n(n+1)}{1+} \frac{(n+1)(n+2)}{1+} \cdots \tag{17.106}$$

and

$$u_n = \frac{n^2+n}{1+u_{n+1}}. \tag{17.107}$$

This shows that $u_n$ is unbounded as $n \to \infty$, and hence $u_n u_{n+1} \approx n^2 + n$ and $u_n \approx n$ for large $n$. Thus, for large $n$ we may write, following Sylvester,

$$\frac{\pi}{2} = 1 + \frac{1}{1+} \frac{2}{1+} \frac{6}{1+} \cdots \frac{n(n-1)}{1+n}. \tag{17.108}$$

This correction by $n$ at the end of the formula improves the $n$th approximant obtained from the continued fraction. Thus, Sylvester noted that the convergents for $n = 4$ and $n = 5$ were

$$\frac{64}{45} = 1.4222 \quad \text{and} \quad \frac{384}{225} = 1.7056,$$

while the corrected values given by (17.108) were

$$\frac{128}{81} = 1.5802 \quad \text{and} \quad \frac{352}{225} = 1.5644.$$

For comparison, note that the actual value of $\frac{\pi}{2}$ to four decimal places is 1.5708; thus, the continued fraction has an advantage over the Wallis product.

## 17.10 Poisson, Jacobi, and Dirichlet: Beta Integrals

The early nineteenth-century mathematicians, striving to better understand how to manipulate definite integrals, used them to give new proofs of already known properties of the gamma and beta integrals. Poisson, Jacobi, and Dirichlet showed how double integrals could be employed to evaluate Euler's beta integral (17.10). It is interesting to see that Poisson made a change of variables in the double integral one variable at a time, while Jacobi changed from one pair of variables to another. Poisson's derivation appeared in papers of 1811 and 1823;[47] Jacobi's proof

---

[47] Poisson (1823); especially see pp. 477–478.

was published in 1833.[48] In 1841, Jacobi wrote his important paper on functional determinants or Jacobians[49] (so called by J. J. Sylvester[50]) from which arose the change of variables formula for $n$-dimensional integrals. Around 1836, the Russian mathematician M. Ostrogradsky gave a derivation of the general change of variables formula based on symbolic manipulation.

In Poisson's evaluation of the beta integral, we first observe that the substitution $t = \frac{1}{1+s}$ gives

$$\int_0^1 t^{p-1}(1-t)^{q-1}dt = \int_0^\infty \frac{s^{q-1}ds}{(1+s)^{p+q}}. \qquad (17.109)$$

Euler knew this, but Poisson also noted that the integrals converged if $p$ and $q$ were real and positive, or if $p$ and $q$ were complex with positive real parts. Poisson started by multiplying the integrals for $\Gamma(p)$ and $\Gamma(q)$ to get

$$\Gamma(p)\Gamma(q) = \int_0^\infty \int_0^\infty e^{-x}e^{-y}x^{p-1}y^{q-1}dxdy. \qquad (17.110)$$

He then substituted $xy$ and $xdy$ for $y$ and $dy$, followed by $\frac{x}{1+y}$ and $\frac{dx}{1+y}$ in place of $x$ and $dx$, to obtain

$$\begin{aligned}
\Gamma(p)\Gamma(q) &= \int_0^\infty \int_0^\infty \frac{e^{-x}x^{p+q-1}y^{q-1}}{(1+y)^{p+q}}dxdy \\
&= \int_0^\infty e^{-x}x^{p+q-1}dx \int_0^\infty \frac{y^{q-1}dy}{(1+y)^{p+q}} \\
&= \Gamma(p+q)\int_0^\infty \frac{y^{q-1}dy}{(1+y)^{p+q}}.
\end{aligned}$$

This proved (17.10). In 1833 Jacobi gave a different substitution. He set $x + y = r$, $x = rw$, so that $r$ ranged from 0 to $\infty$ and $w$ from 0 to 1. He then noted that

$$dx\,dy = rdr\,dw, \qquad (17.111)$$

so that the change of variables from $x, y$ to $r, w$ in (17.110) gave

$$\Gamma(p)\Gamma(q) = \int_0^\infty e^{-r}r^{p+q-1}\,dr \int_0^1 w^{p-1}(1-w)^{q-1}\,dw.$$

He did not explain how he obtained (17.111), probably because he took it to be well known.

We note that Poisson gave the conditions for convergence of the integrals, reflecting the increasing awareness among mathematicians that rigor was important. In fact, the works of Gauss, Cauchy, and Abel on infinite series contain the first significant expressions of this rigor. And Dirichlet was particularly attentive to questions of rigor

---

[48] Jacobi (1833).
[49] Jacobi (1841).
[50] Sylvester (1853a), especially Art. 65.

in his important work on integrals, both in papers and in lectures. For example, he discussed the conditions, such as absolute convergence, for changing the order of integration in a double integral.

Dirichlet's evaluation of Euler's integral and his proof of Gauss's multiplication formula using integrals are both given in the exercises at the end of this chapter. Here we mention a multivariable extension of Euler's integral presented by Dirichlet to the Berlin Academy in 1839:[51]

$$\int \cdots \int x_1^{\alpha_1-1} x_2^{\alpha_2-1} \cdots x_n^{\alpha_n-1} dx_1 dx_2 \cdots dx_n = \frac{\Gamma(\alpha_1)\Gamma(\alpha_2)\cdots\Gamma(\alpha_n)}{\Gamma(1+\alpha_1+\cdots+\alpha_n)}, \quad (17.112)$$

where $\alpha_i > 0$ and the integral is taken over the region $\sum_{i=1}^{n} x_i \le 1$, $x_i \ge 0$, $i = 1, 2, \ldots n$. Note that this formula is an iterated form of Euler's beta integral; in 1941, Selberg discovered a genuine multidimensional generalization of Euler's beta integral.[52] Also note that the use of Euler's beta integral yields a new proof of Euler's reflection formula. For this purpose, take $q = 1 - p$ in the infinite integral (17.109). Then we have

$$\Gamma(p)\Gamma(1-p) = \int_0^\infty \frac{s^{p-1}}{1+s} ds.$$

So if we verify that the value of the integral is $\frac{\pi}{\sin p\pi}$, we have our proof. Euler himself evaluated this integral in 1738 for $p$ a rational number. See Section 12.6 in this connection. Dedekind included an improved and streamlined version of this method in his doctoral thesis of 1852.[53] He let

$$B = \int_0^\infty \frac{x^{\frac{m}{n}-1}}{x+1} dx = n \int_0^\infty \frac{x^{m-1}}{x^n+1} dx. \quad (17.113)$$

Then

$$x^n + 1 = (x - \zeta)(x - \zeta^3)\cdots(x - \zeta^{2n-1}),$$

where $\zeta = e^{\frac{\pi i}{n}}$. By a partial fractions expansion, and applying (12.54), he obtained

$$\frac{x^{m-1}}{x^n+1} = \frac{-1}{n} \sum_{k=1}^{n} \frac{\zeta^{(2k-1)m}}{x - \zeta^{2k-1}}.$$

From this, Dedekind could conclude

$$n \int \frac{x^{m-1}}{x^n+1} dx = -\sum_{k=1}^{n} \zeta^{m(2k-1)} \log(\zeta^{2k-1} - x).$$

The last expression was easy to evaluate at $x = 0$ but not at $x = \infty$. So Dedekind rewrote this expression as

[51] Dirichlet (1839b).
[52] Selberg (1989) vol. 1, pp. 204–213 and vol. 1, pp. 62–73.
[53] Dedekind (1930) vol. I, pp. 1–26.

$$-\sum_{k=1}^{n} \zeta^{m(2k-1)} \log\left(\frac{\zeta^{2k-1}}{x} - 1\right) - \log x \sum_{k=1}^{n} \zeta^{m(2k-1)}, \qquad (17.114)$$

a sum that was zero because

$$\sum \zeta^{m(2k-1)} = \zeta^m \sum (\zeta^{2m})^{k-1} = \zeta^m \frac{\zeta^{2mn} - 1}{\zeta^{2m} - 1} = 0.$$

Thus,

$$n \int \frac{x^{m-1}}{x^n + 1} dx = -\sum_{k=1}^{n} \zeta^{m(2k-1)} \log\left(\frac{\zeta^{2k-1}}{x} - 1\right). \qquad (17.115)$$

He next used this relation at $x = \infty$ and the previous one at $x = 0$ to get

$$B = \sum_{k=1}^{n} \zeta^{m(2k-1)} \log(\zeta^{2k-1}) = \frac{\pi i}{n} \sum_{k=1}^{n} (2k-1)\zeta^{m(2k-1)} = \frac{\pi}{\sin(\frac{m\pi}{n})}.$$

Dedekind remarked that his use of the second relation to evaluate the integral at $\infty$ made his proof shorter than the ones found in integral calculus textbooks. It is interesting to compare Dedekind's derivation with that of Euler, (12.68). Dedekind gave three other evaluations of this integral. One of these used Cauchy's new calculus of residues. Another proof, included in the exercises, employed differential equations, and was an original contribution of Dedekind. In the third proof, given earlier by Euler, Dedekind expressed the integral as a partial fractions expansion.

## 17.11  The Volume of an *n*-Dimensional Ball

Dirichlet applied the integral formula (17.112) to find the volume of a unit ball in $n$ dimensions. However, Dirichlet's proof and his evaluations of other integrals involving surface areas in $n$ dimensions omitted some details. Dirichlet published his 1839 paper in Germany on multiple integrals;[54] he placed an even more brief proof in Liouville's French journal.[55] Liouville responded the same year with a paper in his journal,[56] providing the necessary details; we follow Dirichlet's argument, with details as presented by Liouville.

Taking $n = 1$ in (17.112) gives

$$\int_0^1 x_1^{\alpha_1 - 1} dx_1 = \frac{1}{\alpha_1} = \frac{\Gamma(\alpha_1)}{\Gamma(1 + \alpha_1)}.$$

Assume the result true for $n = k$. For $\alpha_i > 0$ and $\sum_{i=1}^{n} x_i \le 1$, $x_i \ge 0$, $i = 1, 2, \ldots, k$, we then have the integral on this region

54  Dirichlet (1839b).
55  Dirichlet (1839a).
56  Liouville (1839).

$$\int \cdots \int x_1^{\alpha_1-1} x_2^{\alpha_2-1} \cdots x_k^{\alpha_k-1} \, dx_1 \cdots dx_k = \frac{\Gamma(\alpha_1)\Gamma(\alpha_2)\cdots\Gamma(\alpha_k)}{\Gamma(1+\alpha_1+\cdots+\alpha_k)}. \qquad (17.116)$$

Now consider the $k+1$-dimensional integral

$$I \equiv \int \cdots \int x_1^{\alpha_1-1} x_2^{\alpha_2-1} \cdots x_{k+1}^{\alpha_{k+1}-1} \, dx_1 \cdots dx_{k+1}$$

$$= \int_0^1 \int_0^{1-x_1} \cdots \int_0^{1-x_1-x_2-\cdots-x_k} x_1^{\alpha_1-1} x_2^{\alpha_2-1} \cdots x_{k+1}^{\alpha_{k+1}-1} \, dx_{k+1} \, dx_k \cdots dx_1$$

$$= \frac{1}{\alpha_{k+1}} \int_0^1 \int_0^{1-x_1} \cdots \int_0^{1-x_1-\cdots-x_{k-1}} x_1^{\alpha_1-1} x_2^{\alpha_2-1} \cdots x_k^{\alpha_k-1}$$

$$\cdot (1-x_1-x_2-\cdots-x_k)^{\alpha_{k+1}} \, dx_k \, dx_{k-1} \cdots dx_1 \qquad (17.117)$$

and set $x_k = t(1-x_1-\cdots-x_{k-1})$ to find that

$$I = \frac{1}{\alpha_{k+1}} \int_0^1 \int_0^{1-x_1} \cdots \int_0^{1-x_1-\cdots-x_{k-2}} \int_0^1 x_1^{\alpha_1-1} x_2^{\alpha_2-1} \cdots x_{k-1}^{\alpha_{k-1}-1}$$

$$\times t^{\alpha_k-1}(1-x_1-x_2-\cdots-x_{k-1})^{\alpha_k+\alpha_{k+1}}(1-t)^{\alpha_{k+1}} \, dt \, dx_{k-1} \cdots dx_1$$

$$= \frac{1}{\alpha_{k+1}} \frac{\Gamma(\alpha_k)\Gamma(\alpha_{k+1}+1)}{\Gamma(\alpha_k+\alpha_{k+1}+1)} \int_0^1 \int_0^{1-x_1} \cdots$$

$$\int_0^{1-x_1-\cdots-x_{k-2}} x_1^{\alpha_1-1} \cdots x_{k-1}^{\alpha_{k-1}-1}(1-x_1-\cdots-x_{k-1})^{\alpha_k+\alpha_{k+1}} \, dx_{k-1} \cdots dx_1.$$

$$(17.118)$$

Observe that integrating (17.116) once with respect to $x_k$ produces

$$\int_0^1 \cdots \int_0^{1-x_1-\cdots-x_{k-1}} x_1^{x_1-1} \cdots x_{k-1}^{\alpha_{k-1}-1}(1-x_1-\cdots-x_{k-1})^{\alpha_k} \, dx_{k-1} \cdots dx_1$$

$$= \alpha_k \frac{\Gamma(\alpha_1)\cdots\Gamma(\alpha_k)}{\Gamma(1+\alpha_1+\cdots+\alpha_k)}; \qquad (17.119)$$

and from this we conclude that the value of the last integral in (17.118) must be

$$(\alpha_k + \alpha_{k_1}) \frac{\Gamma(\alpha_1)\cdots\Gamma(\alpha_{k-1})\Gamma(\alpha_k+\alpha_{k+1})}{\Gamma(1+\alpha_1+\cdots+\alpha_k+\alpha_{k+1})}. \qquad (17.120)$$

Substituting (17.120) in (17.118) finally gives us

$$I = \frac{1}{\alpha_{k+1}} \frac{\Gamma(\alpha_k)\Gamma(\alpha_{k+1}+1)}{\Gamma(\alpha_k+\alpha_{k+1}+1)} \cdot \frac{(\alpha_k+\alpha_{k+1})\Gamma(\alpha_1)\cdots\Gamma(\alpha_{k-1})\Gamma(\alpha_k+\alpha_{k+1})}{\Gamma(1+\alpha_1+\cdots+\alpha_{k+1})}$$

$$= \frac{\Gamma(\alpha_1)\cdots\Gamma(\alpha_{k+1})}{\Gamma(1+\alpha_1+\cdots+\alpha_{k+1})},$$

completing our proof of (17.112).

Note that the region of integration for the integral in (17.112) is the region enclosed by $x_i \geq 0$ and $\sum_{i=1}^{n} x_i \leq 1$. By applying a change of variables, following Dirichlet and Liouville, we can evaluate the integral:

$$\int \cdots \int_V x_1^{\alpha_1 - 1} \cdots x_n^{\alpha_n - 1}\, dx_1 \cdots dx_n = \frac{\prod_{i=1}^{n} \left(\frac{a_i^{\alpha_i}}{p_i}\right) \Gamma\left(\frac{x_i}{p_i}\right)}{\Gamma\left(1 + \sum_{i=1}^{n} \frac{\alpha_i}{p_i}\right)}, \qquad (17.121)$$

with conditions on V set as

$$x_i \geq 0, \quad a_i \geq 0, \quad p_i \geq 0, \quad i = 1, \ldots, n, \quad \sum_{i=1}^{n} \left(\frac{x_i}{a_i}\right)^{p_1} \leq 1.$$

To reduce the integral in (17.121) to (17.112), let $y_i = \left(\frac{x_i}{a_i}\right)^{p_i}$, so that

$$\frac{dy_i}{y_i} = p_i \frac{dx_i}{x_i}.$$

Substitute in (17.121) to get

$$\frac{a_1^{\alpha_1} a_2^{\alpha_2} \cdots a_n^{\alpha_n}}{p_1 p_2 \cdots p_n} \int \cdots \int_{\overline{V}} y_1^{\frac{\alpha_1}{p_1} - 1} y_2^{\frac{\alpha_2}{p_2} - 1} - y_n^{\frac{\alpha_n}{p_n} - 1}\, dy_1 \cdots dy_n,$$

where $\overline{V}$ is defined by $u_i \geq 0$ and $\sum_{i=1}^{n} y_i \leq 1$; (17.121) is now verified.
Thus, to find the volume $V$ of the region enclosed by

$$x_i \geq 0, \quad \sum_{i=1}^{n} \left(\frac{x_i}{a_i}\right)^{p_1} \leq 1,$$

take $\alpha_i = 1, i = 1, 2, \ldots, n$ in (17.121) to get

$$V = \frac{\prod_{i=1}^{n} \left(\frac{a_i}{p_i}\right) \Gamma\left(\frac{1}{p_i}\right)}{\Gamma\left(1 + \sum_{i=1}^{n} \frac{1}{p_i}\right)}. \qquad (17.122)$$

Formula (17.122) implies that the volume of the $n$-dimensional ellipsoid $\sum_{i=1}^{n} \left(\frac{x}{a_i}\right)^2 \leq 1$ must be

$$\frac{a_1 a_2 \cdots a_n \pi^{\frac{n}{2}}}{\Gamma\left(1 + \frac{n}{2}\right)} \qquad (17.123)$$

and the volume of a $n$-dimensional ball of radius $r$ is

$$\frac{r^n \pi^{\frac{n}{2}}}{\Gamma\left(1 + \frac{n}{2}\right)}. \qquad (17.124)$$

Dirichlet and Liouville also computed the surface area $S$, defined by $x_i \geq 0$, $\sum_{i=1}^{n} x_i = 1$:

$$\int \cdots \int_{S} x_1^{\alpha_1 - 1} x_2^{\alpha_2 - 1} \cdots x_n^{\alpha_n - 1} \, dx_1 \cdots dx_{n-1} = \frac{\Gamma(\alpha_1) \cdots \Gamma(\alpha_n)}{\Gamma(\alpha_1 + \cdots + \Gamma_n)}. \quad (17.125)$$

To sketch a proof, observe that $S$ is defined by $x_n = 1 - x_1 - x_2 - \cdots - x_{n-1}$, where $x_i \geq 0$, $i = 1, 2, \ldots, n$, and $x_1 + x_2 + \cdots + x_{n-1} \leq 1$. Thus, the integral in (17.125) can be written as

$$\int_0^1 \int_0^{1-x_1} \cdots \int_0^{1-x_1 - \cdots - x_{n-2}} x_1^{\alpha_1 - 1} \cdots x_{n-1}^{\alpha_{n-1} - 1}$$

$$\cdot (1 - x_1 - \cdots - x_{n-1})^{\alpha_n - 1} \, dx_{n-1} \cdots dx_1,$$

and by applying (17.119), we get our result.

## 17.12   The Selberg Integral

We have seen that Dirichlet's integral (17.112) is a generalization of Euler's beta integral. In 1941, Selberg presented another generalization:[57]

If $n$ is a positive integer; $\alpha, \beta, \gamma$ are complex numbers such that

$$\operatorname{Re} \alpha > 0, \ \operatorname{Re} \beta > 0$$

and

$$\operatorname{Re} \gamma > -\min \left\{ \frac{1}{n}, \frac{\operatorname{Re} \alpha}{n-1}, \frac{\operatorname{Re} \beta}{n-1} \right\};$$

and if

$$\Delta(x) = \prod_{1 \leq i < j \leq n} (x_i - x_j), \quad (17.126)$$

then

$$S_n(\alpha, \beta, \gamma) = \int_0^1 \cdots \int_0^1 \prod_{i=1}^{n} (x_i^{\alpha - 1}(1 - x_i)^{\beta - 1}) |\Delta(x)|^{2\gamma} \, dx_1 \cdots dx_n$$

$$= \prod_{j=1}^{n} \frac{\Gamma(\alpha + (j-1)\gamma) \, \Gamma(\beta + (j-1)\gamma) \Gamma(1 + j\gamma)}{\Gamma(\alpha + \beta + (n+j-2)\gamma) \Gamma(1 + \gamma)}. \quad (17.127)$$

In 1941, Selberg needed the integral formula (17.127) in a paper on entire functions,[58] in which he wrote that if the formula were new, he would publish

[57]  Selberg (1941).
[58]  Selberg (1944).

a proof elsewhere. And he published the proof, but in a popular mathematics magazine,[59] concerning which he later remarked:[60]

> This paper was published with some hesitation, and in Norwegian, since I was rather doubtful that the results were new. The journal is one which is read by mathematics-teachers in the gymnasium, and the proof was written out in detail so it should be understandable to someone who knew a little bit about analytic functions and analytic continuation.

For his generalization, Selberg required a weak form of Carlson's theorem, mentioned at the end of Section 17.4. Using only Cauchy's integral formula, he proved that if $f(z)$ was analytic and bounded for Re $z \geq 0$ and if $f(z) = 0$ for $z = 0, 1, 2, \ldots$, then $f(z)$ must be identically zero. We are fortunate that Selberg provided this nice proof for his audience of mathematics teachers; if he had been writing for mathematics researchers, he most probably would have simply cited Carlson's theorem. We here give Selberg's proof and later we present Askey's proof of Selberg's integral formula, that does not use Carlson's theorem. We mention that G. W. Anderson in 1991 gave a proof of Selberg's integral formula using Dirichlet's integral (17.112) but not Carlson's theorem.[61]

For Selberg's weak Carlson's theorem proof,[62] note that it is a consequence of Cauchy's integral formula, or Cauchy's residue theorem, that

$$f(a) = \frac{(a-1)(a-2)\cdots(a-n)}{2\pi i} \int_{-i\infty}^{i\infty} \frac{f(z)\,dz}{(z-a)(z-1)\cdots(z-n)} \qquad (17.128)$$

for $n > a > 0$. For $a > 1$, (17.128) implies that

$$|f(a)| \leq \frac{[a]!\,(n-[a])!}{2\pi} \int_{-\infty}^{\infty} \frac{|f(it)|\,dt}{\sqrt{(a^2+t^2)(1+t^2)\cdots(n^2+t^2)}}$$

$$\leq \frac{[a]!\,(n-[a])!}{2\pi n!} \int_{-\infty}^{\infty} \frac{|f(it)|}{1+t^2}\,dt. \qquad (17.129)$$

The integral converges because $f(z)$ is bounded and because $a \geq 1$:

$$\frac{[a]!\,(n-[a])!}{n!} \to 0 \quad \text{as } n \to \infty.$$

Hence, by (17.129), $f(a) = 0$ for $a \geq 1$ and, by analytic continuation, $f(a) = 0$ for $a \geq 0$.

Most mathematicians and physicists were hardly aware of Selberg's integral for at least thirty years. Some researchers in entire functions probably knew it, because Selberg included it in his 1941 paper on Gelfond's theorem, and this paper was referenced in Boas's well-known work on entire functions. In the early 1960s, physicists F. J. Dyson and M. L. Mehta conjectured a limiting case of Selberg's

[59] Selberg (1944).
[60] Selberg (1982) vol. I, p. 212.
[61] Anderson (1991).
[62] Selberg (1944).

integral in their book on the statistical theory of energy levels of complex systems.[63] Then in 1976, when he, Selberg , and Dyson were all at the Institute for Advanced Study at Princeton, Enrico Bombieri was led to consider integrals of the Selberg type in the course of his work on prime number theory. Perceiving his integrals to be similar to those used in physics, Bombieri went to consult Dyson, who referred him to Mehta's book, *Random Matrices*. Bombieri next approached Selberg to discuss his problem on the distribution of primes, and Selberg recognized Bombieri's integral as a more complex version of his generalized beta integral.[64]

Selberg's original proof was complicated. An idea for a simplified proof was found by K. Aomoto in 1987.[65] In fact, Aomoto's method evaluated a slightly more general integral. Thus, following Aomoto, denote the integrand in Selberg's integral by

$$\Phi(x) \equiv \Phi(x; \alpha, \beta, \gamma) \equiv \prod_{i=1}^{n} x_i^{\alpha-1}(1-x_i)^{\beta-1} \prod_{1 \le i < j \le n} |x_i - x_j|^{2\gamma}$$

and set

$$I_k \equiv \int_{C_n} \prod_{i=1}^{k} x_i \Phi(x; \alpha, \beta, \gamma) dx,$$

where $C_n$ is the $n$-dimensional cubic and $dx = dx_1 dx_2 \cdots dx_n$. Aomoto found a relation between $I_k$ and $I_{k-1}$ by observing that, since

$$\frac{d}{dx}|x|^c = c|x|^{c-1} \operatorname{sgn} x = \frac{c|x|^c}{x} \quad \text{if } x \ne 0,$$

we can write

$$0 = \int_{C_n} \frac{\partial}{\partial x_1} \left( (1-x_1) \prod_{i=1}^{k} x_i \Phi(x) \right) dx$$

$$= \alpha \int_{C_n} (1-x_1) \prod_{i=2}^{k} x_i \Phi(x) dx - \beta \int_{C_n} \prod_{i=1}^{k} x_i \Phi(x) dx$$

$$+ 2\gamma \sum_{j=2}^{n} \int_{C_n} \frac{(1-x_1) \prod_{i=1}^{k} x_i \Phi(x) dx}{x_1 - x_j}. \tag{17.130}$$

Aomoto proved two lemmas that revealed how the third integral in (17.130) can be written in terms of $I_k$ and $I_{k-1}$. Lemma 1 states,

$$\int_{C_k} \frac{\prod_{j=1}^{k} x_j \Phi(x)}{x_1 - x_j} dx = \begin{cases} 0, & \text{if } 2 \le j \le k \\ \frac{1}{2} I_{k-1}, & \text{if } k < j \le n. \end{cases}$$

[63] Dyson and Mehta (1963)..
[64] For more history of the Selberg integral, see Forrester and Warnaar (2008).
[65] Aomoto (1987).

In the case $2 \leq j \leq k$, the transposition $x_1 \leftrightarrow x_j$ changes $x_1 - x_j$ to $x_j - x_1$ and $\prod_{j=1}^{k} x_j$ remains unchanged. Hence the value of the integral in this case is zero. On the other hand, when $k < j \leq n$, the same transposition effects the change

$$\frac{x_1}{x_1 - x_j} \rightarrow \frac{x_j}{x_j - x_1} = 1 - \frac{x_1}{x_1 - x_j}$$

so that

$$2 \int_{C_n} \frac{\prod_{i=1}^{k} x_i \, \Phi(x)}{x_1 - x_j} dx = \int_{C_n} \prod_{i=2}^{k} x_i \, \Phi(x) dx = I_{k-1}.$$

Lemma 2 states,

$$\int_{C_n} \frac{x_1 \prod_{i=1}^{k} x_i \, \Phi(x)}{x_1 - x_j} dx = \begin{cases} \frac{1}{2} I_k, & \text{if } 2 \leq j \leq k, \\ I_k & \text{if } k < j \leq n. \end{cases}$$

When $2 \leq j \leq j$, the transposition $x_1 \leftrightarrow x_j$ produces

$$\frac{x_1^2 x_j}{x_1 - x_j} \rightarrow \frac{x_1 x_j^2}{x_j - x_1} = x_1 x_j - \frac{x_1^2 x_j}{x_1 - x_j},$$

proving the first part. For the second part, note that

$$\frac{x_1^2}{x_1 - x_j} = x_1 + \frac{x_1 x_j}{x_1 - x_j}.$$

Note that the last term changes sign in the transposition $x_1 \leftrightarrow x_j$, so this term leads to the value zero for the integral. The first term, $x_1$, produces $I_k$, proving the second lemma.

Applying these two lemmas to (17.130) gave Aomoto the relation

$$0 = \alpha I_{k-1} - (\alpha + \beta) I_k + \gamma(n - k) I_{k-1} - \gamma(zn - k - 1) I_k,$$

or

$$I_k = \frac{\alpha + (n - k)\gamma}{\alpha + \beta + (2n - k - 1)\gamma} I_k. \qquad (17.131)$$

Iteration of (17.131) $k$ times produced

$$\int_{C_n} \prod_{i=1}^{k} x_i \Phi(x) \, dx = \prod_{i=1}^{k} \frac{\alpha + (n - i)\gamma}{\alpha + \beta + (2n - i - 1)\gamma} \int_{C_n} \Phi(x) \, dx. \qquad (17.132)$$

Note that the last integral is Selberg's integral, $S_n(\alpha, \beta, \gamma)$. Taking $k = n$ in (17.132), produced

480

The Gamma Function

$$S_n(\alpha + 1, \beta, \gamma) = \prod_{j=1}^{n} \frac{\alpha + (n-j)\gamma}{\alpha + \beta + (2n-j-1)\gamma} S_n(\alpha, \beta, \gamma) \qquad (17.133)$$

$$= \prod_{j=1}^{n} \frac{\alpha + (j-1)\gamma}{\alpha + \beta + (n+j-2)\gamma} S_n(\alpha, \beta, \gamma).$$

This implied that when $\alpha$ was a positive integer

$$S_n(\alpha, \beta, \gamma) = \prod_{j=1}^{n} \prod_{m=1}^{\alpha-1} \frac{m + (j-1)\gamma}{m + \beta + (n+j-2)\gamma} S_n(1, \beta, \gamma). \qquad (17.134)$$

With $\beta$ also a positive integer, symmetry in $\alpha$ and $\beta$ gave

$$S_n(1, \beta, \gamma) = \prod_{j=1}^{n} \prod_{l=1}^{\beta-1} \frac{l + (j-1)\gamma}{1 + l + (n+j-1)\gamma} S_n(1, 1, \gamma). \qquad (17.135)$$

Observe that

$$\prod_{m=1}^{\alpha-1} \frac{m + (j-1)\gamma}{m + \beta + (n+j-1)\gamma} \prod_{l=1}^{\beta-1} \frac{l + (j-1)\gamma}{1 + l + (n+j-1)\gamma}$$

$$= \frac{\Gamma(\alpha + (j-1)\gamma)\,\Gamma(\beta + (j-1)\gamma)}{\Gamma(\alpha + \beta + (n+j-1)\gamma)} \cdot \frac{\Gamma(2 + (n+j-1)\gamma)}{\left(\Gamma(1 + (n-j)\gamma)\right)^2}. \qquad (17.136)$$

Combining (17.134), (17.135), and (17.136), Aomoto obtained

$$S_n(\alpha, \beta, \gamma) = D_n(\gamma) \prod_{j=1}^{n} \frac{\Gamma(\alpha + (j-1)\gamma)\,\Gamma(\beta + (j-1)\gamma)}{\Gamma(\alpha + \beta + (n+j-1)\gamma)}, \qquad (17.137)$$

where $D_n(\gamma)$ depends only on $\gamma$ and $n$, and not on $\alpha$ and $\beta$.

Formula (17.137) was derived earlier by Selberg, using a different method. From this point on, Aomoto's proof followed Selberg's. To determine the value of $D_n(\gamma)$, Selberg took $\gamma$ to be a positive integer. Observe that with $\alpha = \beta = 1$, (17.137) becomes

$$S_n(1, 1, \gamma) = \int_{C_n} |\Delta(x)|^{2\gamma} dx = D_n(\gamma) \prod_{j=1}^{n} \frac{\Gamma(1 + (j-1)\gamma)\Gamma(1 + (j-1)\gamma)}{\Gamma(2 + (n+j-1)\gamma)}.$$

$$(17.138)$$

Now, following Selberg, take $x_n$ to be the largest of the variables $x_1, x_2, \ldots, x_n$; since any of the $n$ variables could be the largest, it follows that

$$S_n(1,1,\gamma) = n \int_0^1 \left( \int_0^{x_n} \cdots \int_0^{x_n} ((x_n - x_1) \cdots (x_n - x_{n-1}))^{2\gamma} \right.$$
$$\left. \cdot |\Delta(x_1,\ldots,x_{n-1})|^{2\gamma} dx_1 \cdots dx_{n-1} \right) dx_n.$$

Make the change of variables $x_i = x_n t_i$, $i = 1, 2, \ldots, n-1$. Then

$$S_n(1,1,\gamma) = n \int_0^1 x_n^{n(n-1)\gamma + n - 1} dx_n \int_0^1 \cdots \int_0^1 ((1-x_1) \cdots (1 - x_{n-1}))^{2\gamma}$$
$$\cdot |\Delta(x_1,\ldots,x_{n-1})|^{2\gamma} dx_1 \cdots dx_{n-1}$$

$$= \frac{1}{(n-1)\gamma + 1} S_{n-1}(1, 2\gamma + 1, 2\gamma)$$

$$= \frac{D_{n-1}(\gamma)}{(n-1)\gamma + 1} \cdot \prod_{j=1}^{n-1} \frac{\Gamma(1 + (j-1)\gamma)\Gamma(1 + (j+1)\gamma)}{\Gamma(2 + (n+j-1)\gamma)}, \quad (17.139)$$

where equation (17.139) follows by an application of (17.137). Equating (17.138) and (17.139) and simplifying, we arrive at

$$D_n(\gamma) = \frac{\Gamma(1 + n\gamma)}{\Gamma(1 + \gamma)} D_{n-1}(\gamma),$$

and, since $D_1(\gamma) = 1$, we finally have

$$D_n(\gamma) = \prod_{j=1}^n \frac{\Gamma(1 + j\gamma)}{\Gamma(1 + \gamma)}. \quad (17.140)$$

Combining (17.140) with (17.137) produces

$$S_n(\alpha, \beta, \gamma) = \prod_{j=1}^n \frac{\Gamma(\alpha + (j-1)\gamma)\Gamma(\beta + (j-1)\gamma)\Gamma(1 + j\gamma)}{\Gamma(\alpha + \beta + (n+j-1)\gamma)\Gamma(1 + \gamma)}, \quad (17.141)$$

proving (17.127) when $\alpha, \beta, \gamma$ are positive integers. To extend the result to $\mathrm{Re}\,\alpha > 0$, $\mathrm{Re}\,\beta > 0$, $\mathrm{Re}\,\gamma > 0$, Selberg applied Stirling's approximation,

$$\Gamma(\omega) \sim \sqrt{2\pi}\, \omega^{\omega - \frac{1}{2}} e^{-\omega} \quad \text{for } |\omega| \to \infty \text{ and } \mathrm{Re}\,\omega > 0,$$

to the expression–call it $T_j$–inside the product in (17.141) to obtain

$$T_j \sim \sqrt{\frac{2\pi j}{1 + \gamma}} \frac{(j-1)^{\alpha - \frac{1}{2}} (n-j)^{\beta - \frac{1}{2}}}{(n+j-2)^{\alpha + \beta - \frac{1}{2}}} \cdot \left( \frac{j^j (j-1)^{j-1} (n-j)^{n-j}}{(n+j-2)^{n+j-2}} \right)^\gamma. \quad (17.142)$$

Note that as $\alpha$ gets large and $1 \leq j \leq n$, the expression

$$\frac{(j-1)^{\alpha-\frac{1}{2}}}{(n+j-2)^{\alpha-\frac{1}{2}}}$$

is bounded and, similarly, as $\beta$ becomes large, the expression

$$\frac{(n-j)^{\beta-\frac{1}{2}}}{(n+j-1)^{\beta-\frac{1}{2}}}$$

is bounded. Selberg observed that for $1 \leq j \leq n$,

$$\frac{j^j (j-1)^{j-1} (n-j)^{n-j}}{(n+j-2)^{n+j-2}} \leq 1.$$

This is not difficult to show. For $2 \leq j \leq n-1$,

$$j^j (j-1)^{j-1} \leq (2j-2)^{2j-2},$$

and

$$j^j (j-1)^{j-1} (n-j)^{n-j} \leq (2j-2)^{2j-2} (n-j)^{n-j} \leq (n-j+2j-2)^{n-j+2j-2}.$$

Thus, (17.142) is bounded for Re $\gamma \geq 0$ when $\gamma \to \infty$ and the right-hand side of (17.141) is bounded for Re $\alpha > 0$, Re $\beta > 0$, and Re $\gamma > 0$. Since the integral $S_n(\alpha, \beta, \gamma)$ is on a $n$-dimensional cube, it follows that $S_n(\alpha, \beta, \gamma)$ is bounded for Re $\alpha > 0$, Re $\beta > 0$, and Re $\gamma > 0$; moreover, $S_n(\alpha, \beta, \gamma)$ is clearly analytic for these values of $\alpha, \beta, \gamma$. By Carlson's theorem, then, (17.141) must hold for Re $\alpha > 0$, Re $\beta > 0$, and Re $\gamma > 0$.

In his lectures on special functions, delivered around the period 1990–1992 at the University of Wisconsin, Richard Askey derived Selberg's integral using a method different from (17.133); we sketch this method: From (17.133) it follows by symmetry in $\alpha$ and $\beta$ and then by iteration that

$$S_n(\alpha, \beta, \gamma) = \prod_{j=1}^{n} \frac{(\alpha + \beta + (2n - j - 1)\gamma)_k}{(\beta + (n-j)\gamma)_k} S_n(\alpha, \beta + k, \gamma)$$

$$= \prod_{j=1}^{n} \frac{(\alpha + \beta + (2n - j - 1)\gamma)_k}{(\beta + (n-j)\gamma)_k}$$

$$\cdot \int_0^k \cdots \int_0^k \prod_{i=1}^{n} \left(\frac{x_i}{k}\right)^{\alpha-1} \left(1 - \frac{x_i}{k}\right)^{\beta+k-1} \prod_{1 \leq i < j \leq n} \left|\frac{x_i - x_j}{k}\right|^{2\gamma} \frac{dx}{k^n}.$$

Let $k \to \infty$ and apply (17.2) to obtain

$$S_n(\alpha,\beta,\gamma) = \prod_{j=1}^{n} \frac{\Gamma(\beta + (n-j)\gamma)}{\Gamma(\alpha+\beta+(2n-j-1)\gamma)}$$

$$\cdot \int_0^\infty \cdots \int_0^\infty \prod_{i=1}^{n} x_i^{\alpha-1} e^{-x_i} \prod_{1 \le i < j \le n} |x_i - x_j|^{2\gamma} dx. \qquad (17.143)$$

Denote the integral in (17.143) by $M_n(\alpha,\beta)$. Symmetry in $\alpha$ and $\beta$ and (17.143) imply that

$$\frac{M_n(\alpha,\gamma)}{\prod_{j=1}^{n} \Gamma(\alpha+(n-j)\gamma)} = \frac{M_n(\beta,\gamma)}{\prod_{j=1}^{n} \Gamma(\beta+(n-j)\gamma)} \equiv D_n(\gamma),$$

because only if it is independent of $\alpha$ and $\beta$ can a function of $\alpha$ and $\gamma$ be equal to a function of $\beta$ and $\gamma$. Therefore

$$S_n(\alpha,\beta,\gamma) = \prod_{j=1}^{n} \frac{\Gamma(\alpha+(n-j)\gamma)\Gamma(\beta(n-j)\gamma)}{\Gamma(\alpha+\beta+(2n-j-1)\gamma)} D_n(\gamma). \qquad (17.144)$$

This time, we compute $D_n(\gamma)$ by a modification of Selberg's method. Observe that by the symmetry in the variables $x_1, x_2, \ldots, x_n$,

$$\int_{C_n} \Phi(x)dx = n! \int_0^1 \int_{x_n}^1 \cdots \int_{x_2}^1 n\, \Phi(x)dx_1 \cdots dx_n. \qquad (17.145)$$

Now for a differentiable function $f$, integration by parts gives

$$\lim_{\alpha \to 0^+} a \int_0^1 t^{\alpha-1} f(t)dt = f(0),$$

a result also true for a continuous function. So multiply (17.145) by $\alpha$ and let $\alpha \to 0^+$ and use (17.144) to find

$$n \int_{C_{n-1}} \prod_{i=1}^{n-1} (x_i^{2\gamma-1}(1-x_i)^{\beta-1}) \prod_{1 \le i < j \le n-1} |x_i - x_j|^{2\gamma} dx$$

$$= D_n(\gamma) \prod_{j=1}^{n} \frac{\Gamma(\beta+(j-1)\gamma)}{\Gamma(\beta+(n+j-2)\gamma)} \prod_{j=2}^{n} \Gamma((j-1)\gamma). \qquad (17.146)$$

Again by (17.144), the integral on the left-hand side of (17.146) also equals

$$n D_{n-1}(\gamma) \prod_{j=1}^{n-1} \frac{\Gamma(2\gamma + (j-1)\gamma)\Gamma(\beta + (j-1)\gamma)}{\gamma(2\gamma + \beta + (n+j-3)\gamma)}. \tag{17.147}$$

Comparing (17.146) and (17.147), reveals the functional relation

$$D_n(\gamma) = \frac{n\Gamma(n\gamma)}{\Gamma(\gamma)} D_{n-1}(\gamma) = \frac{\Gamma(n\gamma+1)}{\Gamma(\gamma+1)} D_{n-1}(\gamma)$$

and therefore

$$D_n(\gamma) = \prod_{j=1}^{n} \frac{\Gamma(1+j\gamma)}{\Gamma(1+\gamma)},$$

completing Askey's evaluation of the Selberg integral.[66]

Also note that (17.143) implies that for Re $\alpha > 0$ and Re $\gamma > 0$,

$$\int_0^\infty \cdots \int_0^\infty \prod_{i=1}^{n} x_i^{\alpha-1} e^{-x_i} \prod_{1 \le i < j \le n} |x_i - x_j|^{2\gamma} dx$$

$$= \prod_{j=1}^{n} \frac{\Gamma(\alpha + (j-1)\gamma)\Gamma(1+j\gamma)}{\Gamma(1+\gamma)}.$$

Now if we take $\alpha = \beta$ and $x_i = \frac{1}{2}\left(1 + \frac{t_i}{\sqrt{2\alpha}}\right)$, $i = 1, \ldots, n$, and let $\alpha \to \infty$, we get the result, for Re $(\gamma > 0)$,

$$\int_0^\infty \cdots \int_0^\infty \exp\left(-\frac{1}{2}\sum_{i=1}^{n} t_i^2\right) \prod_{1 \le i < j \le n} |t_i - t_j|^{2\gamma} dt = (2\pi)^{\frac{n}{2}} \prod_{j=1}^{n} \frac{\Gamma(1+j\gamma)}{\Gamma(1+\gamma)}.$$

## 17.13   Good's Proof of Dyson's Conjecture

One implication of Selberg's integral formula is an integral that arises in physics, known as Dyson's integral:

$$\int_{-\pi}^{\pi} \cdots \int_{-\pi}^{\pi} \prod_{1 \le j < k \le n} |e^{i\theta_j} - e^{i\theta_k}|^{2\gamma} d\theta_1 \cdots d\theta_n = (2\pi)^n \frac{\Gamma(n\gamma+1)}{(\Gamma(\gamma+1))^n}.$$

In fact, this is a particular case of Selberg's integral. It can be used to find the constant term in the product

$$\prod_{l,j} \left(1 - \frac{z_l}{z_j}\right)^k, \tag{17.148}$$

---

[66] See Andrews et al. (1999) pp. 405–406.

where $k$ is a positive integer and $l \neq j$. To verify this claim, observe that for $z_j = e^{i\theta_j}$,

$$|z_j - z_l|^{2k} = (z_j - z_l)^k (z_j^{-1} - z_l^{-1})^k = \left(1 - \frac{z_l}{z_j}\right)^k \left(1 - \frac{z_j}{z_l}\right)^k.$$

Also note the fact that any power other than 0 of $z_j$ vanishes on integration. So the constant term in (17.148) is given by

$$\frac{\Gamma(nk+1)}{\Gamma(k+1)^n} = \frac{(nk)!}{(k!)^n}.$$

In 1962, Dyson conjectured that if $a_1, a_2, \ldots, a_n$ were nonnegative integers, then the constant term in the product

$$\prod_{\substack{j,l \\ j \neq l}} \left(1 - \frac{x_j}{x_l}\right) \quad \text{was} \quad \frac{(a_1 + a_2 + \cdots + a_n)!}{a_1! \, a_2! \cdots a_n!}.$$

Wilson[67] and Gunson[68] independently proved Dyson's conjecture, and in 1970 Good provided a short proof,[69] starting with the Waring–Lagrange interpolation formula and deriving equation (9.31):

$$\sum_{j=1}^{n} \prod_{\substack{1 \le k \le n \\ j \neq k}} \frac{1}{1 - \frac{x_j}{x_k}} = 1.$$

He then set

$$F_n(a_1, a_2, \ldots, a_n) = \prod_{j \neq l} \left(1 - \frac{x_j}{x_l}\right)^{a_j},$$

and multiplied (9.31) by $F_n(a_1, a_2, \ldots, a_n)$ to arrive at the recurrence relation

$$F_n(a_1, a_2, \ldots, a_n) = \sum_{j=1}^{n} F_n(a_1, \ldots, a_j - 1, \ldots, a_n). \qquad (17.149)$$

He next let C. T. $F_n(a_1, a_2, \cdots, a_n)$ denote the constant term in $F_n(a_1, a_2, \ldots, a_n)$ so that (17.149) implied that

$$\text{C. T. } F_n(a_1, a_2, \cdots, a_n) = \sum_{j=1}^{n} \text{C. T. } F_n(a_1, \ldots, a_j - 1, \ldots, a_n).$$

[67]  Wilson (1962).
[68]  Gunson (1962).
[69]  Good (1970).

Good then observed that C. T. $F_n(0, 0, \ldots, 0) = 1$, and if $a_k = 0$, then

$$\text{C. T. } F_n(a_1, a_2, \cdots, a_n) = \text{C. T. } F_{n-1}(a_1, \ldots, a_{k-1}, a_k, \ldots, a_n),$$

because only the nonpositive powers of $x_k$ appeared in $F_n(a_1, a_2, \cdots, a_n)$. He then noted that, in fact,

$$\frac{(a_1 + a_2 + \cdots + a_n)!}{a_1! a_2! \cdots a_n!}$$

satisfied the same recurrence relation (17.149) and initial conditions, completing his proof.

## 17.14  Bohr, Mollerup, and Artin on the Gamma Function

In 1922, an important Danish language textbook on analysis was published, written by colleagues at the Polytecknisk Lareanstat, Harald Bohr (1887–1951) and Johannes Mollerup (1872–1937). Bohr gained early fame in 1908 as a member of his country's silver-medal Olympic soccer team; he later worked on the Riemann zeta function and did his most original mathematical work in creating the theory of almost periodic functions. Bohr and Mollerup's four-volume work, *Laerebog i Matematisk Analyse* in various editions, had a profound effect on the teaching of analysis in Denmark, greatly raising the standards. In this work,[70] they applied the idea of logarithmic convexity to prove that the gamma integral equaled the infinite product, that is,

$$\int_0^\infty e^{-x} x^{m-1} dx = \lim_{n \to \infty} \frac{n! \, n^{m-1}}{m(m+1) \cdots (m+n-1)}. \tag{17.150}$$

They started with the right-hand side of (17.150) as the definition of $\Gamma(x)$. They relied on a definition of the Danish mathematician Johan Jensen of a convex function as a function $\phi(x)$ with the property that for every pair of numbers $x_1 < x_3$ and $x_2 = \frac{x_1 + x_3}{2}$

$$\phi(x_2) \leq \frac{\phi(x_1) + \phi(x_3)}{2}. \tag{17.151}$$

When $\phi$ was continuous, Jensen noted that (17.151) was equivalent to

$$\phi(tx_1 + (1-t)x_3) \leq t\phi(x_1) + (1-t)\phi(x_3), \, 0 < t < 1 \tag{17.152}$$

for all pairs of numbers $x_1 < x_3$. We note that Jensen's definition of convexity arose from a study of inequalities. Jensen's work is discussed in our Section 6.7.

In proving (17.150), we follow closely the Bohr–Mollerup notation and argument; by contrast, textbooks usually follow the treatment of Artin. The result we now refer to as the Bohr–Mollerup theorem was not explicitly stated by Bohr and Mollerup but

[70] Bohr and Mollerup (1922) vol. 3.

follows from their argument. Bohr and Mollerup denoted the integral in (17.150) by $\Lambda(m)$ and then observed that $\Lambda(x + 1) = x\Lambda(x)$ and $\Lambda(1) = 1$. Moreover, by the Cauchy–Schwarz inequality

$$\left(\Lambda\left(\frac{x_1 + x_3}{2}\right)\right)^2 \leq \Lambda(x_1)\Lambda(x_3), \quad 0 < x_1 < x_3. \tag{17.153}$$

Observe that this result is equivalent to the logarithmic convexity of $\Lambda(x)$ because $\Lambda(x)$ is continuous and (17.153) implies

$$\ln \Lambda\left(\frac{x_1 + x_3}{2}\right) \leq \frac{1}{2}\left(\ln \Lambda(x_1) + \ln \Lambda(x_3)\right).$$

For a definition of logarithmic convexity, see Section 3.2. Following Bohr and Mollerup, write $x_2 = \frac{x_1 + x_3}{2}$. They then set

$$P(x) = \frac{\Lambda(x)}{\Gamma(x)} \tag{17.154}$$

where $\Gamma(x)$ was defined by (17.4). It followed that $P(1) = 1$ and

$$P(x + 1) = P(x) \quad \text{for} \quad x > 0. \tag{17.155}$$

Bohr and Mollerup then used (17.153) to show that $P(x) \equiv 1$. For this purpose, they noted that when $n$ was an integer,

$$\lim_{n\to\infty} \frac{\Gamma(x_1 + n)\Gamma(x_3 + n)}{(\Gamma(x_2 + n))^2} = 1, \tag{17.156}$$

because

$$\lim_{n\to\infty} \frac{\Gamma(n + x)}{n^{x-1}n!} = \lim_{n\to\infty} \frac{(n - 1 + x)(n - 2 + x)\cdots x\Gamma(x)}{n^{x-1}n!} = 1.$$

Now by the periodicity of $P(x)$ given in (17.155), they had

$$\frac{P(x_1)P(x_3)}{(P(x_2))^2} = \frac{P(x_1 + n)P(x_3 + n)}{(P(x_2 + n))^2}.$$

By letting $n \to \infty$ in this equation and using (17.153) and (17.156), they could see that

$$1 \leq \frac{P(x_1)P(x_3)}{(P(x_2))^2}. \tag{17.157}$$

Next, they supposed $P(x)$ was not a constant. Then, because $P(x + 1) = P(x)$, it was possible to choose $x_1 < x_2$ such that $P(x_1) < P(x_2)$. They took the sequence $x_1 < x_2 < x_3 < x_4 < \cdots$ such that the difference between two consecutive numbers was always the same, that is, equal to $x_2 - x_1$. This meant that if $x_{n-1} < x_n < x_{n+1}$ was a part of the above sequence, then $x_n = \frac{x_{n-1} + x_{n+1}}{2}$. By (17.157), they obtained

$$1 < \frac{P(x_2)}{P(x_1)} \le \frac{P(x_3)}{P(x_2)} \le \frac{P(x_4)}{P(x_3)} \le \cdots, \qquad (17.158)$$

implying by induction that

$$\frac{P(x_n)}{P(x_1)} \ge \left(\frac{P(x_2)}{P(x_1)}\right)^n.$$

They could conclude that $P(x_n) \to \infty$ as $n \to \infty$. They noted that $P(x)$ was continuous on [1,2] and hence bounded on that interval. So they got a contradiction by the periodicity of $P$. Thus, $P(x)$ was a constant, necessarily equal to 1, and their proof was complete.

Emil Artin (1898–1962) saw that the Bohr–Mollerup proof of (17.150) could be simplified if (17.152) instead of (17.151) were used for convexity.[71] Artin's argument applied Hölder's inequality to show that

$$\ln(\Lambda(tx_1 + (1-t)x_3)) \le t \ln \Lambda(x_1) + (1-t) \ln \Lambda(x_3), \quad \text{for } 0 < t < 1.$$

He then proved more generally that if $f(0) = 1$, $f(x+1) = xf(x)$, and $\ln f(x)$ satisfied (17.152), then

$$f(x) = \lim_{n\to\infty} \frac{n!\, n^{x-1}}{x(x+1)\cdots(x+n-1)}.$$

Artin's proof was quite short. Note that by the first two conditions, $\ln f(n) = \ln(n-1)!$, when $n$ was a positive integer. Next, let $0 < x \le 1$. With $x_1 = n$, $x_3 = x + n + 1$ and $t = \frac{x}{1+x}$ in (17.152) Artin had

$$(x+1)\ln n! \le x \ln(n-1)! + \ln f(n+1+x).$$

This simplified to

$$\frac{n!\, n^{x-1}}{x(x+1)\cdots(x+n-1)} \cdot \frac{n}{n+x} \le f(x).$$

Similarly, with $x_1 = n+1$, $x_3 = n+2$ and $t = 1-x$, he had, after simplification,

$$f(x) \le \frac{n!\,(n+1)^x}{x(x+1)\cdots(x+n)} = \frac{n!\,n^{x-1}}{x(x+1)\cdots(x+n-1)} \cdot \left(\frac{n+1}{n}\right)^{x-1} \cdot \frac{n+1}{n+x}.$$

The two inequalities yielded the required formula when $n \to \infty$. Artin was a number theorist and algebraist. In algebra, he was a disciple of Emmy Noether (1882–1935) and advocated a very abstract point of view. It is therefore interesting to see him make this contribution to special functions. Some of his other results in this area are mentioned in the exercises. In addition, the reader may refer to Sections 3.2 and 3.4, for the use of logarithmic convexity by Wallis and Stieltjes.

---

[71] Artin (1964) pp. 14–15.

## 17.15   Kummer's Fourier Series for $\ln \Gamma(x)$

By an interesting application of definite integrals, in 1847 Kummer derived the Fourier series for $\ln \Gamma(x)$.[72] This formula is important in number theory, although Kummer's purpose was to obtain a new derivation for Gauss's multiplication formula for the gamma function. The latter can be written as

$$\sum_{k=0}^{n-1} \ln \Gamma\left(x + \frac{k}{n}\right) = \frac{1}{2}(n-1)\ln 2\pi + \frac{1}{2}(1-2nx)\ln n + \ln \Gamma(nx). \quad (17.159)$$

Kummer explained why he thought of the Fourier series in this connection. Suppose

$$f(x) = A_0 + 2\sum_{k=1}^{\infty} A_k \cos 2k\pi x + 2\sum_{k=1}^{\infty} B_k \sin 2k\pi x, \quad \text{for } 0 < x < 1, \quad (17.160)$$

where

$$A_k = \int_0^1 f(x)\cos 2k\pi x\, dx, \quad B_k = \int_0^1 f(x)\sin 2k\pi x\, dx. \quad (17.161)$$

Then

$$\sum_{k=0}^{n-1} f\left(x + \frac{k}{n}\right) = n\left(A_0 + 2\sum_{k=1}^{\infty} A_{nk}\cos 2kn\pi x + 2\sum_{k=1}^{\infty} B_{nk}\sin 2kn\pi x\right). \quad (17.162)$$

Moreover, by denoting

$$F(x) = A_0 + 2\sum_{k=1}^{\infty} A_{nk}\cos 2k\pi x + 2\sum_{k=1}^{\infty} B_{nk}\sin 2k\pi x,$$

the right-hand side of (17.162) was $nF(nx)$. Thus, equation (17.162) was suggestive of Gauss's formula (17.159). So Kummer took $f(x) = \ln \Gamma(x)$ in (17.160). Then by Euler's reflection formula

$$\ln \Gamma(x) + \ln \Gamma(1-x) = \ln 2\pi - \ln(2\sin \pi x)$$

$$= \ln 2\pi - \sum_{k=1}^{\infty} \frac{\cos 2k\pi x}{k}, \quad (17.163)$$

where the last relation followed from Euler's Fourier series for $\ln(2\sin \pi x)$. By (17.160),

$$\ln \Gamma(x) + \ln \Gamma(1-x) = 2A_0 + \sum_{k=1}^{\infty} 4A_k \cos 2k\pi x$$

[72] Kummer (1847).

and hence, by (17.163),

$$A_0 = \frac{1}{2}\ln 2\pi, \quad A_k = \frac{1}{4k}.$$

Kummer had to work harder to find $B_k$. He started with Plana's formula

$$\ln \Gamma(x) = \int_0^1 \left(\frac{1 - t^{x-1}}{1 - t} - x + 1\right)\frac{dt}{\ln t}, \quad x > 0, \tag{17.164}$$

an integrated form of Gauss's formula for $\Psi(x)$. So he had

$$B_k = \int_0^1 \int_0^1 \left(\frac{1 - t^{x-1}}{1 - t} - x + 1\right)\frac{\sin 2k\pi x}{\ln t}dt\, dx. \tag{17.165}$$

Since

$$\int_0^1 \sin 2k\pi x\, dx = 0, \quad \int_0^1 x \sin 2k\pi x\, dx = -\frac{1}{2k\pi}$$

and

$$\int_0^1 t^{x-1} \sin 2kx\, dx = \frac{(1 - t)2k\pi}{t((\ln t)^2 + 4k^2\pi^2)},$$

Kummer reduced (17.165) to

$$B_k = \int_0^1 \left(\frac{-2k\pi}{t((\ln t)^2 + 4k^2\pi^2)} + \frac{1}{2k\pi}\right)\frac{dt}{\ln t}.$$

Then, with $t = e^{-2k\pi u}$,

$$B_k = \frac{1}{2k\pi}\int_0^\infty \left(\frac{1}{1 + u^2} - e^{-2k\pi u}\right)\frac{du}{u}.$$

When $k = 1$,

$$B_1 = \frac{1}{2\pi}\int_0^\infty \left(\frac{1}{1 + u^2} - e^{-2\pi u}\right)\frac{du}{u}.$$

Kummer then employed a result of Dirichlet:

$$-\frac{\gamma}{2\pi} = \frac{1}{2\pi}\int_0^\infty \left(e^{-u} - \frac{1}{1 + u}\right)\frac{du}{u},$$

where $\gamma$ was Euler's constant. See Exercise 3b in this connection. Therefore,

$$B_1 - \frac{\gamma}{2\pi} = \frac{1}{2\pi}\int_0^\infty \frac{e^{-u} - e^{-2\pi u}}{u}du + \frac{1}{2\pi}\int_0^\infty \left(\frac{1}{1 + u^2} - \frac{1}{1 + u}\right)\frac{du}{u}.$$

The first integral ln $2\pi$ and a change of variables $t$ to $\frac{1}{t}$ showed that the value of the second integral was 0. Thus,

$$B_1 = \frac{\gamma}{2\pi} + \frac{1}{2\pi} \ln 2\pi.$$

To find $B_k$, he observed that

$$k B_k - B_1 = \frac{1}{2\pi} \int_0^\infty \frac{e^{-2\pi u} - e^{-2k\pi u}}{u} du = \frac{1}{2\pi} \ln k.$$

Thus,

$$B_k = \frac{1}{2k\pi}(\gamma + \ln 2k\pi), \quad k = 1, 2, 3, \ldots,$$

and Kummer got his Fourier expansion:

$$\ln \Gamma(x) = \frac{1}{2} \ln 2\pi + \sum_{k=1}^\infty \frac{\cos 2\pi kx}{2k} + \frac{1}{\pi} \sum_{k=1}^\infty \frac{\gamma + \ln 2\pi + 2\ln k}{2k} \sin 2k\pi x.$$

## 17.16 Exercises

(1) Show that by taking $k = p + q\sqrt{-1}$, $p > 0$ in

$$\int_0^\infty x^{n-1} e^{-kx} dx = \frac{\Gamma(n)}{k^n},$$

we get

$$\int_0^\infty x^{n-1} e^{-px} \cos qx \, dx = \frac{\Gamma(n) \cos n\theta}{f^n}$$

and

$$\int_0^\infty x^{n-1} e^{-px} \sin qx \, dx = \frac{\Gamma(n) \sin n\theta}{f^n},$$

where $f = (p^2 + q^2)^{\frac{1}{2}}$ and $\tan \theta = \frac{q}{p}$.

Deduce that

$$\int_0^\infty e^{-px} \frac{\cos qx}{\sqrt{x}} dx = \frac{\sqrt{\pi}}{f} \sqrt{\frac{f-p}{2}},$$

$$\int_0^\infty e^{-px} \frac{\sin qx}{\sqrt{x}} dx = \frac{\sqrt{\pi}}{f} \sqrt{\frac{f+p}{2}},$$

$$\int_0^\infty \frac{\cos x}{\sqrt{x}} \, dx = \sqrt{\frac{\pi}{2}}, \quad \int_0^\infty \frac{\sin x}{\sqrt{x}} \, dx = \sqrt{\frac{\pi}{2}},$$

$$\int_0^\infty e^{-px} \frac{\sin qx}{x} \, dx = \theta, \quad \text{and} \quad \int_0^\infty \frac{\sin x}{x} \, dx = \frac{\pi}{2}.$$

All these definite integrals appeared in Euler's paper of 1781, though he had evaluated some of them earlier by other methods. See Eu I-19 pp. 217–227. Euler's deductions were formal and he was the first to make use of complex parameters in this way. Although he initially assumed the parameter $p > 0$, he let $p \to 0$ to obtain the later integrals. He expressed this by setting $p = 0$. He had some reservations about presenting the final integral for $\frac{\sin x}{x}$ but numerical computation convinced him of its correctness. This paper was influential in the development of complex analysis by Cauchy. Legendre and Laplace referred to it when extending its methods to evaluate other integrals and these results motivated Cauchy to begin his work on complex integrals in 1814.

(2) If Euler found a formula interesting, he often evaluated it in more than one way. Complete the details and verify the steps of the three methods he gave to show that the integral $\int_0^{\frac{\pi}{2}} \ln \sin \phi \, d\phi = -\frac{\pi}{2} \ln 2$.

(a) Euler began by setting $x = \sin \phi$ and then $\cos \phi = y$ to get

$$\int_0^1 \frac{\ln x}{\sqrt{1 - x^2}} \, dx = \int_0^1 \frac{\ln \sqrt{(1 - y^2)}}{\sqrt{1 - y^2}} \, dy$$

$$= - \int_0^1 \frac{\left( \frac{y^2}{2} + \frac{y^4}{4} + \frac{y^6}{6} + \cdots \right)}{\sqrt{1 - y^2}} \, dy$$

$$= - \frac{\pi}{2} \left( \frac{1}{2^2} + \frac{1 \cdot 3}{2 \cdot 4^2} + \frac{1 \cdot 3 \cdot 5}{2 \cdot 4 \cdot 6^2} + \frac{1 \cdot 3 \cdot 5 \cdot 7}{2 \cdot 4 \cdot 6 \cdot 8^2} + \cdots \right).$$

Euler showed that the sum of the series was ln 2. For this purpose, he noted that by the binomial expansion

$$\int_0^x \frac{1}{z} \left( \frac{1}{\sqrt{1 - z^2}} - 1 \right) dz = \frac{1}{2^2} x^2 + \frac{1 \cdot 3}{2 \cdot 4^2} x^4 + \frac{1 \cdot 3 \cdot 5}{2 \cdot 4 \cdot 6^2} x^6 + \text{etc.}$$

He applied the substitution $v = \sqrt{1 - z^2}$ to evaluate

$$\int \frac{1}{z\sqrt{1 - z^2}} \, dz = - \ln \frac{1 + \sqrt{1 - z^2}}{z} + C.$$

Hence

$$\int_0^x \frac{1}{z} \left( \frac{1}{\sqrt{1 - z^2}} - 1 \right) dz = \ln \frac{2}{1 + \sqrt{1 - x^2}}.$$

(b) In his second method, Euler started with a divergent series. He applied the addition formula $2 \sin n\theta \sin \theta = \cos(n-1)\theta - \cos(n+1)\theta$ to get

$$\frac{\cos \theta}{\sin \theta} = 2 \sin 2\theta + 2 \sin 4\theta + 2 \sin 6\theta + 2 \sin 8\theta + \text{ etc.}$$

He integrated to obtain

$$\ln \sin \theta = C - \cos 2\theta - \frac{1}{2} \cos 4\theta - \frac{1}{3} \cos 6\theta - \frac{1}{4} \cos 8\theta - \text{ etc.}$$
$$(17.166)$$

Then $\theta = \frac{\pi}{2}$ gave $C = -\ln 2$. Euler integrated again to obtain the required result.

(c) Euler proved the more general formula

$$\int_0^1 x^{p-1} X \ln x \, dx = \int_0^1 x^{p-1} X \, dx \int_0^1 \frac{x^{p-1}(x^m - 1)}{1 - x^n} \, dx, \qquad (17.167)$$

where $X = (1 - x^n)^{\frac{m-n}{n}}$. The result in (a) and (b) would be obtained by setting $n = 2, m = p = 1$. To prove (17.167), Euler set $P = \int_0^1 x^{p-1} X \, dx$. By an argument which gives (17.37), show that

$$P = \frac{n}{m} \cdot \frac{2n}{m+n} \cdot \frac{3n}{m+2n} \cdots \times \frac{p+m}{p} \cdot \frac{p+m+n}{p+n} \cdot \frac{p+m+2n}{p+2n} \cdot \text{etc.}$$

He then let $p$ be the variable and $m, n$ be constants to obtain

$$\frac{1}{P} \frac{dP}{dp} = \frac{1}{m+p} - \frac{1}{p} + \frac{1}{m+p+n} - \frac{1}{p+n}$$
$$+ \frac{1}{m+p+2n} - \frac{1}{p+2n} + \text{ etc.}$$

Prove the result by showing that that this partial fractions expansion equals

$$\int_0^1 \frac{x^{p-1}(x^m - 1)}{1 - x^n} \, dx.$$

Euler actually worked with a product for $\frac{P}{Q}$ where $Q$ was an integral similar to $P$. Lacroix (1819) gave the preceding simplification on p. 437. These results are contained in Eu. I-18, pp. 23–50. This volume is full of ingenious evaluations of definite integrals. Observe that (17.166) is the Fourier series expansion of $\ln \sin \theta$ used by Kummer to obtain (17.163). Also, Euler could have derived (17.166) without using divergent series by expanding $\ln(1 - e^{2i\theta})$. But Euler treated divergent series as very much a part of mathematics, a view validated only in the twentieth century

(3) The following derivation of Gauss's multiplication formula is due to Dirichlet (1969) pp. 274–276. Verify the successive steps for Re $a > 0$:

(a) $\int_0^\infty (e^{-y} - e^{-sy})\frac{dy}{y} = \ln s$.

(b) Next

$$\Gamma'(a) = \int_0^\infty e^{-s} s^{a-1} \ln s \, ds$$

$$= \int_0^\infty \frac{dy}{y} \left( e^{-y} \int_0^\infty e^{-s} s^{a-1} \, ds - \int_0^\infty e^{-(1+y)s} s^{a-1} \, ds \right)$$

$$= \Gamma(a) \int_0^\infty \frac{dy}{y} \left( e^{-y} - \frac{1}{(1+y)^a} \right).$$

$$(17.168)$$

(c) $\frac{d}{da} \ln \Gamma(a) = \int_0^1 \left( e^{1-\frac{1}{x}} - x^a \right) \frac{dx}{x(1-x)}$.

(d) Let

$$S = n \int_0^1 \left( \frac{ne^{1-\frac{1}{x^n}}}{1 - x^n} - \frac{x^{na}}{1-x} \right) \frac{dx}{x}.$$

Then

$$S = \sum_{k=0}^{n-1} \frac{d}{da} \ln \Gamma \left( a + \frac{k}{n} \right).$$

(e) Change $a$ to $na$ in (c) and subtract the result from (d) to see that
$S - \frac{d}{da} \ln \Gamma(na)$ is independent of $a$. Denote this quantity by $p$ and
integrate to get

$$\prod_{k=0}^{n-1} \Gamma \left( a + \frac{k}{n} \right) = q p^a \Gamma(na).$$

(f) Change $a \to a + \frac{1}{n}$ in (e) to deduce that $p = n^{-n}$.

(g) Euler's formula (17.17) implies that $q = (2\pi)^{\frac{n-1}{2}} \sqrt{n}$.

(h) Show that Euler's formula (17.17) is obtained by applying

$$\Gamma(x)\Gamma(1-x) = \frac{\pi}{\sin \pi x}.$$

(4) Show that for suitably chosen constants $a$, $b$, $c$, and $k$,

(a) $\int_0^\infty e^{-(c+z)y} y^{a-1} dy = \frac{\Gamma(a)}{(c+z)^a}$.

(b) $\int_0^\infty e^{-cy} y^{a-1} \left( \int_0^\infty e^{-(k+y)z} z^{b-1} dz \right) dy = \Gamma(a) \int_0^\infty \frac{e^{-kz} z^{b-1}}{(c+z)^a} dz$.

(c) $\Gamma(b) \int_0^\infty \frac{e^{-cy} y^{a-1}}{(k+y)^b} dy = \Gamma(a) \int_0^\infty \frac{e^{-kz} z^{a-1}}{(c+z)^a} dz.$

(d) $\int_0^\infty \frac{y^{a-1}}{(1+y)^{a+b}} dy = \frac{\Gamma(a)\Gamma(b)}{\Gamma(a+b)}.$

See Dirichlet (1969) vol. I, p. 278.

(5) Use Dirichlet's integral formula for $\frac{\Gamma'(a)}{\Gamma(a)}$ (17.168) to show that for $a > 0$

$$\gamma + \frac{d}{da} \ln \Gamma(a+1) = \int_0^1 \frac{1 - y^a}{1 - y} dy$$

$$= \left(1 - \frac{1}{a+1}\right) + \left(\frac{1}{2} - \frac{1}{a+2}\right) + \left(\frac{1}{3} - \frac{1}{a+3}\right) + \cdots.$$

Deduce the infinite product for $\Gamma(a)$:

$$\Gamma(a+1) = a\Gamma(a) = e^{-\gamma a} \frac{e^a}{1+a} \cdot \frac{e^{\frac{a}{2}}}{1 + \frac{a}{2}} \cdot \frac{e^{\frac{a}{3}}}{1 + \frac{a}{3}} \cdots.$$

For details, see Schlömilch (1843).

(6) Show that Dirichlet's integral formula, (17.168) can be obtained from Gauss's (17.173).

(7) For $0 < b < 1$, set $B(b) = \int_0^\infty \frac{t^{b-1}}{1+t} dt$. Show that

$$\int_0^\infty \frac{t^{b-1}}{st+1} dt = Bs^{-b}$$

and

$$\int_0^\infty \frac{t^{b-1}}{t+s} dt = Bs^{b-1}.$$

Deduce that

$$B\frac{(s^{b-1} - s^{-b})}{s - 1} = \int_0^\infty \frac{t^{b-1}(t-1)}{(st+1)(t+s)} dt. \qquad (17.169)$$

Observe that

$$B^2 = \int_0^\infty \frac{1}{s+1} \left(\int_0^\infty \frac{t^{b-1}}{t+s} dt\right) ds$$

$$= \int_0^\infty \frac{t^{b-1} \ln t}{t - 1} dt.$$

Deduce that

$$\int_{1-y}^y B^2 dt = \int_0^\infty \frac{t^{y-1} - t^{-y}}{t - 1} dt.$$

From (17.169) obtain

$$B(b) \int_{1-b}^{b} [B(t)]^2 dt = 2 \int_0^\infty \frac{t^{b-1} \ln t}{1+t} dt = 2B'(b).$$

Observe that $B(b) = B(1-b)$ and deduce that $B'\left(\frac{1}{2}\right) = 0$,

$$\int_{1-b}^{b} [B(t)]^2 dt = 2 \int_{\frac{1}{2}}^{b} [B(t)]^2 dt$$

and

$$B(b) \int_{\frac{1}{2}}^{b} [B(t)]^2 dt = B'(b).$$

Now show that $B$ satisfies the differential equation $BB'' - (B')^2 = B^4$. Solve the differential equation with initial conditions $B\left(\frac{1}{2}\right) = \pi$ and $B'\left(\frac{1}{2}\right) = 0$ to obtain $B = \pi \csc \pi b$. This is Dedekind's evaluation of the Eulerian integral $B$, a part of his doctoral dissertation. See Dedekind (1930) vol. 1, pp. 19–22 and 29–31.

(8) Let $c_1, c_2, \ldots, c_n$ be positive constants and set $f(x) = (x + c_1)(x + c_2) \cdots (x + c_n)$. Show that for $0 < b < n$

$$\int_0^\infty \frac{x^{b-1}}{f(s)} dx = \frac{\pi}{\sin b\pi} \sum_{k=1}^{n} \frac{c_k^{b-1}}{f'(-c_k)},$$

where $f'$ denotes the derivative of $f$. See Dedekind (1930) vol. 1, p. 24.

(9) (a) Suppose that $\phi(x)$ is positive and twice continuously differentiable on $0 < x < \infty$ and satisfies (i) $\phi(x + 1) = \phi(x)$ and (ii) $\phi\left(\frac{x}{2}\right)\phi\left(\frac{x+1}{2}\right) = d\phi(x)$, where $d$ is a constant. Prove that $\phi$ is a constant.

(b) Show that $\Gamma(x)\Gamma(1-x) \sin \pi x$ satisfies the conditions of the first part of the problem. Deduce Euler's formula (17.11). This proof of Euler's reflection formula (17.11) is due to Artin (1964) chapter 4.

(10) Suppose that $f$ is a positive and twice continuously differentiable function on $0 < x < \infty$ and satisfies $f(x + 1) = xf(x)$ and $2^{2x-1}f(x)f\left(x + \frac{1}{2}\right) = \sqrt{\pi} f(2x)$. Show that $f(x) = \Gamma(x)$. See Artin (1964).

(11) Prove the following results of Gauss on the digamma function:

(a) For a positive integer $n$,

$$\Psi(z + n) = \Psi(z) + \frac{1}{z+1} + \frac{1}{z+2} + \cdots + \frac{1}{z+n}.$$

(b) $\Psi(0) = \Gamma'(1) = -\gamma = -0.57721566490153286060653$. Euler computed the constant $\gamma$ correctly to fifteen decimal places by an application of the Euler–Maclaurin summation formula. About twenty years later, in 1790, the Italian mathematician, Lorenzo Mascheroni (1750–1800) computed $\gamma$ to thirty-two decimal places by the same method. To compute

$\gamma$, Gauss gave two asymptotic series for $\Psi(z)$, obtained by taking the derivatives of the de Moivre and Stirling asymptotic series for $\ln(z+1)$. His value differed from Mascheroni's in the twentieth place and so he persuaded F. B. G. Nicolai, a calculating prodigy, to the repeat the computation and to extend it further. Nicolai calculated to forty places, given by Gauss in a footnote, and verified that Gauss's computation was correct.

(c)

$$\Psi(-z) - \Psi(z-1) = \pi \cot \pi z. \tag{17.170}$$

(d)

$$\Psi(x) - \Psi(y) = -\frac{1}{x+1} + \frac{1}{y+1} - \frac{1}{x+2} + \frac{1}{y+2} - \frac{1}{x+3} + \text{etc.}$$

(e)

$$\Psi(z) + \Psi\left(z - \frac{1}{n}\right) + \Psi\left(z - \frac{2}{n}\right) + \cdots + \Psi\left(z - \frac{n-1}{n}\right)$$
$$= n\Psi(nz) - n \ln n.$$

(f)

$$\Psi\left(-\frac{1}{n}\right) + \Psi\left(-\frac{2}{n}\right) + \cdots + \Psi\left(-\frac{n-1}{n}\right) = -(n-1)\gamma - n \ln n.$$

(g) For $n$ an odd integer and $m$ a positive integer less than $n$,

$$\Psi\left(-\frac{m}{n}\right) = -\gamma + \frac{1}{2}\pi \cot \frac{m\pi}{n} - \ln n$$
$$+ \sum_{k=1}^{n-1} \cos \frac{2km\pi}{n} \ln\left(2 - 2\cos \frac{k\pi}{n}\right). \tag{17.171}$$

(h) For $n$ even,

$$\Psi\left(-\frac{m}{n}\right) = \pm \ln 2 - \gamma + \frac{1}{2}\pi \cos \frac{m\pi}{n} - \ln n$$
$$+ \sum_{k=1}^{n-2} \cos \frac{2km\pi}{n} \ln\left(2 - 2\cos \frac{k\pi}{n}\right), \tag{17.172}$$

where the upper sign is taken for $m$ even, and the lower for $m$ odd.

(i)

$$\Psi(t) = \int_0^1 \left(-\frac{1}{\ln x} - \frac{x^t}{1-x}\right) dx, \quad t > -1. \tag{17.173}$$

In unpublished work, Gauss explained how $\Psi(z)$ could be used to express the second independent solution of the hypergeometric equation in certain special circumstances. See Gauss (1868–1927) vol. 3, pp. 154–160.

(12) Prove that

$$\frac{1}{2\pi i}\int_{-i\infty}^{i\infty}\Gamma(a+s)\Gamma(b+s)\Gamma(c-s)\Gamma(d-s)ds$$

$$=\frac{\Gamma(a+c)\Gamma(a+d)\Gamma(b+c)\Gamma(b+d)}{\Gamma(a+b+c+d)},$$

where the path of integration is curved so that the poles of $\Gamma(c-s)\Gamma(d-s)$ lie on the right of the path and the poles of $\Gamma(a+s)\Gamma(b+s)$ lie on the left. This formula is due to Barnes (1908); it played an important role in his theory of the hypergeometric function. It is an extension of Euler's beta integral formula (17.10), as pointed out by Askey. This can be seen by replacing $b$ and $d$ by $b-it$ and $d+it$, respectively, and then setting $s=tx$ and letting $t\to\infty$.

(13) Suppose $F$ is a holomorphic function in the right half complex plane Re $z>0$. Suppose also that $F(1)=1$, $F(z+1)=zF(z)$ and that $F(z)$ is bounded in the vertical strip $1\le$ Re $z<2$. Then $F(z)=\Gamma(z)$ for Re $z>0$. This uniqueness theorem, useful for giving short proofs of several basic results on the gamma function, was proved by Helmut Wielandt in 1939 and published by Konrad Knopp in 1941. See Remmert (1996), who quotes a letter of Wielandt explaining this and gives references.

(14) Prove Dirichlet's formula: Suppose $c_1, c_2, c_3, \ldots$ is a sequence of complex numbers which satisfy $c_{n+k}=c_n$. Suppose $\sum_{n=1}^{\infty}\frac{c_n}{n^s}$ converges absolutely. Then

$$\sum_{n=1}^{\infty}\frac{c_n}{n^s}=\frac{1}{\Gamma(s)}\int_0^1\frac{\sum_{n=1}^{k}c_nx^{n-1}}{1-x^k}\ln^{s-1}\left(\frac{1}{x}\right)dx.$$

This is the formula Dirichlet applied to the problem of primes in arithmetic progressions. See Dirichlet (1969) vol. I, pp. 421–422.

(15) Observe that for $a<\frac{1}{2}$

$$\int_0^1 t^{\frac{1}{2}-a}(1-t)^{\frac{1}{2}-a}dt=\frac{\left(\Gamma(\frac{1}{2}-a)\right)^2}{\Gamma(1-2a)}.$$

Write the integrand as $2^{2a-1}(1-(1-2t)^2)^{a-\frac{1}{2}}$, apply a change of variables, and then use Euler's reflection formula to obtain

$$\sqrt{\pi}\Gamma(a)=2^{1-2a}\cos(a\pi)\Gamma(2a)\cdot\Gamma\left(\frac{1}{2}-a\right).$$

This proof of the duplication formula is Legendre's (1811–1817) vol. 1, p. 284.

(16) Prove that

$$\frac{\int_0^\infty \sinh^{\beta-1} u \cosh^{-a} u e^{-xu}\,du}{(\beta-1)\int_0^\infty \sinh^{\beta-2} u \cosh^{1-a} u e^{-xu}\,du}$$

$$= \frac{1}{x+} \frac{\alpha\beta}{x+} \frac{(\alpha+1)(\beta+1)}{x+} \frac{(\alpha+2)(\beta+2)}{x+} \cdots.$$

See Stieltjes (1993) vol. II, p. 391.

(17) From Stieltjes's formula in the previous exercise, deduce that

$$\frac{\Gamma\left(x-\tfrac{1}{2}a+\tfrac{1}{4}\right)\Gamma\left(x+\tfrac{1}{2}a+\tfrac{3}{4}\right)}{\Gamma\left(x+\tfrac{1}{2}a+\tfrac{1}{4}\right)\Gamma\left(x-\tfrac{1}{2}a+\tfrac{3}{4}\right)}$$

$$= 1 + \frac{2a}{4x-a+} \frac{1^2-a^2}{4x+} \frac{2^2-a^2}{4x+} \frac{3^2-a^2}{4x+} \cdots$$

$$\frac{\Gamma\left(x-\tfrac{1}{2}a+\tfrac{1}{4}\right)\Gamma\left(x+\tfrac{1}{2}a+\tfrac{1}{4}\right)}{\Gamma\left(x-\tfrac{1}{2}a+\tfrac{3}{4}\right)\Gamma\left(x+\tfrac{1}{2}a+\tfrac{3}{4}\right)}$$

$$= \frac{4}{4x+} \frac{1^2-4a^2}{8x+} \frac{3^2-4a^2}{8x+} \frac{5^2-4a^2}{8x+} \frac{7^2-4a^2}{8x+} \cdots.$$

Also show that if

$$\frac{1\cdot3\cdot5\cdots(2n-1)}{2\cdot4\cdot6\cdots2n} = \frac{1}{\sqrt{(\pi(n+\epsilon))}},$$

$$\text{then} \quad \phi(n) = 1 + \frac{2}{8n-1+} \frac{1\cdot3}{8n+} \frac{3\cdot5}{8n+} \frac{5\cdot7}{8n+} \frac{7\cdot9}{8n+} \cdots.$$

See Stieltjes (1993) vol. II, pp. 396–398.

## 17.17 Notes on the Literature

Tweddle (2003) contains the English translation of Stirling's *Methodus Differentialis*. Tweddle has helpfully added 120 pages of notes to clarify and explain Stirling's propositions in modern terms. For more history of the gamma function, see Davis (1959) and Dutka (1991).

# 18

# *The Asymptotic Series for* $\ln \Gamma(x)$

## 18.1  Preliminary Remarks

The answer to the thought-provoking question: "How large is $n!$ for large $n$?" was first given a good approximate answer in about 1730 by the joint efforts of Abraham de Moivre (1667–1754) and James Stirling (1692–1770). The story of their cooperation is fascinating.

Born in France, de Moivre was a victim of religious discrimination there, leading him as a young man to relocate to Britain, where he worked the rest of his life. De Moivre's motivation in developing an easily useable approximation for the magnitude of $n!$ arose from his interest in probability theory, a subject he began cultivating in 1707 at the age of 40. He became familiar with the works of Jakob Bernoulli, Niklaus I Bernoulli, and Pierre Montmort and went on to make very important contributions in that field. De Moivre supported himself as a consultant to gamblers, speculators, and rich patrons, helping them solve problems related to games of chance or the calculation of annuities. He published the first volume of his *Doctrine of Chances* in 1718. This work may have led Sir Alexander Cuming (1690–1775) to consult de Moivre on a problem that arose in the context of gambling. Cuming lived an eventful and long life, becoming a Fellow of the Royal Society of London as well as a Cherokee chief, and yet he died in poverty.[1]

To set up Cuming's problem, first let $p$ be the probability of any toss of a coin resulting in a head; the probability of $x$ heads in $n$ tosses would then be

$$b(x,n,p) = \binom{n}{x} p^x (1-p)^{n-x}.$$

Cuming's problem[2] for de Moivre reduced to the calculation of the mean deviation

$$\sum_{x=0}^{n} |x - np| \, b(x,n,p).$$

---

[1] Stephens and Lee (1908) vol. 5, pp. 294–295.
[2] Hald (1990) p. 470.

For example, given $p = \frac{1}{2}$, he had the value of the sum as

$$\frac{n}{2^{n+1}} \binom{n}{\lfloor \frac{n}{2} \rfloor}.$$

With $n$ an even number, $2m$, the expression reduced to

$$\frac{m}{2^{2m}} \binom{2m}{m}. \tag{18.1}$$

Of course, this step was the easy part. De Moivre's real problem was to obtain a fairly accurate approximation for (18.1) that would also be practical to use by a person such as Cuming. Here recall that we write $a_m \sim b_m$ as $m \to \infty$ if $\lim_{m\to\infty} \frac{a_m}{b_m} = 1$. Briefly, de Moivre first found

$$\frac{1}{2^{2m}} \binom{2m}{m} \sim \frac{2.168}{\sqrt{2m-1}} \left(1 - \frac{1}{2m}\right)^{2m} \quad \text{as } m \to \infty; \tag{18.2}$$

he then noted that

$$\left(1 - \frac{1}{2m}\right)^{2m} \sim \frac{1}{e} \quad \text{as } m \to \infty$$

and obtained

$$\frac{1}{2^{2m}} \binom{2m}{m} \sim \frac{2.168}{e} \frac{1}{\sqrt{2m-1}}. \tag{18.3}$$

Note that the value of $2.168\,e^{-1}$ is approximately 0.7976. To derive (18.3), he took the logarithm of the left-hand side of (18.2), expanded the logarithms as infinite series, and then changed the order of summation and applied Bernoulli's formula for the sum of powers of consecutive integers. We present a fuller account later in this chapter.

In June, 1729, de Moivre received a letter from Stirling and published it in his 1730 work, *Miscellanea Analytica*. Stirling wrote:[3]

> About four years ago, when I informed Mr. *Alex. Cuming* that problems concerning the Interpolation and Summation of series and others of this type which are not susceptible to the commonly accepted analysis, can be solved by Newton's Method of Differences, the most illustrious man replied that he doubted if the problem solved by you some years before about finding the middle coefficient in an arbitrary power of the binomial could be solved by differences. Then, led by curiosity and confident that I would be doing a favour to a most deserving man of Mathematics, I took it up willingly: and I admit that difficulties arose which prevented me from arriving at the desired conclusion rapidly, but I do not regret the labour, if I have in fact finally achieved a solution which is so acceptable to you that you consider it worthy of inclusion in your own writings.

---

[3] Stirling and Tweddle (2003) pp. 285–287, especially p. 285. See Tweedie (1922) p. 46 for the original Latin.

Interestingly, Cuming was the intermediary in connecting Stirling and de Moivre and, in fact, Stirling was inducted into the Royal Society of London on the nomination of Cuming. Stirling's letter gave the formulas,[4] first for the square of the reciprocal of the left-hand side of equation (18.2):

$$
\left( \frac{1}{2^{2m}} \binom{2m}{m} \right)^{-2}
$$

$$
= \pi m \left( 1 + \frac{1}{2^2 (m+1)} + \frac{(1 \cdot 3)^2}{2^4 2! \, (m+1)(m+2)} + \frac{(1 \cdot 3 \cdot 5)^2}{2^6 3! \, (m+1)(m+2)(m+3)} + \cdots \right)
$$
(18.4)

and then for the square of the left-hand side of (18.2)

$$
\left( \frac{1}{2^{2m}} \binom{2m}{m} \right)^2 = \frac{2}{\pi (2m+1)} \times \left( 1 + \frac{1}{2^2 \left(m + \frac{3}{2}\right)} + \frac{(1 \cdot 3)^2}{2^4 2! \left(m + \frac{3}{2}\right)\left(m + \frac{5}{2}\right)} \right.
$$

$$
\left. + \frac{(1 \cdot 3 \cdot 5)^2}{2^6 3! \left(m + \frac{3}{2}\right)\left(m + \frac{5}{2}\right)\left(m + \frac{7}{2}\right)} + \cdots \right). \quad (18.5)
$$

In his letter, Stirling did not give any indication of a proof except to say that he had established these formulas by the method of differences. The details are contained in his book,[5] however, and we present them in the course of the present chapter. In fact, identities (18.4) and (18.5) can also be proved as particular cases of Gauss's formula (17.14). However, it would be eighty years before Gauss would introduce this result.

Observe that by taking the first term of the series from (18.5), one gets a first approximation

$$
\frac{1}{2^{2m}} \binom{2m}{m} \sim \frac{1}{\sqrt{\pi m}}, \quad \text{as } m \to \infty,
$$

where the second and later terms can be neglected when $m$ is large, leading to a more elegant approximation than de Moivre's (18.3). When de Moivre received Stirling's communication, he was astonished to see the appearance of $\pi$ in this context. He searched the literature for a result that would explain this phenomenon and found that Wallis's formula was just what he needed. In fact, Stirling had applied Wallis's formula.

Stirling presented his results with proofs, along with other results, in his 1730 *Methodus Differentialis*. Formulas (18.4) and (18.5) appear as examples of propositions 22 and 23. Proposition 28 discusses the problem of finding the sum of any number of logarithms, whose arguments are in arithmetic progression. As example 2 of this result, Stirling gave the formula: for $s = m + \frac{1}{2}$,

[4] ibid. p. 286.
[5] Stirling (1730).

$$\ln m! = s \ln s + \frac{1}{2} \ln 2\pi - s - \frac{1}{24s} + \frac{7}{2880s^3} - \cdots, \qquad (18.6)$$

whose statement contained a recursive procedure to find the coefficients of $\frac{1}{s^{2n-1}}$ for $n = 1, 2, 3, \ldots$. After seeing this formula, de Moivre observed that the method that yielded him (18.3) would produce an infinite series for $\ln m!$, a series he presented in the supplement to his *Miscellanea Analytica*:[6]

$$\ln m! = \left( m + \frac{1}{2} \right) \ln m + \frac{1}{2} \ln 2\pi - m + \frac{1}{12m}$$

$$- \frac{1}{360m^3} + \frac{1}{1260m^5} - \frac{1}{1680m^7} + \cdots. \qquad (18.7)$$

The method employed by de Moivre would give the coefficient of $\frac{1}{m^{2k-1}}$ as $\frac{B_{2k}}{2k(2k-1)}$, though, following the custom of the seventeenth- and eighteenth-century mathematicians, he did not write the general term.

Observe that the terms of the series in (18.7), after the third, appear to get smaller as $m$ increases. However, by writing out many more terms of the series, we find that this is a misleading impression. In fact, the series diverges so that the series (18.6) and (18.7) cannot converge, but de Moivre was not aware of the exponential growth of Bernoulli numbers. The first dozen values of the Bernoulli numbers might even suggest that they could be bounded. In his *Supplement*, de Moivre wrote that the coefficients of the terms after the fourth do not decrease and in the 1756 edition of his *Doctrine of Chances*, he remarked that the series converged, but slowly. Now in 1740, Euler proved a formula, that we have given as (16.78); we can rewrite it as

$$B_{2k} = \frac{(-1)^{k-1}(2k)!}{2^{2k-1}\pi^{2k}} \left( 1 + \frac{1}{2^{2k}} + \frac{1}{3^{2k}} + \cdots \right).$$

Using (18.8), this implies

$$B_{2k} = O\left( \left( \frac{k}{\pi} \right)^{2k+\frac{1}{2}} \right).$$

To find an approximate value of $\ln m!$, de Moivre added the terms given in our (18.7). This gave a good approximation because the error term had the same order as the last term employed, that is $\frac{1}{m^7}$. The eighteenth-century mathematicians were intuitively aware of this method of summing divergent series of this type, so that they summed the terms of the series as long as the terms decreased, and stopped summing when the terms began to increase. Such divergent series are known as asymptotic series; although such series diverge, they yield very good approximations of the series they represent, as long as as a suitable number of terms of the series is utilized.

[6] de Moivre (1730b).

On this basis, we can see that the first three terms given in (18.7), expressed as

$$\left(m + \frac{1}{2}\right)\ln m + \frac{1}{2}\ln 2\pi - m = \ln\left(\sqrt{2\pi m}\, m^m e^{-m}\right),$$

approximate $\ln m!$ to the order of $\frac{1}{m}$ when $m$ is large. Thus, we have the approximation

$$m! \sim \sqrt{2\pi}\, m^{m+\frac{1}{2}} e^{-m}, \quad m \to \infty. \tag{18.8}$$

Though (18.8) actually follows immediately from de Moivre's (18.7), the first person to point out (18.8) was Euler in a letter to Goldbach dated July 4, 1744.[7] The result (18.8) is now called Stirling's approximation. By using Stirling's series (18.6), we obtain the approximation

$$m! \sim \sqrt{2\pi}\left(\frac{m + \frac{1}{2}}{e}\right)^{m+\frac{1}{2}}. \tag{18.9}$$

In the *Miscellanea Analytica* of 1730, de Moivre first derived an asymptotic series for the logarithm of the left-hand side of (18.2), that in modern notation can be rendered as

$$\ln\left(\frac{1}{2^{2m}}\binom{2m}{m}\right) = \left(2m - \frac{1}{2}\right)\ln(2m - 1) - 2m\ln(2m) + \ln 2 - \frac{1}{2}\ln(2\pi)$$

$$+ 1 - \sum_{k=1}^{\infty} \frac{B_{2k}}{(2k-1)2k}\left(\frac{2}{m^{2k-1}} - \frac{1}{(2m-1)^{2k-1}}\right); \tag{18.10}$$

note the similarity with (18.7). In fact, de Moivre proved both (18.7) and (18.10) by applying three formulas: the Mercator-Newton power series for $\ln(1 + x)$; Jakob Bernoulli's formula for $\sum_{k=1}^{n} k^m$; and Wallis's formula for $\pi$. We give the details later in the chapter. Observe that Stirling's series (18.4) and (18.5) for the square of $\frac{1}{2^{2m}}\binom{2m}{m}$ and its reciprocal are convergent series. By contrast, de Moivre's series for the logarithm of $\frac{1}{2^{2m}}\binom{2m}{m}$ is an asymptotic series, a series that diverges but whose first few terms give a very good approximation.

Soon after the work of de Moivre and Stirling, Euler and Maclaurin independently discovered a very general formula from which (18.6) and (18.7) could be easily derived. In his paper on the gamma function,[8] Gauss referred to Euler's derivation. He noted that though Euler stated the result for $\ln \Gamma(x)$ when $x$ was a positive integer, Euler's method applied to the general case so that for any real $x > 0$

---

[7] Fuss (1968) vol. 1, p. 283.
[8] Gauss (1813) art. 29.

$$\ln \Gamma(x+1) = \left(x+\frac{1}{2}\right)\ln x - x + \frac{1}{2}\ln 2\pi + \frac{B_2}{1\cdot 2x} + \frac{B_4}{3\cdot 4x^3} + \frac{B_6}{5\cdot 6x^5} + \text{etc.}$$

$$(18.11)$$

Note that in his 1813 paper, Gauss denoted the absolute values of the Bernoulli numbers by $A, B, C, D, \ldots$ and used his own notation, $\Pi(x)$, to mean $\Gamma(x+1)$. He also stated that it was important to understand why the first few terms of a divergent series might yield an excellent approximation. He also pointed out that the series (18.6) and (18.7) of Stirling and de Moivre respectively, when put in their more general form for $\ln \Gamma(x+1)$, could be obtained from each other by the duplication formula for the gamma function. Note that the case $n = 2$ of (17.15) gives the duplication formula.

In 1843,[9] Cauchy gave an explanation of the peculiar nature of the series (18.11). He proved that

$$\mu(x) = \ln \Gamma(x) - \left(x-\frac{1}{2}\right)\ln x + x - \frac{1}{2}\ln 2\pi$$

$$= \sum_{k=1}^{m} \frac{B_{2k}}{(2k-1)2kx^{2k-1}} + \frac{\theta B_{2m+2}}{(2m+1)(2m+2)x^{2m+1}}, \text{ where } 0 < \theta < 1. \quad (18.12)$$

This implies that when $m$ terms of the series (18.11) are used, the error is less than the $(m+1)$th term and has the same sign as that term, determined by the sign of $B_{2m+2}$. Thus, the eighteenth-century mathematicians had the judgment to choose an ideal stopping point in the series for their numerical calculations. Although Poisson and Jacobi had already proved Cauchy's result in a more general situation, they did not specifically note this important particular case. It is also possible that Cauchy wished to show how an integral representation for $\mu(x)$ in (18.12), due to his friend Binet, could be used for the proof of (18.12).

Jacques Binet (1786–1856) studied at the École Polytechnique from 1804 to 1806 and returned to the institution as an instructor in 1807. His main interests were astronomy and optics, though he contributed some important papers in mathematics. He was a good friend of Cauchy, and in 1812 the two generalized some results on determinants and took the subject to a higher level of generality. In particular, Binet stated the multiplication theorem in more general terms, so that his work can be taken as an early discussion of the product of two rectangular matrices. In an 1839 paper of over 200 pages,[10] Binet gave two integral representations for $\mu(x)$ in (18.12). These are now called Binet's formulas. In applying integrals to study the gamma function, Binet was following the trend of the 1830s. Thus, he used Euler's formula for the beta integral $\int_0^1 x^{m-1}(1-x)^{n-1}\,dx$ to prove Stirling's formulas (18.4) and (18.5).

Although the asymptotic series (18.11) and similar series were used frequently after 1850, it was in 1886 that Henri Poincaré gave a formal definition. He noted that it was well known to geometers that if $S_n$ denoted the terms of the series for $\ln \Gamma(x+1)$ up to

---

[9] Cauchy (1843b).
[10] Binet (1839).

and including $\left(\frac{B_{2n}}{2n(2n-1)}\right)\frac{1}{x^{2n}}$, then the expression $x^{2n+1}(\ln \Gamma(x+1) - S_n)$ tended to 0 when $x$ increased indefinitely. He then defined an asymptotic series:

I say that a *divergent series*

$$A_0 + \frac{A_1}{x} + \frac{A_2}{x^2} + \cdots + \frac{A_n}{x^n} + \cdots,$$

where the sum of the first $n+1$ terms is $S_n$, *asymptotically* represents a function $J(x)$ if the expression $x^n(J - S_n)$ tends to 0 when $x$ increases indefinitely.

He showed that asymptotic series behaved well under the algebraic operations of addition, subtraction, multiplication, and division. Term-by-term integration of an asymptotic series also worked, but not differentiation. Poincaré noted that the theory remained unchanged if one supposed that $x$ tended to infinity radially (in the complex plane) with a fixed nonzero argument. However, a divergent series could not represent one and the same function $J$ in all directions of radial approach to infinity. He also observed that the same series could represent more than one function asymptotically. Poincaré applied his theory of asymptotic series to the solution of differential equations, though the British mathematician George Stokes developed some of these ideas earlier, in the 1850s and 1860s, in connection with Bessel's equation.

The Dutch mathematician Thomas Joannes Stieltjes (1856–1894) also developed a theory of asymptotic series; following Legendre, he labeled it semiconvergent series. Stieltjes's paper[11] appeared in the same year as Poincaré's, 1886. Then in 1889, Stieltjes extended formula (18.12) to the slit complex plane $\mathbb{C}^- = \mathbb{C} \setminus (-\infty, 0]$. Until then, the formula was known to hold only in the right half-plane. He accomplished this extension by a systematic use of the formula[12]

$$\mu(z) = \int_0^\infty \frac{t - [t] - \frac{1}{2}}{z + t}\, dt, \ z \in \mathbb{C}^-, \tag{18.13}$$

where $\mu$ was defined by (18.12).

## 18.2   De Moivre's Asymptotic Series

De Moivre's derivation of (18.7) in the *Supplement*[13] started with

$$\ln \frac{m^{m-1}}{(m-1)!} = \sum_{k=1}^{m-1} \ln\left(1 - \frac{k}{m}\right)^{-1} = -\sum_{k=1}^{m-1} \ln\left(1 - \frac{k}{m}\right)$$

$$= \sum_{k=1}^{m-1}\sum_{n=1}^{\infty} \frac{k^n}{nm^n} = \sum_{n=1}^{\infty} \frac{1}{nm^n} \sum_{k=1}^{m-1} k^n. \tag{18.14}$$

[11] Stieltjes (1886a).
[12] Stieltjes (1889).
[13] de Moivre (1730b).

We remark that de Moivre did not use the summation or factorial notation. He effected the change in the order of summation by writing the series for $\ln(1 - \frac{k}{m})$ in rows, for some values of $k$, and then summing the columns. He then reproduced Jakob Bernoulli's table for sums of powers of integers and applied it to the inner sum in the last expression of (18.14). In modern notation, Bernoulli's formula, also given as (2.26), may be written as

$$\sum_{k=1}^{m-1} k^n = \frac{(m-1)^{n+1}}{n+1} + \frac{1}{2}(m-1)^n$$

$$+ \binom{n}{1}\frac{B_2}{2}(m-1)^{n-1} + \binom{n}{3}\frac{B_4}{4}(m-1)^{n-3} + \cdots . \qquad (18.15)$$

Thus, de Moivre had the equation

$$\ln\frac{m^{m-1}}{(m-1)!} = \frac{1}{m}\left(\frac{(m-1)^2}{2} + \frac{m-1}{2}\right) \qquad (18.16)$$

$$+ \frac{1}{2mm}\left(\frac{(m-1)^3}{3} + \frac{(m-1)^2}{2} + \frac{m-1}{6}\right)$$

$$+ \frac{1}{3m^3}\left(\frac{(m-1)^4}{4} + \frac{(m-1)^3}{2} + 3B_2\frac{(m-1)^2}{2}\right)$$

$$+ \frac{1}{4m^4}\left(\frac{(m-1)^5}{5} + \frac{(m-1)^4}{2} + 4B_2\frac{(m-1)^3}{2} + 4B_4\frac{(m-1)}{4}\right) + \cdots .$$

He then set $x = \frac{m-1}{m}$ and changed the order of summation to get

$$\ln\frac{m^{m-1}}{(m-1)!} = m\left(\frac{x^2}{2} + \frac{x^3}{6} + \frac{x^4}{12} + \cdots\right) + \frac{1}{2}\left(x + \frac{x^2}{2} + \frac{x^3}{3} + \cdots\right)$$

$$+ \frac{B_2}{2m}\left(x + x^2 + x^3 + \cdots\right) + \frac{B_4}{4m^3}\left(x + \frac{4}{2}x^2 + \frac{5\cdot4}{3\cdot2}x^3 + \cdots\right) + \cdots . \qquad (18.17)$$

The general term, left unexpressed by de Moivre, as was the common practice, would have been

$$\frac{B_{2r}}{2rm^{2r-1}}\left(\binom{2r}{1}\frac{x}{2r} + \binom{2r+1}{2}\frac{x^2}{2r+1} + \binom{2r+2}{3}\frac{x^3}{2r+2} + \cdots\right)$$

$$= \frac{B_{2r}}{2r(2r-1)m^{2r-1}}\left(\binom{2r-1}{1}x + \binom{2r}{2}x^2 + \binom{2r+1}{3}x^3 + \cdots\right)$$

$$= \frac{B_{2r}}{2r(2r-1)m^{2r-1}}\left((1-x)^{-2r+1} - 1\right)$$

$$= \frac{B_{2r}}{2r(2r-1)}\left(1 - \frac{1}{m^{2r-1}}\right). \qquad (18.18)$$

The second series in (18.17) summed to

$$-\ln(1-x) = -\ln\left(1 - \frac{m-1}{m}\right) = \ln m,$$

and the first series turned out to be the integral of this series, equal to

$$(1-x)\ln(1-x) + x = \frac{m-1-\ln m}{m}.$$

This computation involves integration by parts, and it is interesting to see how de Moivre handled it. He used Newton's notation for fluxions, as was only natural since de Moivre worked in England and was Newton's friend. He set

$$v = \ln \frac{1}{1-x} = x + \frac{x^2}{2} + \frac{x^3}{3} + \frac{x^4}{4} + \cdots.$$

Then

$$v\dot{x} = x\dot{x} + \frac{1}{2}x^2\dot{x} + \frac{1}{3}x^3\dot{x} + \frac{1}{4}x^4\dot{x} + \cdots$$

$$F.v\dot{x} = \frac{1}{2}x^2 + \frac{1}{6}x^3 + \frac{1}{12}x^4 + \frac{1}{20}x^5 + \cdots . \quad (\text{F.} = \text{fluent} = \text{integral}).$$

He then set

$$q = vx - F.v\dot{x} = x^2 + \frac{1}{2}x^3 + \frac{1}{3}x^4 + \frac{1}{4}x^5 + \cdots - \left(\frac{1}{2}x^2 + \frac{1}{6}x^3 + \frac{1}{12}x^4 + \cdots\right)$$

so that

$$\dot{q} = \dot{x}\left(2x + \frac{3}{2}x^2 + \frac{4}{3}x^3 + \frac{5}{4}x^4 + \cdots\right) - \dot{x}\left(x + \frac{1}{2}x^2 + \frac{1}{3}x^3 + \frac{1}{4}x^4 + \cdots\right)$$

$$= \frac{\dot{x}x}{1-x} = -\dot{x} + \dot{v}.$$

Therefore $q = -x + v$ and $F.v\dot{x} = vx - q = \frac{m-1-\ln m}{m}$. Using the above simplifications, (18.17) became

$$\ln \frac{m^{m-1}}{(m-1)!} = (m-1) - \ln m + \frac{1}{2}\ln m$$

$$+ \frac{B_2}{2}\left(1 - \frac{1}{m}\right) + \frac{B_4}{3 \cdot 4}\left(1 - \frac{1}{m^3}\right) + \frac{B_6}{5 \cdot 6}\left(1 - \frac{1}{m^5}\right) + \cdots$$

$$= (m-1) - \frac{1}{2}\ln m + \frac{1}{12} - \frac{1}{360} + \frac{1}{1260} - \frac{1}{1680} + \cdots$$

$$- \frac{1}{12m} + \frac{1}{360m^3} - \frac{1}{1260m^5} + \frac{1}{1680m^7} - \cdots .$$

Note that we have substituted the numerical values of Bernoulli numbers:

$$B_2 = \frac{1}{6}, \ B_4 = -\frac{1}{30}, \ B_6 = \frac{1}{42}, \ B_8 = -\frac{1}{30}, \cdots ;$$

de Moivre worked only with these numerical values in his calculations. After adding ln $m$ to each side and rearranging terms, de Moivre had

$$\ln m! = \left(m + \frac{1}{2}\right) \ln m - m + 1 - \frac{1}{12} + \frac{1}{360} - \frac{1}{1260} + \frac{1}{1680} - \cdots$$
$$+ \frac{1}{12m} - \frac{1}{360m^3} + \frac{1}{1260m^5} - \frac{1}{1680m^7} + \cdots . \qquad (18.19)$$

De Moivre remarked that the constant in this equation could be quickly computed by taking $m = 2$. In that case,

$$C = 1 - \frac{1}{12} + \frac{1}{360} - \frac{1}{1260} + \frac{1}{1680} - \cdots$$
$$= 2 - \frac{3}{2} \ln 2 - \frac{1}{12 \times 2} + \frac{1}{360 \times 8} - \frac{1}{1260 \times 32} + \frac{1}{1680 \times 128} - \cdots . \qquad (18.20)$$

As we have noted before, the two series here are divergent but the terms as written down by de Moivre gave a good approximation for $C$. After learning of Stirling's result on the asymptotic value of $\frac{1}{2^{2m}} \binom{2m}{m}$, de Moivre realized that $C = \frac{1}{2} \ln(2\pi)$, and he proved it using Wallis's formula. Stirling and de Moivre's derivations for the value of $C$ were identical; we present the details in the next section.

## 18.3  Stirling's Asymptotic Series

Stirling's *Methodus Differentialis* gave several ingenious applications of difference equations. His derivations of (18.4) and (18.5) were probably his most imaginative use of difference equations. It is obvious from his letter to de Moivre that he was quite proud of his solutions. We give details of this work and derive equation (18.6). Proposition 23 of Stirling's book states the problem: to find the ratio of the middle coefficient to the sum of all coefficients in any power of the binomial. Stirling observed that the sequence $1, 2, \frac{8}{3}, \frac{16}{5}, \frac{128}{35}, \ldots$ or $2^{2m} \div \binom{2m}{m}$, $m = 0, 1, 2, \ldots$ satisfied the relation $T' = \frac{(n+2)T}{n+1}$, where $n = 2m = 0, 2, 4, \ldots$ and $T'$ denoted the term after $T$. So if $T$ was the $n$th term, $T'$ would be obtained by changing $n$ to $n + 2$ in $T$. In modern notation,

$$T_{n+2} = \frac{n+2}{n+1} T_n.$$

After squaring the relation, Stirling rewrote it to get

$$2T'^2 + (n+2)(T^2 - T'^2) - \frac{T'^2}{n+2} = 0, \tag{18.21}$$

the difference equation into which Stirling substituted an inverse factorial series to solve his problem. Since he had so many difference equations from which to choose, it is hard to discern how Stirling was guided to this one; it worked very successfully. Stirling first took

$$
\begin{aligned}
T^2 &= An + \frac{Bn}{n+2} + \frac{Cn}{(n+2)(n+4)} + \frac{Dn}{(n+2)(n+4)(n+6)} + \cdots \\
&= An + B + \frac{C - 2B}{n+2} + \frac{D - 4C}{(n+2)(n+4)} + \cdots .
\end{aligned} \tag{18.22}
$$

Then

$$T'^2 = A(n+2) + B + \frac{C - 2B}{n+4} + \frac{D - 4C}{(n+4)(n+6)} + \cdots , \tag{18.23}$$

so that

$$(n+2)(T^2 - T'^2) = -2A(n+2) + \frac{2C - 4B}{n+4} + \frac{4D - 16C}{(n+4)(n+6)} + \cdots . \tag{18.24}$$

It followed from (18.22) after replacing $T$ by $T'$ and $n$ by $n+2$ that

$$\frac{T'^2}{n+2} = A + \frac{B}{n+4} + \frac{C}{(n+4)(n+6)} + \frac{D}{(n+4)(n+6)(n+8)} + \cdots .$$

He used the three series (18.22), (18.23), and (18.24) in (18.21), and the result was

$$2B - A + \frac{4C - 9B}{n+4} + \frac{6D - 25C}{(n+4)(n+6)} + \frac{8E - 49D}{(n+4)(n+6)(n+8)} + \cdots = 0. \tag{18.25}$$

Since the constant term and the numerators of the other terms must be zero in (18.25), it followed that

$$2B - A = 0, \quad 4C - 9B = 0, \quad 6D - 25C = 0, \quad 8E - 49D = 0,$$

so that, clearly, Stirling had the series in (18.4) except for the value of $A$. As was the practice among the eighteenth-century mathematicians, Stirling computed only the first few coefficients $B, C, D, \ldots$ and gave no expression for the general term. To find $A$, he argued that by (18.22) for large $n$, $T^2 = An$. Stirling then made an application of Wallis's formula, to which he referred in his exposition. For $n = 2m$,

$$\frac{T^2}{n} = \frac{1}{2m}\left(\frac{1}{2^{2m}}\binom{2m}{m}\right)^{-2}$$

$$= \frac{2^2 \times 4^2 \times \cdots \times (2m)^2}{3^2 \times 5^2 \times \cdots \times (2m-1)^2 \times 2m + 1} \cdot \frac{2m+1}{2m} \rightarrow \frac{\pi}{2} \quad \text{as } n \rightarrow \infty.$$

First in his *Miscellanea* and then again in the 1756 edition of his *Doctrine of Chances*, de Moivre praised Stirling for his introduction of $\pi$ in the asymptotic series for $\ln n!$. In the latter work he wrote, "I own with pleasure that this discovery, besides that it saved trouble, has spread a singular Elegancy on the Solution."[14]

Stirling found the series in (18.5) in a similar manner. This time he let $T_{n+2} = \frac{n+1}{n+2}$; after squaring this relation he could thus write

$$(n+1)(n+3)(T_n^2 - T_{n+2}^2) - 2(n+1)T_n^2 - T_{n+1}^2 = 0.$$

Stirling next let

$$T_n^2 = \frac{A}{n+1} + \frac{B}{(n+1)(n+3)} + \frac{C}{(n+1)(n+3)(n+5)}$$

$$+ \frac{D}{(n+1)(n+3)(n+5)(n+7)} + \cdots \quad (18.26)$$

and applied a procedure similar to that by which he solved (18.21) to get, as before:

$$2B - A = 0, \quad 4C - 9B = 0, \quad 6D - 25C = 0, \quad 8E - 49D = 0, \ldots,$$

relations that gave him (18.5).

Now series (18.6) for $\ln m!$ was a corollary of Stirling's main result, contained in proposition 28 of his book. The purpose of the proposition was to find the sum of any number of logarithms, whose arguments were in arithmetic progression. He denoted the progression by $x+n, x+2n, x+3n, \ldots, z-n$, where the logarithms were taken base 10. Since $\log_{10} x = \frac{\ln x}{\ln 10}$, he defined the number $a = \frac{1}{\ln 10}$ and gave the approximate value of $a$ to be 0.43429,44819,03252. To state his result, Stirling began with the series

$$f(z) = \frac{z}{2n}\log_{10} z - \frac{a}{2n}z + aA_1\frac{n}{z} + aA_2\frac{n^3}{z^3} + aA_3\frac{n^5}{z^5} + aA_4\frac{n^7}{z^7} + \cdots, \quad (18.27)$$

where the numbers $A_1, A_2, A_3, \ldots$ were such that

$$\sum_{k=1}^{m}\binom{2m-1}{2k-2}A_k = -\frac{1}{4m(2m+1)}. \quad (18.28)$$

He had the values

$$A_1 = -\frac{1}{12}, \quad A_2 = \frac{7}{360}, \quad A_3 = -\frac{31}{1260}, \quad A_4 = \frac{127}{1680}, \quad A_5 = -\frac{511}{1188}.$$

---

[14] de Moivre (1967) p. 244.

In fact, one can show that

$$A_k = -\frac{(2^{2k-1}-1)B_{2k}}{2k(2k-1)}. \tag{18.29}$$

Stirling may not have recognized this connection with Bernoulli numbers when he discovered his result on $\ln m!$. Later, after reading de Moivre's book, where Bernoulli numbers were explicitly mentioned, Stirling investigated properties of these numbers and discussed them in some unpublished notes. See equation (10.77) for a relation between Bernoulli and Stirling numbers.

Stirling's main theorem was that

$$\log_{10}((x+n)(x+3n)(x+5n)\cdots(z-n)) = f(z) - f(x), \tag{18.30}$$

where $f(z)$ was the series defined in (18.27). His proof consisted in observing that

$$f(z) - f(z-2n) = \log_{10} z - a\left(\frac{n}{z} + \frac{1}{2}\left(\frac{n}{z}\right)^2 + \frac{1}{3}\left(\frac{n}{z}\right)^3 + \cdots\right)$$

$$= \log_{10} z - \log_{10}\left(1 - \frac{n}{z}\right) = \log_{10}(z-n). \tag{18.31}$$

Stirling apparently left the verification of this equation to the reader. He made the remark that the terms in $f(z)$ and $f(z-2n)$ had first to be reduced to the same form. The theorem follows immediately from (18.31):

$$f(z) - f(x) = (f(z) - f(z-2n))$$
$$+ (f(z-2n) - f(z-4n)) + \cdots + (f(x+2n) - f(x))$$
$$= \log_{10}(z-n) + \log_{10}(z-3n) + \cdots + \log_{10}(x+n).$$

Stirling applied his theorem to derive his series for $\log_{10} m!$ by taking $x = \frac{1}{2}, n = \frac{1}{2}$, and $z = m + \frac{1}{2}$. From this he had

$$f\left(m + \frac{1}{2}\right) = \left(m + \frac{1}{2}\right)\log_{10}\left(m + \frac{1}{2}\right) - a\left(m + \frac{1}{2}\right)$$
$$- \frac{a}{24\left(m + \frac{1}{2}\right)} + \frac{7a}{2880\left(m + \frac{1}{2}\right)^3} - \text{etc.}$$

Next, by (18.30), $\log_{10} m! = f\left(m + \frac{1}{2}\right) - f\left(\frac{1}{2}\right)$. Stirling wrote that

$$-f\left(\frac{1}{2}\right) = \frac{1}{2}\log_{10} 2\pi \approx 0.39908,99341,79,$$

but he did not explain how he arrived at $\frac{1}{2}\log_{10} 2\pi$. Perhaps he numerically computed $\log_{10} m! - f\left(m + \frac{1}{2}\right)$ for a large enough value of $m$ and noticed that he had half the value of $\log_{10} 2\pi$; he must have been very familiar with this value, based on his extensive numerical calculations. Recall that he had recognized $\sqrt{\pi}$ from its numerical value when computing $\left(\frac{1}{2}\right)$ !. Of course, he could also have provided a proof using Wallis's formula, as he did in the situation discussed above.

We can prove that the sequence (18.29) satisfies the recurrence relation (18.28) by first setting $C_k$ as the coefficients in the Taylor series expansion of $\csc x$:

$$\frac{\csc x}{x} = \frac{1}{x^2} + \sum_{k=1}^{\infty} (-1)^{k-1} 2C_k \frac{x^{2k-2}}{(2k-2)!}, \quad 0 < x < \pi.$$

From (16.81) we know that

$$C_k = \frac{(2^{2k-1} - 1) B_{2k}}{2k(2k-1)}.$$

Now by considering the Taylor series

$$\sin x = x - \frac{x^3}{3!} + \frac{x^5}{5!} - \frac{x^7}{7!} + \cdots,$$

we see that

$$\left( \frac{1}{x^2} + \sum_{k=1}^{\infty} (-1)^{k-1} 2C_k \frac{x^{2k-2}}{(2k-2)!} \right) \left( \sum_{k=1}^{\infty} (-1)^{k-1} \frac{x^{2k-1}}{(2k-1)!} \right) = \frac{1}{x}. \quad (18.32)$$

Equating the coefficient of $x^{2m-1}$, $m = 1, 2, 3, \ldots$, in (18.32) produces

$$\frac{(-1)^m}{(2m+1)!} + \frac{(-1)^{m-1} 2C_m}{1!\,(2m-2)!} + \frac{(-1)^{m-1} 2C_{m-1}}{3!\,(2m-4)!} + \cdots + \frac{(-1)^{m-1} 2C_1}{(2m-1)!\,0!} = 0;$$

multiply by $(2m-1)!$ to arrive at

$$\sum_{k=1}^{m} \binom{2m-1}{2k-2} C_k = \frac{1}{4m(2m+1)}. \quad (18.33)$$

When we compare (18.33) with (18.28), it is clear that

$$A_k = -C_k = -\frac{(2^{2k-1} - 1) B_{2k}}{2k(2k-1)}.$$

By using Lagrange's symbolic method, we obtain an alternative perspective on $A_k$ as expressed in (18.29). To see this, we invoke Lagrange's formulas from Chapter 21. Taking $D$ to be $\frac{d}{dx}$,

$$e^D f(x) = f(x+1),$$
$$e^{-D} f(x) = f(x-1),$$

$$\frac{1}{e^D - 1} = \frac{1}{D} \left( 1 - \frac{1}{2}D + \sum_{k=1}^{\infty} B_{2k} \frac{D^{2k}}{(2k)!} \right).$$

Now Stirling required the sum $f(x+1) + f(x+3) + \cdots + f(x+2n-1) + \cdots$. Let $S(x+1)$ denote this sum. Then

$$(e^D - e^{-D})\, S(x) = S(x+1) - S(x-1) = f(x-1)$$

and hence

$$S = \frac{1}{e^D - e^{-D}}\, f = \frac{e^D}{e^{2D} - 1}\, f$$

$$= \left( \frac{1}{e^D - 1} - \frac{1}{e^{2D} - 1} \right) f$$

$$= \frac{1}{D}\left( 1 - \frac{1}{2}D + \sum_{k=1}^{\infty} B_{2k} \frac{D^{2k}}{(2k)!} \right) f$$

$$= -\frac{1}{2D}\left( 1 - \frac{1}{2}(2D) + \sum_{k=1}^{\infty} B_{2k}\, 2^{2k} \frac{D^{2k}}{(2k)!} \right) f$$

$$= \frac{1}{2D}\, f + \sum_{k=1}^{\infty} \frac{(1 - 2^{2k-1}) B_{2k}}{(2k)!}\, D^{2k-1} f.$$

Taking $f(x) = \log_{10} x$ and $\frac{1}{2} D^{-1} f = \frac{1}{2} \int f(x)\, dx$, and noting that $D^{2k-1} \ln x = \frac{(2k-2)!}{x^{2k-1}}$, we obtain exactly Stirling's formula for the sum of logarithms (18.27).

## 18.4   Binet's Integrals for $\ln \Gamma(x)$

Binet knew that Stirling's two series (18.4) and (18.5) could be derived by Gauss's summation formula. In addition, he gave an interesting proof of (18.4) using integrals:[15]

$$B\left( m, \frac{1}{2} \right) = \int_0^1 x^{m-1} (1-x)^{-\frac{1}{2}}\, dx = \int_0^1 x^{m-\frac{1}{2}} (1-x)^{-\frac{1}{2}} (1 - (1-x))^{-\frac{1}{2}}\, dx$$

$$= \int_0^1 x^{m-\frac{1}{2}} (1-x)^{-\frac{1}{2}} \left( 1 + \frac{\frac{1}{2}}{1!}(1-x) + \frac{\frac{1}{2} \cdot \frac{3}{2}}{2!}(1-x)^2 + \cdots \right) dx$$

$$= B\left( m + \frac{1}{2}, \frac{1}{2} \right) + \frac{\frac{1}{2}}{1!} B\left( m + \frac{1}{2}, \frac{3}{2} \right) + \frac{\frac{1}{2} \cdot \frac{3}{2}}{2!} B\left( m + \frac{1}{2}, \frac{5}{2} \right) + \cdots.$$

$$(18.34)$$

When Euler's formula, $B(x, y) = \frac{\Gamma(x)\Gamma(y)}{\Gamma(x+y)}$, was applied in (18.34), Stirling's formula (18.4) followed. Binet's proof of (18.5) ran along similar lines. He also gave two integral representations for

---

[15] Binet (1839) pp. 239–241.

$$\mu(x) = \ln \Gamma(x) - \left( x - \frac{1}{2} \right) \ln x + x - \frac{1}{2} \ln 2\pi. \qquad (18.35)$$

These useful representations were:

$$\mu(x) = \int_0^\infty \left( \frac{1}{2} - \frac{1}{t} + \frac{1}{e^t - 1} \right) \frac{e^{-xt}}{t} \, dt, \qquad (18.36)$$

$$\mu(x) = 2 \int_0^\infty \frac{\arctan \left( \frac{t}{x} \right)}{e^{2\pi t} - 1} \, dt. \qquad (18.37)$$

Binet demonstrated the equality of the two expressions for $\mu(x)$ by using the two formulas:[16]

$$\int_0^\infty e^{-sy} \sin (ty) \, dy = \frac{t}{t^2 + s^2}, \qquad (18.38)$$

$$4 \int_0^\infty \frac{\sin (ty)}{e^{2\pi y} - 1} \, dy = \frac{e^t + 1}{e^t - 1} - \frac{2}{t}. \qquad (18.39)$$

He attributed the first of these to Euler and the second to Poisson. He multiplied the second equation by $e^{-st} \, dt$, integrated over $(0, \infty)$, and then used the first integral to get

$$\int_0^\infty e^{-st} \left( \frac{e^t + 1}{e^t - 1} - \frac{2}{t} \right) dt = 4 \int_0^\infty \frac{t \, dt}{(t^2 + s^2)(e^{2\pi t} - 1)}.$$

Binet then integrated both sides of this equation with respect to $s$ over the interval $(x, \infty)$, and changed the order of integration to obtain

$$\int_0^\infty \left( \frac{1}{2} - \frac{1}{t} + \frac{1}{e^t - 1} \right) \frac{e^{-xt}}{t} \, dt = 2 \int_0^\infty \frac{\arctan \left( \frac{t}{x} \right)}{e^{2\pi t} - 1} \, dt.$$

Thus, it was sufficient to prove one of the formulas for $\mu(x)$, and Binet proved the first one, starting from the definition of $\mu(x)$:

$$\mu(x + 1) - \mu(x) = \left( x + \frac{1}{2} \right) \ln \left( 1 - \frac{1}{x+1} \right) + 1$$

$$= -\sum \frac{(n-1)}{2n(n+1)(x+1)^n},$$

or

$$2\mu(x) = 2\mu(x + 1) + \sum \frac{n-1}{n(n+1)(x+1)^n}. \qquad (18.40)$$

---

[16] ibid. pp. 321–323.

By Stirling's approximation, he had $\mu(x) \to 0$ as $x \to \infty$, so that

$$\mu(x) = \sum_{k=0}^{\infty} \left( \mu(x+k) - \mu(x+k+1) \right)$$

and hence, by (18.40),

$$2\mu(x) = \frac{1}{2 \cdot 3} \sum_{k=1}^{\infty} \frac{1}{(x+k)^2} + \frac{2}{3 \cdot 4} \sum_{k=1}^{\infty} \frac{1}{(x+k)^3} + \frac{3}{4 \cdot 5} \sum_{k=1}^{\infty} \frac{1}{(x+k)^4} + \text{ etc.}$$

By Euler's gamma integral

$$\frac{\Gamma(n+1)}{(k+x)^{n+1}} = \int_0^{\infty} t^n e^{-t(k+x)} \, dt, \tag{18.41}$$

and therefore

$$\Gamma(n+1) \sum_{k=1}^{\infty} \frac{1}{(x+k)^{n+1}} = \int_0^{\infty} t^n \left( e^{-t(x+1)} + e^{-t(x+2)} + e^{-t(x+3)} + \cdots \right) dt$$

$$= \int_0^{\infty} \frac{t^n e^{-xt}}{e^t - 1} \, dt.$$

He then wrote

$$2\mu(x) = \int_0^{\infty} \frac{e^{-xt}}{e^t - 1} \left( \frac{t}{2 \cdot 3} + \frac{2t^2}{2 \cdot 3 \cdot 4} + \frac{3t^3}{2 \cdot 3 \cdot 4 \cdot 5} + \cdots \right) dt$$

and an easy calculation showed that the sum of the series inside the parentheses would be

$$(e^t - 1) \left( \frac{1}{t} - \frac{2}{t^2} \right) + \frac{2}{t}.$$

This completes Binet's ingenious proof of his formulas.

De Moivre's form of the asymptotic series for $\ln \Gamma(x)$ can be obtained from Binet's integrals. Start with Euler's generating function for Bernoulli numbers,

$$\frac{t}{e^t - 1} = 1 - \frac{1}{2}t + \sum_{n=1}^{\infty} \frac{B_{2n}}{(2n)!} t^{2n}.$$

It follows easily that the integrand in the first integral for $\mu(x)$ is

$$\frac{1}{t} \left( \frac{1}{e^t - 1} - \frac{1}{t} + \frac{1}{2} \right) = \sum_{n=0}^{\infty} \frac{B_{2n+2}}{(2n+2)!} t^{2n}. \tag{18.42}$$

We substitute this series in the integrand and integrate term by term; an application of (18.41) then yields de Moivre's asymptotic series. Unfortunately, however, this last operation is invalid because the series (18.42) is convergent only for $|t| < 2\pi$, whereas we are integrating on $(0, \infty)$.

## 18.5  Cauchy's Proof of the Asymptotic Character of de Moivre's Series

In an 1843 paper in *Comptes Rendus*,[17] Cauchy wrote that Stirling's series for the logarithm of a product whose factors increase in an arithmetic progression was divergent and therefore it had no sum. However, he maintained, good approximations could be obtained from this and other similar series for the functions represented by the series expansions. Cauchy's main result in this paper was that in the series he denoted as Stirling's series (though he was actually dealing with de Moivre's series) and other similar series, the first neglected term gave the upper bound of the error.

Thus, according to Cauchy's result, if we were to take the terms $\left(m + \frac{1}{2}\right) \ln m - m + \frac{1}{2} \ln 2\pi$ from the series (18.7), representing the function $\ln m!$, the error would have to be less than the first of the neglected terms, $\frac{1}{12m}$. The basic formula Cauchy employed was a finite form of equation (18.42):

$$\frac{1}{t}\left(\frac{1}{e^t - 1} - \frac{1}{t} + \frac{1}{2}\right) = \sum_{k=0}^{m-1} \frac{B_{2k+2}}{(2k+2)!} t^{2k} + \frac{\theta(t) B_{2m+2}}{(2m+2)!} t^{2m}, \quad 0 < \theta < 1. \tag{18.43}$$

He noted that (18.42) converged only when $|t| < 2\pi$, while (18.43) was always true. To prove (18.43), he first derived the partial fraction expansion of its left-hand side:

$$\frac{1}{t}\left(\frac{1}{e^t - 1} - \frac{1}{t} + \frac{1}{2}\right) = \sum_{k=0}^{\infty} \frac{2}{t^2 + (2k\pi)^2}, \tag{18.44}$$

using his own original complex analytic methods. However, the result was originally due to Euler. See Euler's equation (16.28), in which one may set $t = 2i\pi x$ to produce (18.44).

Cauchy next observed that half of each term in the sum on the right-hand side of (18.44) could be expanded as

$$\frac{1}{(2k\pi)^2 + t^2} = \frac{1}{(2k\pi)^2} - \frac{t^2}{(2k\pi)^4} + \cdots \pm \frac{t^{2m-2}}{(2k\pi)^{2m}} \mp \frac{t^{2m}}{(2k\pi)^{2m}\left((2k\pi)^2 + t^2\right)}; \tag{18.45}$$

to see this, one may employ the identity

$$\frac{1 \pm x^{2n}}{1 + x^2} = 1 - x^2 + x^4 - \cdots \pm x^{2n-2}.$$

[17]  Cauchy (1843b).

Concerning the last term in (18.45), Cauchy then noted:

$$\frac{t^{2m}}{(2k\pi)^{2m}\left((2k\pi)^2 + t^2\right)} < \frac{t^{2m}}{(2k\pi)^{2m+2}}. \tag{18.46}$$

Next, from (18.45) and (18.46),

$$\sum_{k=1}^{\infty} \frac{1}{(2k\pi)^2 + t^2} = \frac{1}{(2\pi)^2}\sum_{k=1}^{\infty}\frac{1}{k^2} - \frac{t^2}{(2\pi)^4}\sum_{k=1}^{\infty}\frac{1}{k^4} + \cdots \pm \frac{t^{2m-2}}{(2\pi)^{2m}}\sum_{k=1}^{\infty}\frac{1}{k^{2m}} + R_m(t)$$

$$\tag{18.47}$$

where

$$0 < |R_m(t)| < \frac{t^{2m}}{(2\pi)^{2m+2}}\sum_{k=1}^{\infty}\frac{1}{k^{2m+2}}.$$

Hence

$$|R_m(t)| = \frac{\theta(t)\,t^{2m}}{(2\pi)^{2m+2}}\sum_{k=1}^{\infty}\frac{1}{k^{2m+2}}$$

where

$$0 < \theta(t) = \frac{(2m\pi)^2}{t^2 + (2m\pi)^2} < 1.$$

An application of Euler's formula

$$\sum_{k=1}^{\infty}\frac{1}{k^{2n}} = (-1)^{n-1}\frac{2^{2n-1}B_{2n}\pi^{2n}}{(2n)!}$$

enabled Cauchy to rewrite (18.47) as

$$\sum_{k=1}^{\infty}\frac{2}{(2k\pi)^2 + t^2} = B_2 + \frac{B_4}{4!}t^2 + \cdots + \frac{B_{2m}}{(2m)!}t^{2m-2} + \theta(t)\frac{B_{2m+2}}{(2m+2)!}t^{2m}.$$

$$\tag{18.48}$$

The result (18.48) when substituted in (18.44) produced (18.43); multiplying (18.43) by $e^{-tx}$ and integrating, Cauchy arrived at

$$\int_0^{\infty}\left(\frac{1}{e^t - 1} - \frac{1}{t} + \frac{1}{2}\right)e^{-tx}\frac{dt}{t} = \sum_{k=0}^{m-1}\frac{B_{2k+2}}{(2k+2)!}\int_0^{\infty}t^{2k}e^{-tx}dt$$

$$+ \frac{B_{2m+2}}{(2m+2)!}\int_0^{\infty}\theta(t)\,t^{2m}e^{-tx}dt. \tag{18.49}$$

Since $\theta(t)$ is continuous and since $t^{2m}e^{-tx} \geq 0$ in $(0,\infty)$, with $\theta_1$ a constant between 0 and 1,

$$\int_0^\infty \theta(t)\, t^{2m} e^{-tx}\, dt \le \theta_1 \int_0^\infty t^{2m} e^{-tx}\, dt. \tag{18.50}$$

Cauchy then noted that when $k$ was a positive integer

$$\int_0^\infty t^{2k} e^{-tx}\, dt = \frac{(2k)!}{x^{2k+1}}. \tag{18.51}$$

Thus, by (18.50), he could rewrite (18.49) as

$$\int_0^\infty \left( \frac{1}{e^t - 1} - \frac{1}{t} + \frac{1}{2} \right) e^{-tx} \frac{dt}{t} = \sum_{k=0}^{m-1} \frac{B_{2k+2}}{(2k+1)(2k+2)x^{2k+1}}$$

$$+ \theta_1 \frac{B_{2m+2}}{(2m+1)(2m+2)x^{2m+1}}, \quad 0 < \theta_1 < 1.$$

By using Binet's formula, given by (18.35) and (18.36), Cauchy obtained his final formula:

$$\ln \Gamma(x) = \left( x - \frac{1}{2} \right) \ln x - x + \frac{1}{2} \ln 2\pi + \sum_{k=0}^{m-1} \frac{B_{2k+2}}{(2k+1)(2k+2)x^{2k+1}}$$

$$+ \theta_1 \frac{B_{2m+2}}{(2m+1)(2m+2)x^{2m+1}}, \quad 0 < \theta_1 < 1. \tag{18.52}$$

This completed Cauchy's proof of his contention that if the $m$th term in the series were the first neglected term, then the absolute value of the error would be less than the absolute value of the $m$th term.

Cauchy also indicated a proof of the result first conjectured by de Moivre: For $c > 1$,

$$\frac{1}{n^c} + \frac{1}{(n+2)^c} + \cdots = \frac{1}{(c-1)n^{c-1}} + \frac{1}{2n^c} + \frac{c B_2}{2 \cdot n^{c+1}} + \frac{c(c+1)(c+2) B_4}{2 \cdot 3 \cdot 4 n^{c+3}}$$

$$+ \frac{c(c+1)(c+2)(c+3)(c+4) B_6}{2 \cdot 3 \cdot 4 \cdot 5 \cdot 6 n^{c+5}}$$

$$+ \frac{c(c+1)(c+2)(c+3)(c+4)(c+5)(c+6) B_8}{2 \cdot 3 \cdot 4 \cdot 5 \cdot 6 \cdot 8 n^{c+7}} + \cdots. \tag{18.53}$$

Note that we have discussed the case $c = 2$ in (10.75), in connection with Stirling's observation on the computation of Bernoulli numbers. Note also that the right-hand side of the series (18.53) is an asymptotic series; thus, Cauchy proved it in the finite form:

$$\sum_{k=0}^\infty \frac{1}{(n+k)^c} = \frac{1}{(c-1)n^{c-1}} + \frac{1}{2n^c} + \sum_{k=0}^{m-1} \frac{c(c+1)\cdots(c+2k)}{(2k+2)!} \frac{B_{2k+2}}{n^{c+2k+1}}$$

$$+ \theta_1 \frac{c(c+1)\cdots(c+2m)}{(2m+2)!} \frac{B_{2m+2}}{n^{c+2m+1}}, \quad 0 < \theta_1 < 1. \tag{18.54}$$

In fact, de Moivre conjectured the infinite series form of (18.54), that exemplifies the manner in which it was used in a finite form by him and his contemporaries.

To derive (18.54), Cauchy multiplied (18.43) by $t^c e^{-nt}$, integrated on $(0, \infty)$, and used (18.51) to obtain

$$\int_0^\infty \left( \frac{1}{e^t - 1} - \frac{1}{t} + \frac{1}{2} \right) t^{c-1} e^{-nt} \frac{dt}{t}$$

$$= \sum_{k=0}^{m-1} \frac{B_{2k+2}}{(2k+2)!} \int_0^\infty t^{c+2k} e^{-nt} dt + \frac{B_{2m+2}}{(2m+2)!} \int_0^\infty \theta(t) t^{c+2m} e^{-nt} dt$$

$$= \sum_{k=0}^{m-1} \frac{B_{2k+2}}{(2k+2)!} \frac{\Gamma(c+2k+1)}{n^{c+2k+1}} + \theta_1 \frac{B_{2m+2}}{(2m+2)!} \frac{\Gamma(c+2m+1)}{n^{c+2m+1}}, \quad 0 < \theta_1 < 1.$$

$$(18.55)$$

Cauchy wrote

$$\frac{1}{e^t - 1} - \frac{1}{t} + \frac{1}{2} = -\frac{1}{2} - \frac{1}{t} + \frac{1}{1 - e^{-t}}$$

$$= -\frac{1}{2} - \frac{1}{t} + 1 + e^{-t} + e^{-2t} + e^{-3t} + \cdots.$$

Next, the first integral in equation (18.55) could be evaluated as

$$\int_0^\infty \left( -\frac{1}{2} t^{c-1} e^{-nt} - t^{c-2} e^{-nt} + t^{c-1} e^{-nt} + t^{c-1} e^{-(n+1)t} + \cdots \right)$$

$$= -\frac{\Gamma(c)}{2n^c} - \frac{\Gamma(c-1)}{n^{c-1}} + \frac{\Gamma(c)}{n^c} + \frac{\Gamma(c)}{(n+1)^c} + \frac{\Gamma(c)}{(n+2)^c} + \cdots$$

and when this was used in (18.55), after rearrangement of terms, the result was (18.54).

As discussed in Chapter 20 in connection with the Euler–Maclaurin formula, (18.54) and (18.55) are particular cases of the results of Poisson and Jacobi on the E–M formula with a remainder term. Poisson and Jacobi derived these formulas using different methods in 1826 and 1834 respectively. Cauchy was probably aware of their results, since he published his paper in 1843. Cauchy's work is of interest because it shows how asymptotic series for $\log \Gamma(x)$ and for $\sum_{k=0}^{\infty} \frac{1}{(n+k)^c}$, $c > 1$, could be obtained from explicit forms of these functions.

## 18.6  Exercises

(1) Prove that if $S_m = \sqrt{2\pi} \left( \frac{m + \frac{1}{2}}{e} \right)^{m + \frac{1}{2}}$ and $D_m = \sqrt{2\pi m} \left( \frac{m}{e} \right)^m$, then

$$\lim_{m \to \infty} \frac{S_m - m!}{m! - D_m} = \frac{1}{2}.$$

This shows that (18.9), implied by Stirling's series, is a better approximation for $m!$ than $D_m$, resulting from de Moivre's series but called Stirling's approximation. Note that $S_m$ gives values larger than $m!$, while $D_m$ underestimates $m!$. See Stirling and Tweddle (1984).

(2) Prove that

$$\int_0^1 \ln \Gamma(x + u)\, du = x \ln x - x + \frac{1}{2} \ln 2\pi \qquad (18.56)$$

by the following methods:

(a) Observe that the integral is equal to

$$\lim_{n \to \infty} \frac{1}{n} \left( \ln \Gamma(x) + \ln \Gamma\left(x + \frac{1}{n}\right) + \cdots + \ln \Gamma\left(x + \frac{n-1}{n}\right) \right).$$

Apply Gauss's multiplication formula and Stirling's approximation.

(b) Apply Euler's reflection formula $\Gamma(x)\Gamma(1 - x) = \frac{\pi}{\sin \pi x}$ to compute the limit

$$\int_0^1 \ln \Gamma(u)\, du = \lim_{n \to \infty} \frac{1}{n} \left( \ln \Gamma\left(\frac{1}{n}\right) + \ln \Gamma\left(\frac{2}{n}\right) + \cdots + \ln \Gamma\left(\frac{n-1}{n}\right) \right)$$

$$= \frac{1}{2} \ln 2\pi. \qquad (18.57)$$

Take the derivative of (18.56) with respect to $x$ to show that

$$\int_0^1 \ln \Gamma(u + x)\, du = x \ln x - x + \int_0^1 \ln \Gamma(u)\, du.$$

(c) Denote the integral in (18.57) by $C$ and show that

$$2C = - \int_0^1 f(u)\, du \quad \text{where} \quad f(x) = \ln \left( \frac{\sin \pi u}{u} \right).$$

Show that $C = \frac{1}{2} \ln 2\pi$ by proving (i) $\int_0^1 f(u)\, du = \int_0^1 f\left(\frac{u}{2}\right) du$ and (ii) $f(u) = f\left(\frac{u}{2}\right) + f\left(\frac{1-u}{2}\right) + \ln 2\pi$.

The proofs in (a) and (b) were published by Stieltjes in 1878. See Stieltjes (1993) vol. 1, pp. 114–118. The proof in (c) was attributed to Lerch by Hermite in his 1891 lectures at the École Normale. See Hermite (1891).

(3) Integrate Gauss's formula for $\frac{\Gamma'(1+y)}{\Gamma(1+y)}$ to obtain Plana's formula

$$\ln \Gamma(u) = \int_0^\infty \left( \frac{1 - e^{(1-u)x}}{e^x - 1} + (u - 1)e^{-x} \right) \frac{dx}{x}.$$

Then, by another integration, show that

$$J \equiv \int_a^{a+1} \ln \Gamma(u)\, du = \int_0^\infty \left( \frac{e^{-ax}}{x} + \frac{e^{-x}}{e^{-x}-1} - \left( a - \frac{1}{2} \right) e^{-x} \right) \frac{dx}{x}.$$

Deduce that

$$\ln \Gamma(a) = J - \frac{1}{2} \ln a + \int_0^\infty \left( \frac{1}{2} - \frac{1}{x} + \frac{1}{e^x - 1} \right) \frac{e^{-ax}}{x}\, dx.$$

This is Binet's formula (18.36) after the value of $J$ is substituted from Exercise 2. This proof of Binet's formula is from Hermite (1891).

(4) Prove the formulas used by Binet:

$$\int_0^\infty e^{-xy} \sin(ty)\, dy = \frac{t}{t^2 + x^2},$$

$$4 \int_0^\infty \frac{\sin(ty)}{e^{2\pi} - 1}\, dy = \frac{e^t + 1}{e^t - 1} - \frac{2}{t}.$$

(5) Let $n = 2m$ and $Y_n = \left( \frac{1}{2^{2m}} \binom{2m}{m} \right)^2$. Prove the recurrence relation

$$(n+1)(n+3)(Y_n - Y_{n+2}) - 2(n+1)Y_n - Y_{n+2} = 0, \quad n = 0, 2, 4, 6, \dots.$$

Assume

$$Y_n = \frac{\alpha_1}{n+1} + \frac{\alpha_2}{(n+1)(n+3)} + \frac{\alpha_3}{(n+1)(n+3)(n+5)} + \dots.$$

Then employ the recurrence relation to prove that $(2k-2)\alpha_k = (2k-3)^2 \alpha_{k-1}$, $k = 2, 3, 4, \dots$. Finally, use Wallis's formula to show that $\alpha_1 = \frac{1}{\pi}$. This is Stirling's formal proof of (18.5) and is very similar to his proof of (18.4) given in the text.

(6) Obtain Binet's proof of (18.5) by observing that

$$B\left( m + \frac{1}{2}, \frac{1}{2} \right) = \int_0^1 x^m (1-x)^{-\frac{1}{2}} (1 - (1-x))^{-\frac{1}{2}}\, dx,$$

and following his argument for (18.4), given in the text.

(7) Note that

$$\ln \left( \frac{1}{2^{2m}} \binom{2m}{m} \right) = (-2m+1)\ln 2 + \sum_{k=1}^{m-1} \ln \frac{1 + \frac{k}{m}}{1 - \frac{k}{m}}.$$

Now apply de Moivre's method from the *Miscellanea Analytica*, given in the text, to obtain (18.10).

(8) Prove the formula (18.44) used by Cauchy in his derivation of the remainder in the series for $\mu(n)$. Cauchy started with the infinite product for $\sinh\left(\frac{t}{2}\right)$, due to Euler, and took the logarithmic derivative. In fact, Euler was aware of this result and this proof.

(9) In the *Methodus*, Stirling gave two more formulas for $b_m = \binom{2m}{m}$:

$$\left(\frac{2^{2m}}{b_m}\right)^2 = \frac{\pi}{2}(2m+1)\left(1 - \frac{1^2}{2(2m-3)} + \frac{1^2 \cdot 3^2}{2 \cdot 4(2m-3)(2m-5)} - \text{etc.}\right),$$

$$\left(\frac{b_m}{2^{2m}}\right)^2 = \frac{1}{\pi m}\left(1 - \frac{1^2}{2(2m-2)} + \frac{1^2 \cdot 3^2}{2 \cdot 4(2m-2)(2m-4)} - \text{etc.}\right).$$

In his analysis of Stirling's work, Binet pointed out that these formulas were incorrect and should be replaced by

$$\left(\frac{2^{2m}}{b_m}\right)^2 = \frac{\pi}{2}(2m+1)\sum_{k=0}^{\infty} \frac{\left(\frac{1}{2}\right)_k \left(-\frac{1}{2}\right)_k}{k!\,(m+1)_k},$$

$$\left(\frac{b_m}{2^{2m}}\right)^2 = \frac{1}{\pi m}\sum_{k=0}^{\infty} \frac{\left(\frac{1}{2}\right)_k \left(-\frac{1}{2}\right)_k}{k!\,\left(m+\frac{1}{2}\right)_k},$$

where $(a)_k = a(a+1)\cdots(a+k-1)$. Prove Binet's formulas. See Binet (1839) pp. 319–320. For an analysis of Stirling's results, see Stirling and Tweddle (2003).

(10) From de Moivre's version (18.11), use Legendre's duplication formula $\sqrt{\pi}\,\Gamma(2x) = 2^{2x-1}\Gamma(x)\Gamma\left(x+\frac{1}{2}\right)$ to obtain Stirling's series

$$\ln\Gamma(x+1) = \left(x+\frac{1}{2}\right)\ln\left(x+\frac{1}{2}\right) - \left(x+\frac{1}{2}\right)$$

$$+ \frac{1}{2}\ln 2\pi - \frac{B_2}{4\left(x+\frac{1}{2}\right)} + \frac{7B_4}{96\left(x+\frac{1}{2}\right)^3} - \text{etc.} \qquad (18.58)$$

See Gauss (1863–1927) vol. III, p. 152.

## 18.7 Notes on the Literature

Hald (1990), a book on the history of probability and statistics, devotes two chapters to the work of de Moivre and he also discusses the work of de Moivre and Stirling; see especially pages 480–489. Schneider (1968) gives a thorough analysis of the totality of de Moivre's mathematics.

# 19

---

# Fourier Series

## 19.1 Preliminary Remarks

The problem of representing functions by trigonometric series has played as significant a role in the development of mathematics and mathematical physics as that of representing functions as power series. Trigonometric series take the form

$$\frac{1}{2}a_0 + a_1 \cos x + b_1 \sin x + a_2 \cos 2x + b_2 \sin 2x + \cdots, \qquad (19.1)$$

and these series naturally made their appearance in eighteenth-century works on astronomy, a subject dealing with periodic phenomena. Now series (19.1) is called a Fourier series if, for some function $f(x)$ defined on $(0, 2\pi)$, the coefficients $a_n$ and $b_n$ can be expressed as

$$a_n = \frac{1}{\pi} \int_0^{2\pi} f(t) \cos nt \, dt; \quad b_n = \frac{1}{\pi} \int_0^{2\pi} f(t) \sin nt \, dt. \qquad (19.2)$$

Moreover, if (19.1) converges to some integrable function $f(x)$ and can be integrated term by term, then the coefficients $a_n$ and $b_n$ will take the form (19.2). Thus, Fourier series have very wide applicability.

In connection with investigations on the vibratory motion of a stretched string, trigonometric series of the type (19.1) were used, although the coefficients were not explicitly written as integrals. These researches led to controversy among the principal investigators, d'Alembert, Euler, Bernoulli, and Lagrange, as to whether an 'arbitrary' function could be represented by such series. This dispute began with d'Alembert's 1746 discovery, published in 1747, of the wave equation describing the motion of the vibrating string:[1]

$$\sigma \frac{\partial^2 y}{\partial t^2} = T \frac{\partial^2 y}{\partial x^2} \quad \text{or} \quad \frac{1}{c^2} \frac{\partial^2 y}{\partial t^2} = \frac{\partial^2 y}{\partial x^2}, \quad c^2 \equiv \frac{T}{\sigma}, \qquad (19.3)$$

---

[1] d'Alembert (1747). This paper was followed in the same year by another in the same journal, with pages consecutive to the earlier one.

where $\sigma$ and $T$ were constants and $y$ was the displacement of the string. The derivation was based on the work of Taylor dating from 1715.[2] D'Alembert showed that the general solution of equation (19.3) would be of the form

$$y = \Phi(ct + x) + \Psi(ct - x),$$

but the initial and boundary conditions implied a relation between $\Phi$ and $\Psi$. For example, at $x = 0$ and $x = l$, the string would be fixed and hence $y = 0$ for all $t$ at these points. This implied that for all $u$

$$0 = \Phi(u) + \Psi(u), \quad \text{or} \quad \Psi(u) = -\Phi(u) \tag{19.4}$$

and

$$0 = \Phi(u + l) + \Psi(u - l). \tag{19.5}$$

By (19.4), the general solution took the form $y = \Psi(ct + x) - \Psi(ct - x)$, and by (19.5), $\Psi$ was periodic: $\Psi(u + 2l) = \Psi(u)$. Interestingly, d'Alembert's paper also gave the first instance of the use of separation of variables to solve partial differential equations. He set

$$\Psi(ct + x) - \Psi(ct - x) = f(t)\, g(x) \tag{19.6}$$

and by differentiation obtained the relation

$$\frac{1}{c^2} \frac{f''}{f} = \frac{g''}{g} = A.$$

Since $f$ was independent of $x$ and $g$ of $t$, $A$ was a constant and the expressions for $f$ and $g$ could be obtained from their differential equations. Note that from the boundary conditions, it can be shown that $f$ and $g$ are sine and cosine functions. D'Alembert, however, saw these solutions as special cases of the general solution.

Euler responded to d'Alembert's work by publishing his ideas on the matter within a few months.[3] Essentially, he and d'Alembert disagreed on the meaning of the function $\Phi(u)$. D'Alembert thought that $\Phi$ had to be an analytic expression, whereas Euler was of the view that $\Phi$ was an arbitrary graph defined only by the periodicity condition

$$\Phi(u + 2l) = \Phi(u).$$

On this view, $\Phi$ could be defined by different expressions in different intervals; in our terms, $\Phi$ would be continuous but its derivative could be piecewise continuous. The functions allowed by Euler as solutions of $\Phi$ would now be called weak solutions of the equation, while d'Alembert required the solutions to be twice differentiable. And while Euler allowed all possible initial conditions on $\Phi$, d'Alembert ruled out certain initial conditions.

[2] Taylor (1715).
[3] Eu. II-10 pp. 50–62. E 119.

Euler also criticized Taylor's contention that an arbitrary initial vibration would eventually settle into a sinusoidal one. He argued from the equation of motion that higher frequencies would also be involved and that the solution could have the form

$$\sum A_n \sin \frac{n\pi x}{l} \cos \frac{n\pi ct}{l}$$

with the initial shape

given by $\sum A_n \sin \frac{n\pi x}{l}$. According to Truesdell, Euler was therefore "*the first to publish formulae for the simple modes of a string* and to observe that they can be combined simultaneously with arbitrary amplitudes."[4] However, Euler did not regard these trigonometric series as the most general solutions of the problem.

At this point, Daniel Bernoulli entered the discussion by presenting in 1747 and in 1748 two memoirs to the Petersburg Academy, eventually published by the Berlin Academy in 1753,[5] in which he explained on physical grounds that the trigonometric solutions found by Euler were in fact the most general possible. Bernoulli's ideas had been developing for more than a decade; moreover, he explained that were based on the work of Taylor, who had observed that the basic shapes of the vibrating string of length $a$ were given by the functions

$$\sin \frac{\pi x}{a}, \ \sin \frac{2\pi x}{a}, \ \sin \frac{3\pi x}{a}, \dots .$$

Bernoulli argued that the general form of the curve for the string would be obtained by linear superposition:

$$y = \alpha \sin \frac{\pi x}{a} + \beta \sin \frac{2\pi x}{a} + \gamma \sin \frac{3\pi x}{a} + \delta \sin \frac{4\pi x}{a} + \text{ etc.}$$

Unlike Euler, Bernoulli thought that all possible curves assumed by the vibrating string could be obtained in this way. It is interesting to note that in 1728 Bernoulli solved a linear difference equation by taking linear combinations of certain special solutions.[6] However, he was unable to extend this idea to the solutions of ordinary linear differential equations; Euler did this around 1740, as we mention in Chapter 14. Finally in 1748, Bernoulli once again proposed this idea to solve a linear partial differential equation, but this time he gave a physical argument. He apparently saw no need here for the differential equation and thought that the mathematics only obscured the main ideas. This led to further discussion and controversy, mainly involving d'Alembert and Bernoulli.

It seems that these discussions led Euler to further ponder on the problem of expanding functions in terms of trigonometric series. In a paper written around 1752, Euler started with the divergent series[7]

---

[4] Truesdell (1960) p. 250.
[5] D. Bernoulli (1753a) and (1753b).
[6] Bernoulli (1982–1996) vol. 2, pp. 49–64.
[7] Eu. I-14 p. 584. E 246.

$$\cos x + \cos 2x + \cos 3x + \cdots = -\frac{1}{2}$$

and after integration obtained the formulas

$$\frac{1}{2}x = \sin x - \frac{1}{2}\sin 2x + \frac{1}{3}\sin 3x - \frac{1}{4}\sin 4x + \cdots,$$

$$\frac{1}{12}\pi^2 - \frac{1}{4}x^2 = \cos x - \frac{1}{2^2}\cos 2x + \frac{1}{3^2}\cos 3x - \frac{1}{4^2}\cos 4x + \cdots.$$

He gave no range of validity for the formulas, but two decades later D. Bernoulli observed[8] that these results were true only in the interval $-\pi < x < \pi$. Euler also continued the integration process to obtain similar formulas with polynomials of degree 3, 4 and 5; obviously, the process could be continued. The polynomials occurring in this situation were the (Jakob) Bernoulli polynomials. Neither Bernoulli nor Euler seems to have noticed this. It appears that the Swiss mathematician Joseph Raabe (1801–1859) was the first to show this explicitly, around 1850.[9] Recall that the Fourier expansion of the Bernoulli polynomials also follows when Poisson's remainder in the Euler–Maclaurin formula, dating from the 1820s, is set equal to the remainder derived by Jacobi in the 1830s.

In 1759, Lagrange wrote a paper on the vibrating string problem[10] in which he attempted to obtain Euler's general solution with arbitrary functions by first finding the explicit solution for the loaded string and then taking the limit. The equations of motion in the latter case were

$$M\frac{d^2 y_k}{dt^2} = c^2 (y_{k+1} - 2y_k + y_{k-1}), \qquad k = 1, 2, \ldots, n, \qquad (19.7)$$

and were first obtained by Johann Bernoulli in 1727. Euler studied them in a slightly different context in 1748 and obtained solutions by setting

$$y_k = A_k \cos \frac{2c}{\sqrt{M}} pt$$

and finding

$$p = \sin \frac{r\pi}{2(n+1)}, \qquad r = 1, \ldots, n$$

and the value of $A_k$ from the corresponding second-order difference equation. Lagrange solved (19.7) by writing the equations as the first-order system

$$\frac{dy_k}{dt} = v_k, \qquad \frac{dv_k}{dt} = c^2 (y_{k+1} - 2y_k + y_{k-1}), \quad k = 1, 2, \ldots. \qquad (19.8)$$

---

[8] D. Bernoulli (1982–1996) vol. 2, pp. 119–121.
[9] Raabe (1850).
[10] Lagrange (1867–1892) vol. 1, pp. 72–90.

In the course of this work, Lagrange came close to deriving the Fourier coefficients in the expansion of a function as a series of sines. Instead, he took a different course, since his aim was to derive Euler's general solution rather than a trigonometric series.

Surprisingly, as early as 1757, while studying the perturbations created by the sun, Alexis-Claude Clairaut gave the Fourier coefficients in the case of a cosine series expansion.[11] He viewed the question of finding the coefficients $A_0, A_1, A_2, \ldots$ in

$$f(x) = A_0 + 2 \sum_{m=1}^{\infty} A_m \cos mx \qquad (19.9)$$

as an interpolation problem, given that values of $f$ were known at $x = \frac{2\pi}{k}, \frac{4\pi}{k}, \frac{6\pi}{k}, \ldots$. He found

$$A_0 = \frac{1}{k} \sum_{m=1}^{\infty} f\left(\frac{2m\pi}{k}\right), \quad A_n = \frac{1}{k} \sum_{m=1}^{\infty} f\left(\frac{2m\pi}{k}\right) \cos \frac{2mn\pi}{k}$$

and then let $k \to \infty$, to get

$$A_n = \frac{1}{2\pi} \int_0^{2\pi} f(x) \cos nx \, dx. \qquad (19.10)$$

Twenty years later, Euler derived (19.10) directly by multiplying (19.9) by $\cos nx$ and using the orthogonality of the cosine function.[12]

Joseph Fourier (1768–1830) lost his parents as a child; he was then sent by the bishop of Auxerre to a military college run by the Benedictines. Fourier's earliest researches were in algebraic equations, and he went to Paris in 1789 to present his results to the Academy. He soon became involved in revolutionary activities and gained a reputation as an orator. In 1795, Fourier began studying with Gaspard Monge; he soon published his first paper and announced plans to present a series of papers on algebraic equations. But Monge selected him to join Napoleon's scientific expedition to Egypt. When Fourier returned to France in 1801, Napoleon appointed him an administrator in Isre. Fourier ably executed his duties, but found time to successfully carry out his difficult researches in heat conduction, presented to the Academy in 1807. His work was reviewed by Lagrange, Laplace, Lacroix, and Monge; Lagrange opposed its publication. Perhaps to make up for this, the Academy then set a prize problem in the conduction of heat, won by Fourier in 1812.

It is not clear whether Euler thought that $f(x)$ in (19.10) was an arbitrary function. But in his 1807 work on heat conduction, Fourier took this view explicitly.[13]

---

[11]  Clairaut (1759) p. 545.
[12]  Eu. I-16 pp. 311–332. E 703.
[13]  See Gratta-Guinness (2003).

He translated a physics problem into the mathematical one of finding a function $v$ such that

$$\frac{d^2v}{dx^2} + \frac{d^2v}{dy^2} = 0,$$

and $v(0,y) = v(r,y) = 0$, $v(x,0) = f(x)$. By a separation of variables, Fourier found $v$ to be given by the series

$$v = a_1 e^{-\frac{\pi y}{r}} \sin \frac{\pi x}{r} + a_2 e^{-\frac{2\pi y}{r}} \sin \frac{2\pi x}{r} + a_3 e^{-\frac{3\pi y}{r}} \sin \frac{3\pi x}{r} + \cdots,$$

with the coefficients $a_1, a_2, a_3, \ldots$ to be obtained from

$$f(x) = a_1 \sin \frac{\pi x}{r} + a_2 \sin \frac{2\pi x}{r} + a_3 \sin \frac{3\pi x}{r} + \cdots. \tag{19.11}$$

Fourier discussed three methods for deriving these coefficients. In one approach, he converted equation (19.11) into a system of infinitely many equations in infinitely many unknowns $a_1, a_2, a_3, \ldots$. He also considered problems that reduced to cosine series and to series with sines as well as cosines. In his 1913 monograph on such systems, *Les systèmes d'équations linéaires* the Hungarian mathematician Frigyes Riesz (1880–1956) wrote that Fourier was the first to deal with linear equations in infinitely many unknowns.[14] Fourier gave two other methods for determining the coefficients. One method depended on the discrete orthogonality of the sine function, and the other on its continuous orthogonality. Dirichlet later gave a brief exposition of Fourier's discrete orthogonality method, explaining why the integral representation for $a_n$ was plausible.

Fourier regarded the use of an infinite system of equations in infinitely many unknowns as important enough to first discuss a particular case. He expanded a constant function as an infinite series of cosines:

$$1 = a \cos y + b \cos 3y + c \cos 5y + d \cos 7y + \text{etc.} \tag{19.12}$$

To see briefly how he determined the coefficients $a, b, c, d, \ldots$,[15] we write the equation in the form

$$1 = \sum_{m=1}^{\infty} a_m \cos(2m-1)y.$$

He took derivatives of all orders of this equation and set $y = 0$ to obtain

---

[14] Riesz (1913) pp. 2–8.
[15] Fourier (1955) pp. 137–143.

$$1 = \sum_{m=1}^{\infty} a_m, \quad 0 = \sum_{m=1}^{\infty} (2m-1)^2 a_m,$$

$$0 = \sum_{m=1}^{\infty} (2m-1)^4 a_m, \quad \text{etc.}$$

He considered the first $n$ equations with $n = 1, 2, 3, \ldots$, and replaced these $n$ equations with a new set of $n$ equations, taking $a_m = 0$ for $m > n$. This new system could be regarded as $n$ equations in the $n$ unknowns $a_1^{(n)}, a_2^{(n)}, \ldots, a_n^{(n)}$. By using the well-known formula now known as Cramer's rule, Fourier calculated the Vandermonde determinants appearing in this situation to find that

$$a_1^{(n)} = \frac{3 \cdot 3}{2 \cdot 4} \cdot \frac{5 \cdot 5}{4 \cdot 6} \cdots \frac{(2n-1)(2n-1)}{(2n-2)(2n)},$$

with a similar formula for $a_m^{(n)}$. He assumed that $a_m^{(n)} \to a_m$ as $n \to \infty$. This, by Wallis's formula, gave him $a_1 = \frac{4}{\pi}$ and in general $a_m = (-1)^{m-1} \frac{4}{(2m-1)\pi}$. By substituting these $a_m$ back in (19.12), he obtained

$$\frac{\pi}{4} = \cos x - \frac{1}{3}\cos 3x + \frac{1}{5}\cos 5x - \frac{1}{7}\cos 7x + \frac{1}{9}\cos 9x - \text{etc.}$$

Fourier did not discuss the validity of his method. Interestingly, according to Riesz, the question of the justification of this process was first considered by Henri Poincaré (1854–1912) in 1885. Poincaré's attention was drawn to this problem by a paper of Paul Appell in which Appell applied Fourier's method to obtain the coefficients of a cosine expansion of an elliptic function. Poincaré gave a simple theorem justifying Appell's calculations.[16] A year later he wrote another paper on the subject, "Sur les Determinants d'ordre Infini."

The term infinite determinant was introduced by the American astronomer and mathematician G. W. Hill (1838–1914) in an 1877 paper on lunar theory. In this paper, he solved the equation

$$\frac{d^2w}{dt^2} + \left( \sum_{n=-\infty}^{\infty} \theta_n e^{int} \right) w = 0$$

by making the substitution

$$w = \sum_{n=-\infty}^{\infty} b_n e^{i(n+c)t}$$

and determining $b_n$ from the infinite system of equations

$$\sum_{k=-\infty}^{\infty} \theta_{n-k} b_k - (n+c)^2 b_n = 0, \quad n = -\infty, \ldots, +\infty.$$

---

[16] See Riesz (1913) pp. 20–24 for references.

Hill employed a procedure similar to that of Fourier and once again Poincaré developed the necessary theorems to justify Hill's result.[17] Poincaré's work was generalized a decade later by the Swedish mathematician Niels Helge von Koch (1870–1924) and this was the starting point for his countryman Ivar Fredholm's (1866–1927) theory of integral equations. Fredholm in turn provided the basis for the pioneering work in the development of functional analysis by David Hilbert and then Riesz, with significant contributions from others such as Erhard Schmidt (1876–1959).[18] They created the ideas and techniques by which linear equations in infinitely many variables could be treated by general methods. The valuable 1913 book by Riesz, one of the earliest monographs on functional analysis, contains an interesting history of the topic. Surely Fourier could not have foreseen that his idea would see such beautiful development. On the other hand, he must have considered it worthy of attention, since he included the long derivation of the formula for the Fourier coefficients by this method when he was well aware of the much shorter method using term-by-term integration.

## 19.2 Euler: Trigonometric Expansion of a Function

In a very interesting paper presented in the Petersburg Academy in 1750, "De Serierum Determinatione seu Nova Methodus Inveniendi Terminos Generales Serierum,"[19] Euler used symbolic calculus to expand a function as a trigonometric series. He also applied the discoveries he had made a decade earlier on solving differential equations with constant coefficients. Given a function $X$, his problem was to determine $y(x)$ such that

$$y(x) - y(x-1) = X(x). \tag{19.13}$$

He viewed this as a differential equation of infinite order:

$$\frac{dy}{dx} - \frac{1}{1 \cdot 2}\frac{d^2 y}{dx^2} + \frac{1}{1 \cdot 2 \cdot 3}\frac{d^3 y}{dx^3} - \cdots = X(x).$$

He noted that if $d^n y/dx^n$ was replaced by $z^n$, then the left-hand side could be expressed as

$$z - \frac{z^2}{1 \cdot 2} + \frac{z^3}{1 \cdot 2 \cdot 3} - \cdots = 1 - e^{-z}.$$

He observed that the factors of $1 - e^{-z}$ were $z$ and $z^2 + 4kk\pi\pi$ for $k = 1, 2, 3, \ldots$. Hence, $dy/dx$ and $d^2 y/dx^2 + 4k^2\pi^2 y$ were factors of the differential equation. The solution of the differential equation corresponding to $dy/dx$ was given by $y = \int X \, dx$, while the solution corresponding to $d^2 y/dx^2 + 4kk\pi\pi$ was given by

---

[17] ibid.
[18] See Dieudonné (1981) pp. 75–120.
[19] Eu. I-14 pp. 463–515. E 189.

$$y = 2(\cos 2k\pi \cos 2k\pi x - \sin 2k\pi \sin 2k\pi x) \int X \cos 2k\pi x \, dx$$

$$+ 2(\cos 2k\pi \sin 2k\pi x + \sin 2k\pi \cos 2k\pi x) \int X \sin 2k\pi x \, dx,$$

and since $\sin 2k\pi = 0$, $\cos 2k\pi = 1$, Euler could write the complete solution

$$y = \int X \, dx + 2\cos 2\pi x \int X \cos 2\pi x \, dx + 2\cos 4\pi x \int X \cos 4\pi x \, dx + \cdots$$

$$+ 2\sin 2\pi x \int X \sin 2\pi x \, dx + 2\sin 4\pi x \int X \sin 4\pi x \, dx + \cdots .$$

$$(19.14)$$

Although (19.14) appears to be a Fourier series, the integrals are indefinite, so it is not. If the integrals were on the interval $[0, 1]$, then the right-hand side would be the Fourier series of $X$, not of $y$.

### 19.3   Lagrange on the Longitudinal Motion of the Loaded Elastic String

In his study of the vibrating string, Lagrange considered the situation in which the masses were assumed to be at a discrete set of points so that he could express the rate of change with respect to $x$ in terms of finite differences.[20] He wrote the equations in the form

$$\frac{dy_k}{dt} = v_k, \quad \frac{dv_k}{dt} = C^2(y_{k+1} - 2y_k + y_{k-1}), \qquad (19.15)$$

where $k = 1, 2, \ldots, m - 1$ and $y_0 \equiv y_m \equiv 0$. His idea was to determine constants $M_k$, $N_k$ and $R$ such that

$$\sum_{k=1}^{m-1} (M_k \, dv_k + N_k \, dy_k) = \sum_{k=1}^{m-1} \left(N_k v_k + C^2 M_k(y_{k+1} - 2y_k + y_{k-1})\right) dt$$

would be reduced to $dz = Rz \, dt$. This required that

$$R(M_k v_k + N_k y_k) = N_k v_k + C^2 M_k(y_{k+1} - 2y_k + y_{k-1}), \quad k = 1, 2, \ldots, m - 1,$$

$$(19.16)$$

or

$$RM_k = N_k, \quad RN_k = C^2(M_{k+1} - 2M_k + M_{k-1}).$$

[20] Lagrange (1867–1892) vol. 1, pp. 72–90.

This meant that $M_k$ satisfied the equation

$$M_{k+1} - \left(\frac{R^2}{C^2} + 2\right) M_k + M_{k-1} = 0. \tag{19.17}$$

Lagrange set $M_k = Aa^k + Bb^k$, so that $a$ and $b$ were roots of

$$x^2 - \left(\frac{R^2}{C^2} + 2\right) x + 1 = 0.$$

Thus,

$$ab = 1 \quad \text{and} \quad a + b = \frac{R^2}{C^2} + 2.$$

Note that because of the restriction on $y_0$ and $y_m$, Lagrange could assume without loss of generality that $M_0 \equiv M_m \equiv 0$. He also set $M_1 \equiv 1$. From this it followed that $A + B = 0$ and $Aa + Bb = 1$. With these initial values, he could find the constants $A, B$ in $M_k$. Thus, he could write

$$M_k = \frac{a^k - b^k}{a - b} \quad \text{and} \quad \frac{a^m - b^m}{a - b} = 0,$$

yielding him $m - 1$ pairs of values $a_n = e^{\frac{n\pi i}{m}}$ and $b_n = e^{-\frac{n\pi i}{m}}$ for $n = 1, 2, \ldots, m - 1$. Corresponding to these were $m - 1$ values of $M$ and $R$:

$$M_{kn} = \frac{e^{\frac{kn\pi i}{m}} - e^{-\frac{kn\pi i}{m}}}{e^{\frac{n\pi i}{m}} - e^{-\frac{n\pi i}{m}}} = \frac{\sin\left(\frac{kn\pi}{m}\right)}{\sin\left(\frac{n\pi}{m}\right)}, \tag{19.18}$$

$$R_n = \pm 2iC \sin\left(\frac{n\pi}{2m}\right), \quad n = 1, 2, \ldots, m - 1. \tag{19.19}$$

For these values he had the corresponding equations

$$dz_n = R_n z_n \, dt \quad \text{where} \quad z_n = \sum_{k=1}^{m-1} (M_{kn} v_k + R_n M_{kn} y_k). \tag{19.20}$$

The solution of the differential equation for $z_n$ yielded

$$z_n = F_n e^{R_n t}$$

with $F_n$ a constant. Next he set

$$Z_n = \sum_{k=1}^{m-1} M_{kn} y_k, \tag{19.21}$$

so that $\frac{dy_k}{dt} = v_k$ implied that

$$\frac{dZ_n}{dt} + R_n Z_n = F_n e^{R_n t}. \tag{19.22}$$

Lagrange expressed the constant $F_n$ as $2R_n K_n$ so that he could solve this differential equation in the form

$$Z_n = K_n e^{R_n t} + L_n e^{-R_n t}, \tag{19.23}$$

where $L_n$ was a constant of integration. Here recall that (19.22) can be solved by multiplying it by the integrating factor $e^{R_n t}$. By substituting the value of $R_n$ from (19.19), he could write $Z_n$ in terms of the sine and cosine functions

$$Z_n = P_n \cos\left(2Ct \sin\left(\frac{n\pi}{2m}\right)\right) + Q_n \frac{\sin\left(2Ct \sin\left(\frac{n\pi}{2m}\right)\right)}{2C \sin\left(\frac{n\pi}{2m}\right)}. \tag{19.24}$$

Then, $Z_n$ being known, the problem was to determine $y_k$ from (19.21); after substituting the value of $M_{kn}$ from (19.18), (19.21) took the form

$$Z_n \sin\frac{n\pi}{m} = \sum_{k=1}^{m-1} y_k \sin\frac{kn\pi}{m}, \qquad m = 1, 2, \ldots, m-1. \tag{19.25}$$

The next step was to obtain the $m-1$ unknowns $y_1, y_2, \ldots, y_{m-1}$ from these $m-1$ equations. Several years before Lagrange, in 1748, Euler had encountered this system of equations in his study of the loaded elastic cord.[21] He was able to write the solution in general after studying the special cases where $m \leq 6$. He saw that the result followed from the discrete orthogonality relation for the sine function:

$$\sum_{k=1}^{m-1} \sin\frac{kn\pi}{m} \sin\frac{kp\pi}{m} = \frac{1}{2}m\delta_{np} \tag{19.26}$$

for which he did not provide a complete proof. But Lagrange gave an ingenious proof of (19.26) and obtained

$$y_j = \frac{2}{m} \sum_{n=1}^{m-1} Z_n \sin\frac{n\pi}{m} \sin\frac{nj\pi}{m}, \tag{19.27}$$

by multiplying (19.25) by $\sin\left(\frac{nj\pi}{m}\right)$, summing over $n$ and applying (19.26). In a later paper, Lagrange observed that the analysis involved in moving from (19.25) to (19.27) also solved an interpolation problem related to trigonometric polynomials. Specifically, given the $m-1$ values

---

[21] Eu. 2-10 pp. 98–131. E 136.

$$f\left(\frac{\pi}{m}\right), f\left(\frac{2\pi}{m}\right), \ldots, f\left(\frac{(m-1)\pi}{m}\right)$$

of a function $f(x)$, the problem was to find a trigonometric polynomial

$$a_1 \sin x + a_2 \sin 2x + \cdots + a_{m-1} \sin (m-1)x \qquad (19.28)$$

passing through $m-1$ points $\left(\frac{k\pi}{m}, f\left(\frac{k\pi}{m}\right)\right), k = 1, 2, \ldots, m-1$. By an application of (19.26), it was clear that

$$ma_n = 2\sin \frac{n\pi}{m} f\left(\frac{\pi}{m}\right) + 2\sin \frac{2n\pi}{m} f\left(\frac{2\pi}{m}\right)$$
$$+ \cdots + 2\sin \frac{(m-1)n\pi}{m} f\left(\frac{(m-1)\pi}{m}\right), \qquad (19.29)$$

and one obtained the coefficients of the trigonometric polynomial interpolating $f(x)$.

We reproduce Dirichlet's 1837 proof[22] of the orthogonality relation (19.26), since it is more illuminating than Lagrange's complicated though clever proof. The same method was clearly described in Fourier's 1822 book on heat. The idea was to apply the addition formula for the sine function; note that this addition formula is also used for the integral analog of (19.26). First note that by the addition formula

$$2\sin \frac{kn\pi}{m} \sin \frac{kp\pi}{m} = \cos \frac{k(n-p)\pi}{m} - \cos \frac{k(n+p)\pi}{m}.$$

So when $n \neq p$, twice the sum in (19.26) is given by

$$\sum_{k=1}^{m-1} \left(\cos \frac{k(n-p)\pi}{m} - \cos \frac{k(n+p)\pi}{m}\right) \qquad (19.30)$$
$$= \frac{\sin \left(m-\frac{1}{2}\right)(n-p)\frac{\pi}{m}}{2\sin (n-p)\frac{\pi}{2m}} - \frac{\sin \left(m-\frac{1}{2}\right)(n+p)\frac{\pi}{m}}{2\sin (n+p)\frac{\pi}{2m}} = 0,$$

since each expression is either $-\frac{1}{2}$ or 0, according as $n-p$ is even or odd. To sum the series in (19.30) Dirichlet employed the formula

$$1 + 2\cos 2\theta + 2\cos 4\theta + \cdots + 2\cos 2s\theta = \frac{\sin (2s+1)\theta}{\sin \theta},$$

also provable by the addition formula for the sine function. Dirichlet pointed out that (19.28) and (19.29) strongly suggested that a function $f(x)$ could be expanded as a Fourier series. Observe that $a_n$ can be expressed as

[22] Dirichlet (1969) vol. 1, pp. 139–142.

$$\frac{2}{\pi}\left[\frac{\pi}{m}\sin\frac{0n\pi}{m}\ f\left(\frac{0\pi}{m}\right)+\frac{\pi}{m}\sin\frac{n\pi}{m}\ f\left(\frac{\pi}{m}\right)+\cdots\right.$$

$$\left.+\frac{\pi}{m}\sin\frac{(m-1)n\pi}{m}\ f\left(\frac{(m-1)\pi}{m}\right)\right],$$

and when $m\to\infty$, the right-hand side tends to

$$\frac{2}{\pi}\int_0^\pi\sin nx\ f(x)\,dx.$$

Thus, $f(x)=a_1\sin x+a_2\sin 2x+\cdots+a_n\sin nx+\cdots$, with

$$a_n=\frac{2}{\pi}\int_0^\pi\sin nx\ f(x)\,dx.$$

Observe here that Lagrange missed this opportunity to discover Fourier series, partly because he was focused on obtaining the results of d'Alembert and Euler and partly because he did not think that functions could be represented by such series.

## 19.4   Euler on Fourier Series

In 1777, Euler submitted a paper to the Petersburg Academy containing a derivation of the Fourier coefficients of a cosine series.[23] This was the first derivation of the coefficients using the orthogonality of the sequence of functions $\cos nx$, $n=1,2,\ldots$. Euler's paper was published in 1798, but its contents did not become generally known until much later; a half century afterward, Riemann thought that Fourier was the first to give such a derivation. Euler expanded a function $\Phi$ as a trigonometric series, $\Phi=A+B\cos\phi+C\cos 2\phi+\cdots$ and gave the coefficients as

$$A=\frac{1}{\pi}\int_0^\pi\Phi\,d\phi,\quad B=\frac{2}{\pi}\int_0^\pi\Phi\,d\phi\cos\phi,\quad C=\frac{2}{\pi}\int_0^\pi\Phi\,d\phi\cos 2\phi,\ \ldots.$$

$$(19.31)$$

His argument was that since $\int d\phi\cos i\phi=\frac{1}{i}\sin i\phi=0$, on integration from $\phi=0$ to $\phi=\pi$, he would get

$$\int_0^\pi\Phi\,d\phi=A\pi.$$

Next, by the addition formula for the cosine function, when $i\neq\lambda$,

$$d\phi\cos i\phi\cos\lambda\phi=\frac{1}{2}d\phi(\cos(i-\lambda)\phi+\cos(i+\lambda)\phi),$$

$$\int_0^\pi d\phi\cos i\phi\cos\lambda\phi=\frac{\sin(i-\lambda)\phi}{2(i-\lambda)}+\frac{\sin(i+\lambda)\phi}{2(i+\lambda)}\Bigg]_0^\pi$$

$$=0.$$

[23] Eu. I-16 pp. 311–332. E 703.

And when $i = \lambda$,

$$\int_0^\pi d\phi\,(\cos i\phi)^2 = \frac{1}{2}\phi + \frac{1}{4i}\sin(2i\phi)\Big]_0^\pi = \frac{1}{2}\pi;$$

for

$$(\cos i\phi)^2 = \frac{1}{2} + \frac{1}{2}\cos(2i\phi).$$

Hence the coefficients $A, B, C, D, \ldots$ were as given in (19.31).

In this paper, Euler included a proof of the well-known recurrence relation

$$\int_0^\pi d\phi\,(\cos\phi)^\lambda = \frac{\lambda-1}{\lambda}\int_0^\pi d\phi\,(\cos\phi)^{\lambda-2}.$$

We mention that Euler wrote $\cos\phi^\lambda$ for $(\cos\phi)^\lambda$. Observe that, though he was well aware of the integration by parts formula in the standard form $\int P\,dQ = PQ - \int Q\,dP$, he usually worked it out in a slightly different way. For example, to prove the recurrence formula, Euler started with

$$\int d\phi\,\cos^\lambda\phi = f\sin\phi\,\cos^{\lambda-1}\phi + g\int d\phi\,\cos^{\lambda-2}\phi,$$

where $f$ and $g$ had to be determined. He differentiated to get

$$\cos^\lambda\phi = f\cos^\lambda\phi - f(\lambda-1)\sin^2\phi\,\cos^{\lambda-2}\phi + g\cos^{\lambda-2}\phi.$$

Since $\sin^2\phi = 1 - \cos^2\phi$,

$$\cos^\lambda\phi = \lambda f\cos^\lambda\phi - f(\lambda-1)\cos^{\lambda-2}\phi + g\cos^{\lambda-2}\phi,$$

for which Euler required that $f = \frac{1}{\lambda}$ and $g = f(\lambda-1)$ or $g = \frac{\lambda-1}{\lambda}$. Hence

$$\int d\phi\,\cos^\lambda\phi = \frac{1}{\lambda}\sin\phi\,\cos^{\lambda-1}\phi + \frac{\lambda-1}{\lambda}\int d\lambda\,\cos^{\lambda-2}\phi.$$

Finally, he took the integral from 0 to $\pi$. In this paper, as in some others, Euler used the notation $\partial\phi$ instead of $d\phi$.

## 19.5 Fourier and Linear Equations in Infinitely Many Unknowns

Fourier rediscovered Euler's derivation of the Fourier coefficients. Nevertheless, Fourier sought alternative derivations to convince the mathematical community of the correctness of the Fourier expansion. He also presented several derivations of Fourier expansions of specific functions. In spite of these efforts, his theory encountered a certain amount of opposition, mainly from the older generation. In section 6 of the third chapter of his famous book, *Théorie analytique de la chaleur*, Fourier considered the problem of determining the coefficients in the sine expansion of an odd function.[24] He reduced this problem to that of solving an infinite system of equations in infinitely many unknowns. It is interesting to see how beautifully Fourier carried out the computations; he started with a sine series expansion of an odd function

$$\phi(x) = a \sin x + b \sin 2x + c \sin 3x + d \sin 4x + \cdots . \qquad (19.32)$$

He let $A = \phi'(0)$, $B = -\phi'''(0)$, $C = \phi^{(5)}(0)$, $D = -\phi^{(7)}(0)$, ..., so that by repeatedly differentiating (19.32), he had

$$
\begin{aligned}
A &= a + 2b + 3c + 4d + 5e + \cdots, \\
B &= a + 2^3 b + 3^3 c + 4^3 d + 5^3 e + \cdots, \\
C &= a + 2^5 b + 3^5 c + 4^5 d + 5^5 e + \cdots, \\
D &= a + 2^7 b + 3^7 c + 4^7 d + 5^7 e + \cdots, \\
E &= a + 2^9 b + 3^9 c + 4^9 d + 5^9 e + \cdots,
\end{aligned}
\qquad (19.33)
$$

and so on. He broke up this system into the subsystems

$$a_1 = A_1$$

$$
\begin{aligned}
a_2 + 2b_2 &= A_2, \\
a_2 + 2^3 b_2 &= B_2,
\end{aligned}
$$

$$
\begin{aligned}
a_3 + 2b_3 + 3c_3 &= A_3, \\
a_3 + 2^3 b_3 + 3^3 c_3 &= B_3, \\
a_3 + 2^5 b_3 + 3^5 c_3 &= C_3,
\end{aligned}
\qquad (19.34)
$$

$$
\begin{aligned}
a_4 + 2b_4 + 3c_4 + 4d_4 &= A_4, \\
a_4 + 2^3 b_4 + 3^3 c_4 + 4^3 d_4 &= B_4, \\
a_4 + 2^5 b_4 + 3^5 c_4 + 4^5 d_4 &= C_4, \\
a_4 + 2^7 b_4 + 3^7 c_4 + 4^7 d_4 &= D_4,
\end{aligned}
\qquad (19.35)
$$

$$
\begin{aligned}
a_5 + 2b_5 + 3c_5 + 4d_5 + 5e_5 &= A_5, \\
a_5 + 2^3 b_5 + 3^3 c_5 + 4^3 d_5 + 5^3 e_5 &= B_5, \\
a_5 + 2^5 b_5 + 3^5 c_5 + 4^5 d_5 + 5^5 e_5 &= C_5, \\
a_5 + 2^7 b_5 + 3^7 c_5 + 4^7 d_5 + 5^7 e_5 &= D_5, \\
a_5 + 2^9 b_5 + 3^9 c_5 + 4^9 d_5 + 5^9 e_5 &= E_5,
\end{aligned}
\qquad (19.36)
$$

and so on. Fourier's strategy was to solve the first equation for $a_1$, the second for $2b_2$, the third for $3c_3$, and so on. He wrote that the equations could be solved by inspection, meaning that they could be obtained by Cramer's rule, since the determinants in the equations were Vandermonde determinants. He also established the recursive relations

---

[24] See Fourier (1955) pp. 168–185.

connecting $a_{j-1}$ with $a_j$, $b_{j-1}$ with $b_j$, $c_{j-1}$ with $c_j$, etc., and similarly with the right-hand sides of the equations, $A_j$, $B_j$, $C_j$, etc. He assumed that as $j \to \infty$,

$$a_j \to a,\, b_j \to b,\, c_j \to c, \text{etc.,}$$

and

$$A_j \to A,\, B_j \to B,\, C_j \to C, \text{etc.}$$

To find the recursive relations, Fourier eliminated $e_5$ from the last five equations to get

$$a_5(5^2 - 1^2) + 2b_5(5^2 - 2^2) + 3c_5(5^2 - 3^2) + 4d_5(5^2 - 4^2) = 5^2 A_5 - B_5,$$
$$a_5(5^2 - 1^2) + 2^3 b_5(5^2 - 2^2) + 3^3 c_5(5^2 - 3^2) + 4^3 d_5(5^2 - 4^2) = 5^2 B_5 - C_5,$$
$$a_5(5^2 - 1^2) + 2^5 b_5(5^2 - 2^2) + 3^5 c_5(5^2 - 3^2) + 4^5 d_5(5^2 - 4^2) = 5^2 C_5 - D_5,$$
$$a_5(5^2 - 1^2) + 2^7 b_5(5^2 - 2^2) + 3^7 c_5(5^2 - 3^2) + 4^7 d_5(5^2 - 4^2) = 5^2 D_5 - E_5.$$
$$(19.37)$$

Fourier then argued that for this system to coincide with the system of four equations in (19.35), he must have

$$a_4 = (5^2 - 1^2)a_5, \quad b_4 = (5^2 - 2^2)b_5, \quad c_4 = (5^2 - 3^2)c_5, d_4 = (5^2 - 4^2)d_5,$$
$$(19.38)$$

$$A_4 = 5^2 A_5 - B_5, \quad B_4 = 5^2 B_5 - C_5, \quad C_4 = 5^2 C_5 - D_5, \quad D_4 = 5^2 D_5 - E_5.$$
$$(19.39)$$

We remark that he wrote out all his equations in this manner, noting that this reasoning would apply in general to the $m \times m$ system of equations. We now write his formulas in shorter form. From the relations (19.38) and (19.39), it is evident that

$$a_{j-1} = a_j(j^2 - 1^2), \quad j = 2, 3, 4, \ldots,$$
$$b_{j-1} = b_j(j^2 - 2^2), \quad j = 3, 4, 5, \ldots,$$
$$c_{j-1} = c_j(j^2 - 3^2), \quad j = 4, 5, 6, \ldots,$$
$$d_{j-1} = d_j(j^2 - 4^2), \quad j = 5, 6, 7, \ldots$$
$$(19.40)$$

and also

$$A_{j-1} = j^2 A_j - B_j, \quad j = 2, 3, 4, \ldots,$$
$$B_{j-1} = j^2 B_j - C_j, \quad j = 3, 4, 5, \ldots,$$
$$C_{j-1} = j^2 C_j - D_j, \quad j = 4, 5, 6, \ldots,$$
$$D_{j-1} = j^2 D_j - E_j, \quad j = 5, 6, 7, \ldots.$$
$$(19.41)$$

As we noted before, Fourier assumed that $a_j \to a$, $b_j \to b, \ldots, A_j \to A, \ldots$ as $j \to \infty$. So from (19.40), he could conclude that

$$a = \frac{a_1}{(2^2 - 1^2)(3^2 - 1^2)(4^2 - 1^2)\cdots},$$

$$b = \frac{b_2}{(3^2 - 2^2)(4^2 - 2^2)(5^2 - 2^2)\cdots}, \qquad (19.42)$$

$$c = \frac{c_3}{(4^2 - 3^2)(5^2 - 3^2)(6^2 - 3^2)\cdots},$$

and so on. Similarly, a repeated application of (19.41) gave him

$$A_1 = A_2 2^2 - B_2, \quad A_1 = A_3 2^2 \cdot 3^2 - B_3(2^2 + 3^2) + C_3,$$

$$A_1 = A_4 2^2 \cdot 3^2 \cdot 4^2 - B_4(2^2 \cdot 3^2 + 2^2 \cdot 4^2 + 3^2 \cdot 4^2) + C_4(2^2 + 3^2 + 4^2) - D_4, \text{ etc.}$$

To understand Fourier's next step, one may divide the first value of $A_1$ by $2^2$, the second value of $A_1$ by $2^2 \cdot 3^2$, the third by $2^2 \cdot 3^2 \cdot 4^2$, and consider the form of the right-hand side. So by dividing the ultimate equation $A_1$ by $2^2 \cdot 3^2 \cdot 4^2 \cdot 5^2 \cdots$, Fourier obtained by equations (19.34) and (19.42)

$$\frac{A_1 (= a_1)}{2^2 \cdot 3^2 \cdot 4^2 \cdot 5^2 \cdots} = \frac{a(2^2 - 1)(3^2 - 1)(4^2 - 1)(5^2 - 1)\cdots}{2^2 \cdot 3^2 \cdot 4^2 \cdot 5^2 \cdots}$$

$$= A - B\left(\frac{1}{2^2} + \frac{1}{3^2} + \frac{1}{4^2} + \cdots\right) + C\left(\frac{1}{2^2 \cdot 3^2} + \frac{1}{2^2 \cdot 4^2} + \frac{1}{3^2 \cdot 4^2} + \cdots\right)$$

$$- D\left(\frac{1}{2^2 \cdot 3^2 \cdot 4^2} + \frac{1}{2^2 \cdot 3^2 \cdot 4^2} + \frac{1}{3^2 \cdot 4^2 \cdot 5^2} + \cdots\right) + \cdots$$

$$= A - BP_1 + CQ_1 - DR_1 + ES_1 - \cdots.$$

$$(19.43)$$

We note that by $P_1, Q_1, R_1, \ldots$ Fourier meant the sums of products of $\frac{1}{2^2}, \frac{1}{3^2}, \frac{1}{4^2}, \cdots$ taken one, two, three, ... at a time. This gave him the value of $a$ in terms of $A$, $B$, $C$, $D$ etc. To find the values of $b, c, d, \ldots$ in a similar manner, Fourier solved the second system and third systems in (19.34) to find $2b_2$ and $3b_3$. Similarly, he solved (19.35) for $4b_4$ etc. to arrive at the solutions:

$$A - BP_2 + CQ_2 - DR_2 + \cdots = 2b_2 \frac{(1^2 - 2^2)}{1^2 \cdot 3^2 \cdot 4^2 \cdot 5^2 \cdots},$$

$$A - BP_3 + CQ_3 - DR_3 + \cdots = 3c_3 \frac{(1^2 - 3^2)(2^2 - 3^2)}{1^2 \cdot 2^2 \cdot 4^2 \cdot 5^2 \cdot 6^2 \cdots}, \qquad (19.44)$$

$$A - BP_4 + CQ_4 - DR_4 + \cdots = 4d_4 \frac{(1^2 - 4^2)(2^2 - 4^2)(3^2 - 4^2)}{1^2 \cdot 2^2 \cdot 3^2 \cdot 5^2 \cdot 6^2 \cdots},$$

and so on. The starting points for deriving these equations were

$$2b_2(1^2 - 2^2) = A_2 1^2 - B_2,$$

$$3c_3(1^2 - 3^2)(2^2 - 3^2) = A_3 1^2 \cdot 2^2 - B_3(1^2 + 2^2) + C_3,$$

$$4d_4(1^2 - 4^2)(2^2 - 4^2)(3^2 - 4^2)$$
$$= A_4 1^2 \cdot 2^2 \cdot 3^2 - B_4(1^2 \cdot 2^2 + 1^2 \cdot 3^2 + 2^2 \cdot 3^2) + C_4(1^2 + 2^2 + 3^2) - D_4,$$
$$\tag{19.45}$$

and so on. As before, these relations were continued by repeated use of (19.41). Once again Fourier applied (19.40) to express $b_2, c_3, d_4, \ldots$ in terms of $b, c, d, \ldots$. Recall that $b_j \to b$, $c_j \to c$, $d_j \to d$, etc. He then had

$$A - BP_1 + CQ_1 - DR_1 + c \cdots = a \left(1 - \frac{1^2}{2^2}\right)\left(1 - \frac{1^2}{3^2}\right)\left(1 - \frac{1^2}{4^2}\right) \cdots,$$

$$A - BP_2 + CQ_2 - DR_2 + \cdots = 2b \frac{(1^2 - 2^2)(3^2 - 2^2)(4^2 - 2^2) \cdots}{1^2 \cdot 3^2 \cdot 4^3 \cdot 5^2 \cdots}$$

$$= 2b\left(1 - \frac{2^2}{1^2}\right)\left(1 - \frac{2^2}{3^2}\right)\left(1 - \frac{2^2}{4^2}\right) \cdots,$$

$$A - BP_3 + CQ_3 - DR_3 + \cdots = 3c\left(1 - \frac{3^2}{1^2}\right)\left(1 - \frac{3^2}{2^2}\right)\left(1 - \frac{3^2}{4^2}\right) \cdots,$$

$$A - BP_4 + CQ_4 - DR_4 + \cdots = 4d\left(1 - \frac{4^2}{1^2}\right)\left(1 - \frac{4^2}{2^2}\right)\left(1 - \frac{4^2}{3^2}\right)\left(1 - \frac{4^2}{5^2}\right) \cdots,$$
$$\tag{19.46}$$

and so on. To compute the values of the products on the right-hand side of (19.46), and the values of $P_j, Q_j, R_j, S_j, \ldots$, observe that

$$\frac{\sin \pi x}{\pi x} = \left(1 - \frac{x^2}{1^2}\right)\left(1 - \frac{x^2}{2^2}\right)\left(1 - \frac{x^2}{3^2}\right) \cdots. \tag{19.47}$$

Fourier did not write down the details of the evaluations of the products on the right-hand sides of (19.46), but they are fairly simple. Note that the first product has a factor $\left(1 - \frac{1^2}{1^2}\right)$ missing, the second $\left(1 - \frac{2^2}{2^2}\right)$, the third $\left(1 - \frac{3^2}{3^2}\right)$, etc. Now the value of the product with a missing $1 - \frac{j^2}{j^2}$ can be evaluated by (19.47) to be

$$\lim_{\epsilon \to 0} \frac{\sin \pi(j + \epsilon)}{\pi(j + \epsilon)\left(1 - \frac{(j+\epsilon)^2}{j^2}\right)} = \lim_{\epsilon \to 0} \frac{j^2(-1)^{j-1} \sin \pi\epsilon}{\pi(j + \epsilon)(2j + \epsilon)\epsilon} = \frac{(-1)^{j-1}}{2}. \tag{19.48}$$

To find $P_j, Q_j, R_j, \ldots$, he expanded the product on the right-hand side as a series

$$1 - Px^2 + Qx^4 - Rx^6 + \cdots, \tag{19.49}$$

so that $P$, $Q$, $R$, ... were sums of products of $1, \frac{1}{2^2}, \frac{1}{3^2}, \ldots$ taken one, two, three, ... (respectively) at a time. He could then equate this series with the known power series for $\frac{\sin(\pi x)}{\pi x}$:

$$1 - \frac{x^2 \pi^2}{3!} + \frac{x^4 \pi^4}{5!} - \frac{x^6 \pi^6}{7!} + \cdots$$

to get

$$P = \frac{\pi^2}{3!}, \quad Q = \frac{\pi^4}{5!}, \quad R = \frac{\pi^6}{7!}, \quad \ldots \tag{19.50}$$

Moreover, it is easy to see that

$$\left(1 - \frac{y}{j^2}\right)\left(1 - P_j y + Q_j y^2 - R_j y^3 + \cdots\right) = 1 - Py + Qy^2 - Ry^3 + \cdots,$$

$$P_j + \frac{1}{j^2} = P, \quad Q_j + \frac{1}{j^2}P_j = Q, \quad R_j + \frac{1}{j^2}Q_j = R, \ldots.$$

Hence

$$P_j = P - \frac{1}{j^2} = \frac{\pi^2}{3!} - \frac{1}{j^2}, \quad Q_j = \frac{\pi^4}{5!} - \frac{1}{j^2}\frac{\pi^2}{3!} + \frac{1}{j^4}, \quad \ldots. \tag{19.51}$$

Using these expressions in (19.46), he obtained the relations

$$\frac{1}{2}a = A - B\left(\frac{\pi^2}{3!} - \frac{1}{1^2}\right) + C\left(\frac{\pi^4}{5!} - \frac{1}{1^2}\frac{\pi^2}{3!} + \frac{1}{1^4}\right)$$

$$- D\left(\frac{\pi^6}{7!} - \frac{1}{1^2}\frac{\pi^4}{5!} + \frac{1}{1^4}\frac{\pi^2}{3!} - \frac{1}{1^6}\right) + \cdots,$$

$$-\frac{1}{2}2b = A - B\left(\frac{\pi^2}{3!} - \frac{1}{2^2}\right) + C\left(\frac{\pi^4}{5!} - \frac{1}{2^2}\frac{\pi^2}{3!} + \frac{1}{2^4}\right) \tag{19.52}$$

$$- D\left(\frac{\pi^6}{7!} - \frac{1}{2^2}\frac{\pi^4}{5!} + \frac{1}{2^4}\frac{\pi^2}{3!} - \frac{1}{2^6}\right) + \cdots,$$

$$\frac{1}{2}3c = A - B\left(\frac{\pi^2}{3!} - \frac{1}{3^2}\right) + C\left(\frac{\pi^4}{5!} - \frac{1}{3^2}\frac{\pi^2}{3!} + \frac{1}{3^4}\right) - \cdots,$$

etc. Now recall that $A = \phi'(0)$, $-B = \phi'''(0)$, $C = \phi^{(5)}(0)$, ...; one may use the expressions for $a, b, c, \ldots$ in equation (19.42) to get

$$\frac{1}{2}\phi(x) = \sin x \left\{\phi'(0) + \phi'''(0)\left(\frac{\pi^2}{3!} - \frac{1}{1^2}\right) + \phi^{(5)}(0)\left(\frac{\pi^4}{5!} - \frac{1}{1^2}\frac{\pi^2}{3!} + \frac{1}{1^4}\right) + \cdots\right\}$$

$$- \frac{1}{2}\sin 2x \left\{\phi'(0) + \phi'''(0)\left(\frac{\pi^2}{3!} - \frac{1}{2^2}\right) + \phi^{(5)}(0)\left(\frac{\pi^4}{5!} - \frac{1}{2^2}\frac{\pi^2}{3!} + \frac{1}{2^4}\right) + \cdots\right\}$$

$$+ \frac{1}{3}\sin 3x \left\{\phi'(0) + \phi'''(0)\left(\frac{\pi^2}{3!} - \frac{1}{3^2}\right) + \phi^{(5)}(0)\left(\frac{\pi^4}{5!} - \frac{1}{3^2}\frac{\pi^2}{3!} + \frac{1}{3^4}\right) + \cdots\right\}$$

.......

Fourier noted that the expression in the first set of chain brackets was the Maclaurin series for

$$\frac{1}{\pi}\left\{\phi(\pi) - \frac{1}{1^2}\phi''(\pi) + \frac{1}{1^4}\phi^{(4)}(\pi) - \frac{1}{1^6}\phi^{(6)}(\pi) + \cdots\right\}.$$

Similarly, the expressions in the second and third brackets were

$$\frac{1}{\pi}\left\{\phi(\pi) - \frac{1}{2^2}\phi''(\pi) + \frac{1}{2^4}\phi^{(4)}(\pi) - \frac{1}{2^6}\phi^{(6)}(\pi) + \cdots\right\};$$

$$\frac{1}{\pi}\left\{\phi(\pi) - \frac{1}{3^2}\phi''(\pi) + \frac{1}{3^4}\phi^{(4)}(\pi) - \frac{1}{3^6}\phi^{(6)}(\pi) + \cdots\right\}.$$

To sum the expressions in chain brackets, Fourier observed that

$$s(x) = \phi(x) - \frac{1}{m^2}\phi''(x) + \frac{1}{m^4}\phi^{(4)}(x) - \cdots$$

satisfied the differential equation

$$\frac{1}{m^2}s''(x) + s(x) = \phi(x).$$

He noted that the general solution of this differential equation was

$$s(x) = C_1 \cos mx + C_2 \sin mx + m \sin mx \int_0^x \phi(t) \cos mt\, dt$$
$$- m \cos mx \int_0^x \phi(t) \sin mt\, dt.$$

Because $\phi(x)$ was an odd function, its even-order derivatives were also odd functions, making $s(x)$ an odd function. This meant that $C_1 = 0$. Hence,

$$s(\pi) = (-1)^{m+1}m \int_0^\pi \phi(t) \sin mt\, dt,$$

and this in turn implied

$$a_m = \frac{2}{\pi}\int_0^\pi \phi(t) \sin mt\, dt.$$

Thus, Fourier found the "Fourier" coefficients.

## 19.6 Dirichlet's Proof of Fourier's Theorem

Fourier's work clearly demonstrated the tremendous significance of trigonometric series in the study of heat conduction and more generally in solving partial differential equations with boundary conditions. As we have seen, Fourier offered many arguments for the validity of his methods. But when the work of Gauss, Cauchy, and Abel on convergence of series became known in the 1820s, Fourier's methods were perceived to be nonrigorous. Lejeune Dirichlet (1805–1859) studied in France with Fourier and Poisson, who introduced him to problems in mathematical physics. At the same time, Dirichlet became familiar with the latest ideas on the rigorous treatment of infinite power series. Dirichlet's first great achievement was to treat infinite trigonometric series with equal rigor, thereby vindicating Fourier, who had befriended him.

In 1829, Dirichlet published his famous paper on Fourier series, "Sur la convergence des séries trigonométriques qui servent a représenter une fonction arbitraire entre des limites données"[25] in the newly founded *Crelle's Journal*. Eight years later he published the same paper in the Berlin Academy journal with further computational details and a more careful analysis of convergence. We follow the 1829 paper, whose title indicates that Dirichlet's aim was to obtain conditions on an arbitrary function so that the corresponding Fourier series would converge to the function. Dirichlet started his paper by observing that Fourier began a new era in analysis by applying trigonometric series in his researches on heat. However, he noted that only one paper, published by Cauchy in 1823, had discussed the validity of this method. Dirichlet noted, moreover, that the results of Cauchy's paper were inconclusive because they were based on the false premise that if the series with $n$th term $v_n = \frac{A \sin nx}{n}$ converged, the series $\sum u_n$ also converged when $\frac{u_n}{v_n}$ had 1 as a limit. Dirichlet produced examples of two series with $n$th terms

$$\frac{(-1)^n}{\sqrt{n}} \quad \text{and} \quad \left(1 + \frac{(-1)^n}{\sqrt{n}}\right)\frac{(-1)^n}{\sqrt{n}}.$$

Dirichlet pointed out that the ratio of the $n$th terms approached 1 as $n$ tended to infinity, but the first series converged and the second diverged.

Now in article 235 of his book on heat, Fourier gave the formula

$$\phi(x) = \frac{1}{\pi} \int \phi(\alpha)\, d\alpha \left(\frac{1}{2} + \sum \cos i(x - \alpha)\right). \tag{19.53}$$

Briefly, Fourier arrived at (19.53) by starting with his result

$$\phi(x) = \frac{1}{2}a_0 + a_1 \cos x + a_2 \cos 2x + \cdots + b_1 \sin x + b_2 \sin 2x + \cdots, \tag{19.54}$$

where

$$a_i = \frac{1}{\pi} \int_{-\pi}^{\pi} f(\alpha) \cos i\alpha\, d\alpha, \quad b_i = \frac{1}{\pi} \int_{-\pi}^{\pi} f(\alpha) \sin i\alpha\, d\alpha.$$

Observing that

$$a_i \cos ix + b_i \sin ix = \frac{1}{\pi} \int_{-\pi}^{\pi} f(\alpha)\left(\cos ix \cos i\alpha + \sin ix \sin i\alpha\right)d\alpha$$

$$= \frac{1}{\pi} \int_{-\pi}^{\pi} f(\alpha) \cos i(x - \alpha)d\alpha, \tag{19.55}$$

---

[25] Dirichlet (1829b).

Fourier substituted (19.55) into (19.54) to obtain (19.53).

Dirichlet analyzed the partial sums of this series under the assumption that the function $f(x)$ was piecewise monotonic. Taking $n + 1$ terms of the series and using

$$\frac{1}{2} + \cos(\alpha - x) + \cos 2(\alpha - x) + \cdots + \cos n(\alpha - x) = \frac{\sin\left(n + \frac{1}{2}\right)(\alpha - x)}{2\sin\frac{1}{2}(\alpha - x)},$$

Dirichlet represented the partial sum by

$$s_n(x) = \frac{1}{\pi}\int_{-\pi}^{\pi}\phi(\alpha)\frac{\sin\left(n + \frac{1}{2}\right)(\alpha - x)}{2\sin\frac{1}{2}(\alpha - x)}\,d\alpha.$$

He proved that this integral converged to

$$\frac{f(x+0) + f(x-0)}{2},$$

when $f$ satisfied certain conditions. For this purpose, he first demonstrated the theorem: For any function $f(\beta)$, continuous and monotonic in the interval $(g, h)$, where $0 \le g < h \le \frac{\pi}{2}$, the integral (for $0 \le g < h$)

$$\int_{g}^{h} f(\beta)\frac{\sin i\beta}{\sin\beta}\,d\beta$$

converges to a limit as $i$ tends to infinity. The limit is zero except when $g = 0$, in which case the limit is $\frac{\pi}{2}f(0)$. We present Dirichlet's argument in a slightly condensed form, for the most part using his notation. First note that we can write

$$\int_{0}^{\infty}\frac{\sin x}{x}\,dx = \frac{\pi}{2} \qquad (19.56)$$

as

$$\int_{0}^{\pi} + \int_{\pi}^{2\pi} + \cdots + \int_{n\pi}^{(n+1)\pi} + \cdots \frac{\sin x}{x}\,dx = \frac{\pi}{2}.$$

Note that the integral (19.56) was evaluated by Euler in a paper of 1781. See Exercise 1 in Chapter 17. Since $\sin x$ changes signs in the successive intervals $[0, \pi]$, $[\pi, 2\pi]$, ... and the integrand is decreasing, we can write the sum as

$$k_1 - k_2 + k_3 - \cdots + (-1)^{n-1}k_n + \cdots = \frac{\pi}{2}, \qquad (19.57)$$

$$\text{where} \quad k_{n+1} = \left|\int_{n\pi}^{(n+1)\pi}\frac{\sin x}{x}\,dx\right|.$$

The series converges and hence $k_n \to 0$ as $n \to \infty$. Now consider the integral

$$I = \int_{0}^{h}\frac{\sin i\beta}{\sin\beta}f(\beta)\,d\beta,$$

where $f(\beta)$ is decreasing and positive. Divide the interval $[0, h]$ by the points

$$0 < \frac{\pi}{i} < \frac{2\pi}{i} < \cdots < \frac{r\pi}{i} < h,$$

where $r$ is the largest integer for which the last inequality holds. $I$ is the sum of the integrals on these $r + 1$ subintervals. On comparing two of the consecutive subintervals, we see that for $v < r$

$$\left| I_v = \int_{(v-1)\frac{\pi}{i}}^{\frac{v\pi}{i}} \frac{\sin i\beta}{\sin \beta} f(\beta) \, d\beta \right| \geq \left| I_{v+1} = \int_{\frac{v\pi}{i}}^{\frac{(v+1)\pi}{i}} \frac{\sin i\beta}{\sin \beta} f(\beta) \, d\beta \right|.$$

Verify this by changing $\beta$ to $\frac{\pi}{i} + \beta$, so that the second integral can be written as

$$-\int_{(v-1)\frac{\pi}{i}}^{\frac{v\pi}{i}} \frac{\sin i\beta}{\sin (\beta + \frac{\pi}{i})} f\left(\beta + \frac{\pi}{i}\right) d\beta.$$

Also, $f(\beta)$ is decreasing so that

$$\frac{f(\beta)}{\sin \beta} > \frac{f\left(\beta + \frac{\pi}{i}\right)}{\sin \left(\beta + \frac{\pi}{i}\right)}.$$

Thus,

$$I = I_1 - I_2 + I_3 - I_4 + \cdots \pm I_r \mp I_h,$$

where $I_h$ is defined over the interval $(\frac{r\pi}{i}, h)$ so that the $I_j$ are positive and decreasing right up to the last term $I_h$. Next, let

$$K_v = \left| \int_{(v-1)\frac{\pi}{i}}^{\frac{v\pi}{i}} \frac{\sin i\beta}{\sin \beta} \, d\beta \right|$$

$$= \left| \int_{(v-1)\pi}^{v\pi} \frac{\sin \gamma}{i \sin (\frac{\gamma}{i})} \, d\gamma \right|.$$

Observe that the last integral is obtained by the change of variables $\gamma = i\beta$. As $i \to \infty$, this integral tends to $\int_{(v-1)\pi}^{v\pi} \frac{\sin \gamma}{\gamma} \, d\gamma = k_v$. Next fix a number $m$, assumed for convenience to be even, and let $r$ be greater than $m$. Let $\rho_v$ be such that

$$f\left(\frac{(v-1)\pi}{i}\right) \leq \rho_v \leq f\left(\frac{v\pi}{i}\right) \quad \text{and} \quad I_v = \rho_v K_v.$$

Then

$$I = (K_1\rho_1 - K_2\rho_2 + K_3\rho_3 - \cdots - K_m\rho_m)$$
$$+ (K_{m+1}\rho_{m+1} - K_{m+2}\rho_{m+2} + \cdots) = I(m) + I',$$

where $I(m)$ consists of the $m$ terms inside the first set of parentheses and $I'$ represents the remaining terms, inside the second set of parentheses. Therefore, the sum $I(m)$ as $i \to \infty$ converges to

$$f(0)(k_1 - k_2 + k_3 - \cdots - k_m) = s_m f(0).$$

This means that the sums $I(m)$ and $s_m f(0)$ can be made less than a positive number $w$ no matter how small. The sum $I'$ is an alternating series with decreasing terms and hence is less than $K_{m+1} \rho_{m+1}$; note that this converges to $k_{m+1} f(0)$. Thus, by (19.57), $|I'| < k_{m+1} |f(0)| + w'$, where $w'$ can be made arbitrarily small. Moreover,

$$\left| \frac{\pi}{2} - s_m \right| < k_{m+1} \quad \text{and so} \quad \left| I - \frac{\pi}{2} f(0) \right| < w + w' + 2f(0)k_{m+1}. \tag{19.58}$$

This proves the theorem for $f$ positive and $g = 0$. If $g > 0$, then

$$\int_g^h \frac{\sin i\beta}{\sin \beta} f\beta \, d\beta = \int_0^h \frac{\sin i\beta}{\sin \beta} f\beta \, d\beta - \int_0^g \frac{\sin i\beta}{\sin \beta} f\beta \, d\beta.$$

At this point, one may conclude that both these integrals tend to $\frac{\pi}{2} f(0)$ as $i \to \infty$. So $I_g \to 0$ as $i \to \infty$. This proves the theorem for positive decreasing $f$. If $f$ also assumes negative values, then choose a constant $C$ large enough that $C + f$ is positive. If $f$ is increasing, $-f$ is decreasing, taking care of that case, and the theorem is proved.

Dirichlet noted that if $f$ was discontinuous at 0, then by the previous argument, $f(0)$ could be replaced by $f(\epsilon)$ where $\epsilon$ was an infinitely small positive number. In his 1837 paper, he denoted $f(x + \epsilon)$ by $f(x + 0)$, the right-hand limit of $f(t)$ as $t \to x$. This is now standard notation.

To prove Fourier's theorem, break up the integral for $s_n(x)$ into two parts, one taken from $-\pi$ to $x$ and the other from $x$ to $\pi$. If $\alpha$ is replaced by $x - 2\beta$ in the first integral and by $x + 2\beta$ in the second, then we have

$$s_n(x) = \int_0^{\frac{\pi+x}{2}} \frac{\sin(2n+1)\beta}{\sin \beta} \phi(x - 2\beta) \, d\beta + \int_0^{\frac{\pi-x}{2}} \frac{\sin(2n+1)\beta}{\sin \beta} \phi(x + 2\beta) \, d\beta.$$

Suppose $x \neq -\pi$ or $\pi$ and $\beta - x < \pi$. The function $\phi(x+2\beta)$ in the second integral may be discontinuous at several points between $\beta = 0$ and $\beta = \frac{\pi - x}{2}$, and it may also have several external points in this interval. Denote these points by $l, l', l'', \ldots, l^v$ in ascending order, and decompose the second integral over the intervals $(0, l), (l, l'), \ldots$. By the theorem, the first of these $v+1$ integrals has the limit $\phi(x+\epsilon)\frac{\pi}{2}$ (i.e., $\phi(x+0)\frac{\pi}{2}$) and the others have the limit zero as $n \to \infty$. If in the first integral for $s_n(x)$ we have $\beta + x \geq \pi$, then write it as

$$\int_0^{\frac{\pi}{2}} \frac{\sin(2n+1)\beta}{\sin \beta} \phi(x - 2\beta) \, d\beta + \int_{\frac{\pi}{2}}^{\frac{\beta+x}{2}} \frac{\sin(2n+1)\beta}{\sin \beta} \phi(x - 2\beta) \, d\beta.$$

The first integral tends to $\phi(x - \epsilon)\frac{\pi}{2}$ as $n \to \infty$. A similar argument shows that

$$\int_0^{\frac{\pi+x}{2}} \frac{\sin(2n+1)\beta}{\sin \beta} \phi(x - 2\beta)\, d\beta \quad \text{tends to} \quad \phi(x+\epsilon)\frac{\pi}{2}.$$

### 19.7   Dirichlet: On the Evaluation of Gauss Sums

In 1835, Dirichlet presented a paper to the Berlin Academy[26] explaining how the definite integral $\int_{-\infty}^{\infty} e^{ix^2}\, dx$ could be applied to evaluate the Gauss sum $\sum_{k=0}^{n-1} e^{\frac{2\pi i k^2}{n}}$. He gave a slightly simpler version of this proof in his famous 1840[27] paper in *Crelle's Journal*, on the applications of infinitesimal analysis to the theory of numbers. In this paper, Dirichlet also derived the class number formula for quadratic forms and proved his well-known theorem on primes in arithmetic progression.

He started his evaluation of the Gauss sum by first proving a finite form of the Poisson summation formula. Note that in 1826, Poisson had used such a finite formula to deduce the Euler–Maclaurin summation. Dirichlet began with a continuous function $g(x)$ in $[0, \pi]$ expandable as a Fourier series:

$$\pi g(x) = c_0 + 2 \sum_{s=1}^{\infty} c_s \cos sx, \tag{19.59}$$

where

$$c_s = \int_0^{\pi} g(x) \cos sx\, dx. \tag{19.60}$$

It followed for $x = 0$ that

$$c_0 + 2 \sum_{s=1}^{\infty} c_s = \pi g(0). \tag{19.61}$$

He then set

$$g(x) = f(x) + f(2\pi - x) + f(2\pi + x) + \cdots + f(2(h-1)\pi + x) + f(2h\pi - x), \tag{19.62}$$

where $f(x)$ was continuous on $[0, 2h\pi]$. He observed that

$$c_s = \int_0^{\pi} g(x) \cos sx\, dx = \int_0^{2h\pi} f(x) \cos sx\, dx. \tag{19.63}$$

By using the value of $g(0)$ from (19.62), he could rewrite (19.61) in the form

[26] Dirichlet (1969) vol. 1, pp. 237–256.
[27] ibid. pp. 410–496, especially pp. 473–479.

$$c_0 + 2 \sum_{s=1}^{\infty} c_s = \pi \left( f(0) + f(2h\pi) + 2 \sum_{s=1}^{h-1} f(2s\pi) \right), \qquad (19.64)$$

where $c_s$ was given by (19.63). This was the finite form of the Poisson summation formula employed by Dirichlet. He then considered the integral

$$\int_{-\infty}^{\infty} \cos x^2 \, dx = a,$$

where $a$ was some number to be determined. Since Euler had evaluated this integral in 1781, Dirichlet knew its exact value. See Exercise 1 in Chapter 17. Note here that Dirichlet's method of evaluating Gauss sums was such that it also determined the value of $a$. He set

$$x = \frac{z}{2} \sqrt{\frac{n}{2\pi}},$$

where $n$ was a positive integer divisible by 4, transforming the integral to

$$\int_{-\infty}^{\infty} \cos \left( \frac{n}{8\pi} z^2 \right) dz = 2a \sqrt{\frac{2\pi}{n}}. \qquad (19.65)$$

Dirichlet then rewrote the last integral as a sum:

$$\sum_{s=-\infty}^{\infty} \int_{2s\pi}^{2(s+1)\pi} \cos \left( \frac{n}{8\pi} z^2 \right) dz = \sum_{s=-\infty}^{\infty} \int_{0}^{2\pi} \cos \frac{n}{8\pi} (2s\pi + z)^2 \, dz. \qquad (19.66)$$

He observed that since $n$ was divisible by 4,

$$\cos \frac{n}{8\pi} (2s\pi + z)^2 = \cos \frac{n}{8\pi} (4s^2\pi^2 + 4s\pi z + z^2) = \cos \left( \frac{snz}{2} + \frac{n}{8\pi} z^2 \right).$$

Then, by the addition formula, he had

$$\cos \left( \frac{snz}{2} + \frac{n}{8\pi} z^2 \right) + \cos \left( -\frac{snz}{2} + \frac{n}{8\pi} z^2 \right) = 2 \cos \left( \frac{snz}{2} \right) \cos \left( \frac{n}{8\pi} z^2 \right).$$

Hence, (19.65) and (19.66) could be expressed as

$$\int_{0}^{2\pi} \cos \left( \frac{n}{8\pi} z^2 \right) dz + 2 \sum_{s=1}^{\infty} \int_{0}^{2\pi} \cos \left( \frac{n}{8\pi} z^2 \right) \cos \left( s \frac{nz}{2} \right) dz = 2a \sqrt{\frac{2\pi}{n}}.$$

Dirichlet substituted $nz = 2x$ in this formula to obtain

$$\int_{0}^{n\pi} \cos \left( \frac{x^2}{2n\pi} \right) dx + 2 \sum_{s=1}^{\infty} \int_{0}^{n\pi} \cos \left( \frac{x^2}{2n\pi} \right) \cos sx \, dx = a \sqrt{2n\pi}. \qquad (19.67)$$

Since $n$ was an even number expressible as $2h$, the sum on the left-hand side coincided with the sum on the left-hand side of (19.64) when $f(x) = \cos(\frac{x^2}{2n\pi})$. By combining (19.64) and (19.67), Dirichlet arrived at the formula

$$\cos 0 + \cos\left(\frac{n}{2}\right)^2 \frac{2\pi}{n} + 2\sum_{s=1}^{\frac{n}{2}-1}\cos s^2\frac{2\pi}{n} = a\sqrt{\frac{2n}{\pi}}. \qquad (19.68)$$

He then observed that

$$\cos s^2\frac{2\pi}{n} = \cos(n-s)^2\frac{2\pi}{n},$$

and therefore (19.68) could be expressed in the simpler form

$$\sum_{s=0}^{n-1}\cos s^2\frac{2\pi}{n} = a\sqrt{\frac{2n}{\pi}}. \qquad (19.69)$$

Dirichlet next remarked that the value of $a$ was independent of $n$ so that by choosing $n = 4$, he could write

$$2 = 2a\sqrt{\frac{2}{\pi}}, \quad \text{or} \quad a = \sqrt{\frac{\pi}{2}},$$

and therefore he could express the Gauss sum as

$$\sum_{s=0}^{n-1}\cos\frac{s^2 2\pi}{n} = \sqrt{n}. \qquad (19.70)$$

Operating in the same manner with $\int_{-\infty}^{\infty}\sin x^2\,dx$, he arrived at

$$\sum_{s=0}^{n-1}\sin\frac{s^2 2\pi}{n} = \sqrt{n}. \qquad (19.71)$$

Dirichlet pointed out that the sums (19.70) and (19.71) could be similarly evaluated for $n$ of the form $4\mu+1$, $4\mu+2$, and $4\mu+3$. However, it was possible to obtain these sums in a different way. For that purpose he defined, for positive integers $m$ and $n$,

$$\sum_{s=0}^{n-1}e^{\frac{2ms^2\pi i}{n}} = \phi(m,n).$$

He then wrote

$$\phi(m,n) = \phi(m',n) \quad \text{when } m \equiv m'(\text{mod } n); \qquad (19.72)$$

$$\phi(m,n) = \phi(c^2m,n) \quad \text{when } c \text{ was prime to } n; \qquad (19.73)$$

$$\phi(m,n)\phi(n,m) = \phi(1,mn) \quad \text{when } m \text{ and } n \text{ were coprime.} \qquad (19.74)$$

Dirichlet proved (19.74) by observing that

$$\phi(m,n)\phi(n,m) = \sum_{s=0}^{n-1}\sum_{t=0}^{m-1} e^{\frac{2ms^2\pi i}{n}} e^{\frac{2nt^2\pi i}{m}}$$

$$= \sum_{s=0}^{n-1}\sum_{t=0}^{m-1} e^{\frac{(m^2s^2+n^2t^2)2\pi i}{mn}}$$

$$= \sum_{s=0}^{n-1}\sum_{t=0}^{m-1} e^{\frac{(ms+nt)^2 2\pi i}{mn}}.$$

Since $m$ and $n$ were chosen relatively prime, Dirichlet argued that $ms+nt$ assumed all the residues (mod $mn$) as $s$ and $t$ ranged over the values $0, 1\ldots, n-1$ and $0, 1, \ldots, m-1$, respectively. Therefore,

$$\phi(m,n)\phi(n,m) = \sum_{s=0}^{mn-1} e^{\frac{2\pi s^2 i}{mn}}, \tag{19.75}$$

and Dirichlet's proof of (19.74) was complete. Note that Gauss gave a similar argument in 1801, though it was published later.[28] Dirichlet then observed that for $n$, a multiple of 4, (19.70) and (19.71) implied

$$\phi(1,n) = (1+i)\sqrt{n}. \tag{19.76}$$

And for odd $n$, (19.74) and (19.76) gave

$$\phi(4,n)\phi(n,4) = \phi(1,4n) = 2(1+i)\sqrt{n}. \tag{19.77}$$

Moreover, by (19.73), $\phi(4,n) = \phi(1,n)$ when $n$ was odd, and by (19.72) $\phi(n,4) = \phi(1,4)$ or $\phi(3,4)$, depending on whether $n = 4\mu + 1$ or $n = 4\mu + 3$. Since

$$\phi(1,4) = 2(1+i), \quad \text{and} \quad \phi(3,4) = 2(1-i),$$

Dirichlet could conclude that

$$\phi(1,n) = \sqrt{n}, \ n = 4\mu + 1; \quad \phi(1,n) = i\sqrt{n}, \ n = 4\mu + 3. \tag{19.78}$$

Finally, when $n = 4\mu + 2$, he argued that $\frac{n}{2}$ and 2 were relatively prime, so that by (19.74)

$$\phi\left(2,\frac{n}{2}\right)\phi\left(\frac{n}{2},2\right) = \phi(1,n) \quad \text{and} \quad \phi\left(\frac{n}{2},2\right) = \phi(1,2) = 0.$$

---

[28] Gauss (1863–1927) vol. 2, pp. 11–45.

Thus,

$$\phi(1,n) = 0, \quad n = 4\mu + 2.$$

Gauss gave a proof of the quadratic reciprocity theorem by using the values of the Gauss sums. Dirichlet repeated these arguments in his papers, though in a simpler form. In fact, in his papers and lectures Dirichlet presented many number theoretic ideas of Gauss within an easily understandable approach. For example, he published a one-page proof of a theorem of Gauss on the biquadratic character of 2. It is interesting to note that the British number theorist, H. J. S. Smith (1826–1883), presented this result in the first part of his report on number theory published in 1859; he wrote in a footnote:[29]

> The death of this eminent geometer in the present year (May 5, 1859) is an irreparable loss to the science of arithmetic. His original investigations have probably contributed more to its advancement than those of any other writer since the time of Gauss; if, at least, we estimate results rather by their importance than by their number. He has also applied himself (in several of his memoirs) to give an elementary character to arithmetical theories which, as they appear in the work of Gauss, are tedious and obscure; and he has thus done much to *popularize* the theory of numbers among mathematicians – a service which it is impossible to appreciate too highly.

Noting Smith's remark on the importance, rather than the number, of Dirichlet's results, we observe that Gauss made a similar comment when he recommended Dirichlet for the order *pour le mérite* in 1845:[30] "The same [Dirichlet] has – as far as I know – not yet published a big work, and also his individual memoirs do not yet comprise a big volume. But they are jewels, and one does not weigh jewels on a grocer's scales."

## 19.8   Schaar: Reciprocity of Gauss Sums

In 1850, the Belgian mathematician Mathias Schaar (1817–1867) derived a remarkable reciprocity formula for Gauss sums:[31]

$$\sqrt{\frac{q}{2p}}\, e^{\frac{\pi i}{4}} \sum_{k=0}^{2p-1} e^{-\frac{\pi q k^2 i}{2p}} = \sum_{k=0}^{q-1} e^{\frac{2\pi p k^2 i}{q}}, \tag{19.79}$$

where $2p$ and $q$ are relatively prime integers. This formula contains the value of the Gauss sum (19.76) and also implies the law of quadratic reciprocity. We present a streamlined version of Schaar's argument. In this context it is interesting to observe that in his evaluation of the Gauss sum, instead of working with

$$f(x) = \cos\frac{x^2}{2n\pi} \quad \text{and} \quad f(x) = \sin\frac{x^2}{2n\pi}$$

[29]  Smith (1965b) p. 72.
[30]  Duke and Tschinkel (2005) p. 18.
[31]  Schaar (1850).

separately, Dirichlet could have used the function

$$f(x) = e^{\frac{ix^2}{2n\pi}} \quad \text{or, indeed, the function} \quad f(x) = e^{\frac{2\pi i x^2}{n}}.$$

The right-hand side of (19.79) suggests that one should consider the function $e^{\frac{2\pi i x^2}{n}}$. Thus, instead of (19.59) and (19.60), take a continuous function $g(x)$ on $[0,1]$ to find that

$$g(x) = \sum_{k=-\infty}^{\infty} c_k e^{-2\pi ikx}, \quad \text{where} \quad c_k = \int_0^1 g(x)e^{2\pi ikx} \, dx;$$

when $x = 0$,

$$g(0) = \sum_{k=-\infty}^{\infty} c_k.$$

Now if we take $f(x)$ continuous on $[0,q]$, and $g(x) = f(x) + f(x+1) + \cdots + f(x+q-1)$, then

$$c_k = \int_0^1 g(x)e^{2\pi ikx} \, dx = \int_0^q f(x)e^{2\pi ikx} \, dx.$$

Moreover, (19.64) gives us

$$\sum_{k=-\infty}^{\infty} \int_0^q f(x)e^{2\pi ikx} \, dx = \sum_{s=0}^{q-1} f(s).$$

Taking $f(x) = e^{\frac{2\pi i p x^2}{q}}$, we then get

$$S \equiv \sum_{k=0}^{q-1} e^{\frac{2\pi i p k^2}{q}}$$

$$= \sum_{v=-\infty}^{\infty} \int_0^q e^{\frac{2\pi i p x^2}{q} + 2\pi i v x} \, dx$$

$$= q \sum_{v=-\infty}^{\infty} \int_0^1 e^{2\pi i p q t^2 + 2\pi i v q t} \, dt$$

$$= q \sum_{v=-\infty}^{\infty} \int_0^1 e^{2\pi i p q (t^2 + \frac{vt}{p})} \, dt$$

$$= q \sum_{v=-\infty}^{\infty} e^{-\frac{\pi i q v^2}{2p}} \int_0^1 e^{2\pi i p q (t + \frac{v}{2p})^2} \, dt. \qquad (19.80)$$

Applying the change of variables $s = t + \frac{v}{2p}$ in (19.80), we see that

$$S = q \sum_{v=-\infty}^{\infty} e^{-\frac{\pi i q v^2}{2p}} \int_{\frac{v}{2p}}^{1+\frac{v}{2p}} e^{2\pi i p q s^2} \, ds. \tag{19.81}$$

Schaar next broke up the sum in (19.81) into $2p$ sums by summing over $v \equiv k \bmod 2p, k = 0, 1, 2, \ldots, 2p - 1$ to obtain

$$S = q \sum_{v=-\infty}^{\infty} \sum_{k=0}^{2p-1} e^{-\frac{\pi i q (2vp+k)^2}{2p}} \int_{v+\frac{k}{2p}}^{v+1+\frac{k}{2p}} e^{2\pi i p q s^2} \, ds$$

$$= q \sum_{k=0}^{2p-1} e^{-\frac{\pi i q k^2}{2p}} \sum_{v=-\infty}^{\infty} \int_{v+\frac{k}{2p}}^{v+1+\frac{k}{2p}} e^{2\pi i p q s^2} \, ds$$

$$= q \sum_{k=0}^{2p-1} e^{-\frac{\pi i q k^2}{2p}} \int_{-\infty}^{\infty} e^{2\pi i p q s^2} \, ds. \tag{19.82}$$

Finally, we see that since

$$\int_{-\infty}^{\infty} e^{2\pi i p q s^2} \, ds = \frac{e^{\frac{\pi i}{4}}}{\sqrt{2pq}},$$

(19.82) implies

$$\sum_{k=0}^{q-1} e^{\frac{2\pi i p k^2}{q}} = S = \sqrt{\frac{q}{2p}} \, e^{\frac{i\pi}{4}} \sum_{k=0}^{2p-1} e^{-\frac{\pi i q k^2}{2p}},$$

completing our summary of Schaar's proof of the reciprocity formula for Gauss sums.

## 19.9   Exercises

(1) Solve the equation $\frac{d^2v}{dx^2} + \frac{d^2v}{dy^2} = 0$ by assuming $v = F(x)f(y)$ to obtain $F(x) = e^{-mx}$, $f(y) = \cos my$. Let $v = \phi(x, y)$ and assume the boundary conditions $\phi\left(x, \pm \frac{\pi}{2}\right) = 0$ and $\phi(0, y) = 1$. Show that

$$\phi(x, y) = \frac{4}{\pi}\left(e^{-x} \cos y - \frac{1}{3} e^{-3x} \cos 3y + \frac{1}{5} e^{-5x} \cos 5y - \cdots\right).$$

See Fourier (1955) pp. 134–144.

(2) Show that $\frac{\pi}{2} = \arctan u + \arctan\left(\frac{1}{u}\right)$. Let $u = e^{ix}$ and expand $\arctan u$ and $\arctan\left(\frac{1}{u}\right)$ as series to obtain

$$\frac{\pi}{4} = \cos x - \frac{1}{3} \cos 3x + \frac{1}{5} \cos 5x - \cdots.$$

See Fourier (1955) p. 154.

(3) Let $a$ denote a quadratic residue modulo $p$, a prime, and let $b$ denote a quadratic nonresidue. Show that

$$\phi(1, p) = 1 + 2 \sum e^{\frac{a2\pi i}{p}} = i^{\frac{(p-1)^2}{4}} \sqrt{p},$$

where the sum is over all the residues $a$. Show also that

$$\phi(m, p) = \left(\frac{m}{p}\right) \phi(1, p) = 1 + 2 \sum e^{\frac{a2m\pi i}{p}},$$

where $\left(\frac{m}{p}\right)$ denotes the Legendre symbol. Deduce that

$$\sum e^{\frac{2m\pi i a}{p}} - \sum e^{\frac{2m\pi i b}{p}} = \left(\frac{m}{p}\right) i^{\frac{(p-1)^2}{4}} \sqrt{p}.$$

See Dirichlet (1969) pp. 478–479.

(4) Suppose $p$ and $q$ are primes. Use $\phi(p, q)\phi(q, p) = \phi(1, pq)$ and the results in the previous exercise to prove the law of quadratic reciprocity:

$$\left(\frac{p}{q}\right)\left(\frac{q}{p}\right) = (-1)^{\frac{p-1}{2} \cdot \frac{q-1}{2}}.$$

This proof originates with Gauss's 1808 paper, "Summatio Quarundam Serierum Singularium." For the derivation discussed in this exercise, see Dirichlet and Dedekind (1999) pp. 206–207.

## 19.10 Notes on the Literature

Truesdell (1960) is a detailed history of the mechanics of flexible or elastic bodies from 1638 to 1788. He includes a discussion of those aspects of the works of Euler, d'Alembert, D. Bernoulli, and Lagrange that led to the consideration of trigonometric series in such problems. For Wiener's treatment of Euler's difference equation (19.13), see Wiener (1979) vol. 2, pp. 443–453. Wiener's paper contributed to the effort to make operational calculus rigorous. Yushkevich (1971) and Bottazzini (1986) chapter 1 deal with the development of the concept of a function in connection with trigonometric series. Fourier's 1807 memoir on heat conduction was never published by the French Academy; Fourier's famous book of 1822 was a reworking of this memoir. However, Grattan-Guinness (1972) has helpfully presented us with the original 1807 memoir. For a mathematical biography of Dirichlet, see Merzbach (2018) and for a more brief biography, see Elstrodt (2005) in Duke and Tschinkel (2005).

# 20

## The Euler–Maclaurin Summation Formula

### 20.1  Preliminary Remarks

The Euler–Maclaurin summation formula is among the most useful and important formulas in all of mathematics, independently discovered by Euler and Maclaurin in the early 1730s.

The Euler–Maclaurin summation formula arose out of efforts to find approximate values for finite and infinite series. During the 1720s, the series $\zeta(2) = \sum_{n=1}^{\infty} \frac{1}{n^2}$ received a good deal of attention. Since the exact evaluation of this series appeared to be out of reach at that time, several mathematicians devised methods to compute approximations for this series. Stirling found some ingenious methods for transforming this and similar slowly convergent series to more rapidly convergent series. See, for example, (10.54) for one such method. In his *Methodus Differentialis* of 1730, Stirling computed $\zeta(2)$ by three different methods, one of which gave the correct value to sixteen decimal places. Around 1727, Daniel Bernoulli and Goldbach also showed a passing interest in the problem by computing $\zeta(2)$ to a few decimal places. This may have caused Euler, their colleague at the St. Petersburg Academy, to study this problem. In a paper of 1731,[1] Euler used integration to derive the formula

$$\sum_{n=1}^{\infty} \frac{1}{n^2} = \sum_{n=0}^{\infty} \frac{1}{2^n} \cdot \frac{1}{(n+1)^2} + (\ln 2)^2. \tag{20.1}$$

Euler derived (20.1), or (15.52), as an immediate consequence of (15.53), but his derivation of (15.53) was complicated. For this reason, in Chapter 15 we gave Abel's derivation.

Observe that the series on the right-hand side of (20.1) was evidently much more rapidly convergent than the original series for $\zeta(2)$, and Euler determined that $\zeta(2) \approx 1.644934$. But Euler had a result more general than (20.1), namely (15.53), that involved the dilogarithmic function defined by the series $\sum_{n=1}^{\infty} \frac{x^n}{n^2}$. Perhaps Euler's work on the dilogarithm led him to apply calculus to the problem of the summation of

---

[1] E20 § 22. Eu. I-14 pp. 25–41.

the general series $\sum_{k=1}^{n} f(k)$. The result was a paper he presented to the Academy in 1732[2] (published in 1738) in which he briefly mentioned the Euler–Maclaurin formula in the form

$$\sum_{k=1}^{n} t(k) = \int t\, dn + \alpha t + \beta \frac{dt}{dn} + \gamma \frac{d^2 t}{dn^2} + \delta \frac{d^3 t}{dn^3} + \cdots, \qquad (20.2)$$

where $\alpha, \beta, \gamma, \ldots$ were computed from the equations

$$\alpha = \frac{1}{2}, \ \beta = \frac{1}{1\cdot 2}\alpha - \frac{1}{1\cdot 2\cdot 3}, \ \gamma = \frac{1}{1\cdot 2}\beta - \frac{1}{1\cdot 2\cdot 3}\alpha + \frac{1}{1\cdot 2\cdot 3\cdot 4};$$

$$\delta = \frac{1}{1\cdot 2}\gamma - \frac{1}{1\cdot 2\cdot 3}\beta + \frac{1}{1\cdot 2\cdot 3\cdot 4}\alpha - \frac{1}{1\cdot 2\cdot 3\cdot 4\cdot 5}, \ldots.$$

He gave an application of (20.2) to the summation of the very simple example, $\sum_{k=1}^{n}(k^2 + 2k)$, and then proceeded to discuss other types of series. He wrote a longer paper[3] on the subject three years later, in 1735, in which he explicitly evaluated $\sum_{k=1}^{n} k^r$ as polynomials in $n$ for $r = 1, 2, \ldots, 16$. This should have alerted Euler to the fact that $\alpha, \beta, \gamma, \delta, \ldots$ were closely related to the Bernoulli numbers. Jakob Bernoulli defined his numbers in exactly this context, except that in his published work he gave the polynomials up to $r = 10$. Euler was perhaps not aware of Bernoulli's work on sums of powers of integers at this stage. By 1727, he had certainly studied at least portions of Bernoulli's *Ars Conjectandi*.[4] But it was only in 1755[5] that Euler, by then fully aware of Bernoulli's contributions, followed de Moivre in adopting the term Bernoulli numbers. In a highly interesting paper,[6] written in 1740, that we discuss in Chapter 2, Euler explained that the generating function for the numbers $\alpha, \beta, \gamma, \ldots$ was given by

$$S = 1 + \alpha z + \beta z^2 + \gamma z^3 + \cdots = \frac{1}{1 - \frac{z}{1\cdot 2} + \frac{z^2}{1\cdot 2\cdot 3} - \frac{z^3}{1\cdot 2\cdot 3\cdot 4} + \frac{z^4}{1\cdot 2\cdot 3\cdot 4\cdot 5} - \cdots}$$

$$= \frac{z e^z}{e^z - 1}. \qquad (20.3)$$

Euler also offered an explanation for the appearance of the Bernoulli numbers in two such very different situations: in the values of $\zeta(2n), n = 1, 2, 3, \ldots$ and in the Euler–Maclaurin formula. In rough terms, his explanation was that the generating function for both cases was the same. In his 1735 paper, he also computed $\zeta(n) = \sum_{k=1}^{\infty} \frac{1}{k^n}$ to fifteen decimal places for $n = 2, 3, 4$, making use of (20.2). The series on the right-hand side of (20.2) in these cases were asymptotic series and Euler manipulated them exactly as Stirling and de Moivre had done in a different context,

---

[2] E25 § 2. Eu. I-14 pp. 42–72.
[3] Eu. I-14 pp. 108–123. E 47 § 23.
[4] Calinger (2016) p. 24.
[5] Eu. I-10 p. 335. E 212 Part II, § 122.
[6] Eu. I-14 pp. 407–462. E 130, § 22.

using the first few terms of the asymptotic series, up to the point where the terms started getting large.

Maclaurin's results related to (20.2) appeared in his influential book *Treatise of Fluxions*. This work was published in 1742 in two volumes, although the first volume, containing the statements of the Euler–Maclaurin formula, was already typeset in 1737. Colin Maclaurin (1698–1746) studied at the University of Glasgow, Scotland, but the mathematician who had the greatest formative influence on him was Newton, whom he met in 1719. Much of Maclaurin's work on algebra, calculus, and dynamics arose directly from topics on which Newton had published results. Maclaurin was professor of mathematics at the University of Edinburgh from 1726 to 1746, having been recommended to the position by Newton. Maclaurin was probably inspired to discover the Euler–Maclaurin formula by the results of de Moivre and Stirling on the asymptotic series for $\sum_{k=1}^{n} \ln k$. Newton had a result on the sum $\sum_{k=1}^{n} \frac{1}{a+kb}$, giving the first terms of the Euler–Maclaurin formula for this particular case. Newton gave this result in a letter of July 20, 1671, to Collins,[7] but Maclaurin was most likely unaware of it.

It is a curious fact that Euler and Maclaurin learned of each other's works even before they were published. This was a result of the brief correspondence between Euler and Stirling. In June 1736, Euler wrote Stirling[8] about his formula and mentioned applications to the summation of $\sum_{k=1}^{\infty} \frac{1}{k^2}$ and $\sum_{k=1}^{n} \frac{1}{k}$. He wrote the latter result as

$$1 + \frac{1}{2} + \frac{1}{3} + \cdots + \frac{1}{x}$$

$$= C + \ln x + \frac{1}{2x} - \frac{1}{12x^2} + \frac{1}{120x^4} - \frac{1}{252x^6} + \frac{1}{240x^8} - \frac{1}{132x^{10}} + \frac{691}{32760x^{12}} - \text{etc.}$$
(20.4)

We can see that the value of $C$, Euler's constant $\gamma$, would be

$$\lim_{x \to \infty} \left( \sum_{k=1}^{x} \frac{1}{k} - \ln x \right),$$

and Euler gave this value as 0.5772156649015329 in his 1735 paper and in his letter. Note that the series on the right-hand side of (20.4), from the fourth term onward, can be written as

$$-\sum_{n=1}^{\infty} \frac{B_{2n}}{2n} \cdot \frac{1}{x^{2n}}.$$

Then in 1737, Stirling received from Maclaurin the galley proofs of some portions of the first volume of Maclaurin's treatise, containing two formulations of the Euler-Maclaurin formula. Because of some business preoccupations, Stirling did not reply

[7] Newton (1959–60) vol. 1, pp. 68–70.
[8] Tweddle (1988) pp. 141–144.

to Euler's letter until April 1738. He then informed Euler about Maclaurin's work and about his communications with Maclaurin on Euler's work. Stirling also told Euler that Maclaurin had promised to acknowledge Euler's work in his book. And indeed Maclaurin did so.[9] Concerning this point, Euler wrote a reply to Stirling on July 27, 1738:[10]

> But in this matter I have very little desire for anything to be detracted from the fame of the celebrated Mr Maclaurin since he probably came upon the same theorem for summing series before me, and consequently deserved to be named as its first discoverer. For I found that theorem about four years ago, at which time I also described its proof and application in greater detail to our Academy.

Unfortunately, Euler uncharacteristically forgot to mention Maclaurin in his differential calculus book of 1755, where he discussed this formula.

Maclaurin presented four formulas and he understood these to be variations of the same result.[11] In modern notation, two of these can be given as

$$\sum_{k=0}^{n-1} f(a+k) = \int_{a}^{a+n} f(x)\,dx + \frac{1}{2}(f(a) - f(a+n)) + \frac{1}{12}(f'(a) - f'(a+n))$$

$$- \frac{1}{720}(f'''(a) - f'''(a+n)) + \frac{1}{30240}(f^v(a) - f^v(a+n)) - \cdots ,$$
(20.5)

$$\sum_{k=0}^{n} f(a+k) = \int_{a-\frac{1}{2}}^{a-\frac{1}{2}+n} f(x)\,dx + \frac{1}{24}\left( f'\left(a - \frac{1}{2}\right) - f'\left(a - \frac{1}{2} + n\right)\right)$$

$$- \frac{7}{5760}\left( f'''\left(a - \frac{1}{2}\right) - f'''\left(a - \frac{1}{2} + n\right)\right)$$

$$+ \frac{31}{967680}\left( f^v\left(a - \frac{1}{2}\right) - f^v\left(a - \frac{1}{2} + n\right)\right) - \cdots . \quad (20.6)$$

Note that the coefficients appearing after the second term on the right-hand side of (20.5) are given by

$$\frac{B_2}{2!}, \frac{B_4}{4!}, \frac{B_6}{6!}, \cdots$$

and the coefficients appearing after the first term on the right-hand side of (20.6) are given by

$$\frac{(2-1)B_2}{2!\,2}, \frac{(2^3-1)B_4}{4!\,2^3}, \frac{(2^5-1)B_6}{6!\,2^5}, \cdots .$$

[9] Maclaurin (1742) p. 691, footnote.
[10] Tweddle (1988) p. 146.
[11] For the original statement of these formulas, see Maclaurin (1742) articles 352–353, pp. 292–293. For the proofs, see articles 828–832, pp. 672–677.

The remaining two formulas were for cases in which the series on the left-hand side were infinite, and where it was assumed that $f(x)$ and its derivatives tended to zero as $x \to \infty$. Maclaurin derived the de Moivre and Stirling forms of the approximations for $n!$ by taking $f(x) = \ln x$ in (20.5) and (20.6), respectively.[12] He also applied his results to obtain Jakob Bernoulli's formula for sums of powers of integers as well as approximations of $\zeta(n)$ for some values of $n$. In addition, he obtained some formulas for approximate integration, such as the three-eighths rule.

In 1772,[13] Lagrange gave a formal expression for the Taylor series as a basis for an interesting derivation of the Euler–Maclaurin summation formula and of some extensions involving sums of sums. Suggestive of important analytical applications, Lagrange's formula for the Taylor series was

$$f(x + h) = f(x) + hf'(x) + \frac{h^2}{2!} f''(x) + \cdots$$

$$= \left( 1 + hD + \frac{h^2 D^2}{2!} + \cdots \right) f(x) = e^{hD} f(x). \qquad (20.7)$$

Here $D$ represents the differential operator $\frac{d}{dx}$. We discuss this in detail in Chapter 21.

Clearly, Lagrange charted out a new approach with his algebraic conception of the derivative, and yet this algebraic perspective can be traced back to the work of Leibniz. Leibniz had been struck by the formal analogy between the differential operator and algebraic quantities; Lagrange implemented this idea by going a step further and identifying the derivative operator with an algebraic quantity. This formal method was used by some French and British mathematicians of the first half of the nineteenth century, leading to significant mathematical developments.

The eighteenth-century mathematicians made very effective use of the Euler–Maclaurin formula but did not seem too concerned about the reasons for this effectiveness, especially where divergent asymptotic series were involved. Gauss, with his interest in rigor, was the first mathematician to express the need for an investigation into this question. He did this in his 1813 paper on the gamma function and hypergeometric series[14] and again in 1816 in a paper on the fundamental theorem of algebra. Interestingly, the rigor needed for the careful discussion of the Euler–Maclaurin summation formula was provided by the French mathematical physicist, S. D. Poisson.

Siméon Denis Poisson (1781–1840) studied at the École Polytechnique in Paris where he came under the influence of Laplace and Lagrange. The latter lectured on analytic functions at the Polytechnique, where he introduced the remainder term for the Taylor series. In the 1820s, Cauchy lectured at the Polytechnique on the application of this remainder term to a rigorous discussion of the power series representation of functions. Poisson's contribution was to derive the remainder term for the Euler–Maclaurin series.[15] His motivation for this 1826 work was to explain an

[12] ibid. articles 838–854, pp. 678–692.
[13] Lagrange (1867–1892) vol. 3, pp. 441–476.
[14] Gauss (1813).
[15] Poisson (1826).

apparent paradox in Legendre's use of the Euler–Maclaurin formula to numerically evaluate the elliptic integral

$$\int_0^{\frac{\pi}{2}} \sqrt{1 - k^2 \sin^2 \theta} \, d\theta. \tag{20.8}$$

The sum on the left-hand side of (20.5), after a small modification to allow for non-integer division points of the interval $(a, a + n)$, can be used to approximate the integral on the right-hand side. Now the integrand in (20.8), $f(x) = \sqrt{1 - k^2 \sin^2 x}$, is such that its odd order derivatives vanish at 0 and $\frac{\pi}{2}$. Thus, the series on the right-hand side of (20.5) vanishes, since it involves only the odd order derivatives. This implies the absurd result that the sum on the left-hand side remains unchanged no matter how many division points are chosen in the interval. It was to resolve this paradox, rather than to explain the effectiveness of the asymptotic series, that Poisson developed the remainder term for the Euler–Maclaurin formula.

Another peculiar feature of Poisson's work was that he used Fourier series instead of Taylor series to find the remainder term. Poisson learned the technique of Fourier series from Fourier's long 1807 paper on heat conduction, on which Poisson wrote a brief summary in 1807.[16] From 1811 on, Poisson published several papers on Fourier series and was very familiar with its techniques. In particular, he applied the result now known as the Poisson summation formula (originally due to Cauchy) to a number of problems, including the present one. He obtained the remainder after $q$ terms as an integral whose integrand had the form

$$\left( \sum_{n=1}^{\infty} \frac{1}{n^{2q}} \cos 2\pi n x \right) f^{(2q)}(x). \tag{20.9}$$

In his 1826 paper, Poisson also applied the Euler–Maclaurin formula to the derivation of a result he attributed to Laplace. Laplace arrived at his result as he attempted to approximate an integral by a sum during his study of the variations of the elements of the orbit of a comet. It is a remarkable fact that James Gregory communicated just this result to Collins in a letter dated November 23, 1670.[17]

Jacobi, who was aware of Poisson's paper, gave a different derivation of Euler–Maclaurin using Taylor series;[18] we state the formula he derived in a slightly more convenient format:

$$\sum_{k=m}^{n} f(k) = \int_m^n f(x) \, dx + \frac{1}{2}(f(m) + f(n))$$

$$+ \sum_{s=1}^{q} \frac{B_{2s}}{(2s)!} \left( f^{(2s-1)}(n) - f^{(2s-1)}(m) \right) + R_q(f), \tag{20.10}$$

---

[16] Poisson (1807).
[17] Turnbull (1939) pp. 118–122.
[18] Jacobi (1834).

where $R_q(f)$ is the remainder term:

$$R_q(f) = \frac{-1}{(2q)!} \int_m^n B_{2q}(x - [x]) f^{(2q)}(x)\, dx. \tag{20.11}$$

The $B_{2s}$ are the Bernoulli numbers; the Bernoulli polynomial $B_q(t)$ is defined by

$$B_q(t) = \sum_{k=0}^q \binom{q}{k} B_k t^k. \tag{20.12}$$

Note that since $B_{2k+1} = 0$ for $k \geq 1$, only odd order derivatives appear in the sum (20.10). It can therefore be shown, applying integration by parts, that changing every $2q$ to $2q + 1$, while also changing the $-$ to $+$, does not effect a change in $R_q$.

Jacobi defined, but did not name, the even Bernoulli polynomials $B_{2n}(x)$ and gave their generating function. In the 1840s, Raabe gave them the name Bernoulli polynomials.[19] Jacobi also proved the important result that $B_{4m+2}(x) - B_{4m+2}$ was positive while $B_{4m}(x) - B_{4m}$ was negative in the interval $(0, 1)$. From this he was able to give sufficient conditions on $f$ that the remainder term had the same sign and the magnitude of at most the first omitted term in the series on the right-hand side of (20.5). One set of sufficient conditions was that the sign of $f^{(2m)}(x)$ did not change for $x > a$ and that the product $f^{(2m)}(x) f^{(2m+2)}(x)$ was positive. Since this was clearly true for $f(x) = \ln x$, Jacobi's result actually implied that de Moivre's and Stirling's series were asymptotic. Thus, though Jacobi did not explicitly mention it, he had resolved the problem raised by Gauss.

The papers of Poisson and Jacobi show that the Euler–Maclaurin formula follows from the Poisson summation formula. Thus, these two extremely important formulas are essentially equivalent. Moreover, by comparing the remainders in the formulas of Poisson and Jacobi, we observe that the Bernoulli polynomials $B_n(x)$ restricted to $0 \leq x \leq 1$ have Fourier series expansions. Surprisingly, Euler and D. Bernoulli were aware of this fact and in the 1770s, Euler gave a very interesting derivation of this result by starting with a divergent series; for a full discussion of this, see Section 20.5.

In 1823, the Norwegian mathematician Niels Abel (1802–1829) found another summation formula, called the Plana–Abel formula.[20] There was little mathematical instruction at Abel's alma mater, University of Christiania, so he independently studied the works of Euler, Lagrange, and Laplace. Thus, even before doing his great work on algebraic equations and elliptic and Abelian functions, Abel made some interesting discoveries as a student. For example, he found an integral representation for the Bernoulli numbers, and he substituted this in the Euler–Maclaurin formula to write:

$$\sum \phi(x) = \int \phi(x)\, dx - \frac{1}{2}\phi(x) + \int_0^\infty \frac{\phi\left(x + \frac{t}{2}\sqrt{-1}\right) - \phi\left(x - \frac{t}{2}\sqrt{-1}\right)}{2\sqrt{-1}} \frac{dt}{e^{\pi t} - 1}.$$

---

[19] Raabe (1848).
[20] Abel (1965) vol. I, pp. 11–27, especially p. 23.

Interestingly, the Italian astronomer and mathematician Giovanni Plana (1781– 1864) discovered this result[21] three years before Abel. Plana studied with Lagrange at the École Polytechnique; both Lagrange and Fourier supported Plana in the course of his long and illustrious career.

## 20.2 Euler on the Euler–Maclaurin Formula

Euler's problem was to sum $\sum_{k=1}^{x} t(k) = S(x)$, where he assumed that $t(x)$ and $S(x)$ were analytic functions for $x > 0$. Naturally, he did not state such conditions, but his calculations imply them. His procedure for solving this problem was almost reckless.[22] He expanded $S(x - 1)$ as a Taylor series:

$$S(x - 1) = S(x) - S'(x) + \frac{1}{2!}S''(x) - \frac{1}{3!}S'''(x) + \cdots .$$

Following his notation except for the factorials, Euler then had

$$t(n) = S(n) - S(n - 1) = \frac{dS}{dn} - \frac{1}{2!}\frac{d^2 S}{dn^2} + \frac{1}{3!}\frac{d^3 S}{dn^3} - \frac{1}{4!}\frac{d^4 S}{dn^4} + \cdots . \qquad (20.13)$$

To determine $S$ from this equation, he assumed

$$S = \int t\, dn + \alpha t + \beta\frac{dt}{dn} + \gamma\frac{d^2 t}{dn^2} + \delta\frac{d^3 t}{dn^3} + \cdots . \qquad (20.14)$$

Next, he substituted this series for $S$ on the right-hand side of (20.13) and equated coefficients. He called this a well-known method, probably referring to the method of undetermined coefficients. In his first paper on this topic, he merely noted the values of $\alpha, \beta, \gamma, \delta, \cdots$ obtained when this substitution was carried out. In the second paper he observed that he got

$$t = \left(t + \alpha\frac{dt}{dn} + \cdots\right) - \frac{1}{2!}\left(\frac{dt}{dn} + \alpha\frac{d^2 t}{dn^2} + \cdots\right) + \frac{1}{3!}\left(\frac{d^2 t}{dn^2} + \alpha\frac{d^3 t}{dn^3} + \cdots\right) - \cdots .$$

The term $t$ on both sides cancelled, and thus the coefficients of $\frac{dt}{dn}, \frac{d^2 t}{dn^2}, \ldots$ had to be zero. This gave him

$$\alpha = \frac{1}{2}, \quad \beta = \frac{\alpha}{2} - \frac{1}{6}, \quad \gamma = \frac{\beta}{2} - \frac{\alpha}{6} + \frac{1}{24}, \ldots,$$

so that

$$\alpha = \frac{1}{2}, \quad \beta = \frac{1}{12}, \quad \gamma = 0, \quad \delta = -\frac{1}{720}.$$

---

[21] Plana (1820).
[22] Eu. I-14 pp. 108–123. E 47.

Finally, Euler could write the Euler–Maclaurin formula as

$$S = \int t\, dn + \frac{1}{2} t + \frac{1}{12} \frac{dt}{dn} - \frac{1}{720} \frac{d^3 t}{dn^3} + \frac{1}{30240} \frac{d^5 t}{dn^5} - \cdots.$$

In fact, he calculated the terms up to the fifteenth derivative.

In later papers and in his 1755 book on the differential calculus, Euler stated this formula in terms of Bernoulli numbers:[23]

$$Sz = \int z\, dz + \frac{1}{2} z + \frac{B_2}{2!} \frac{dz}{dx} + \frac{B_4}{4!} \frac{d^3 z}{dx^3} + \frac{B_6}{6!} \frac{d^5 z}{dx^5} + \cdots. \tag{20.15}$$

Here $z$ was a function of $x$ and $Sz$ represented the sum of a series whose last term was $z$. For example, for the case $z = \frac{1}{x^n}$:

$$Sz = \frac{1}{1^n} + \frac{1}{2^n} + \cdots + \frac{1}{x^n}, \tag{20.16}$$

or when $z = \ln x$:

$$Sz = \ln 1 + \ln 2 + \cdots + \ln x = \ln x!. \tag{20.17}$$

Recall from Chapter 18 that de Moivre gave an asymptotic formula for $\ln x!$; in his differential calculus,[24] Euler derived this formula from (20.15) by taking $z = \ln x$, and $C$ as the constant of integration, obtaining

$$\ln x! = C + x \ln x - x + \frac{1}{2} \ln x + \frac{B_2}{1 \cdot 2} \frac{1}{x} + \frac{B_4}{3 \cdot 4} \frac{1}{x^3} + \frac{B_6}{5 \cdot 6} \frac{1}{x^5} + \cdots. \tag{20.18}$$

He next set $x = 1$ to find

$$C = 1 - \frac{B_2}{1 \cdot 2} - \frac{B_4}{3 \cdot 4} - \frac{B_6}{5 \cdot 6} - \frac{B_8}{7 \cdot 8} - \cdots, \tag{20.19}$$

a series he knew to be divergent. In fact, he remarked that this excessively divergent series was unfit to be used to find an approximate value of $C$. So Euler employed Wallis's formula, as had de Moivre, to calculate $C$. Recall Wallis's formula

$$\frac{\pi}{2} = \frac{2 \cdot 2 \cdot 4 \cdot 4 \cdot 6 \cdot 6 \cdot 8 \cdot 8 \cdot \ \text{ etc.}}{1 \cdot 3 \cdot 3 \cdot 5 \cdot 5 \cdot 7 \cdot 7 \cdot 9 \cdot \ \text{ etc.}};$$

to apply this, Euler started by taking $x = \infty$ in (20.18) to obtain

$$\sum_{k=1}^{x} \ln k = C + \left( x + \frac{1}{2} \right) \ln x - x, \tag{20.20}$$

$$\sum_{k=1}^{2x} \ln k = C + \left( 2x + \frac{1}{2} \right) \ln 2x - 2x. \tag{20.21}$$

[23] Eu. 1- 10$_2$ § 140. E 212.
[24] ibid. § 157—159

He then added $\ln 2$ to each term on the left-hand side of (20.20) and $x \ln 2$ to the right-hand side to write

$$\sum_{k=1}^{x} \ln(2k) = C + \left(x + \frac{1}{2}\right) \ln x - x \ln 2 - x. \tag{20.22}$$

Subtracting (20.22) from (20.21), he arrived at

$$\sum_{k=1}^{x} \ln(2k - 1) = x \ln x + \left(x + \frac{1}{2}\right) \ln 2 - x. \tag{20.23}$$

Still taking $x$ to be "infinitely" large, Euler took the logarithm of each side of Wallis's formula and then applied (20.22) and (20.23) to find that

$$\ln \frac{\pi}{2} = 2 \sum_{k=1}^{x} \ln(2k) - \ln(2x) - 2 \sum_{k=1}^{x} \ln(2k - 1)$$
$$= 2C + (2x + 1) \ln x + 2x \ln 2 - 2x - \ln 2 - 2x \ln x - (2x + 2) \ln 2 + 2x$$
$$= 2C - 2 \ln 2$$

or

$$C = \frac{1}{2} \ln(2\pi).$$

Euler then gave the decimal value of $\frac{1}{2} \ln(2\pi)$ as

$$0.9189385332046727417803297,$$

and wrote that the value of the divergent series (20.19) was $\frac{1}{2} \ln(2\pi)$. Recall that Euler was of the view that divergent series had a sum; he believed that, depending on the type of series, certain divergent series could be used to find the approximate values of the functions being expanded and some could not be so used.

In Chapter 18, we saw that de Moivre conjectured the asymptotic formula for $c > 1$

$$\sum_{k=0}^{\infty} \frac{1}{(n + k)^c} = \frac{1}{(c - 1)n^{c-1}} + \frac{1}{2n^c} + \sum_{k=0}^{\infty} \frac{B_{2k} \, c(c + 1) \cdots (c + 2k - 2)}{n^{c+2k-1}}; \tag{20.24}$$

he also conjectured that

$$\frac{1}{n} + \frac{1}{n+1} + \frac{1}{n+2} + \cdots + \frac{1}{a-1}$$
$$= \ln \frac{a}{n} + \frac{1}{2n} + \frac{B_2}{2n^2} + \frac{B_4}{4n^4} + \cdots - \frac{1}{2a} - \frac{B_2}{2a^2} - \frac{B_4}{4a^4} - \cdots. \tag{20.25}$$

Now de Moivre may have obtained (20.25) by taking the term-by-term derivative of the asymptotic series for $\ln x!$ and then obtained (20.24) by further differentiation.

Although de Moivre did not give proofs for these results, Stirling specifically noted the case $c = 2$:[25]

$$\sum_{k=0}^{\infty} \frac{1}{(n+k)^2} = \frac{1}{n} + \frac{1}{2n^2} + \sum_{k=1}^{\infty} \frac{B_{2k}}{n^{2k+1}};$$

(20.26)

Euler gave a proof of this case,[26] taking $z = \frac{1}{x^2}$ in (20.15) to obtain

$$1 + \frac{1}{2^2} + \cdots + \frac{1}{x^2} = C - \frac{1}{x} + \frac{1}{2x^2} - \frac{B_2}{x^3} - \frac{B_4}{x^5} - \frac{B_6}{x^7} - \frac{B_8}{x^9} - \frac{B_{10}}{x^{11}} - \cdots.$$

(20.27)

Note that $C = \sum_{k=1}^{\infty} \frac{1}{n^2}$ so that (20.27) is equivalent to (20.26). To find an approximate value for $C = \frac{\pi^2}{6}$, Euler set $x = 10$ in (20.27); he calculated

$$\sum_{k=1}^{10} \frac{1}{k^2} = 1.549767731166540690.$$

He then took ten terms of the series on the right-hand side of (20.27) and obtained

$$C = \frac{\pi^2}{6} = 1.644934066848226430.$$

Note that although (20.27) was a divergent series, it yielded good approximations to the value of $C$ because of its asymptotic character. Although Euler had not fully fathomed the nature of such a series, to be revealed later by the work of Poisson and Jacobi, his mathematical experience and intuition allowed him to utilize it effectively for correct results.

## 20.3 Maclaurin's Derivation of the Euler–Maclaurin Formula

Maclaurin's proof of the Euler–Maclaurin formula is similar to that of Euler, though the procedure appears to be more rigorous. Maclaurin described his results in geometric terms, but his arguments were mostly analytic. However, to enter Maclaurin's geometric mode of thought, we start with his proof of the integral test, usually attributed to Cauchy, who proved it in his lectures of 1828. The Euler–Maclaurin formula may be viewed as a refinement of the integral test.

---

[25] Tweddle (1988) pp. 15–16.
[26] ibid. § 148–49.

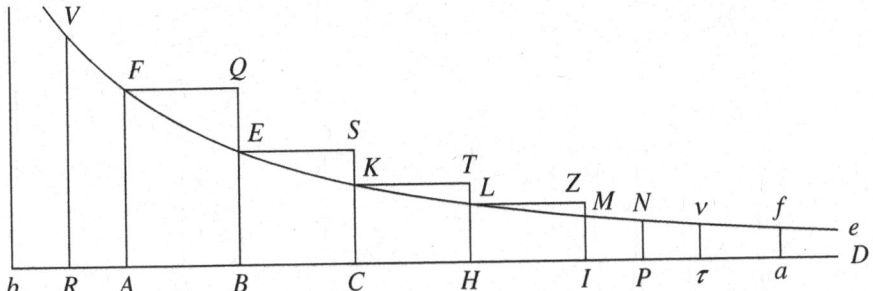

Figure 20.1 Maclaurin's geometric statement of his formula.

Referring to Figure 20.1, in section 350 of his treatise, Maclaurin wrote:[27]

Let the terms of any progression be represented by the perpendiculars $AF, BE, CK, HL, \&c.$ that stand upon the base $AD$ at equal distances; and let $PN$ be any ordinate of the curve $FNe$ that passes through the extremities of those perpendiculars. Suppose $AP$ to be produced; and according as the area $APNF$ has a limit which it never amounts to, or may be produced till it exceed any give space, there is a limit which the sum of the progression never amounts to, or it may be continued till its sum exceed any given number. For let the rectangles $FB, EC, KH, LI, \&c.$ be completed, and, the area $APNF$ being continued over the same base, it is always less than the sum of all those rectangles, but greater than the sum of all the rectangles after the first. Therefore the area $APNF$ and the sum of those rectangles either both have limits, or both have none; and it is obvious, that the same is to be said of the sum of the ordinates $AF, BE, CK, HL, \&c.$ and of the sum of the terms of the progression that are represented by them.

Maclaurin's derivation of the Euler–Maclaurin formula followed a slightly less dangerous path than Euler's. His initial description was geometric, but once he had defined his terms with the help of a picture, his argument was analytic. The Maclaurin series for $f(x) = \int_0^x y(t)\, dt$ was

$$f(x) = xy(0) + \frac{x^2}{2!}y'(0) + \frac{x^3}{3!}y''(0) + \cdots .$$

Thus,

$$\int_0^1 y\, dx = y(0) + \frac{1}{2!}y'(0) + \frac{1}{3!}y''(0) + \frac{1}{4!}y'''(0) + \cdots ,$$

or

$$y(0) = \int_0^1 y\, dx - \frac{1}{2!}y'(0) - \frac{1}{3!}y''(0) - \frac{1}{4!}y'''(0) - \cdots . \qquad (20.28)$$

[27] Maclaurin (1742) pp. 289–290.

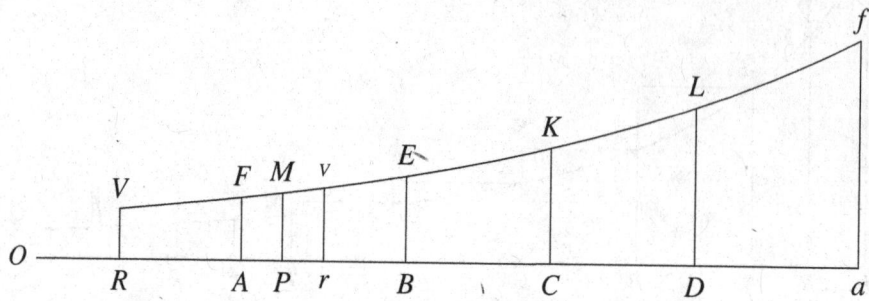

Figure 20.2 Maclaurin's representation of E–M formula.

Similarly

$$y'(0) = \int_0^1 y'\,dx - \frac{1}{2!}y''(0) - \frac{1}{3!}y'''(0) - \frac{1}{4!}y^{iv}(0) - \cdots,$$

$$y''(0) = \int_0^1 y''\,dx - \frac{1}{2!}y'''(0) - \frac{1}{3!}y^{iv}(0) - \cdots,$$

$$y'''(0) = \int_0^1 y'''\,dx - \frac{1}{2!}y^{iv}(0) - \cdots.$$

Maclaurin used these equations to eliminate $y'(0), y''(0), \ldots$ in (20.28), obtaining

$$y(0) = \int_0^1 y\,dx - \frac{1}{2}\int_0^1 y'\,dx + \frac{1}{12}\int_0^1 y''\,dx - \frac{1}{720}\int_0^1 y^{(4)}\,dx + \cdots. \quad (20.29)$$

Thus, he obtained another form of the Euler–Maclaurin formula:

$$y(0) + y(1) + \cdots + y(n-1)$$

$$= \int_0^n y\,dx - \frac{1}{2}\int_0^n y'\,dx + \frac{1}{12}\int_0^n y''\,dx - \frac{1}{720}\int_0^n y^{iv}\,dx + \cdots. \quad (20.30)$$

Maclaurin also explained how the coefficients were obtained. The reader should now have little trouble in following Maclaurin's proof of (20.29) from section 828 of his book, while referring to Figure 20.2.[28]

Suppose the base $AP = z$, the ordinate $PM = y$, and, the base being supposed to flow uniformly, let $\dot{z} = 1$. Let the first ordinate $AF$ be represented by $a$, $AB = 1$, and the area $ABEF = A$. As $A$ is the area generated by the ordinate $y$, so let $B, C, D, E, F, \&c.$ represent the areas upon the same base $AB$ generated by the respective ordinates $\dot{y}, \ddot{y}, \dddot{y}, \ddddot{y}, \&c.$ Then $AF = a = A - \frac{B}{2} + \frac{C}{12} - \frac{E}{720} + \frac{G}{30240} - \&c.$ For, by art. 752, $A = a + \frac{\dot{a}}{2} + \frac{\ddot{a}}{6} + \frac{\dddot{a}}{24} + \frac{\ddddot{a}}{120} + \&c.$, whence we have the equation $(Q)\ a = A - \frac{\dot{a}}{2} - \frac{\ddot{a}}{6} - \frac{\dddot{a}}{24} - \frac{\ddddot{a}}{120} - \&c.$ In like manner, $\dot{a} = B - \frac{\ddot{a}}{2} - \frac{\dddot{a}}{6} - \frac{\ddddot{a}}{24} - \&c.\ \ddot{a} = C - \frac{\dddot{a}}{2} - \frac{\ddddot{a}}{2} - \&c.\ \dddot{a} = D - \frac{\ddddot{a}}{2} - \&c.\ \dddot{a} = E - \&c.$ by which latter equations, if we exterminate

28 Maclaurin (1742) pp. 672–673.

$\dot{a}, \ddot{a}, \dddot{a}, \ddddot{a}, \&c.$ from the value of $a$ in the equation $Q$, we find that $a = A - \frac{B}{2} + \frac{C}{12} - \frac{E}{720} + \&c.$ The coefficients are continued thus: let $k, l, m, n, \&c.$ denote the respective coefficients of $\dot{a}, \ddot{a}, \dddot{a}, \&c.$ in the equation $Q$; that is, let $k = \frac{1}{2}, l = \frac{1}{6}, m = \frac{1}{120}, \&c.$; suppose $K = k = \frac{1}{2}$, $L = kK - l = \frac{1}{12}, M = kL - KK + m = 0, N = kM - lL + mK - n = -\frac{1}{720}$, and so on; then $a = A - KB + LC - MD + NE - \&c.$ where the coefficients of the alternate areas $D, F, H, \&c.$ vanish.

## 20.4  Poisson's Remainder Term

Poisson entitled his 1826 paper on the Euler–Maclaurin formula "Sur le calcul numérique des intégrales définies."[29] This paper applied Fourier series, and the Poisson summation formula in particular, to a variety of problems. Poisson began with a brief sketch of the proof, that he had given in an earlier paper, that the Abel means of a Fourier series of a given function converged to the function at a point of continuity. Note that the definition of an Abel mean is given in Chapter 32. Since the Abel mean of a convergent series converges to the same value as the series, Poisson mistakenly assumed that his proof sufficed to show that the Fourier series of a function converged to the function. Poisson used his result on Fourier series to prove the Euler–Maclaurin formula, undermining the proof. However, the work of Fejér some seventy years later filled in the gap and rescued Poisson's proof and its expression for the remainder term. Fejér's proof is discussed in Section 31.2.

Poisson started his proof of the Euler–Maclaurin formula by partitioning the interval $[-a, a]$ into $2n$ equal parts with $a = n\omega$. The partition points were given, then, by

$$-n\omega < -n\omega + \omega < \cdots < 0 < \omega < \cdots < n\omega.$$

He applied the trapezoidal rule, given in our Section 9.7, on each subinterval to a function $f$, a function he implicitly assumed to be differentiable $2m$ times. Thus

$$\int_{-n\omega+k\omega}^{-n\omega+(k+1)\omega} f(t)dt \approx \frac{\omega}{2}\big(f(-n\omega + k\omega) + f(-n\omega + (k+1)\omega)\big),$$

$$\int_{-a}^{a} f(t) = \sum_{k=0}^{2n-1} \int_{-n\omega+k\omega}^{-n\omega+(k+1)\omega} f(t)dt$$

$$\approx \omega P_n,$$

where

$$P_n = \frac{1}{2}f(-n\omega) + f(-n + \omega) + f(-n\omega + 2\omega) + \cdots$$

$$+ f(n\omega - 2\omega) + f(n\omega - \omega) + \frac{1}{2}f(n\omega). \quad (20.31)$$

[29] Poisson (1826).

He then gave the equation

$$\int_{-a}^{a} f(t)dt = \omega P_n + Q_n, \tag{20.32}$$

where $Q_n$ denoted the error in the approximation $\omega P_n$ for the integral. To find an expression for $Q_n$, though he did not refer to Fourier, he employed Fourier's formula (19.53) rescaled to the interval $(-a, a)$:

$$f(x) = \frac{1}{2a}\int_{-a}^{a} f(t)\,dt + \frac{1}{a}\int_{-a}^{a}\sum_{k=1}^{\infty}\cos\frac{k\pi(x-t)}{a} f(t)\,dt, \tag{20.33}$$

when $-a < x < a$. When $x = \pm a$, the left-hand side was replaced by $\frac{1}{2}(f(a) + f(-a))$. He took $x = n\omega, (n-1)\omega, \ldots, 0, -\omega, -2\omega, \ldots, -(n-1)\omega$ successively in (20.33) and added to get

$$P_n = \frac{2n}{2n\,\omega}\int_{-a}^{a} f(t)\,dt + \frac{1}{n\,\omega}\int_{-a}^{a}\sum_{j=-(n-1)}^{n}\sum_{k=1}^{\infty}\cos\frac{k\pi(j\,\omega - t)}{n\,\omega} f(t)\,dt. \tag{20.34}$$

It is easy to check that the inner sum in (20.34), after changing the order of summation, would be

$$\sum_{j=-(n-1)}^{n}\cos\frac{k\pi(j\,\omega - t)}{n\,\omega} = \begin{cases} 2n\cos\frac{k\pi t}{n\,\omega} & \text{when } k = 2nl, \\ 0 & \text{otherwise.} \end{cases}$$

So he obtained, with $a = n\,\omega$,

$$\omega P_n = \int_{-a}^{a} f(t)\,dt + 2\int_{-a}^{a}\sum_{l=1}^{\infty}\cos\frac{2l\pi t}{\omega} f(t)\,dt$$

$$= \int_{-a}^{a} f(t)\,dt - Q_n. \tag{20.35}$$

To find another expression for $Q_n$, Poisson applied integration by parts repeatedly to get

$$\int_{-a}^{a}\cos\frac{2l\pi t}{\omega} f(t)\,dt = \frac{\omega}{2\pi l}\sin\frac{2l\pi t}{\omega} f(t)\Big|_{-a}^{a} - \int_{-a}^{a}\frac{\omega}{2\pi l}\sin\frac{2l\pi t}{\omega} f'(t)\,dt$$

$$= \frac{\omega^2}{4\pi^2 l^2}\cos\frac{2l\pi t}{\omega} f'(t)\Big|_{-a}^{a} - \int_{-a}^{a}\frac{\omega^2}{4\pi^2 l^2}\cos\frac{2l\pi t}{\omega} f''(t)\,dt$$

$$= \frac{\omega^2}{4\pi^2 l^2}\left(f'(a) - f'(-a)\right) - \frac{\omega^2}{4\pi^2}\cdot\int_{-a}^{a}\frac{1}{l^2}\cos\frac{2l\pi t}{\omega} f''(t)\,dt.$$

Thus, after $m$ repetitions of this process,

$$Q_n = -\frac{2\omega^2}{4\pi^2} \sum_{l=1}^{\infty} \frac{1}{l^2} \left( f'(a) - f'(-a) \right) + \frac{2\omega^4}{(2\pi)^4} \sum_{l=1}^{\infty} \frac{1}{l^4} \left( f'''(a) - f'''(-a) \right)$$

$$- \frac{2\omega^6}{(2\pi)^6} \sum_{l=1}^{\infty} \frac{1}{l^6} \left( f^{v}(a) - f^{v}(-a) \right) + \cdots$$

$$+ \frac{2\omega^{2m}}{(2\pi)^{2m}} \sum_{l=1}^{\infty} \frac{1}{l^{2m}} \left( f^{(2m-1)}(a) - f^{(2m-1)}(-a) \right) + 8R_m \qquad (20.36)$$

where

$$R_m = -2(-1)^m \left( \frac{\omega}{2\pi} \right)^{2m} \int_{-a}^{a} \sum_{l=1}^{\infty} \frac{1}{l^{2m}} \cos \frac{2l\pi t}{\omega} f^{(2m)}(t) \, dt. \qquad (20.37)$$

Now to determine $\sum \frac{1}{l^{2m}}$, Poisson took $f(x) = e^x$ to obtain

$$\int_{-a}^{a} f(t) dt = e^a - e^{-a}$$

$$f^{(2j-1)}(a) - f^{(2j-1)}(-a) = e^a - e^{-a}.$$

He then let $n = 1$, so that setting $\omega = a$ in (20.35) and (20.31) gave him

$$P_1 = \frac{1}{2} e^{-a} + e^0 + \frac{1}{2} e^a = \frac{1}{2} \left( e^{\frac{1}{2}a} + e^{-\frac{1}{2}a} \right)^2.$$

Substituting these values in (20.35) and (20.36) and dividing by $e^a - e^{-a}$, Poisson obtained

$$\frac{a}{2} \cdot \frac{e^{\frac{1}{2}a} + e^{-\frac{1}{2}a}}{e^{\frac{1}{2}a} - e^{-\frac{1}{2}a}} = 1 + \frac{2a^2}{(2\pi)^2} \sum_{l=1}^{\infty} \frac{1}{l^2} - \frac{2a^4}{(2\pi)^4} \sum_{l=1}^{\infty} \frac{1}{l^4} + \frac{2a^6}{(2\pi)^6} \sum_{l=1}^{\infty} \frac{1}{l^6} - \cdots .$$

$$(20.38)$$

Now note that the left-hand side of (20.38) can be rewritten as

$$\frac{a}{2} \cdot \frac{e^a + 1}{e^a - 1} = \frac{a}{2} \left( 1 + \frac{2}{e^a - 1} \right) = \frac{a}{e^a - 1} + \frac{a}{2};$$

though Poisson does not refer to it as such, by (2.36), this represents the generating function for the even Bernoulli numbers:

$$\frac{a}{e^n - 1} + \frac{a}{2} = 1 + \frac{B_2}{2!} a^2 + \frac{B_4}{4!} a^4 + \cdots . \qquad (20.39)$$

Equating the coefficients in the powers of $a$ in (20.38) and (20.39), Poisson succeeded in rederiving Euler's formula (2.35):

$$1 + \frac{1}{2^{2m}} + \frac{1}{3^{2m}} + \cdots = \frac{(-1)^{m-1} B_{2m} 2^{2m-1} \pi^{2m}}{(2m)!}. \tag{20.40}$$

By substituting these values in (20.36), Poisson had

$$Q_n = \frac{B_2}{2!}(f'(a) - f'(-a))\omega^2 + \frac{B_4}{4!}(f'''(a) - f'''(-a))\omega^4 + \cdots$$
$$+ \frac{B_{2m}}{(2m)!}(f^{(2m-1)}(a) - f^{(2m-1)}(-a))\omega^{2m} + R_m. \tag{20.41}$$

Substituting the expression for $Q_n$ from (20.41) into (20.32), with $R_m$ given by (20.37), Poisson obtained the Euler–Maclaurin formula:

$$\int_{-a}^{a} f(t)dt = \omega \sum_{k=-n+1}^{n-1} f(k\omega) + \frac{\omega}{2}(f(-a) + f(a))$$
$$+ \sum_{k=1}^{m} \frac{B_{2k}}{(2k)!}(f^{(2k-1)}(a) - f^{(2k-1)}(-a))\omega^{2k} + R_m.$$

Poisson also observed that the term in the integrand of (20.37), given by

$$\sum_{l=1}^{\infty} \frac{1}{l^{2m}} \cos \frac{2l\pi t}{\omega},$$

is less in absolute value than

$$\sum_{l=1}^{\infty} \frac{1}{l^{2m}}.$$

Thus, if $f^{(2m)}(t)$ does not change sign in the interval $[-a, a]$, we may write the absolute value of the integral (20.37) as:

$$\left| 2\left(\frac{\omega}{2\pi}\right)^{2m} \int_{-a}^{a} \sum_{l=1}^{\infty} \frac{1}{l^{2m}} \cos \frac{2\pi lt}{\omega} f^{(2m)}(t)dt \right|$$
$$\leq \frac{\omega^{2m}}{(2m)!} |B_{2m}| \left| f^{(2m-1)}(a) - f^{(2m-1)}(-a) \right|. \tag{20.42}$$

Thus, the right-hand side of (20.42) gives the absolute value of the last term before $R_m$ in (20.36); under the given condition on the function, the error $R_m$ was less in absolute value than the last term of the series $Q_n$, explaining why the first few terms of the series for $\ln \Gamma(x)$, although it diverged, yielded a good approximation of the function. Recall from Section 18.10 that Cauchy gave a simpler but less general derivation of this result seventeen years after Poisson.

## 20.5 Jacobi's Remainder Term

The first rigorous treatment of the Euler–Maclaurin summation formula was given by Jacobi in a paper dated June 2, 1834.[30] He employed the Taylor series with remainder, a topic treated rigorously by Cauchy in the 1820s in his lectures at the École Polytechnique; in this connection, see our Section 11.9. In his paper, Jacobi obtained the Euler–Maclaurin formula with remainder for positive integers $a$ and $x$ in the form

$$\sum_{k=a}^{x} f(k) = \int_{a}^{x} f(t)dt + \frac{1}{2}(f(a) + f(x))$$
$$+ \sum_{k=1}^{m+1} \frac{B_{2k}}{(2k)!}\left(f^{(2k-1)}(x) - f^{(2k-1)}(a)\right) + R_{2m+2} \quad (20.43)$$

where

$$R_{2m+2}(f) = \frac{-1}{(2m+2)!} \int_{a}^{x} B_{2m+2}(t - [t]) f^{(2m+2)}(t)\, dt. \quad (20.44)$$

Note that $B_n(x)$ represents the $n$th Bernoulli polynomial, defined by the generating function

$$\frac{te^{xt}}{e^t - 1} = \sum_{n=0}^{\infty} B_n(x) \frac{t^n}{n!},$$

and $B_n$ denotes the $n$th Bernoulli number, defined by the value of the $n$th Bernoulli polynomial at $x = 0$, that is, $B_n = B_n(0)$. For a discussion of the generating function of the Bernoulli polynomials, see Section 2.10. Jacobi used the generating function for Bernoulli numbers to show that for $m \geq 1$

$$\frac{1}{(2m+1)!} + B_1\frac{1}{(2m)!} + \frac{B_2}{2!}\frac{1}{(2m-1)!} + \frac{B_4}{4!}\frac{1}{(2m-3)!} + \cdots + \frac{B_{2m}}{(2m)!} = 0, \quad (20.45)$$

$$\frac{1}{(2m+2)!} + B_1\frac{1}{(2m+1)!} + \frac{B_2}{2!}\frac{1}{(2m)!} + \frac{B_4}{4!}\frac{1}{(2m-2)!} + \cdots + \frac{B_{2m}}{(2m)!\,2!} = 0, \quad (20.46)$$

two relations that, for $m \geq 1$, are equivalent to:

$$B_{m+1}(1) - B_{m+1} = 0, \quad (20.47)$$

an equation we proved in Section 2.10.

Observe that for the case in which $m+1$ is odd, so that $B_{m+1} = 0$, equation (20.47) is equal to (20.45); otherwise (20.47) is equal to (20.46). Also note that equations

[30] Jacobi (1834).

(20.45) and (20.46) are identical with those employed by Seki and Bernoulli to define Bernoulli numbers.

Jacobi proceeded to define the functions

$$\psi(x) = \int_a^x f(t)dt \quad \text{and} \quad \psi(x) - \psi(x - h) = \Theta(x), \quad h > 0. \tag{20.48}$$

Taking $x - a$ to be an integer multiple of $h$, he denoted

$$\sum_a^x \Theta(x) = \Theta(a + h) + \Theta(a + 2h) + \cdots + \Theta(x) = \Phi(x) - \Phi(a) = \Phi(x). \tag{20.49}$$

Now since $\Theta(x) = \Phi(x) - \Phi(x - h)$, Taylor's theorem with integral remainder gave

$$\Theta(x) = \Phi'(x)h - \Phi''(x)\frac{h^2}{2!} + \cdots + (-1)^{n-1}\Phi^{(n)}(x)\frac{h^n}{n!}$$
$$+ (-1)^n \int_0^h \frac{(h - t)^n}{n!} \Phi^{(n+1)}(x - t)\,dt. \tag{20.50}$$

Because $\Phi'(x) = f(x)$, Jacobi had $\Phi^{(m+1)}(x) = f^{(m)}(x)$; then taking

$$I_n = \int_0^h \frac{(h - t)^n}{n!} f^{(n)}(x - t)\,dt,$$

(20.50) could be rewritten as

$$\Theta(x) = f(x)h - f'(x)\frac{h^2}{2!} + f''(x)\frac{h^3}{3!} - \cdots + (-1)^{(n-1)}f^{(n-1)}(x)\frac{h^n}{n!} + (-1)^n I_n. \tag{20.51}$$

The next steps Jacobi delineated were to replace $n$ by $n - 1$ in (20.51); take its derivative and multiply that derivative by $\frac{1}{2}h$; then change $n$ to $n - 2$, $n$ to $n - 4, \ldots, n$ to $n - 2m$; correspondingly take the second derivative and multiply it by $\frac{B_2}{2!}h^2$, take the fourth derivative and multiply it by $\frac{B_4}{4!}h^4, \ldots$, and, finally, take the $(2m)$th derivative and multiply it by $\frac{B_{2m}}{(2m)!}h^{2m}$. Though he did not write it out, note that the result of this process is

$$\frac{1}{2}\Theta'(x)h = f'(x)\frac{h^2}{1!\,2} - f''(x)\frac{h^3}{2!\,2} + \cdots$$
$$+ (-1)^n f^{(n-1)}(x)\frac{h^n}{(n-1)!\,2} + (-1)^{n-1}\frac{h}{2} I_{n-1}, \tag{20.52}$$

$$\frac{B_2}{2!}\,\Theta''(x)k^2 = f''(x)\frac{B_2h^3}{2!} + \cdots$$

$$+ (-1)^{n-1} f^{(n-1)}(x)\,\frac{B_2h^n}{(n-2)!\,2!} + (-1)^{n-2}\frac{B_2h^2}{2!}\,I_{n-2} \quad (20.53)$$

$$\cdots \quad \cdots \quad \cdots \ .$$

Unlike Euler, Jacobi omitted many details in his explications, since by that time many received and known results could be assumed. Expanding, then, Jacobi's terse argument, note that when (20.50), (20.51), (20.52), (20.53), and so on were added, the sum would be

$$\Theta(x) + \frac{1}{2}\,\Theta'(x)h + \frac{B_2}{2!}\,\Theta''(x)h^2 + \frac{B_4}{4!}\,\Theta^{iv}(x)h^4 + \cdots + \frac{B_{2m}}{(2m)!}\,\Theta^{(2m)}(x)h^{2m}$$

$$= f(x)h + \int_0^h T_m\, f^{(2m+2)}(x-t)\,dt \quad (20.54)$$

where, since $B_1 = -\frac{1}{2}$,

$$T_m = \frac{(h-t)^{2m+2}}{(2m+2)!} + B_1\frac{(h-t)^{2m+1}h}{(2m+1)!} + B_2\frac{(h-t)^{2m}h^2}{(2m)!} + \cdots + B_{2m}\frac{(h-t)^2h^{2m}}{2!}$$

$$= \frac{h^{2m+2}}{(2m+2)!}\left(B_{2m+2}\left(1-\frac{t}{h}\right) - B_{2m+2}\right). \quad (20.55)$$

Observe that the terms involving $f'(x)h^2$, $f''(x)h^3$, ... vanish. For example, the coefficient of $f''(x)h^3$ in the sum (20.54) is

$$\frac{1}{3!} + \frac{B_1}{2!\,2} + \frac{B_2}{2!},$$

that becomes zero when we take $m = 1$ in (20.45). Next note that the coefficient of $f'''(x)h^4$ is

$$\frac{1}{4!} + \frac{B_1}{3!} + \frac{B_2}{2!\,2!} + \frac{B_4}{4!},$$

a sum that is zero by (20.46). Thus, following this pattern, all the powers of $h$ up to $h^{2m}$ reduce to zero, as Jacobi mentions, explaining (20.54) and (20.55). How did Jacobi perceive, after such a complicated calculation, that the terms in (20.54) would vanish? It would appear that the clue might have been that, in terms of symbolic calculus discussed in our Chapter 21,

$$\Theta(x) = \psi(x) - \psi(x-h)$$
$$= \psi(x) - e^{hD}\,\psi(x).$$

Thus

$$\Theta(x) = (1 - e^{-hD})\psi$$

or

$$\psi' = \frac{e^{hD}}{e^{hD} - 1}\Theta' + \frac{1}{h}\frac{hD}{e^{hD} - 1}\Theta',$$

$$= \Theta' + \frac{1}{h}\sum_{n=0}^{\infty}\frac{B_n}{n!}\Theta^{(n)}h^n$$

or

$$hf = h\Theta + \frac{1}{2}\Theta'h + \frac{B_2}{2!}\Theta''h^2 \cdots . \tag{20.56}$$

Jacobi surely saw that the sum on the right-hand side of (20.56), up to $2m$ terms, was identical with the sum on the left-hand side of (20.54) up to $2m$ terms, so that the terms would cancel, leaving $hf$. Jacobi thus stated the E–M formula with remainder:

$$\sum_a^x \left( \frac{\Theta(x)}{h} + \frac{1}{2}\Theta'(x) + \frac{B_2}{2!}\Theta''(x)h + \cdots + \frac{B_{2m}}{(2m)!}\Theta^{(2m)}(x)h^{2m} \right)$$

$$= \int_a^x \left( \frac{f(t)}{h} + \frac{1}{2}f'(t) + \frac{B_2}{2!}f''(t)h + \cdots + \frac{B_{2m}}{(2m)!}f^{(2m)}(t)h^{2m-1} \right) dt$$

$$= \sum_a^x f(x) + \int_0^h T_m \sum_a^x f^{(2m+2)}(x - t)\,dt, \tag{20.57}$$

with $T_m$ defined by (20.55). Thus, (20.57) gives Jacobi's Euler–Maclaurin formula for the case in which $x$ and $a$ are not necessarily integers, from which we can derive the modern form of the Euler–Maclaurin formula, (20.43) and (20.44).

To derive the modern form from Jacobi's result, take $a$ and $x$ to be positive integers and take $h = 1$. Then

$$\int_a^x \left( f(t) + \frac{1}{2}f'(t) + \frac{B_2}{2!}f''(t) + \cdots + \frac{B_{2m}}{(2m)!}f^{(2m)}(t) \right) dt$$

$$= \int_a^x f(t)dt + \frac{1}{2}(f(x) - f(a)) + \frac{B_2}{2!}(f'(x) - f'(a)) + \cdots$$

$$+ \frac{B_{2m}}{(2m)!}(f^{(2m-1)}(x) - f^{(2m-1)}(a)). \tag{20.58}$$

Moreover, with $u = 1 - t$,

$$\int_0^1 T_m \sum_a^x f^{(2m+2)}(x - t)\,dt$$

$$= \frac{1}{(2m+2)!} \int_0^1 \left(B_{2m+2}(1-t) - B_{2m+2}\right)\left(f^{(2m+2)}(a+1-t)\right.$$

$$\left. + f^{(2m+2)}(a+2-t) + \cdots\right)dt$$

$$= \frac{1}{(2m+2)!} \int_0^1 \left(B_{2m+2}(u) - B_{2m+2}\right)$$

$$\cdot \left(f^{(2m+2)}(a+u) + f^{(2m+2)}(a+1+u) + \cdots\right)du. \quad (20.59)$$

Now with $s$ a positive integer such that $a + s + u = v$, we have

$$\int_0^1 B_{2m+2}(u) f^{(2m+2)}(a+s+u)\,du$$

$$= \int_{a+s}^{a+s+1} B_{2m+2}\left(v - (a+s)\right) f^{(2m+2)}(v)\,dv$$

and since $a + s \le v < a + s + 1$, so that $\lfloor v \rfloor = a + s$, we may write

$$= \int_{a+s}^{a+s+1} B_{2m+2}\left(v - \lfloor v \rfloor\right) f^{(2m+2)}(v)\,dv.$$

Denoting $B_{2m+2}\left(v - \lfloor v \rfloor\right) - B_{2m+2}$ by $\overline{B}_{2m+2}(v)$, the right-hand side of (20.59) may be rewritten

$$\frac{1}{(2m+2)!}\left(\int_a^{a+1} \overline{B}_{2m+2}(v) f^{(2m+2)}(v)\,dv + \int_{a+1}^{a+2} \overline{B}_{2m+2}(v) f^{(2m+2)}(v)\,dv\right.$$

$$\left. + \cdots + \int_{x-a}^x \overline{B}_{2m}(v) f^{(2m+2)}(v)\,dv\right)$$

$$= \frac{1}{(2m+2)!} \int_a^x \left(B_{2m+2}(v - \lfloor v \rfloor) - B_{2m+2}\right) f^{(2m+2)}(v)\,dv. \quad (20.60)$$

By taking (20.57), (20.58), and (20.60) together, we obtain (20.43) and (20.44), the modern form of the Euler–Maclaurin formula with remainder term.

In his 1834 paper, Jacobi also proved the theorem: For $0 < x < 1$, the polynomial $B_{2m}(x) - B_{2m}$ is always positive when $m$ is odd and always negative when $m$ is even.

To understand Jacobi's proof, recall from Chapter 2 the generating function for $2\big(B_{2m}(x) - B_{2m}\big) \equiv 2\overline{B}_{2m}(x)$:

$$\left(\frac{t(e^{xt}-1)}{e^t-1} + \frac{t(e^{-xt}-1)}{1-e^{-t}}\right) = t\left(\frac{1-e^{xt}}{1-e^t} - \frac{1-e^{-xt}}{1-e^{-t}}\right). \quad (20.61)$$

Jacobi set $x' = 1 - x$ and rewrote (20.61) as

$$t\left(\frac{1-e^{xt}}{1-e^t} + \frac{e^t - e^{x't}}{1-e^t}\right) = t\,\frac{(1-e^{xt})(1-e^{x't})}{1-e^t}$$

$$= -t\,\frac{(e^{\frac{xt}{2}} - e^{-\frac{xt}{2}})(e^{\frac{x't}{2}} - e^{-\frac{x't}{2}})}{e^{\frac{t}{2}} - e^{-\frac{t}{2}}}. \quad (20.62)$$

Jacobi next employed Euler's product formula (15.18) to express (20.62), the generating function of $2\overline{B}_{2m}(x)$, as an infinite product, obtaining

$$-t^2 xx' \prod_{n=1}^{\infty} \frac{\left(1 + \frac{x^2 t^2}{4n^2\pi^2}\right)\left(1 + \frac{x'^2 t^2}{4n^2\pi^2}\right)}{\left(1 + \frac{x^2}{4n^2\pi^2}\right)} = 2\sum_{n=1}^{\infty} \overline{B}_{2m}(x)\,t^{2m}. \quad (20.63)$$

He next set $y = -\frac{t^2}{4n^2\pi^2}$, and observed that the $n$th term in the infinite product on the left-hand side of (20.63) could be written as

$$\frac{(1-x^2 y)(1-x'y)}{1-y} = 1 + (1 - x^2 - x'^2)y + \frac{(1-x^2)(1-x'^2)y^2}{1-y}$$

$$= 1 + 2xx'y + xx'(2+xx')\,\frac{y^2}{1-y}$$

$$= 1 + 2xx'y + xx'(2+xx')y^2\,(1 + y + y^2 + \cdots). \quad (20.64)$$

In view of the given condition that $0 < x < 1$, so that $0 < x' < 1$, Jacobi could conclude that the coefficients of the powers of $y$ (or powers of $-t^2$) were all positive. Therefore, in the infinite product on the left-hand side of (20.63), the coefficients of the odd powers of $-y$ (or $t^2$) were necessarily negative for $0 < x < 1$. This fact in turn implied that $\overline{B}_{2m}(x)$ in (20.63) was negative with $m$ odd and positive for $m$ even. And this completed the proof of his theorem.

From this theorem, Jacobi deduced that if $f^{(2m)}(t)$ and $f^{(2m+2)}(t)$ had the same sign in the interval $(a, x)$, then

$$\int_a^x \overline{B}_{2m+2}\big(t - \lfloor t\rfloor\big)\,f^{(2m+2)}(t)\,dt = \theta(-1)^m\,B_{2m+2}\big(f^{(2m+1)}(x) - f^{(2m+1)}(a)\big),$$

where $0 \le \theta \le 1$. This implied that the error incurred in taking $m + 1$ terms of the Euler–Maclaurin series must be of the same order as the term just preceding the first neglected term.

## 20.6  Bernoulli Polynomials

As we have seen, the expressions found by Jacobi and Poisson for the remainder term in the Euler–Maclaurin formula took quite different forms. When these expressions are equated, we obtain the trigonometric expansions of the Bernoulli polynomials:

$$B_{2m}(t - [t]) = 2(-1)^{m-1}(2m)! \sum_{l=0}^{\infty} \frac{\cos 2\pi\, lt}{(2\pi\, l)^{2m}}, \quad m \geq 1, \ 0 \leq t \leq 1 \quad (20.65)$$

and taking the derivative, as we showed in equation (16.126), gives us

$$B_{2m-1}(t - [t]) = 2(-1)^{m}(2m - 1)! \sum_{l=0}^{\infty} \frac{\sin 2\pi\, lt}{(2\pi\, l)^{2m-1}}, \quad m \geq 2, \ 0 \leq t \leq 1,$$

$$(20.66)$$

an equation that holds for $m = 1$ where $0 < t < 1$. Note that the series on the right-hand sides of equations (20.65) and (20.66) can be shown to be the Fourier series of the functions on the left-hand sides.

Recall that we saw equations (20.65) and (20.66) in Chapter 16, connected with the 1809 work of Spence; Euler found results for $B_m(x - [x])$, $n = 1, 2, \ldots$, equivalent to the initial cases of (20.65) and (20.66) in his 1753 paper, "Subsidium calculi sinuum."[31] By setting $x = ae^{i\phi}$ in the equation $x + x^2 + \cdots = \frac{x}{1-x}$, he got

$$(\cos\phi + i\sin\phi) + a(\cos 2\phi + i\sin 2\phi) + \cdots = \frac{e^{i\phi}}{1 - ae^{i\phi}} = \frac{e^{i\phi}(1 - ae^{-i\phi})}{1 - 2a\cos\phi + a^2}.$$

$$(20.67)$$

On taking the real part of (20.67), he obtained

$$\cos\phi + a\cos 2\phi + a^2\cos 3\phi + \cdots = \frac{\cos\phi - a}{1 - 2a\cos\phi + a^2}; \quad (20.68)$$

he then set $a = -1$ to arrive at

$$\cos\phi - \cos 2\phi + \cos 3\phi - \cdots = \frac{1}{2}. \quad (20.69)$$

Of course, $-1$ was not a legitimate value of $a$, because (20.68) is correct only for $|a| < 1$. Euler had a knack for handling divergent series so as to obtain useful results; in fact, his method here was later legitimized by the idea of an Abel mean, a concept discussed in Chapter 32. Note that $\frac{1}{2}$ is the Abel mean if the series $\cos\phi - \cos 2\phi + \cdots$; we also mention here that the Abel mean of the series $\sum_{n=0}^{\infty} a_n$ is given by $\lim_{x \to 1^-}$.

---

[31] Eu. 1-14 pp. 542–584. E 246.

Euler integrated equation (20.69) to obtain[32]

$$\sin\phi - \frac{\sin 2\phi}{2} + \frac{\sin 3\phi}{3} - \cdots = \frac{\phi}{2},$$ (20.70)

where the constant of integration can be seen to be zero. Integrating again, he arrived at

$$\cos\phi - \frac{\cos 2\phi}{2^2} + \frac{\cos 3\phi}{3^2} - \cdots = \frac{\pi^2}{12} - \frac{\phi^2}{4},$$ (20.71)

since the constant of integration was

$$C = 1 - \frac{1}{2^2} + \frac{1}{3^2} - \cdots = \frac{\pi^2}{12}.$$ (20.72)

Now observe that if we set $\phi = \pi - \theta$ in (20.70) and (20.71), we will get the series for $B_1\left(\frac{\theta}{2\pi}\right)$ and $B_2\left(\frac{\theta}{2\pi}\right)$, series that hold for $0 < \frac{\theta}{2\pi} < 1$. In a 1773 paper, "Theoria Elementaria serierum,"[33] Daniel Bernoulli started with (20.68), took $a = 1$, and repeatedly integrated to obtain the trigonometric series for $B_1\left(\frac{\theta}{2\pi}\right), B_2\left(\frac{\theta}{2\pi}\right), \ldots, B_6\left(\frac{\theta}{2\pi}\right)$. A year later, Euler followed up with a paper, "Nova methodus quantitates integrales determinandi,"[34] dealing with the integral and series expressions for $B_m\left(\frac{\theta}{2\pi}\right)$. In section 40 of his paper, he integrated the formula

$$\cos\phi + \cos 2\phi + \cos 3\phi + \cdots = -\frac{1}{2}$$

to arrive at

$$\sin\phi + \frac{\sin 2\phi}{2} + \frac{\sin 2\phi}{3} + \cdots = A - \frac{1}{2}\phi.$$ (20.73)

He observed that he could not determine the value of $A$ by setting $\phi = 0$, reasoning that when $\phi$ was small, then $\frac{\sin m\phi}{m} = \phi$, so that the left-hand side of (20.73) became $\phi + \phi + \phi + \cdots$, and this could not equal the right-hand side. However, he noted that he could set $\phi = \pi$, because when $\phi = \pi + \omega$, where $\omega$ was an infinitesimal, he got the left-hand side as $\omega - \omega + \omega - \omega + \cdots$. Taking $\phi = \pi$ gave him $0 = A - \frac{\pi}{2}$ or $A = \frac{\pi}{2}$, so that (20.73) could be written as

$$\sin\phi + \frac{\sin 2\phi}{2} + \frac{\sin 3\phi}{3} + \cdots = \frac{\pi - \phi}{2}, \quad 0 < \phi < 2\pi;$$ (20.74)

after integration, Euler obtained

$$\cos\phi + \frac{\cos 2\phi}{2^2} + \frac{\cos 3\phi}{3^2} + \cdots = B - \frac{\pi}{2}\phi + \frac{1}{4}\phi^2.$$ (20.75)

[32] ibid. § 53.
[33] Bernoulli (1982–1996) vol. 2, pp. 119–134, especially pp. 120–121.
[34] Eu.1-17 pp. 421–457. E 464.

He set $\phi = 0$ and $\phi = \pi$ in (20.75) to get the two equations

$$\sum_{n=1}^{\infty} \frac{1}{n^2} = B \quad \text{and} \quad \sum_{n=1}^{\infty} \frac{(-1)^n}{n^2} = B - \frac{\pi^2}{4};$$

adding, he obtained

$$2 \sum_{n=1}^{\infty} \frac{1}{(2n)^2} = 2B - \frac{\pi^2}{4},$$

$$\frac{1}{2} \sum_{n=1}^{\infty} \frac{1}{n^2} = \frac{1}{2}B = 2B - \frac{\pi^2}{4},$$

$$B = \frac{\pi^2}{6}.$$

Thus

$$\sum_{n=1}^{\infty} \frac{\cos n\phi}{n^2} = \frac{\pi^2}{6} - \frac{\pi}{2}\phi + \frac{1}{4}\phi^2, \quad 0 \le \phi \le 2\pi. \tag{20.76}$$

Another integration produced

$$\sum_{n=1}^{\infty} \frac{\sin n\phi}{n^3} = \frac{\pi^2 \phi}{6} - \frac{\pi \phi^2}{4} + \frac{\phi^3}{12}, \quad 0 \le \phi \le 2\pi$$

and yet another integration yielded

$$\sum_{n=1}^{\infty} \frac{\cos n\phi}{n^4} = \frac{\pi^4}{90} - \frac{\pi^2 \phi^2}{12} + \frac{\pi \phi^3}{12} - \frac{\phi^4}{48}, \quad 0 \le \phi \le 2\pi.$$

Euler found the constant by evaluating $\phi = 0$ and $\phi = \pi$ as in the derivation of (20.76); he noted that this method of calculating the constants was contained in Bernoulli's 1773 paper. Euler calculated the next two cases in a similar manner:

$$\sum_{n=1}^{\infty} \frac{\sin n\phi}{n^5} = \frac{\pi^4 \phi}{90} - \frac{\pi^2 \phi^3}{36} + \frac{\pi \phi^4}{48} - \frac{\phi^5}{240},$$

$$\sum_{n=1}^{\infty} \frac{\cos n\phi}{n^6} = \frac{\pi^6}{945} - \frac{\pi^4}{90}\frac{\phi^2}{2} + \frac{\pi^2}{6}\frac{\phi^4}{24} - \frac{\pi}{2}\frac{\phi^3}{120} + \frac{1}{2}\frac{\phi^6}{720}. \tag{20.77}$$

Apparently, Euler did not observe the connection between polynomials such as the right-hand side of equation (20.77) and the polynomials obtained by Jakob Bernoulli when he expressed the sum $1^k + 2^k + \cdots + (n-1)^k$ as a polynomial in $n$. For example, Bernoulli's table for sums of powers, given in our Chapter 2, has

$$1^5 + 2^5 + \cdots + (n-1)^5 = \frac{1}{6}n^6 - \frac{1}{2}n^5 + \frac{5}{12}n^4 - \frac{1}{12}n^2$$

$$= \frac{1}{6}\left(n^6 - 3n^5 + \frac{5}{2}n^4 - \frac{1}{2}n^2\right).$$

Now if we set $\phi = 2\pi t$ in (20.77), we have

$$2(6!) \sum_{n=1}^{\infty} \frac{\cos 2n\pi t}{n^6} = t^6 - 3t^5 + \frac{5}{2}t^4 - \frac{1}{2}t^2 + \frac{1}{42}, \quad 0 \le t \le 1; \tag{20.78}$$

had Euler written (20.78), rather than (20.77), he might have noticed this relationship. As it is, both Euler and Daniel Bernoulli succeeded in deriving the trigonometric series expansions for the Bernoulli polynomials.

The main purpose of Euler's 1774 "Nova methodus"[35] paper was to express the Bernoulli polynomials as integrals. Euler denoted by $P$ and $Q$ the real and imaginary parts of equation (20.67):

$$P \equiv a\cos\phi + a^2\cos 2\phi + a^3\cos 3\phi + \cdots = \frac{a\cos\phi - a^2}{1 - 2a\cos\phi + a^2}, \tag{20.79}$$

$$Q \equiv a\sin\phi + a^2\sin 2\phi + a^3\sin 3\phi + \cdots = \frac{a\sin\phi}{1 - 2a\cos\phi + a^2}. \tag{20.80}$$

In addition he defined:

$$P_n \equiv \int_0^a \frac{P_{n-1}}{a}da, \quad P_0 \equiv P,$$

$$Q_n \equiv \int_0^a \frac{Q_{n-1}}{a}da, \quad Q_0 \equiv Q.$$

Actually, Euler wrote out only the series for $P_1, Q_1, P_2, Q_2, P_3, Q_3, P_4, Q_4$; he did not define the general case. Thus, he had:

$$P_1(a) = \frac{a\cos\phi}{1} + \frac{a^2\cos 2\phi}{2} + \frac{a^3\cos 3\phi}{3} + \cdots,$$

$$Q_1(a) = \frac{a\sin\phi}{1} + \frac{a^2\sin 2\phi}{2} + \frac{a^3\sin 3\phi}{3} + \cdots,$$

$$P_2(a) = \frac{a\cos\phi}{1^2} + \frac{a^2\cos 2\phi}{2^2} + \frac{a^3\cos 3\phi}{3^2} + \cdots,$$

$$Q_2(a) = \frac{a\sin\phi}{1^2} + \frac{a^2\sin 2\phi}{2^2} + \frac{a^3\sin 3\phi}{3^2} + \cdots,$$

and so on; observe that if we set $a = 1$ in these series, then $Q_1(1)$ is essentially $B_1\left(\frac{\phi}{2\pi}\right)$, $P_2(1)$ is essentially $B_2\left(\frac{\phi}{2\pi}\right)$, and so on. Euler observed that

[35] ibid, especially sections 32–39.

$$P_2(1) = \int_0^1 \frac{P_1(a)}{a}\, da = \ln a\, P_1(a)\Big|_0^1 - \int_0^1 \frac{P(a)}{a} \ln a\, da$$

$$= -\int_0^1 \frac{P(a)}{a} \ln a\, da$$

and similarly

$$Q_2(1) = -\int_0^1 \frac{Q(a)}{a} \ln a\, da$$

and, in general,

$$P_n(1) = \frac{(-1)^{n-1}}{(n-1)!} \int_0^1 \frac{P(a)}{a} (\ln a)^{n-1}\, da,$$

$$Q_n(1) = \frac{(-1)^{n-1}}{(n-1)!} \int_0^1 \frac{Q(a)}{a} (\ln a)^{n-1}\, da.$$

Euler's formulas for $P_n(1)$ and $Q_n(1)$ can also be expressed in terms of Bernoulli polynomials.

$$B_{2n}(\phi) = (-1)^{n-1} \int_0^1 \frac{(\cos 2\pi\phi - a)(\ln a)^{2n-1}\, da}{1 - 2a \cos 2\pi\phi + a^2},$$

$$B_{2n+1}(\phi) = (-1)^n \int_0^1 \frac{(\sin 2\pi\phi - a)(\ln a)^{2n}\, da}{1 - 2a \cos 2\pi\phi + a^2}.$$

Now Euler obtained (20.74) by integrating a divergent series. We may ask, then, did he have another method of obtaining that result? Indeed, Euler himself noted that by term by term integration of (20.80),

$$Q_1(1) = \frac{\sin\phi}{1} + \frac{\sin 2\phi}{2} + \frac{\sin 3\phi}{3} + \cdots$$

$$= \int_0^1 \frac{a \sin\phi\, da}{1 - 2a \cos\phi + a^2} = \arctan \frac{a \sin\phi}{1 - a \cos\phi}\Big|_0^1$$

$$= \arctan \frac{\sin\phi}{1 - \cos\phi} = \arctan \cot \frac{\phi}{2} = \frac{\pi - \phi}{2}, \quad 0 < \phi < 2\pi.$$

Of course, the term-by-term integration is the most problematic step to justify. However, since a power series is involved, one does not require the concept of uniform convergence for this step.[36] We note that later mathematicians, such as Abel and Dedekind, dealt with exactly these formulas of Euler and their convergence problems. In addition to this approach, Euler offered yet another method, perhaps more easily

---

[36] See, for example, Nevanlinna and Paatero (2007) pp. 103–105.

justified, in his 1773 paper "Summatio progressionum."[37] In example 2, contained in the final portion of the paper, Euler considered the formula

$$-\log(1-z) = z + \frac{z^2}{2} + \frac{z^3}{3} + \cdots,$$

in which he set $z = xe^{i\phi}$ and $z = xe^{-i\phi}$ to arrive at

$$2\sqrt{-1}\left(x\sin\phi + x^2\frac{\sin 2\phi}{2} + \cdots\right) = -\log(1 - xe^{i\phi}) + \log(1 - xe^{-i\phi})$$

$$= \log\frac{1 - x\cos\phi + ix\cos\phi}{1 - x\cos\phi - ix\cos\phi}$$

$$= 2\sqrt{-1}\arctan\frac{x\sin\phi}{1 - x\cos\phi}, \quad \phi \neq 0.$$

Taking $x = 1$ produced the result. To justify this step, we may apply Abel's continuity theorem, discussed in Chapter 4. First we must verify the convergence of the series

$$e^{i\phi} + \frac{e^{2i\phi}}{2} + \frac{e^{3i\phi}}{3} + \cdots, \quad 0 < \phi < 2\pi. \tag{20.81}$$

For this purpose, we apply a procedure due to Dirichlet, based on Abel's method of summation by parts. In fact, in a paper of 1877,[38] Dedekind applied exactly this procedure to prove that this series and more general series converged. Let $s_n = e^{i\phi} + e^{i2\phi} + \cdots + e^{in\phi}$ and let $s$ denote the series (20.81). Then

$$s = \frac{s_1}{1} + \frac{s_2 - s_1}{2} + \frac{s_3 - s_2}{3} + \cdots$$

$$= s_1\left(1 - \frac{1}{2}\right) + s_2\left(\frac{1}{2} - \frac{1}{3}\right) + s_3\left(\frac{1}{3} - \frac{1}{4}\right) + \cdots. \tag{20.82}$$

Observe that

$$s_n = \frac{e^{i\phi}(1 - e^{in\phi})}{1 - e^{i\phi}} = \frac{-(1 - e^{i\phi})(1 - e^{in\phi})}{2 - 2\cos\phi}.$$

Thus, for a fixed $\phi \neq 0$, $|s_n| \leq \frac{1}{\sin^2\frac{\phi}{2}}$. It then follows from (20.82) that $S$ converges absolutely, since

$$|s_1|\left(1 - \frac{1}{2}\right) + |s_2|\left(\frac{1}{2} - \frac{1}{3}\right) + |s_3|\left(\frac{1}{3} - \frac{1}{4}\right) + \cdots$$

$$\leq \frac{1}{\sin n^2\frac{\phi}{2}}\left(\left(1 - \frac{1}{2}\right) + \left(\frac{1}{2} - \frac{1}{3}\right) + \cdots\right) \leq \frac{1}{\sin^2\frac{\phi}{2}}.$$

[37] Eu. I-15 pp. 168–184. E 447.
[38] Dedekind (1877).

Thus, the series $s$ converges when $0 < \phi < 2\pi$. Now Abel's continuity theorem states that if the series $a_1 + a_2 + a_3 + \cdots$ converges to $A$, then

$$\lim_{x \to 1^-} (a_1 x + a_2 x^2 + a_3 x^3 + \cdots) = A;$$

for more on this theorem, see Section 4.5.

Taking $a_n = \frac{e^{in\phi}}{n}$, we have

$$\frac{x e^{i\phi}}{1} + \frac{x^2 e^{2i\phi}}{2} + \frac{x^3 e^{3i\phi}}{3} + \cdots = \log(1 - x e^{i\phi}), \quad 0 < \phi < 2\pi, \ 0 \le x < 1.$$

Since $\log(1 - x e^{i\phi})$, $0 < \phi < 2\pi$, is a continuous function in $x$, it follows that

$$\lim_{x \to 1^-} \log(1 - x e^{i\phi}) = \log(1 - e^{i\phi}).$$

Moreover, Abel's theorem implies that

$$\lim_{x \to 1^-} \left( x e^{i\phi} + \frac{x^2 e^{2i\phi}}{2} + \cdots \right) = e^{i\phi} + \frac{e^{2i\phi}}{2} + \cdots.$$

This completes the proof that

$$e^{i\phi} + \frac{e^{2i\phi}}{2} + \frac{e^{3i\phi}}{3} + \cdots = -\log(1 - e^{i\phi}).$$

Finally, we note that equation (20.78) and a similar equation involving the sine series can be generalized:[39]

$$2(-1)^{k-1}(2k)! \sum_{n=1}^{\infty} \frac{\cos 2\pi nt}{(2\pi n)^{2k}} = B_{2k}(t), \quad k \ge 1, \ 0 \le t \le 1 \qquad (20.83)$$

and

$$2(-1)^k (2k-1)! \sum_{n=1}^{\infty} \frac{\sin 2\pi nt}{(2\pi n)^{2k-1}} = B_{2k-1}(t), \quad k \ge 1, \ 0 < t < 1, \qquad (20.84)$$

results apparently first stated and proved by Raabe, who recognized that the polynomials on the right-hand sides of equations (20.83) and (20.84) were Bernoulli polynomials; these equations in turn imply

$$B_n(t) = (-1)^n B_n(1 - t)$$

and

$$|B_{2n}(t)| \le |B_{2n}(0)| = |B_{2n}|.$$

[39] Raabe (1850).

## 20.7  Number Theoretic Properties of Bernoulli Numbers

Recall that Bernoulli numbers, $B_{2k}$, are rational numbers. Take their reduced form to be $B_{2k} = \frac{N_{2k}}{D_{2k}}$, with $N_{2k}$ and $D_{2k}$ relatively prime. Thomas Clausen and Karl von Staudt independently published a complete determination of $D_{2k}$ in 1840. Clausen first published the theorem without proof in the *Astronomische Nachrichten*,[40] with a promise to publish the proof. Soon thereafter, von Staudt published a proof,[41] ending his paper by remarking that many years earlier he had communicated the result to Gauss and later, upon seeing Clausen's one-paragraph announcement, he (von Staudt) decided to publish his proof. Clausen did not publish his proof, assumedly because of von Staudt's paper.

The theorem in question states that for primes $p$, $B_{2k} + \sum_{p-1|2k} \frac{1}{p}$ is an integer. In particular, $D_{2k}$ is the product of primes $p$ such that: $p - 1|2k$ and $6|D_{2k}$.

In 1911, Ramanujan published his first paper,[42] and it contained a suggestion for a proof of the Clausen-von Staudt result. The topic of this paper was Bernoulli numbers, and much of it was rediscovered material, but Ramanujan took a fresh perspective in approaching the topic. He wrote that several number theoretic properties of Bernoulli numbers would follow from properties of specific series. In this connection, he observed that if it could be proved that

$$2\left(\frac{1}{x+2} - \frac{1}{x+4} + \frac{1}{x+6} - \cdots\right)$$

could be expanded in ascending powers of $\frac{1}{x}$ with integer coefficients, then it would follow that $2^{2k}(2^{2k} - 1)\frac{B_{2k}}{2k}$ was an integer.

In fact, Ramanujan stated the Clausen-von Staudt formula as a result on series. However, this result had already been proved by K. Schwering, a student of Weierstrass and Kummer at Berlin. Schwering's proof involved infinite series.[43] He first showed that

$$\frac{1}{x} + \frac{1}{2x^2} + \frac{B_2}{x^3} + \frac{B_4}{x^5} + \frac{B_6}{x^7} + \frac{B_8}{x^9} + \cdots = \sum_{k=1}^{\infty} \frac{(k-1)!}{kx(x+1)(x+2)\cdots(x+k-1)},$$

$$(20.85)$$

derivable by combining (10.54) due to Stirling with (10.75) due to de Moivre. Note that in (20.85), the series on the left-hand side is an asymptotic series and the series on the right-hand side is an inverse factorial series that converges for $x > 0$. When the latter series is expanded in powers of $\frac{1}{x}$, the result is an asymptotic series and the different powers of $\frac{1}{x}$ can be equated. Unlike Stirling and de Moivre, Schwering

[40]  Clausen (1840).
[41]  von Staudt (1840).
[42]  Ramanujan (1911).
[43]  Schwering (1899).

regarded the first Bernoulli number as $-\frac{1}{2}$, as we do today. Thus, the formula of Schwering corresponding to (20.85) had $-\frac{1}{2x^2}$ as the second term on the left-hand side, so that the denominator on the right-hand side was $k(x+1)\cdots(x+k)$. Of course, this difference does not affect the calculation.

At this point, Schwering considered the term on the right-hand side of (20.85) (in Stirling's notation),

$$\frac{(k-1)!}{kx(x+1)(x+2)\cdots(x+k-1)}. \tag{20.86}$$

He observed that if $k$ were a composite integer greater than 4, then $k$ would divide $(k-1)!$. Thus, (20.86) could be expanded in powers of $\frac{1}{x}$ with integer coefficients, because (20.86) could be expressed as

$$
\frac{(k-1)!}{k} \cdot \frac{1}{x^k} \left(1+\frac{1}{x}\right)^{-1} \left(1+\frac{2}{x}\right)^{-1} \cdots \left(1+\frac{k-1}{x}\right)^{-1}
$$
$$
= \frac{(k-1)!}{kx^k} \left(1-\frac{1}{x}+\frac{1}{x^2}-\cdots\right) \cdots \left(1-\frac{k-1}{x}+\frac{(k-1)^2}{x^2}-\cdots\right).
$$

The idea of Schwering's proof of the Clausen-von Staudt formula was that since $B_{2k}$ was the coefficient of $\frac{1}{x^{2k+1}}$ on the left-hand side of (20.85), it was sufficient to show that, modulo the integers and for primes $p$, the coefficient of the odd power $\frac{1}{x^{2k+1}}$ was $\sum_{p-1|2k} \frac{1}{p}$.

For $k=4$, Schwering noted that the term in (20.86) became

$$\frac{3}{2x(x+1)(x+2)(x+3)}; \tag{20.87}$$

also note that

$$\frac{1}{x(x+1)(x+2)(x+3)} \equiv \frac{1}{x^2(x^2-1)} = \frac{1}{x^4}\left(1+\frac{1}{x^2}+\frac{1}{x^4}+\cdots\right) \text{ mod } 2. \tag{20.88}$$

Thus, the term (20.87) did not contribute a fraction to the odd powers of $\frac{1}{x}$ because these powers were all even integers that cancelled with the 2 in the denominator to produce an integer. We see that for the case in which $k$ was composite, Schwering did not have to consider the terms in (20.86). For the case $k=p$, a prime, he noted that the congruence

$$x(x+1)(x+2)\cdots(x+p-1) \equiv x(x^{p-1}-1) \text{ mod } p \tag{20.89}$$

and Wilson's theorem

$$(p-1)! \equiv -1 \text{ mod } p$$

implied that

$$\frac{(p-1)!}{px(x+1)(x+2)\cdots(x+p-1)} \equiv \frac{-1}{px(x^{p-1}-1)}. \qquad (20.90)$$

Note that Lagrange's proof of (20.89) and Wilson's theorem appear in Section 10.8. We may now write

$$\frac{-1}{px(x^{p-1}-1)} = \frac{-1}{px}\left(\frac{1}{x^{p-1}} + \frac{1}{x^{2(p-1)}} + \frac{1}{x^{3(p-1)}} + \cdots\right).$$

Therefore, if $p-1|2k$, then the term (20.86) contributes $-\frac{1}{p}$+ integer to the coefficient of $\frac{1}{x^{2k+1}}$. Thus, since $B_{2k}$ is the coefficient of $\frac{1}{x^{2k+1}}$ on the left-hand side of (20.85),

$$B_{2k} = -\sum_{p-1|2k}\frac{1}{p} + \text{integer},$$

completing the proof.

In his 1911 paper, Ramanujan wrote that the theorem now credited to Clausen and von Staudt, could be deduced from the proposition: The series

$$\frac{1}{2x^2} + \frac{1}{(x+1)^2} + \frac{1}{(x+2)^2} + \cdots - \frac{1}{x} - \frac{1}{6(x^3-x)}$$
$$+ \frac{1}{5(x^5-x)} + \frac{1}{7(x^7-x)} + \cdots,$$

where $5, 7, 11, 13, \ldots$ are primes greater than 3, can be expanded in ascending powers of $\frac{1}{x}$ with integral coefficients; note that this proposition can be inferred from Schwering's work. Samuel Wagstaff's paper[44] gives a full discussion of this matter and other results that Ramanujan rediscovered, with a complete list of references.

Ramanujan indicated a proof that $2^{2k}(2^{2k}-1)\frac{B_{2k}}{2k}$, with $k = 1, 2, 3\ldots$, were integers. Given his familiarity with series, Ramanujan most probably knew that these were tangent numbers, and thus integers, a fact first proved by Euler as we saw in equation (16.84). However, it appears that here Ramanujan wished to indicate a proof along the same lines as the proof of the Clausen-von Staudt theorem. He briefly stated that, based on the asymptotic series for $\frac{\Gamma'(x)}{\Gamma(x)}$, one could derive the equation

$$\frac{1}{x+2} - \frac{1}{x+4} + \frac{1}{x+6} - \frac{1}{x+8} + \frac{1}{x+10} - \cdots$$
$$= \frac{1}{2x} - 2(2^2-1)\frac{B_2}{2x^2} - 2^3(2^4-1)\frac{B_4}{4x^4} - 2^5(2^6-1)\frac{B_6}{6x^6} - \cdots. \qquad (20.91)$$

[44] Wagstaff (1981).

From Boole's formula, originally due to Euler, we may derive (20.91); in this connection see Section 21.8 on Boole's symbolic method. In Boole's formula,

$$\sum_{k=0}^{\infty} (-1)^k f(x+k) = \frac{1}{2} f(x) + \sum_{n=1}^{\infty} \frac{(1-2^{2n})B_{2n}}{(2n)!} \frac{d^{2n-1} f}{dx^{2n-1}},$$

take $f(x) = \frac{1}{x}$ and then subtract $\frac{1}{x}$ from each side; finally, change $x$ to $\frac{x}{2}$ to arrive at (20.91).

Next, express two times the left-hand side of (20.91) as an inverse factorial series to show that it can be expanded in ascending powers of $\frac{1}{x}$ with integer coefficients. To find this expression in terms of inverse factorials, we can follow a method that Ramanujan himself often employed, for example in his paper, "A series for Euler's constant $\gamma$."[45] Now since

$$\int_0^1 t^{\frac{x}{2}} dt = \frac{2}{x+2},$$

we have

$$2\left(\frac{1}{x+2} - \frac{1}{x+4} + \frac{1}{x+6} - \cdots\right) = \int_0^1 \left(t^{\frac{x}{2}} - t^{\frac{x}{2}+1} + t^{\frac{x}{2}+2} - \cdots\right) dt$$

$$= \int_0^1 \frac{t^{\frac{x}{2}}}{1+t} dt. \qquad (20.92)$$

Using integration by parts or from (17.24), we may write

$$\frac{1}{2} \int_0^1 t^{\frac{x}{2}} \left(\frac{1-t}{2}\right)^n dt = \frac{n!}{(x+2)(x+4)\cdots(x+2n+2)}. \qquad (20.93)$$

Applying (20.93) and the expansion

$$1 + \frac{1-t}{2} + \left(\frac{1-t}{2}\right)^2 + \cdots = \frac{1}{1 - \frac{1-t}{2}} = \frac{2}{1+t}, \quad 0 < t < 1,$$

we see that the right-hand side of (20.92) may be given as

$$\int_0^1 \frac{t^{\frac{x}{2}}}{1+t} dt = \frac{1}{2} \int_0^1 t^{\frac{x}{2}} \sum_{n=0}^{\infty} \left(\frac{1-t}{2}\right)^n dt$$

$$= \sum_{n=0}^{\infty} \frac{n!}{(x+2)(x+4)\cdots(x+2n+2)}.$$

[45] Ramanujan (1917).

Thus, we have shown that the left-hand side of (20.92) can be expanded as an inverse factorial series:

$$2\left(\frac{1}{x+2} - \frac{1}{x+4} + \frac{1}{x+6} - \cdots\right) = \sum_{n=0}^{\infty} \frac{n!}{(x+2)(x+4)\cdots(x+2n+2)}. \tag{20.94}$$

Moreover, since

$$\frac{n!}{(x+2)(x+4)\cdots(x+2n+2)} = \frac{n!}{x^{n+1}}\left(1+\frac{2}{x}\right)^{-1}\cdots\left(1+\frac{2n+2}{x}\right)^{-1},$$

we see that the right-hand side of (20.94) can be expanded in powers of $\frac{1}{x}$ with integer coefficients, thus confirming that the tangent numbers $2^{2k}(2^{2k}-1)\frac{B_{2k}}{2k}, k = 1, 2, 3, \ldots$ are integers.

Ramanujan also remarked that $2(2^{2k}-1)B_{2k}$ could be shown to be an integer if it could be proved that

$$\frac{1}{(x+1)^2} - \frac{1}{(x+2)^2} + \frac{1}{(x+3)^2} - \cdots \tag{20.95}$$

was expandable in powers of $\frac{1}{x}$ with integer coefficients. He was able to show that the expression (20.95) was equal to

$$\frac{1}{2x^2} - \sum_{k=1}^{\infty}(2^{2k}-1)\frac{B_{2k}}{x^{2k+1}}, \tag{20.96}$$

starting with de Moivre's formula (10.75), taken in the form

$$\frac{1}{(x+1)^2} + \frac{1}{(x+2)^2} + \frac{1}{(x+3)^2} + \cdots = \frac{1}{x} + \frac{1}{2x^2} + \sum_{k=1}^{\infty}\frac{B_{2k}}{x^{2k+1}}. \tag{20.97}$$

He then changed $x$ to $\frac{x}{2}$ in (20.95) and subtracted half of the resulting equation from (20.97) to accomplish this. Note that this result also follows from Boole's formula.

The next step would be to show that the series (20.95) can be expanded as an inverse factorial series with integer coefficients, and this can be done by taking the derivative of (20.94), so that

$$\sum_{k=1}^{\infty}\frac{(-1)^{k-1}}{(x+k)^2} = \sum_{n=0}^{\infty}\frac{n!}{(x+2)(x+4)\cdots(x+2n+2)}\left(\frac{1}{x+2} + \cdots + \frac{1}{x+2n+2}\right), \tag{20.98}$$

a step Ramanujan did not present. However, he made similar arguments in his paper on Euler's constant. We can now see that the right-hand side of (20.98) can be expanded in ascending powers of $\frac{1}{x}$ with integer coefficients. For example, we have

$$\frac{n!}{(x+2)(x+4)\cdots(x+2n+2)} = \frac{n!}{x^{2n+2}}\left(1+\frac{2}{x}\right)^{-2}\cdots\left(1+\frac{2n+2}{x}\right)^{-1},$$

and each of the $\left(1+\frac{2}{x}\right)^{-2}, \left(1+\frac{4}{x}\right)^{-1}, \cdots, \left(1+\frac{2n+2}{x}\right)^{-1}$ can be expanded in a series of the appropriate type. This proves that $2(2^{2k}-1)B_{2k}$ is an integer, a result that actually also follows from the Clausen-von Staudt theorem because we can verify that if $p-1|2k$, then $p|2^{2k}-1$. Thus, $2(2^2k-1)$ cancels every prime in the denominator of $B_{2k}$. Ramanujan was surely aware of this, but here he clearly wished to use purely series considerations to demonstrate that $2(2^{2k}-1)B_{2k}$ was an integer.

Recall that Euler had shown that $2(2^{2k}-1)B_{2k}, k = 1,2,3,\ldots$ were coefficients in the power series expansion of $\csc x$ and were therefore integers.[46] The reader may also see the discussion in Section 16.8 and, in particular, equation (16.86).

## 20.8 Exercises

(1) In his 1671 letter to Collins, Newton expressed his result on $\sum_{k=1}^{p}\frac{a}{b+kc}$ as follows:

> Any musical progression $\frac{a}{b}\cdot\frac{a}{b+c}\cdot\frac{a}{b+2c}\cdot\frac{a}{b+3c}\cdot\frac{a}{b+4c}$ etc. being propounded whose last term is $\frac{a}{d}$: for ye following operation choose any convenient number $e$ (whither whole broken or surd) which intercedes these limits $\frac{2mn}{b+d}$ and $\sqrt{mn}$; supposing $b-\frac{1}{2}c$ to bee $m$, and $b+\frac{1}{2}c$ to bee $n$. And this proportion will give you the aggregate of the terms very near the truth.
>
> As ye Logarithm of $\frac{e+\frac{1}{2}c}{e-\frac{1}{2}c}$ to ye Logarithm of $\frac{n}{m}$, so is $\frac{a}{e}$ to ye desired summe.

Verify Newton's approximation, stated in modern terminology: Let $m = b-\frac{1}{2}c$ and $n = d+\frac{1}{2}c$ and $d = b+(p-1)c$. Then

$$\sum_{k=1}^{p}\frac{a}{b+kc} \approx \frac{a\ln(\frac{n}{m})}{e\ln\left(\frac{e+\frac{1}{2}c}{e-\frac{1}{2}c}\right)},$$

where $\frac{2mn}{m+n} < e < \sqrt{mn}$. See Newton (1959–1960) vol. 1, pp. 68–70. The editorial notes on p. 70 are helpful.

(2) Suppose $p = \int\frac{\partial x}{x}\ln y$ and $q = \int\frac{\partial y}{y}\ln x$, where the symbols $\partial x, \partial y$ denote partial differentiation.

(a) Show that $p+q = \ln x\ln y + C$.

(b) Take $y = 1-x$, $0 < x < 1$. Show that

$$p = -\sum_{n=1}^{\infty}\frac{x^n}{n^2} \quad \text{and} \quad q = -\sum_{n=1}^{\infty}\frac{y^n}{n^2}.$$

---

[46] Eu. I-10$_2$ § 223. E 212.

Let $x \to 1$ to get $C = -\sum_{n=1}^{\infty} \frac{1}{n^2} = -\frac{\pi^2}{6}$. Thus,

$$\text{Li}_2(x) + \text{Li}_2(y) = \frac{\pi^2}{6} - \ln x \ln y, \qquad (20.99)$$

where $\text{Li}_2(x) = \sum_{n=1}^{\infty} \frac{x^n}{n^2}$.

(c) Take $y = x - 1$. Observe that

$$\ln y = \ln x + \ln\left(1 - \frac{1}{x}\right) = \ln x - \frac{1}{x} - \frac{1}{2x^2} - \frac{1}{3x^3} - \frac{1}{4x^4} - \text{etc.}$$

Show that

$$p = \frac{1}{2}(\ln x)^2 + \text{Li}_2\left(\frac{1}{x}\right) \text{ and } q = -\text{Li}_2(-y).$$

Deduce that

$$p + q = \frac{\pi^2}{6} + \ln x \ln \frac{y}{\sqrt{x}}.$$

(d) Deduce from (c) that for $a = \frac{\sqrt{5}-1}{2}$

$$\text{Li}_2(a) - \text{Li}_2(-a) = \frac{\pi^2}{6} - \ln a \ln(a\sqrt{a}).$$

The results and methods above are from Euler's paper, written in 1779 and published in 1811, devoted entirely to the topic of the dilogarithm. By using partial derivatives, he made the proof of (20.99) somewhat shorter than his 1729 proof of the same result. In 1735 Euler had also been able to evaluate $\zeta(2)$; he made use of this in his 1779 paper. See Eu. I-16-2 pp. 117–138.

(3) Let $s = 1^n + 2^n + 3^n + \cdots + x^n$. Use Euler–Maclaurin summation to prove

$$s = \frac{x^{n+1}}{n+1} + \frac{x^n}{2} + \frac{1}{2}\binom{n}{1}\frac{x^{n-1}}{6} - \frac{1}{4}\binom{n}{3}\frac{x^{n-3}}{30} + \frac{1}{6}\binom{n}{5}\frac{x^{n-5}}{42}$$

$$- \frac{1}{8}\binom{n}{7}\frac{x^{n-7}}{30} + \frac{1}{10}\binom{n}{9}\frac{5x^{n-9}}{66} - \frac{1}{12}\binom{n}{11}\frac{691x^{n-11}}{2730}$$

$$+ \frac{1}{14}\binom{n}{13}\frac{7x^{n-13}}{6} - \frac{1}{16}\binom{n}{15}\frac{3617x^{n-15}}{510} + \text{etc.}$$

Euler gave this formula in his 1736 paper (Eu. I-14 pp. 108–123. E 47.) and specifically listed the sums for $n = 1, 2, \ldots, 16$. Note the explicit appearance of the Bernoulli numbers, $\frac{1}{6}, -\frac{1}{30}, \frac{1}{42}, \ldots, -\frac{3617}{510}$ in Euler's presentation of the formula. Naturally, Euler wrote

$$\frac{1}{k+1}\binom{n}{k} \quad \text{as} \quad \frac{n(n-1)\cdots(n-k+1)}{1\cdot 2\cdots k(k+1)}.$$

(4) Verify Euler's computations leading to the value of $\gamma$ to fifteen decimal places. He took $n = 10$ in (20.4) and determined that

$$1 + \frac{1}{2} + \frac{1}{3} + \cdots + \frac{1}{10} = 2.9289682539682539.$$

He also knew that $\ln 10 = 2.302585092994045684$. He used precisely the terms in (20.4) so that he had to calculate

$$\frac{1}{20} - \frac{1}{1200} + \frac{1}{1200000} - \cdots + \frac{691}{32760 \times 10^{12}} - \frac{1}{12 \times 10^{14}}.$$

From this he obtained

$$\text{Const.} = \gamma = \lim_{n\to\infty}\left(\sum_{k=1}^{n}\frac{1}{k} - \ln n\right) = 0.5772156649015329.$$

This is just one example of the kind of numerical calculation that Euler undertook on a regular, if not daily, basis.

In this connection, also prove the inequalities of Mengoli:

$$\frac{1}{n+1} + \frac{1}{n+2} + \cdots + \frac{1}{np} < \ln p < \frac{1}{n} + \frac{1}{n+1} + \cdots + \frac{1}{np-1}. \quad (20.100)$$

Mengoli proved these in his *Geometria Speciosa* of 1659. See Hofmann's article on the Euler–Maclaurin formula, in Hofmann (1990) vol. 1, pp. 233–240, especially p. 237.

(5) Use the Euler–Maclaurin formula to show that

$$\sum_{k=1}^{\infty}\frac{1}{k^3} = 1.202056903159594,$$

$$\sum_{k=1}^{\infty}\frac{1}{k^4} = 1.0823232337110824.$$

These results and the evaluation of $\gamma$ in the previous exercise are in Euler's 1736 paper (Eu. I-14 pp. 118–121).

(6) Verify Euler's formal computations to obtain a formula for the alternating series

$$s(x) = f(x) - f(x+b) + f(x+2b) - \text{etc.}$$

From $s(x + 2b) - s(x) = -f(x) + f(x + b)$, deduce that

$$f(x + b) - f(x) = \frac{2b}{1!}\frac{ds}{dx} + \frac{4b^2}{2!}\frac{d^2s}{dx^2} + \text{etc.}$$

Thus,

$$\frac{b}{1!}\frac{df}{dx} + \frac{b^2}{2!}\frac{d^2f}{dx^2} + \text{etc.} = \frac{2b}{1!}\frac{ds}{dx} + \frac{4b^2}{2!}\frac{d^2s}{dx^2} + \text{etc.} \tag{20.101}$$

Assume with Euler that

$$\frac{ds}{dx} = \alpha\frac{df}{dx} + \beta\frac{d^2f}{dx^2} + \gamma\frac{d^3f}{dx^3} + \text{etc.}$$

and substitute in (20.101), equating coefficients, to get

$$2s = \text{Const.} + f(x) + \frac{b}{2}\frac{df}{dx} - \frac{b^3}{4!}\frac{d^3f}{dx^3} + \frac{3b^5}{6!}\frac{d^5f}{dx^5} - \frac{17b^7}{8!}\frac{d^7f}{dx^7}$$

$$+ \frac{155b^9}{10!}\frac{d^9f}{dx^9} - \frac{2073b^{11}}{12!}\frac{d^{11}f}{dx^{11}} + \frac{38227b^{13}}{14!}\frac{d^{13}f}{dx^{13}} - \text{etc.} \tag{20.102}$$

Euler used this formula to compute the series

$$\frac{1}{x} - \frac{1}{x+b} + \frac{1}{x+2b} - \frac{1}{x+3b} + \text{etc.}$$

He applied it to

$$\frac{\pi}{4} = 1 - \frac{1}{3} + \frac{1}{5} - \frac{1}{7} + \cdots$$

by taking $x = 25, b = 2$ and then computing $1 - \frac{1}{3} + \cdots - \frac{1}{23}$ separately. In this way he obtained a value of $\frac{\pi}{4}$ correct to eleven decimal places. The summation formula (20.102) is often called the Boole summation formula, though Euler had the result a hundred years before Boole. This work is in Eu. I-14 pp. 128–130.

(7) Using the method of Exercise 6, show that

$$f(x) - f(x + 1) + f(x + 2) - \cdots \tag{20.103}$$

$$= \frac{1}{2}f(x) + \sum_{n=1}^{\infty} \frac{(1 - 2^{2n})B_{2n}}{(2n)!}\frac{d^{2n-1}f}{dx^{2n-1}}. \tag{20.104}$$

Although this is not Euler's notation, he pointed out the connection between the terms of the series and the Bernoulli numbers. The paper appeared in 1788, though it was presented to the St. Petersburg Academy in 1776. See Eu. I-16-1 p. 57.

(8) Show formally that when $m$ is a positive integer

$$a^m - (a+b)^m + (a+2b)^m - (a+3b)^m + \cdots$$

$$= \frac{a^m}{2} - \frac{(2^2-1)B_2}{2}\binom{m}{1}a^{m-1}b - \frac{(2^4-1)B_4}{4}\binom{m}{3}a^{m-3}b^3 - \cdots$$

$$- \frac{(2^{m+1}-1)B_{m+1}}{m+1}b^m.$$

For $a=0, b=1$ this gives

$$0^m - 1^m + 2^m - 3^m + 4^m - \cdots = -\frac{B_{m+1}(2^{m+1}-1)}{m+1}.$$

This also follows from the formula (20.104) above. Euler used this obviously divergent series to prove the functional relation for the zeta function. See Eu. I-15 p. 76.

## 20.9 Notes on the Literature

Hardy (1949), a work on divergent series, gives an excellent treatment of the work of Jacobi and Poisson on the Euler–Maclaurin. Cohen (2007) presents a modern treatment of the E–M and its extensions, particularly to sums involving periodic arithmetic functions.

# 21

## *Operator Calculus and Algebraic Analysis*

### 21.1   Preliminary Remarks

The operator or operational calculus, the method of treating differential operators as algebraic objects, was once thought to have originated with the English physicist and electrical engineer Oliver Heaviside (1850–1925). Indeed, Heaviside revived and brilliantly applied this method to problems in mathematical physics. But the basic ideas can actually be traced back to Leibniz and Lagrange, who must be given credit as the founders of the operational method. With his notation for the differential and integral, Leibniz was able to regard some results on derivatives and integrals as analogs of algebraic results. The later insight of Lagrange was to extend this analogy to infinite series of differentials so that, in particular, he could write the Taylor expansion as an exponential function of a differential operator. In fact, this formal approach to infinite series appeared in the work of Newton himself. For Newton, infinite series in algebra served a purpose analogous to infinite decimals in arithmetic: They were necessary to carry out the algebraic operations to their completion. Newton's insightful algorithms using formal power series were of very wide applicability in analysis, algebra, and algebraic geometry; their power lay precisely in their formal nature. Thus, the algebraic analysis of the eighteenth century can trace its origins to Newton's genius. A branch of algebraic analysis focusing on the combinatorial aspects of power series was developed by the eighteenth-century German combinatorial school.

In a letter of May 1695 to Johann Bernoulli,[1] Leibniz pointed out the formal resemblance between the expression for the $n$th derivative of a product $xy$ and the binomial expansion of $(x + y)^n$. For $n = 2$, for example, Leibniz wrote

$$(x + y)^2 = 1x^2 + 2xy + 1y^2, \; d^2.xy = 1yddx + 2dydx + 1xddy. \qquad (21.1)$$

He made similar remarks in a September 1695 letter to l'Hôpital,[2] in which he used the symbol $p$ for the power (or exponent) so that the analogy would be even more evident. Thus, he denoted $x^n$ by $p^n x$, so that he could write

---

[1] Leibniz (1971) vol. 3/1, pp. 174–179, especially p. 175.
[2] Leibniz (1971) vol. I, pp. 297–303, especially pp. 301–302.

$$p^e(x+y) = p^e x \cdot p^0 y + \frac{e}{1} p^{e-1} x \cdot p^1 y + \frac{e \cdot e - 1}{1 \cdot 2} p^{e-2} x \cdot p^2 y + \cdots;$$

$$d^e(xy) = d^e x d^0 y + \frac{e}{1} d^{e-1} x \cdot d^1 y + \frac{e \cdot e - 1}{1 \cdot 2} d^{e-2} x \cdot d^2 y + \cdots.$$

The exponent $e$ could be a positive or negative integer; when $e = -n$ was negative, $d^{-n}$ denoted an $n$-fold integral. He mentioned that the $e = -1$ case was also noted by Johann Bernoulli. In fact, the formula in that case would be equivalent to Taylor's formula. Finally in 1710, Leibniz published a paper[3] on this symbolic analogy. Later, Lagrange, inspired by this paper, extended the scope of this analogy by treating the symbol $d$, denoting the differential operator, as an algebraic object. In a paper of 1772 presented to the Berlin Academy,[4] he gave the Taylor series formula as

$$u(x + \xi) = u + \frac{du}{dx}\xi + \frac{d^2u}{dx^2}\frac{\xi^2}{2} + \frac{d^3u}{dx^3}\frac{\xi^3}{2 \cdot 3} + \cdots = e^{\frac{du}{dx}\xi}, \qquad (21.2)$$

where $\left(\frac{du}{dx}\xi\right)^n$ in the expansion of the exponential was understood to be $\frac{d^n u}{dx^n}\xi^n$. We observe that this point of view was not foreign to Euler. In a brilliant paper of 1750,[5] he suggested replacing the $n$th derivative by $z^n$ in order to solve a differential equation of infinite order.

The generation of mathematicians after Lagrange chose, for clarity, to separate the symbol $\frac{d}{dx}$ from the function $u$ upon which it acted. Lacroix, for example, in his influential work summarizing the eighteenth-century discoveries in calculus, wrote $e^{\frac{du}{dx}\xi}$ as $e^{\xi \frac{d}{dx}} u$. It appears that the French mathematician L. F. Arbogast (1759–1803), the collector and preserver of important mathematical works, was the first to separate the operator from the object on which it operated. Arbogast's method, published in 1800, so impressed the English mathematician Charles Babbage that he wrote,[6]

Arbogast, in the 6th article of his, "Calcul des derivations," where, by a peculiarly elegant mode of separating the symbols of operation from those of quantity, and operating *upon them* as upon analytical symbols; he derives not only these, but many other much more general theorems with unparalleled conciseness.

Returning to Lagrange's paper, we note that he observed that the difference operator could be expressed as

$$\Delta u(x) = u(x + \xi) - u(x) = e^{\xi \frac{du}{dx}} - 1. \qquad (21.3)$$

In Arbogast's notation, write $\Delta u = \left(e^{\xi \frac{d}{dx}} - 1\right)u$. Lagrange applied this formula to obtain a formal though very simple derivation of the Euler–Maclaurin summation, and he extended this to situations involving sums of sums. Laplace used Lagrange's operational method in his work on difference equations; he also attempted to give a

[3] Leibniz (1971) vol. 5, pp. 377–382.
[4] Lagrange (1772).
[5] Eu. I-14 pp. 463–515. E 189.
[6] Babbage and Herschel (1813) p. xi.

rigorous derivation of Lagrange's formulas. In his work of 1800, Arbogast applied this technique to numerous problems, including the solutions of differential equations.

Then in 1808, Barnabé Brisson (1777–1828) independently applied this method to differential equations. A graduate of the École Polytechnique, Brisson published his paper in its journal. In the 1810s, other French mathematicians such as F. J. Servois (1768–1847) and J. F. Français (1775–1833) applied the methods of Lagrange and Arbogast to obtain some results on series. Servois also considered the logical foundation of the operator method. However, after the 1821 publication of Cauchy's *Analyse algébrique*, effectively establishing the limit concept at the foundation of analysis, the operational methods ceased to be developed in France. But in 1826, Cauchy presented a justification of the operational calculus using Fourier transforms. Interestingly, in the early twentieth century, integral transforms were applied to rigorize the operator methods employed by the physicist Heaviside. In 1926, Norbert Wiener created generalized harmonic analysis and one of his motivations was to provide rigorous underpinnings for the operational method.

During the 1830s and 1840s, important work in the operational calculus was done in Britain. Robert Murphy (1806–1843), Duncan Gregory (1813–1844), and George Boole (1815–1864) applied the methods to somewhat more difficult problems than those considered by Franais and Servois. Much of the British work was done without full awareness of the earlier Continental work, so that even as late as 1851, William F. Donkin (1814–1869) published a paper in the *Cambridge and Dublin Mathematical Journal* giving an exposition of Arbogast's method of derivations. Thus, the British work was not a direct continuation of the work of Arbogast, Franais, and Servois; its origins and motivations lay in a more formal and/or symbolic mathematical approach.

To understand the historical background of the British operational calculus, note that Britain produced a number of outstanding mathematicians in the first half of the eighteenth century, including Cotes, de Moivre, Taylor, Stirling, and Maclaurin. A large part of their work elaborated on or continued the study of topics opened up by Newton. There were also some good textbook writers such as Thomas Simpson and Edmund Stone who explained these developments to a larger audience. In the second half of the century, there was a swift decline in the development of mathematics in Britain. Mathematics was sustained at Cambridge by the almost solitary figure of Edward Waring, whose main interests were algebra and combinatorics, but he had few followers or students and little influence. Also, John Landen did interesting work in analysis, making a significant contribution to elliptic integrals.

British mathematicians had long paid scant attention to the major mathematical advances in continental Europe: the calculus of several variables and its applications to problems of mathematical physics developed by Euler, Fontaine, Clairaut, d'Alembert, Lagrange, and Laplace; major works in algebra produced by Euler, Lagrange, Vandermonde, and Ruffini; and also the brilliant progress in number theory made by Euler and Lagrange. In the early nineteenth century, Robert Woodhouse (1773–1827) appears to be one of the first British mathematicians to attempt to expand the focus of mathematics at Cambridge. He leaned strongly toward a formal or symbolic approach and his main interests lay in the foundations of calculus and the appropriate notation for its development. He also wrote expository works in subjects such as the calculus of variations and gravitation, and his efforts brought continental work on these topics to the notice of the British. In 1803, Woodhouse wrote *The Principles of Analytical Calculation*, a polemical work on the foundation of calculus. He reviewed the foundational ideas of his predecessors: Newton, Leibniz, d'Alembert, Landen, and Lagrange. He rejected the limits of Newton and d'Alembert as well as the infinitesimals of Leibniz as inconsistent and inadequate, advocating instead the algebraic approach of Lagrange and Arbogast, though he disputed specific details. In the preface of his book, he wrote,[7]

> I regard the rule for the multiplication of algebraic symbols, by which addition is compendiously exhibited, as the true and original basis of that calculus, which is equivalent to the fluxionary or differential calculus; on the direct operations of multiplication, are founded the reverse operations of division and extraction of roots, ... they are still farther comprehended under a general formula, called the expansion, or development of a function: from the second term of this expansion, the fluxion or differential of a quantity may immediately be deduced, and in a particular application, it appears to represent the velocity of a body in a motion.

Concerning the equal sign, $=$, Woodhouse maintained that in the context of series, this sign did not denote numerical equality but the result of an operation. So if $\frac{1}{1+x}$ denoted the series obtained by dividing 1 by $1 + x$, then

$$\frac{1}{1+x} = 1 - x + x^2 - x^3 + \cdots .$$

On the other hand, if $\frac{1}{x+1}$ represented the series obtained when 1 was divided by $x+1$, then

$$\frac{1}{x+1} = \frac{1}{x} - \frac{1}{x^2} + \frac{1}{x^3} - \frac{1}{x^4} + \cdots .$$

Woodhouse remarked with reference to the two series that the equality $\frac{1}{1+x} = \frac{1}{x+1}$ could not be affirmed.

Woodhouse wrote other articles and books advocating his formal point of view. In his 1809 textbook, *A Treatise on Plane and Spherical Trigonometry*, he defined the trigonometric functions by their series expansions and made arguments for the advantages of the analytic approach over the geometric approach of Newton's *Principia*. Though his 1809 treatise acquired some popularity and went into several editions, Woodhouse was unable to convert the Cambridge dons. Progress in introducing the analytic approach into the curriculum was achieved mainly through the efforts of his students: Edward Ffrench Bromhead (1789–1855), Charles Babbage (1791–1871), George Peacock (1791–1858), and John Herschel (1792–1871). As students at Cambridge, they formed the Analytical Society in 1812 to promote and practice analytical mathematics; they decided to publish a journal called the *Memoirs of the Analytical Society*, though Babbage had wished to name it *The Principles of Pure D-ism* in opposition to the Dot-age of the university. Only one volume was

---

[7] Woodhouse (1803) p. II.

published; it appeared in 1813 and contained one article by Babbage on functional equations and two by Herschel, on trigonometric series and on finite difference equations. As the members scattered, the Analytical Society ceased to meet, though many of its members became fellows or professors at Cambridge. In any case, Babbage, Peacock, and Herschel remained friends. They translated an elementary text by Lacroix on differential and integral calculus and in 1820 published a supplementary collection of examples on calculus, difference equations, and functional equations. Their efforts gradually influenced the teaching of mathematics at Cambridge, leading to the acceptance of Continental methods.

Even though Babbage succeeded Woodhouse as Lucasian Professor of Mathematics in 1828, he spent very little time at Cambridge. Consistent with his formal approach, he became interested in the mechanization of computation and spent the rest of his life on the problems associated with that. He first developed plans to construct a "difference engine," complete with printing device; he hoped it would eventually compute up to twenty decimal places using sixth-order differences. A Swedish engineer, Georg Schentz, used Babbage's description to build a machine with a printer capable of computing eight decimal places using fourth-order differences. Instead of actually building a machine, Babbage himself went on to design a more elaborate computer, called the "analytical engine," inspired by the study of Jacquard's punched cards weaving machine.

John Herschel lost interest in pure mathematics and became a professor of astronomy at Cambridge. So that left George Peacock to carry out the reform or modernization of the teaching of mathematics at Cambridge and more generally in England. In 1830, he first published an algebra textbook, later published in two volumes[8] in which he attempted to put the theory of negative and complex numbers on a firm foundation by dividing algebra into two parts, arithmetical and symbolical. The symbols of arithmetical algebra represented positive numbers, whereas the domain of the symbols in symbolical algebra was extended by the principle of the permanence of equivalent forms. This abstract principle implied, according to Peacock, that any formula in symbolical algebra would yield a formula in arithmetical algebra if the variables were properly chosen. Note that this approach excluded the possibility of a noncommutative algebra.

Ironically, this algebraic approach to calculus taken by the British mathematicians of the 1820s and 1830s stood in contrast with the rigorous methods contemporaneously introduced in Europe by Gauss, Cauchy, Abel, and Dirichlet. The next generation of British mathematicians, including Duncan Gregory, Robert Murphy, George Boole, Leslie Ellis (1817–1858) and others, were aware of the continental approach and yet they felt that their own methods had legitimacy. Early death prevented the talented Duncan Gregory from preparing a new foundation for this method. However, the symbolic method, even if lacking in rigor, had significant influence. The origin of some aspects of modern operational calculus and of the theory of distributions can be seen in the symbolic methods of Gregory and Boole.

---

[8] Peacock (1843–1845).

Moreover, some of the methods themselves were put on a more solid foundation through G. C. Rota's twentieth-century work on umbral calculus.[9]

The British symbolic approach served as the starting point for some significant developments: the symbolic logic of Boole and Augustus De Morgan (1806–1871) and the invariant theory of Boole, Cayley, and Sylvester. Consider, for example, Boole's remarks in the introduction to his 1847 work, *The Mathematical Analysis of Logic*:[10]

> They who are acquainted with the present state of the theory of Symbolical Algebra, are aware, that the validity of the processes of analysis does not depend upon the interpretation of the symbols which are employed, but solely upon the laws of their combination. Every system of interpretation which does not affect the truth of the relations supposed, is equally admissible.

G. H. Hardy wrote in his book *Divergent Series* that the British symbolical mathematicians had the spirit but not the accuracy of the twentieth-century algebraists.[11] Nevertheless, there is at least one example of abstract algebraic work consistent with the standards of today: Hamilton's theory of couples and of quaternions. The former laid a rigorous algebraic basis for complex numbers. And Hamilton reported that his 1843 discovery of quaternions was guided by a determination for consistency, so that he left open the possibility of an algebra with zero divisors or with noncommutativity. It is noteworthy that around 1819, Gauss composed a multiplication table for quaternions,[12] though apparently he did not develop this further.

Papers by Murphy, Duncan Gregory, and Boole published between 1835 and 1845 provided important steps toward the creation of concepts laying the groundwork for the eventual construction of abstract algebraic theories. Murphy, the son of a shoemaker-parish clerk in Cork County, Ireland, studied mathematics on his own, and his talent soon became known. In 1819, Mr. Mulcahy, a teacher in Cork County, published mathematical problems in the local newspaper; he soon began to receive original solutions from an anonymous reader. He was surprised to discover that his correspondent was a boy of 13. After this, Murphy began to receive encouragement and financial assistance to continue his studies. In 1825, some of his work was brought to the attention of Woodhouse; consequently, Murphy was admitted to Gonville and Caius College, Cambridge, from which he graduated in 1828. In an 1835 paper on definite integrals, Murphy introduced the idea of orthogonal functions, giving them the name reciprocal functions. In his 1837 paper,[13] "First Memoir on the Theory of Analytical Operations," he defined what he called linear operations and showed that their sums and products, obtained by composition, were also linear operations, though the products were not necessarily commutative. He stated a binomial theorem for noncommutative operations, and went on to consider inverses of operations, proving that the inverse of the product of two operations $A$ and $B$ was $B^{-1}A^{-1}$. Murphy also defined the kernel of an operation, naming it the appendage of the operation. He applied his theory mainly to three operations: the differential operator, the difference

---

[9] See, for example, Mullin and Rota (1970).
[10] Boole (1847) p. 3.
[11] Hardy (1949) p. 18.
[12] Gauss (1863–1927) vol. 8, pp. 357–361.
[13] Murphy (1837).

operator $\Delta$, and the operator transforming a function $f(x)$ to $f(x + h)$. Thus, in his paper Murphy isolated and defined some basic ideas of a system of abstract algebra.

Around this time, Duncan F. Gregory, descendent of the great James Gregory, also began to develop his mathematical ideas. Gregory was born at Edinburgh, Scotland, and graduated from Trinity College, Cambridge, in 1837. Even as a student, Gregory was interested in mathematical research and in encouraging British mathematicians to take up this activity. As a step in this direction, in 1837 he helped found the *Cambridge Mathematical Journal,* of which he was the editor until a few months before his premature death in February 1844. R. Leslie Ellis, who served as editor after this, wrote that Gregory was particularly well qualified for this position for "his acquaintance with mathematical literature was very extensive, while his interest in all subjects connected with it was not only very strong, but also singularly free from the least tinge of jealous or personal feeling. That which another had done or was about to do, seemed to give him as much pleasure as if he himself had been the author of it, and this even when it related to some subject which his own researches might seem to have appropriated."[14] In addition, D. F. Gregory encouraged undergraduates to publish and permitted authors to publish anonymously so that they need not fear for their reputations. This journal was later renamed *Cambridge and Dublin Mathematical Journal*; it then evolved into the *Quarterly Journal of Pure and Applied Mathematics,* with editors including William Thomson and J. W. L. Glaisher. In fact, most British mathematicians of the period contributed papers to the *CMJ,* including Augustus De Morgan, J. J. Sylvester, George Gabriel Stokes, Arthur Cayley, George Boole, and William Thomson.

By about 1845, research in operational calculus was no longer widely pursued. In the 1890s, however, Heaviside revived operational methods in order to solve differential equations occurring in electrical engineering problems. Heaviside may or may not have independently devised these methods, but he made at least one important new contribution: The use of the step function $H(t) = 0$ with $t$ negative, and $H(t) = 1$ with $t$ nonnegative. By taking the derivative of this function, he obtained the Dirac delta function. In some situations, Heaviside also used the derivative of the delta function. Because these methods were so successful in solving problems in electrical engineering, mathematicians such as Wiener, Carson, Doetsch, and van der Pol made substantial progress toward putting them on a rigorous footing.

Toward the end of the eighteenth century, algebraic analysis was taken in a different direction by the German combinatorial school founded by C. F. Hindenburg (1741–1808), Professor at Leipzig. Since combinatorial considerations were important in probability computations as well as in deriving formulas for higher derivatives of products of functions and of compositions of functions, Hindenburg saw that he could find relations between/among series through the use of combinatorial concepts. This school took as its starting point and inspiration Euler's extensive use of series to tackle various mathematical problems, as set forth in his *Introductio in Analysin Infinitorum* of 1748. The combinatorial school played a significant role

---

[14] Ellis (1845).

in the overall development of mathematics in Germany;[15] it included among its early members Christian Kramp (1760–1826), Gauss's thesis supervisor J. F. Pfaff (1765–1825), and H. A. Rothe (1773–1842). Later, H. F. Scherk (1798–1885), Franz Ferdinand Schweins (1780–1856), August Leopold Crelle (1780–1855), Weierstrass's teacher Christoph Gudermann (1798–1852), and Moritz A. Stern (1807–1894) made contributions to this tradition, and many of them were active in instituting educational reforms in Germany. In fact, it is very likely that Weierstrass chose to make power series the fundamental object in his study of analysis because of his early contact with Gudermann. Also, Riemann's earliest research on fractional derivatives and infinite series was done while he was a student under Stern, though Riemann eventually took a completely different route as a result of his later association with Dirichlet and Gauss. The combinatorial school produced some interesting results useful even today, and their approach to infinite series is not without significance in modern mathematical research.

Hindenburg believed, and his colleagues agreed, that his most important work was the polynomial formula he proved in 1779. A power series raised to an exponent is another power series:

$$(1 + a_1x + a_2x^2 + a_3x^3 + \cdots)^m = 1 + A_1x + A_2x^2 + A_3x^3 + \cdots.$$

Hindenburg's formula expressed $A_n$ in terms of $a_1, a_2, \ldots, a_n$. De Moivre had already done this[16] with $m$ a positive integer in a paper of 1697. Leibniz and Johann Bernoulli also considered this case in letters exchanged in 1695. This particular case is quite useful; for example, it can be applied to give a short proof of Faà di Bruno's formula, giving the $n$th derivative of a composition of two functions. Faà di Bruno stated this in the mid-nineteenth century without referring to the earlier proofs;[17] Arbogast offered a proof in 1800.[18]

Using Newton's binomial theorem, Hindenburg extended his formula to fractional and negative $m$ by expanding $(1 + y)^m$, where $y = a_1x + a_2x^2 + a_3x^3 + \cdots$, and $y^n$ was obtained from the polynomial theorem for positive integral $n$. Part of Hindenburg's achievement was to clarify the combinatorial content of the formula. De Moivre had given only the recursive rule for the calculation of $A_{n+1}$ from $A_n$. In the notation presented by B. F. Thibaut in his 1809 textbook *Grundriss der Allgemeinen Arithmetik*,[19] Hindenburg's formula would be expressed as

$$A_n = \sum_{h=1}^{n} \binom{m}{h} p^{n^h} C.$$

The symbol $n^h C$ represented the sum of all products of $h$ factors taken from $a_1, a_2, \ldots, a_n$, so that the sum of the indices in each summand was $n$. The symbol $p$

[15] See Jahnke (1993).
[16] de Moivre (1697).
[17] Faà di Bruno (1857).
[18] Arbogast (1800) pp. 30–31, 310–313.
[19] Jahnke (1993).

stood for the coefficient associated with each summand, each summand consisting of $h$ factors, and this coefficient gave the number of different permutations of the $h$ factors. Thus, $^{6^3}C$ stood for $a_1^2 a_4 + a_1 a_2 a_3 + a_2^3$ and the number of terms in the sum was the number of partitions of 6 with exactly three parts. Therefore, $p^{6^3}C$ represented

$$\frac{3!}{2!\,1!} a_1^2 a_4 + \frac{3!}{(1!)^3} a_1 a_2 a_3 + a_2^3.$$

The combinatorial school set great importance on this formula, overestimating its potential. Still, Hindenburg's formula is useful in power series manipulation.

In 1793, H. A. Rothe used Hindenburg's formula to state the reversion of series formula as a combinatorial relation. Two years later, Rothe and Pfaff showed the equivalence of Rothe's formula with the Lagrange inversion formula. In modern times, Lagrange's formula has been regarded as more combinatorial than analytic in nature; in this respect, the combinatorialists were on the right track. Rothe also found one important terminating version of the $q$-binomial theorem, published in the preface of his 1811 book.[20] In this formula, the coefficients of the powers of $x$ are $q$-extensions of the binomial coefficients, now called Gaussian polynomials. It is possible that Rothe may have discovered these polynomials even before Gauss's work of 1805, published in 1811. It would be nice to know Rothe's combinatorial interpretation of these polynomials; he gave no proof or comment. In order to get an insight into the combinatorialists' mathematical style, consider comment given by Thomas Muir in his monumental *The Theory of Determinants in the Historical Order of Development*:[21]

> Rothe was a follower of Hindenburg, knew Hindenburg's preface to Rüdiger's *Specimen Analyticum*, and was familiar with what had been done by Cramer and Bézout . . . . His memoir is very explicit and formal, proposition following definition, and corollary following proposition, in the most methodical manner.

Christian Kramp taught mathematics, chemistry, and experimental physics at École Centrale in Cologne and in 1809 he became professor of mathematics and dean of the faculty of science at Strasbourg. He was a follower of Hindenburg and contributed articles to various journals edited by Hindenburg. In a paper of 1796,[22] he derived some interesting properties of Stirling numbers. One object of interest for him was what he termed a factorial:

$$a^{n|d} = a(a+d)(a+2d) \cdots (a+(n-1)d).$$

He expanded this as a polynomial in $a$ and $d$, and obtained formulas for the coefficients involving Stirling numbers. Denoting the Stirling numbers of the first and second kinds by $s(n,k)$ and $S(n,k)$, Kramp proved that[23]

20   Rothe (1811).
21   Muir (1960) vol. I, p. 55.
22   Kramp (1796).
23   See Knuth (1992).

$$|s(n+1, n+1-k)| = \sum \binom{n+1}{k+l} \frac{(k+l)!}{j_1! \, 2^{j_1} j_2! \, 3^{j_2} j_3! \, 4^{j_3} \cdots},$$

$$S(n+k, n) = \sum \binom{n+k}{k+l} \frac{(k+l)!}{j_1! \, (2!)^{j_1} j_2! \, (3!)^{j_2} j_3! \, (4!)^{j_3} \cdots},$$

where the sums were over all nonnegative $j$ such that $j_1 + 2j_2 + 3j_3 + \cdots + kj_k = k$, and where $l = j_1 + j_2 + j_3 + \cdots + j_k$. Kramp also introduced the factorial notation, $n!$.

## 21.2 Euler's Solution of a Difference Equation

In a very interesting paper of 1750,[24] "De serierum determinatione seu nova methodus inveniendi terminos generales serierum," Euler used symbolic calculus to solve the difference equation

$$y(x) - y(x-1) = X(x). \tag{21.4}$$

In his solution, Euler applied to differential equations of infinite order the method applicable to differential equations of finite order; thus, his work is difficult to confirm. However, it is interesting to compare Euler's results with the results obtained by conventional methods. By applying the Taylor series to expand $y(x-1)$, he had

$$y(x-1) = y(x) - y'(x) + \frac{1}{2!} y''(x) \cdots,$$

and he could write equation (21.4) in the form

$$\frac{dy}{dx} - \frac{1}{2!} \frac{d^2 y}{dx^2} + \frac{1}{3!} \frac{d^3 y}{dx^3} - \cdots = X(x). \tag{21.5}$$

Before working with equation (21.4), Euler took the case $X \equiv 0$ and obtained the solution of (21.5) as a series $\sum a_n \sin nx$. We can omit discussion of this portion of the paper, as the solution for this case is contained within the solution in general of equation (21.5). Euler viewed (21.5) as a nonhomogeneous equation of infinite order with constant coefficients and, in a highly unorthodox way, he applied the same technique that he had earlier applied to nonhomogeneous equations equations of finite order with constant coefficients. His method of solving such differential equations is discussed in our Section 14.6. We restate that result: Suppose $\alpha_1, \alpha_2, \ldots, \alpha_n$ are the roots of the $n$th-degree polynomial $P(x)$. Then with $D = \frac{d}{dx}$, the equation

$$P(D)y = X$$

has the solution

$$y = \frac{e^{\alpha_1 x}}{P'(\alpha_1)} \int e^{-\alpha_1 x} X dx + \cdots + \frac{e^{\alpha_n x}}{P'(\alpha_n)} \int e^{-\alpha_n x} X dx. \tag{21.6}$$

[24] Eu. I-14 pp. 463–515. E 189.

Euler observed that if $\frac{d^n y}{dx^n}$ on the left-hand side of (21.5) were replaced by $z^n$, $n = 1, 2, 3\ldots$, then the result would be

$$z - \frac{z^2}{2!} + \frac{z^3}{3!} - \cdots = 1 - e^{-z}.$$

The roots of $P(z) = 1 - e^{-z}$ were given by $2\pi in$, where $n$ was an integer. Thus, $P'(e^{2\pi in}) = 1$; the solution of (21.5), an equation of infinite order, would be given by

$$y = \int X\,dx + \sum_{k=1}^{\infty} \left( e^{2\pi ikx} \int e^{-2\pi ikx} X\,dx + e^{-2\pi ikx} \int e^{2\pi ikx} X\,dx \right). \quad (21.7)$$

Euler combined the two terms in parentheses within the summation in (21.7) to obtain

$$y = \int X\,dx + 2\sum_{k=1}^{\infty} \left( \cos(2\pi kx) \int \cos(2\pi kx) X\,dx + \sin(2\pi kx) \int \sin(2\pi kx) X\,dx \right);$$

$$(21.8)$$

here note that

$$e^{\pm 2\pi ikx} = \cos(2\pi kx) \pm \sin(2\pi kx).$$

Observe that because its integrals are indefinite, (21.8) is not a Fourier series, though it may resemble one. Euler applied (21.8) to the difference equation

$$y(x) - y(x-1) = \ln x, \quad y(0) = 0. \quad (21.9)$$

Note that a solution of this equation is $y(x) = \ln\Gamma(x+1)$. Substituting $X = \ln x$ in (21.8) gave Euler

$$y = \int \ln x\,dx + 2\sum_{k=1}^{\infty} \left( \cos(2\pi kx) \int \cos(2\pi kx) X\,dx \right.$$

$$\left. + \sin(2\pi kx) \int \sin(2k\pi x) \ln x\,dx \right). \quad (21.10)$$

Euler calculated the integrals

$$\int \ln x\,dx = x \ln x - x, \quad (21.11)$$

$$\int \ln x \cos mx\,dx = \frac{1}{m}\sin mx \left( \ln x + \frac{1}{m^2 x^2} - \frac{3!}{m^4 x^4} + \frac{5!}{m^6 x^6} - \cdots \right)$$

$$+ \frac{1}{m}\cos mx \left( \frac{1}{mx} - \frac{2!}{m^3 x^3} + \frac{4!}{m^5 x^5} - \frac{6!}{m^7 x^7} + \cdots \right), \quad (21.12)$$

$$\int \ln x \sin mx \, dx = -\frac{1}{m} \cos mx \left( \ln x + \frac{1}{m^2 x^2} - \frac{3!}{m^4 x^4} + \frac{5!}{m^6 x^6} - \cdots \right)$$

$$+ \frac{1}{m} \sin mx \left( \frac{1}{mx} - \frac{2!}{m^3 x^3} + \frac{4!}{m^5 x^5} - \frac{6!}{m^7 x^7} + \cdots \right).$$

$$(21.13)$$

He did not explain his method in obtaining (21.12) and (21.13), but certainly one way of deriving them is to use integration by parts. For example,

$$\int \ln x \cos mx \, dx = \frac{\ln x}{m} \sin mx - \frac{1}{m} \int \sin mx \, \frac{dx}{x}$$

$$= \frac{\ln x}{m} \sin mx + \frac{\cos mx}{m^2 x} - \frac{1}{m^2} \int \cos mx \, \frac{dx}{x^2}$$

and so on. Euler noted that (21.12) and (21.13) implied that

$$2 \cos mx \int \ln x \cos mx \, dx + 2 \sin mx \int \ln x \sin mx \, dx$$

$$= \frac{2}{m^2 x} \left( 1 - \frac{2!}{m^2 x^2} + \frac{4!}{m^4 x^4} - \frac{6!}{m^6 x^6} + \text{etc.} \right) + a_m \cos mx + b_m \sin mx, \quad (21.14)$$

where $a_m$ and $b_m$ were constants of integration. He next substituted (21.11) and (21.14) in (21.10) to obtain, with $P$ a constant of integration,

$$y = P + x \ln x - x + \sum_{k=1}^{\infty} \left( a_k \cos(2k\pi x) + b_k \sin(2k\pi x) \right) + \sum_{k=1}^{\infty} \frac{B_{2k}}{2k(2k-1)x^{2k-1}},$$

$$(21.15)$$

where he applied (16.31) to obtain the infinite series with Bernoulli numbers $B_{2k}$. Since $y(1) = 0$, he set $x = 1$ in (21.15) to obtain

$$0 = P - 1 + \sum_{k=1}^{\infty} a_k + C, \quad C = \sum_{k=1}^{\infty} \frac{B_{2k}}{2k(2k-1)}.$$

Euler noted that he had found $C = \frac{1}{2} \ln 2\pi$ in his differential calculus book. For Euler's derivation of $C$, see (20.19) and the subsequent discussion.

Euler was thus able to rewrite (21.15) as

$$y = x \ln x - x + \frac{1}{2} \ln 2\pi + \sum_{k=1}^{\infty} \left( a_k \cos(2k\pi x) + b_k \sin(2k\pi x) \right)$$

$$+ \sum_{k=1}^{\infty} \frac{B_{2k}}{2k(2k-1)x^{2k-1}}.$$

$$(21.16)$$

Note that (21.16) is the equation for $\ln \Gamma(x+1)$; it has an indeterminacy in that $a_k$ and $b_k$ are not known. Of course, this is understandable because $\ln \Gamma(x)$ is not completely determined by the two conditions given in (21.9):

$$\ln \Gamma(x+1) = \ln x + \ln \Gamma(x), \qquad \ln \Gamma(1) = 0.$$

Recall that the Bohr–Mollerup–Artin theorem required a third condition: that $\ln \Gamma(x)$ be a convex function. Note also that (21.16) implies that

$$\Gamma(x+1) = \sqrt{2\pi}\, \frac{x^x}{e^x}\, e^{\sum (a_n \cos nx + b_n \sin nx)} \cdot e^{\frac{1}{12x} - \frac{1}{360x^3} + \cdots},$$

$$\Gamma(x+1) = \sqrt{2\pi}\, x^x\, e^{-x}\, e^{\sum (a_n \cos nx + b_n \sin nx)} \cdot e^{\frac{1}{12x} - \frac{1}{360x^3} + \cdots} \tag{21.17}$$

and since

$$\Gamma(x+1) \sim \sqrt{2\pi}\, x^{x+\frac{1}{2}} e^{-x} \quad \text{as } x \to \infty,$$

by (21.17) we have

$$e^{\sum (a_n \cos nx + b_n \sin nx)} \sim \sqrt{x} \quad \text{as } x \to \infty. \tag{21.18}$$

Thus, the convexity of $\ln \Gamma(x)$ has somehow allowed us to obtain (21.18).

We add the observation that in 1923, the Danish mathematician Niels Erik Nörlund gave the solution to the equation $y(x+1) - y(x) = X(x)$ in an alternate form.[25] His result may be stated:

$$y(x) = \lim_{t \to 0} \left( \int_a^\infty X(x) e^{-tx}\, dx - \sum_{s=0}^\infty X(x+s)\, e^{-t(x+s)} \right), \tag{21.19}$$

providing the limit exists. Sufficient conditions for its existence are that $X(x)$, for all $x > 0$, has a continuous derivative of order $m$ that tends to 0 as $x \to \infty$; that the integral $\int_0^\infty B_m(-t + [-t])\, X^{(m)}(x+t)\, dt$ converges uniformly over $x \in (0,1)$, where $B_m(t)$ represents the $m$th Bernoulli polynomial.

## 21.3  Lagrange's Extension of the Euler–Maclaurin Formula

In his 1772 "Sur une nouvelle espèce de calcul," Lagrange set out to create a new symbolic method in calculus.[26] As a first step, he expressed the Taylor series of a function $u(x, y, z, \dots)$ of several variables as

$$u(x + \xi, y + \psi, z + \zeta, \dots) = e^{\xi \frac{du}{dx} + \psi \frac{du}{dy} + \zeta \frac{du}{dz} + \cdots}.$$

[25] Nörlund (1923). Also see Nörlund (1924).
[26] Lagrange (1772).

In his symbolic notation, Lagrange understood the numerator of the $n$th term of $e^{\frac{du}{dx}\xi+\cdots}$ to represent $\left(\xi\frac{d}{dx}+\psi\frac{d}{dy}+\cdots\right)^{n}u$ rather than $\left(\xi\frac{du}{dx}+\psi\frac{du}{dy}+\cdots\right)^{n}$. In effect, Lagrange was treating the derivative operator as an algebraic quantity. The later notation of Arbogast makes this approach clearer,[27] allowing us to write $e^{\xi\frac{d}{dx}}u$ for Lagrange's $e^{\frac{d}{dx}\xi}$ and $\left(e^{\xi\frac{d}{dx}}-1\right)^{\lambda}u$ for his $\left(e^{\frac{du}{dx}\xi}-1\right)^{\lambda}$. It is easy to see that the last expression is the symbolic form of the $\lambda$th difference $\Delta^{\lambda}u$, since we may write

$$\Delta u = u(x+\xi)-u(x) = \left(e^{\xi\frac{d}{dx}}-1\right)u. \tag{21.20}$$

It follows that the difference operator $\Delta$ can be identified with the operator $e^{\xi\frac{d}{dx}}-1$ and the repeated application of these operations yields

$$\Delta^{\lambda}u = \left(e^{\xi\frac{d}{dx}}-1\right)^{\lambda}u. \tag{21.21}$$

Following Leibniz, Lagrange noted that, given the derivative operator $d$, he could write

$$d^{-1}=\int,\ d^{-2}=\int^{2},\cdots$$

and

$$\Delta^{-1}=\Sigma,\ \Delta^{-2}=\Sigma^{2},\ldots.$$

Here $\int^{2}$ stood for an iterated integral and $\Sigma^{2}$ for an iterated sum. Lagrange applied his symbolic method to a generalization of the Euler–Maclaurin summation by expanding the expression in (21.21). He assumed the series expansion

$$(e^{\omega}-1)^{\lambda}=\omega^{\lambda}(1+A\omega+B\omega^{2}+C\omega^{3}+D\omega^{4}+\cdots)$$

and took the logarithm to obtain

$$\lambda\ln(e^{\omega}-1)-\lambda\ln\omega = \ln(1+A\omega+B\omega^{2}+C\omega^{3}+D\omega^{4}+\cdots).$$

By differentiation, he found

$$\lambda\left(\frac{e^{\omega}}{e^{\omega}-1}-\frac{1}{\omega}\right)=\frac{A+2B\omega+3C\omega^{2}+4D\omega^{3}+\cdots}{1+A\omega+B\omega^{2}+C\omega^{3}+D\omega^{4}+\cdots}. \tag{21.22}$$

Since

$$\frac{e^{\omega}}{e^{\omega}-1}=\frac{1}{1-e^{-\omega}}=\frac{1}{\omega-\frac{\omega^{2}}{2}+\frac{\omega^{3}}{2\cdot3}-\frac{\omega^{4}}{2\cdot3\cdot4}+\cdots},$$

he obtained the equation

---

[27] Arbogast (1800) p. 350.

$$\lambda\left(\frac{1}{2} - \frac{\omega}{2\cdot 3} + \frac{\omega^2}{2\cdot 3\cdot 4} - \cdots\right)(1 + A\omega + B\omega^2 + C\omega^3 + \cdots)$$

$$= \left(1 - \frac{\omega}{2} + \frac{\omega^2}{2\cdot 3} - \frac{\omega^3}{2\cdot 3\cdot 4} + \cdots\right)(A + 2B\omega + 3C\omega^2 + \cdots). \quad (21.23)$$

Finally, by equating the coefficients of the powers of $w$, Lagrange found that

$$A = \frac{\lambda}{2}, \ 2B = \frac{(\lambda+1)A}{2} - \frac{\lambda}{2\cdot 3}, \ 3C = \frac{(\lambda+2)}{2}B - \frac{(\lambda+1)}{2\cdot 3}C + \frac{\lambda}{2\cdot 3\cdot 4},$$

$$4D = \frac{(\lambda+3)}{2}C - \frac{(\lambda+2)}{2\cdot 3}B + \frac{(\lambda+1)}{2\cdot 3\cdot 4}A - \frac{\lambda}{2\cdot 3\cdot 4\cdot 5} \ \text{etc.} \quad (21.24)$$

Now because $\Delta^{-\lambda} = \Sigma^{\lambda}$, Lagrange could replace $\lambda$ by $-\lambda$ in (21.21) to get his extension of the Euler–Maclaurin formula:

$$\Sigma^{\lambda}u = \frac{\int^{\lambda} u\, dx^{\lambda}}{\xi^{\lambda}} + \alpha\frac{\int^{\lambda-1} u\, dx^{\lambda-1}}{\xi^{\lambda-1}} + \beta\frac{\int^{\lambda-2} u\, dx^{\lambda-2}}{\xi^{\lambda-2}} + \cdots. \quad (21.25)$$

We observe that when $\lambda$ was changed to $-\lambda$ in (21.24), Lagrange denoted the changed values of $A, B, C, \ldots$ by $\alpha, \beta, \gamma, \ldots$. The case $\lambda = 1$ in (21.25) is immediately distinguishable as the Euler–Maclaurin formula. In the same paper, Lagrange then proceeded to derive a formula for repeated integrals in terms of sums. He rewrote
(21.20) as

$$\xi\frac{du}{dx} = \ln(1 + \Delta)u$$

or in Arbogast's notation,

$$\xi\frac{d}{dx} = \ln(1 + \Delta) \quad (21.26)$$

and more generally

$$\xi^{\lambda}\frac{d^{\lambda}}{dx^{\lambda}} = [\ln(1 + \Delta)]^{\lambda}. \quad (21.27)$$

By expanding the right-side of (21.27) as a series in powers of $\Delta$, Lagrange obtained the coefficients of the expansion by a method similar to the one that gave him (21.24). Once again, he replaced $\lambda$ by $-\lambda$ to obtain

$$\frac{\int^{\lambda} u\, dx^{\lambda}}{\xi} = \sum^{\lambda} u + \mu\sum^{\lambda-1} u + \nu\sum^{\lambda-2} u + \omega\sum^{\lambda-3} u + \cdots, \quad (21.28)$$

where

$$\mu = \frac{\lambda}{2}, \ 2\nu = \frac{(\lambda-1)\mu}{2} - \frac{\lambda}{2\cdot 3}, \ 3\omega = \frac{(\lambda-2)\nu}{2} - \frac{(\lambda-1)\mu}{2\cdot 3} + \frac{\lambda}{3\cdot 4},$$

$$4\chi = \frac{(\lambda - 3)\,\omega}{2} - \frac{(\lambda - 2)v}{2 \cdot 3} + \frac{(\lambda - 1)\mu}{3 \cdot 4} - \frac{\lambda}{4 \cdot 5}, \text{ etc.}$$

When $\lambda = 1$ in (21.28), Lagrange got the value of the integral in terms a sum involving finite differences:

$$\frac{\int u\,dx}{\xi} = \sum u + \mu u + v\Delta u + \omega\Delta^2 u + \chi\Delta^3 u + \cdots. \tag{21.29}$$

This is exactly the formula communicated by James Gregory to Collins in November 1670. Recall that Gregory most probably discovered this formula by integrating the Gregory–Newton interpolation formula. Lagrange may not have been aware of Gregory's work, but he referred to Cotes, Stirling, and others who used similar, though not identical, results. Laplace may have found this result independently; he used it in his astronomical work. Indeed, this formula was sometimes attributed to Laplace, in particular by Poisson.

Lagrange employed (21.29) to derive an inverse factorial series for $\ln\left(1 + \frac{1}{x}\right)$, taking $u = \frac{1}{x}$ and $\xi = 1$ to obtain

$$\ln x = \sum \frac{1}{x} + \frac{\mu}{x} + v\Delta\frac{1}{x} + \omega\Delta^2\frac{1}{x} + \chi\Delta^3\frac{1}{x} + \cdots,$$

where

$$\sum \frac{1}{x} = \frac{1}{x-1} + \frac{1}{x-2} + \frac{1}{x-3} + \cdots,$$

$$\Delta\frac{1}{x} = \frac{1}{x+1} - \frac{1}{x} = -\frac{1}{x(x+1)}, \tag{21.30}$$

$$\Delta^2\frac{1}{x} = \frac{2}{x(x+1)(x+2)}, \tag{21.31}$$

$$\Delta^3\frac{1}{x} = -\frac{2 \cdot 3}{x(x+1)(x+2)(x+3)}, \text{ etc.} \tag{21.32}$$

Changing $x$ to $x + 1$, Lagrange obtained a similar series for $\ln(x + 1)$. He then subtracted the series for $\ln x$ to obtain the desired result.

Lagrange's heuristic method was immediately welcomed as a powerful tool in discovering interesting and useful formulas. Laplace's papers on finite differences in the 1770s discussed and used Lagrange's symbolic method. Laplace thought that Lagrange's formula $\Delta^n u = \left(e^{h\frac{d}{dx}} - 1\right)^n u$ could be rigorously established by the use of formal power series. He observed that

$$\Delta^n u = \frac{d^n u}{dx^n}h^n + A_1\frac{d^{n+1}u}{dx^{n+1}}h^{n+1} + A_2\frac{d^{n+2}u}{dx^{n+2}}h^{n+2} + \cdots \tag{21.33}$$

for constant $A_1, A_2, \ldots$. He believed the problem could be reduced to proving that these coefficients were identical with the coefficients of the powers of $h$ in the expansion of $\left(e^h - 1\right)^n$. He noted that the constants $A_1, A_2, \ldots$ were the same for all functions $u$, so he took $u = e^x$. Then $\Delta e^x = e^{x+h} - e^x = e^x \left( e^h - 1 \right)$ and, more generally, $\Delta^n e^x = e^x \left( e^h - 1 \right)^n$. Thus,

$$\left( e^h - 1 \right)^n = h^n + A_1 h^{n+1} + A_2 h^{n+2} + \cdots , \tag{21.34}$$

completing his argument.

In 1807, John Brinkley (1766-1835), professor of mathematics at the University of Dublin and a mentor to Hamilton, presented in the *Philosophical Transactions* an interesting expression for the constants $A_1, A_2, \ldots$ .[28] He noted that

$$
\begin{aligned}
\left( e^h - 1 \right)^n &= e^{nh} - \binom{n}{1} e^{(n-1)h} + \binom{n}{2} e^{(n-2)h} - \cdots \\
&= \left( 1 - \binom{n}{1} + \binom{n}{2} - \cdots \right) \\
&\quad + \left( n - \binom{n}{1}(n-1) + \binom{n}{2}(n-2) - \cdots \right) \frac{h}{1!} \\
&\quad + \left( n^2 - \binom{n}{1}(n-1)^2 + \binom{n}{2}(n-2)^2 - \cdots \right) \frac{h^2}{2!} + \cdots \\
&\quad + \left( n^m - \binom{n}{1}(n-1)^m + \binom{n}{2}(n-2)^m - \cdots \right) \frac{h^m}{m!} + \cdots . \tag{21.35}
\end{aligned}
$$

Furthermore, it was clear from the formula

$$\Delta^n f(0) = f(n) - \binom{n}{1} f(n-1) + \binom{n}{2} f(n-2) - \cdots \tag{21.36}$$

that the coefficient of $\frac{h^m}{m!}$ in (21.35) was $\Delta^n 0^m$. Thus, Brinkley had

$$A_k = \frac{\Delta^n 0^{n+k}}{(n+k)!}. \tag{21.37}$$

Note that, by (10.67), these numbers are related to Stirling numbers of the second kind:

$$(n+k)! \, A_k = n! \, S(n+k, n). \tag{21.38}$$

Brinkley (c. 1763–1835) studied at Cambridge, graduating senior wrangler in 1788. He was the first Royal Astronomer of Ireland and later became Bishop of Cloyne. It was to Brinkley that the 17-year-old Hamilton communicated his work on geometrical optics. Brinkley encouraged Hamilton by presenting his work to the Irish Academy with the legendary assertion, "This young man, I do not say will be, but is, the

---

[28] Brinkley (1807). See also Lacroix, Babbage, and Herschel (1816) p. 478.

first mathematician of his age."[29] Fittingly, Hamilton succeeded Brinkley as Royal Astronomer.

Finally, we note some very interesting connections with the results discussed in this section. First, when the coefficients of $A_k$ in (21.33), given by Brinkley's formula as

$$A_k = \frac{n!\, S(n+k,n)}{(n+k)!},$$
(21.39)

are substituted in (21.34), the result is the exponential generating function for the Stirling numbers of the second kind:

$$\frac{(e^x - 1)^n}{n!} = \sum_{m=n}^{\infty} \frac{S(m,n)x^m}{m!}.$$
(21.40)

We may also derive the generating function for Stirling numbers of the first kind from results given in this section, combined with Stirling's (10.50) and (10.57). When Lagrange's formulas (21.30) through (21.32), of which Stirling was well aware, are substituted in Stirling's (10.50), we arrive at

$$-D\left(\frac{1}{z}\right) = -\Delta \frac{1}{z} + \frac{1}{2}\Delta^2 \frac{1}{z} - \frac{1}{3}\Delta^3 \frac{1}{z} + \cdots,$$
(21.41)

where $D$ denotes the derivative with respect to $z$. Next, (21.41) may be written as

$$D = \log(1 + \Delta),$$

an expression equivalent to (21.26). Thus, we can see that Stirling's (10.51) and (10.52) give the formulas for

$$\frac{\left(\log(1 + \Delta)\right)^2}{2!} \quad \text{and} \quad \frac{\left(\log(1 + \Delta)\right)^3}{3!}$$

as series in powers of $\Delta$. We may then generalize these formulas by using (10.57), with $n = \lambda$, and apply

$$\Delta^m \left(\frac{1}{z}\right) = \frac{(-1)^m\, m!}{z(z+1)\cdots(z+m)}$$
(21.42)

to get

$$(-1)^\lambda D^\lambda \left(\frac{1}{z}\right) = (-1)^\lambda \frac{\left(\log(1+\Delta)\right)^\lambda}{\lambda!}\left(\frac{1}{z}\right) = \sum_{m=\lambda}^{\infty} c(m,\lambda)(-1)^m \frac{\Delta^m \left(\frac{1}{z}\right)}{m!}.$$
(21.43)

---

[29] This remark appears in many places, including Robert Perceval Graves's article about Hamilton in the *Dublin University Magazine* of 1842, vol. 19, pp. 94–110.

Furthermore, since by (10.60) $s(m,\lambda) = (-1)^{m-\lambda} c(m,\lambda)$, (21.43) implies that

$$\frac{\left(\log(1+x)\right)^{\lambda}}{\lambda!} = \sum_{m=\lambda}^{\infty} s(m,\lambda) \frac{x^m}{m!}, \tag{21.44}$$

giving us the exponential generating function for the Stirling numbers of the first kind.

Lagrange's calculations produce another relation: between Bernoulli numbers and Stirling numbers of the second kind. To see this, note that the left-hand side of (21.22) may be expanded in terms of Bernoulli numbers:

$$\frac{e^{\omega}}{e^{\omega}-1} - \frac{1}{\omega} = \frac{1}{\omega}\left(\frac{\omega e^{\omega}}{e^{\omega}-1} - 1\right)$$

$$= \frac{1}{\omega}\left(\sum_{n=0}^{\infty} B_n(1)\frac{\omega^n}{n!} - 1\right)$$

$$= \sum_{n=1}^{\infty} B_n(1)\frac{\omega^{n-1}}{n!}$$

$$= \frac{1}{2}\sum_{n=1}^{\infty} B_{n+1}\frac{\omega^n}{(n+1)!}.$$

Now $B_1(1) = \frac{1}{2}$ and $B_n(1) = B_n$ when $n \geq 2$; therefore, (21.23) can be rewritten as

$$\lambda\left(\frac{1}{2} + \sum_{k=1}^{\infty} B_{k+1}\frac{\omega^k}{(k+1)!}\right)\left(1 + \sum_{k=1}^{\infty} A_k \omega^k\right) = \sum_{k=1}^{\infty} k A_k \omega^{k-1},$$

where $A_k$ is given by (21.39). Equating the coefficients of $\omega^{n-1}$ on both sides yields the required result:

$$\frac{n S(\lambda+n,\lambda)}{\lambda(\lambda+1)\cdots(\lambda+n)} = A_{n-1} + \sum_{k=1}^{n} \frac{B_k A_{n-k}}{k!}$$

where

$$A_k = \frac{S(\lambda+k,\lambda)}{(\lambda+1)\cdots(\lambda+k)}. \tag{21.45}$$

## 21.4   Français's Method of Solving Differential Equations

Jacques F. Français, whose mathematical work incorporated some results from the notebooks of his late brother François, based his solution of ordinary differential equations with constant coefficients[30] on the relation between the Arbogast operator $E$ and the differential operator $D$:

$$E \phi(x) = \phi(x+1) = e^D \phi, \tag{21.46}$$

where $D = \frac{d}{dx}$. We note that Français's notation had $\delta$ for $D$. Now if $\phi$ were a solution of the equation

$$\frac{d\phi}{dx} - a\phi = 0, \quad \text{or} \quad (D-a)\phi = 0, \tag{21.47}$$

then Français had $D - a = 0$ by the separation of the operator. By (21.46) he had $E = e^D = e^a$ or $E^k = e^{ak}$ and hence $1 = e^{ak} E^{-k}$. He then used this relation to solve (21.47):

$$\phi(x) = 1\phi(x) = e^{ak} E^{-k} \phi(x) = e^{ak} \phi(x-k)$$

or

$$\phi(k) = \phi(0)e^{ak}.$$

Thus, $\phi(x) = Ce^{ax}$, where $C$ was a constant, and the differential equation (21.47) was solved. To solve the general homogeneous differential with constant coefficients

$$D^n \phi + a_1 D^{n-1} \phi + \cdots + a_n \phi = 0,$$

Français separated the operator and factored the $n$th degree polynomial in $D$ to obtain

$$(D - \alpha_1)(D - \alpha_2) \cdots (D - \alpha_n) = 0.$$

This gave him the $n$ equations:

$$D - \alpha_1 = 0, \; D - \alpha_2 = 0, \ldots, \; D - \alpha_n = 0$$

whose solutions he expressed as $e^{\alpha_1 x}$, $e^{\alpha_2 x}$, $\ldots, e^{\alpha_n x}$. Note that this is an exhaustive list of the independent solutions, under the condition that $\alpha_1, \alpha_2, \ldots, \alpha_n$ are all distinct. Français also applied the operational method for the summation of series. He found a series for $\pi$, reminiscent of a result obtained by the Kerala school. In a paper of 1811, he started with Euler's series

$$\frac{\pi}{4}\alpha = \sin \alpha - \frac{1}{3^2} \sin 3\alpha + \frac{1}{5^2} \sin 5\alpha - \frac{1}{7^2} \sin 7\alpha + \cdots.$$

[30] Français (1812–1813).

He rewrote this as

$$\frac{\pi}{2}\alpha\sqrt{-1}$$

$$= \left(e^{\alpha\sqrt{-1}} - e^{-\alpha\sqrt{-1}}\right) - \frac{1}{3^2}\left(e^{3\alpha\sqrt{-1}} - e^{-3\alpha\sqrt{-1}}\right) + \frac{1}{5^2}\left(e^{5\alpha\sqrt{-1}} - e^{-5\alpha\sqrt{-1}}\right) - \cdots .$$

$$(21.48)$$

He then set $\alpha\sqrt{-1} = D$, $e^D = E$ to obtain

$$\frac{\pi}{2}D = (E - E^{-1}) - \frac{1}{3^2}(E^3 - E^{-3}) + \frac{1}{5^2}(E^5 - E^{-5}) - \cdots .$$

Français next applied this operator equation to $\phi(x)$, so that

$$\frac{\pi}{2}\phi'(x) = \phi(x+1) - \phi(x-1) - \frac{1}{3^2}\left(\phi(x+3) - \phi(x-3)\right) + \cdots .$$

Recall that $E\phi(x) = \phi(x+1)$. Taking $\phi(x) = x$, he obtained Leibniz's formula. For $\phi(x) = \frac{1}{x}$, he found

$$\frac{\pi}{4}\cdot\frac{1}{x^2} = \left(\frac{1}{x^2-1}\right) - \frac{1}{3}\left(\frac{1}{x^2-3^2}\right) + \frac{1}{5}\left(\frac{1}{x^2-5^2}\right) - \frac{1}{7}\left(\frac{1}{x^2-7^2}\right) + \cdots .$$

$$(21.49)$$

Putting $\frac{1}{x^2} = -a$, he could rewrite as

$$\frac{\pi}{4} = \frac{1}{1+a} - \frac{1}{3}\cdot\frac{1}{1+3^2a} + \frac{1}{5}\cdot\frac{1}{1+5^2a} - \frac{1}{7}\cdot\frac{1}{1+7^2a} + \cdots .$$

Then again, by taking $\phi(x) = \ln x$, Français obtained the formula

$$\frac{\pi}{2}\cdot\frac{1}{x} = \ln\left(\frac{x+1}{x-1}\right) - \frac{1}{3^2}\ln\left(\frac{x+3}{x-3}\right) + \frac{1}{5^2}\ln\left(\frac{x+5}{x-5}\right) - \cdots .$$

Finally, to derive another interesting series, he set $a\sqrt{-1} = \frac{1}{x}$ in (21.49) and then integrated, obtaining

$$\frac{\pi}{4}a = \arctan a - \frac{1}{3^2}\arctan 3a + \frac{1}{5^2}\arctan 5a - \cdots .$$

## 21.5  Herschel: Calculus of Finite Differences

In the appendix to their English translation of Lacroix's book,[31] Babbage and Herschel included a large number of examples on functional and difference equations, some of which were original. Like his followers, Herschel showed much manipulative ability

---

[31] Lacroix, Babbage, and Herschel (1816).

of an algebraic kind. To see some of Herschel's work on difference equations, consider the equation

$$u_{x+1}u_x - a(u_{x+1} - u_x) + 1 = 0,$$

first given in a paper of Laplace but for which Herschel found a new solution. He differentiated and rewrote it as

$$(a + u_{x+1})\frac{du_x}{dx} - (a - u_x)\frac{du_{x+1}}{dx} = 0.$$

He solved for $a$ in the original equation and used this value in the second equation to find, after simplification

$$\frac{du_{x+1}}{1 + u_{x+1}^2} - \frac{du_x}{1 + u_x^2} = 0, \text{ or } \Delta \int \frac{du_x}{1 + u_x^2} = A,$$

where $A$ was a constant depending on $a$. Solving this simple difference equation, he had

$$\int \frac{du_x}{1 + u_x^2} = Ax + C,$$

where $C$ was an arbitrary constant. After computing the integral, he got $u_x = \tan(Ax + C)$ and therefore

$$u_{x+1} = \tan(Ax + C + A) = \frac{u_x + \tan A}{1 - u_x \tan A}.$$

At this point he rewrote the original difference equation as

$$u_{x+1} = \frac{u_x + \frac{1}{a}}{1 - u_x \cdot \frac{1}{a}},$$

so he could conclude that $\tan A = \frac{1}{a}$ or $A = \tan^{-1}\frac{1}{a}$. Thus, Herschel obtained the result

$$u_x = \tan\left(x \tan^{-1}\frac{1}{a} + C\right).$$

To see an example of Herschel's symbolic approach, take an analytic function $f(x)$, and let

$$f(e^t) = A_0 + A_1 t + A_2 t^2 + \cdots + A_x t^x + \cdots.$$

Herschel wished to find an expression for $A_x$; he started with the Taylor expansion

$$f(e^t) = f(1) + \frac{f'(1)}{1}(e^t - 1) + \frac{f''(1)}{1 \cdot 2}(e^t - 1)^2 + \cdots,$$

and noted that for $x \geq 1$, the coefficient of $t^x$ in $f(1)$ was 0; in

$$\frac{f'(1)}{1}(e^t - 1),$$

it was

$$\frac{f'(1)}{1} \cdot \frac{1}{1 \cdot 2 \cdots x} = \frac{f'(1)}{1} \cdot \frac{\Delta 0^x}{1 \cdot 2 \cdots x};$$

and in

$$\frac{f''(1)}{1 \cdot 2}(e^t - 1)^2,$$

it was

$$\frac{f''(1)}{1 \cdot 2} \cdot \frac{\Delta^2 0^x}{1 \cdot 2 \cdots x},$$

and so on. He could conclude that

$$A_x = \frac{1}{1 \cdot 2 \cdots x}\left(f(1) \cdot 0^x + \frac{f'(1)}{1}\Delta 0^x + \frac{f''(1)}{1 \cdot 2}\Delta^2 0^x + \cdots\right).$$

He then wrote, "Let the symbols of operation be separated from those of quantity, and we get

$$A_x = \frac{1}{1 \cdot 2 \cdots x}\left(f(1) + \frac{f'(1)}{1}\Delta + \frac{f''(1)}{1 \cdot 2}\Delta^2 + \cdots\right)0^x = \frac{f(1 + \Delta)0^x}{1 \cdot 2 \cdots x}. \text{ "}$$

Herschel apparently saw that taking a function of an operator was somewhat problematic, commenting that it should be understood to have no other meaning than its development, of which it is a mere abbreviated expression."

## 21.6  Murphy's Theory of Analytical Operations

Murphy began his 1837 paper[32] on analytical operations with the statement, "The elements of which every distinct analytical process is composed are three, namely, first the *Subject*, that is, the symbol on which a certain notified operation is to be performed; secondly, the *Operation* itself, represented by its own symbol; thirdly, the *Result*, which may be connected with the former two by the algebraic sign of equality." He defined several operations. For example, he denoted by $\Psi$ the operation changing $x$ to $x+h$, and by $\Delta$ the operation subtracting the subject from the result of changing $x$ to $x + h$ in the subject. He wrote these operations as

$$[f(x)]\Psi = f(x + h), \quad [f(x)]\Delta = f(x + h) - f(x).$$

[32] Murphy (1837).

The operations themselves could be algebraically combined. Thus, $\Psi = \Delta + 1$, where 1 was the operation under which the subject remained the same. Murphy defined the linearity of an operation:

$$[f(x) + \phi(x)]\Psi = [f(x)]\Psi + [\phi(x)]\Psi.$$

He called two operations fixed or free, depending on whether they were noncommutative or commutative in the given situation. Thus,

$$[x^n]x\Psi = [x^{n+1}]\Psi = (x + h)^{n+1},$$
$$[x^n]\Psi x = [(x + h)^n]x = x(x + h)^n,$$

so that $x\Psi \neq \Psi x$; but for a constant $a$

$$[x^n]a\Psi = [ax^n]\Psi = a(x + h)^n,$$
$$[x^n]\Psi a = [(x + h)^n]a = a(x + h)^n,$$

so that $a\Psi = \Psi a$.

He also stated a noncommutative binomial theorem: When $\theta$ and $\theta'$ were fixed operations,

$$(\theta + \theta')^n = \theta^{(n)} + \theta^{(n-1)}\theta' + \theta^{(n-2)}\theta'^{(2)} + \cdots + \theta\theta'^{(n-1)} + \theta'^{(n)}.$$

Here the term $\theta^{(n-1)}\theta'$ represented the sum of $n$ terms formed by placing $\theta'$ at the beginning, at the end, and in all the $n - 2$ intermediate positions of the expression $\theta \cdot \theta \cdots \theta = \theta^{n-1}$. Similarly, $\theta^{(n-2)}\theta'^{(2)}$ signified a similar sum of $\frac{n(n-1)}{2}$ terms and so on.

Murphy carefully defined some important algebraic concepts, such as the inverse and the kernel. Concerning the inverse: "Suppose $\theta$ to represent any operation which performed on a subject $[u]$ gives $y$ as the result, then the inverse operation is denoted by $\theta^{-1}$, and is such that when $[y]$ is made the subject $u$ becomes the result." The kernel was called the appendage, denoted by $[0]\theta^{-1}$. Murphy showed, for example, that if $d_x$ denoted the derivative with respect to $x$, then $[0]d_x^{-1}$ consisted of all the constants. To prove this, he took $[0]d_x^{-1} = \phi(x)$, implying that $[\phi(x)]d_x = 0$, and hence

$$[\phi(x)]d_x^2 = 0, [\phi(x)]d_x^3 = 0, \ldots.$$

Murphy then employed Taylor's theorem,

$$\phi(x + h) = \phi(x) + h\frac{d\phi}{dx} + \frac{h^2}{1 \cdot 2}\frac{d^2\phi}{dx^2} + \cdots,$$

to obtain $\phi(x + h) = \phi(x)$, meaning $\phi(x)$ was a constant. Here Murphy assumed without comment that $\phi$ was analytic. To find the kernel of $d_x^{-n}$, he observed that

$$[0]d_x^{-2} = [0]d_x^{-1}d_x^{-1} = [c]d_x^{-1} + [0]d_x^{-1} = cx + c',$$

and, more generally,

$$[0]d_x^{-n} = A_1 x^{n-1} + A_2 x^{n-2} + \cdots + A_n.$$

In another interesting example, Murphy let the subject be $f(x + y)$; let $\Psi_x$ be the operation under which $x$ received an increment $h$; and let $\Psi_y$ be the operation under which $y$ received an increment of $h$. Then, obviously, $[f(x + y)](\Psi_x - \Psi_y) = 0$, and therefore $f(x + y)(\Delta_y - \Delta_x) = 0$, so that $f(x + y)$ was a value in $[0](\Delta_y - \Delta_x)^{-1}$. He explained that $(\Delta_y - \Delta_x)^{-1}$ could be expanded as

$$(\Delta_y - \Delta_x)^{-1} = \Delta_y^{-1} + \Delta_y^{-2}\Delta_x + \Delta_y^{-3}\Delta_x^2 + \Delta_y^{-4}\Delta_x^3 + \cdots,$$

so that

$$[0](\Delta_y - \Delta_x)^{-1} = [0]\left(\Delta_y^{-1} + \Delta_y^{-2}\Delta_x + \Delta_y^{-3}\Delta_x^2 + \cdots\right). \tag{21.50}$$

Murphy then explained how to derive the Gregory–Newton interpolation formula from (21.50). He noted that $[0]\Delta_y^{-1}$ was a function independent of $y$, so he had $[0]\Delta_y^{-1} = \phi(x)$, where $\phi$ was an arbitrary function of $x$. Similarly, $[0]\Delta_y^{-2} = \phi(x) \cdot \frac{y}{h}$, where the appendage was omitted without loss of generality. Then again,

$$[0]\Delta_y^{-3} = \phi(x) \cdot \frac{y(y - h)}{1 \cdot 2 \cdot h^2}$$

for

$$[y(y - h)]\Delta_y = (y + h)y - y(y - h) = 2hy.$$

By a similar argument,

$$[0]\Delta_y^{-4} = \phi(x) \cdot \frac{y(y - h)(y - 2h)}{1 \cdot 2 \cdot 3 \cdot h^3},$$

and so on. In this way, Murphy obtained the relation

$$[0](\Delta_y - \Delta_x)^{-1} = \phi(x) + \frac{y}{h} \cdot \Delta\phi(x) + \frac{y(y - h)}{1 \cdot 2 \cdot \cdot h^2} \cdot \Delta^2\phi(x)$$
$$+ \frac{y(y - h)(y - 2h)}{1 \cdot 2 \cdot 3 \cdot h^3} \cdot \Delta^3\phi(x) + \cdots,$$

and "since $f(x + h)$ is included in this general expression, the particular form to be assigned to the arbitrary $\phi(x)$ is known by making $y = 0$, which gives $\phi(x) = f(x)$." Thus, Murphy had the Gregory–Newton interpolation formula,

$$f(x + y) = f(x) + \frac{y}{h}\Delta f(x) + \frac{y(y - h)}{1 \cdot 2 \cdot h^2}\Delta^2 f(x) + \frac{y(y - h)(y - 2h)}{1 \cdot 2 \cdot 3 \cdot h^3}\Delta^3 f(x) + \cdots.$$

Note that this is equivalent to (9.7). From this he derived the binomial theorem as Cotes had done, and perhaps James Gregory before him. Murphy took $f(x) = (1 + b)^x$,

$h = 1$ and observed that $\Delta^n f(x) = (1+b)^x b^n$ so that the binomial theorem followed after dividing both sides of the equation by $(1+b)^x$:

$$(1+b)^y = 1 + yb + \frac{y(y-1)}{1 \cdot 2} b^2 + \frac{y(y-1)(y-2)}{1 \cdot 2 \cdot 3} b^3 + \cdots .$$

## 21.7 Duncan Gregory's Operational Calculus

Duncan Gregory published many papers on operational calculus, illustrating the power of the method by elegant derivations of known results.[33] Gregory's proof of Leibniz's formula for the $n$th derivative of a product of two functions began with the observation that Euler's proof of the binomial theorem

$$(a+b)^n = a^n + na^{n-1}b + \frac{n(n-1)}{1 \cdot 2} a^{n-2}b^2 + \frac{n(n-1)(n-2)}{1 \cdot 2 \cdot 3} a^{n-3}b^3 + \cdots$$

required that $n$ was a fraction. More importantly, Gregory wrote that $a, b$ should satisfy

(1) The commutative law, $\quad ab = ba$,

(2) The distributive law, $\quad c(a+b) = ca + cb$,

(3) The index law, $\quad a^m \cdot (a^n) = a^{m+n}$.

Gregory added, "Now, since it can be shown that the operations both in the Differential Calculus and the Calculus of Finite Differences are subject to these laws, the Binomial Theorem may be at once assumed as true with respect to them, so that it is not necessary to repeat the demonstration of it for each case." To prove Leibniz's theorem, Gregory observed that

$$\frac{d}{dx}(uv) = u \frac{dv}{dx} + v \frac{du}{dx}.$$

He then rewrote this equation, as had Arbogast:

$$\frac{d}{dx}(uv) = \left( \frac{d'}{dx} + \frac{d}{dx} \right) uv,$$

where $\frac{d'}{dx}$ acted on $v$ but not on $u$ and $\frac{d}{dx}$ acted on $u$ but not on $v$. Since these operations were independent of each other, they commuted, so that

$$\left( \frac{d}{dx} \right)^n (uv) = \left( \frac{d'}{dx} + \frac{d}{dx} \right)^n uv$$

$$= \left( \left( \frac{d'}{dx} \right)^n + n \left( \frac{d'}{dx} \right)^{n-1} \frac{d}{dx} + \cdots \right) uv$$

$$= u \frac{d^n v}{dx^n} + n \frac{d^{n-1}v}{dx^{n-1}} \frac{du}{dx} + \frac{n(n-1)}{1 \cdot 2} \frac{d^2 u}{dx^2} \frac{d^{n-2}v}{dx^{n-2}} + \cdots .$$

[33] Gregory (1865) pp. 14–27, 108–123.

. Gregory remarked that this result was true with $n$ negative or fractional, or "in the cases of integration and general differentiation." He then took $v = 1$ and $n = -1$ to obtain Bernoulli's formula,

$$\int u \, dx = xu - \frac{x^2}{1 \cdot 2} \frac{du}{dx} + \frac{x^3}{1 \cdot 2 \cdot 3} \frac{d^3 u}{dx^3} - \cdots .$$

Using Arbogast's $E$ operator, Gregory derived a proof of the Newton–Montmort transformation, given by Euler in his 1755 differential calculus book. Suppose

$$S = ax + a_1 x^2 + a_2 x^3 + a_3 x^4 + \cdots$$

and $a_1 = Ea, a_2 = E^2 a, a_3 = E^3 a, \ldots$. We write $E$ instead of Gregory's $D$, since $D$ might be confused with the derivative. Recall that $E = 1 + \Delta$, where $\Delta$ is the difference operator; thus, Gregory derived the Newton–Montmort transformation:

$$
\begin{aligned}
S &= (x + x^2 E + x^3 E^2 + \cdots)a \\
&= x(1 - xE)^{-1} a = x(1 - x - x\Delta)^{-1} a \\
&= \frac{x}{1-x} \left( 1 - \frac{x}{1-x} \Delta \right)^{-1} a \\
&= \frac{x}{1-x} \left( 1 + \frac{x}{1-x} \Delta + \frac{x^2}{(1-x)^2} \Delta^2 + \cdots \right) a \\
&= \frac{ax}{1-x} + \Delta a \left( \frac{x}{1-x} \right)^2 + \Delta^2 a \left( \frac{x}{1-x} \right)^3 + \cdots .
\end{aligned}
$$

Recall that we have discussed this formula earlier, as (10.3).

Gregory also found an operational method for solving linear ordinary differential equations with constant coefficients. He began with the equation

$$\frac{d^n y}{dx^n} + A \frac{d^{n-1} y}{dx^{n-1}} + B \frac{d^{n-2} y}{dx^{n-2}} + \cdots + R \frac{dy}{dx} + Sy = X,$$

where $X$ was a function of $x$. After separating "the signs of operation from those of quantity," the equation became

$$\left( \frac{d^n}{dx^n} + A \frac{d^{n-1}}{dx^{n-1}} + B \frac{d^{n-2}}{dx^{n-2}} + \cdots + R \frac{d}{dx} + S \right) y = X.$$

Note that this can also be written as

$$f \left( \frac{d}{dx} \right) y = X,$$

where $f$ is a polynomial. Gregory's problem was to find

$$y = \left\{ f \left( \frac{d}{dx} \right) \right\}^{-1} X$$

and he first worked out the simplest case, where $f(x) = 1 + x$ and $X = 0$. Gregory calculated

$$y = \left(1 + \frac{d}{dx}\right)^{-1} 0$$

$$= \left(1 + \frac{d^{-1}}{dx^{-1}}\right)^{-1} \frac{d^{-1}}{dx^{-1}} 0 = \left(1 + \frac{d^{-1}}{dx^{-1}}\right)^{-1} C$$

$$= \left(1 - \frac{d^{-1}}{dx^{-1}} + \frac{d^{-2}}{dx^{-2}} - \cdots\right) C$$

$$= C\left(1 - x + \frac{x^2}{1 \cdot 2} - \frac{x^3}{1 \cdot 2 \cdot 3} + \cdots\right) = Ce^{-x}.$$

He noted that $\frac{d^{-1}}{dx^{-1}} = \int dx$. Now note that if $f(x) = a + x$, one would get $y = ce^{-ax}$. Gregory then observed that

$$\left(\frac{d}{dx} \pm a\right)^n X = e^{\mp ax} \left(\frac{d}{dx}\right)^n e^{\pm ax} X,$$

provable by means of the binomial theorem. Gregory finally considered the general case:

$$\left(\frac{d}{dx} - a_1\right)\left(\frac{d}{dx} - a_2\right)\left(\frac{d}{dx} - a_3\right) \cdots \left(\frac{d}{dx} - a_n\right) y = X.$$

He applied $\left(\frac{d}{dx} - a_1\right)^{-1}$ to both sides of the equation to find

$$\left(\frac{d}{dx} - a_2\right)\left(\frac{d}{dx} - a_3\right) \cdots \left(\frac{d}{dx} - a_n\right) y = \left(\frac{d}{dx} - a_1\right)^{-1} X$$

$$= e^{a_1 x} \int e^{-a_1 x} X \, dx.$$

Similarly,

$$\left(\frac{d}{dx} - a_3\right) \cdots \left(\frac{d}{dx} - a_n\right) y = \left(\frac{d}{dx} - a_2\right)^{-1} e^{a_1 x} \int e^{-a_1 x} X dx$$

$$= e^{a_2 x} \int e^{(a_1 - a_2)x} \left(\int e^{-a_1 x} X \, dx\right) dx$$

$$= \frac{e^{a_1 x} \int e^{-a_1 x} X \, dx}{a_1 - a_2} + \frac{e^{a_2 x} \int e^{-a_2 x} X \, dx}{a_2 - a_1},$$

using integration by parts in the last step. Thus, Gregory's final formula took the form

$$y = \frac{e^{a_1 x} \int e^{-a_1 x} X \, dx}{(a_1 - a_2)(a_1 - a_3) \cdots (a_1 - a_n)} + \cdots + \frac{e^{a_n x} \int e^{-a_n x} X \, dx}{(a_n - a_1)(a_n - a_2) \cdots (a_n - a_{n-1})}.$$

In 1811, Français had used the same method, going a step beyond Gregory by using partial fractions to decompose $\left\{ f\left(\frac{d}{dx}\right) \right\}^{-1}$. By this technique,

$$y = \left\{ f\left(\frac{d}{dx}\right) \right\}^{-1} X = \sum_{i=1}^{n} \frac{N_i}{\frac{d}{dx} - a_i} X,$$

where Français assumed that the roots $a_1, a_2, \ldots, a_n$ of $f(x) = 0$ were distinct. When the value of $N_i$ was substituted, the result was the same as Gregory's. Français also showed that his method could be extended to the case of repeated roots.

## 21.8  Boole's Operational Calculus

In his 1844 paper "On a General Method in Analysis,"[34] Boole extended Murphy and Gregory's symbolic method to treat problems on linear ordinary and partial differential equations with variable coefficients, linear difference equations, summation of series, and the computation of multiple integrals. He started his paper by stating several general propositions on functions of commutative and noncommutative operators. He made frequent use of some special cases and noted that they were already known: Let $x = e^{\theta}$. Then $x \frac{d}{dx} = \frac{d}{d\theta} = D$, so

$$f(D)e^{m\theta}u = e^{m\theta} f(D + m)u, \qquad (21.51)$$

$$f(D)e^{m\theta} = f(m)e^{m\theta}, \qquad (21.52)$$

$$D(D - 1) \cdot (D - n + 1)u = x^n \left(\frac{d}{dx}\right)^n u. \qquad (21.53)$$

Though Boole did not explicitly say so, $f(x)$ is a function expandable as a series. Relation (21.51) can be verified from the particular case $f(D) = D^n$. In this case, by Leibniz's formula for the $n$th derivative of a product, we can see that

$$\frac{d^n}{d\theta^n}(e^{m\theta}u) = e^{m\theta} \left(\frac{d}{d\theta} + m\right)^n u.$$

The other two cases can be easily verified.

Boole's fundamental theorem of development was given by the formula

$$f_0(D)u + f_1(D)e^{\theta}u + f_2(D)e^{2\theta}u + \cdots$$

$$= \sum \left\{ (f_0(m)u_m + f_1(m)u_{m-1} + f_2(m)u_{m-2} + \cdots)e^{m\theta} \right\}, \qquad (21.54)$$

---

[34] Boole (1844b).

where $u = \sum u_m e^{m\theta}$. He verified (21.54) by substituting the series for $u$ on the left-hand side of the equation and applying (21.51). Boole applied his development theorem to the summation of series, noting that if the coefficients of $u$ satisfied a linear recurrence relation

$$f_0(m)u_m + f_1(m)u_{m-1} + \cdots = 0,$$

then (21.54) yielded a differential equation satisfied by $u$. In cases where this differential equation could be solved in closed form, he had the sum of the series $\sum u_n x^n$. This method allowed him to use the recurrence relation satisfied by the coefficients of the series in order to quickly find the differential equation satisfied by the series.

As an example of a series summation, Boole considered for any real $n$,

$$u = 1 - \frac{n^2}{1 \cdot 2}x^2 + \frac{n^2(n^2 - 2^2)}{1 \cdot 2 \cdot 3 \cdot 4}x^4 - \frac{n^2(n^2 - 2^2)(n^2 - 4^2)}{1 \cdot 2 \cdots 6}x^6 + \cdots . \qquad (21.55)$$

In this case,

$$u_m = -\frac{n^2 - (m-2)^2}{m(m-1)}u_{m-2}.$$

He could then immediately write the differential equation satisfied by the series as

$$u - \frac{(D-2)^2 - n^2}{D(D-1)}e^{2\theta}u = 1,$$

or

$$D(D-1)u - \left((D-2)^2 - n^2\right)e^{2\theta}u = 0. \qquad (21.56)$$

Applying (21.51) and (21.53),

$$(D-2)^2 e^{2\theta}u = e^{2\theta}D^2 u = e^{2\theta}\left(D(D-1) + D\right)u$$

$$= x^2\left(x^2\frac{d^2}{dx^2} + x\frac{d}{dx}\right)u.$$

Thus, (21.56) was simplified to

$$(1 - x^2)\frac{d^2u}{dx^2} - x\frac{du}{dx} + n^2 u = 0. \qquad (21.57)$$

Boole then substituted $\sqrt{1 - x^2}\frac{d}{dx} = \frac{d}{dy}$, or $y = \sin^{-1} x$, to convert the differential equation (21.57) to $\frac{d^2u}{dy^2} + n^2 u = 0$. This gave Boole the solution

$$u = c_1 \cos(ny) + c_2 \sin(ny) = c_1 \cos(n \sin^{-1} x) + c_2 \sin(n \sin^{-1} x)$$

with the constants $c_1$ and $c_2$ equal to 1 and 0, respectively. He therefore had

$$\cos(n \sin^{-1} x) = 1 - \frac{n^2}{1 \cdot 2} x^2 + \frac{n^2(n^2 - 2^2)}{1 \cdot 2 \cdot 3 \cdot 4} x^4 - \cdots$$

or

$$\cos(ny) = 1 - \frac{n^2}{2!} \sin^2 y + \frac{n^2(n^2 - 2^2)}{4!} \sin^4 y - \frac{n^2(n^2 - 2^2)(n^2 - 4^2)}{6!} \sin^6 y + \cdots .$$

Similarly, Boole noted

$$\sin(ny) = n \sin y - \frac{n(n^2 - 1^2)}{3!} \sin^3 y + \frac{n(n^2 - 1^2)(n^2 - 3^2)}{5!} \sin^5 y + \cdots .$$

Recall that Newton discovered this series (8.16) and communicated it to Leibniz in his first letter of 1676. The series was afterward employed by Gauss to prove that $\Gamma(x)\Gamma(1 - x) = \frac{\pi}{\sin \pi x}$. Boole also used his method to solve some linear differential equations with variable coefficients and considered even more complex equations requiring a somewhat more elaborate technique. A simple example may explain his basic method. Boole set out to solve a differential equation with variable coefficients, commenting that it occurred in the theory of the "Earth's figure":

$$\frac{d^2 u}{dx^2} + q^2 u - \frac{6u}{x^2} = 0. \tag{21.58}$$

In his solution, Boole employed the general proposition:[35]

The equation $u + \phi(D)e^{r\theta} u = U$ will be converted into the form $v + \psi(D)e^{r\theta} v = V$, by the relations

$$u = P_r \frac{\phi(D)}{\psi(D)} v, \quad U = P_r \frac{\phi(D)}{\psi(D)} V, \tag{21.59}$$

wherein $P_r \frac{\phi(D)}{\psi(D)}$ denotes the infinite symbolical product $\frac{\phi(D)\phi(D-r)\phi(D-2r)\cdots}{\psi(D)\psi(D-r)\psi(D-2r)\cdots}$.

Boole proved this by assuming $u = f(D)v$ and substituting in the first equation in (21.59) to get

$$f(D)v + \phi(D)e^{r\theta} f(D)v = U.$$

By (21.53) this became

$$f(D)v + \phi(D)f(D - r)e^{r\theta} v = U$$

or

$$v + \frac{\phi(D)f(D - r)}{f(D)} e^{r\theta} v = (f(D))^{-1} U.$$

---

[35] ibid.

Thus

$$\psi(D) = \frac{\phi(D)f(D-r)}{f(D)}$$

or

$$f(D) = \frac{\phi(D)}{\psi(D)}f(D-r) = \frac{\phi(D)\phi(D-r)}{\psi(D)\psi(D-r)}f(D-2r) = \cdots.$$

In general, Boole attempted to choose $v$ such that it satisfied the equation $\frac{d^n v}{dx^n} \pm q^n v = X$. Boole applied his general proposition to rewrite (21.58) as

$$u + \frac{q^2}{(D+2)(D-3)}e^{2\theta}u = 0.$$

Now Boole required the equation for $v$ to be $\frac{d^2 v}{dx^2} + q^2 v = X$, or

$$v + \frac{q^2}{D(D-1)}e^{2\theta}v = V.$$

Here

$$\phi(D) = \frac{q^2}{(D+2)(D-3)}, \quad \psi(D) = \frac{q^2}{D(D-1)},$$

so that, by the general proposition,

$$P_2\frac{\phi(D)}{\psi(D)} = \frac{D-1}{D+2}.$$

Thus, $u = \frac{D-1}{D+2}v$ and $0 = \frac{D-1}{D+2}V$. Boole could then take $V = 0$ and $v = c\sin(qx+c_1)$. Finally,

$$u = \frac{D-1}{D+2}v = (1 - 3(D+2)^{-1})c\sin(qx+c_1)$$

$$= c(1 - 3e^{-2\theta}D^{-1}e^{2\theta})\sin(qx+c_1)$$

$$= c\left(1 - \frac{3}{x^2}\left(x\frac{d}{dx}\right)^{-1}x^2\right)\sin(qx+c_1)$$

$$= c\sin(qx+c_1) - \frac{3}{x^2}\int dx\, x\sin(qx+c_1)$$

$$= c\left(\left(1 - \frac{3}{q^2x^2}\right)\sin(qx+c_1) + \frac{3}{qx}\cos(qx+c_1)\right).$$

In 1860, Boole wrote a book on the calculus of finite differences,[36] in which he discussed operators methods at length. In particular, he employed the operator $e^D$ to express the series

$$\phi(x) - \phi(x+1) + \phi(x+2) - \phi(x+3) + \cdots$$

as an asymptotic series using Bernoulli numbers.[37] He observed that

$$\phi(x) - \phi(x+1) + \phi(x+2) - \phi(x+3) + \cdots$$
$$= (1 - e^D + e^{2D} - e^{3D} + \cdots)\phi(x)$$
$$= \frac{1}{e^D + 1}\phi(x).$$

He next noted that

$$\frac{1}{e^D + 1} = \frac{1}{e^D - 1} - \frac{2}{e^{2D} - 1}$$

$$= \frac{1}{D} - \frac{1}{2} + \frac{B_2}{2!}D + \frac{B_4}{4!}D^3 + \cdots - 2\left(\frac{1}{2D} - \frac{1}{2} + \frac{B_2}{2!}2D + \frac{B_4}{4!}2^4 D^3 + \cdots\right)$$

$$= \frac{1}{2} - \frac{B_2}{2!}(2^2 - 1)D + \frac{B_4}{4!}(2^4 - 1)D^3 - \text{etc.}$$

Thus, the formula he proved, now known as Boole's formula, may be written as

$$\sum_{j=0}^{\infty}(-1)^j \phi(x+j) = \frac{\phi(x)}{2} + \sum_{k=1}^{\infty}(-1)^k \frac{B_{2k}(2^{2k} - 1)}{(2k)!}\phi^{(2k-1)}(x).$$

As we have mentioned before, Boole's formula was originally derived by Euler in a paper of 1736[38] and then presented in several later papers[39] and in his differential calculus book of 1755.[40]

## 21.9 Jacobi and the Symbolic Method

In 1847, Jacobi wrote an interesting paper[41] using the operational method to derive two results on transformations of series. He applied the second of these to the derivation of a result in the theory of hypergeometric series knows as Pfaff's transformation. Jacobi apparently wished to bring attention to Pfaff's important result. This wish was finally fulfilled in about 1970, when Richard Askey read Jacobi's paper and made Pfaff's work known to the mathematical community. Jacobi did not explain why he chose to explore the operational method. The work of the British

---

[36] Boole (2003).
[37] Boole (2003) pp. 101–102.
[38] Eu. I-14 pp. 124–137, especially § 9–11. E 55.
[39] Boole's result in almost the same form: Eu. I-14 pp. 124–137, especially § 14. E 55. Also E 617.
[40] Eu. I-10₂ § 179–183. E 212.
[41] Jacobi (1847).

mathematicians may have appealed to his algorithmic style; note that he had visited Britain in 1842. Both of these transformations had been earlier presented in Euler's differential calculus book. Euler's second formula (10.15) stated that if

$$f(x) = a + bx + cx^2 + dx^3 + \cdots$$

then

$$aA_0 + bA_1x + cA_2x^2 + dA_3x^3 + \cdots$$

$$= A_0 f(x) + \Delta A_0 x \frac{df}{dx} + \frac{\Delta^2 A_0}{1 \cdot 2} x^2 \frac{d^2 f}{dx^2} + \frac{\Delta^3 A_0}{1 \cdot 2 \cdot 3} x^3 \frac{d^3 f}{dx^3} + \cdots.$$

Jacobi's proof of this formula was similar to Duncan Gregory's proof of the Newton–Montmort formula, discussed earlier. Recall the Arbogast operator $E$ used by Gregory: $E^k A_0 = A_k$. The formal steps of the argument were then

$$aA_0 + bA_1x + cA_2x^2 + dA_3x^3 + \cdots$$

$$= (a + bxE + cx^2E^2 + dx^3E^3 + \cdots)A_0$$

$$= f(xE)A_0 = f(x + x(E-1))A_0 = f(x + x\Delta)A_0$$

$$= \left( f(x) + f'(x)x\Delta + \frac{f''(x)}{1 \cdot 2}x^2\Delta^2 + \cdots \right) A_0$$

$$= A_0 f(x) + \Delta A_0 x \frac{df}{dx} + \frac{\Delta^2 A_0}{1 \cdot 2}x^2 \frac{d^2 f}{dx^2} + \cdots.$$

To obtain Pfaff's transformation, Jacobi specialized the sequence $A_0, A_1, A_2, A_3, \ldots$ to

$$1, \frac{\beta}{\gamma}, \frac{\beta(\beta+1)}{\gamma(\gamma+1)}, \frac{\beta(\beta+1)(\beta+2)}{\gamma(\gamma+1)(\gamma+2)}, \cdots$$

and noted that the first and second differences were

$$\frac{\beta-\gamma}{\gamma}, \frac{\beta-\gamma}{\gamma}\frac{\beta}{\gamma+1}, \frac{\beta-\gamma}{\gamma}\frac{\beta(\beta+1)}{(\gamma+1)(\gamma+2)}, \frac{\beta-\gamma}{\gamma}\frac{\beta(\beta+1)(\beta+2)}{(\gamma+1)(\gamma+2)(\gamma+3)}, \cdots$$

and

$$\frac{(\beta-\gamma)(\beta-\gamma-1)}{\gamma(\gamma+1)}, \frac{(\beta-\gamma)(\beta-\gamma-1)}{\gamma(\gamma+1)}\frac{\beta}{\gamma+2},$$

$$\frac{(\beta-\gamma)(\beta-\gamma-1)}{\gamma(\gamma+1)}\frac{\beta(\beta+1)}{(\gamma+2)(\gamma+3)}, \cdots.$$

In general, he observed,

$$\Delta^m A_n = \frac{(\beta-\gamma)(\beta-\gamma-1)\cdots(\beta-\gamma-m+1)\cdot\beta(\beta+1)\cdots(\beta+n-1)}{\gamma(\gamma+1)(\gamma+2)\cdots(\gamma+m+n-1)}.$$

In particular, when $n = 0$, he got

$$\Delta^m A_0 = \frac{(\beta - \gamma)(\beta - \gamma - 1) \cdots (\beta - \gamma - m + 1)}{\gamma(\gamma + 1) \cdots (\gamma + m - 1)}.$$

Jacobi then set

$$f(x) = (1 - x)^{-\alpha} = 1 + \alpha x + \frac{\alpha(\alpha + 1)}{1 \cdot 2} x^2 + \frac{\alpha(\alpha + 1)(\alpha + 2)}{1 \cdot 2 \cdot 3} x^3 + \cdots;$$

so that Euler's transformation reduced to Pfaff's transformation:

$$1 + \frac{\alpha \cdot \beta}{1 \cdot \gamma} x + \frac{\alpha(\alpha + 1) \cdot \beta(\beta + 1)}{1 \cdot 2 \cdot \gamma(\gamma + 1)} x^2 + \frac{\alpha(\alpha + 1)(\alpha + 2) \cdot \beta(\beta + 1)(\beta + 2)}{1 \cdot 2 \cdot 3 \cdot \gamma(\gamma + 1)(\gamma + 2)} x^3 + \cdots$$

$$= \frac{1}{(1 - x)^\alpha} \left( 1 + \frac{\alpha(\beta - \gamma)}{1 \cdot \gamma} \frac{x}{1 - x} + \frac{\alpha(\alpha + 1) \cdot (\beta - \gamma)(\beta - \gamma - 1)}{1 \cdot 2 \cdot \gamma(\gamma + 1)} \frac{x^2}{(1 - x)^2} + \cdots \right).$$

## 21.10  Cartier: Gregory's Proof of Leibniz's Rule

In his 2000 paper "Mathemagics," Pierre Cartier gives a rigorous version[42] of Arbogast and Gregory's argument for Leibniz's rule. Cartier makes use of the tensor product $V \otimes W$ of vector spaces $V$ and $W$. The vector space $V \otimes W$ consists of all finite sums $\sum \lambda_i (v_i \otimes w_i)$, where $\lambda_i$ are scalars, $v_i \in V$ and $w_i \in W$, and where $v \otimes w$ is bilinear in $v$ and $w$. For the purpose at hand, let $I$ be an interval on the real line and $C^\infty(I)$ be the vector space of infinitely differentiable functions on $I$. Define the operators $D_1$ and $D_2$ on $C^\infty(I) \otimes C^\infty(I)$ by

$$D_1(f \otimes g) = Df \otimes g, \quad D_2(f \otimes g) = f \otimes Dg.$$

The two operators commute, that is, $D_1 D_2 = D_2 D_1$. Now define

$$\overline{D}(f \otimes g) = Df \otimes g + f \otimes Dg,$$

so that $\overline{D} = D_1 + D_2$; we can then conclude that

$$\overline{D}^n(f \otimes g) = \sum_{k=0}^{n} \binom{n}{k} D^k f \otimes D^{n-k} g.$$

We can convert the tensor product to an ordinary product by observing that $f \cdot g$ is bilinear in $f$ and $g$ and hence there is a linear map

$$\mu : C^\infty(I) \otimes C^\infty(I) \to C^\infty(I)$$

---

[42] Cartier (2000).

such that $\mu(f \otimes g) = f \cdot g$. The proof can now be completed:

$$D^n(fg) = D^n(\mu(f \otimes g)) = \mu(\overline{D}^n(f \otimes g))$$

$$= \mu\left(\sum_{k=0}^{n} \binom{n}{k} D^k f \otimes D^{n-k} g\right)$$

$$= \sum_{k=0}^{n} \binom{n}{k} \mu(D^k f \otimes D^{n-k} g)$$

$$= \sum_{k=0}^{n} \binom{n}{k} D^k f \cdot D^{n-k} g.$$

Observe that Cartier succeeds in resolving the problem in Gregory's presentation, that the operators $D_1$ and $D_2$ do not apply to both $f$ and $g$.

## 21.11 Hamilton's Algebra of Complex Numbers and Quaternions

We have noted a strong algebraic spirit in the work of the British mathematicians of 1830–1850 and have observed important modern algebraic concepts in the work of Murphy. However, William R. Hamilton's (1805–1865) algebraic work was thoroughly modern in its structure and presentation. In 1826, Hamilton's friend J. T. Graves communicated to him some results on imaginary logarithms. This led Hamilton to formulate the theory of algebraic couples as the proper logical foundation for complex numbers. He finally presented this to the British Association in 1834.[43] Gauss also got these ideas around the same time. Hamilton defined complex numbers as a set of pairs of real numbers, called couples, with addition and multiplication defined in a special way. More generally, he determined the necessary and sufficient conditions for a set of couples to form a commutative and associative division algebra. Hamilton first defined the sum and scalar multiplication of couples:

$$(b_1, b_2) + (a_1, a_2) = (b_1 + a_1, b_2 + a_2);$$
$$a \times (a_1, a_2) = (aa_1, aa_2).$$

He took the last equation as the first step toward the definition of the product of two couples by identifying the real number with a couple $(a, 0)$ to get

$$(a, 0) \times (a_1, a_2) = (a, 0)(a_1, a_2) = (a_1, a_2)(a, 0) = (aa_1, aa_2).$$

His aim was to define multiplication in order to satisfy the two conditions

$$(b_1 + a_1, b_2 + a_2)(c_1, c_2) = (b_1, b_2)(c_1, c_2) + (a_1, a_2)(c_1, c_2), \qquad (21.60)$$

[43] Hamilton (1835).

$$(c_1, c_2)(b_1 + a_1, b_2 + a_2) = (c_1, c_2)(b_1, b_2) + (c_1, c_2)(a_1, a_2). \qquad (21.61)$$

Now for this type of multiplication to be possible, he had to have

$$\begin{aligned}
(c_1, c_2)(a_1, a_2) &= (c_1, 0)(a_1, a_2) + (0, c_2)(a_1, a_2) \\
&= (c_1 a_1, c_1 a_2) + (0, c_2)(a_1, 0) + (0, c_2)(0, a_2) \\
&= (c_1 a_1, c_1 a_2) + (0, c_2 a_1) + (0, c_2)(0, a_2) \\
&= (c_1 a_1, c_1 a_2 + c_2 a_1) + (0, c_2)(0, a_2). \qquad (21.62)
\end{aligned}$$

It remained to define the product $(0, c_2)(0, a_2) = c_2 a_2 (0, 1)(0, 1)$ contained in the last step. Hamilton set

$$(0, 1)(0, 1) = (\gamma_1, \gamma_2) \qquad (21.63)$$

and determined the necessary and sufficient condition on $\gamma_1$ and $\gamma_2$ so that the two conditions (21.60) and (21.61) would hold: He supposed $(b_1, b_2)$ to be the result of the product on the left-hand side of (21.62); then by equation (21.63) he had

$$b_1 = c_1 a_1 + \gamma_1 a_2 c_2,$$
$$b_2 = c_1 a_2 + c_2 a_1 + \gamma_2 a_2 c_2.$$

Now to be able to solve these equations for $a_1, a_2$ when $c_1 c_2 \neq 0$, the necessary and sufficient condition was the nonvanishing of the determinant, where the determinant was given as

$$c_1(c_1 + \gamma_2 c_2) - \gamma_1 c_2^2 = \left( c_1 + \frac{1}{2}\gamma_2 c_2 \right)^2 - \left( \gamma_1 + \frac{1}{4}\gamma_2^2 \right) c_2^2.$$

This expression was nonvanishing for all $c_1 c_2 \neq 0$ if

$$\gamma_1 + \frac{1}{4}\gamma_2^2 < 0.$$

The case in which $\gamma_1 = -1, \gamma_2 = 0$ gave Hamilton the usual multiplication rule for complex numbers:

$$(b_1, b_2)(a_1, a_2) = (b_1, b_2) \times (a_1, a_2) = (b_1 a_1 - b_2 a_2, b_2 a_1 + b_1 a_2).$$

Further developing the theory of complex numbers, Hamilton showed that the principal square root of $(-1, 0)$ was $(0, 1)$, and since $(-1, 0)$ could be replaced by $-1$ for brevity, he obtained

$$\sqrt{-1} = (0, 1).$$

He then wrote[44]

In the THEORY OF SINGLE NUMBERS, the symbol $\sqrt{-1}$ is *absurd*, and denotes an IMPOSSIBLE EXTRACTION, or a merely IMAGINARY NUMBER; but in the THEORY OF COUPLES, the same symbol $\sqrt{-1}$ is *significant*, and denotes a POSSIBLE EXTRACTION, or a REAL COUPLE, namely (as we have just now seen) the *principal square-root of the couple* $(-1, 0)$. In the latter theory, therefore, though not in the former, this sign $\sqrt{-1}$ may properly be employed; and we may write, if we choose, for any couple $(a_1, a_2)$ whatever,

$$(a_1, a_2) = a_1 + a_2\sqrt{-1}, \ldots.$$

Hamilton next attempted to extend his work to triples, or triplets. His motivation was to obtain an algebra applicable to three-dimensional geometry and physics. He was well aware of the geometrical interpretation of complex numbers as vectors in two dimensions. Under this interpretation, the parallelogram law determined addition; moreover, the length (or modulus) of the product of two complex numbers turned out to be the product of the lengths of the two numbers. In October 1843, Hamilton described his train of thought as he worked toward his October 16 discovery of quaternions. He explained that he considered triplets of the form $x + iy + jz$ representing points $(x, y, z)$ in space. Here $j$ was "another sort of $\sqrt{-1}$, perpendicular to the plane itself."[45] Addition and subtraction of triplets was a simple matter, but multiplication turned out to be a challenge: In a letter of 1865 to his son, Hamilton recalled:[46]

Every morning in the early part of the above-cited month, on my coming down to breakfast, your brother William Edwin and yourself used to ask me, 'Well, Papa, can you multiply triplets?' Whereto I was always obliged to reply, with a sad shake of the head, 'No, I can only add and subtract them.'

In his 1843 description of his discovery, Hamilton recounted his dilemma:[47] In order to multiply triplets, term-by-term multiplication had to be possible and the modulus of the product was required to equal the product of the moduli. He observed that

$$(a + iy + jz)(x + iy + jz) = ax - y^2 - z^2 + i(a + x)y + j(a + x)z + (ij + ji)yz \tag{21.64}$$

and that

$$(a^2 + y^2 + z^2)(x^2 + y^2 + z^2) = (ax - y^2 - z^2)^2 + (a + x)^2(y^2 + z^2).$$

So the rule for the moduli implied that the last term in (21.64) should be zero, or, $ij + ji = 0$. He was sufficiently audacious to consider the possibility that $ij = ji = 0$;

---

[44] ibid. 127–128.
[45] Hamilton (1945).
[46] Graves (1885) pp. 434–435.
[47] Hamilton (1945).

in modern terminology, this meant that $i$ and $j$ would be zero divisors. However, when he examined the general case

$$(a + ib + jc)(x + iy + jz) = ax - by - cz + i(ay + bx)$$
$$+ j(az + cx) + ij(bx - cy)$$

and the corresponding formula for the moduli

$$(a^2 + b^2 + c^2)(x^2 + y^2 + z^2) = (ax - by - cz)^2 + (ay + bx)^2$$
$$+ (az + cx)^2 + (bz - cy)^2,$$

he saw that the coefficient of $ij$, $bz - cy$, could not be dropped and hence $ij$ could not be zero. Put even more simply, the moduli of the product $ij$ had to be 1 and not 0.

When he reached this result, it dawned on Hamilton that to multiply triplets, he must admit in some sense a fourth dimension, and he described this realization in a letter to his friend J. T. Graves, written the day after he discovered quaternions. By a remarkable coincidence, after completing this letter he came across the May 1843 issue of the *Cambridge Mathematical Journal* containing a paper by Cayley on analytical geometry of $n$ dimensions. In a postscript to his letter to Graves, Hamilton noted that he did not yet know whether or not his ideas were similar to Cayley's.

Continuing his description, Hamilton saw that he had to introduce a new imaginary $k$ such that $ij = k$. Thus, he discovered quaternions! Moreover, $ji = -ij = -k$. He wondered whether $k^2 = 1$. But this produced the equation

$$(a + ib + jc + kd)(\alpha + i\beta + j\gamma + k\delta)$$
$$= a\alpha - b\beta - c\gamma + d\delta + i(a\beta + \cdots) + j(a\gamma + \cdots) + k(a\delta + d\alpha + \cdots),$$

implying that

$$(a^2 + b^2 + c^2 + d^2)(\alpha^2 + \beta^2 + \gamma^2 + \delta^2)$$
$$= (a\alpha - b\beta - c\gamma + d\delta)^2 + (a\beta + \cdots)^2 + (a\gamma + \cdots)^2 + (a\delta + d\alpha + \cdots)^2.$$

Of course, this relation could not possibly hold, because the term $2a\alpha d\delta$ in the first square on the right-hand side would not cancel the same term in the last square. So Hamilton took $k^2 = -1$. He then supposed that associativity would probably hold true and hence $-j = (ii)j = i(ij) = ik$. Similarly, $j(ii) = (ji)i = -ki$ or $j = ki$, and so $ik = -ki$. In this way he obtained the basic relations:

$$i^2 = j^2 = k^2 = -1, \ ij = k, \ jk = i, \ ki = j, \ ji = -k, \ kj = -i, \ ik = -j.$$

The product of two quaternions then emerged as

$$(a + ib + jc + kd)(\alpha + i\beta + j\gamma + k\delta)$$
$$= a\alpha - b\beta - c\gamma - d\delta + i(a\beta + b\alpha + c\delta - d\gamma)$$
$$+ j(a\gamma - b\delta + c\alpha + d\beta) + k(a\delta + b\gamma - c\beta + d\alpha).$$

And of course, with this definition,

$$(a^2 + b^2 + c^2 + d^2)(\alpha^2 + \beta^2 + \gamma^2 + \delta^2)$$
$$= (a\alpha - b\beta - c\gamma - d\delta)^2 + (a\beta + b\alpha + c\delta - d\gamma)^2$$
$$+ (a\gamma - b\delta + c\alpha + d\beta)^2 + (a\delta + b\gamma - c\beta + d\alpha)^2.$$

Thus, the modulus of the product equaled the product of the moduli! Interestingly, Euler also knew this identity in 1748, in connection with the representation of a number as a sum of four squares.[48] Hamilton concluded his description of this discovery:[49]

Hence we may write, on the plan of my theory of couples,

$$(a, b, c, d)(\alpha, \beta, \gamma, \delta) =$$

$$(a\alpha - b\beta - c\gamma - d\delta, \; a\beta + b\alpha + c\delta - d\gamma, \; a\gamma - b\delta + c\alpha + d\beta, \; a\delta + b\gamma - c\beta + d\alpha).$$

Hence

$$(a, b, c, d)^2 = (a^2 - b^2 - c^2 - d^2, 2ab, 2ac, 2ad).$$

Thus

$$(0, x, y, z)^2 = -(x^2 + y^2 + z^2); \quad (0, x, y, z)^3 = -(x^2 + y^2 + z^2)(0, x, y, z);$$

$$(0, x, y, z)^4 = +(x^2 + y^2 + z^2)^2; \& c.$$

Therefore

$$e^{(0,x,y,z)} = e^{(ix+jy+kz)} = 1 + \frac{ix + jy + kz}{1} - \frac{x^2 + y^2 + z^2}{1 \cdot 2} - \& c;$$

$$= \cos \sqrt{x^2 + y^2 + z^2} + \frac{ix + jy + kz}{\sqrt{x^2 + y^2 + z^2}} \sin \sqrt{x^2 + y^2 + z^2}$$

and the *modulus* of $e^{(0,x,y,z)} = 1$. [Like the modulus of $e^{(0,x)}$ or $e^{\sqrt{-1}x}$] Let

$$\sqrt{x^2 + y^2 + z^2} = \rho, \; x = \rho \cos \phi, \; y = \rho \sin \phi \cos \psi, \; z = \rho \sin \phi \sin \psi;$$

then

$$e^{\rho(i \cos \phi + j \sin \phi \cos \psi + k \sin \phi \sin \psi)}$$

$$= \cos \rho + (i \cos \phi + j \sin \phi \cos \psi + k \sin \phi \sin \psi) \sin \rho;$$

a theorem, which when $\phi = 0$, becomes the well-known equation

$$e^{i\rho} = \cos \rho + i \sin \rho, \; i = \sqrt{-1}.$$

---

[48] Fuss (1968) vol. 1, p. 452. See also Eu. I-2 pp. 338–372, especially pp. 368–369. E 242 § 93.
[49] Hamilton (1945).

Hamilton's letter led John T. Graves in December 1843 to produce an eight-dimensional division algebra, the algebra of octaves or octonions. The law of moduli was maintained within this system, so that

$$(a_1^2 + a_2^2 + \cdots + a_8^2)(b_1^2 + b_2^2 + \cdots + b_8^2) = c_1^2 + c_2^2 + \cdots + c_8^2.$$

Hamilton observed that while associativity held for quaternions, it failed to hold for octonions. Graves did not publish his work, though Cayley rediscovered and published it in 1845. Octonions are therefore called Cayley numbers.

The German mathematician A. Hurwitz wrote that Hamilton's two requirements, that term-by-term multiplication be valid and that the product of the moduli be equal to the moduli of the product of $n$-tuples $(x_1, x_2, \ldots, x_n)$, in fact held only for $n = 1, 2, 4, 8$. This explains why Hamilton was unable to discover a way of multiplying triplets. In the 1870s, C. S. Peirce and Frobenius gave another explanation for Hamilton's failure to work out a three-dimensional division algebra, i.e., an algebra of triplets. They proved that the only real finite-dimensional associative division algebras were: the real numbers, the complex numbers, and the quaternions.

We have seen that Hamilton was initially hesitant to move to the fourth dimension and was struck by Cayley's work outlining a geometry of $n$ dimensions. Note Felix Klein's telling remark on George Green's 1835 paper concerning the attraction of an ellipsoid, "This investigation merits special mathematical interest ... because it is carried out for $n$ dimensions, long before the development of $n$-dimensional geometry in Germany began."[50] Such was the influence of the formal algebraic approach taken by British mathematicians of the early 1800s, that even the applied mathematician George Green was willing to consider the novel concept of an $n$-dimensional space.

## 21.12  Exercises

(1) Solve the differential equation

$$0 = y - \frac{1}{2!}\frac{d^2 y}{dx^2} + \frac{1}{4!}\frac{d^4 y}{dx^4} - \frac{d^6 y}{dx^6} + \cdots$$

using Euler's method, by which he obtained

$$y = \sum \left( a_k e^{\frac{(2n+1)\pi x}{2}} + b_k e^{-\frac{(2n+1)\pi x}{2}} \right).$$

See section 50 of Euler's paper E 62.

---

[50] Klein (1979) p. 217.

(2) Prove Faà di Bruno's formula for the $m$th derivative of a composition of two functions:

$$\frac{d^m}{dt^m} g(f(t))$$

$$= \sum \frac{m!}{b_1! b_2! \cdots b_m!} g^{(k)}(f(t)) \left(\frac{f'(t)}{1!}\right)^{b_1} \left(\frac{f''(t)}{2!}\right)^{b_2} \cdots \left(\frac{f^{(m)}(t)}{m!}\right)^{b_m},$$

where the sum is over different solutions of $b_1 + 2b_2 + \cdots + mb_m = m$ and $k = b_1 + b_2 + \cdots + b_m$. Faà di Bruno gave the right hand side in the form of a determinant:

$$\begin{vmatrix} \binom{m-1}{0} f' g & \binom{m-1}{1} f'' g & \binom{m-1}{2} f''' g & \cdots & \binom{m-1}{m-2} f^{(m-1)} g & \binom{m-1}{m-1} f^{(m)} g \\ -1 & \binom{m-2}{0} f' g & \binom{m-2}{1} f'' g & \cdots & \binom{m-2}{m-3} f^{(m-2)} g & \binom{m-2}{m-2} f^{(m-1)} g \\ 0 & -1 & \binom{m-3}{0} f' g & \cdots & \binom{m-3}{m-4} f^{(m-3)} g & \binom{m-3}{m-3} f^{(m-2)} g \\ \cdot & \cdot & -1 & \cdots & \cdot & \cdot \\ \cdot & \cdot & \cdot & \cdots & \cdot & \cdot \\ \cdot & \cdot & \cdot & \cdots & \cdot & \cdot \\ 0 & 0 & 0 & \cdots & \binom{1}{0} f' g & \binom{1}{1} f'' g \\ 0 & 0 & 0 & \cdots & -1 & \binom{0}{0} f' g \end{vmatrix}$$

where $f^{(i)} \equiv f^{(i)}(t)$ and $g^k = g^{(k)}(f(t))$. Faà di Bruno published this formula without proof or reference in 1855 and then again in 1857. The formulation as a determinant appears to be original with Faà di Bruno, who may also be the only mathematician to be beatified by the Catholic Church. The papers of Craik (2005) and Johnson (2002) contain a detailed history of Faà di Bruno's formula.

(3) Solve the difference equation

$$u_{x+1}^2 - 4u_x^2(u_x^2 + 1) = 0.$$

Herschel's hint for the solution is to set $u_x = \sqrt{-1} \sin v_x$. See Herschel (1820) p. 34.

(4) Solve the difference equation

$$u_{x+1} u_x - a_x(u_{x+1} - u_x) + 1 = 0.$$

Herschel remarked that this was a slight generalization of the equation worked out in our text. See Herschel (1820) pp. 36–37.

(5) Sum the series $u = \frac{4x^2}{1 \cdot 2 \cdot 3} + \frac{5x^4}{2 \cdot 3 \cdot 4} + \frac{6x^5}{3 \cdot 4 \cdot 5} + \cdots$. See Boole (1844b) p. 264.

(6) Using Boole's notation given in the text, prove his proposition: The equation

$$u + \phi(D)e^{r\theta} u = U$$

will be converted to the form

$$v + \psi(D)e^{r\theta}v = V,$$

by the relations $u = e^{n\theta}v$ and $U = e^{n\theta}V$. See Boole (1844b) p. 247.

(7) Let $d_x$ denote the derivative with respect to $x$. Note that $[f(x + y)](d_y - d_x) = 0$. Apply Murphy's method for the difference operator to the differential in order to obtain the Taylor series for $f(x + y)$. See Murphy (1837) p. 196.

(8) Sum the series

$$\sum_{n=1}^{\infty} \arctan\left(\frac{1}{1 + n + n^2}\right).$$

See Herschel (1820) p. 57.

## 21.13   Notes on the Literature

Friedelmeyer (1994) gives an extensive discussion of Arbogast and his work. Babbage (1961), edited by Morrison and Morrison, contains papers of Babbage, including one discussing the Analytical Society. Enros (1983) is a nice discussion of the Analytical Society. Becher (1980) is an interesting article on Woodhouse, Babbage, and Peacock. For the early nineteenth-century work on operational calculus in Britain, see Allaire and Bradley (2002).

See also articles by E. L. Ortiz and of S. E. Despeaux in Gray and Parshall (2007) and the paper of Koppelman (1971) for the role of the operational method in the development of abstract algebra. For remarks on the influence of the German combinatorial school on Gudermann and Weierstrass, see Manning (1975).

# 22

## Trigonometric Series after 1830

### 22.1 Preliminary Remarks

At the end of his 1829 paper on Fourier series, Dirichlet pointed out that the concept of the definite integral required further investigation if the theory of Fourier series were to include functions with an infinite number of discontinuities.[1] In this connection, he gave the example of a function $\phi(x)$ defined as a fixed constant for rational $x$ and another fixed constant for irrational $x$. Such a function could not be integrated by Cauchy's definition of an integral. Dirichlet stated his plan to publish a paper on this topic at the foundation of analysis, but he never presented any results on it, though he gave important applications of Fourier series to number theory.

Bernhard Riemann (1826–1866), a student of Dirichlet, took up this question as he discussed trigonometric series in his *Habilitation* paper of 1854.[2] The first part of the paper gave a brief history of Fourier series, a topic Riemann studied with Dirichlet's help. In the later portion, Riemann briefly considered a new definition of the integral and then went on to study general trigonometric series of the form

$$\frac{1}{2}a_0 + \sum_{n=1}^{\infty}(a_n \cos nx + b_n \sin nx),$$

where the coefficients $a_n$ and $b_n$ were not necessarily defined by the Euler–Fourier integrals. Using these series, he could represent nonintegrable functions in terms of trigonometric series. Here he introduced methods still not superseded, though they have been further developed. He associated with the trigonometric series a continuous function $F(x)$ obtained by twice formally integrating the series. Riemann then defined the generalized second, or Riemann–Schwarz, derivative of $F(x)$ as

$$\lim_{h \to 0} \frac{\Delta^2 F(x-h)}{h^2}$$

[1] Dirichlet (1969) vol. 1, pp. 117–132, especially p. 132.
[2] Riemann (1990) pp. 259–296.

and proved that if the trigonometric series converged to some $f(x)$, then the Riemann–Schwarz derivative was equal to $f(x)$. In addition, he proved that if $a_n$ and $b_n$ tended to zero, then

$$\lim_{h \to 0} \frac{\Delta^2 F(x - h)}{h} = 0. \tag{22.1}$$

Riemann's paper also contained a number of very interesting examples, raising important questions. Perhaps he did not publish the paper because he was unable to answer these questions; Dedekind had it published in 1867 after Riemann's premature death.

The publication of this paper of Riemann led Heinrich Heine (1821–1881) to ask whether more than one trigonometric series could represent the same function. He applied Weierstrass's result that when a series converged uniformly, term-by-term integration was possible. Weierstrass taught this theorem in his lectures in Berlin starting in the early 1860s, though he had discovered it two decades earlier. From this result, Heine concluded that a uniformly convergent trigonometric series was a Fourier series; he defined a generally uniformly convergent series to cover the case of the series of continuous functions converging to a discontinuous function. Such series converged uniformly on the intervals obtained after small neighborhoods around the discontinuities had been removed. In a paper of 1870, Heine stated and proved that a function could not be represented by more than one generally uniformly convergent trigonometric series.[3]

When Georg Cantor (1845–1918) joined Heine at the University of Halle in 1869, Heine awakened his interest in this uniqueness question. Cantor had studied at the University of Berlin under Kummer, Kronecker, and Weierstrass and wrote his thesis on quadratic forms under Kummer. At Heine's suggestion, Cantor studied Riemann's paper containing the observation, without proof, that if

$$a_n \cos nx + b_n \sin nx \to 0 \quad \text{as} \quad n \to \infty$$

for all $x$ in an interval, then $a_n \to 0$ and $b_n \to 0$ as $n \to \infty$. Cantor's first paper on trigonometric series, published in 1870,[4] provided a proof of this important assertion, now known as the Cantor-Lebesgue theorem. This was the first step in Cantor's proof of the uniqueness theorem that if two trigonometric series converge to the same sum in $(0, 2\pi)$ except for a finite number of points, then the series are identical. Note that Henri Lebesgue (1875–1941) later proved the theorem in a more general context.

To prove his theorem, Cantor needed to show that if the generalized second, or Riemann–Schwarz, derivative of a continuous function was zero in an interval, then the function was linear in that interval. So in a letter of February 17, 1870, he asked his friend Hermann Schwarz for a proof of this result. Schwarz had received his

[3] Heine (1870).
[4] Cantor (1870a).

doctoral degree a few years before Cantor, but they had both studied under Kummer and Weierstrass at Berlin. Schwarz left the University of Halle in 1869 and went to Zurich, but they corresponded often. In fact, Schwarz wrote to Cantor on February 25, 1870, "The fact that I wrote to you at length yesterday is no reason why I should not write again today." In this letter Schwarz gave what he said was the first rigorous proof of the theorem that if a function had a zero derivative at every value in an interval, then the function was a constant in that interval.[5]

Schwarz provided a proof of the result Cantor needed for his uniqueness theorem. Cantor next studied the case with exceptional points, at which the series was not known to converge to zero. Was the value of every coefficient still zero? He supposed $c$ to be an exceptional point in an interval $(a, b)$ so that the series converged to zero in the intervals $(a, c)$ and $(c, b)$. Now Riemann's second theorem, given by (22.1), implied that the slopes of the two lines had to be the same, and hence $F(x)$ was linear in $(a, b)$, and the uniqueness theorem followed. Clearly, the argument could be extended to a finite number of exceptional points. When Cantor realized this, he asked whether there could be an infinite number of exceptional points; he soon understood that even if the exceptional points were infinite in number, as long as they had only a finite number of limit points, his basic argument would still be effective.

Leopold Kronecker (1823–1891) was initially quite interested in the work of Cantor on the uniqueness of trigonometric series. After the publication of Cantor's first paper, Kronecker explained to him that the proof of the Cantor-Lebesgue theorem could be simplified by means of an idea contained in Riemann's paper. However, as Cantor's work progressed and he began to use increasingly intricate infinite sets, Kronecker lost sympathy with Cantor's ideas and became a passionate critic of the theory of infinite sets. Cantor, on the other hand, abandoned the study of trigonometric series and after 1872 became more and more intrigued by infinite sets, at that time completely unexplored territory. Luckily, Cantor found an understanding and kindred spirit in Dedekind, who had himself done some work on infinite sets. Cantor started a correspondence with Dedekind in 1872 that continued off and on for several years.[6] Dedekind helped Cantor write up a concise proof of the countability of the set of algebraic numbers, and in 1874 this theorem appeared in Cantor's first paper on infinite sets.

Though there was some opposition to Cantor's theory, it was directly and indirectly successful as sets became basic objects in the language of mathematics. Without this concept, such early twentieth-century innovations as measure theory and the Lebesgue integral would hardly have been possible. These advances in turn had consequences for the theory of trigonometric series and the theory of uniqueness of such series. As an example, consider the noteworthy theorem of W. H. Young from a 1909 paper: "If the values of a function be assigned at all but a countable set of points, it can be expressed as a trigonometric series in at most one way."[7]

---

[5] For an English translation, see Meschkowski (1964), pp. 87–89.
[6] See Ferreirós (1993) for references.
[7] Young (1909).

## 22.2   The Riemann Integral

In his 1854 paper on trigonometric series,[8] Riemann observed that since the Euler–Fourier coefficients were defined by integrals, he would begin his study of Fourier series with a clarification of the concept of an integral. To understand his definition, let $f$ be a bounded function defined on an interval $(a,b)$ and let $a = x_0 < x_1 < x_2 < \cdots < x_{n-1} < x_n = b$. Denote the length of the subinterval $x_k - x_{k-1}$ by $\delta_k$ where $k = 1, 2, \ldots, n$. Let $0 \le \epsilon_k \le 1$ and set

$$s = \delta_1 f(a + \epsilon_1 \delta_1) + \delta_2 f(x_1 + \epsilon_2 \delta_2) + \delta_3 f(x_2 + \epsilon_3 \delta_3) + \cdots + \delta_n f(x_{n-1} + \epsilon_n \delta_n).$$

Riemann noted that the value of the sum $s$ depended on $\delta_k$ and $\epsilon_k$, but if it approached infinitely close to a fixed limit $A$ as all the $\delta$s became infinitely small, then this limit would be denoted by $\int_a^b f(x)\,dx$. On the other hand, if the sum $s$ did not have this property, then $\int_a^b f(x)\,dx$ had no meaning.

Riemann then extended the definition of an integral to include unbounded functions, as Cauchy had done. Thus, if $f(x)$ was infinitely large at a point $c$ in $(a,b)$, then

$$\int_a^b f(x)\,dx = \lim_{\alpha_1 \to 0} \int_a^{c-\alpha_1} f(x)\,dx + \lim_{\alpha_2 \to 0} \int_{c+\alpha_2}^b f(x)\,dx.$$

Riemann next raised the question: When was a function integrable? He gave his answer in terms of the variations of the function within subintervals. He let $D_k$ denote the difference between the largest and the smallest values of the function in the interval $(x_{k-1}, x_k)$ for $k = 1, 2, \ldots, n$. He then argued that if the function was integrable, the sum

$$\sum = \delta_1 D_1 + \delta_2 D_2 + \cdots + \delta_n D_n$$

must become infinitely small as the values of $\delta$ became small. He next observed that for $\delta_k \le d$ $(k = 1, \ldots, n)$, this sum would have a largest value, $\Delta(d)$. Moreover, $\Delta(d)$ decreased with $d$ and $\Delta(d) \to 0$ as $d \to 0$. He noted that if $s$ were the total length of those intervals in which the function varied more than some value $\sigma$, then the contribution of those intervals to $\sum$ was $\ge \sigma s$. Thus, he arrived at

$$\sigma s \le \delta_1 D_1 + \delta_2 D_2 + \cdots + \delta_n D_n \le \Delta$$

or

$$s \le \frac{\Delta}{\sigma}.$$

From this inequality, he concluded that for a given $\sigma$, $\frac{\Delta}{\sigma}$ could be made arbitrarily small by a suitable choice of $d$ and hence the same was true for $s$. Riemann could then state that a bounded function $f(x)$ was integrable only if the total length of the

---

[8]  Riemann (1990) pp. 259–296.

intervals in which the variations of $f(x)$ were $> \sigma$ could be made arbitrarily small by a suitable choice of $d$. He also gave a short argument proving the converse. Riemann's proof omitted some details necessary to make it completely convincing. In fact, in an 1875 paper presented to the London Mathematical Society, H. J. S. Smith formulated a clearer definition of integrability and a modified form of Riemann's theorem.[9]

Riemann gave a number of interesting examples of applications of this theorem, remarking that they were quite novel. For instance, he considered the function defined by the series

$$f(x) = \frac{(x)}{1} + \frac{(2x)}{4} + \frac{(3x)}{9} + \cdots = \sum_{n=1}^{\infty} \frac{(nx)}{n^2}, \qquad (22.2)$$

where $(x)$ was the difference between $x$ and the closest integer; in the ambiguous case when $x$ was at the midpoint between two successive integers, $(x)$ was taken to be zero. He showed that for $x = \frac{p}{2n}$ where $p$ and $n$ were relatively prime,

$$f(x+0) = f(x) - \frac{1}{2nn}\left(1 + \frac{1}{9} + \frac{1}{25} + \cdots\right) = f(x) - \frac{\pi\pi}{16nn},$$

$$f(x-0) = f(x) + \frac{1}{2nn}\left(1 + \frac{1}{9} + \frac{1}{25} + \cdots\right) = f(x) + \frac{\pi\pi}{16nn};$$

at all other values of $x$, $f(x)$ was continuous. Riemann applied his theorem to show that, although (22.2) had an infinite number of discontinuities, it was integrable over $(0, 1)$.

## 22.3 Smith: Revision of Riemann and Discovery of the Cantor Set

Henry Smith did his most notable work in number theory and elliptic functions, but his 1875 paper "On the Integration of Discontinuous Functions" also obtained some important results later found by Cantor. Though continental mathematicians did not notice this paper, it anticipated by eight years Cantor's construction of a ternary set.[10] In order to reformulate Riemann's definition and theorem on integrability, Smith efficiently set up the modern definition of the Riemann integral in terms of the upper and lower Riemann sums, politely pointing out the gap in Riemann's work:[11]

> Riemann, in his Memoir . . . , has given an important theorem which serves to determine whether a function $f(x)$ which is discontinuous, but not infinite, between the finite limits $a$ and $b$, does or does not admit of integration between those limits, the variable $x$, as well as the limits $a$ and $b$, being supposed real. Some further discussion of this theorem would seem to be desirable, partly because, in one particular at least, Riemann's demonstration is wanting in formal accuracy, and partly because the theorem itself appears to have been misunderstood, and to have been made the basis of erroneous inferences.

[9] Smith (1875).
[10] Smith (1965a) vol. 2, pp. 94–95.
[11] ibid. pp. 86–89.

Let $d$ be any given positive quantity, and let the interval $b - a$ be divided into any *segments* whatever, $\delta_1 = x_1 - a$, $\delta_2 = x_2 - x_1$, ..., $\delta_n = b - x_{n-1}$, subject only to the condition that none of these segments surpasses $d$. We may term $d$ the *norm* of the division; it is evident that there is an infinite number of different divisions having a given norm; and that a division appertaining to any given norm, appertains also to every greater norm. Let $\epsilon_1, \epsilon_2, \ldots, \epsilon_n$ be positive proper fractions; if, when the norm $d$ is diminished indefinitely, the sum

$$S = \delta_1 f(a + \epsilon_1 \delta_1) + \delta_2 f(x_1 + \epsilon_2 \delta_2) + \cdots + \delta_n f(x_{n-1} + \epsilon_n \delta_n)$$

converges to a definite limit, whatever be the mode of division, and whatever be the fractions $\epsilon_1, \epsilon_2, \ldots, \epsilon_n$, that limit is represented by the symbol $\int_a^b f(x)dx$, and the function $f(x)$ is said to admit of integration between the limits $a$ and $b$. We shall call the values of $f(x)$ corresponding to the points of any segment the *ordinates* of that segment; by the *ordinate difference* of a segment we shall understand the difference between the greatest and least ordinates of the segment. For any given division $\delta_1, \delta_2, \ldots, \delta_n$, the greatest value of $S$ is obtained by taking the maximum ordinate of each segment, and the least value of $S$ by taking the minimum ordinate of each segment; if $D_i$ is the ordinate difference of the segment $d_i$, the difference $\theta$ between those two values of $S$ is

$$\theta = \delta_1 D_1 + \delta_2 D_2 + \cdots + \delta_n D_n.$$

But, for a given norm $d$, the greatest value of $S$, and the least value of $S$, will in general result, not from one and the same division, but from two different divisions, each of them having the given norm. Hence the difference $\Theta$ between the greatest and least values that $S$ can acquire for a given norm, is, in general, greater than the greatest of the differences $\theta$. To satisfy ourselves, in any given case, that $S$ converges to a definite limit, when $d$ is diminished without limit, we must be sure that $\Theta$ diminishes without limit; and it is not enough to show (as the form of Riemann's proof would seem to imply) that $\theta$ diminishes without limit, even if this should be shown for every division having the norm $d$.

With this revised definition of the integral, Smith was in a position to restate Riemann's condition for integrability:

Let $\sigma$ be any given quantity, however small; if, in every division of norm $d$, the sum of the segments, of which the ordinate differences surpass $\sigma$, diminishes without limit, as $d$ diminishes without limit, the function admits of integration; and, *vice versa*, if the function admits of integration, the sum of these segments diminishes without limit with $d$.

Recall that Cantor was led to his theory of infinite sets through his researches in trigonometric series, and these in turn had their origins in Riemann's paper. This paper also inspired mathematicians to investigate the possibility of other peculiar or pathological functions and to construct infinite sets with apparently strange properties. In 1870 Hermann Hankel, a student of Riemann, constructed infinite nowhere-dense sets and he gave a flawed proof that a function with discontinuities only on a nowhere dense set was integrable. However, Hankel succeeded in proving that the set of points of continuity of an integrable function was dense.

Smith was the first to notice the mistake in Hankel's proof; to begin to tackle this problem, he divided the interval $(0, 1)$ into $m \geq 2$ equal parts where the last segment was not further divided. The remaining $m - 1$ segments were again divided into $m$ equal parts with the last segments of each left undivided. This process was continued ad infinitum to obtain the set $P$ of division points. Smith proved that $P$ was nowhere dense; he called them points "in loose order." The union of the set $P$ and its limit points is now called a Cantor set since in 1883 Cantor constructed such a set with $m = 3$. Smith showed that after $k$ steps, the total length of the divided segments was $\left(1 - \frac{1}{m}\right)^k$; so that as $k$ increased indefinitely, the points of $P$ were located on segments occupying only an infinitesimal portion of the interval $(0, 1)$. He then applied Riemann's criterion for integrability to show that bounded functions whose discontinuities occur only at points in $P$ would be integrable.

With a slight modification of this construction, Smith showed that there existed nowhere dense sets of positive measure. The first step in his modification was the same as before. In the second step he divided the $m - 1$ divided segments into $m^2$ parts, but did not further divide the last segment of each of these. The $(m - 1)(m^2 - 1)$ remaining segments were divided into $m^3$ parts, and so on. After $k$ steps, Smith found the total length of the divided segments to be $\left(1 - \frac{1}{m}\right)\left(1 - \frac{1}{m^2}\right) \cdots \left(1 - \frac{1}{m^k}\right)$. He noted that the limit $\prod_{k=1}^{\infty} \left(1 - \frac{1}{m^k}\right)$ was not equal to zero. He again proved that the set of division points $Q$ was nowhere dense but that in this case a function with discontinuities at the points in $Q$ was not integrable. Smith then noted,[12] "The result obtained in the last example deserves attention, because it is opposed to a theory of discontinuous functions, which has received the sanction of an eminent geometer, Dr. Hermann Hankel, whose recent death at an early age is a great loss to mathematical science."

In his thesis of 1902,[13] Lebesgue proved that a bounded function was Riemann integrable if and only if the set of its discontinuities was of measure zero.[14] Smith would perhaps not have been surprised at this result.

## 22.4 Riemann's Theorems on Trigonometric Series

In his 1854 paper, after defining the integral, Riemann also investigated the question of whether a function could be represented by a trigonometric series without assuming any specific properties of the function, such as whether the function was integrable. Of course, if a function is not integrable it cannot have a Fourier series. Thus, Riemann focused on series of the form

$$\Omega = \frac{1}{2}a_0 + (a_1 \cos x + b_1 \sin x) + (a_2 \cos 2x + b_2 \sin 2x) + \cdots$$
$$= A_0 + A_1 + A_2 + \cdots,$$

where $A_0 = \frac{a_0}{2}$ and for $n > 0$

$$A_n = a_n \cos nx + b_n \sin nx.$$

[12] ibid. p. 95.
[13] Lebesgue (1902).
[14] See Hawkins (1975) p. 127.

He assumed that $A_n \to 0$ as $n \to \infty$ and he associated with $\Omega$ a function $F(x)$ obtained by twice formally integrating the series $\Omega$. Thus, he set

$$C + C'x + A_0 \frac{xx}{2} - A_1 - \frac{A_2}{4} - \frac{A_3}{9} - \cdots = F(x) \qquad (22.3)$$

and proved $F(x)$ continuous by showing that the series was uniformly convergent, though he did not use this terminology. He then stated his first theorem on $F(x)$:

If the series $\Omega$ converges,

$$\frac{F(x+\alpha+\beta) - F(x+\alpha-\beta) - F(x-\alpha+\beta) + F(x-\alpha-\beta)}{4\alpha\beta}, \qquad (22.4)$$

converges to the same value as the series if $\alpha$ and $\beta$ become infinitely small in such a way that their ratio remains finite (bounded).

By using the addition formula for sine and cosine, Riemann saw that expression (22.4) reduced to

$$A_0 + A_1 \frac{\sin\alpha}{\alpha} \frac{\sin\beta}{\beta} + A_2 \frac{\sin 2\alpha}{2\alpha} \frac{\sin 2\beta}{2\beta} + A_2 \frac{\sin 3\alpha}{3\alpha} \frac{\sin 3\beta}{3\beta} + \cdots .$$

When $\alpha = \beta$, he had the equation

$$\frac{F(x+2\alpha) - 2F(x) + F(x-2\alpha)}{4\alpha\alpha} = A_0 + A_1 \left(\frac{\sin\alpha}{\alpha}\right)^2 + A_2 \left(\frac{\sin 2\alpha}{2\alpha}\right)^2 + \cdots . \qquad (22.5)$$

Riemann first proved this theorem for the $\alpha = \beta$ case and then deduced the general case. Observe that as $\alpha \to 0$, the series (22.5) converges termwise to $\Omega$. Thus, Riemann's task was essentially to show that (22.5) converged uniformly with respect to $\alpha$. We follow Riemann in detail, keeping in mind that Riemann did not use absolute values as we would today. Suppose that the series $\Omega$ converges to a function $f(x)$. Write

$$A_0 + A_1 + \cdots + A_{n-1} = f(x) + \epsilon_n \qquad (22.6)$$

so that

$$A_0 = f(x) + \epsilon_1 \quad \text{and} \quad A_n = \epsilon_{n+1} - \epsilon_n. \qquad (22.7)$$

Riemann noted that, because of convergence, for any positive number $\delta$, there existed an integer $m$ such that $\epsilon_n < \delta$ for $n > m$. By (22.7) and using summation by parts, he concluded that

$$\sum_{n=0}^{\infty} A_n \left(\frac{\sin n\alpha}{n\alpha}\right)^2 = f(x) + \sum_{n=1}^{\infty} \epsilon_n \left(\left(\frac{\sin(n-1)\alpha}{(n-1)\alpha}\right)^2 - \left(\frac{\sin n\alpha}{n\alpha}\right)^2\right). \qquad (22.8)$$

He then took $\alpha$ to be sufficiently small, so that $m\alpha < \pi$ and let $s$ be the largest integer in $\frac{\pi}{\alpha}$. He divided the last sum into three parts:

$$\sum_{n=1}^{m} + \sum_{n=m+1}^{s} + \sum_{n=s+1}^{\infty}.$$

The first sum was a finite sum of continuous functions, and it could be made arbitrarily small by taking $\alpha$ sufficiently small. In the second sum, the factor multiplying $\epsilon_n$ was positive and hence the sum could be written

$$< \delta \left( \left( \frac{\sin m\alpha}{m\alpha} \right)^2 - \left( \frac{\sin s\alpha}{s\alpha} \right)^2 \right).$$

Note that Riemann assumed $\frac{\pi}{2} \le \alpha \le \pi$ for any $n$ in the second sum, although he did not explicitly mention this. To show that the third sum could be made arbitrarily small, he rewrote the general term as the sum of

$$\epsilon_n \left( \left( \frac{\sin(n-1)\alpha}{(n-1)\alpha} \right)^2 - \left( \frac{\sin(n-1)\alpha}{n\alpha} \right)^2 \right) \quad \text{and}$$

$$\epsilon_n \left( \left( \frac{\sin(n-1)\alpha}{n\alpha} \right)^2 - \left( \frac{\sin n\alpha}{n\alpha} \right)^2 \right) = -\epsilon_n \frac{\sin(2n-1)\alpha \sin\alpha}{(n\alpha)^2}.$$

It was then clear that the general term in the third sum was less than

$$\delta \left( \frac{1}{(n-1)^2 \alpha\alpha} - \frac{1}{nn\alpha\alpha} \right) + \delta \frac{1}{nn\alpha}.$$

Thus, the third sum was less than

$$\delta \left( \frac{1}{(s\alpha)^2} + \frac{1}{s\alpha} \right).$$

Then, for infinitely small $\alpha$, this expression became

$$\delta \left( \frac{1}{\pi\pi} + \frac{1}{\pi} \right).$$

Riemann concluded that the infinite series on the right-hand side of (22.8) could not be greater than

$$\delta \left( 1 + \frac{1}{\pi} + \frac{1}{\pi^2} \right),$$

so that the theorem was proved. Riemann's argument can be shortened by observing that the second and third sums, in absolute value, are together less than

$$\delta \sum_{n=m+1}^{\infty} \int_{(n-1)\alpha}^{n\alpha} \left| \frac{d}{dt}\left(\frac{\sin^2 t}{t^2}\right) \right| dt < \delta \int_0^{\infty} \left| \frac{d}{dt}\left(\frac{\sin^2 t}{t^2}\right) \right| dt.$$

Since the last integral is convergent, the result follows. To prove the general result, when $\alpha \neq \beta$, Riemann set

$$F(x+\alpha+\beta) - 2F(x) + F(x-\alpha-\beta) = (\alpha+\beta)^2(f(x)+\delta_1),$$
$$F(x+\alpha-\beta) - 2F(x) + F(x-\alpha+\beta) = (\alpha-\beta)^2(f(x)+\delta_2),$$

so that

$$\frac{F(x+\alpha+\beta) - F(x+\alpha-\beta) - F(x-\alpha+\beta) + F(x-\alpha-\beta)}{4\alpha\beta}$$

$$= f(x) + \frac{(\alpha+\beta)^2}{4\alpha\beta}\delta_1 - \frac{(\alpha-\beta)^2}{4\alpha\beta}\delta_2.$$

The special case $\alpha = \beta$ implied that $\delta_1$ and $\delta_2$ became small as $\alpha$ and $\beta$ got small. Moreover, the factors $\frac{(\alpha+\beta)^2}{4\alpha\beta}$ and $\frac{(\alpha-\beta)^2}{4\alpha\beta}$ remained bounded when $\frac{\beta}{\alpha}$ was bounded. This proved the general case. Observe that the limit

$$\lim_{h\to 0} \frac{F(x+h) + F(x-h) - 2F(x)}{h^2}$$

is called the Schwarz, or Riemann–Schwarz, derivative of $F$. Riemann called this the "second differential quotient." Note that

$$F(x+h) + F(x-h) - 2F(x) = \Delta^2 F(x-h),$$
$$\text{where } \Delta F(x-h) = F(x) - F(x-h).$$

In general, $F(x)$ is continuous, as Riemann proved, but not necessarily differentiable. So here we have an instance of a generalized second derivative, although Riemann did not express himself in those terms.

Riemann's second theorem stated that when $A_n \to 0$ as $n \to \infty$, then

$$\frac{F(x+2\alpha) + F(x-2\alpha) - 2F(x)}{2\alpha}$$

tends to 0 as $\alpha$ tends to 0. In his terse style, Riemann gave a succinct argument for this, along lines similar to his proof of his first theorem. In his *The Apprenticeship of a Mathematician*, André Weil wrote that both he and his sister Simone found great value in the works of great minds and that he was very lucky to start off his mathematical reading of the greats with Riemann; he found that Riemann's works "are not hard to read, as long as one realizes that every word is loaded with meaning; there is perhaps no other mathematician whose writing matches Riemann's for density."[15]

---

[15]  Weil (1992) p. 40.

## 22.5 The Riemann–Lebesgue Lemma

The Riemann–Lebesgue lemma states that if $f(x)$ is integrable over $(a,b)$, then as $t \to \infty$,

$$\int_a^b f(x) \cos tx \, dx \to 0, \quad \text{and} \quad \int_a^b f(x) \sin tx \, dx \to 0.$$

Note that this result implies that the $n$th Fourier coefficients of an integrable function tend to zero as $n \to \infty$. Riemann derived his lemma from his integrability condition in an interesting way. Again in his 1854 paper, he began by writing

$$\int_0^{2\pi} f(x) \sin nx \, dx = \sum_{k=1}^n \int_{\frac{2(k-1)\pi}{n}}^{\frac{2k\pi}{n}} f(x) \sin nx \, dx.$$

He noted that $\sin nx$ was positive in the first half of the subinterval $\left( \frac{2(k-1)\pi}{n}, \frac{2k\pi}{n} \right)$ and negative in the second half. He supposed that in the whole subinterval he had $m_k \leq f(x) \leq M_k$, where $M_k$ was taken to be the largest value of $f(x)$ in the subinterval and $m_k$ the least. We may assume these to be the least upper bound and greatest lower bound, respectively. Thus, in the first half of the subinterval,

$$\int_{\frac{2(k-1)\pi}{n}}^{\frac{(2k-1)\pi}{n}} f(x) \sin nx dx \leq M_k \int_{\frac{2(k-1)\pi}{n}}^{\frac{(2k-1)\pi}{n}} \sin nx dx = \frac{2M_k}{n}.$$

Similarly, in the second half of the subinterval, the integral would be less than $-\frac{2m_k}{n}$. It followed that

$$\int_{\frac{2(k-1)\pi}{n}}^{\frac{2k\pi}{n}} f(x) \sin nx dx \leq \frac{2}{n}(M_k - m_k)$$

and hence

$$\left| \int_0^{2\pi} f(x) \sin nx \, dx \right| \leq \sum_{k=1}^n \frac{2}{n}(M_k - m_k) = \frac{1}{\pi} \sum_{k=1}^n \delta_k D_k,$$

where $\delta_k$ was the length of the $k$th subinterval and $D_k$ was the variation of $f(x)$ on that interval. By his own definition of integrability of $f(x)$, the sum $\sum \delta_k D_k$ had to become infinitely small as $n$ became infinitely large. This proved the theorem. Observe that the definition of integrability was perfect for obtaining this result on the Fourier coefficients, leading some to speculate that Riemann fashioned the definition with this result in mind.

## 22.6 Schwarz's Lemma on Generalized Derivatives

Recall that in connection with his work on trigonometric series, Cantor in 1870 asked Schwarz whether the following result was true: If $F(x)$ is continuous in an interval $a \leq x \leq b$ and

$$\lim_{\alpha \to 0} \frac{F(x+\alpha) - 2F(x) + F(x-\alpha)}{\alpha\alpha} = 0 \qquad (22.9)$$

for all $x$ in the interval, then $F(x)$ is a linear function. Schwarz replied to this letter that this was indeed a theorem and provided a proof, republished twenty years later in his collected mathematical works.[16] Cantor used the theorem and gave the proof, credited to Schwarz, in his 1870 paper.

Now note that if $F$ is twice differentiable, then its second derivative and generalized second derivative are identical; moreover, by (22.9), $F(x)$ is linear. Briefly, Schwarz's proof of the general case began by setting

$$\phi(x) = \left| F(x) - F(a) - \frac{x-a}{b-a}(F(a) - F(b)) \right| - \frac{1}{2}k(x-a)(b-x), \qquad (22.10)$$

where $k$ was a positive quantity to be chosen later. Schwarz did not employ the absolute value sign, instead using an $\epsilon$, equal to plus or minus 1, as a factor to maintain a positive value. Observe that $\phi(a) = 0$ and $\phi(b) = 0$. If the expression inside the absolute value sign in (22.10) is zero for all $x$ in $a \leq x \leq b$, then $F(x)$ is a linear function. Suppose the value of the expression is not zero. Since $\phi(x)$ is continuous, it has a maximum at some point $x_0$. Take $k$ sufficiently small that the value of $\phi(x_0)$ is positive. By the definition of maximum,

$$\phi(x_0 + \alpha) - \phi(x_0) \leq 0 \quad \text{and} \quad \phi(x_0 - \alpha) - \phi(x_0) \leq 0;$$

thus,

$$\phi(x_0 + \alpha) - 2\phi(x_0) + \phi(x_0 - \alpha) \leq 0.$$

But

$$\lim_{\alpha \to 0} \frac{\phi(x_0 + \alpha) - 2\phi(x_0) + \phi(x_0 - \alpha)}{\alpha\alpha}$$

$$= \lim_{\alpha \to 0} \left( \frac{F(x_0 + \alpha) - 2F(x_0) + F(x_0 + \alpha)}{\alpha\alpha} + k \right) = k > 0.$$

This contradiction implies that $F(x)$ is a linear function. Note that Weierstrass is credited with the 1841 invention of the absolute value sign we use today.

## 22.7  Cantor's Uniqueness Theorem

Cantor first stated his uniqueness theorem in 1870, though he later gave generalizations.[17] His first theorem stated that if a trigonometric series

$$\frac{1}{2}a_0 + \sum_{n=1}^{\infty}(a_n \cos nx + b_n \sin nx)$$

[16] Schwarz (1972) vol. 2, pp. 341–343.
[17] Cantor (1870a).

converged to zero at every point of the interval $(-\pi, \pi)$, then $a_0 = 0$ and $a_n = b_n = 0$ for $n \geq 1$. To prove this, Cantor first used the convergence of the trigonometric series to produce a tedious proof that $a_n \to 0$ and $b_n \to 0$ as $n \to \infty$. Later on, Kronecker helped Cantor realize that he could greatly streamline his proof by working with a different series. But we continue with Cantor's original proof based on this result. Observe that he could apply Riemann's second theorem so that the second Riemann–Schwarz derivative of

$$F(x) = \frac{1}{4} a_0 x^2 - \sum_{n=1}^{\infty} \frac{1}{n^2} (a_n \cos nx + b_n \sin nx)$$

was zero in $(-\pi, \pi)$. By Schwarz's lemma, $F(x)$ was a linear function $ax + b$, and he had

$$\sum_{n=1}^{\infty} \frac{1}{n^2} (a_n \cos nx + b_n \sin nx) = \frac{1}{4} a_0 x^2 - ax - b.$$

Since the left-hand side was periodic, $a_0$ and $a$ had to be zero. Because the series was uniformly convergent, Cantor could multiply by $\cos mx$ and $\sin mx$ and integrate term by term to obtain

$$\frac{\pi a_m}{m^2} = -b \int_{-\pi}^{\pi} \cos mx \, dx = 0, \quad \frac{\pi b_m}{m^2} = -b \int_{-\pi}^{\pi} \sin mx \, dx = 0,$$

for $m \geq 1$. This concludes Cantor's original proof. Observe that as a student of Weierstrass, he was quite familiar with uniform convergence and its connection with integration, but at that time the concept of uniform convergence was not well known.

To take care of the first step concerning $a_n$ and $b_n$, Kronecker pointed out that it was not necessary to prove that these coefficients tended to zero. Instead, he called the trigonometric series in the theorem $f(x)$ and defined a new function in terms of $u$:

$$g(u) = \frac{1}{2} \big( f(x + u) + f(x - u) \big)$$

$$= \frac{1}{2} a_0 + \sum_{n=1}^{\infty} (a_n \cos nx + b_n \sin nx) \cos nu = \frac{1}{2} a_0 + \sum_{n=1}^{\infty} A_n \cos nu.$$

Since the series $f(x)$ converged, $g(u)$ also converged and therefore $A_n = a_n \cos nx + b_n \sin nx \to 0$ as $n \to \infty$ for all $x$ in $(-\pi, \pi)$. With this new first step, Riemann's second theorem could then be applied, using $g(u)$ instead of $f(x)$, yielding $A_n = 0$ for $n \geq 1$. Thus, $a_n = 0$ and $b_n = 0$ for $n \geq 1$, so he also had $a_0 = 0$. Though Kronecker assisted Cantor with this argument, the germ of the idea was already in Riemann's paper.

Cantor extended the uniqueness theorem in an 1871 paper[18] by requiring convergence to zero of $\frac{1}{2} a_0 + \sum_{n=1}^{\infty} A_n$ at all but a finite number of points in $(-\pi, \pi)$.

---

[18] Cantor (1871).

He supposed $x_\nu$ to be a point at which the series did not converge. Now by Cantor's first proof, on the left-hand side of $x_\nu$, $F(x) = k_\nu x + l_\nu$ for some constants $k_\nu$ and $l_\nu$, whereas on the right-hand side, $F(x) = k_{\nu+1}x + l_{\nu+1}$. Now because $F(x)$ was continuous, $k_\nu x_\nu + l_\nu = k_{\nu+1}x_\nu + l_\nu$ and by Riemann's second theorem

$$\lim_{\alpha \to 0} \frac{F(x_\nu + \alpha) - 2F(x_\nu) + F(x_\nu - \alpha)}{\alpha}$$
$$= \lim_{\alpha \to 0} \frac{x_\nu(k_{\nu+1} - k_\nu) + l_{\nu+1} - l_\nu + \alpha(k_{\nu+1} - k_\nu)}{\alpha} = 0.$$

This implied that $k_{\nu+1} = k_\nu$ and $l_{\nu+1} = l_\nu$; therefore, $F(x)$ was defined by the same linear function in the whole interval $(-\pi, \pi)$.

Cantor then extended the argument to an infinite set with a finite number of limit points. Summarizing his argument, suppose $x_1, x_2, x_3, \ldots$ to be a sequence with one limit point $x$. Then, by the previous argument, the isolated points $x_1, x_2, x_3, \ldots$ can be removed, and then finally, after an infinite number of steps, $x$ is isolated and can be removed. Kronecker was horrified at this mode of argument, involving the completion of an infinite number of steps; he suggested to Cantor that he refrain from publishing his paper. But to Cantor's way of thinking, this kind of reasoning was quite legitimate, since he subscribed to the concept of a completed infinity. Cantor gave further extensions of his uniqueness theorem to more general infinite sets. The enterprise led him to turn his attention toward set theory rather than analysis, and he spent the rest of his life creating and developing the theory of infinite sets.

## 22.8 Exercises

(1) A solution of an equation $a_0 x^n + a_1 x^{n-1} + \cdots + a_n = 0$ where $a_0, a_1, \ldots, a_n$ are integers is called an algebraic number. Let $|a_0| + |a_1| + \cdots + |a_n| + n$ be the height of the equation. Show that there exist only a finite number of equations of a given height. Use this theorem to prove Dedekind's result that the set of algebraic numbers is countable, that is, the set can be put in one-to-one correspondence with the set of natural numbers. This theorem and this proof appeared in print in an 1874 paper of Cantor. Dedekind had communicated the proof to Cantor in November 1873. Uncharacteristically, Cantor did not mention Dedekind's contribution. See Ferreirós (1993) for references and a possible explanation.

(2) Read Wilbraham (1848); this paper contains the first discussion of the Gibbs phenomenon, dealing with overshoot in the convergence of the partial sums of certain Fourier series in the neighborhood of a discontinuity of the function. See Hewitt and Hewitt (1980) for a detailed discussion and history of the topic.

(3) In his paper, Riemann gave the function $f(x) = \frac{d}{dx}\left(x^\nu \cos \frac{1}{x}\right)$, where $0 < \nu < \frac{1}{2}$ as an example of an integrable function, not representable as a Fourier series and having an infinite number of maxima and minima. Analyze this claim.

(4) Show that the series $\sum_{n=1}^{\infty} \frac{\sin n^2 x}{n^2}$ converges to a continuous function. Prove that the function does not have a derivative at $\zeta\pi$ if $\zeta$ is irrational; prove the same if $\zeta = \frac{2A}{4B+1}$ or $\frac{2A+1}{2B}$ for integers $A$ and $B$. Show that when $\zeta = \frac{2A+1}{2B+1}$, the function has a derivative $= -\frac{1}{2}$. In 1916, Hardy proved the nondifferentiability portion of this above result; in 1970, J. Gerver proved the differentiability portion. In his lectures, Riemann discussed this series, apparently without stating the theorem. Weierstrass was of the opinion that Riemann may have intended this to be an example of a continuous but nondifferentiable function. Unable to prove this, Weierstrass constructed a different example, given in the next exercise. See Segal (1978).

(5) Show that the function $f(x) = \sum_{n=0}^{\infty} b^n \cos(a^n\pi x)$, where $0 < b < 1$; $a$ is an odd integer; $ab > 1 + \frac{3\pi}{2}$, and the function $g(x) = \sum_{n=1}^{\infty} \frac{\cos(n!x)}{n!}$ are both continuous and everywhere nondifferentiable. Weierstrass presented the first example in his lectures, and Paul du Bois-Reymond published it in 1875. G. Darboux published the second example in 1879. See Weierstrass (1894–1927) vol. 2, pp. 71–74.

(6) Let $t = \sum \frac{c_n}{2^n}$, with $c_n = 0$ or 1, be the binary expansion of $0 \le t \le 1$. Set $f(t) = \sum \frac{a_n}{2^n}$, where $a_n$ denotes the number of zeros among $c_1, c_2, \ldots, c_n$ if $c_0 = 0$; if $c_0 = 1$, then $a_n$ denotes the number of ones. Prove that $f(t)$ is continuous and single-valued for $0 \le t \le 1$ and that $f(t)$ is not differentiable for any $t$. See Takagi (1990), pp. 5–6. Teiji Takagi (1875–1960) graduated from the University of Tokyo and then studied under Schwarz, Frobenius, and Hilbert in Berlin and Göttingen 1898–1901. Even before going to Germany, Takagi studied Hilbert's 1897 *Zahlbericht*. His thesis proved the statement from Kronecker's *Jugendtraum* that all the abelian extensions of the number field $Q(\sqrt{-1})$ can be obtained by the division of the lemniscate. Takagi did his most outstanding work in class field theory; he was one of the first Japanese mathematicians to begin his career after the transition to Western mathematics in Japan, and he was instrumental in establishing a tradition of algebraic number theory there. See Miyake (1994) and Sasaki (1994). These two papers, along with other papers of interest, are contained in Sasaki, Sugiura, and Dauben (1994).

(7) For Bolzano's example of a continuous nowhere differentiable function, dating from about 1830, read Strichartz (1995) pp. 403–406. He gives a graphical presentation and points out that it has close connections with fractals.

(8) Show that the series $\sum_{n=2}^{\infty} \frac{\sin nx}{\ln n}$ converges to a function not integrable in any interval containing the origin. Then derive the conclusion that this trigonometric series is not a Fourier series. This example is due to P. Fatou and is referred to in Lebesgue (1906) p. 124.

(9) Prove W. H. Young's theorem that if $q_0 \ge q_1 \ge \cdots$ form a monotone descending sequence with zero as limit, and their decrements also form a monotone descending sequence, viz., $q_0 - q_1 \ge q_1 - q_2 \ge \cdots$, then the trigonometric series

$$\frac{1}{2}q_0 + \sum_{n=1}^{\infty} q_n \cos nx$$

is the Fourier series of a positive summable function. Use this to prove that

$$\sum_{n=2}^{\infty} \frac{\cos nx}{(\ln n)^c},$$

where $c > 0$ is a Fourier series. For this and the next exercise, see G. C. Young and W. H. Young (2000) pp. 449–478.

(10) Prove that if $q_1 \geq q_2 \geq \cdots$ form a monotone descending sequence of constants with zero as limit and $\sum_{n=1}^{\infty} n^{-1}q_n$ converges, then

$$\sum_{n=1}^{\infty} q_n \sin nx$$

is the Fourier series of a summable function bounded below for positive values of $x$ and bounded above for negative values of $x$. See Exercise 9.

(11) Prove that if $f \in L^1(-\pi,\pi)$, then the Poisson integral

$$\frac{1}{2\pi}\int_{-\pi}^{\pi} f(t)\frac{1-r^2}{1-2r\cos(t-x)+r^2}dt$$

converges almost everywhere (a.e.) to $f(x)$ as $r \to 1^-$. See Fatou (1906).

(12) Given a series $\sum_{n=1}^{\infty} A_n$, $A_n = a_n \cos nx + b_n \sin nx$, define its conjugate as the series $\sum_{n=1}^{\infty} B_n$, where $B_n = -b_n \cos nx + a_n \sin nx$. Suppose then that $\sum_{n=1}^{\infty}(a_n^2 + b_n^2) < \infty$. Prove the Riesz-Fischer theorem that there exist functions $f, g \in L^2(-\pi,\pi)$ such that $f \sim \sum A_n$ and $g \sim \sum B_n$. Show also Lusin's result that

$$\frac{1}{2\pi}\int_{-\pi}^{\pi} f(t)\frac{1-r^2}{1-2r\cos(t-x)+r^2}dt$$
$$= \frac{1}{\pi}\int_{-\pi}^{\pi} g(t)\frac{r\sin(t-x)}{1-2r\cos(t-x)+r^2}dt = f(x) \text{ a.e.}$$

Next, deduce the formula for the Cauchy principal value integral:

$$\lim_{\epsilon \to 0^+}\frac{1}{\pi}\int_{\epsilon \leq |t| \leq \pi} g(x+t)\frac{dt}{2\tan\frac{t}{2}} = f(x) \text{ a.e.} \qquad (22.11)$$

For an arbitrary function $g$, the conjugate $\tilde{g}$ is defined by the negative of the principal value integral in (22.11). If $g \in L^1$, then in general $\tilde{g}$ might or might not be in $L^1$, but $\tilde{g} \in L^p$ for $0 < p < 1$. If $g \in L^p$, for $p > 1$, then $\tilde{g} \in L^p$. Note that Lusin proved the last result when $p = 2$. See Lusin (1913). Nikolai Lusin (1883–1950) was a student of Dmitri Egorov (1869–1931) at Moscow

University and he founded an important school of mathematics there with students such as Kolmogorov, Menshov, and Privalov. They developed what is now called the complex method in Fourier analysis.

(13) Concerning $S_n f$, the $n$th partial sum of the Fourier series of $f$, given by $\sum_{k=1}^{n} A_k$, show that

$$S_n f(x) = \sum_{k=1}^{n} (a_k \cos kx + b_k \sin kx)$$

$$= \frac{1}{\pi} \int_{-\pi}^{\pi} g(x+t) \left( \frac{1}{2 \tan \frac{t}{2}} - \frac{\cos \left(n + \frac{1}{2}\right)t}{2 \sin \frac{t}{2}} \right) dt,$$

where the two integrals on the right-hand side should be taken as Cauchy principal values. Combine this with the result in Exercise 12 to show that $\lim_{n \to \infty} S_n f(x) = f(x)$ a.e. if and only if the principal value integral satisfies

$$\lim_{n \to \infty} \int_{-\pi}^{\pi} g(x+t) \frac{\cos nt}{t} dt = 0 \quad \text{a.e.}$$

From (22.11) it follows that the principal value integral $\int_{-\pi}^{\pi} \frac{g(x+t)}{t} dt$ exists a.e. for $g \in L^2$. Lusin also had an example of a continuous function $g$ with $\int_{-\pi}^{\pi} \left| \frac{g(x+t)}{t} \right| dt = \infty$ on a set of positive measure. Note that in order for this principal value integral to converge, there must have been a good deal of cancellation. Lusin conjectured the almost everywhere convergence of the Fourier series of square integrable functions because he thought that the cancellation in the principal value integral was the reason for the convergence of the series. Kolmogorov (1923) contains an example of an integrable, but not square-integrable, function whose Fourier series diverged everywhere. Lennart Carleson proved Lusin's conjecture in 1966, and Richard Hunt soon extended Carleson's theorem to $L^p$ functions with $p > 1$. One of the important concepts needed in the Carleson and Hunt proofs was that of maximal functions. For a locally integrable function $f$, the Hardy–Littlewood maximal function is defined by

$$Mf(x) = \sup_{h>0} \frac{1}{h} \int_{x-h}^{x+h} |f(t)| dt.$$

(14) Prove that if $f \in L^p(-\pi,\pi)$ for $1 < p < \infty$, then $\tilde{f} \in L^p$ and $\|\tilde{f}\|_p \leq C_p \|f\|_p$. This theorem is due to Marcel Riesz (1928). Also deduce that $\|S_n f\|_p \leq C_p \|f\|_p$.

(15) Show that if $f \in L^p$ for $1 < p < \infty$, then

$$\|Mf\|_p \leq C_p \|f\|_p.$$

Show also that if $f(r,x)$ is the Poisson integral of $f$, then

$$\sup_{0 \leq r < 1} |f(r,x)| \leq C \, Mf(x),$$

and hence $\left\| \sup_{0 \leq r < 1} |f(r,x)| \right\|_p \leq C_p \, \|f\|_p.$

These results were published in 1930 by Hardy and Littlewood; see Hardy (1966–1979) pp. 509–544, especially pp. 530–538.

## 22.9  Notes on the Literature

For Carleson's proof of the convergence theorem, mentioned in Exercise 13, see Carleson (1966). Hunt's extension can be found in Haimo (1968), pp. 235–255. For historical background on the convergence of Fourier series, see Hunt's paper in Butzer and Sz.-Nagy (1974).

Laugwitz (1999) presents a lively account of Riemann's life and mathematical work, including trigonometric series and complex variables. Riemann (2004) contains an English translation of his papers and lectures. Hawkins (1975) presents a detailed but very readable account of the development of integration theory from Riemann to Lebesgue. For a modern discussion of Riemann integrability, see Bressoud (2007), p. 251 and for more on Lebesgue, see Bressoud (2008).

Purkert and Ilgauds (1985) contains the correspondence between Cantor and Schwarz. Cantor (1932) contains his work on the uniqueness of trigonometric series. See Dauben (1979) for a discussion of the development of Cantor's mathematical thought. Cooke (1993) is an interesting history of the work on the uniqueness of trigonometric series, and it also surveys recent contributions. The article by Zygmund in Ash (1976) contains some insightful remarks on the development of Fourier series.

# 23

# The Hypergeometric Series

## 23.1 Preliminary Remarks

The hypergeometric series and associated functions are among the most important in mathematics, partly because they cover a large class of valuable special functions as either particular cases or as limiting cases. More importantly, because they have the appropriate degree of generality, very useful transformation formulas and other relations can be proved about them. The hypergeometric series is defined by

$$F(a,b,c,x) = {}_2F_1\left(\begin{matrix} a,b \\ c \end{matrix}; x\right) = 1 + \frac{a \cdot b}{1 \cdot c} x + \frac{a(a+1) \cdot b(b+1)}{1 \cdot 2 \cdot c(c+1)} x^2 + \cdots . \quad (23.1)$$

The expressions involved can be written more briefly if we adopt the modern notation for the shifted factorial:

$$(a)_n = a(a+1)\cdots(a+n-1) \quad \text{for } n \geq 1, \ (a)_0 = 1. \quad (23.2)$$

Thus

$$F(a,b,c,x) = {}_2F_1\left(\begin{matrix} a,b \\ c \end{matrix}; x\right) = \sum_{n=0}^{\infty} \frac{(a)_n (b)_n}{n! \, (c)_n} x^n. \quad (23.3)$$

The subscript notation in $F$ was introduced in the twentieth century when similar series with varying numbers of parameters, such as $a,b,c$, were considered. Note the following examples of hypergeometric series in Gauss's notation:

$$(1-x)^{-\alpha} = F(\alpha, 1, 1, x); \quad \log\frac{1+x}{1-x} = 2x \, F\left(\frac{1}{2}, 1, \frac{3}{2}, x^2\right);$$

$$e^x = \lim_{a \to \infty} F\left(1, 1, 1, \frac{x}{a}\right); \quad J_\alpha(x) = \frac{\left(\frac{x}{2}\right)^\alpha}{\Gamma(\alpha+1)} \lim_{a,b \to \infty} F\left(a, b, \alpha+1, -\frac{x^2}{4ab}\right).$$

Historically, hypergeometric series occurred not only in the study of power series but also as inverse factorial series in finite difference theory. James Stirling, in particular, employed them in the approximate summation of series and in this connection also discovered special cases of important transformation formulas. However, in 1778, Euler first introduced the hypergeometric series in the form (23.1). He proved[1] that the series satisfied the second-order differential equation:

$$x(1-x)\frac{d^2 F}{dx^2} + (c - (a+b+1)x)\frac{dF}{dx} - abF = 0, \qquad (23.4)$$

and then used this equation to prove an important transformation formula:

$$F(a,b,c,x) = (1-x)^{c-a-b} F(c-a, c-b, c, x). \qquad (23.5)$$

The binomial factor can be moved to the left-hand side, as $(1-x)^{a+b-c}$. When this is expanded as a series and multiplied by the hypergeometric function on the left-hand side, the coefficients of $x^n$ on the two sides give the identity

$$\sum_{k=0}^{n} \frac{(a)_k (b)_k (a+b-c)_{n-k}}{k! \, (c)_k (n-k)!} = \frac{(c-a)_n (c-b)_n}{n! \, (c)_n} \qquad (23.6)$$

or

$$\sum_{k=0}^{n} \frac{(-n)_k (a)_k (b)_k}{k! \, (c)_k (1+a+b-c-n)_k} = \frac{(c-a)_n (c-b)_n}{(c)_n (c-a-b)_n}, \qquad (23.7)$$

or in the following modern notation, whose meaning is obvious from (23.7):

$$_3F_2\left(\begin{matrix} -n, a, b \\ c, 1+a+b-c-n \end{matrix}; 1\right) = \frac{(c-a)_n (c-b)_n}{(c)_n (c-a-b)_n}. \qquad (23.8)$$

Observe that this identity is formally equivalent to (23.5).

In 1797, Johann Friedrich Pfaff (1765–1825) proved Euler's transformation (23.5) by giving an inductive proof of (23.8).[2] Pfaff was among the leading mathematicians in Germany during the late eighteenth and early nineteenth centuries; he was the formal thesis advisor for Gauss. His results on second-order differential equations were inspired by Euler, whose work on this topic appeared in his three volumes on the integral calculus.[3] Euler's work on series provided the starting point for the German combinatorial school founded by C. F. Hindenburg (1741–1808), of which Pfaff was a member. Pfaff's formula (23.8) is very useful for evaluating certain types of sums of products of binomial coefficients occurring in combinatorial problems. No one seems to have taken notice of this work; in order to save this identity and some other of Pfaff's results from oblivion, Jacobi referred to it in a paper of 1847.[4] We remark

---

[1] Eu. I-16$_2$ pp. 41–55. E 710.
[2] Pfaff (1797b).
[3] Eu. I-11, 12, 13. E 342, E 366, E 385.
[4] Jacobi (1969) vol. 6, pp. 174–182, especially p. 178.

that Jacobi was interested in the history of mathematics and consistently attempted to give credit to the original discoverer of a concept or formula. In spite of Jacobi's efforts, this identity was forgotten for many years. In 1890, it was finally rediscovered and published by L. Saalschütz,[5] with whose name it was associated for many years. In the 1970s, Askey noticed Jacobi's reference and renamed it the so that it is now called the Pfaff–Saalschütz identity.[6] Pfaff could not have foreseen that in the 1990s, his method of proving (23.8) would become the foundation of George Andrews's general method for proving hypergeometric identities useful in computer algebra systems. Pfaff also found the terminating form of another important hypergeometric transformation:

$$F(a,b,c,x) = (1-x)^{-a} F\left(a, c-b, c, \frac{x}{x-1}\right). \tag{23.9}$$

Note that Pfaff took the parameter $a$ to be a negative integer so that the series on both sides were finite. Pfaff derived this formula from a study of the differential equation

$$x^2(a+bx^n)\frac{d^2y}{dx^2} + x(c+ex^n)\frac{dy}{dx} + (f+gx^n)y = X,$$

where $X$ was a function of $x$. Euler earlier discussed the homogeneous form of this equation in his book on the integral calculus. Note that Newton's transformation (10.4) is a particular case of (23.9), obtained by taking $a=1$, $b=\frac{1}{2}$, $c=\frac{3}{2}$, and $x=-t^2$. Stirling's formula (10.13), obtained by equating the series in (10.31) and (10.32), can also be derived from (23.9) by taking $a=-1$ and $x=\frac{1}{m}$. It is possible that Pfaff was motivated to study the series in (23.9) by Hindenburg's 1778 work[7] on the following problem: For given numbers $\alpha$ and $\beta$, transform a series $ay+by^2+cy^3+\cdots$ to a series of the form

$$\frac{Ay}{\alpha+\beta y} + \frac{By^2}{(\alpha+\beta y)^2} + \frac{Cy^3}{(\alpha+\beta y)^3} + \cdots;$$

thus, determine $A, B, C, \ldots$ in terms of $a, b, c, \ldots$.

Gauss was the first mathematician to undertake a systematic and thorough study of the hypergeometric function. His treatment of the subject appeared in a paper of 1813. It is possible that Gauss was introduced to the topic when he visited Helmstedt in 1799 to use the university library and rented a room in Pfaff's home. One imagines this to be very likely, since Gauss and Pfaff took walks together every evening and discussed mathematics. Gauss does not refer to earlier work on hypergeometric series so it is hard to determine what he had learned from others. The two most notable features of Gauss's contributions to hypergeometric series were his use of contiguous relations to derive the basic formulas and his determination of the conditions for the convergence of the series. Some of his unpublished work shows that he wanted to

[5] Saalschütz (1890).
[6] Askey (1975) p. 62; Andrews et al. (1999) p. 69.
[7] Hindenburg (1778).

build the foundation of analysis on a rigorous theory of limits, for which purpose he carefully defined the concepts of superior and inferior limits of sequences.

Gauss defined functions contiguous to $F(a,b,c,x)$ as those functions arising from it when the first, second or third parameter $a,b,c$ was increased or diminished by one while the other two remained the same. Gauss may have seen the importance of contiguous functions by reading Stirling's 1730 *Methodus*. He found that there was a linear relation between $F(a,b,c,x)$ and any two contiguous functions; such an equation is now called a contiguous relation. Clearly there would be $\binom{6}{2} = 15$ such relations, and Gauss listed all of them in the first section of his 1813 paper. From these relations he derived continued fractions expansions of ratios of hypergeometric functions, his fundamental summation formula for $F(a,b,c,1)$, and the differential equation for $F(a,b,c,x)$. He derived the latter in the second (unpublished) part of his paper. In this part, Gauss derived transformation formulas in the same manner as Euler before him, except that he also gave examples of quadratic transformations. For example:

$$F\left(a,b,a+b+\frac{1}{2},4x-4x^2\right) = F\left(2a,2b,a+b+\frac{1}{2},x\right). \qquad (23.10)$$

Gauss treated $a,b,c$, and $x$ as complex variables and in this connection he pointed out that it was necessary to exercise care when dealing with values of $x$ outside the circle of convergence of the series. Thus, when $x$ was changed to $1-x$ in (23.10), the left-hand side would remain unchanged, leading to the evidently contradictory result that

$$F\left(2a,2b,a+b+\frac{1}{2},x\right) = F\left(2a,2b,a+b+\frac{1}{2},1-x\right). \qquad (23.11)$$

Gauss called this result a paradox and his explanation, from the unpublished portion of his paper, is highly interesting, showing that as early as 1812 he was thinking of analytic continuation of functions:[8]

> To explain this, it ought to be remembered that proper distinction should be made between the two significations of the symbol $F$, viz., whether it represents the function whose nature is expressed by the differential equation [(23.4)], or simply the sum of an infinite series. The latter is always a perfectly determinate quantity so long as the fourth element lies between $-1$ and $+1$, and care must be taken not to exceed these limits for otherwise it is entirely without any meaning. On the other hand, according to the former signification, it [$F$] represents a general function which always varies subject to the law of continuity if the fourth element vary continuously whether you attribute real values or imaginary values to it, provided you always avoid the values 0 and 1. Hence it is evident that in the latter sense, the function may for equal values of the fourth element (the passage or rather the return being made through imaginary quantities) attain unequal values of which that which the *series* $F$ represents is only one, so that it is not at all contradictory that while some *one* value of the function $F(a,b,a+b+\frac{1}{2},4y-4yy)$ is equal to $F(2a,2b,a+b+\frac{1}{2},y)$ the *other* value should be equal to $F(2a,2b,a+b+\frac{1}{2},1-y)$ and it would be just as absurd to deduce thence the equality of these values as it would be to conclude, that since Arc. $\sin\frac{1}{2} = 30^o$,

[8] Gauss (1863–1927) vol. 3, pp. 226–227. For the English translation, see Kikuchi (1891) pp. 144–145.

Arc. sin $\frac{1}{2} = 150^o, 30^o = 150^o$. – But if we take $F$ in the less general sense, viz. simply as the sum of the series $F$, the arguments by which we have deduced (23.10), necessarily suppose $y$ to increase from the value 0 only up to the point when $x[= 4y - 4yy]$ becomes $= 1$, i.e. up to $y = \frac{1}{2}$. At this point, indeed, the *continuity* of the series $P = F(a, b, a + b + \frac{1}{2}, 4y - 4yy)$ is interrupted, for evidently $\frac{dP}{dy}$ jumps suddenly from a positive (finite) value to a negative. Thus in this sense equation (23.10) does not admit of being extended outside the limits $y = \frac{1}{2} - \sqrt{\frac{1}{2}}$ up to $y = \frac{1}{2}$. If preferred, the same equation can also be put thus: –

$$F\left(a, b, a + b + \frac{1}{2}, x\right) = F\left(2a, 2b, a + b + \frac{1}{2}, \frac{1 - \sqrt{1 - x}}{2}\right).$$

Again, Gauss's letter of December 18, 1811, to his friend F. W. Bessel (1784–1846) shows how far he had advanced in developing a theory of functions of complex variables:[9]

What should we make of $\int \phi x.dx$ for $x = a + bi$? Obviously, if we're to proceed from clear concepts, we have to assume that $x$ passes, via infinitely small increments (each of the form $\alpha + i\beta$), from that value at which the integral is supposed to be 0, to $x = a + bi$ and that then all the $\phi x.dx$ are summed up. In this way the meaning is made precise. But the progression of $x$ values can take place in infinitely many ways: Just as we think of the realm of all real magnitudes as an infinite straight line, so we can envision the realm of all magnitudes, real and imaginary, as an infinite plane wherein every point which is determined by an abscissa $a$ and an ordinate $b$ represents as well the magnitude $a + bi$. The continuous passage from one value of $x$ to another $a + bi$ accordingly occurs along a curve and is consequently possible in infinitely many ways. But I maintain that the integral $\int \phi x.dx$ computed via two different such passages always gets the same value as long as $\phi x = \infty$ never occurs in the region of the plane enclosed by the curves describing these two passages. This is a very beautiful theorem, whose not-so-difficult proof I will give when an appropriate occasion comes up. It is closely related to other beautiful truths having to do with developing functions in series. The passage from point to point can always be carried out without ever touching one where $\phi x = \infty$. However, I demand that those points be avoided lest the original basic conception of $\int \phi x.dx$ lose its clarity and lead to contradictions. Moreover it is also clear from this how a function generated by $\int \phi x.dx$ could have several values for the same values of $x$, depending on whether a point where $\phi x = \infty$ is gone around not at all, once, or several times. If, for example, we define log $x$ via $\int \frac{1}{x} dx$ starting at $x = 1$, then arrive at log $x$ having gone around the point $x = 0$ one or more times or not at all, every circuit adds the constant $+2\pi i$ or $-2\pi i$; thus the fact that every number has multiple logarithms becomes quite clear.

Thus in 1811, Gauss had a clear conception of complex integration and had discovered Cauchy's integral theorem, published by Cauchy in 1825. He had also begun to understand the reason for a function being multivalued; this understanding informed Gauss's comments on (23.11). It is possible that Gauss was motivated to study quadratic transformations by his discovery during the mid-1790s of the connection between the arithmetic-geometric mean and the complete elliptic integral. This integral is defined by

---

[9] Gauss (1863–1927) vol. 8, pp. 90–92. For the English translation of this portion of the letter, see Remmert (1991) pp. 167–168.

$$K(k) = \int_0^{\frac{\pi}{2}} \frac{d\theta}{\sqrt{1 - k^2 \sin^2 \theta}} = \frac{\pi}{2} F\left(\frac{1}{2}, \frac{1}{2}, 1, k^2\right).$$

In his unpublished paper,[10] Gauss also computed two independent solutions of the hypergeometric equation in the neighborhood of 0, 1, and $\infty$. He obtained explicit formulas linearly relating a solution in the neighborhood of one of these points with two independent solutions in the neighborhood of another point. As an example, consider Gauss's result

$$F(a,b,c,x) = \frac{\Gamma(c)\Gamma(b-a)}{\Gamma(c-a)\Gamma(b)}(-x)^{-a} F\left(a, a+1-c, a+1-b, \frac{1}{x}\right)$$
$$+ \frac{\Gamma(c)\Gamma(a-b)}{\Gamma(a)\Gamma(c-b)}(-x)^{-b} F\left(b, b+1-c, b+a-a, \frac{1}{x}\right).$$

The functions on the right-hand side were solutions in the neighborhood of infinity. Gauss also considered the case where the parameter $c$ was an integer so that the second independent solution involved a logarithmic term. Euler was also aware of this situation. Gauss went further by showing that the digamma function, $\psi(x) \equiv \frac{\Gamma'(x)}{\Gamma(x)}$, defined in the first part of his paper, could be employed to obtain an expression for the second solution.

Gauss's paper was quite influential, especially among German mathematicians, who produced much important research on this topic in the next three or four decades. In 1833, as part of his doctoral dissertation, P. Vorsselman de Heer gave the integral representation[11]

$$F(a,b,c,x) = \frac{\int_0^1 t^{b-1}(1-t)^{c-b-1}(1-xt)^{-a}dt}{\int_0^1 t^{b-1}(1-t)^{c-b-1}dt}. \tag{23.12}$$

Note that the integral in the denominator is the beta integral, evaluated by Euler, equal to $\frac{\Gamma(b)\Gamma(c-b)}{\Gamma(c)}$. This integral representation of $F(a,b,c,x)$ was independently found by Kummer and published a few years later in his long memoir on hypergeometric functions.[12] However, in a posthumous paper, Jacobi attributed this formula to Euler,[13] though it seems that it does not appear explicitly in Euler's work. However, Euler did give an integral representation of a solution of a differential equation closely related to the hypergeometric equation; this may have been Jacobi's reason for the attribution.

In 1828, the Danish mathematician Thomas Clausen (1801–1885) obtained a significant result of a different kind.[14] Clausen was born to poor farming people and did not learn to read or write until the age of 12. He encountered many difficulties due to his humble origins. But Gauss thought highly of him, and Clausen's

[10] Gauss (1863–1927) vol. 3, pp. 207–229. For an English translation of this paper, see Kikuchi (1891) pp. 121–149.
[11] Vorselman de Heer (1833).
[12] Kummer (1836) § 27.
[13] Jacobi (1859) p. 149.
[14] Clausen (1828).

abundant mathematical talent was eventually recognized. He considered the square of a hypergeometric series and found

$$(F(a,b,c,x))^2 \equiv \left({}_2F_1\left({a,b \atop c};x\right)\right)^2 = {}_3F_2\left({2a,2b,a+b \atop 2a+2b,a+b+\frac{1}{2}};x\right), \quad (23.13)$$

where

$$\sum_{n=0}^{\infty} \frac{(a)_n(b)_n(c)_n}{n!\,(d)_n(e)_n}x^n \equiv {}_3F_2\left({a,b,c \atop d,c};x\right).$$

In 1836, Ernst Kummer (1810–1893) published the first major work on hypergeometric functions after Gauss.[15] He rediscovered much of the material in the unpublished portion of Gauss's paper, including quadratic transformations. In fact, these transformations are implicitly contained in Gauss's published paper. Kummer also found some results for ${}_3F_2$ functions, including the existence of three-term contiguous relations when $x = 1$. Kummer was trained as a high school teacher; he taught at that level 1831–1841. In 1834, while serving a year in the army, he communicated some papers in analysis to Jacobi who is reported by E. Lampe to have commented: "There we are; now the Prussian musketeers even enter into competition with the professors by way of mathematical works."[16] However, Jacobi was impressed with the work done by Kummer under difficult circumstances and wrote in his reply, "If you think that I could be of any help with obtaining an academic position, I would be happy to offer my humble services – less because I think that you would need them, or that they would be significant, but as a token of my great respect for your talent and your works."[17] Dirichlet and Jacobi worked to find Kummer a university position. Kummer became a professor at Breslau in 1842 and moved to Berlin in 1855, when Dirichlet vacated his chair there to take up the position at Göttingen left open by Gauss's death.

In the 1840s, Jacobi wrote some interesting results on hypergeometric series. In the posthumous paper mentioned earlier, he showed that the sequence of hypergeometric polynomials $F(-n,b,c,x)$ where $n = 0, 1, 2, \ldots$, were orthogonal with respect to a suitable distribution. Following Euler, he also worked out how definite integrals could be employed to study solutions of the hypergeometric equation. In another paper, discussed in Section 21.9, he applied the symbolic method to obtain some known transformation formulas for hypergeometric functions.

In a paper of 1857, Bernhard Riemann took a very different approach to hypergeometric functions as part of his new theory of functions of a complex variable.[18] Riemann gave the foundation of this theory in his famous doctoral dissertation of 1851.[19] An important idea first given in this work and later applied to the theory of

---

[15] Kummer (1836).
[16] Kummer (1975) vol. 1, p. 18.
[17] See Pieper (2007) pp. 214–215.
[18] Riemann (1990) pp. 99–115.
[19] ibid. pp. 35–75.

abelian functions, hypergeometric functions, and the zeta function was that a complex analytic function was to a large extent determined by the nature and location of its singularities. The singularities of the hypergeometric equation are at 0, 1, and $\infty$. In his 1857 paper, Riemann considered the more general case where the singularities of a function were at three distinct values $a, b$, and $c$. He axiomatically defined a set of functions, called $P$-functions, satisfying certain properties in the neighborhood of the three singularities, but without reference to the hypergeometric function or equation. Riemann showed that $P$-functions were solutions of a second-order differential equation reducible to the hypergeometric equation when the singular points were 0, 1, and $\infty$. He also developed a very simple transformation theory for $P$-functions by means of which one could derive a large number of relations among hypergeometric functions with little calculation.

We have seen that Gauss emphasized the fact that the hypergeometric series represented a hypergeometric function in only a small part of the domain of definition of the function. Moreover, the function was multivalued. Perhaps unable to develop a theory of complex variables to treat the hypergeometric function to his satisfaction, Gauss held back publication of the second part of his paper on the subject. Riemann saw Gauss's full paper in 1855, after Gauss's death. Surely this problem left pending by Gauss provided Riemann with great motivation for his landmark 1857 paper. Riemann also had a strong interest in mathematical physics; as he mentioned in the introduction to his paper, the hypergeometric function had numerous applications in physical and astronomical researches. After 1857, Riemann continued his investigations on the theory of ordinary differential equations with algebraic coefficients. His lectures and writings on the topic were published posthumously and eventually led to the formulation of what is now known as the Riemann–Hilbert problem.

Felix Klein (1849–1925) was one of the earliest mathematicians to understand and propagate the ideas of Riemann. In 1893, he gave a course of lectures on Riemann's theory of hypergeometric functions.[20] Interestingly, a decade later, the English mathematician E. W. Barnes (1874–1953) presented an alternative development of the hypergeometric function, based on the complex analytic technique of the Mellin transform, making use of Cauchy's calculus of residues.[21]

R. H. Mellin (1854–1935) was a Finnish mathematician who studied analysis first under Mittag-Leffler in Stockholm and then with Weierstrass in Berlin. He started teaching in 1884 at what was later named the Technical University of Finland. He founded a tradition of research in complex function theory in Finland, continued by mathematicians such as Ernst Lindelöf, Frithiof and Rolf Nevanlinna, and Lars V. Ahlfors. Mellin gave a general formulation of the Mellin transform in an 1895 treatise on the gamma and hypergeometric functions. For a function $f(x)$ integrable on $(0, \infty)$, the Mellin transform is defined by

$$F(s) = \int_0^\infty x^{s-1} f(x) dx. \tag{23.14}$$

---

[20] These lectures were published in 1933. See Klein (1933).
[21] Barnes (1908).

If $f(x) = O(x^{-a+\epsilon})$ as $x \to 0+$ and $f(x) = O(x^{b-\epsilon})$ as $x \to +\infty$, for $\epsilon > 0$ and $a < b$, then the integral converges absolutely and defines an analytic function in the strip $a < \operatorname{Re} s < b$. Mellin gave the inversion formula:

$$f(x) = \frac{1}{2\pi i} \int_{c-\infty i}^{c+\infty i} x^{-s} F(s) ds, \quad a < c < b. \tag{23.15}$$

In particular, we have the pair of formulas (stated without convergence conditions) very useful in analytic number theory:

$$\Gamma(s) = \int_0^\infty x^{s-1} e^{-x} dx \quad \text{and} \quad e^{-x} = \frac{1}{2\pi i} \int_{c-\infty i}^{c+\infty i} \Gamma(s) x^{-s} ds. \tag{23.16}$$

In fact, Riemann had already used the Mellin transform in his famous paper on the distribution of primes.[22] Other particular cases of the transform were derived by others, including Mellin himself, before he stated the general formula. The second formula in (23.16) was apparently first discovered by the French mathematician Eugène Cahen in 1894.[23] His thesis on the Riemann zeta function and its analogs contains several interesting results on Dirichlet series, though some of these were not rigorously proved until more than a decade later. Cahen followed Riemann in taking the Mellin transforms of a function analogous to the theta function to obtain functional equations for the corresponding Dirichlet series. He considered some analogs of the theta function:

$$\sum_{n=1}^\infty \left(\frac{n}{p}\right) e^{-\frac{n^2\pi x}{p}}, \quad \sum_{n=1}^\infty n\left(\frac{n}{p}\right) e^{-\frac{n^2\pi x}{p}}, \quad \sum_{n=1}^\infty \frac{\sigma_1(n)}{n} e^{-2n\pi x}$$

where $\left(\frac{n}{p}\right)$ denoted the Legendre symbol and $\sigma_1(n)$ the sum of the divisors of $n$. Cahen employed the first sum when $p \equiv 1 \pmod 4$, and the second when $p \equiv 3 \pmod 4$.

E. W. Barnes studied at Trinity College, Cambridge, from 1893 to 1896. Most of his mathematical work was done in the period 1897–1910 on the double gamma function, hypergeometric functions and Mellin transforms, and the theory of entire functions. In 1915 Barnes left Cambridge to pursue his second career. He was ordained in 1922 and appointed to the Bishopric of Birmingham in 1924, an office he held until 1952.

Barnes's starting point was the observation that from Euler's integral representation, and by expanding $(1 - xt)^{-a}$ as a series, the Mellin transform of the hypergeometric function would be

$$\int_0^\infty x^{s-1} F(a,b,c,-x) dx = \frac{\Gamma(c)}{\Gamma(a)\Gamma(b)} \frac{\Gamma(s)\Gamma(a-s)\Gamma(b-s)}{\Gamma(c-s)}, \tag{23.17}$$

---

[22] Riemann (1990) pp. 177–185.
[23] Cahan (1894).

for min $(\operatorname{Re} a, \operatorname{Re} b) > \operatorname{Re} s > 0$. This suggested the integral representation for the hypergeometric function:

$$\frac{\Gamma(a)\Gamma(b)}{\Gamma(c)} F(a,b,c,x) = \frac{1}{2\pi i} \int_{k-\infty i}^{k+\infty i} \frac{\Gamma(s)\Gamma(a-s)\Gamma(b-s)}{\Gamma(c-s)} (-x)^{-s} ds, \quad (23.18)$$

where min $(\operatorname{Re} a, \operatorname{Re} b) > k > 0$ and $c \neq 0, -1, -2, \dots$. This is Barnes's integral for the hypergeometric function and provides the basis for an alternative development of these functions. A precise statement of the integral formula requires conditions on the path of integration.

## 23.2   Euler's Derivation of the Hypergeometric Equation

We follow Euler's notation as it is easy to understand and his derivation is quite short and straightforward.[24] Euler let $s$ denote the hypergeometric series (23.1). Then

$$\partial(x^c \partial s) = ab\, x^{c-1} + \frac{ab}{1 \cdot c}(a+1)(b+1)x^c + \cdots$$

$$\partial(x^a s) = ax^{a-1} + \frac{ab}{1 \cdot c}(a+1)x^a + \cdots .$$

Note that, for the sake for brevity, he frequently suppressed $\partial x$. Now

$$\partial(x^{b+1-a}\partial(x^a s)) = abx^{b-1} + \frac{ab}{1 \cdot c}(a+1)(b+1)x^b + \cdots$$
$$= x^{b-c}\partial(x^c \partial s),$$

or

$$\partial(ax^b s + x^{b+1}\partial s) = x^{b-c}(cx^{c-1}\partial s + x^c \partial\partial s)$$

or

$$a(bx^{b-1}s + x^b \partial s) + (b+1)x^b \partial s + x^{b+1}\partial\partial s = cx^{b-1}\partial s + x^b \partial\partial s.$$

Dividing by $x^{b-1}$, he got the hypergeometric equation

$$x(1-x)\partial\partial s + (c - (a+b+1)x)\,\partial s - abs = 0. \qquad (23.19)$$

Euler gave an equally simple proof of the transformation formula. He showed that $s = (1-x)^n z$ also satisfied a second-order differential equation with the hypergeometric form when $n = c - a - b$. He started by taking the logarithmic derivative of $s$ to obtain

---

[24] Eu. I-16$_2$ pp. 41–55. E 710.

$$\frac{\partial s}{s} = \frac{\partial z}{z} - \frac{n\partial x}{1-x}. \tag{23.20}$$

The derivative of this equation was

$$\frac{\partial\partial s}{s} - \frac{(\partial s)^2}{s^2} = \frac{\partial\partial z}{z} - \frac{(\partial z)^2}{zz} - \frac{n(\partial x)^2}{(1-x)^2}. \tag{23.21}$$

We remark that Euler wrote $\partial s^2$ for $(\partial s)^2$. He then squared (23.20) to get

$$\frac{(\partial s)^2}{s^2} = \frac{(\partial z)^2}{z^2} - \frac{2n\partial x\partial z}{z(1-x)} + \frac{nn(\partial x)^2}{(1-x)^2}.$$

He added this equation to (23.21) to get

$$\frac{\partial\partial s}{s} = \frac{\partial\partial z}{z} - \frac{2n\partial x\partial z}{z(1-x)} + \frac{n(n-1)(\partial x)^2}{(1-x)^2}. \tag{23.22}$$

When (23.20) and (23.22) were applied to the hypergeometric equation (23.19), he could write

$$x(1-x)\frac{\partial\partial z}{z} - \frac{2nx\partial x\partial z}{z} + (c-(a+b+1)x)\frac{\partial z}{z}$$

$$+ \frac{n(n-1)x(\partial x)^2}{1-x} - \frac{n\,(c-(a+b+1)x)\,\partial x}{1-x} - ab = 0. \tag{23.23}$$

Next, the two terms with $1-x$ in the denominator, the second of which had a suppressed $\partial x$, combined to form

$$\frac{n\,((n+a+b)x - c)}{1-x}.$$

When $n+a+b = c$, the factor $1-x$ cancelled. For this $n$, (23.23) was reduced to

$$x(1-x)\partial\partial z + [c+(a+b-2c-1)x]\partial z - (c-a)(c-b)z = 0, \tag{23.24}$$

an equation of the hypergeometric type. Thus,

$$z = F(c-a, c-b, c, x) = (1-x)^{a+b-c}F(a, b, c, x). \tag{23.25}$$

This proved Euler's transformation (23.5).

## 23.3  Pfaff's Derivation of the $_3F_2$ Identity

We have already noted that equation (23.25) is equivalent to Pfaff's identity (23.7). Pfaff gave a very interesting proof of this,[25] given here in modern notation using shifted factorials. Let

---

[25] Pfaff (1797b) pp. 51–52.

$$S_n(a,b,c) = \sum_{j=0}^{n} \frac{(-n)_j(a)_j(b)_j}{j!\,(c)_j(1-n+a+b-c)_j}.$$

Then, by a simple calculation,

$$S_n(a,b,c) - S_{n-1}(a,b,c)$$

$$= \sum_{j=0}^{n} \left( \frac{(-n)_j(a)_j(b)_j}{j!\,(c)_j(1-n+a+b-c)_j} - \frac{(1-n)_j(a)_j(b)_j}{j!\,(c)_j(2-n+a+b-c)_j} \right)$$

$$= \frac{-(1+a+b-c)ab}{c(1+a+b-c-n)(2+a+b-c-n)} S_{n-1}(a+1,b+1,c+1). \qquad (23.26)$$

By induction, the recurrence (23.26), combined with the initial value $S_0(a,b,c) = 1$, uniquely determines $S_n(a,b,c)$. Pfaff could easy verify that

$$\sigma_n(a,b,c) = \frac{(c-a)_n(c-b)_n}{(c)_n(c-a-b)_n}$$

satisfied the same recurrence relation and initial condition, proving his formula (23.7).

This formula is quite useful and important, though this does not seem to have been realized until the twentieth century when it found applications to the evaluation of combinatorial sums of products of binomial coefficients. In this connection, the Chinese mathematician Li Shanlan (1811–1882) is of historical interest. He was trained in the Chinese mathematical tradition, though later in life he came to learn about Western works on algebra, analytic geometry, and calculus.[26] At the age of 8, he studied the ancient Chinese text *Jiuzhang Suanshu*, and six years later he read a Chinese translation of the first six books of Euclid's *Elements*. Soon after that, he studied Chinese works on algebra and trigonometry. Eventually he became interested in the summation of finite series. He made some interesting discoveries involving Stirling numbers, Euler numbers and other numbers and series of combinatorial significance, contained in his work *Duoji Bilei*. This may be translated as "Heaps Summed Using Analogies;" heaps refer to finite sums. In this work, Li Shanlan developed and generalized the concepts and formulas of earlier researchers such as Wang Lai (1768–1813) and Dong Youcheng (1791–1823). Li Shanlan presented the following summation formula:

$$\sum_{j=0}^{k} \binom{k}{j}^2 \binom{n+2k-j}{2k} = \binom{n+k}{k}^2. \qquad (23.27)$$

This formula was brought to the notice of the Hungarian mathematician Paul Turán (1910–1976) in 1937. He gave a proof using Legendre polynomials, published

---

[26] Martzloff (1997) pp. 341–350.

in 1954.[27] This aroused the curiosity of other mathematicians, and it was established that the combinatorial sum (23.27) could be written as

$$\binom{n+2k}{2k} {}_3F_2\left(\begin{matrix} -k, \ -k, \ -n \\ 1, \ -n-2k \end{matrix}; 1\right),$$

and therefore (23.27) could be derived from Pfaff's formula. Jacobi's perceptive effort to prevent this formula from being forgotten provides further evidence of his insight into formulas and his stature as an algorist.

As another application of Pfaff's identity, note that it can be written as

$$\sum_{k=0}^{n} \frac{(-n)_k (a)_k (b)_k}{k!\,(c)_k (-n+1+a+b-c)_k} = \frac{(c-a)_n}{n!\,n^{c-a-1}} \cdot \frac{(c-b)_n}{n!\,n^{c-b-1}} \cdot \frac{n!\,n^{c-1}}{(c)_n} \cdot \frac{n!\,n^{c-a-b-1}}{(c-a-b)_n}.$$

When $n \to \infty$ and $\mathrm{Re}(c-a-b) > 0$, by (17.4), we obtain Gauss's ${}_2F_1$ summation mentioned earlier as (17.14):

$$F(a,b,c,1) = \frac{\Gamma(c)\Gamma(c-a-b)}{\Gamma(c-a)\Gamma(c-b)}, \qquad (23.28)$$

though we do not know whether Gauss was aware of this derivation. Note that for $a = -m$, a negative integer, (23.28) reduces to the Vandermonde identity (Chu–Vandermonde identity), discussed in Section 25.10.

## 23.4 Gauss's Contiguous Relations and Summation Formula

The contiguous relations can be given in compact form if we use the following notation for contiguous functions:

$$F = F(a,b,c,x), \quad F(a+) = F(a+1,b,c,x), \quad \text{etc.}$$

Gauss wrote down all of the fifteen contiguous relations connecting $F$ with two functions contiguous to it.[28] Here we give four examples:

$$(c-2a-(b-a)x)F + a(1-x)F(a+) - (c-a)F(a-) = 0, \qquad (23.29)$$

$$(c-a-b)F + a(1-x)F(a+) - (c-b)F(b-) = 0, \qquad (23.30)$$

$$(c-a-1)F + aF(a+) - (c-1)F(c-) = 0, \qquad (23.31)$$

$$c(c-1-(2c-a-b-1)x)F + (c-a)(c-b)xF(c+)$$
$$-c(c-1)(1-x)F(c-) = 0. \qquad (23.32)$$

[27] Turán (1990) vol. 1, pp. 743–747.
[28] Gauss (1813) § 7.

From the fifteen relations, one may obtain other relations in which more than one parameter is changed by one or more; we give a relation presented by Gauss, where our notation has the obvious meaning.

$$F(b+,c+) - F = \frac{a(c-b)x}{c(c+1)} F(a+,b+,c+2). \tag{23.33}$$

Gauss proved relations (23.30) and (23.31): First, let

$$M = \frac{(a+1)_{n-1}(b)_{n-1}}{n!\,(c)_n}.$$

Then the coefficients of $x^n$ in $F, F(b-), F(a+), F(c-)$, and $xF(a+)$ would be

$$a(b+n-1)M, \quad a(b-1)M, (a+n)(b+n-1)M,$$

$$\frac{a(b+n-1)(c+n-1)M}{c-1}, \quad n(c+n-1)M,$$

respectively. To obtain (23.31), it was therefore sufficient for him to check that

$$a(c-a-1)(b+n-1) + a(a+n)(b+n-1) - a(b+n-1)(c+n-1) = 0.$$

Equation (23.30) can be proved in a similar manner; equation (23.33) can also be proved by the direct method. Gauss found his formula (23.28) for $F(a,b,c,1)$ by taking $x = 1$ in (23.4) to obtain

$$F(a,b,c,1) = \frac{(c-a)(c-b)}{c(c-a-b)} F(a,b,c+1,1). \tag{23.34}$$

Note that he proved the convergence of the series for $\mathrm{Re}(c-a-b) > 0$; thus, the series on the right-hand side also converged. By repeated application of this equation he got

$$F(a,b,c,1) = \frac{(c-a)_n(c-b)_n}{(c)_n(c-a-b)_n} F(a,b,c+n,1). \tag{23.35}$$

Gauss could then express the right-hand side of the equation in terms of the gamma function, just as we obtained (23.28); he then let $n \to \infty$ to get the result.

## 23.5 Gauss's Proof of the Convergence of $F(a,b,c,x)$ for $c-a-b>0$

Gauss's proof of this important result was based on the formula

$$(\beta - \alpha - 1)\left(1 + \frac{\alpha}{\beta} + \frac{\alpha(\alpha+1)}{\beta(\beta+1)} + \cdots + \frac{(\alpha)_k}{(\beta)_k}\right) = \beta - 1 - \frac{(\alpha)_{k+1}}{(\beta)_k}. \tag{23.36}$$

This summation formula follows immediately from the following relation; although Gauss did not state it explicitly, he knew it well from his numerous calculations with hypergeometric series.

$$\frac{(\alpha)_k}{(\beta)_{k-1}} - \frac{(\alpha)_{k+1}}{(\beta)_k} = (\beta - \alpha - 1)\frac{(\alpha)_k}{(\beta)_k}. \tag{23.37}$$

A simple algebraic calculation is sufficient to check this relation. The idea was to write a hypergeometric term as a difference of two terms. It is interesting that in 1978, Bill Gosper showed the tremendous effectiveness of this approach in the summation of series of hypergeometric type. Gosper's method is now one of the fundamental algorithms used to sum such series.

Now note that the ratio of the $(n + 1)$th term over the $n$th term of the series $F(a,b,c,x)$ (omitting $x$) is

$$\frac{(a + n)(b + n)}{(1 + n)(c + n)} = \frac{n^2 + (a + b)n + ab}{n^2 + (c + 1)n + c}. \tag{23.38}$$

We take $a$, $b$, $c$ real, though the argument also applies to complex values. Gauss proved, more generally,[29] that if the ratio of the consecutive terms in a series was

$$\frac{n^\lambda + An^{\lambda-1} + Bn^{\lambda-2} + Cn^{\lambda-3} + \cdots}{n^\lambda + an^{\lambda-1} + bn^{\lambda-2} + cn^{\lambda-3} + \cdots} \tag{23.39}$$

and $A - a$ was a negative quantity with absolute value greater than unity, then the series converged. And when this result is applied to the special case of $F(a,b,c,x)$, it follows from (23.38) that the hypergeometric series converges for $c + 1 - a - b > 1$ or $c - a - b > 0$. To prove the theorem, write the series, for which the ratio of terms is given by (23.39), as $M_1 + M_2 + M_3 + \cdots$. We remark that Gauss did not use subscripts; he wrote the series as $M + M' + M'' + \cdots$. Now since $a > A + 1$, there is a sufficiently small number $h$ such that $a - h > A + 1$, or $a - h - 1 > A$. Now observe that if the fraction (23.39) is multiplied by $\frac{n}{n-1-h}$, we have

$$\frac{n}{n - 1 - h}\frac{M_{n+1}}{M_n} = \frac{n^{\lambda+1} + An^\lambda + \cdots}{n^{\lambda+1} + (a - h - 1)n^\lambda + \cdots}.$$

If $n$ is large enough, the last ratio is less than 1. Suppose this true for $n \geq N$. Then

$$|M_{N+1}| < \frac{N - 1 - h}{N}|M_n|,$$

$$|M_{N+2}| < \frac{N - h}{N + 1}|M_{n+1}| < \frac{(N - h - 1)(N - h)}{N(N + 1)}|M_N|,$$

$$\cdots$$

$$|M_{N+k}| < \frac{(N - h - 1)(N - h)\cdots(N - h - 1 + k - 1)}{N(N + 1)\cdots(N + k - 1)}|M_N|.$$

---

[29] Gauss (1813) § 16.

Hence,

$$|M_N| + |M_{N+1}| + \cdots + |M_{N+k}|$$

$$= |M_N| \left( 1 + \frac{N-h-1}{N} + \frac{(N-h-1)(N-h)}{N(N+1)} + \cdots + \frac{(N-h-1)_k}{(N)_k} \right)$$

$$= \frac{|M_N|}{h} \left( N - 1 - \frac{(N-h-1)_{k+1}}{(N)_k} \right),$$

following from (23.36). The term $\frac{(N-h-1)_{k+1}}{(N)_k}$ tends to zero as $k \to \infty$ because

$$\lim_{k \to \infty} \frac{(N-h-1)_{k+1}}{(N)_k} = \lim_{k \to \infty} \left( \frac{k! \, k^N}{(N)_k} \right) \left( \frac{(N-h)_k}{k! \, k^{N-h}} \right) \frac{N-h-1}{k^h}. \qquad (23.40)$$

Now the two expressions in parentheses on the right-hand side of (23.40) have the limit $\frac{\Gamma(N)}{\Gamma(N-h)}$, while

$$\lim_{k \to \infty} \frac{N-h-1}{k^h} = 0.$$

Thus, Gauss proved that

$$\sum_{k=N}^{\infty} |M_k| < \frac{N-1}{h} |M_N|,$$

and the convergence of $\sum_{n=1}^{\infty} M_n$ followed.

Observe that Gauss's method leads to a great refinement of the ratio test.

## 23.6  Raabe's Test for Convergence

Joseph Raabe refined Gauss's convergence result into a general test for convergence of a series. In a paper of 1832,[30] he stated his convergence test:

Let $a_0 + a_1 + a_2 + \cdots$ be a series of positive terms. Suppose $\lim_{n \to \infty} n \left( \frac{a_n}{a_{n+1}} - 1 \right) = k$. Then the series converges when $k > 1$ and diverges when $k < 1$.

Raabe proved this theorem in section 11 of his paper. In section 1, he proved the integral test and in section 2 he applied it to show that the series

$$1 + \frac{1}{2^m} + \frac{1}{3^m} + \cdots$$

converged for $m > 1$ and diverged when $m \leq 1$. Then in section 7 he applied this result to prove that the series

---

[30] Raabe (1832) pp. 63–64.

$$\frac{1}{1+m} + \frac{1}{(1+m)\left(1+\frac{m}{2}\right)} + \frac{1}{(1+m)\left(1+\frac{m}{2}\right)\left(1+\frac{m}{3}\right)} + \cdots$$

$$= \frac{1}{1+m} + \frac{1\cdot 2}{(1+m)(2+m)} + \frac{1\cdot 2\cdot 3}{(1+m)(2+m)(3+m)} + \cdots \quad (23.41)$$

converged when $m > 1$ and diverged when $m \le 1$. To verify this result, he noted that if $u_n$ denoted the $n$th term of (23.41), then

$$u_n n^m = \frac{n!\, n^m}{(m+1)_n}.$$

Next, according to (17.5),

$$\lim_{n\to\infty} u_n n^m = \Gamma(m+1),$$

so Raabe could conclude that $\sum_{n=1}^{\infty} u_n$ would behave as the series $\Gamma(m+1)\sum_{n=1}^{\infty} \frac{1}{n^m}$, so that the result was verified.

To prove his main theorem, Raabe observed that for $m > k$ and $N$ large enough,

$$a_{N+1} + a_{N+2} + \cdots > a_N \left( \frac{1}{1+\frac{m}{N}} + \frac{1}{\left(1+\frac{m}{N}\right)\left(1+\frac{m}{N+1}\right)} + \cdots \right)$$

and for $m < k$ and $N$ large enough,

$$a_{N+1} + a_{N+2} + \cdots < a_N \left( \frac{1}{1+\frac{m}{N}} + \frac{1}{\left(1+\frac{m}{N}\right)\left(1+\frac{m}{N+1}\right)} + \cdots \right).$$

The convergence of (23.41) for $m > 1$ and its divergence for $m < 1$ completed the proof of Raabe's convergence theorem.

In section 12 of his paper, Raabe deduced Gauss's convergence result, given in Section 23.5 of this chapter. He noted that

$$\frac{a_n}{a_{n+1}} = \frac{n^h + a_1 n^{h-1} + \cdots + a_n}{n^h + A_1 n^{h-1} + \cdots + A_n},$$

implied, with $\omega = \frac{1}{n}$, that

$$n\left( \frac{a_n}{a_{n+1}} - 1 \right) = \frac{(a_1 - A_1) + (a_2 - A_2)\omega + \cdots + (a_h - A_h)\omega^{h-1}}{1 + A_1\omega + \cdots + A_h\omega^h}.$$

Hence

$$\lim_{n\to\infty} \left( \frac{a_n}{a_{n+1}} - 1 \right) = a_1 - A_1,$$

confirming that the series converged when $a_1 - A_1 > 1$ and diverged when $a_1 - A_1 < 1$.

### 23.7  Gauss's Continued Fraction

Gauss derived an important continued fraction from the contiguous relation (23.33).[31] He set

$$G(a,b,c,x) = \frac{F(a,b+1,c+1,x)}{F(a,b,c,x)},$$

so that

$$\frac{F(a+1,b,c+1,x)}{F(a,b,c,x)} = \frac{F(b,a+1,c+1,x)}{F(b,a,c,x)} = G(b,a,c,x).$$

Then, dividing (23.33) by $F(a,b+1,c+1,x)$, he obtained

$$1 - \frac{1}{G(a,b,c,x)} = \frac{a(c-b)}{c(c+1)} x G(b+1,a,c+1,x),$$

or

$$G(a,b,c,x) = \frac{1}{1 - \frac{a(c-b)}{c(c+1)} x G(b+1,a,c+1,x)}. \tag{23.42}$$

This process could be continued:

$$G(b+1,a,c+1,x) = \frac{1}{1 - \frac{(b+1)(c+1-a)}{(c+1)(c+2)} x G(a+1,b+1,c+2,x)},$$

and thus

$$G(a,b,c,x) = \frac{1}{1-} \frac{\alpha_0 x}{1-} \frac{\beta_1 x}{1-} \frac{\alpha_1 x}{1-} \frac{\beta_2 x}{1-} \cdots, \tag{23.43}$$

where

$$\alpha_n = \frac{(a+n)(c+n-b)}{(c+2n)(c+2n+1)} \quad \text{and} \quad \beta_n = \frac{(b+n)(c+n-a)}{(c+2n-1)(c+2n)}. \tag{23.44}$$

Gauss mentioned an important particular case: when $b = 0$. In that case,

$$G(a,0,c-1,x) = F(a,1,c,x), \tag{23.45}$$

and the formulas in (23.44) took the form

$$\alpha_n = \frac{(a+n)(c+n-1)}{(c+2n-1)(c+2n)} \quad \text{and} \quad \beta_n = \frac{n(c+n-1-a)}{(c+2n-2)(c+2n-1)}. \tag{23.46}$$

---

[31]  Gauss (1813) § 12.

For $a = 1$ and $c = \frac{3}{2}$ and $x = t^2$, Gauss had

$$\log \frac{1+t}{1-t} = \frac{2t}{1-} \frac{\frac{1}{3}t^2}{1-} \frac{\frac{2 \cdot 2}{3 \cdot 5}t^2}{1-} \frac{\frac{3 \cdot 3}{5 \cdot 7}t^2}{1-} \cdots .$$

This continued fraction played a fundamental role in Gauss's theory of numerical integration.

## 23.8 Gauss: Transformations of Hypergeometric Functions

Gauss found solutions of the hypergeometric equation other than $F(\alpha, \beta, \gamma, x)$ and also used the hypergeometric equation to obtain transformation formulas,[32] just as Euler had done. Note that Gauss used the symbols $\alpha, \beta$, and $\gamma$ and employed $a, b$ for variables in a different context. We shall follow that practice here. He set $x = 1 - y$ in the hypergeometric equation to get

$$(y - yy) \frac{dd P}{dy^2} + (\alpha + \beta + 1 - \gamma - (\alpha + \beta + 1)y) \frac{dP}{dy} - \alpha \beta P = 0.$$

Clearly, $P = F(\alpha, \beta, \alpha + \beta + 1 - \gamma, y)$ was a solution of this equation and hence $F(\alpha, \beta, \alpha + \beta + 1 - \gamma, 1 - x)$ would be an independent solution of the hypergeometric equation. Gauss noted that any solution of the hypergeometric equation must be a linear combination of these two. He then looked for solutions of the form $P = x^\mu P'$ by substituting this expression for $P$ in the equation. He observed that the equation for $P'$ was of the hypergeometric form when $\mu = 0$ or $\mu = 1 - \gamma$. In the latter case, the equation for $P'$ was

$$(x - xx) \frac{dd P'}{dx^2} + (2 - \gamma - (\alpha + \beta + 3 - 2\gamma)x) \frac{dP'}{dx}$$
$$- (\alpha + 1 - \gamma)(\beta + 1 - \gamma) P' = 0.$$

Thus,

$$P = x^{1-\gamma} F(\alpha + 1 - \gamma, \beta + 1 - \gamma, 2 - \gamma, x)$$
$$= (1 - x)^{\gamma - \alpha - \beta} x^{1-\gamma} F(1 - \alpha, 1 - \beta, 2 - \gamma, x)$$

would be another solution of the original hypergeometric equation. Observe that the last step followed from an application of Euler's transformation (23.5). It then followed that there existed constants $M$ and $N$ such that

$$F(\alpha, \beta, \alpha + \beta + 1 - \gamma, 1 - x) = M F(\alpha, \beta, \gamma, x)$$
$$+ N x^{1-\gamma}(1 - x)^{\gamma - \alpha - \beta} F(1 - \alpha, 1 - \beta, 2 - \gamma, x).$$

---

[32] Gauss (1863–1927) vol. 3, pp. 208–223. For an English translation, see Kikuchi (1891) pp. 122–137.

Gauss determined after three pages of interesting calculations that

$$M = \frac{\Gamma(\alpha + \beta + 1 - \gamma)\Gamma(1 - \gamma)}{\Gamma(\alpha + 1 - \gamma)\Gamma(\beta + 1 - \gamma)} \quad \text{and} \quad N = \frac{\Gamma(\alpha + \beta + 1 - \gamma)\Gamma(\gamma - 1)}{\Gamma(\alpha)\Gamma(\beta)}.$$

We observe that the case in which $\alpha$ is a negative integer was given by Pfaff in 1797.[33] In this case, the second term is zero because $\Gamma(\alpha)$ appears in the denominator. Gauss remarked that this formula was useful for computational purposes. Clearly, a series would converge more rapidly for $x$ between 0 and $\frac{1}{2}$ than for $x$ between $\frac{1}{2}$ and 1. A formula of this type could be applied to convert a slowly convergent series to two more rapidly convergent ones. But Gauss cautioned that this formula would not be applicable if the series to be transformed was such that the third parameter minus the sum of the first two turned out to be an integer. He then went on to show that if this occurred, the formula could be modified by the use of his $\Psi$ function and the logarithm. He explicitly worked out the formula for the elliptic integral $F\left(\frac{1}{2}, \frac{1}{2}, 1, 1 - x\right)$.

Gauss also found solutions at infinity. He set $x = \frac{1}{y}$ and then $P = y^\mu P'$ and observed that $P'$ was hypergeometric when $\mu = \alpha$ or $\beta$. Thus, he obtained $P$ as

$$x^{-\alpha} F\left(\alpha, \alpha + 1 - \gamma, \alpha + 1 - \beta, \frac{1}{x}\right) \quad \text{or} \quad x^{-\beta} F\left(\beta, \beta + 1 - \gamma, \beta + 1 - \alpha, \frac{1}{x}\right).$$

He then expressed $F(\alpha, \beta, \gamma, x)$ as a linear combination of these solutions.

Gauss derived another general transformation formula by taking $x = \frac{y}{y-1}$ in the hypergeometric equation and then $P = (1 - y)^\mu P'$, so that another hypergeometric equation would be obtained when $\mu = \alpha$ or $\beta$. This gave the necessary result:

$$F(\alpha, \beta, \gamma, x) = (1 - y)^\alpha F(\alpha, \gamma - \beta, \gamma, y)$$
$$= (1 - x)^{-\alpha} F\left(\alpha, \gamma - \beta, \gamma, \frac{x}{x - 1}\right).$$

In 1797 Pfaff published this result for the case in which $\alpha$ was a negative integer.[34] Gudermann proved the generalization in 1830. Three years later, P. Vorsselman de Heer noted in his thesis[35] that Euler's transformation could be obtained when the preceding transformation was applied to itself; Kummer also observed this fact.

In the published part of his 1813 paper,[36] Gauss found the values of the coefficients in the expansion

$$(aa + bb - 2ab\cos\phi)^{-n} = A + 2A'\cos\phi + 2A''\cos 2\phi + 2A'''\cos 3\phi + \cdots$$
$$(23.47)$$

[33] Pfaff (1797a).
[34] Pfaff (1797a).
[35] Vorselman de Heer (1833).
[36] Gauss (1813) § 6.

in terms of hypergeometric series. He noted that

$$A^{(p)} = \frac{1}{a^{2n}} \binom{n+p-1}{p} \left(\frac{b}{a}\right)^p F\left(n, n+p, p+1, \frac{bb}{aa}\right)$$

$$= \frac{1}{(aa+bb)^n} \binom{n+p-1}{p} \left(\frac{ab}{aa+bb}\right)^p$$

$$\times F\left(\frac{n+p}{2}, \frac{n+p+1}{2}, p+1, \frac{4aabb}{(aa+bb)^2}\right)$$

$$= \frac{1}{(a \pm b)^{2n}} \binom{n+p-1}{p} \left(\frac{ab}{(a \pm b)^2}\right)^p F\left(n+p, p+\frac{1}{2}, n+\frac{1}{2}, \frac{\pm 4ab}{(a \pm b)^2}\right).$$

$$(23.48)$$

Note that Euler studied the series (23.47) in a 1749 memoir on the perturbation of planetary orbits and in 1766 Lagrange found the first series for $A^{(p)}$ in (23.48).[37] This series and its coefficients have been studied intensively, both analytically and numerically, and Gauss's interest in them was evident. If we take $a = 1$ and $x = b^2$, the second equation in (23.48) gives

$$(1+x)^{n+p} F(n, n+p, p+1, x) = F\left(\frac{n+p}{2}, \frac{n+p+1}{2}, p+1, \frac{4x}{(1+x)^2}\right).$$

$$(23.49)$$

This is an example of a quadratic transformation because the variable on one side is $x$, or it could be a fractional linear transformation of $x$, while the variable on the right-hand side involves $x^2$. It is very likely that equation (23.49) led Gauss to study such transformations in the second (unpublished) part of his paper. He set $x = \frac{4y}{(1+y)^2}$ in the hypergeometric equation and then $P = (1+y)^{2\alpha} Q$ to find that the equation satisfied by $Q$ was

$$(1+y)(y-y^2)\frac{d^2 Q}{dy^2} + (\gamma - (4\beta - 2\gamma)y + (\gamma - 4\alpha - 2)y^2)\frac{dQ}{dy}$$

$$- 2\alpha(2\beta - \gamma + (2\alpha + 1 - \gamma)y)Q = 0.$$

Now note that when $\beta = \alpha + \frac{1}{2}, 1 + y$ is a common factor in this equation. We remark that equation (23.49) guided Gauss in the substitutions for $x, P$ and $\beta$. We next have $Q = F(2\alpha, 2\alpha + 1 - \gamma, \gamma, y)$ and finally

$$(1+y)^{2\alpha} F(2\alpha, 2\alpha + 1 - \gamma, \gamma, y) = F\left(\alpha, \alpha + \frac{1}{2}, \gamma, \frac{4y}{(1+y)^2}\right). \quad (23.50)$$

---

[37] Dutka (1984) p. 22.

## 23.9   Kummer's 1836 Paper on Hypergeometric Series

Kummer independently rediscovered Gauss's unpublished results on hypergeometric functions, including the quadratic transformations. Of course, he was familiar with Gauss's published paper and with the work of Euler, Pfaff, Jacobi, and Gudermann on this topic. Kummer took a general approach. He set out to determine all functions of $z$ and $w$ of $x$ such that $y = wF(\alpha',\beta',\gamma',z)$ satisfied the equation

$$y'' + (\gamma - (\alpha+\beta+1)x)y' - \alpha\beta y = 0$$

and $\alpha',\beta',\gamma'$ were linear combinations of $\alpha,\beta,\gamma$.[38] He found $z$ to be a fractional linear transformation $\frac{ax+b}{cx+d}$ and that $w$ could be taken to be

$$x^{1-\gamma},\ (1-x)^{\gamma-\alpha-\beta},\ x^{1-\gamma}(1-x)^{\gamma-\alpha-\beta},\ \text{ or } 1.$$

Specifically, $z$ could be any one of the six fractional linear transformations serving to permute the values 0, 1, and $\infty$. These would be

$$z=x,\ z=1-x,\ z=\frac{1}{x},\ z=\frac{1}{1-x},\ z=\frac{x}{x-1},\ z=\frac{x-1}{x}.$$

When $z=x$, he obtained the four forms

$$F(\alpha,\beta,\gamma,x),\quad (1-x)^{\gamma-\alpha-\beta}F(\gamma-\alpha,\gamma-\beta,\gamma,x),$$

$$x^{1-\gamma}F(\alpha-\gamma+1,\beta-\gamma+1,2-\gamma,x),$$

$$x^{1-\gamma}(1-x)^{\gamma-\alpha-\beta}F(1-\alpha,1-\beta,2-\gamma,x).$$

Thus, he obtained twenty-four solutions of the hypergeometric equation and determined the linear relation among any three of them.

Kummer may have become interested in quadratic transformations after studying Gauss's published equation (23.48). His interest in elliptic integrals may have provided him with further motivation to study these transformations. It was clear to Kummer, as it was to Gauss, that quadratic transformations existed when the parameters $\alpha,\beta,\gamma$ in $F(\alpha,\beta,\gamma,x)$ satisfied certain relations. So Kummer considered the linear relations among the parameters leading to such transformations. In this way, he rediscovered Gauss's results as well as new ones. For example, he obtained

$$F(\alpha,\beta,2\beta,x) = (1-x)^{\beta-\alpha}\left(1-\frac{x}{2}\right)^{\alpha-2\beta}$$
$$\cdot F\left(\beta-\frac{\alpha}{2},\frac{2\beta-\alpha+1}{2},\beta+\frac{1}{2},\left(\frac{x}{2-x}\right)^2\right).$$

[38] Kummer (1836).

Note that by applying Euler's transformation to the right-hand side, we get the simpler form

$$F(\alpha, \beta, 2\beta, x) = \left(1 - \frac{x}{2}\right)^{-\alpha} F\left(\frac{\alpha}{2}, \frac{\alpha+1}{2}, \beta + \frac{1}{2}, \left(\frac{x}{2-x}\right)^2\right).$$

This and Gauss's transformation (23.50) are the two basic quadratic transformations; from these, the others can be obtained by using fractional linear transformations or the three term relations among the different solutions of the hypergeometric equation. At the end of his paper, Kummer commented on the more general hypergeometric series

$$1 + \frac{\alpha \cdot \beta \cdot \lambda}{1 \cdot \gamma \cdot \nu} x + \frac{\alpha(\alpha+1) \cdot \beta(\beta+1) \cdot \lambda(\lambda+1)}{1 \cdot 2 \cdot \gamma(\gamma+1) \cdot \nu(\nu+1)} x^2 + \cdots.$$

He wrote that he was unable to obtain general transformation formulas for this function, although he had several for the case $x = 1$. As an example, he presented

$$\sum_{k=0}^{\infty} \frac{(\alpha)_k (\beta)_k (\lambda)_k}{k! \, (\gamma)_k (\nu)_k} = \frac{\Gamma(\nu)\Gamma(\nu+\gamma-\alpha-\beta-\lambda)}{\Gamma(\nu-\lambda)\Gamma(\nu+\gamma-\alpha-\beta)} \sum_{k=0}^{\infty} \frac{(\gamma-\alpha)_k(\gamma-\beta)_k(\lambda)_k}{k! \, (\gamma)_k(\nu+\gamma-\alpha-\beta)_k}.$$
(23.51)

He observed that, in general, this series could not be summed in terms of the gamma function, but when $\lambda = 1$ and $\nu = 2(\alpha + \beta - \gamma + 1)$, then its value would be

$$\frac{(\alpha+\beta-\gamma+1)(\gamma-1)}{(\alpha-\gamma+1)(\beta-\gamma+1)} \left(\frac{\Gamma(\gamma-1)\Gamma(\alpha+\beta-\gamma+1)}{\Gamma(\alpha)\Gamma(\beta)} - 1\right). \qquad (23.52)$$

Recall that Stirling discovered a particular case of Kummer's transformation where $\lambda = 1$ and $\nu = \beta + 1$; see (10.41).

## 23.10 Jacobi's Solution by Definite Integrals

Euler gave a method of solving differential equations using definite integrals. He applied it to solve several second-order differential equations, including one related to the hypergeometric equation. Jacobi worked out the specific details of the method for the hypergeometric equation and showed how to obtain the twenty-four solutions of Kummer.[39] Jacobi started with the observation that for

$$V = u^{\beta-1}(1-u)^{\gamma-\beta-1}(1-xu)^{-\alpha}, \qquad (23.53)$$

$$x(1-x)\frac{d^2V}{dx^2} + (\gamma - (\alpha+\beta+1)x)\frac{dV}{dx} - \alpha\beta V = -\alpha \frac{d}{du}\left(\frac{u(1-u)}{1-xu}V\right)$$
$$= -\alpha u^{\beta}(1-u)^{\gamma-\beta}(1-xu)^{-\alpha-1}.$$

---

[39] Jacobi (1859).

Hence, $y = \int_0^1 V \, du$ would be a solution of the hypergeometric equation for $\beta > 0$ and $\gamma - \beta > 0$ because

$$-\alpha \int_0^1 d\left(\frac{u(1-u)}{1-xu}V\right) = -\alpha\frac{u(1-u)}{1-xu}V\Big]_0^1 = -\alpha u^\beta(1-u)^{\gamma-\beta}(1-xu)^{-\alpha-1}\Big]_0^1 = 0.$$

The expression $u^\beta(1-u)^{\gamma-\beta}(1-xu)^{-\alpha-1}$ also vanished at $u = \pm\infty$ when $\gamma - \alpha - 1 < 0$. So if $g$ and $h$ were a pair of the values $0, 1, \pm\infty$, then, Jacobi observed, the integral $y = \int_g^h V \, du$ would be a solution of the hypergeometric equation under suitable conditions on $\alpha, \beta, \gamma$.

Jacobi also considered a solution of the form $y = \int_g^{\frac{\epsilon}{x}} V \, du$ where $\epsilon$ was a constant. When this $y$ was substituted in the hypergeometric equation, Jacobi obtained

$$-(\gamma - \beta - 1)\epsilon^\beta(1-\epsilon)^{1-\alpha}x^{1-\gamma}(x-\epsilon)^{\gamma-\beta-2} + \alpha g^\beta(1-g)^{\gamma-\beta}(1-xg)^{-\alpha-1}.$$

The expression involving $\epsilon$ vanished for $\epsilon = 1$ when $1 - \alpha > 0$, so for $y = \int_g^{\frac{1}{x}} V \, du$ to be a solution, Jacobi required that $1 - \alpha > 0$. Taking $x$ to be positive, Jacobi had the six solutions:

- $y = \int_0^1 V \, du$, when $\beta$ and $\gamma - \beta$ were positive;
- $y = \int_0^{-\infty} V \, du$, when $\beta$ and $\alpha + 1 - \gamma$ were positive;
- $y = \int_1^\infty V \, du$, when $\gamma - \beta$ and $\alpha + 1 - \gamma$ were positive;
- $y = \int_0^{\frac{1}{x}} V \, du$, when $\beta$ and $1 - \alpha$ were positive;
- $y = \int_{\frac{1}{x}}^\infty V \, du$, when $\alpha + 1 - \gamma$ and $1 - \alpha$ were positive;
- $y = \int_1^{\frac{1}{x}} V \, du$, when $\gamma - \beta$ and $1 - \alpha$ were positive.

Jacobi then noted that the integral $\int_0^1 u^\lambda(1-u)^\mu(1-au)^\nu \, du$ was in fact a constant times the series $F(-\nu, \lambda + 1, \lambda + \mu + 2, a)$. This series could be derived by expanding $(1 - au)^\nu$ by the binomial expansion and then performing term-by-term integration. Also, note that the last five integrals could actually be obtained by a suitable substitution in the first one. For example, to go from $\int_0^1 V \, du$ to $\int_0^{-\infty} V \, du$, set $u = \frac{v-1}{v}$. Then

$$y = \int_0^{-\infty} V \, du = (-1)^\beta x^{-\alpha} \int_0^1 v^{\alpha-\gamma}(1-v)^{\beta-1}\left(1 - v\frac{x-1}{x}\right)^{-\alpha} dv.$$

The corresponding hypergeometric series would be

$$F\left(\alpha, \alpha + 1 - \gamma, \alpha + \beta + 1 - \gamma, \frac{x-1}{x}\right).$$

In this manner, Jacobi represented the six integrals as hypergeometric functions:

- $F(\alpha,\beta,\gamma,x)$, substitution $u = v$;
- $x^{-\alpha} F\left(\alpha,\alpha+1-\gamma,\alpha+\beta+1-\gamma,\frac{x-1}{x}\right)$, substitution $u = \frac{v-1}{v}$;
- $x^{-\alpha} F\left(\alpha,\alpha+1-\gamma,\alpha+1-\beta,\frac{1}{x}\right)$, substitution $u = \frac{1}{v}$;
- $x^{-\beta} F\left(\beta,\beta+1-\gamma,\beta+1-\alpha,\frac{1}{x}\right)$, substitution $u = \frac{v}{x}$;
- $x^{1-\gamma} F(\alpha+1-\gamma,\beta+1-\gamma,2-\gamma,x)$, substitution $u = \frac{1}{xv}$;
- $x^{\alpha-\gamma}(1-x)^{\gamma-\alpha-\beta} F\left(\gamma-\alpha,1-\alpha,\gamma+1-\alpha-\beta,\frac{x-1}{x}\right)$,
  substitution $u = \frac{1}{x+(1-x)v}$.

Jacobi then observed that, other than the identity, the fractional linear transformations mapping 0, 1 to itself could be given as

$$u = 1 - v, \quad u = \frac{v}{1-x+vx}, \quad u = \frac{1-v}{1-vx}.$$

Then $V\, du$ was, respectively,

$$(1-x)^{-\alpha} v^{\gamma-\beta-1}(1-v)^{\beta-1}\left(1-\frac{xv}{x-1}\right)^{-\alpha} dv,$$

$$(1-x)^{-\beta} v^{\beta-1}(1-v)^{\gamma-\beta-1}\left(1-\frac{xv}{x-1}\right)^{\alpha-\gamma} dv,$$

$$(1-x)^{\gamma-\alpha-\beta} v^{\gamma-\beta-1}(1-v)^{\beta-1}(1-vx)^{\alpha-\gamma} dv.$$

Observe that we have $y = \int_0^1 V\, du$ as a constant times each of the four expressions:

$$F(\alpha,\beta,\gamma,x) = (1-x)^{-\alpha} F\left(\alpha,\gamma-\beta,\gamma,\frac{x}{x-1}\right)$$

$$= (1-x)^{-\beta} F\left(\gamma-\alpha,\beta,\gamma,\frac{x}{x-1}\right)$$

$$= (1-x)^{\gamma-\alpha-\beta} F(\gamma-\alpha,\gamma-\beta,\gamma,x).$$

Similarly, there are four expressions with each of the six integral solutions, yielding Kummer's twenty-four solutions.

## 23.11 Riemann's Theory of Hypergeometric Functions

Kummer showed that the twenty-four solutions of the hypergeometric equation could be expressed as hypergeometric series in $x, 1-x, \frac{1}{x}, 1-\frac{1}{x}, \frac{1}{1-x}, \frac{1}{1-\frac{1}{x}}$ multiplied by suitable powers of $x$ and/or $1-x$. He also gave the relations among any three overlapping solutions. In a paper of 1857,[40] Riemann reversed this process, starting with a set of functions with three properties; these properties in turn uniquely determined the functions up to a constant factor, as well as the differential equation of which these functions were the complete set of solutions. He denoted by

$$
P \left\{ \begin{matrix} a & b & c \\ \alpha & \beta & \gamma & x \\ \alpha' & \beta' & \gamma' \end{matrix} \right\}
$$

any function satisfying the three properties:

- For all values of $x$ except $a$, $b$, $c$, called branch points, $P$ was single valued and finite.
- Between any three branches $P'$, $P''$, $P'''$ of this function, there was a linear homogeneous relation with constant coefficients,

$$
C'P' + C''P'' + C'''P''' = 0.
$$

- The function could be written in the form

$$
C_\alpha P^\alpha + C_{\alpha'} P^{\alpha'}, \ C_\beta P^{(\beta)} + C_{\beta'} P^{(\beta')}, \ C_\gamma P^{(\gamma)} + C_{\gamma'} P^{(\gamma')},
$$

where $C_\alpha$, $C_{\alpha'}$, $\cdots C_{\gamma'}$ were constants and

$$
(x-a)^{-\alpha} P^{(\alpha)}, \ (x-a)^{-\alpha'} P^{(\alpha')}
$$

were single valued near $x = a$ and nonvanishing and finite at $x = a$; a similar requirement would hold for

$$
(x-b)^{-\beta} P^{(\beta)}, \ (x-b)^{-\beta'} P^{(\beta')} \quad \text{at } x = b
$$

and for

$$
(x-c)^{-\gamma} P^{(\gamma)}, \ (x-c)^{-\gamma'} P^{(\gamma')} \quad \text{at } x = c.
$$

Moreover, $\alpha - \alpha'$, $\beta - \beta'$, $\gamma - \gamma'$ were not integers and $\alpha + \alpha' + \beta + \beta' + \gamma + \gamma' = 1$.

It follows immediately from the definition of $P$ that if $x'$ is a fractional linear transformation of $x$ mapping $a$, $b$, $c$ to $a'$, $b'$, $c'$, then

$$
P \left\{ \begin{matrix} a & b & c \\ \alpha & \beta & \gamma & x \\ \alpha' & \beta' & \gamma' \end{matrix} \right\} = P \left\{ \begin{matrix} a' & b' & c' \\ \alpha & \beta & \gamma & x' \\ \alpha' & \beta' & \gamma' \end{matrix} \right\} \tag{23.54}
$$

Here recall that every conformal mapping of $\mathbb{C} \cup \{\infty\}$ is of the form

$$
x' = \frac{\lambda x + \mu}{\delta x + \nu}, \quad \text{where} \quad \lambda \nu - \mu \delta = 1.
$$

[40] Riemann (1990) pp. 99–115.

We can therefore choose $a'$, $b'$, $c'$ to be 0, $\infty$, 1. It is also clear from the definition of the Riemann $P$-function that

$$\left(\frac{x-a}{x-b}\right)^{\delta} P \left\{ \begin{matrix} a & b & c \\ \alpha & \beta & \gamma & x \\ \alpha' & \beta' & \gamma' \end{matrix} \right\} = P \left\{ \begin{matrix} a & b & c \\ \alpha+\delta & \beta-\delta & \gamma & x \\ \alpha'+\delta & \beta'-\delta & \gamma' \end{matrix} \right\};$$

$$x^{\delta}(1-x)^{\epsilon} P \left\{ \begin{matrix} 0 & \infty & 1 \\ \alpha & \beta & \gamma & x \\ \alpha' & \beta' & \gamma' \end{matrix} \right\}$$

$$= P \left\{ \begin{matrix} 0 & \infty & 1 \\ \alpha+\delta & \beta-\delta-\epsilon & \gamma+\epsilon & x \\ \alpha'+\delta & \beta'-\delta-\epsilon & \gamma'+\epsilon \end{matrix} \right\}. \qquad (23.55)$$

Following Riemann, we write $P \begin{pmatrix} \alpha & \beta & \gamma \\ \alpha' & \beta' & \gamma' \end{pmatrix} x$ when the first row is 0 $\infty$ 1. We may immediately write the relations

$$P \begin{pmatrix} 0 & a & 0 \\ 1-c & b & c-a-b \end{pmatrix} x = (1-x)^{-a} P \begin{pmatrix} 0 & a & 0 & x \\ 1-c & c-b & b-a & x-1 \end{pmatrix}$$

$$= x^{-a} P \begin{pmatrix} 0 & a & 0 & \frac{1}{x} \\ b-a & 1-c+b & c-a-b \end{pmatrix}$$

$$= (1-x)^{c-a-b} P \begin{pmatrix} 0 & c-a & 0 \\ 1-c & c-b & a+b-c \end{pmatrix} x. \qquad (23.56)$$

Riemann also studied contiguous relations satisfied by the $P$-functions. Following Gauss, he used these relations to find the differential equation satisfied by $P$. In fact, Riemann worked out the details only for the case $\gamma = 0$, sufficient for his purpose. Felix Klein's student Erwin Papperitz (1857–1938) presented the general case in 1889. Riemann found that the equation satisfied by $P \begin{pmatrix} \alpha & \beta & 0 \\ \alpha' & \beta' & \gamma' \end{pmatrix} x$ was

$$(1-x)\frac{d^2 y}{d \log x^2} - (\alpha + \alpha' + (\beta + \beta')x)\frac{dy}{d \log x} + (\alpha\alpha' - \beta\beta' x)y = 0,$$

from which he quite easily showed that

$$F(a,b,c,x) = \text{const.} \, P \begin{pmatrix} 0 & a & 0 \\ 1-c & b & c-a-b \end{pmatrix} x. \qquad (23.57)$$

Moreover, the Pfaff and Euler transformations follow from (23.56) and (23.57).

Riemann's work on the hypergeometric equation led to important developments in the theory of linear differential equations. Riemann himself foresaw some of these developments, though he did not publish his ideas. In 1904, James Pierpont wrote about this aspect of nineteenth-century mathematics:[41] "A particular class of linear differential equations of great importance is the hypergeometric equation; the results obtained by Gauss, Kummer, Riemann, and Schwarz relating to this equation have had the greatest influence on the development of the general theory. The great extent of the theory of linear differential equations may be estimated when we recall that within its borders it embraces not only almost all the elementary functions, but also the modular and automorphic functions."

## 23.12 Exercises

(1) Verify (23.52).

(2) Show that $y = \frac{(\arcsin x)^2}{2}$ satisfies the differential equation

$$(1 - x^2)\frac{d^2 y}{dx^2} - x\frac{dy}{dx} - 1 = 0.$$

Deduce Takebe's formula $\dfrac{1}{2}(\arcsin x)^2 = \displaystyle\sum_{n=0}^{\infty} \frac{2^{2n}(n!)^2}{(2n+2)!} x^{2n+2}.$

Prove Clausen's 1828 observation that this formula is a particular case of his formula (23.13). See Clausen (1828). Also see Eu. 14 pp. 156–186 and the correspondence of Euler and Johann Bernoulli on this topic: Eu. 4A-2 pp. 161–262.

(3) Prove the following examples mentioned in Gauss's 1813 paper on hypergeometric series.

$$\sin nt = n \sin t \, F\left(\frac{1}{2}n + \frac{1}{2}, -\frac{1}{2}n + \frac{1}{2}, \frac{3}{2}, \sin^2 t\right);$$

$$\sin nt = n \sin t \cos t \, F\left(\frac{1}{2}n + 1, -\frac{1}{2}n + 1, \frac{3}{2}, \sin^2 t\right);$$

$$\cos nt = F\left(\frac{1}{2}n, -\frac{1}{2}n, \frac{1}{2}, \sin^2 t\right);$$

$$\cos nt = \cos t \, F\left(\frac{1}{2}n + \frac{1}{2}, -\frac{1}{2}n + \frac{1}{2}, \frac{1}{2}, \sin^2 t\right).$$

See Gauss (1863–1927) vol. 3, p. 127.

(4) Show that

$$e^t = \frac{1}{1-}\, \frac{t}{1+}\, \frac{\frac{1}{2}t}{1-}\, \frac{\frac{1}{6}t}{1+}\, \frac{\frac{1}{6}t}{1-}\, \frac{\frac{1}{10}t}{1+}\, \frac{\frac{1}{10}t}{1-} \cdots$$

---

[41] Pierpont (2000).

$$t = \cfrac{\sin t \cos t}{1-} \cfrac{\frac{1\cdot2}{1\cdot3}\sin^2 t}{1-} \cfrac{\frac{1\cdot2}{5\cdot7}\sin^2 t}{1-} \cfrac{\frac{3\cdot4}{5\cdot7}\sin^2 t}{1-} \cfrac{\frac{3\cdot4}{7\cdot9}\sin^2 t}{1-}.$$

See Gauss (1863–1927) vol. 3, pp. 136–137.

(5) Show that

$$F\left(2,4,\frac{9}{2},x\right) = (1-x)^{-\frac{3}{2}} F\left(\frac{5}{2},\frac{1}{2},\frac{9}{2},x\right).$$

Gauss stated this without proof in the *Ephemeridibus Astronomicis Berolinensibus* 1814, p. 257. He gave a proof in his unpublished second part of his paper. See Gauss (1863–1927) vol. 3, p. 209.

(6) Set $x = 1 - y$ in the hypergeometric differential equation and from its form deduce that $F(a,b,a+b+1-c,1-x)$ is another solution of the hypergeometric equation. See Gauss (1863–1927), vol. 3, p. 208.

(7) Show that when $x = \frac{y}{y-1}$, the hypergeometric equation changes to

$$(1-y)(y-y^2)\frac{d^2 F}{dy^2} + (1-y)(c+(a+b-c-1)y)\frac{dF}{dy} + abF = 0.$$

In this equation set $F = (1-y)^\mu G$ to show that $G$ satisfies

$$(1-y)(y-y^2)\frac{d^2 G}{dy^2} + (1-y)(c+(a+b-c-1)y - 2\mu y)\frac{dG}{dy}$$

$$+\big((ab-\mu(a+b-c-1)y) + (\mu^2-\mu)y\big)G = 0.$$

Show that when $\mu = a$ or $\mu = b$, then the coefficient of $G$ is divisible by $(1-y)$ and thus deduce that

$$F(a,b,c,x) = (1-x)^{-a} F\left(a,c-b,c,\frac{x}{x-1}\right).$$

This is Gauss's proof of Pfaff's transformation. See Gauss (1863–1927) vol. 3, pp. 217–218.

(8) Set $x = 4y - 4y^2$. Show that the hypergeometric equation takes the form

$$(y-y^2)\frac{d^2 F}{dy^2} + (c-(4a+4b+2)y + (4a+4b+2)y^2)$$

$$\times \frac{1}{1-2y}\frac{dF}{dy} - 4abF = 0.$$

Next, show that the fraction in the middle term is removed by putting $c = a+b+\frac{1}{2}$. Also deduce that

$$F\left(a,b,a+b+\frac{1}{2},4y-4y^2\right) = F\left(2a,2b,a+b+\frac{1}{2},y\right).$$

It was by this example that Gauss illustrated the multivaluedness of the hypergeometric function. See Gauss (1863–1927) vol. 3, pp. 225–227.

(9) Prove Kummer's transformation (23.51) and its corollary (23.52). Kummer stated these formulas without proof at the end of his 1836 paper. See Kummer (1975) vol. 2, pp. 75–166.

(10) Prove Euler's continued fraction formula

$$\frac{\beta x}{\gamma} \frac{{}_2F_1(-\alpha, \beta + 1; \gamma + 1, -x)}{{}_2F_1(-\alpha; \beta; \gamma; -x)}$$

$$= \frac{\beta x}{\gamma - (\alpha + \beta + 1)x} + \frac{(\beta + 1)(\alpha + \gamma + 1)x}{\gamma + 1 - (\alpha + \beta + 2)x}$$

$$+ \frac{(\beta + 2)(\alpha + \gamma + 2)x}{\gamma + 2 - (\alpha + \beta + 3)x} + \cdots .$$

See Eu. I-14 pp. 291–349.

(11) Prove Ramanujan's integral formula

$$\int_0^\infty x^{s-1}(\phi(0) - \phi(1)x + \phi(2)x^2 - \cdots) \, dx = \frac{\pi}{\sin s\pi}\phi(-s).$$

See Berndt (1985–1998) part I, pp. 295–307. See also Hardy (1978) pp. 186–190; he relates this formula of Ramanujan with a 1914 interpolation theorem of F. Carlson, useful in proving hypergeometric formulas.

(12) Use the following outline to determine when the square of a hypergeometric function

$$y = \sum_{n=0}^\infty \frac{(\alpha)_n(\beta)_n}{n!\,(\gamma)_n} x^n \quad \text{takes the form} \quad z = \sum_{n=0}^\infty \frac{(\alpha')_n(\beta')_n(\delta')_n}{n!\,(\gamma')_n(\epsilon')_n} x^n.$$

(a) Show that when the hypergeometric equation is multiplied by $x$ and then differentiated, the result is

$$(x^3 - x^2)\frac{d^3y}{dx^3} + ((\alpha + \beta + 4)x^2 - (y + 2)x)\frac{d^2y}{dx^2}$$

$$+ ((2\alpha + 2\beta + \alpha\beta + 2)x - \gamma)\frac{dy}{dx} + \alpha\beta y = 0.$$

(b) Show that $z$ satisfies the differential equations

$$(x^3 - x^2)\frac{d^3z}{dx^3} + ((3 + \alpha' + \beta' + \delta')x^2 - (1 + \gamma' + \epsilon')x)\frac{d^2z}{dx^2}$$

$$+ (1 + \alpha' + \beta' + \delta' + \alpha'\beta' + \alpha'\delta' + \beta'\delta')x - \gamma'\epsilon')\frac{dz}{dx} + \alpha'\beta'\delta'z = 0.$$

(c) Show that if $z = y^2$, then the equation in (b) becomes

$$(x^3 - x^2)2y\frac{d^3y}{dx^3} + ((3 + a')x^2 - (1 + d')x)2y\frac{d^2y}{dx^2}$$

$$+ ((1 + a' + b')x - e')2y\frac{dy}{dx} + c'y^2$$

$$+ 6(x^3 - x^2)\frac{dy}{dx}\cdot\frac{d^2y}{dx^2} + 2((3 + a')x^2 - (1 + d')x)\left(\frac{dy}{dx}\right)^2 = 0,$$

where $a' = \alpha' + \beta' + \delta'$, $b' = \alpha'\beta' + \alpha'\delta' + \beta'\delta'$, $c' = \alpha'\beta'\delta'$, $d' = \epsilon' + \gamma'$, $e' = \epsilon'\gamma'$.

(d) Multiply the hypergeometric equation by $\frac{2yA}{x} + B\frac{dy}{dx}$ and equation (a) by $2y$ and add the two equations. Compare the resulting equation with (c), and deduce that

$$\gamma = \alpha + \beta + \frac{1}{2}, A = 2\alpha + 2\beta - 1, B = b', a' = 3\alpha + 3\beta,$$

$$b' = 2\alpha^2 + 8\alpha\beta + 2\beta^2, c' = 4(\alpha + \beta)\alpha\beta, d' = 3\gamma - 1, e' = (2\gamma - 1)\gamma.$$

(e) Deduce that $\alpha' = 2\alpha$, $\beta' = 2\beta$, $\delta' = \alpha + \beta$, $\gamma' = \gamma$, $\epsilon' = 2\gamma - 1$.

(f) Conclude that

$$\left({}_2F_1\left(\begin{matrix}\alpha, \beta, \\ \alpha + \beta + \frac{1}{2}\end{matrix}; x\right)\right)^2 = {}_3F_2\left(\begin{matrix}2\alpha, 2\beta, \alpha + \beta \\ \alpha + \beta + \frac{1}{2}, 2\alpha + 2\beta\end{matrix}; x\right).$$

See Clausen (1828).

## 23.13 Notes on the Literature

For a history of the hypergeometric series, see Dutka (1984). The reader may read more on Li Shanlan and other Chinese mathematicians in Martzloff (1997), pp. 341–350. Discussions of Gauss's convergence test are available in Bressoud (2007) and Knopp (1990). In 1859, perhaps influenced by the work of Jacobi, Riemann gave a course of lectures in which he defined the $P$-function by means of a complex integral. See Riemann (1990), pp. 667–691.

# 24

## Orthogonal Polynomials

### 24.1 Preliminary Remarks

Orthogonal polynomials played an important role in the nineteenth-century development of continued fractions, hypergeometric series, numerical integration, and approximation theory; in the twentieth century, they additionally contributed to progress in the moment problem and in functional analysis. However, orthogonal polynomials may not have received recognition proportional to their significance, leading Barry Simon to dub them "the Rodney Dangerfield of analysis."[1][2] Nevertheless, when Paul Nevai edited the proceedings of a 1989 conference on this subject, he stamped on the dedication page, "I love orthogonal polynomials."[3]

A sequence of polynomials $p_n(x)$, $n = 0, 1, 2, \ldots$, is said to be orthogonal with respect to a weight function $w(x)$ over an interval $(a, b)$ where $-\infty \le a < b \le \infty$, if

$$\int_a^b p_n(x) p_m(x) w(x) \, dx = A_n \delta_{mn}, \tag{24.1}$$

where $A_n \ne 0$. In a paper on probability written in the early 1770s, Lagrange defined a sequence of polynomials containing as special cases the Legendre polynomials. Denoted by $P_n(x)$, the Legendre polynomials are obtained when $a = -1$, $b = 1$ and $w(x) \equiv 1$, in (24.1). Lagrange gave a three-term recurrence relation for his sequence of polynomials; for the particular case of Legendre polynomials, this recurrence amounted to

$$(2n + 1)x P_n(x) = (n + 1) P_{n+1}(x) + n P_{n-1}(x), \tag{24.2}$$

$$n = 1, 2, 3, \ldots, \qquad P_0(x) = 1, \ P_1(x) = x.$$

In a paper of 1785[4] on the attraction of spheroids of revolution, Legendre defined the polynomials now bearing his name by the expansion

[1] Simon (2005).
[2] Simon (2005).
[3] Nevai (1990).
[4] Legendre (1785).

$$(1 - 2\cos\theta y + y^2)^{-\frac{1}{2}} = 1 + P_1(\cos\theta)y + P_2(\cos\theta)y^2 + P_3(\cos\theta)y^3 + \cdots.$$
$$(24.3)$$

In this memoir, Legendre needed only the polynomials of even degree and he explicitly presented $P_2(\cos\theta)$, $P_4(\cos\theta)$, $P_6(\cos\theta)$, and $P_8(\cos\theta)$. We note his first two examples:

$$P_2(\cos\theta) = \frac{3}{2}\cos^2\theta - \frac{1}{2}; \quad P_4(\cos\theta) = \frac{5\cdot7}{2\cdot4}\cos^4\theta - \frac{3\cdot5}{2\cdot4}2\cos^2\theta + \frac{1\cdot3}{2\cdot4}.$$
$$(24.4)$$

In the second volume of his *Exercices de calcul intégral* of 1817,[5] Legendre gave the orthogonality relation and an expression for the general $P_n(x)$. Legendre polynomials played an important role in the celestial mechanics of Laplace, Legendre, and others.

Gauss used Legendre polynomials in his paper on numerical integration, presented in Göttingen in 1814,[6] extending the work of Newton and Cotes. But Gauss did not refer to the earlier work on these polynomials; rather, he conceived of Legendre polynomials as an outgrowth of his work in hypergeometric series. The groundbreaking approach and methodology taken by Gauss in this paper led to important advances in nineteenth-century numerical analysis. Briefly summarizing Gauss, we suppose $\int_c^d y(x)\,dx$ is to be computed. Let points $a(=a_0), a_1, \ldots, a_n$ be chosen in $[c,d]$ and let the corresponding values of $y$ at these points be $y_0, y_1, \ldots, y_n$. Set

$$f(x) = (x - a)(x - a_1)(x - a_2)\ldots(x - a_n). \qquad (24.5)$$

Note that the $n$th degree Lagrange–Waring polynomial

$$z_n(x) = \sum_{k=0}^{n} \frac{f(x)y_k}{f'(a_k)(x - a_k)} \qquad (24.6)$$

passes through $(a_k, y_k)$, $k = 0, 1, \ldots, n$, and therefore interpolates $y(x)$; thus, we may write $y(x) = z_n(x) + r_n(y)$. Then

$$\int_c^d y(x)\,dx = \sum_{k=0}^{n} \lambda_k y_k + R_n(y), \qquad (24.7)$$

where

$$\lambda_k = \int_c^d \frac{f(x)\,dx}{f'(a_k)(x - a_k)} \quad \text{and} \quad R_n(y) = \int_c^d r_n(y)\,dx. \qquad (24.8)$$

It is clear that if $y(x)$ is a polynomial of degree $\leq n$, then $y(x) = z_n(x)$ and hence $R_n(y) = 0$. In the Newton–Cotes scheme, the points $a_0, a_1, \ldots, a_n$ were

---

[5] Legendre (1811–1817) vol. 2, pp. 249–250.
[6] Gauss (1863–1927) vol. 3, pp. 163–196.

equally spaced. Gauss considered whether he could prove $R_n(y) = 0$ for a larger class of polynomials by varying the nodes $a, a_1, \ldots, a_n$. Since there were $n+1$ points to be varied, he wanted the class of polynomials, for which $R_n(y) = 0$, to consist of all polynomials of degree $\leq 2n+1$; indeed, he succeeded in proving this. In short, his argument began with the observation that

$$R_n\left(\frac{1}{t-x}\right) = R_n\left(\frac{1}{t} + \frac{x}{t^2} + \frac{x^2}{t^3} + \cdots\right) = \sum_{k=0}^{\infty} \frac{R_n(x^k)}{t^{k+1}}. \qquad (24.9)$$

His problem was to choose $a, a_1, \ldots, a_n$ so that $R_n(x^k) = 0$ for $k = 0, 1, \ldots, 2n+1$; then he could write

$$R_n\left(\frac{1}{t-x}\right) = O\left(\frac{1}{t^{2n+3}}\right), \quad t \to \infty. \qquad (24.10)$$

When $[c, d] = [-1, 1]$, from his results on hypergeometric functions, Gauss had

$$\int_{-1}^{1} \frac{dx}{t-x} = \ln \frac{1+\frac{1}{t}}{1-\frac{1}{t}} = \frac{2}{t-} \frac{\frac{1}{3}}{t-} \frac{2 \cdot \frac{2}{3 \cdot 5}}{t-} \frac{3 \cdot \frac{3}{5 \cdot 7}}{t-} \cdots. \qquad (24.11)$$

He also knew that the $(n+1)$th convergent of this continued fraction was a rational function $\frac{S_n(t)}{P_{n+1}(t)}$, where $S_n$ was of degree $n$ and $P_{n+1}$ of degree $n+1$. Moreover, this rational function approximated the continued fraction up to the order $t^{-2n-3}$. So Gauss factorized $P_{n+1}(t)$ and wrote $\frac{S_n(x)}{P_{n+1}(x)}$ as a sum of partial fractions:

$$\frac{S_n(x)}{P_{n+1}(x)} = \sum_{k=0}^{n} \frac{\lambda_k}{x-a_k}. \qquad (24.12)$$

He then easily showed that by using these $a, a_1, \ldots, a_n$ and $\lambda, \lambda_1, \ldots, \lambda_n$, he would obtain the result. Gauss explicitly wrote down the polynomials $P_{n+1}(x)$ for $n = 0, 1, 2, \ldots, 6$; we can see they are Legendre polynomials of degrees 1 to 7, although Gauss did not make this observation. Instead, he gave the hypergeometric representations of the polynomials $P_{n+1}$ and of the remainder

$$\ln \frac{1+\frac{1}{t}}{1-\frac{1}{t}} - \frac{S_n(t)}{P_{n+1}(t)}.$$

At the end of the paper, he computed the zeros of the Legendre polynomials of degree seven and less with the corresponding $\lambda$. He used these results to compute the integral $\int \frac{dx}{\ln x}$ over the interval $x = 100000$ to $x = 200000$. Note that Gauss was well aware that $\int_2^x \frac{dt}{\ln t}$ gave a good approximation for the number of primes less than $x$.

In a paper of 1826, Jacobi pointed out that Gauss's proof ultimately depended on the orthogonality of $P_{n+1}(x)$.[7] To see this, suppose $y(x)$ is a polynomial of degree at

---

[7] Jacobi (1826).

most $2n + 1$. Then $y(x) = q(x)P_{n+1}(x) + r(x)$, where $q(x)$ and $r(x)$ are polynomials of degree at most $n$. Next note that

$$\int_{-1}^{1} y(x)dx = \int_{-1}^{1} q(x)P_{n+1}(x)dx + \int_{-1}^{1} r(x)dx,$$

where the first integral on the right-hand side vanishes by the orthogonality of $P_{n+1}(x)$, and the second integral can be exactly computed by the Newton–Cotes method because the degree of $r(x)$ is not greater than $n$. In fact, Jacobi did not start his reasoning process with the Legendre polynomial; at that time he may not have known of the earlier work of Legendre, Laplace, and others on Legendre polynomials. His argument produced these polynomials, their orthogonality, and the byproduct that

$$P_n(x) = \frac{1}{2^n n!} \frac{d^n}{dx^n} (x^2 - 1)^n. \tag{24.13}$$

Interestingly enough, Rodrigues and Ivory had already independently discovered this useful and important formula for Legendre polynomials.[8] In 1808, Olinde Rodrigues enrolled in the Lycée Impérial, later named Lycée Louis-LeGrand and where Galois also studied. After graduating in 1812, he was admitted to the Université de Paris, submitting a doctoral thesis on the attraction of spheroids in 1815. Unfortunately, the haphazard journal in which his memoir on this subject was published produced only three volumes from 1814 to 1816. This partly explains why Rodrigues's work and the formula for Legendre polynomials, in particular, were not noticed. For several decades, the result was referred to as the formula of Ivory and Jacobi. In 1865, Hermite finally pointed out Rodrigues's paper; Cayley referred to it in a different context in 1858.

James Ivory (1765–1842) was an essentially self-taught Scottish mathematician whose interest was mainly in applied areas. He received much recognition, but perhaps suffered from depression, curtailing his career; in a letter to MacVey Napier he declared, "I believe on the whole I am the most unlucky person that ever existed."[9] Most of Ivory's inspiration was drawn from the work of the French mathematicians Laplace, Legendre, and Lagrange; his papers contain several references to Laplace's book on celestial mechanics. Ivory published (24.13) in a 1824 paper on the shape of a revolving homogeneous fluid mass in equilibrium; he derived the formula from a result in his earlier 1812 paper on the attraction of a spheroid. In his 1824 paper, Ivory remarked on the formula, "From this very simple expression, the most remarkable properties of the coefficients of the expansion of $\frac{1}{f}$, are very readily deduced."[10] Here $f$ refers to the expression $(1 - 2\cos\theta y + y^2)^{\frac{1}{2}}$.

The Irish mathematician Robert Murphy (1806–1843), mentioned in Chapter 21, had a brief mathematical career during which he published papers on integral equations, operator theory, and algebraic equations. He was perhaps the first to

[8] Rodrigues (1816); Ivory (1824).
[9] Craik (2000).
[10] Ivory (1824).

understand the significance of orthogonality; in a series of papers on integral equations in the early 1830s, Murphy considered the following problem: Suppose

$$\phi(x) = \int_0^1 t^x f(t)\,dt, \quad x = 0, 1, 2, \ldots.$$

Determine the function $f(t)$ from the function $\phi(x)$. One of the simplest results he stated in this connection was that if $\phi(x)$ was of the form $\frac{A}{x} + \frac{B}{x^2} + \frac{C}{x^3} + \cdots$, then $f(t)$ would be given by $\frac{1}{t}$ multiplied by the coefficient of $\frac{1}{x}$ in $\phi(x) \cdot t^{-x}$. As an extension of the previously stated problem, Murphy considered the determination of $f(t)$ from a knowledge of $\phi(x)$ for a finite number of values of $x$, say $x = 0, 1, \ldots, n-1$. The simplest case is when $\phi(x) = 0$ and this leads to the Legendre polynomials as solutions for $f(t)$. Murphy called such functions reciprocal rather than orthogonal. He also considered cases where $t$ was replaced by $\ln t$, and this led him to the Laguerre polynomials

$$T_n(u) = \frac{e^{-u}}{n!} \cdot \frac{d^n}{du^n}(u^n e^u), \quad n = 0, 1, 2, \ldots.$$

He proved their orthogonality and found their generating function by applying the Lagrange inversion formula. Recall that a century before this, D. Bernoulli and Euler had studied these Laguerre polynomials; Bernoulli computed the zeros of several of them by his method of recurrent series. See Exercise 3 in Chapter 14.

It may be fair to say that Pafnuty Chebyshev was the creator of the theory of orthogonal polynomials and its applications. In an important paper of 1855, Chebyshev introduced and studied discreet orthogonal polynomials.[11] This and later papers were associated with the areas of continued fractions, least squares approximations, interpolation, and approximate quadrature. Later in his career, Chebyshev's excellent students, including A. A. Markov and E. I. Zolotarev, continued his work in these and other areas.

## 24.2 Legendre's Proof of the Orthogonality of His Polynomials

In his *Exercices*, Legendre used the generating function for Legendre polynomials to offer an elegant and short proof of their orthogonality.[12] Note that Legendre denoted $P_n(x)$ by $X^n$, but we use the more modern notation. He started with the generating function

$$(1 - 2xy + y^2)^{-\frac{1}{2}} = \sum_{n=0}^{\infty} P_n(x) y^n,$$

[11] Chebyshev (1899–1907) vol. 1, pp. 203–230.
[12] Legendre (1811–1817) vol. 2, pp. 224–232.

where $|x| \leq 1$ and $|y| < 1$, and considered the integral

$$I = \int_{-1}^{1} \left( \sum_{n=0}^{\infty} P_n(x) r^n y^n \right) \left( \sum_{m=0}^{\infty} \frac{P_m(x) y^m}{r^m} \right) dx \qquad (24.14)$$

$$= \int_{-1}^{1} \frac{dx}{\sqrt{(1 - 2xry + r^2 y^2)(1 - \frac{2xy}{r} + \frac{y^2}{r^2})}}.$$

He set

$$x = \frac{1 + r^2 y^2 - z^2}{2ry}$$

to obtain

$$I = -\frac{1}{2} \int_{1+ry}^{1-ry} \frac{dz}{\sqrt{(z^2 - 1 + r^2 + y^2 - r^2 y^2)}}$$

$$= \ln \left( -z + \sqrt{(z^2 - 1 + r^2 + y^2 - r^2 y^2)} \right) \Big|_{1+ry}^{1-ry}$$

$$= \ln(-1 + ry + r - y) - \frac{1}{2} \ln(-1 - ry + r + y)$$

$$= \ln \frac{1+y}{1-y} = 2 + \frac{2}{3} y^2 + \frac{2}{5} y^4 + \frac{2}{7} y^6 + \cdots + \frac{2}{2n+1} y^{2n} + \cdots .$$

Comparing this expression with the integral (24.14), he obtained orthogonality:

$$\int_{-1}^{1} P_n(x) P_m(x) \, dx = \frac{2}{2n+1} \delta_{mn}. \qquad (24.15)$$

He also used the generating function to obtain

$$P_n(x) = \frac{1 \cdot 3 \cdot 5 \cdots (2n-1)}{1 \cdot 2 \cdot 3 \cdots n} x^n - \frac{1 \cdot 3 \cdots (2n-3)}{1 \cdot 2 \cdot (n-2)} \frac{x^{n-2}}{2} + \cdots . \qquad (24.16)$$

## 24.3 Gauss on Numerical Integration

Gauss started his 1815 paper[13] with a discussion of the Newton–Cotes method for numerical integration. Let $[0, 1]$ be the interval and let $a, a_1, a_2, \ldots, a_n$ be $n+1$ points in that interval. Set

$$f(x) = (x - a)(x - a_1)(x - a_2) \ldots (x - a_n)$$

$$= x^{n+1} + c_1 x^n + c_2 x^{n-1} + \cdots + c_{n+1}. \qquad (24.17)$$

[13] Gauss (1863–1927) vol. 3, pp. 163–196.

Let $y$ be a function to be integrated over $[0, 1]$ and let $y(= y_0), y_1, y_2, \ldots, y_n$ be its values at $a = a_0, a_1, a_2, \ldots, a_n$, respectively. The Lagrange–Waring interpolating polynomial of degree $n$ for $y$ is then given by

$$g(x) = \frac{f(x)y}{f'(a)(x - a)} + \sum_{k=1}^{n} \frac{f(x)y_k}{f'(a_k)(x - a_k)} = \sum_{k=0}^{n} \frac{f(x)y_k}{f'(a_k)(x - a_k)}. \qquad (24.18)$$

The Newton–Cotes method then consists in integrating the interpolating polynomial:

$$\int_0^1 y \, dt = \sum_{k=0}^{n} \lambda_k y_k + R_n(y) \qquad (24.19)$$

where

$$\lambda_k = \int_0^1 \frac{f(x) \, dx}{f'(a_k)(x - a_k)}, \quad k = 0, 1, \ldots n, \qquad (24.20)$$

and $R_n(y)$ is the remainder. This remainder is zero when $y$ is a polynomial of degree at most $n$. Gauss asked whether it was possible to choose $a, a_1, \ldots, a_n$ in such a way that the remainder would be zero for polynomials of degree at most $2n + 1$, with the points $a, a_1, \ldots, a_n$ no longer equally spaced as in the Newton–Cotes procedure. Gauss observed that since
$f(a) = 0$, he had

$$\frac{f(x)}{x - a} = \frac{x^{n+1} - a^{n+1} + c_1(x^n - a^n) + \cdots + c_n(x - a)}{x - a}$$
$$= x^n + x^{n-1}a + x^{n-2}a^2 + \cdots + a^n$$
$$+ c_1 x^{n-1} + c_1 x^{n-2} a + \cdots + c_1 a^{n-1}$$
$$+ \cdots.$$

Hence, after rearranging terms,

$$\int_0^1 \frac{f(x)}{x - a} \, dx = a^n + c_1 a^{n-1} + c_2 a^{n-2} + \cdots + c_n$$
$$+ \frac{1}{2}(a^{n-1} + c_1 a^{n-2} + \cdots + c_{n-1})$$
$$+ \frac{1}{3}(a^{n-2} + c_1 a^{n-3} + \cdots + c_{n-2})$$
$$+ \cdots$$
$$+ \frac{1}{n}(a + c_1)$$
$$+ \frac{1}{n+1}. \qquad (24.21)$$

Here Gauss noted that the nonnegative powers of $x$ in the product

$$(x^{n+1} + c_1 x^n + \cdots + c_{n+1}) \left( \frac{1}{x} + \frac{1}{2x^2} + \frac{1}{3x^3} + \cdots \right) = -f(x) \ln \left( 1 - \frac{1}{x} \right)$$

$$(24.22)$$

gave the terms on the right-hand side of (24.21) when $x = a$. So he could write

$$-f(x) \ln \left( 1 - \frac{1}{x} \right) = T_1(x) + T_2(x), \qquad (24.23)$$

where $T_1(x)$ was the polynomial or principal part of $-f(x) \ln \left( 1 - \frac{1}{x} \right)$; then, by (24.20),

$$T_1(a_k) = \int_0^1 \frac{f(x)}{x - a_k} \, dx = \lambda_k f'(a_k), \quad k = 0, 1, \ldots, n. \qquad (24.24)$$

Denoting $R_n(x^m)$ by $k_m$, Gauss used (24.19) to obtain

$$\sum_{k=0}^{n} \lambda_k a_k^m = \frac{1}{m+1} - k_m, \quad m = 0, 1, 2, \ldots. \qquad (24.25)$$

It followed that

$$\sum_{k=0}^{n} \frac{\lambda_k}{x - a_k} = \sum_{k=0}^{n} \frac{\lambda_k}{x} \left( 1 - \frac{a_k}{x} \right)^{-1} = \sum_{k=0}^{n} \left( \frac{\lambda_k}{x} + \frac{\lambda_k a_k}{x^2} + \frac{\lambda_k a_k^2}{x^3} + \cdots \right)$$

$$= (1 - k_0) \frac{1}{x} + \left( \frac{1}{2} - k_1 \right) \frac{1}{x^2} + \left( \frac{1}{3} - k_2 \right) \frac{1}{x^3} + \left( \frac{1}{4} - k_3 \right) \frac{1}{x^4} + \cdots$$

$$= -\ln \left( 1 - \frac{1}{x} \right) - \left( \frac{k_{n+1}}{x^{n+2}} + \frac{k_{n+2}}{x^{n+3}} + \cdots \right), \qquad (24.26)$$

where Gauss used the fact that $k_j = 0$ for $j = 0, 1, \ldots, n$. Gauss then had to determine conditions on $f(x)$ so that $k_j = 0$ for $j = n+1, n+2, \ldots, 2n$ as well. By (24.23) and (24.24), he could deduce that

$$T_1(x) = f(x) \sum_{k=0}^{n} \frac{\lambda_k}{x - a_k}.$$

This was possible because both sides were polynomials of degree $n$, equal for $n+1$ values of $x$, given by $a_0, a_1, a_2, \ldots, a_n$. Thus, by (24.23) and (24.26), it followed that

$$f(x) \left( \frac{k_{n+1}}{x^{n+2}} + \frac{k_{n+2}}{x^{n+3}} + \frac{k_{n+3}}{x^{n+4}} + \cdots \right) = T_2(x). \qquad (24.27)$$

Gauss used this analysis to find $f(x)$ of small degrees. For example, when $n = 0$, $f(x) = x + c_1$, he had to consider

$$(x + c_1)\left(\frac{1}{x} + \frac{1}{2x^2} + \frac{1}{3x^3} + \cdots\right).$$

For the coefficient of $\frac{1}{x}$ to be zero, he required that $c_1 + \frac{1}{2} = 0$ or $c_1 = -\frac{1}{2}$. For $n = 1$, $f(x) = x^2 + c_1 x + c_2$, and the coefficients of $\frac{1}{x}$ and $\frac{1}{x^2}$ in the expansion $-f(x)\ln(1 - \frac{1}{x})$ then had to be zero. He could then write the equations for $c_1$ and $c_2$:

$$c_2 + \frac{1}{2}c_1 + \frac{1}{3} = 0 \quad \text{and} \quad \frac{1}{2}c_2 + \frac{1}{3}c_1 + \frac{1}{4} = 0.$$

Thus, $c_1 = -1$ and $c_2 = \frac{1}{6}$ and so the polynomial was $x^2 - x + \frac{1}{6}$. Gauss then changed the variable so that the interval of integration became $[-1, 1]$. In this case, he had to choose the polynomial $U(x)$ of degree $n + 1$ so that

$$\frac{1}{2}U(x)\ln\frac{1 + \frac{1}{x}}{1 - \frac{1}{x}} = U(x)\left(\frac{1}{x} + \frac{1}{3x^2} + \frac{1}{5x^3} + \cdots\right) = U_1(x) + U_2(x)$$

had appropriate negative powers of $x$ with zero coefficients. In fact, the zeros $u$ of $U$ were related to zeros $a$ of $f$ by $u = 2a - 1$; $U_1$ and $U_2$ corresponded to $T_1$ and $T_2$ of equation (24.23). With this change of variables, the polynomials for $n = 1, 2, 3$ were $x$, $x^2 - \frac{1}{3}$, and $x^3 - \frac{3x}{5}$. Note that these are the Legendre polynomials of the first three degrees, normalized so that they are monic.

Gauss proceeded to give a method using continued fractions in order to quickly determine the polynomials $f(x)$. From his paper on hypergeometric functions, he had the expression

$$\phi(x) = \frac{1}{2}\ln\frac{x + 1}{x - 1} = \frac{1}{x-}\frac{\frac{1^2}{3}}{x-}\frac{\frac{2^2}{3\cdot 5}}{x-}\frac{\frac{3^2}{5\cdot 7}}{x-}\cdots. \qquad (24.28)$$

He then showed that if the $n$th convergent of his continued fraction was $\frac{P_n(x)}{Q_n(x)}$, then

$$\phi(x) - \frac{P_n(x)}{Q_n(x)} = O\left(\frac{1}{x^{2n+1}}\right).$$

From this he could conclude that if $Q_n(x)$ was monic, then

$$Q_{n+1}(x) = U(x) \quad \text{and} \quad P_{n+1}(x) = U_1(x).$$

In this manner, Gauss completely solved his problem. Observe that the points $a_k$, $k = 0, 1, \ldots, n$ are the zeros of the Legendre polynomials and that the numbers $\lambda_k$ could be obtained from

$$\frac{P_{n+1}(x)}{Q_{n+1}(x)} = \sum_{k=0}^{n} \frac{\lambda_k}{x - a_k},$$

so that $\lambda_k = \frac{P_{n+1}(a_k)}{Q'_{n+1}(a_k)}$.

## 24.4 Jacobi's Commentary on Gauss

In the introduction to his 1826 paper[14] on Gauss's new method of approximate quadrature, Jacobi remarked that the simplicity and elegance of Gauss's results led him to believe that there was a simple and direct way of deriving them. The object of his paper was to present such a derivation, making use of his work from his doctoral dissertation on the Waring–Lagrange interpolation formula. Jacobi proceeded, in his usual lucid style, to show that Gauss's numerical integration method was effective because of its use of orthogonal polynomials. Abbreviating Jacobi's work for convenience, suppose $\phi(x) = \prod_{k=1}^{n}(x - x_k)$, where the $x_i$ are distinct and suppose $f(x)$ is a polynomial of degree $\leq n - 1$. Then

$$\frac{f(x)}{\phi(x)} = \frac{A_1}{x - x_1} + \frac{A_2}{x - x_2} + \cdots + \frac{A_n}{x - x_n},$$

where

$$A_k = \lim_{x \to x_k} \frac{(x - x_k)f(x)}{\phi(x)} = \lim_{x \to x_k} \frac{(x - x_k)f(x)}{\phi(x) - \phi(x_k)} = \frac{f(x_k)}{\phi'(x_k)}.$$

So if $x_1, x_2, \ldots, x_n$ are the interpolation points, then any polynomial $f(x)$ of degree at most $n - 1$ can be expressed by the formula

$$f(x) = \sum_{k=1}^{n} \frac{f(x_k)\phi(x)}{\phi'(x_k)(x - x_k)},$$

attributed by Jacobi to Lagrange. The integral of such a polynomial is given exactly by the Newton–Cotes formula. On the other hand, if the degree of $f$ is greater than $n - 1$, then divide $f(x)$ by $\phi(x)$ to get

$$\frac{f(x)}{\phi(x)} = V(x) + \frac{U(x)}{\phi(x)},$$

where $U(x)$ and $V(x)$ are polynomials and the degree of $U$ is less than or equal to $n - 1$. Now assume with Jacobi that

$$f(x) = a + a_1 x + a_2 x^2 + \cdots + a_n x^n + a_{n+1} x^{n+1} + \cdots + a_{2n} x^{2n} + \cdots$$

[14] Jacobi (1826).

and

$$\frac{1}{\phi(x)} = \frac{A_1}{x^n} + \frac{A_2}{x^{n+1}} + \cdots + \frac{A_{n+1}}{x^{2n}} + \cdots.$$

Then

$$V(x) = a_n A_1 + a_{n+1}(A_1 x + A_2) + a_{n+2}(A_1 x^2 + A_2 x + A_3) + \cdots$$
$$+ a_{2n-1}(A_1 x^{n-1} + A_2 x^{n-2} + \cdots + A_n) + \cdots.$$

Jacobi observed that according to Newton's method, to compute $\int_{-1}^{1} f(x)\,dx$, one would substitute $U(x)$ for $f(x)$ and the error would be

$$\Delta = \int f(x)\,dx - \int U\,dx = \int \phi(x) V\,dx.$$

He then noted that the expression for $V$ did not involve $a_1, a_2, \ldots, a_{n-1}$ and hence the error, $\Delta$, would be independent of these coefficients of $f$. The question was whether $\phi$ could be chosen so that the error would be independent of $a_n, a_{n+1}$, etc. Clearly, if $\int \phi x^k = 0$, for $k = 0, 1, \ldots, l$, then $\Delta$ would also be independent of $a_n, a_{n+1}, \ldots, a_{n+l-1}$. Since $\int (\phi(x))^2 dx > 0$, the value of $l$ could be at most $n - 1$. Thus if, $\int \phi x^k = 0$, for $k = 0, 1, \ldots, n-1$, then $\int f(x)\,dx$ was exact for polynomials of degree $\leq 2n-1$. This meant that $\phi(x)$ should be a constant multiple of the Legendre polynomial of degree $n$ and Jacobi had succeeded in showing that orthogonality lay at the root of the Gaussian method of numerical integration.

## 24.5 Murphy and Ivory: The Rodrigues Formula

Robert Murphy's discussion of orthogonal polynomials appeared in his two publications of 1833 and 1835 on the inverse method of definite integrals, written in 1832 and 1833, and in his 1833 treatise on physics.[15] He considered the integral $\phi(x) = \int_0^1 f(t) t^x dt$ and determined the form of the polynomial $f(t)$ such that $\phi(x)$ was zero for $x = 0, 1, \ldots, n - 1$. He let

$$f(t) = 1 + A_1 t + A_2 t^2 + \cdots + A_n t^n,$$

so that

$$\phi(x) = \frac{1}{x+1} + \frac{A_1}{x+2} + \frac{A_2}{x+3} + \cdots + \frac{A_n}{x+n+1} = \frac{P}{Q}, \tag{24.29}$$

where $Q = (x+1)(x+2)\cdots(x+n+1)$ and $P$ was a polynomial of degree at most $n$. To find an expression for $f(t)$ when $\phi(x) = 0$ for $x = 0, 1, \ldots, n - 1$, Murphy argued that $P$ would have the form $cx(x-1)\cdots(x-n+1)$. Thus,

[15] Murphy (1833a), (1833c) and (1835).

$$\frac{1}{x+1} + \frac{A_1}{x+2} + \frac{A_2}{x+3} + \cdots + \frac{A_n}{x+n+1} = \frac{cx(x-1)\cdots(x-n+1)}{(x+1)(x+2)\cdots(x+n+1)}.$$

Multiplying both sides by $x+k$ and then setting $x = -k$ for $k = 1, \ldots, n+1$, he got the result

$$c = (-1)^n, \quad A_1 = -\frac{n}{1}\cdot\frac{n+1}{1}, \quad A_2 = \frac{n(n-1)}{1\cdot 2}\cdot\frac{(n+1)(n+2)}{1\cdot 2}, \text{ etc.}$$

Hence,

$$f(t) = 1 - \frac{n}{1}\cdot\frac{n+1}{1}t + \frac{n(n-1)}{1\cdot 2}\cdot\frac{(n+1)(n+2)}{1\cdot 2}t^2 - \cdots$$

$$= \frac{\frac{d^n}{dt^n}\left(t^n\left(1 - nt + \frac{n(n-1)}{1\cdot 2}t^2 - \cdots\right)\right)}{1\cdot 2\cdot 3\cdots n}$$

$$= \frac{1}{1\cdot 2\cdots n}\frac{d^n}{dt^n}\left(t(1-t)\right)^n,$$

completing Murphy's proof of the Rodrigues formula.

Now Ivory's proof involved differential equations. He showed that $P_k(x)$, the $k$th Legendre polynomial, satisfied the equation

$$(k-n)(k+n+1)(1-x^2)^n\frac{d^n P_k}{dx^n} + \frac{d}{dx}\left((1-x^2)^{n+1}\frac{d^{n+1}}{dx^{n+1}}P_k\right) = 0.$$

Ivory presented this result in his 1812 paper. Twelve years later, unaware of Rodrigues's earlier work, he observed that by a repeated use of this equation, he could obtain the Rodrigues formula. He set

$$\phi_n = (1-x^2)^n\frac{d^n}{dx^n}P_k \quad \text{and} \quad \phi_0 = P_k,$$

from which he had

$$\phi_0 + \frac{1}{k(k+1)}\frac{d}{dx}\phi_1 = 0,$$

$$\phi_1 + \frac{1}{(k-1)(k+2)}\frac{d}{dx}\phi_2 = 0,$$

$$\cdots$$

$$\phi_{k-1} + \frac{1}{1\cdot 2k}\frac{d}{dx}\phi_k = 0.$$

Thus,

$$\phi_0 = \frac{(-1)^k}{1\cdot 2\cdot 3\cdots 2k}\frac{d^k}{dx^k}\left((1-x^2)^k\frac{d^k}{dx^k}P_k\right).$$

Now, from (24.16), he could deduce $\frac{d^k}{dx^k} P_k = 1 \cdot 3 \cdot 5 \cdots (2k-1)$, and therefore he could write the required result,

$$P_k(x) = \frac{(-1)^k}{2 \cdot 4 \cdot 6 \cdots 2k} \frac{d^k}{dx^k} (1-x^2)^k.$$

Also mentioned in Chapter 25 on $q$-series, Olinde Rodrigues employed differential equations to obtain his formula in 1815; in fact, he referred to Ivory's 1812 paper. Still, note that Rodrigues has priority in this matter, since Ivory did not work out his final result until 1824.

## 24.6 Liouville's Proof of the Rodrigues Formula

In 1837, Ivory and Jacobi published a joint paper in Liouville's journal, containing a proof of the Rodrigues formula, using Lagrange inversion.[16] They were both unaware that the French mathematician Rodrigues had already published his result in 1815, albeit in an obscure journal. This interesting collaboration between Ivory and Jacobi took place at the suggestion of Jacobi, who wrote to Ivory that, since they had independently obtained the Rodrigues formula, they could publish a joint paper to broadcast this result in France, where it was unknown. In the same issue of his journal, Liouville published an alternate, more transparent proof,[17] in fact similar to one published by Jacobi almost ten years earlier. Liouville started by reproducing Legendre's result that

$$\int_{-1}^{1} x_m x_n \, dx = \frac{2}{2n+1} \delta_{mn},$$

where $x_m$ denoted the Legendre polynomial of degree $m$. Liouville then observed that $x_n$ was a polynomial of exact degree $n$, and hence any $n$th degree polynomial had to be a linear combination of the polynomials $x_0, x_1, \ldots, x_n$. He let $y$ be any polynomial of degree $n-1$, so that for some constants $A_0, A_1, \ldots, A_{n-1}$

$$y = A_0 + A_1 x_1 + A_2 x_2 + \cdots + A_{n-1} x_{n-1}.$$

From this, he had

$$d^n y = 0, \qquad \int_{-1}^{1} y x_n \, dx = 0.$$

Since $d^n y = 0$, repeated integration by parts yielded

$$\int_{-1}^{x} y x_n \, dt = y \int_{-1}^{x} x_n \, dt - y' \int_{-1}^{x} \int_{-1}^{t} x_n \, dt_1 \, dt + y'' \int_{-1}^{x} \int_{-1}^{t} \int_{-1}^{t_1} x_n \, dt_2 \, dt_1 \, dt$$

$$+ \cdots + (-1)^{n-1} y^{(n-1)} \int_{-1}^{x} \int_{-1}^{t} \cdots \int_{-1}^{t_{n-2}} x_n \, dt_{n-1} \, dt_{n-2} \cdots dt.$$

[16] Ivory and Jacobi (1837).
[17] Liouville (1837a).

Because the left-hand side was zero for $x = 1$, and $y$ was an arbitrary polynomial of degree $n - 1$, for $x = 1$ he obtained

$$\int_{-1}^{x} x_n \, dt = 0, \quad \int_{-1}^{x} \int_{-1}^{t} x_n \, dt_1 \, dt = 0, \cdots,$$

$$\int_{-1}^{x} \int_{-1}^{t} \cdots \int_{-1}^{t_{n-2}} x_n \, dt_{n-1} \, dt_{n-2} \cdots dt = 0.$$

Liouville denoted the polynomial of degree $2n$ in the last equation by $\phi(x)$, or,

$$\phi(x) = \int_{-1}^{x} \int_{-1}^{t} \cdots \int_{-1}^{t_{n-2}} x_n \, dt_{n-1} \, dt_{n-2} \cdots dt.$$

Then

$$\phi(1) = \phi'(1) = \cdots = \phi^{(n-1)}(1) = 0,$$

implying that $(x - 1)^n$ was a factor of $\phi(x)$. Since it was obvious that $\phi(-1) = \phi'(-1) = \cdots = \phi^{(n-1)}(-1) = 0$, he could conclude that $(x + 1)^n$ was also a factor; hence, $(x^2 - 1)^n$ was a factor of $\phi(x)$. Also, $\phi(x)$ was of degree $2n$, so, clearly, $\phi(x) = D(x^2 - 1)^n$ for a constant $D$. Therefore, for some constant $H_n$,

$$x_n = H_n \frac{d^n}{dx^n}(x^2 - 1)^n$$

$$= H_n \left( (x + 1)^n \frac{d^n}{dx^n}(x - 1)^n + \frac{n}{1} \cdot \frac{d}{dx}(x + 1)^n \frac{d^{n-1}}{dx^{n-1}}(x - 1)^n + \cdots \right).$$

Observe that Liouville applied Leibniz's formula for the $n$th derivative of a product. Now note that, except for the first, every term in this expression was zero at $x = 1$, so that he could write

$$x_n(1) = 1 \cdot 2 \cdot 3 \cdots n \cdot 2^n H_n.$$

Note also that, for $x = 1$, the generating function of the Legendre polynomials is

$$(1 - 2xz + z^2)^{-\frac{1}{2}} = (1 - z)^{-1} = (1 + z + z^2 + \cdots).$$

So Liouville could conclude that $x_n(1) = 1$, and $H_n = \frac{1}{n!2^n}$, proving the result. Liouville also proved that a function $f(x)$ could be expanded in terms of $x_n$.[18] He first set

$$F(x) = \sum_{n=0}^{\infty} \frac{2n + 1}{2} \cdot x_n \cdot \int_{-1}^{1} f(x)x_n \, dx.$$

[18] Liouville (1837b).

Multiplying both sides by $x_n$ and integrating over $(-1, 1)$, he obtained

$$\int_{-1}^{1} (F(x) - f(x))x_n \, dx = 0,$$

and therefore

$$\int_{-1}^{1} \big(F(x) - f(x)\big) y \, dx = 0$$

for an arbitrary polynomial $y$. Liouville took $y = x^n$ to get

$$\int_{-1}^{1} (F(x) - f(x))x^n \, dx = 0, \ n = 0, 1, 2, \ldots.$$

He then concluded that $f(x) = F(x)$ and the result was proved.

To show that his conclusion was justified, Liouville also derived the additional theorem that if $f(x)$ was continuous and finite on $[a, b]$ and $\int_a^b x^n f(x) \, dx = 0$ for $n = 0, 1, 2, \ldots$, then $f(x) = 0$ on $[a, b]$. His proof applied only to those functions having a finite number of changes of sign in the interval $[a, b]$, though he failed to remark on this. He began his proof by assuming the geometrically evident proposition that if $f(x)$ was always nonnegative in $[a, b]$ and $\int_a^b f(x) \, dx = 0$, then $f(x)$ had to be identically zero. We remark that Cauchy's ideas on integrals and continuity from the 1820s can be applied to provide an effective proof of this assumption. Next, Liouville supposed that $f(x)$ changed sign at the values $x_1, x_2, \ldots, x_n$ inside $[a, b]$. He let $\psi(x) = (x - x_1)(x - x_2) \cdots (x - x_n)$, and noted that $f(x)\psi(x)$ would have no changes of sign in $[a, b]$ and that $\int_a^b \psi(x)f(x) \, dx = 0$. He could conclude $f(x)\psi(x) \equiv 0$ and $f(x) \equiv 0$, giving him the required result. A modern proof of the proposition might use the Weierstrass approximation theorem, but that was not stated until some decades later. Moreover, observe that ideas such as uniform convergence had not been discovered in Liouville's time. In fact, he did not even clarify the type of interval he was working with and had to explicitly state that $f(x)$ was finite, or bounded, at all points.

## 24.7 The Jacobi Polynomials

In his significant 1859 posthumously published paper, "Untersuchungen über die Differentialgleichung der hypergeometrischen Reihe,"[19] edited by Heine, Jacobi used the hypergeometric differential equation to derive a Rodrigues-type formula for hypergeometric polynomials. These polynomials are now referred to as Jacobi polynomials, and Jacobi further showed them to be orthogonal with respect to

---

[19] Jacobi (1859).

the beta distribution. In this paper, Jacobi also obtained the generating function for Jacobi polynomials by an application of the Lagrange inversion formula. Note that Jacobi polynomials are in fact generalizations of Legendre polynomials. It is hard to determine exactly when Jacobi discovered these polynomials. In the 1840s, he published some papers on hypergeometric functions and related topics, but remarks of Kummer indicate that Jacobi had studied these functions even earlier than that. Jacobi started his investigations with the observation that if $y$ satisfied the hypergeometric differential equation, then

$$x(1-x)y^{(2)} + (c - (a+b+1)x)y' - aby = 0,$$

$$x(1-x)y^{(3)} + (c+1-(a+b+3)x)y^{(2)} - (a+1)(b+1)y' = 0,$$

$$x(1-x)y^{(4)} + (c+2-(a+b+5)x)y^{(3)} - (a+2)(b+2)y^{(2)} = 0,$$

$$\cdots$$

$$x(1-x)y^{(n+1)} + \left(c+n-1-(a+b+2n-1)x\right)y^{(n)}$$
$$-(a+n-1)(b+n-1)y^{(n-1)} = 0. \quad (24.30)$$

To understand why these equations follow one after the other, note that, in Gauss's notation, if

$$y = F(a,b,c,x), \quad \text{then} \quad y' = \left(\frac{ab}{c}\right) F(a+1, b+1, c+1, x).$$

In more modern notation, one might write $y$ as $_2F_1\left(\begin{matrix} a,b \\ c \end{matrix}; x\right)$. The parameters $a$, $b$, $c$ change to $a+1$, $b+1$, $c+1$, respectively, when one takes the derivative of a hypergeometric function. Following Jacobi, multiply (24.30) by

$$x^{c+n-2}(1-x)^{a+b-c+n-1}$$

and rewrite as

$$\frac{d}{dx}\left(x^n(1-x)^n My^{(n)}\right) = (a+n-1)(b+n-1)x^{n-1}(1-x)^{n-1}My^{(n-1)},$$

where $M = x^{c-1}(1-x)^{a+b-c}$. By iteration, he had

$$\frac{d^n}{dx^n}\left(x^n(1-x)^n My^{(n)}\right) = a(a+1)\cdots(a+n-1)b(b+1)\cdots(b+n-1)My.$$

Next Jacobi took $b = -n$, so that $y = F(-n,a,c,x)$ would be a polynomial of degree $n$; then $y^{(n)}$ was a constant and the equation became

$$F(-n,a,c,x) = \frac{x^{1-c}(1-x)^{c+n-a}}{c(c+1)\cdots(c+n-1)}\frac{d^n}{dx^n}[x^{c+n-1}(1-x)^{a-c}].$$

Replacing $a$ by $a+n$, he obtained the Rodrigues-type formula

$$X_n = F(-n,a+n,c,x) = \frac{x^{1-c}(1-x)^{c-a}}{c(c+1)\cdots(c+n-1)}\frac{d^n}{dx^n}[x^{c+n-1}(1-x)^{a+n-c}].$$

Jacobi used the Lagrange inversion formula to find the generating function of the polynomials $X_n$; for $\xi = 1 - 2x$ and $(c)_n$ denoting the shifted factorial:

$$\sum_{n=0}^{\infty}\frac{(c)_n}{n!}h^n X_n =$$

$$\frac{x^{1-c}(1-x)^{c-a}\left(h-1+\sqrt{1-2h\xi+h^2}\right)^{c-1}\left(h+1-\sqrt{1-2h\xi+h^2}\right)^{a-c}}{(2h)^{a-1}\sqrt{1-2h\xi+h^2}}.$$

He then used the hypergeometric differential equation to prove the orthogonality relation for $X_n$ when $c > 0$ and $a + 1 - c > 0$. Observe that the latter conditions were necessary for the convergence of the integrals. He then let

$$J_{m,n} = \int_0^1 X_m X_n x^{c-1}(1-x)^{a-c}\,dx.$$

Since $X_n$ satisfied the differential equation

$$x(1-x)X_n'' + (c-(a+1)x)X_n' = -n(n+a)X_n,$$

he could deduce that

$$-n(n+a)J_{m,n} = \int_0^1 X_m\frac{d}{dx}[x^c(1-x)^{a+1-c}X_n']\,dx$$

$$= \int_0^1 X_n\frac{d}{dx}[x^c(1-x)^{a+1-c}X_m']\,dx = -m(m+a)J_{m,n}.$$

Thus, taking $m \neq n$, he had $J_{m,n} = 0$. When $m = n$, then integration by parts yielded

$$n(n+a)J_{n,n} = \int_0^1 X_n' X_n' x^c(1-x)^{a+1-c}\,dx.$$

Since $X_m'$ and $X_n'$ were again hypergeometric polynomials, this relation implied that

$$(n-1)(n+a+1)\int_0^1 X_m' X_n' x^c(1-x)^{a+1-c}\,dx = \int_0^1 X_m'' X_n'' x^{c+1}(1-x)^{a+2-c}\,dx.$$

Now $X_n^{(n)}$ was a constant so a repeated application of this formula finally produced a beta integral, computable in terms of gamma functions. The eventual result was then

$$J_{n,n} = \frac{n!}{a+2n} \frac{(\Gamma(c))^2 \Gamma(a-c+n+1)}{\Gamma(a+n)\Gamma(c+n)}.$$

The polynomials $X_n$ are the Jacobi polynomials, except for a constant factor. In more modern notation, taking $c = \alpha + 1$ and $a = \alpha + \beta + 1$ in the expression for $X_n$, the Jacobi polynomials may be expressed as

$$P^{(\alpha,\beta)}(\xi) := \frac{(\alpha+1)_n}{n!} F\left(-n, n+\alpha+\beta+1, \alpha+1, \frac{1-\xi}{2}\right)$$

$$= (-1)^n \frac{(1-\xi)^{-\alpha}(1+\xi)^{-\beta}}{2^n} \frac{d^n}{d\xi^n}[(1-\xi)^{n+\alpha}(1+\xi)^{n+\beta}].$$

Thus, observe that the orthogonality relation would hold over $[-1, 1]$ with respect to the beta distribution $(1-x)^\alpha(1+x)^\beta$ for $\alpha, \beta > -1$. Also we note that it can be shown, by using the hypergeometric differential equation, that the Jacobi polynomial $P_n^{(\alpha,\beta)}(x)$ is a solution of the differential equation

$$(1-x^2)\, y'' + \left(\alpha - \beta + (\alpha+\beta+1)x\right) y' + n(n+\alpha+\beta+1)\, y = 0. \qquad (24.31)$$

Jacobi briefly noted that for $x = \frac{1-\xi}{2}$,

$$(1 - 2h\xi + h^2)^{-c} = \sum_{n=0}^{\infty} h^n Y_n, \qquad (24.32)$$

where

$$Y_n = \frac{2c(2c+1)\cdots(2c+n-1)}{n!} F\left(-n, 2c+n, \frac{2c+1}{2}, x\right)$$

$$= \frac{4^n c(c+1)\cdots(c+n-1)}{(2c+n)(2c+n+1)\cdots(2c+2n-1)} \frac{[x(1-x)]^{\frac{1}{2}(1-2c)}}{n!}$$

$$\times \frac{d^n}{dx^n}[x(1-x)]^{\frac{1}{2}(2c+2n-1)}.$$

We now designate the $Y_n$ as ultraspherical or Gegenbauer polynomials:

$$Y_n := C_n^c(\xi) = \frac{(2c)_n}{(c+\frac{1}{2})_n} P_n^{(c-\frac{1}{2}, c-\frac{1}{2})}(\xi),$$

where $(a)_n$ denotes the shifted factorial $a(a+1)\cdots(a+n-1)$. Gegenbauer polynomials,[20] named after the Austrian mathematician Leopold Gegenbauer (1849–1903), student of Weierstrass and Kronecker, are special cases of Jacobi polynomials.

---

[20] Gegenbauer (1884).

They occur when the parameters $\alpha$ and $\beta$ are equal, and they are of great independent interest. Note that the generating function for the ultraspherical polynomials (24.32) is different from the generating function obtained from the one for the Jacobi polynomials.

## 24.8   Stieltjes: Zeros of Jacobi Polynomials

In an 1886 paper published in *Acta Mathematica*,[21] Stieltjes presented an interesting interpretation of the zeros of Jacobi polynomials by viewing them in terms of the positions of some point masses distributed on $[-1, 1]$, attracted to one another with a force inverse to the distance between them.

Thomas Jan Stieltjes (1856–1894) entered the Polytechnical School in Delft in 1873. He spent nearly all his time studying the works of Gauss and Jacobi, as a result of which he was unable to pass his examinations in spite of repeated attempts. Thus, he found himself unable to find a suitable position in the Netherlands, except as a calculator at an observatory. In 1882, he began a correspondence with Hermite, by whose assistance in 1886 he defended a thesis on asymptotic series at the Sorbonne and eventually obtained a professorship at Toulouse in France.

In 1894, Stieltjes published his famous paper[22] on the convergence of continued fractions of the form

$$\frac{1}{a_1 z+} \; \frac{1}{a_2 z+} \; \frac{1}{a_3 z+} \; \frac{1}{a_4 z+} \cdots, \quad a_n \geq 0, \quad n = 1, 2, 3, \ldots, \quad z \in \mathbb{C}.$$

This groundbreaking paper devotes one chapter to the Stieltjes integral. Since we will utilize this integral repeatedly in this book, we present Stieltjes's definition at the end of this section. The paper contained many important results,[23] as Poincaré's praise of the work[24] indicates:

> Therefore Stieltjes' work is one of the most remarkable Memoires in Analysis which have been written in the past years; it adds to many others which have placed their author in an eminent rank within the science of our period.... The committee takes pride in proposing the Academy award Mr. Stieltjes the highest evidence of his approval by ordering the insertion of his Memoire "Sur les fractions continues" into the Collection of foreign Scholars (in the Academy) and the committee expresses the wish that a prize could be awarded him from the Lecomte foundation.

In his 1886 *Acta Mathematica* paper,[25] Stieltjes took masses of mass $b$ fixed at $-1$ and of mass $a$ fixed at 1, along with unit masses located at $x_1 > x_2 > \cdots > x_n$.

---

[21] Stieltjes (1886b), especially pp. 387–388.
[22] Stieltjes (1894).
[23] See W. Van Assche in Stieltjes (1993) vol. I, pp. 6–11.
[24] Stieltjes (1993) vol. I, p. 4.
[25] Stieltjes (1886b).

He assumed that the masses were in equilibrium, meaning that the total force on the unit mass at each point $x_i$, $i = 1, 2, \ldots, n$, was zero:

$$\frac{a}{1 + x_i} - \frac{b}{1 - x_i} + \sum_{j \neq i} \frac{1}{x_i - x_j} = 0, \quad i = 1, 2, \ldots, n. \tag{24.33}$$

Now in a paper of 1885,[26] Stieltjes had already shown that a condition more general than (24.33) would imply that the polynomial

$$y = f(x) = (x - x_1)(x - x_2) \cdots (x - x_n) \tag{24.34}$$

satisfied a second-order linear differential equation. Thus, Stieltjes could conclude[27] from (24.33) that $y = f(x)$ would satisfy the equation

$$(1 - x^2)\, y'' + 2\big(a - b - (a + b)x\big)\, y' + Cy = 0, \tag{24.35}$$

where $C = n(n + 2a + 2b - 1)$. But to prove (24.35), following Stieltjes's method of 1885,[28] observe that the logarithmic derivative of (24.34) produces

$$\frac{y'}{y} - \frac{1}{x - x_i} = \frac{1}{x - x_1} + \cdots + \frac{1}{x - x_{i-1}} + \frac{1}{x - x_{i+1}} + \cdots + \frac{1}{x - x_n}; \tag{24.36}$$

next, following Stieltjes, find the limit of (24.36) as $x \to x_i$ by an application of L'Hôpital's rule, to get

$$\frac{y''(x_i)}{2y'(x_i)} = \sum_{j \neq i} \frac{1}{x_i - x_j}. \tag{24.37}$$

Combine (24.35) and (24.37) to get

$$\frac{1}{2} \frac{y''(x_i)}{y'(x_i)} + \frac{a}{1 + x_i} - \frac{b}{1 - x_i} = 0, \quad i = 1, 2, \ldots, n$$

or

$$(1 - x_i^2)\, y''(x_i) + 2\big(a - b - (a + b)x_i\big)\, y'(x_i) = 0, \quad i = 1, \ldots, n.$$

Now since $y = f(x)$ is a polynomial of degree $n$, the expression

$$(1 - x^2)\, y''(x) + 2\big(a - b - (a + b)x\big)\, y'(x)$$

is a polynomial of degree $\leq n$ and is equal to zero for $x = x_i$, $i = 1, 2, \ldots, n$; thus it is a constant multiple of $y$ and, finally, $y$ satisfies the differential equation

[26] Stieltjes (1885a).
[27] Stieltjes (1886b).
[28] Stieltjes (1885a) p. 324.

$$(1 - x^2) y'' + 2(a - b - (a + b)x) y' + Cy = 0. \qquad (24.38)$$

As for the constant $C$, Stieltjes found it by equating the coefficient of $x^n$ in (24.38):

$$-n(n - 1) - 2(a + b)n + C = 0$$

or

$$C = n(n + 2a + 2b - 1).$$

Now if we set $a = \frac{\alpha+1}{2}$ and $b = \frac{\beta+1}{2}$, then (24.38) becomes

$$(1 - x^2) y'' + (\alpha - \beta + (\alpha + \beta + 1) x) y' + n(n + \alpha + \beta + 1) y = 0.$$

Comparison with (24.31) shows that the polynomial

$$f(x) = (x - x_1)(x - x_2) \cdots (x - x_n)$$

is in fact the Jacobi polynomial $P_n^{(\alpha, \beta)}(x)$, except for a constant factor. Thus Stieltjes had actually proved: Let $1 > x_1 > x_2 > \cdots > x_n > -1$ and let $x_i$ satisfy the conditions given in (24.33); then the $x_i$, $i = 1, 2, \ldots, n$, are the zeros of the Jacobi polynomial $P_n^{(2a-1, 2b-1)}(x)$.

Also note that we can reformulate this theorem of Stieltjes in terms of the potential energy of the system with mass $b$ fixed at $-1$, mass $a$ fixed at 1 and unit masses at $x_1 > x_2 > \cdots > x_n$, between $-1$ and 1. The logarithmic potential energy can be given as

$$T(x_1, x_2, \ldots, x_n) = -a \sum_{i=1}^{n} \ln|1 + x_i| - b \sum_{i=1}^{n} \ln|1 - x_i| - \sum_{1 \le i < j \le n} \ln|x_i - x_j|.$$

$$(24.39)$$

The system is in equilibrium when $\frac{\partial T}{\partial x_i} = 0$, $i = 1, 2, \ldots, n$, equations identical with (24.33). Thus, we can say that when the logarithmic potential energy (24.39) is at a minimum, then $x_1, x_2, \ldots, x_n$ are the zeros of the Jacobi polynomial $P_n^{(2a-1, 2b-1)}(x)$.

Turning to Stieltjes's definition of his integral,[29] he began with an increasing function $\phi(x)$ on an interval $[a, b]$. He showed that $\phi$ must have a left- and right-hand limit at each $x \in (a, b)$, denoting these limits by $\phi^-(x)$ and $\phi^+(x)$ respectively. He observed that there would be a jump discontinuity of $\phi^+(x) - \phi^-(x)$ at $x$ when $\phi^+(x) \ne \phi^-(x)$. Since the sum of the jump discontinuities in $[a, b]$ had to be less than $\phi(b) - \phi(a)$, he was able to prove that the number of points of discontinuity for $\phi$ was countable. But the number of points in $[a, b]$ was uncountable, so that every subinterval of $[a, b]$, no matter how small, had to contain points of continuity of $\phi$.

---

[29] Stieltjes (1993) vol. 2, pp. 665–668.

He next considered the set of points

$$a = x_0 < x_1 < x_2 < \cdots < x_{n-1} < x_n = b$$

and took numbers $\xi_k$, $k = 1, 2, \ldots, n$ such that $x_{k-1} \le \xi_k \le x_k$. Stieltjes then wrote that the limit, if it existed, of the sum

$$f(\xi_1)\big(\phi(x_1) - \phi(x_0)\big) + f(\xi_2)\big(\phi(x_2) - \phi(x_1)\big) + \cdots + f(\xi_n)\big(\phi(x_n) - \phi(x_{n-1})\big)$$

was denoted by

$$\int_a^b f(u)\, d\phi(u). \tag{24.40}$$

His clear meaning here was that the limit, as a norm $\|P\| = \max\{|x_i - x_{i-1}|,$ $i = 1, \ldots, n\}$ of the partition $P(x_0, x_1, \ldots, x_n)$, tended to zero. Note that in his definition of the integral, Stieltjes did not use the upper and lower sums, considered by Poincaré's thesis advisor, Gaston Darboux.[30]

Finally, Stieltjes noted that if $f$ was continuous on $[a, b]$, then the integral (24.40) would exist. Luxemburg has remarked[31] that the Stieltjes integral went unnoticed by mathematicians until 1909 when Frigyes Riesz employed it to state his famous theorem on continuous linear functionals on the space of continuous functions $[a, b]$.[32]

## 24.9 Askey: Discriminant of Jacobi Polynomials

In 1983–84, Richard Askey observed that he could find the discriminant of the Jacobi polynomial by a new method and he presented this derivation in his lectures. First, take $f(x)$ to be a polynomial of degree $n$ such that

$$f(x) = a_0 x^n + a_1 x^{n-1} + \cdots + a_n = a_0(x - x_1)(x - x_2) \cdots (x - x_n).$$

The discriminant $D_n(f)$ of $f(x)$ is defined by

$$D_n(f) = a_0^{2n-2} \prod_{1 \le i < j \le n} (x_i - x_j)^2. \tag{24.41}$$

Now observe that the condition (24.39), where $a > 0$, $b > 0$, implies that the maximum of

$$H(x_1, x_2, \ldots, x_n) := \prod_{i=1}^n x_i^a (1 - x_i)^b \prod_{1 \le i < j \le n} |x_i - x_j| \tag{24.42}$$

occurs when $x_1, x_2, \ldots, x_n$ are the zeros of the Jacobi polynomial (with change of variables)

---

[30] Darboux (1875).
[31] Stieltjes (1993) vol. I, pp. 60–65.
[32] Riesz (1909). For a more modern treatment, see Douglas (1972) pp. 18–20.

$$P_n^{(2a-1)(2b-1)}(1-2x).$$

Askey perceived that by maximizing (24.42) and using the Selberg integral, he could arrive at a new derivation for the discriminant of the Jacobi polynomial. For this purpose, he used a result contained in the book of his colleague Walter Rudin:[33]

Let $\mu$ be a positive measure on a measure space $X$; let

$$\|f\|_k = \left(\int_X |f|^k \, d\mu\right)^{\frac{1}{k}} < \infty$$

for some $k$ in $0 < k < \infty$, with $\|f\|_\infty > 0$. Then

$$\|f\|_k \to \|f\|_\infty \quad \text{as } k \to \infty.$$

Apply Rudin's result to the situation

$$d\mu = dx_1 dx_2 \cdots dx_n, \quad X = [0,1]^n,$$

$$f(x_1, x_2, \ldots, x_n) = \prod_{i=1}^n x_i^{2a}(1-x_i)^{2b} \prod_{1 \le i < j \le n} (x_i - x_j)^2 \tag{24.43}$$

and observe that since $f$ is continuous on $[0,1]^n$,

$$\|f\|_\infty = \max_{x \in [0,1]^n} f(x).$$

Askey next took Selberg's integral formula (17.127)

$$\int_X \prod_{i=1}^n x_i^{\alpha-1}(1-x_i)^{\beta-1} \prod_{1 \le i < j \le b} |x_i - x_j|^{2\gamma} \, d\mu$$

$$= \prod_{j=1}^n \frac{\Gamma(\alpha + (j-1)\gamma)\Gamma(\beta + (j-1)\gamma)\Gamma(1+j\gamma)}{\Gamma(\alpha + \beta + (n+j-2)\gamma)\Gamma(1+\gamma)},$$

in which he set $\alpha - 1 = 2ak$, $\beta - 1 = 2bk$, and $\gamma = k$ to obtain

$$\left(\int_X f^k \, d\mu\right)^{\frac{1}{k}} = \left(\prod_{j=1}^n \frac{\Gamma((2a+j-1)k+1)\Gamma((2b+j-1)k+1)\Gamma(jk+1)}{\Gamma((2a+2b+n+j-2)k+2)\Gamma(k+1)}\right)^{\frac{1}{k}}.$$

He next let $k \to \infty$ and applied Stirling's approximation to arrive at

$$\max_{x \in [0,1]^n} f(x) = \prod_{j=1}^n \frac{(2a+j-1)^{2a+j-1}(2b+j-1)^{2b+j-1}j^j}{(2a+2b+n+j-2)^{2a+2b+n+j-2}}, \tag{24.44}$$

---

[33] Rudin (1966) p. 70.

implying that if $x_1, x_2, \ldots, x_n$ were zeros of the Jacobi polynomial $P_n^{(2a, 2b)}(1 - 2x)$, then

$$\prod_{1 \le i < j \le n} (x_i - x_j)^2 = \prod_{j=1}^{n} \frac{(2a + j - 1)^{2a+j-1}(2b + j - 1)^{2b+j-1} j^j}{x_j^{2a}(1 - x_j)^{2b}(2a + 2b + n + j - 2)^{2a+2b+n+j-2}}.$$

$$(24.45)$$

But since

$$P_n^{(2a, 2b)}(1 - 2x) = \frac{(2a + 1)_n}{n!} \, {}_2F_1\left(\begin{matrix} -n, n + 2a + 2b + 1 \\ 2a + 1 \end{matrix}; x\right)$$

$$= \frac{(-1)^n(2 + 2b + n + 1)_n}{n!}(x - x_1)(x - x_2)\cdots(x - x_n),$$

$$(24.46)$$

Askey could write

$$\prod_{j=1}^{n} x_j^{2a} = \left(\frac{(2a + 1)_n}{(2a + 2b + n + 1)_n}\right)^{2a}$$

$$(24.47)$$

and

$$\prod_{j=1}^{n} (1 - x_j)^{2b} = \left(\frac{(-1)^n(2a + 1)_n}{(2a + 2b + n + 1)_n} \, {}_2F_1\left(\begin{matrix} -n, n + 2a + 2b + 1 \\ 2a + 1 \end{matrix}; 1\right)\right)^{2b}.$$

$$(24.48)$$

Moreover, he could sum the ${}_2F_1$ in (24.48) by the Vandermonde, or Chu–Vandermonde, identity discussed in Section 25.10. Thus, he found that

$$\prod_{j=1}^{n} (1 - x_j)^{2b} = \left(\frac{(2b + 1)_n}{(2a + 2b + n + 1)_n}\right)^{2b}.$$

$$(24.49)$$

Using (24.46), the value of $a_0$ in (24.41) must be $\frac{(-1)^n(2a+2b+n+1)_n}{n!}$. Thus, by calling upon (24.45), (24.47), and (24.49), and replacing $2a - 1, 2b - 1$ by $\alpha, \beta$, Askey finally got

$$D_n\left(P_n^{(\alpha, \beta)}(x)\right) = 2^{-n(n-1)} \prod_{j-1}^{n} j^{j-2n+2}(j + \alpha)^{j-1}(j + \beta)^{j-1}(\alpha + \beta + n + j)^{n-j},$$

the discriminant of a Jacobi polynomial that was apparently first discovered by Stieltjes.[34]

---

[34] Stieltjes (1885b). Also see Szegő (1975) pp. 142–143.

## 24.10   Chebyshev: Discrete Orthogonal Polynomials

P. L. Chebyshev introduced discrete orthogonal polynomials into mathematics in his 1855 article, "Sur les fractions continues."[35] His work in this area, like many of his other efforts, was motivated by practical problems for which he sought effective solutions. Chebyshev made frequent use of orthogonal polynomials, continuous as well as discrete, and he was probably the first mathematician to emphasize their importance and applicability to problems in both pure and applied mathematics. Chebyshev was greatly influenced in this connection by the papers of Gauss and Jacobi on numerical integration. Chebyshev studied at Moscow University from 1837 to 1841 where N. D. Brashman instructed him in practical mechanics, motivating some of Chebyshev's later work. In 1846, Chebyshev wrote a master's thesis on a topic in probability; this subject also became his lifelong interest. Chebyshev's 1855 paper laid the foundation for his work on orthogonal polynomials. In presenting his work, we at times follow the notation given by N. I. Akhiezer in his article on Chebyshev's work. This notation more clearly reveals the dependence of certain quantities on the given variables.

Chebyshev began his paper by stating the problem in rather general and vague terms: Suppose $F(x)$ is approximately known for $n+1$ values $x = x_0, x_1, \ldots, x_n$, and that $F(x)$ can be represented by a polynomial of degree $m \leq n$,

$$a + bx + cx^2 + \cdots + gx^{m-1} + hx^m.$$

Find the value of $F(x)$ at $x = X$ so that the errors in $F(x_0)$, $F(x_1)$, ..., $F(x_n)$ have minimal influence on $F(X)$. From a practical standpoint, the problem makes good sense. For example, the values of some function $y = F(x)$ may be obtained by observation for $x = x_0, x_1, \ldots, x_n$. These values would have experimental errors so that $y_i \simeq F(x_i)$. Thus, $F(x)$ is a polynomial of degree $m \leq n$, and the problem is to determine $F(x)$ in such a way that the errors of observation have the least influence. In more specific terms, Chebyshev stated the problem:

Find a polynomial $F(x)$ of the form

$$F(x) = \mu_0\lambda_0(x)y_0 + \mu_1\lambda_1(x)y_1 + \cdots + \mu_n\lambda_n(x)y_n, \qquad (24.50)$$

where $\lambda_i(x)$ are unknown polynomials of degree $\leq m$ and $\mu_i > 0$ are weights associated with observed values $y_i$ subject to the following two conditions: The identity

$$f(X) = \mu_0\lambda_0(X)f(x_0) + \mu_1\lambda_1(X)f(x_1) + \cdots + \mu_n\lambda_n(X)f(x_n) \qquad (24.51)$$

must hold for any polynomial of degree at most $m$; and one must minimize the sum

$$W(X) = \mu_0\big(\lambda_0(X)\big)^2 + \mu_1\big(\lambda_1(X)\big)^2 + \cdots + \mu_n\big(\lambda_n(X)\big)^2. \qquad (24.52)$$

---

[35]  Chebyshev (1899–1907) vol. 1, pp. 203–230.

Thus, $W(X)$ had to be minimized with respect to the constraints of (24.51), equivalent to the $m + 1$ conditions

$$X^k = \mu_0 \lambda_0(X) x_0^k + \mu_1 \lambda_1(X) x_1^k + \cdots + \mu_n \lambda_n(X) x_n^k, \qquad (24.53)$$

for $k = 0, 1, \ldots, m$. Then Chebyshev applied the method of Lagrange multipliers with $\lambda_0, \lambda_1, \ldots, \lambda_n$ as the variables and with $l_0(X), l_1(X), \ldots, l_m(X)$ as the $m + 1$ multipliers. This gave him the $n + 1$ relations

$$\frac{\partial W}{\partial \lambda_i} - \frac{\partial}{\partial \lambda_i} \sum_{k=0}^{m} l_k(X) \left( \mu_0 \lambda_0 x_0^k + \cdots + \mu_n \lambda_n x_n^k - X^k \right) = 0,$$

or

$$2\lambda_i(X) = l_0(X) + l_1(X) x_i + \cdots + l_m(X) x_i^m, \qquad (24.54)$$

for $i = 0, 1, \ldots, m + 1$. Chebyshev wrote that the whole difficulty boiled down to solving this system of equations. He denoted the polynomial on the right-hand side of equation (24.54) by $2K_m(X, x_i)$, obtaining

$$K_m(X, x) = \frac{1}{2} \sum_{k=0}^{m} l_k(X) x^k. \qquad (24.55)$$

Thus, Chebyshev's problem was to find an expression for $\lambda_i(X) = K_m(X, x_i)$, $i = 0, 1, \ldots, n$. Note that the constraints (24.53) could be written as

$$\sum_{i=0}^{n} \mu_i K_m(X, x_i) x_i^k = X^k, \quad k = 0, 1, \ldots, m. \qquad (24.56)$$

These relations implied that the polynomials $K_m(X, x_i)$ should be such that, for some function $A(X)$,

$$\sum_{i=0}^{n} \frac{\mu_i K_m(X, x_i)}{x - x_i} - \frac{1}{x - X} = \frac{A(X)}{x^{m+2}} + \cdots . \qquad (24.57)$$

Note that this relation could be rewritten as

$$K_m(X, x) \sum_{i=0}^{n} \frac{\mu_i}{x - x_i} - N(X, x) - \frac{1}{x - X} = \frac{A(X)}{x^{m+2}} + \cdots \qquad (24.58)$$

with $N(X, x)$ a polynomial of degree $m - 1$ in $x$. Of course, this was made possible by the elementary relation that if $g(x)$ is a polynomial of degree $m$, then there is a polynomial $h(x)$ of degree $m - 1$ such that

$$\frac{g(x)}{x - x_i} = h(x) + \frac{g(x_i)}{x - x_i}.$$

Here Chebyshev considered an additional but related problem: Find a polynomial $\psi_m(x)$ of degree $m$ and a polynomial $\pi_m(x)$ of degree at most $m - 1$ so that

$$\psi_m(x) \sum_{i=0}^{n} \frac{\mu_i}{x - x_i} - \pi_m(x) = O\left(\frac{1}{x^{m+2}}\right). \tag{24.59}$$

Chebyshev's study of Gauss's paper on numerical integration showed him that the answer lay in the continued fraction expansion

$$\sum_{i=0}^{n} \frac{\mu_i}{x - x_i} = \frac{1}{q_1+} \frac{1}{q_2+} \frac{1}{q_3+} \cdots \frac{1}{q_{n+1}}, \tag{24.60}$$

where $q_m = A_m x + B_m$ were linear functions for $m = 1, 2, \ldots$. In fact, the $m$th ($m \leq n + 1$) convergent was the rational function $\frac{\pi_m(x)}{\psi_m(x)}$, producing the polynomials required in (24.59). Chebyshev proved that

$$\lambda_i(x) = K_m(x, x_i) = (-1)^m \frac{\psi_{m+1}(x)\psi_m(x_i) - \psi_m(x)\psi_{m+1}(x_i)}{x - x_i}. \tag{24.61}$$

He then derived another relation for $\lambda_i(x)$ by using the three-term relation for $\psi_m(x)$ obtained from the continued fraction

$$\psi_{m+1}(x) = q_{m+1}\psi_m(x) + \psi_{m-1}(x)$$
$$= (A_{m+1}x + B_{m+1})\psi_m(x) + \psi_{m-1}(x).$$

When this was substituted in (24.61) he could obtain, after simplification,

$$(-1)^m \lambda_i(x) = \left(A_{m+1}\psi_m(x)\psi_m(x_i) - \frac{\psi_m(x)\psi_{m-1}(x_i) - \psi_m(x_i)\psi_{m-1}(x)}{x - x_i}\right).$$

Repeating this process $m$ times, he got

$$(-1)^m \lambda_i(x) = \sum_{j=0}^{m} (-1)^{m-j} A_{j+1}\psi_j(x_i)\psi_j(x)$$
$$= \frac{\psi_{m+1}(x)\psi_m(x_i) - \psi_m(x)\psi_{m+1}(x_i)}{x - x_i}. \tag{24.62}$$

This important relation is usually called the Christofell–Darboux formula; they obtained it in a similar way, but Chebyshev published the formula more than a decade earlier. When he substituted the value of $\lambda_i(x)$ in (24.62) in (24.50), Chebyshev had

$$F(x) = \sum_{i=0}^{n} \left(\sum_{j=0}^{m} (-1)^j A_{j+1}\psi_j(x_i)\psi_j(x)\right) \mu_i F(x_i). \tag{24.63}$$

He then set $F(x) = \psi_m(x)$ and equated the coefficients of $\psi_k(x)$ on both sides to obtain the orthogonality relation

$$(-1)^k A_{k+1} \sum_{i=0}^{n} \mu_i \psi_k(x_i) \psi_m(x_i) = \delta_{km}, \tag{24.64}$$

and in particular

$$A_{k+1} = \frac{(-1)^k}{\sum_{i=0}^{n} \mu_i \psi_k^2(x_i)}, \tag{24.65}$$

where $\psi_k(x)$ was the sequence of discrete orthogonal polynomials. So his final result was expressed as

$$\lambda_i(x) = \frac{\sum_{k=0}^{m} \mu_k \psi_k(x_i) \psi_k(x)}{\sum_{i=0}^{n} \mu_k \psi_k^2(x_i)}. \tag{24.66}$$

Chebyshev concluded his paper by stating and proving two results on least squares. For the first result, he supposed $V$ to be a polynomial of degree $m$ with the coefficient of $x^m$ the same as that of $\psi_m(x)$. He then showed that the sum $\sum_{i=0}^{n} \mu_i V^2(x_i)$ had the least value when $V = \psi_m(x)$. To prove this, Chebyshev set

$$V = A_0 \psi_0(x) + \cdots + A_{m-1} \psi_{m-1}(x) + \psi_m(x)$$

and then

$$\sum_{i=0}^{n} \mu_i V^2(x_i) = \sum_{i=0}^{n} \mu_i \big( A_0 \psi_0(x_i) + \cdots + A_{m-1} \psi_{m-1}(x_i) + \psi_m(x_i) \big)^2.$$

For a minimum, the derivatives with respect to the $A_j$ should be zero. Thus,

$$2 \sum_{i=0}^{n} \mu_i \psi_j(x_i) \big( A_0 \psi_0(x_i) + A_{m-1} \psi_{m-1}(x_i) + \psi_m(x_i) \big) = 0, \quad j = 0, \ldots, m-1.$$

An application of the orthogonality relation (24.64) gave

$$A_j \sum_{i=0}^{n} \mu_{i|} \psi_j^2(x_i) = 0, \quad j = 0, 1, \ldots, m-1.$$

This implied $A_j = 0$, $j = 0, 1, \ldots, m-1$, and hence $V = \psi_m(x)$. In his second result, Chebyshev proved that

$$\sum_{i=0}^{n} \mu_i \left( F(x_i) - \sum_{j=0}^{m} A_j \psi_j(x_i) \right)^2 \tag{24.67}$$

was a minimum when

$$A_j = \frac{\sum_{i=0}^{n} \mu_i \psi_j(x_i) F(x_i)}{\sum_{i=0}^{n} \mu_i \psi_j^2(x_i)}.$$

He once more took the derivatives of (24.67) with respect to $A_j$, $j = 0, \ldots, m$ to obtain

$$2 \sum_{i=0}^{n} \mu_i \psi_j(x_i) \left( F(x_i) - \sum_{j=0}^{m} A_j \psi_j(x_i) \right) = 0, \quad j = 0, 1, \ldots, m.$$

Again using the orthogonality relation (24.64), these equations reduced to

$$\sum_{i=0}^{n} \mu_i \psi_j(x_i) F(x_i) - A_j \sum_{i=0}^{n} \mu_i \psi_j^2(x_i) = 0, \quad j = 0, 1, \ldots, m,$$

implying the required result.

## 24.11  Chebyshev and Orthogonal Matrices

In his 1855 paper, before the theory of matrices was formally developed, Chebyshev gave a very interesting construction of an orthogonal matrix, noting that in a paper of 1771, Euler also constructed such squares.[36] However, after 1855, Chebyshev did not develop this topic further. Chebyshev defined the function

$$\Phi_k(x_i) = \sqrt{\alpha_i}\, \psi_k(x_i), \quad i, k = 0, 1, \ldots, n$$

where

$$\alpha_i = \frac{\mu_i}{\sum_{j=0}^{n} \mu_j \psi_j^2(x_i)}.$$

He then considered the square tableau

$$\begin{matrix}
\Phi_0(x_0) & \Phi_0(x_1) & \cdots & \Phi_0(x_n) \\
\Phi_1(x_0) & \Phi_1(x_1) & \cdots & \Phi_1(x_n) \\
\vdots & \vdots & & \vdots \\
\Phi_n(x_0) & \Phi_n(x_1) & \cdots & \Phi_n(x_n).
\end{matrix} \qquad (24.68)$$

From the orthogonality relation (24.64), he deduced that the sum of the squares of the terms in each row and in each column was one. Also, in any two rows or columns, the sum of the products of their corresponding terms would be zero.

## 24.12  Chebyshev's Discrete Legendre and Jacobi Polynomials

In his 1864 paper "Sur l'interpolation," Chebyshev took $\mu_i = 1$ and $x_i = i$ for $i = 0, 1, \ldots, n - 1$ with $\mu$ as already defined.[37] See equations (24.59) and (24.60).

---

[36] Eu. I-6 pp. 287–315. E 407.
[37] Chebyshev (1899–1907) vol. 1, pp. 541–560.

Then the polynomials $\psi_0(x)$, $\psi_1(x)$, $\psi_2(x)$, ... were the denominators in the continued fraction expansion of

$$\frac{1}{x} + \frac{1}{x-1} + \frac{1}{x-2} + \cdots + \frac{1}{x-n+1}.$$

By (24.64), these in turn satisfied the relations

$$\sum_{i=0}^{n} \psi_l(i)\psi_m(i) = 0 \quad \text{for} \quad m < l. \tag{24.69}$$

Chebyshev found a Rodrigues-type formula for $\psi_k(x)$, where the differential operator was replaced by the finite difference operator. His two-step approach was exactly the discrete analog of the method employed by Jacobi in his paper on numerical integration. In the first step, Chebyshev proved that if there was a polynomial $f(x)$ of degree $l$ such that

$$\sum_{i=0}^{n} f(i)\psi_m(i) = 0 \quad \text{for} \quad m < l,$$

then there existed a constant $C$ such that $f(x) = C\psi_l(x)$. For the second step, he showed that the polynomial of degree $l$ given by

$$f(x) = \Delta^l x(x-1)\cdots(x-l+1)(x-n)(x-n-1)\cdots(x-n-l+1)$$

satisfied the required condition. Thus, he had the Rodrigues-type formula $\psi_l(x) = C_l f(x)$, where $C_l$ was a constant.

Chebyshev also gave an interesting interpolation formula in terms of $\psi_l(x)$. He supposed $u_0, u_1, \ldots, u_{n-1}$ to be the values of a function $u$ at $x = 0, 1, \ldots, n-1$. The interpolation formula would then be expressed as

$$u = \frac{\sum_{i=0}^{n-1} u_i}{n} + \frac{3\sum_{i=0}^{n-1}(i+1)(n-i-1)\Delta u_i}{1^2 n(n^2-1^2)} \Delta x(x-n)$$

$$+ 5\sum_{i=0}^{n-1} \frac{(i+1)(i+2)(n-i-1)(n-i-2)\Delta^2 u_i}{(2!)^2 n(n^2-1^2)(n^2-2^2)}$$

$$\times \Delta^2 x(x-1)(x-n)(x-n-1) + \cdots. \tag{24.70}$$

Interestingly, in an 1858 paper, "Sur une nouvelle série,"[38] Chebyshev took the points as $x_1 = h$, $x_2 = 2h$, ..., $x_n = nh$ such that the orthogonal polynomials could be expressed in the form

$$\psi_l(x) = C_l \Delta^l (x-h)(x-2h)\cdots(x-lh)(x-nh-h)\cdots(x-nh-lh).$$

---

[38] Chebyshev (1858).

The formula corresponding to (24.70) would then be written

$$u = \frac{1}{n}\sum u_i + \frac{3\sum i(n-i)\Delta u_i}{1^2 n(n^2-1^2)h^2}\Delta(x-h)(x-nh-h)$$

$$+ 5\frac{\sum i(i+1)(n-i)(n-i-1)\Delta^2 u_i}{(2!)^2 n(n^2-1^2)(n^2-2^2)h^4}$$

$$\times \Delta(x-h)(x-2h)(x-nh-h)(x-nh-2h)+\cdots. \qquad (24.71)$$

Chebyshev observed that if he set $h = \frac{1}{n}$ in his interpolation formula (24.71) and let $n \to \infty$, he obtained a series in terms of Legendre polynomials. On the other hand, if he set $h = \frac{1}{n^2}$ and let $n \to \infty$, he arrived at the Maclaurin series expansion! At the end of the paper, Chebyshev made the insightful remark that one might use discrete orthogonal polynomials to approximate the sum $\sum_{i=1}^{n} F(ih)$, just as Gauss had used Legendre polynomials for numerical integration.

Chebyshev appears to have independently discovered the Jacobi polynomials.[39] He gave the generating function and proved orthogonality for these polynomials in an 1870 paper based on the work of Legendre. Chebyshev there proved that if

$$F(s,x) = \frac{(1+s+\sqrt{1-2sx+s^2})^{-\gamma}(1-s+\sqrt{1-2sx+s^2})^{-\mu}}{\sqrt{1-2sx+x^2}}$$

$$= \sum_{n=0}^{\infty} T_n(x)s^n,$$

then

$$\int_{-1}^{1} F(s,x)F(t,x)(1-x)^{\mu}(1+x)^{\gamma}\, dx$$

was purely a function of $st$. This gave the orthogonality of $T_n(x)$ with respect to the beta distribution $(1-x)^{\mu}(1+x)^{\gamma}$. Note that the $T_n(x)$ are the Jacobi polynomials, generalizing the Legendre polynomials. On the basis of this work, in 1875 Chebyshev defined the discrete Jacobi polynomials. He also showed that his interpolation formulas could be applied to problems in ballistics.

## 24.13  Exercises

(1) Show that

$$\int_{-1}^{1}\frac{1}{x-u}\frac{du}{\sqrt{1-u^2}} = \frac{\pi}{\sqrt{x^2-1}} = \frac{\pi}{x-}\frac{1}{2x-}\frac{1}{2x-}\cdots.$$

[39] Chebyshev (1899–1907) vol. 2, pp. 1–8.

Show also that the denominators of the convergents of the continued fractions are $\cos\phi$, $\cos 2\phi$, ..., where $x = \cos\phi$. See Chebyshev (1899–1907) vol. 1, pp. 501–508, especially p. 502.

(2) Let $d$ be the greatest integer in $\frac{n}{2}$, where $n$ is a positive integer. Show that the ultraspherical polynomials $C_n^\lambda$ satisfy the relation

$$C_n^\lambda = \sum_{k=0}^{d} \frac{(\lambda)_{n-k}(\lambda - \mu)_k (n + \mu - 2k)}{k!\,\mu(\mu + 1)_{n-k}} C_{n-2k}^\mu(x),$$

where for any quantity $a$, $(a)_k$ denotes the shifted factorial $a(a + 1)\cdots(a + k - 1)$. This is known as the Gegenbauer–Hua formula. See Gegenbauer (1884).

(3) Show that

$$\int_x^\infty \frac{e^{-t}}{t}\,dt = \frac{e^{-x}}{x + 1-}\ \frac{1}{x + 3-}\ \frac{\frac{1}{2^2}}{\frac{x+5}{4}-}\ \frac{\frac{1}{3^2}}{\frac{x+7}{9}-}\ \frac{1}{\frac{x+9}{16}-}\cdots.$$

Denote the $m$th convergent by $\frac{e^{-x}\phi_m(x)}{f_m(x)}$. Then show that

$$xf_n'(x) = nf_n(x) - n^2 f_{n-1}(x),$$

$$f_{n+1}(x) = (x + 2n + 1)f_n(x) - n^2 f_{n-1}(x),$$

$$\int_{-\infty}^0 e^x f_n(x) f_m(x)\,dx = (n!)^2 \delta_{mn}.$$

See Laguerre (1972) vol. 1, pp. 431–435.

(4) Show that

$$\frac{1}{(n - k)!}\frac{d^{n-k}}{dx^{n-k}}\left(x^2 - 1\right)^n = \frac{(x^2 - 1)^k}{(n + k)!}\frac{d^{n+k}}{dx^{n+k}}\left(x^2 - 1\right)^n.$$

See Rodrigues (1816).

(5) Show that if $z = \cos x$, then

$$\frac{d^{i-1}(1 - z^2)^{\frac{2i-1}{2}}}{dz^{i-1}} = (-1)^{i-1}3\cdot 5\cdots(2i - 1)\frac{\sin ix}{i};$$

$$\frac{d^i(1 - z^2)^{\frac{2i-1}{2}}}{dz^i}dz = (-1)^{i-1}3\cdot 5\cdots(2i - 1)\cos ix\,dx.$$

See Jacobi (1969) vol. 6, pp. 90–91.

## 24.14　Notes on the Literature

Goldstine's (1977) very interesting book on the history of numerical analysis gives a thorough and readable account of Gauss's work on numerical integration; see pp. 224–232. *Liouville's Journal*, founded in 1836 and the second-oldest mathematics journal after *Crelle's Journal*, gave Liouville the opportunity to review many papers before they appeared in print, and then react to them. For example, the paper of Ivory and Jacobi stimulated Liouville to write his two short notes (1837a) and (1837b) in the same volume of his journal. Altmann and Ortiz (2005) is completely devoted to the work of Rodrigues in and outside of mathematics; see particularly the articles by Grattan-Guinness and Askey.

See N. I. Akhiezer's readable and insightful commentary on Chebyshev's work on continued fractions in Kolmogorov and Yushkevich (1998). Steffens (2006) gives a fairly comprehensive history of Chebyshev and his students' contributions to orthogonal polynomials and approximation theory. Nevai (1990) is a collection of interesting papers on orthogonal polynomials and their numerous applications. Roy (1993) gives more details on Chebyshev's work in orthogonal polynomials.

# Bibliography

Abel, N. 2007. *Abel on Analysis: Papers of N.H. Abel on Abelian and Elliptic Functions and the Theory of Series*. Heber City, UT: Kendrick Press. Translated by Phillip Horowitz from the second edition of Abel's *Oeuvres*.

Abel, N.H. 1826. Untersuchungen über die Reihe $1 + (m/1)x + (m(m - 1)/2)x^2 + \cdots$. *J. Reine Angew. Math.*, **1**, 311–339.

Abel, N.H. 1965. *Oeuvres complètes*. New York: Johnson Reprint. Edited by L. Sylow and S. Lie.

Abhyankar, S.S. 1976. Historical ramblings in algebraic geometry and related algebra. *Am. Math. Monthly*, **83**, 409–448.

Acosta, D.J. 2003. Newton's rule of signs for imaginary roots. *Am. Math. Monthly*, **110**, 694–706.

Ahlfors, L.V. 1982. *Collected Papers*. Boston: Birkhäuser. Asst. Editor: R.M. Shortt.

Ahlgren, S., and Ono, K. 2001. Addition and counting: The arithmetic of partitions. *Notices A.M.S.*, **48**, 978–984.

Alder, H.L. 1969. Partition identities — from Euler to the present. *Am. Math. Monthly*, **76**, 733–746.

Alexander, J.W. 1915. Functions which map the interior of the unit circle upon simple regions. *Ann. of Math.*, **17**, 12–22.

Allaire, P., and Bradley, R.E. 2004. Symbolical algebra as a foundation for calculus: D.F. Gregory's contribution. *Hist. Math.*, **29**, 395–426.

Almkvist, G., and Berndt, B. 1988. Gauss, Landen, Ramanujan, the arithmetic-geometric mean, ellipses, $\pi$, and the Ladies Diary. *Am. Math. Monthly*, **95**, 585–607.

Altmann, S., and Ortiz, E.L. 2005. *Olinde Rodrigues and His Times*. Providence: A.M.S.

Anderson, G.W. 1991. A short proof of Selberg's generalized beta formula. *Forum. Math.*, **3**, 415–417.

Anderson, M., Katz, V., and Wilson, R. (eds). 2004. *Sherlock Holmes in Babylon*. Washington, D.C.: M.A.A.

Anderson, M., Katz, V., and Wilson, R. (eds). 2009. *Who Gave You the Epsilon?* Washington, D.C.: M.A.A.

Andrews, G. 1981. Ramanujan's "lost" notebook. III. The Rogers-Ramanujan continued fraction. *Adv. Math.*, **41**, 186–208.

Andrews, G. 1986a. Eureka! num $= \Delta + \Delta + \Delta$. *J. Num. Theory*, **23**, 285–293.

Andrews, G. 1986b. *q-Series: Their Development and Application in Analysis, Number Theory, Combinatorics, Physics, and Computer Algebra*. Providence: A.M.S.

Andrews, G. 1998. *The Theory of Partitions*. Cambridge University Press.

Andrews, G., and Garvan, F. 1988. Dyson's crank of a partition. *Bull. A.M.S.*, **18**, 167–171.

Andrews, G., Askey, R., Berndt, B., Ramanathan, K., and Rankin, R. (eds). 1988. *Ramanujan Revisited*. Boston: Acad. Press

Andrews, G., Askey, R., and Roy, R. 1999. *Special Functions*. Cambridge: Cambridge University Press.

Aomoto, K. 1987. Jacobi polynomials associated with Selberg's integral. *SIAM J. Math. Phys.*, **18**, 545–549.

Arakawa, T., Ibukiyama, T., and Kaneko, M. 2014. *Bernoulli Numbers and Zeta Functions*. New York: Springer.

Arbogast, L. 1800. *Du calcul des dérivations*. Strasbourg: Levrault.

Archimedes, and Heath, T.L. 1953. *The Works of Archimedes*. New York: Dover. Translated with commentary by T.L. Heath; originally published in 1897.

Arnold, V.I. 1990. *Huygens and Barrow, Newton and Hooke*. Boston: Birkhäuser. Translated by E.J.F. Primrose.

Arnold, V.I. 2007. *Yesterday and Long Ago*. New York: Springer.

Artin, E. 1964. *The Gamma Function*. New York: Holt, Reinhart and Winston. Translated by Michael Butler.

Ash, J.M. 1976. *Studies in Harmonic Analysis*. Washington, D.C.: M.A.A.

Askey, R. 1975. *Orthogonal Polynomials and Special Functions*. Philadelphia: SIAM.

Askey, R., and Gasper, G. 1976. Positive Jacobi polynomial sums, II. *Amer. J. Math*, **98**, 709–737.

Askey, R., and Ismail, M. 1980. The Rogers $q$-ultraspherical polynomials. Pages 175–182 of: *Approximation Theory III*. New York: Acad. Press. Edited by E. W. Cheney.

Askey, R., and Ismail, M. 1983. A generalization of untraspherical polynomials. Pages 55–78 of: *Studies in Pure Mathematics, to the Memory of Paul Turán*. Basel: Birkhäuser. Edited by P. Erdös.

Askey, R., and Wilson, J. 1985. *Some Basic Hypergeometric Orthogonal Polynomials That Generalize Jacobi Polynomials*. Providence: Memoirs of the A.M.S.

Atkin, A.O.L. 1967. Proof of a Conjecture of Ramanujan. *Glasgow Math. J.*, **8**, 14–32.

Atkin, A.O.L. 1968. Multiplicative congruence properties. *Proc. London Math. Soc.*, **18**, 563–576.

Atkin, A.O.L., and Lehner, J. 1970. Hecke Operators on $\Gamma_0(m)$. *Math. Ann.*, **185**, 134–160.

Atkin, A.O.L., and Swinnerton-Dyer, P. 1954. Some properities of partititons. *Proc. London Math. Soc.*, **4**, 84–106.

Babbage, C., and Herschel, J. 1813. *Memoirs of the Analytical Society*. Cambridge: Cambridge University Press.

Bäcklund, R. 1918. Über die Nullstellen der Riemannschen Zetafunktion. *Acta Math.*, **41**, 345–375.

Baernstein, A. (ed). 1986. *The Bieberbach Conjecture*. Providence: A.M.S.

Bag, A.K. 1966. Trigonometrical series in the Karanapaddhati and the probable date of the text. *Indian J. Hist. of Sci.*, **1**, 98–106.

Baillaud, B., and Bourget, H. (eds). 1905. *Correspondance d'Hermite et de Stieltjes*. Paris: Gauthier-Villars.

Baker, A. 1988. *New Advances in Transcendence Theory*. New York: Cambridge University Press.

Baker, A., and Masser, D.W. 1977. *Transcendence Theory*. New York: Academic Press.

Barnes, E.W. 1908. A new development of the theory of the hypergeometric function. *Proc. London Math. Soc.*, **6**(2), 141–177.

Baron, M.E. 1987. *The Origins of the Infinitesimal Calculus*. New York: Dover.

Barrow, I. 1735. *Geometrical Lectures*. London: Austen. Translated by E. Stone.

Barrow, I. 1916. *Geometrical Lectures of Isaac Barrow*. Chicago: Open Court. Translated with comprehensive introduction by J.M. Child.

Bateman, H. 1907. The correspondence of Brook Taylor. *Bibliotheca Math.*, **7**, 367–371.

Bateman, P.T., and Diamond, H.G. 1996. A hundred years of prime numbers. *Am. Math. Monthly*, **103**(9), 729–741. Reprinted in Anderson, Katz, and Wilson (2009), pp. 328–336.

Baxter, R.J. 1980. Hard hexagons: exact solution. *J. Phys.*, **A 13**, L61–L70. Letter to the editor.

Becher, H.W. 1980. Woodhouse, Babbage, Peacock, and modern algebra. *Hist. Math.*, **7**, 389–400.

Bell, E.T. 1937. *Men of Mathematics*. New York: Simon and Schuster.

Berggren, L., Borwein, J., and Borwein, P. (eds). 1997. *Pi: A Source Book*. New York: Springer-Verlag.

Berndt, B. 1985–1998. *Ramanujan's Notebooks*. New York: Springer-Verlag.

Berndt, B. 1998. *Gauss and Jacobi sums*. New York: Wiley.

Berndt, B., and Ono, K. 2001. Ramanujan's Unpublished Manuscript on the Partition and Tau Functions with Proof and Commentary. Pages 39–110 of: Foata, D., and Han, G.-N. (eds), *The Andrews Festschrift*. New York: Springer.

Bernoulli, D. 1753a. Réflexions et éclairissements sur les nouvelles vibrations des cordes exposées dans les Mémoires de l'Académie de 1747 et 1748. *Hist. Acad. Sci. Berlin*, **9**, 147–172.

Bernoulli, D. 1753b. Sur le mélange de plusieurs especes de vibrations simples isochrones, qui peuvent coexister dans un même système de corps. *Hist. Acad. Sci. Berlin*, **9**, 173–195.

Bernoulli, D. 1982–1996. *Die Werke von Daniel Bernoulli*. Basel: Birkhäuser.

Bernoulli, Ja. 1744. *Jacobi Bernoulli Basileensis, Opera*. Geneva: Cramer and Philibert.

Bernoulli, Ja. 1993–1999. *Die Werke von Jakob Bernoulli*. Basel: Birkhäuser.

Bernoulli, Ja., and Sylla, E.D. 2006. *The Art of Conjecturing, Translation of Ars Conjectandi*. Baltimore: Johns Hopkins University Press. Translated with comprehensive introduction by E.D. Sylla.

Bernoulli, Joh. 1696. Curvatura radii in diaphanis non uniformiformibus. *Acta Erud.*, **16**, 206–211. Reprinted in Bernoulli (1968) vol. 1, pp. 187–193.

Bernoulli, Joh. 1742. *Opera Omnia*. Lausanne; Geneva: Bousquet.

Bernoulli, Joh. 1968. *Opera omnia*. Hildesheim, Germany: G. Olms Verlag.

Bernoulli, N. 1738. Inquisitio in summam seriei $1 + \frac{1}{4} + \frac{1}{9} + \frac{1}{16} + \frac{1}{25} + \frac{1}{36} +$ etc. *Comment. Petropolitanae*, **10**, 19–21.

Bers, L. 1998. *Selected Works of Lipman Bers*. Providence: A.M.S. Edited by I. Kra and B. Maskit.

Bertrand, J. 1845. Mémoire sur le nombre de valeurs que peut prendre une fonction quand on y permute les lettres qu'elle renferme. *J. École Poly.*, **18**, 123–140.

Beukers, F., Brownawell, W.D., and Heckman, G. 1988. Siegel normality. *Ann. of Math.*, **127**, 279–308.

Bézout, É. 2006. *General Theory of Algebraic Equations*. Princeton: Princeton University Press. Translated by E. Feron.

Bhaskaracharya. 2018. *Translation of the Surya Siddhanta and of the Siddhanta Siromani*. London: Forgotten Books. First published 1861. Translated with commentary by B.D. Sastri and L. Wilkinson.

Bieberbach, L. 1916. Über die Koeffizienten derjenigen Potenzreihen, ... *S.B. Preuss. Akad. Wiss.*, **138**, 940–955.

Binet, J. 1839. Mémoire sur les intégrales définies Eulériennes. *J. École Poly.*, **16**, 123–343.

Bissell, C. C. 1989. Cartesian geometry: The Dutch contribution. *Math. Intelligencer*, **9**(4), 38–44.

Boas, R.P. 1954. *Entire Functions*. New York: Academic Press.

Boehle, K. 1933. Über die Transzendenz von Potenzen mit algebraischen Exponenten. *Math. Ann.*, **108**, 56–74.

Bogolyubov, N.N., Mikhaĭlov, G.K., and Yushkevich, A.P. (eds). 2007. *Euler and Modern Science*. Washington, D.C.: M.A.A.

Bohr, H., and Mollerup, J. 1922. *Laerebog i Matematisk Analyse*. Copenhagen: Jul. Gjellerups Forlag.

Bolibruch, A.A., Osipov, Yu.S., and Sinai, Ya.G. (eds). 2006. *Mathematical Events of the Twentieth Century*. New York: Springer.

Bolzano, B. 1930. *Functionenlehre*. Prague: Roy. Bohemian Acad. Sci. Edited by K. Rychlik.

Bolzano, B. 1980. A translation of Bolzano's paper on the intermediate value theorem. *Hist. Math.*, **7**, 156–185. Translated by S.B. Russ.

Bombieri, E., and Gubler, W. 2006. *Heights in Diophantine Geometry*. New York: Cambridge University Press.

Boole, G. 1839. Researches in the theory of analytical transformations, with a special application to the reduction of the general equation of the second order. *Cambridge Math. J.*, **2**, 34–78.

Boole, G. 1841. Exposition of a general theory of linear transformations, Parts I and II. *Cambridge Math. J.*, **3**, 1–20, 106–111.

Boole, G. 1844a. Notes on linear transformations. *Cambridge Math. J.*, **4**, 167–71.

Boole, G. 1844b. On a general method in analysis. *Phil. Trans. Roy. Soc. London*, **124**, 225–282.

Boole, G. 1845. Notes on linear transformations. *Cambridge Math. J.*, **6**, 106–113.

Boole, G. 1847. *The Mathematical Analysis of Logic*. London: George Bell.

Boole, G. 1877. *A Treatise on Differential Equations*. London: Macmillan.

Borel, É. 1896. Démonstration élémenatire d'un théorème de M. Picard sur les fonctions entières. *Comptes Rendus*, **122**, 1045–1048.

Borel, É. 1900. *Leçons sur les fonctions entières*. Paris: Gauthier-Villars.

Bornstein, M. 1997. *Symbolic Integration*. New York: Springer-Verlag.

Boros, G., and Moll, V. 2004. *Irresistible Integrals*. New York: Cambridge University Press.

Borwein, J., Bailey, D., and Girgensohn, R. 2004. *Experimentation in Mathematics: Computational Paths to Discovery*. Natick, MA: Peters.

Bos, H.J.M. 1974. Differentials, higher-order differentials and the derivative in the Leibnizian calculus. *Archive Hist. Exact Sci.*, **14**, 1–90.

Bos, H.J.M. 1996. Johann Bernoulli on Exponential Curves. *Nieuw Archief Wisk.*, **14**, 1–19.

Bottazzini, U. 1986. *The Higher Calculus*. New York: Springer-Verlag. Translated by W. Van Egmond.

Bourbaki, N. 1994. *Elements of the History of Mathematics*. New York: Springer-Verlag. Translated by J. Meldrum.

Boyer, C.B. 1943. Pascal's formula for the sums of the powers of integers. *Scripta Math.*, **9**, 237–244.

Boyer, C.B., and Merzbach, U.C. 1991. *A History of Mathematics*. New York: Wiley.

Bradley, R.E., and Sandifer, C.E. (eds) 2007. *Leonhard Euler: Life, Work and Legacy*. Amsterdam: Elsevier.

Bradley, R.E., and Sandifer, C.E. (eds) 2009. *Cauchy's Cours d'analyse*. New York: Springer. Translated with commentary by Bradley and Sandifer.

Brahmagupta. 1817. *Algebra, with arithmetic and mensuration, from the Sanscrit of Brahmegupta and Bháscara*. London: Murray. Translated with notes by H.T. Colebrooke.

Bressoud, D. 2002. Was calculus invented in India? *College Math. J.*, **33**(1), 2–13. Reprinted in Anderson, Katz, and Wilson (2004), 131–137.

Bressoud, D. 2007. *A Radical Approach to Real Analysis, second edition*. Washington, D.C.: M.A.A.

Bressoud, D. 2008. *A Radical Approach to Lesbesgue's Theory of Integration*. Cambridge: Cambridge University Press.

Bressoud, D., and Zeilberger, D. 1982. A short Rogers–Ramanunan bijection. *Discrete Math.*, **38**, 313–315.

Brezinski, C. 1991. *History of Continued Fractions and Padé Approximations*. New York: Springer.

Briggs, H. 1624. *Arithmetica Logarithmica*. London: W. Jones.

Briggs, H. 1633. *Trigonometria Britannica*. Gouda: Pierre Rammasen.

Brinkley, J. 1807. An investigation of the general term of an important series in the inverse method of finite differences. *Phil. Trans.*, **97**, 114–132.

Briot, C., and Bouquet, J.C. 1856. Recherches sur les propriétés des fonctions définies par des équations différentielles. *J. École Poly.*, **t. 21, cahier 36**, 133–198.

Briot, C., and Bouquet, J.C. 1859. *Théorie des fonctions doublement périodiques et, en particulier, des fonctions elliptiques*. Paris: Mallet-Bachelier.

Bronstein, M. 1997. *Symbolic Integration*. Heidelberg: Springer-Verlag.

Bronwin, B. 1849. On the determination of the coefficients in any series of sines and cosines of multiples of a variable angle from particular values of that series. *Phil. Magazine*, **34**, 260–268.

Brouncker, W. 1668. The squaring of the hyperbola, by an infinite series of rational numbers, together with its demonstration. *Phil. Trans.*, **3**, 645–649.

Browder, F.E. (ed). 1976. *Mathematical Developments Arising from Hilbert Problems*. Providence: A.M.S.

Buchler, J. 1955. *The Philosophy of Peirce: Selected Writings*. New York: Dover.

Budan de Boislaurent, F. 1822. *Nouvelle méthode pour la résolution des équations numériques ...* Paris: Dondey-Dupré.

Bühler, W.K. 1981. *Gauss: A Biographical Study*. New York: Springer-Verlag.

Bunyakovski, V. 1859. Sur quelques inégalités concernant les intégrales ordinaires et les intégrales aux différences finies. *Mém. de Acad. Sci. St.-Pétersbourg*, **1**, 1–18.

Burn, R.P. 2001. Alphose Antonio de Sarasa and logarithms. *Hist. Math.*, **28**, 1–17.

Burnside, W.S., and Panton, A.W. 1960. *The Theory of Equations*. New York: Dover.

Butzer, P.L., and Sz.-Nagy, B. 1974. *Linear Operators and Approximation II*. Basel: Birkhäuser.

Cahen, E. 1894. Sur la fonction $\zeta(s)$ de Riemann et sur des fonctions analogues. *Ann. Sci. École Norm. Sup.*, **11**, 75–164.

Cajori, F. 1913. *A History of Mathematics*. New York: Macmillan.

Cajori, F. 1993. *A history of mathematical notations*. New York: Dover. Two volumes bound as one.

Campbell, G. 1728. A method of determining the number of impossible roots in affected aequations. *Phil. Trans. Roy. Soc.*, **35**, 515–531.

Campbell, P.J. 1978. The origin of "Zorn's Lemma" *Hist. Math.*, **5**, 77–89.

Cannon, J.T., and Dostrovsky, S. 1981. *The Evolution of Dynamics: Vibration Theory from 1687 to 1742*. New York: Springer-Verlag.

Cantor, G. 1870a. Beweis, dass eine für jeden reellen Werth von $x$ durch eine trigonometrische Reihe gegebene Function $f(x)$ sich nur auf eine einzige Weise in dieser Form darstellen lässt. *J. Reine Angew. Math*, **72**, 139–142.

Cantor, G. 1870b. Über einen die trigonometrischen Reihen betreffenden Lehrsatz. *J. Reine Angew. Math*, **72**, 130–138. Reprinted in Cantor (1932) pp. 71–79.

Cantor, G. 1871. Notiz zu dem vorangehenden Aufsatze. *J. Reines Angew. Math*, **73**, 294–296.

Cantor, G. 1872. Über die Ausdehnung eines Satzes aus der Theorie der trigonometrischen Reihen. *Math. Ann.*, **5**, 123–132.

Cantor, G. 1932. *Gesammelte Abhandlungen*. Berlin: Springer. Edited by E. Zermelo.

Carathéodory, C. 1912. Untersuchungen über die konformen Abbildungen von festen und veränderlichen Gebieten. *Math. Ann.*, **72**, 107–144.

Cardano, G. 1993. *Ars Magna or the Rules of Algebra*. New York: Dover. Translated by T.R. Witmer.

Carleson, L. 1966. On convergence and growth of partial sums of Fourier series. *Acta. Math.*, **116**, 135–157.

Cartier, P. 2000. Mathemagics. Pages 6–67 of: Planat, M. (ed), *Lecture Notes in Physics, vol. 550*. Berlin: Springer.

Cauchy, A.-L. 1823. *Résumé des leçons données à l'École Royale Polytechnique sur le calcul infinitésimal*. Paris: De Bure.

Cauchy, A.-L. 1827. Mémoire sur les intégrales définies. *Mém. Acad. Roy. Sci.*, **1**, 601–799. First presented to the Academy in 1814. Reprinted in Cauchy's *Oeuvres complètes* (1), vol. 1, pp. 329–506.

Cauchy, A.-L. 1829. *Calcul différentiel*. Paris: De Bure.

Cauchy, A.-L. 1840–1841. *Exercices d'analyse et de physique mathématique*. Paris: Bachelier.

Cauchy, A.-L. 1843a. Mémoire sur les fonctions dont plusieurs valuers sont liées entre elles par une équation linéaire, et sur diverses ... *Comptes Rendus*, **17**, 523–531. Reprinted in Cauchy (1882–1974), sér. 1, vol. 8, pp. 42–50.

Cauchy, A.-L. 1843b. Sur l'emploi légitime des séries divergentes. *Comptes Rendus*, **17**, 370–376.

Cauchy, A.-L. 1853. Note sur les séries convergentes dont les divers termes sont des fonctions continues ... *Comptes Rendus*, **36**, 454–459.

Cauchy, A.-L. 1882–1974. *Oeuvres complètes*. Paris: Gauthier-Villars.

Cauchy, A.-L. 1989. *Analyse algébrique*. Paris: Gabay.

Cayley, A. 1843. On the theory of determinants. *Trans. Cambridge Phil. Soc.*, **8**, 1–16. Reprinted in Cayley (1889–1898) vol. 1, pp. 63–79.

Cayley, A. 1845a. Mémoire sur les fonctions doublement périodiques. *J. Math. Pures Appl.*, **10**, 385–410.

Cayley, A. 1845b. On the theory of linear transformations. *Cambridge Math. J.*, **1**, 193–209. Reprinted in Cayley (1889–1898) vol. 1, pp. 80–94.

Cayley, A. 1846. On linear transformations. *Cambridge and Dublin Math. J.*, **1**, 104–122. Reprinted in Cayley (1889–1898) vol. 1, pp. 95–112.

Cayley, A. 1854. An introductory memoir upon quantics. *Phil. Trans. Roy. Soc. London*, **144**, 244–258. Reprinted in Cayley (1889–1898) vol. 2, pp. 221–234.

Cayley, A. 1855. A second memoir upon quantics. *Phil. Trans. Roy. Soc. London*, **146**, 101–126. Reprinted in Cayley (1889–1898) vol. 2, pp. 250–275.

Cayley, A. 1856. A third memoir upon quantics. *Phil. Trans. Roy. Soc. London*, **146**, 627–647.

Cayley, A. 1889–1898. *Collected Mathematical Papers*. Cambridge: Cambridge University Press.

Cayley, A. 1895. *Elliptic Functions*. London: Bell.

Cesàro, E. 1890. Sur la multiplication des séries. *Bull. Sci. Math.*, **14**, 114–20.

Chabert, J.-L. 1999. *A History of Algorithms: From the Pebble to the Microchip*. New York: Springer. Translated by C. Weeks.

Chakravarti, G. 1932. Growth and development of permutations and combinations in India. *Bull. Calcutta Math. Soc.*, **24**, 79–88.

Chandrasekhar, S. 1995. *Newton's Principia for the Common Reader*. Oxford: Oxford University Press.

Charzynski, Z., and Schiffer, M. 1960. A new proof of the Bieberbach conjecture for the fourth coefficient. *Arch. Rational Mech. Anal.*, **5**, 187–193.

Chebyshev, P. 1848. Sur la fonction qui détermine la totalité des nombres premiers inférieurs à une limite donnée. *Mém. savants étrangers Acad. Sci. St. Petersbourg*, **6**, 1–19.

Chebyshev, P. 1850. Mémoire sur nombres premiers. *Mém. savants étrangers Acad. Sci. St. Petersbourg*, **1**, 17–33.

Chebyshev, P. 1858. Sur une nouvelle série. *Bull. Phys. Math. Acad. St. Petersburg*, **17**, 257–261.

Chebyshev, P. L. 1899–1907. *Oeuvres de P.L. Tchebychef*. St. Petersburg: Acad. Impériale Sci.

Cheney, E. (ed). 1980. *Approximation Theory III*. New York: Academic Press.

Cherry, W., and Ye, Z. 2001. *Nevanlinna Theory of Value Distribution*. New York: Springer.

Chowla, S. 1934. Congruence properties of partitions. *J. London Math. Soc.*, **9**, 247.

Christoffel, E.B. 1867. Sul problema delle temperature stazionarie e la rappresentazione di una superficie. *Annali Mat. Pura Appl.*, **1**, 89–103. Translated by C. Formenti.

Chudnovsky, D.V., and Chudnovsky, G.V. 1988. Approximations and complex multiplication according to Ramanujan. Pages 375–472 of: Andrews, G., Askey, R., Berndt, B., Ramanathan, K.G., and Rankin, R. (eds), *Ramanujan Revisited: Proceedings of the Ramanujan Centenary Conference held at the University of Illinois, Urbana-Champaign, Illinois, June 1–5, 1987*. Boston: Academic Press.

Clairaut, A.-C. 1734. Solution de plusieurs problèmes où il s'agit de trouver des courbes ... *Hist. Acad. Roy. Sci.*, 196–215.

Clairaut, A.-C. 1739. Recherches générales sur le calcul intégral. *Mém. Acad. Roy. Sci.*, **1**, 425–436.

Clairaut, A.-C. 1740. Sur l'intégration ou la construction des équations différentielles du premier ordre. *Mém. Acad. Roy. Sci.*, **2**, 293–323.

Clairaut, A.-C. 1754. Sur l'oribit apparente du Soleil autour de la terre, ... *Mém Hist. Acad. Sci. Paris*, **9**, 521–565.

Clarke, F. M. 1929. *Thomas Simpson and his Times*. Baltimore: Waverly Press.

Clausen, T. 1828. Ueber die Fälle, wenn die Reihe von der Form $y = 1 + \cdots$ etc. ein Quadrat von der Form $z = 1 + \cdots$ etc. hat. *J. Reine Angew. Math.*, **3**, 89–91.

Clausen, T. 1832. Über die Function $\sin\phi + \frac{1}{2^2}\sin 2\phi + \frac{1}{3^2}\sin 3\phi +$ etc. *J. Reine Angew. Math.*, **8**, 298–300.

Clausen, T. 1840. Theorem. *Astronom. Nach.*, **17**, 351–352.

Clausen, T. 1858. Beweiss des von Schlömilch ... *Archive Math. Phys.*, **30**, 166–169.

Cohen, H. 2007. *Number Theory, Volume II: Analytic and Modern Tools*. New York: Springer.

Cooke, R. 1984. *The Mathematics of Sonya Kovalevskaya*. New York: Springer-Verlag.

Cooke, R. 1993. Uniqueness of trigonometric series and descriptive set theory 1870–1985. *Archive Hist. Exact Sci.*, **45**, 281–334.

Cooper, S. 2006. The quintuple product identity. *Int. J. Number Theory*, **2**, 115–161.

Corry, L. 2004. *Modern Algebra and the Rise of Mathematical Structures*. Basel: Birkhäuser.

Cotes, R. 1714. Logometria. *Phil. Trans.*, **29**, 5–45.

Cotes, R. 1722. *Harmonia Mensurarum*. Cambridge: Cambridge University Press. Edited by Robert Smith.

Cox, D.A. 1984. The arithmetic-geometric mean of Gauss. *L'enseignement math.*, **30**, 275–330.

Cox, D.A. 2004. *Galois Theory*. Hoboken: Wiley.

Craig, J. 1685. *Methodus figurarum lineis rectis et curvis*. London: Pitt.

Craik, A.D.D. 2000. James Ivory, F.R.S., mathematician: "The most unlucky person that ever existed." *Notes and Records Roy. Soc. London*, **54**, 223–247.

Craik, A.D.D. 2005. Prehistory of Faà di Bruno's formula. *Am. Math. Monthly*, **112**, 119–130.

Crilly, T. 2006. *Arthur Cayley*. Baltimore: Johns Hopkins University Press.

d'Alembert, J. 1747. Recherches sur la courbe que forme une corde tenduë mise en vibration. *Hist. Acad. Roy. Sci. Belles-Lettres, Berlin*, **3**, 214–219, 220–249.

d'Alembert, J. 1761–1780. *Opuscules mathématiques*. Paris: David.

Darboux, G. 1875. Mémoire sur les fonctions discontinues. *Ann. École Norm. Sup.*, **4**, 57–112.

Datta, B., and Singh, A.N. 1962. *History of Hindu Mathematics*. Bombay: Asia Pub. House.

Datta, B.B. 1929. The Jaina school of mathematics. *Calcutta Math. Soc.*, **21**, 115–145.

Dauben, J. 1979. *Georg Cantor*. Princeton: Princeton University Press.

Davenport, H. 1980. *Multiplicative Number Theory*. New York: Springer-Verlag.

Davenport, H., and Hasse, H. 1935. Die Nullstellen der Kongruenz-zetafunktionen in gewissen zyklischen Fällen. *J. Reine Angew. Math.*, **172**, 151–182.

Davis, P.J. 1959. Leonhard Euler's Integral: A historical profile of the gamma function. *Am. Math. Monthly*, **66**, 849–869.

de Beaune, F., Girard, A., and Viète, F. 1986. *The Early Theory of Equations: On Their Nature and Constitution*. Annapolis, MD: Golden Hind Press. Translated by R. Schmidt.

de Branges, L. 1985. A proof of the Bieberbach conjecture. *Acta Math.*, **157**, 137–162.

de Moivre, A. 1697. A method of raising an infinite multinomial to any given power, or extracting any given root of the same. *Phil. Trans.*, **19**, 619–625.

de Moivre, A. 1698. A method of extracting the root of an infinite equation. *Phil. Trans.*, **20**, 190–193.

de Moivre, A. 1707. Aequationum quarundam potestatis tertiae, ... *Phil. Trans.*, **25**, 2368–2371.

de Moivre, A. 1730a. *Miscellanea analytica de seriebus et quadraturis*. London: Tonson and Watts.

de Moivre, A. 1730b. *Miscellaneis analyticis supplementum*. London: Tonson and Watts.

de Moivre, A. 1967. *The Doctrine of Chances*. New York: Chelsea.

Dedekind, R. 1857. Abriss einer Theorie der höheren Kongruenzen in Bezug auf einen reellen Primzahl-Modulus. *J. Reine Angew. Math.*, **54**, 1–26. Reprinted in Dedekind (1930) vol. 1, pp. 40–67.

Dedekind, R. 1872. *Stetigkeit and irrationale Zahlen*. Braunschweig: Vieweg.

Dedekind, R. 1877. Schreiben an Herr Borchardt über die Theorie der elliptischen Modulfunktionen. *J. Reine Angew. Math.*, **83**, 265–292. Reprinted in Dedekind's *Werke* (1930) vol. 1 pp. 174–201.

Dedekind, R. 1930. *Gesammelte Mathematische Werke*. Braunschweig: F. Vieweg. Edited by R. Fricke, E. Noether and Ø. Ore.

Dedekind, R. 1963. *Essays on the Theory of Numbers*. New York: Dover. Translated by W.W. Beman.

Dedekind, R., and Weber, H. 2012. *Theory of Algebraic Functions of One Variable*. Providence: A.M.S. Translated by J. Stillwell.

Delone, B.N. 2005. *The St. Petersburg School of Number Theory*. Providence: A.M.S.

Descartes, R. 1897–1913. *Oeuvres*. Paris: Léopold Cerf. edited by C. Adam and P. Tannery.

Descartes, R. 1954. *La Géométrie*. New York: Dover. Translated by D.E. Smith and M.L. Latham.

Dieudonné, J. 1931. Sur les fonctions univalentes. *Comptes Rendus*, **192**, 1148–1150.

Dieudonné, J. 1981. *History of Functional Analysis*. Amsterdam: Elsevier.

Dini, U. 1878. *Fondamenti per la teorica delle funzione de variabili reali.* Pisa: Nistri.

Dirichlet, P.G.L. 1829a. Note sur les intégrales définies. *J. Reine Angew. Math,* **4**, 94–98.

Dirichlet, P.G.L. 1829b. Sur la convergence des séries trigonométriques qui servent à représenter une fonction arbitraire entres des limites données. *J. Reine Angew. Math.,* **4**, 157–169. Reprinted in Dirichlet's *Werke* vol. 1, pp. 117–132.

Dirichlet, P.G.L. 1837. Über die Darstellung ganz willkürlicher Functionen durch Sinus- und Cosinusreihen. *Reper. der Physik,* **1**, 152–174. Reprinted in Dirichlet's *Werke,* vol. 1, pp. 133–160.

Dirichlet, P.G.L. 1839a. Sur une nouvelle méthode pour la détermination des intégrales multiples. *J. Math. Pures Appl.,* **4**, 164–168. Reprinted in Dirichlet's *Werke* vol. 1, pp. 377–380.

Dirichlet, P.G.L. 1839b. Ueber eine neue Methode zur Bestimmung vielfacher Integrale. *Akad. Wiss. Berlin von 1839,* 18–25. Reprinted in Dirichlet's *Werke* vol. 1, pp. 381–390; see pp. 393–410 for later expanded version from 1841.

Dirichlet, P.G.L. 1840. Recherches sur diverses applications de l'analyse infinitésimale à la théorie des nombres. *J. Reine Angew. Math.,* **19, 21**, (19): 324–369, (21): 1–12, 134–155. Reprinted in Dirichlet's *Werke* vol. 1, pp. 411–496.

Dirichlet, P.G.L. 1863. Démonstration d'un théorème d'Abel. *J. Math. Pures App.,* **7**(2), 253–255. Also in *Werke* vol. I, pp. 305–306.

Dirichlet, P.G.L. 1969. *Mathematische Werke.* New York: Chelsea.

Dirichlet, P.G.L., and Dedekind, R. 1999. *Lectures on Number Theory.* Providence: A.M.S. Translated by John Stillwell.

Dörrie, H. 1965. *100 Great Problems of Elementary Mathematics.* New York: Dover. Translated by D. Antin.

Douglas, R.G. 1972. *Banach Algebra Techniques in Operator Theory.* New York: Academic Press.

Doxiadis, A., and Mazur, B. (eds). 2012. *Circles Disturbed.* Princeton: Princeton University Press.

Dugac, P. 1973. Éléments d'analyse de Karl Weierstrass. *Archive Hist. Exact. Sci.,* **10**, 41–176.

Duke, W., and Tschinkel, Y. (eds). 2005. *Analytic Number Theory: A Tribute to Gauss and Dirichlet.* Providence: A.M.S.

Dunham, W. 1990. *Journey Through Genius.* New York: Wiley.

Dunnington, G. 2004. *Gauss: Titan of Science.* Washington, D.C.: M.A.A.

Dupré, A. 1859. *Examen d'une proposition de Legendre relative à la théorie des nombres.* Paris: MalletBachelier.

Duren, P.L. 1983. *Univalent Functions.* New York: Springer-Verlag.

Dutka, J. 1984. The early history of the hypergeometric series. *Archive Hist. Exact Sci.,* **31**, 15–34.

Dutka, J. 1991. The early history of the factorial function. *Archive Hist. Exact Sci.,* **43**, 225–249.

Dyson, F. 1944. Some guesses in the theory of partitions. *Eureka (Cambridge),* **8**, 10–15.

Dyson, F. 1962. Statistical theory of the energy levels of complex systems. *J. Math. Phys.,* **3**, 140–156.

Edwards, A.W.F. 1986. A quick route to sums of powers. *Am. Math. Monthly,* **93**, 451–455.

Edwards, A.W.F. 2002. *Pascal's Arithmetical Triangle.* Baltimore: Johns Hopkins University Press.

Edwards, H.M. 1977. *Fermat's Last Theorem.* Berlin: Springer-Verlag.

Edwards, H.M. 1984. *Galois Theory.* New York: Springer-Verlag.

Edwards, H.M. 2001. *Riemann's Zeta Function.* New York: Dover.

Edwards, J. 1954a. *An Elementary Treatise on the Differential Calculus.* London: Macmillan.

Edwards, J. 1954b. *Treatise on Integral Calculus.* New York: Chelsea.

Eie, M. 2009. *Topics In Number Theory.* Singapore: World Scientific.

Eisenstein, G. 1844. Beiträge zur Kreistheilung. *J. Reine Angew. Math.,* **27**, 269–278.

Eisenstein, G. 1975. *Mathematische Werke.* New York: Chelsea.

Elliott, E.B. 1964. *An Introduction to the Algebra of Quantics.* New York: Chelsea.

Ellis, D.B., Ellis, R., and Nerurkar, M. 2001. The topological dynamics of semigroup actions. *Trans. A.M.S.,* **353**, 1279–1320.

Ellis, R.L. 1845. Memoir to D.F. Gregory. *Cambridge and Dublin Math. J.,* **4**, 145–152.

Elstrodt, J. 2005. The Life and Work of Gustav Lejeune Dirichlet (1805–1859). In: Duke, W., and Tschinkel, Y. (eds), *Analytic Number Theory: A Tribute to Gauss and Dirichlet.* Providence: A.M.S.

Emch, G.G., Sirdhara, R., and Srinivas, M.D. (eds). 2005. *Contributions to the History of Indian Mathematics*. New Delhi: Hindustan Book Agency and Springer.

Eneström, G. 1897. Sur la découverte de l'intégrale complète des équations différentielles linéaires à coefficients constants. *Bib. Math.*, **11**, 43–50.

Eneström, G. 1905. Der Briefwechsel zwischen Leonhard Euler und Johann I Bernoulli. *Bib. Math.*, **6**, 16–87.

Engelsman, S.B. 1984. *Families of Curves and the Origins of Partial Differentiation*. Amsterdam: North-Holland.

Enros, P. 1983. The analytical society (1812–1813). *Hist. Math.*, **10**, 24–47.

Erdős, P. 1932. Beweis eines Satzes von Tschebyschef. *Acta. Sci. (Szeged)*, **5**, 194–198.

Erdős, P. 1949. On a new method in elementary number theory which leads to an elementary proof of the prime number theorem. *Proc. Nat. Acad. Sci. USA*, **35**, 374–384.

Erman, A. (ed). 1852. *Briefwechsel zwischen Olbers und Bessel*. Leipzig: Avenarius and Mendelssohn.

Euler, L. 1911–2000. *Leonhardi Euleri Opera omnia*. Berlin: Teubner.

Euler, L. 1985. An essay on continued fractions. *Math. Syst. Theory*, **18**, 295–328. Translated by M. Wyman and B. Wyman.

Euler, L. 1988. *Introduction to Analysis of the Infinite*. New York: Springer-Verlag. Translated by J.D. Blanton.

Euler, L. 2000. *Foundations of Differential Calculus. English Translation of First Part of Euler's Institutiones calculi differentialis*. New York: Springer-Verlag. Translated by J.D. Blanton.

Faà di Bruno, F. 1857. Note sur une nouvelle formule de calcul différentiel. *Quarterly J. Pure Appl. Math.*, **1**, 359–360.

Fagnano, G.C. 1911. *Opere matematiche*. Rome: Albrighi.

Farkas, H.M., and Kra, I. 2001. *Theta Constants, Riemann Surfaces and the Modular Group*. Providence: A.M.S.

Fatou, P. 1906. Séries trigonométriques et séries de Taylor. *Acta Math.*, **30**, 335–400.

Favard, J. 1935. Sur les polynômes de Tchebicheff. *Comptes Rendus*, **200**, 2052–2053.

Feigenbaum, L. 1985. Taylor and the method of increments. *Archive Hist. Exact Sci.*, **34**, 1–140.

Feingold, M. 1990. *Before Newton*. Cambridge: Cambridge University Press.

Feingold, M. 1993. Newton, Leibniz and Barrow too. *Isis*, **84**, 310–338.

Fejér, L. 1900. Sur les fonctions bornées et intégrables. *Comptes Rendus*, **131**, 984–987.

Fejér, L. 1904. Untersuchungen über Fouriersche Reihen. *Math. Ann.*, **58**, 51–69.

Fejér, L. 1908. Sur le développement d'une fonction arbitraire suivant les fonctions de Laplace. *Comptes Rendus*, **146**, 224–227.

Fejér, L. 1970. *Gesammelte Arbeiten*. Basel: Birkhäuser. Edited by Paul Turán.

Fekete, M., and Szegő, G. 1933. Eine Bemerkung über ungerade schlichte Funktionen. *J. London Math. Soc.*, **8**, 85–89.

Feldheim, E. 1941. Sur les polynômes généralisés de Legendre. *Bull. Acad. Sci. URSS. Ser. Math. [Izvestia Acad. Nauk SSSR]*, **5**, 241–254.

Ferraro, G. 2004. Differentials and differential coefficients in the Eulerian foundations of the calculus. *Hist. Math.*, **31**, 34–61.

Ferreirós, J. 1993. On the relations between Georg Cantor and Richard Dedekind. *Hist. Math.*, **20**, 343–63.

FitzGerald, C.H. 1972. Quadratic inequalities and coefficient estimates for schlicht functions. *Arch. Rational Mech. Anal.*, **46**, 356–368.

FitzGerald, C.H. 1985. The Bieberbach conjecture: Retrospective. *Notices A.M.S.*, **32**, 2–6.

Foata, D., and Han, G.-N. 2001. The triple, quintuple and sextuple product identities revisited. Pages 323–334 of: Foata, D., and Han, G.-N. (eds), *The Andrews Festschrift: Seventeen Papers on Classical Number Theory and Combinatorics*. New York: Springer.

Fomenko, O.M., and Kuzmina, G.V. 1986. The last 100 days of the Bieberbach conjecture. *Math. Intelligencer*, **8**, 40–47.

Forrester, P.J., and Warnaar, S.O. 2008. The Importance of the Selberg Integral. *Bull. A.M.S.*, **45**, 498–534.

Fourier, J. 1955. *The Analytical Theory of Heat*. New York: Dover. Translated by A. Freeman.

Français, J.F. 1812–1813. Analise transcendante. Memoire tendant à démontrer la légitimité de la séparation des échelles de différentiation et d'intégration des fonctions qu'elles affectent. *Annales Gergonne*, **3**, 244–272.

Frei, G. 2007. The unpublished section eight: On the way to function fields over a finite field. Pages 159–198 of: Goldstein, C., Schappacher, N., and Schwermer, J. (eds), *The Shaping of Arithmetic after C.F. Gauss's Disquisitiones Arithmeticae*. New York: Springer.

Friedelmeyer, J.P. 1994. *Le calcul des dérivations d'Arbogast dans le projet d'algébrisation de l'analyse à fin du xviiie siècle*. Nantes, France: U. Nantes.

Frobenius, G. 1878. Über lineare Substitutionen und bilineare Formen. *J. Reine Angew. Math*, **84**, 1–63.

Frobenius, G. 1880. Ueber die Leibnizsche Reihe. *J. Reine Angew. Math.*, **89**, 262–264.

Fuss, P.H. (ed). 1968. *Correspondance mathématique et physique*. New York: Johnson Reprint.

Galois, É. 1830. Sur la théorie des nombres. *Bull. Sci. Math. Phys. Chem.*, **13**, 428–435.

Galois, É. 1897. *Oeuvres mathématiques*. Paris: Gauthier-Villars.

Garabedian, P., and Schiffer, M. 1955. A proof of the Bieberbach conjecture for the fourth coefficient. *J. Rational Mech. Anal.*, **4**, 427–465.

Gårding, L. 1994. *Mathematics in Sweden before 1950*. Providence: A.M.S.

Garsia, A.M., and Milne, S.C. 1981. A Rogers-Ramanujan bijection. *J. Combin. Theory*, **31**, 289–339.

Gauss, C.F. 1813. *Disquisitiones generales circa–seriem infinitam*. Gottingen: Comm. Soc. Reg. Gott: II. Reprinted in *Werke*, vol. 3, pp. 123–162.

Gauss, C.F. 1815. *Methodus nova integralium valores per approximationem inveniendi*. Göttingen: Dieterich. This monograph was presented to the Göttingen Society in 1814. It was reprinted in Gauss's *Werke* vol. 3, pp. 163–196.

Gauss, C.F. 1863–1927. *Werke*. Leipzig: Teubner.

Gauss, C.F. 1965. *Disquisitiones Arithmeticae (An English Translation)*. New Haven, Conn.: Yale University Press. Translated by A.A. Clarke.

Gauss, C.F. 1981. *Arithmetische Untersuchungen*. New York: Chelsea.

Gautschi, W. 1986. My involvement in de Branges's proof. Pages 205–211 of: Baernstein, A. (ed), *The Bieberbach Conjecture*. Providence: A.M.S.

Gegenbauer, L. 1884. Zur Theorie der Functionen $C_n^\nu(x)$. *Denkschriften Akad. Wiss. Wien, Math. Naturwiss. Klasse*, **48**, 293–316.

Gelfand, I.M. 1988. *Collected Papers*. New York: Springer-Verlag.

Gelfand, I.M., Kapranov, M.M., and Zelevinsky, A.V. 1994. *Discriminants, Resultants, and Multidimensional Determinants*. Boston: Birkhäuser.

Gelfond, A.O. 1934. Sur le septième problème de Hilbert. *Dok. Akad. Nauk SSSR*, **2**, 1–6.

Gelfond, A.O. 1939. In Russian: On the approximation by algebraic numbers of the ratio of the logarithms of two algebraic numbers. *Izvestia Akad. Nauk. SSSR*, **5–6**, 509–518.

Gelfond, A.O. 1960. *Transcendental and Algebraic Numbers*. New York: Dover. Translated by L. Boron.

Gelfond, A.O., and Linnik, Yu.V. 1966. *Elementary Methods in the Analytic Theory of Numbers*. Cambridge, MA: MIT Press. Translated by D.E. Brown.

Georgiadou, M. 2004. *Constantine Carathéodory*. New York: Springer-Verlag.

Gerhardt, K.I. (ed). 1899. *Der Briefwechsel von Gottfried Wilhelm Leibniz mit Mathematikern*. Berlin: Mahler and Müller.

Girard, A. 1884. *Invention nouvelle en l'algebre*. Leiden: Haan.

Glaisher, J.W.L. 1878. Series and products for $\pi$ and powers of $\pi$. *Messenger Math.*, **7**, 75–80.

Glaisher, J.W.L. 1883. A theorem in partitions. *Messenger Math.*, **12**, 158–170.

Goethe, N., Beeley, P., and Rabouin, D. (eds). 2015. *G.W. Leibniz, Interrelation between Mathematics and Philosophy*. Dordrecht: Springer.

Goldstein, C., Schappacher, N., and Schwermer, J. (eds). 2007. *The Shaping of Arithmetic after C.F. Gauss's Disquisitiones Arithmeticae*. New York: Springer.

Goldstein, L. J. 1973. A history of the prime number theorem. *Am. Math. Monthly*, **80**, 599–615. Correction, p. 1115. Reprinted in Anderson, Katz, and Wilson (2009), pp. 318–327.

Goldstine, H.H. 1977. *A History of Numerical Analysis*. New York: Springer-Verlag.

Gong, S. 1999. *The Bieberbach Conjecture*. Providence: A.M.S.

Gonzalez-Velasco, E.A. 2011. *Journey through Mathematics*. New York: Springer.

Good, I.J. 1970. Short proof of a conjecture of Dyson. *J. Math. Phys.*, **11**, 1884.

Gordan, P. 1868. Beweis, dass jede Covariante und Invariante einer binären Form eine ganze Function mit numerischen Coefficienten einer endlichen Anzahl solcher Formen ist. *J. Reine Angew. Math.*, **69**, 323–354.

Gordon, B. 1961. A combinatorial generalization of the Rogers-Ramanujan identities. *Amer. J. Math.*, **83**, 393–399.

Gouvêa, F.Q. 1994. A marvelous proof. *Am. Math. Monthly*, **101**, 203–222.

Gowing, R. 1983. *Roger Cotes*. New York: Cambridge University Press.

Grabiner, J.V. 1981. *The Origins of Cauchy's Rigorous Calculus*. Cambridge, MA: MIT Press.

Grabiner, J.V. 1990. *The Calculus as Algebra*. New York: Garland Publishing.

Grace, J.H., and Young, A. 1965. *The Algebra of Invariants*. New York: Chelsea.

Graham, G., Rothschild, B.L., and Spencer, J.H. 1990. *Ramsey Theory*. New York: Wiley.

Grattan-Guinness, I. 1972. *Joseph Fourier 1768–1830*. Cambridge, MA: MIT Press.

Grattan-Guinness, I. (ed). 2005. *Landmark Writings in Western Mathematics*. Amsterdam: Elsevier.

Graves, R.P. 1885. *Life of Sir William Rowan Hamilton*. London: Longmans.

Gray, J. 1986. *Linear Differential Equations and Group Theory from Riemann to Poincaré*. Boston: Birkhäuser.

Gray, J. 1994. On the history of the Riemann mapping theorem. *Rendiconti Circolo Matematico Palermo*, **34**, 47–94.

Gray, J., and Parshall, K.H. (eds). 2007. *Episodes in the History of Modern Algebra (1800–1950)*. Providence: A.M.S.

Green, G. 1970. *Mathematical Papers*. New York: Chelsea. Edited by N.M. Ferrers.

Greenberg, J.L. 1995. *The Problem of the Earth's Shape from Newton to Clairaut*. Cambridge: Cambridge University Press.

Gregory, D.F. 1865. *The Mathematical Writings of Duncan Farquharson Gregory*. Cambridge: Deighton, Bell. Edited by W. Walton.

Gregory, J. 1668. *Exercitationes Geometricae*. London: Godbid.

Grinshpan, A.Z. 1972. Logarithmic coefficients of functions in the class $S$, English translation. *Siberian Math J.*, **13**, 793–801.

Gronwall, T.H. 1913. On the degree of convergence of Laplace's series. *Trans. A.M.S.*, **15**, 1–30.

Gronwall, T.H. 1914. Some remarks on conformal representation. *Ann. Math.*, **16**, 72–76.

Grootendorst, A.W., and van Maanen, J.A. 1982. Van Heuraet's letter (1659) on the rectification of curves. *Nieuw Archief Wiskunde*, **30**(3), 95–113.

Grunsky, H. 1939. Koeffizientenbedingungen für schlicht abbildende meromorphe Funktionen. *Math. Zeit.*, **45**, 29–61.

Gudermann, C. 1838. Theorie der Modular-Functionen und der Modular-Integrale. *J. Reine Angew. Math.*, **18**, 220–258.

Guicciardini, N. 1989. *The Development of Newtonian Calculus in Britain 1700–1800*. Cambridge: Cambridge University Press.

Gunson, J. 1962. Proof of a conjecture of Dyson in the statistical theory of energy levels. *J. Math. Physics*, **3**, 752–753.

Gupta, R.C. 1969. Second order interpolation in Indian mathematics up to the fifteenth century. *Ind. J. Hist. Sci.*, **4**, 86–98.

Gupta, R.C. 1977. Paramesvara's rule for the circumradius of a cyclic quadrilateral. *Hist. Math.*, **4**, 67–74.

Hadamard, J. 1893. Étude sur les propriétés des fonctions entières et en particulier d'une fonction considerée par Riemann. *J. Math. Pures Appl.*, **4**, 171–215.

Hadamard, J. 1896. Sur la distribution des zéros de la fonction $\gamma(s)$ et ses conséquences arithmétiques. *Bull. Soc. Math. France*, **24**, 199–220.

Hadamard, J. 1899. Théorème sur séries entières. *Acta Math.*, **22**, 55–64.

Haimo, D.T. 1968. *Orthogonal Expansions and Their Continuous Analogues*. Carbondale: Southern Illinois University Press.

Hald, A. 1990. *A History of Probability and Statistics and Their Applications Before 1750*. New York: Wiley.

Halley, E. 1695. A most compendious method for constructing the logarithms … *Phil. Trans.*, **19**, 58–67.

Hamel, G. 1905. Eine Basis aller Zahlen und die unstetiggen Lösungen der Funktionalgleichung: $f(x + y) = f(x) + f(y)$. *Math. Ann.*, **60**, 459–462.

Hamilton, W.R. 1835. *Theory of Conjugate Functions or Algebraic Couples*. Dublin: Philip Dixon Hardy.

Hamilton, W.R. 1945. Quaternions. *Proc. Roy. Irish Acad.*, **50**, 89–92.

Hankel, H. 1864. Die Eulerschen Integrale bei unbeschränkter Variabilität des Arguments. *Zeit. Math. Phys.*, **9**, 1–21.

Hardy, G.H. 1905. *The Integration of Functions of a Single Variable*. Cambridge: Cambridge University Press.

Hardy, G.H. 1929. Prolegomena to a chapter on inequalities. *J. London Math. Soc.*, **1**, 61–78.

Hardy, G.H. 1937. *A Course in Pure Mathematics*. Cambridge: Cambridge University Press.

Hardy, G.H. 1949. *Divergent Series*. Oxford: Clarendon.

Hardy, G.H. 1966–79. *Collected Papers*. Oxford: Clarendon.

Hardy, G.H. 1978. *Ramanujan*. New York: Chelsea.

Hardy, G.H., and Heilbronn, H. 1914. Edmund Landau. *J. London Math. Soc.*, **13**, 302–310.

Hardy, G.H., and Littlewood, J.E. 1913. Tauberian theorems concerning power series of positive terms. *Messenger Math.*, **42**, 191–192.

Hardy, G.H., and Littlewood, J.E. 1914. Tauberian theorems concerning power series and Dirichlet series whose coefficients are positive. *Proc. London Math. Soc.*, **13**, 174–191.

Hardy, G.H., and Littlewood, J.E. 1918. Contributions to the theory of the Riemann zeta-function and the theory of the distribution of primes. *Acta Math.*, **41**, 119–196.

Hardy, G.H., and Littlewood, J.E. 1921. On a Tauberian theorem for Lambert's series, and some fundamental theorems in the analytic theory of numbers. *Proc. London Math. Soc.*, **19**, 21–29.

Hardy, G.H., and Littlewood, J.E. 1922. Some problems of 'partitio numerorum' IV. *Math. Zeit.*, **12**, 161–188.

Hardy, G.H., Littlewood, J., and Pólya, G. 1967. *Inequalities*. Cambridge: Cambridge University Press.

Harkness, J., and Morley, F. 1898. *Introduction to the Theory of Analytic Functions*. London: Macmillan.

Harriot, T., and Stedall, J. 2003. *The Greate Invention of Algebra: Thomas Harriot's Treatise on Equations*. Oxford: Oxford University Press. Introduction and commentary by J. Stedall.

Harriot, T., Beery, J., and Stedall, J. 2009. *Thomas Harriot's Doctrine of Triangular Numbers: 'The Magisteria Magna.'* Zürich: European Math. Soc. Extensive background and commentary by Beery and Stedall.

Hasse, H., and Davenport, H. 2014. *Manuskripte Hasse-Davenport*. Heidelberg: Heidelberg University Communication between Davenport and Hasse: heidelberg.de.

Hawking, S. 2005. *God Created the Integers*. Philadelphia: Running Press.

Hawkins, T. 1975. *Lebesgue Theory*. New York: Chelsea.

Hayman, W.K. 1964. *Meromorphic Functions*. Oxford: Clarendon.

Hayman, W.K. 1994. *Multivalent Functions*. Cambridge: Cambridge University Press.

Heegner, K. 1952. Diophantische Analysis und Modulfunktionen. *Math. Zeit.*, **56**, 227–253.

Heine, E. 1847. Untersuchungen über die Reihe … *J. Reine Angew. Math.*, **34**, 285–328.

Heine, E. 1870. Über trigonometrische Reihen. *J. Reine Angew. Math.*, **71**, 353–365.

Heine, E. 1872. Die Elemente der Functionenlehre. *J. Reine Angew. Math.*, **74**, 172–188.

Heine, E. 1878. *Handbuch der Kugelfunctionen.* Berlin: Reimer.

Hellegourarch, Y. 2002. *Invitation to the Mathematics of Fermat-Wiles.* London: Acad. Press. Translated by Leila Schneps.

Hermann, J. 1716. De variationibus chordarum tensarum disquisitio, ... *Acta Erud.*, 370–377.

Hermann, J. 1719. Solution duorum problematum ... *Acta Erud.*, **August, 1719**, 351–361.

Hermite, C. 1848. Note sur la théorie des fonctions elliptiques. *Cam. and Dub. Math. J.*, **3**, 54–56.

Hermite, C. 1873. *Cours d'analyse.* Paris: Gauthier-Villars.

Hermite, C. 1891. *Cours de M. Hermite (rédigé en 1882 par M. Andoyer).* Ithaca: Cornell University Library Reprint.

Hermite, C. 1905–1917. *Oeuvres.* Paris: Gauthier-Villars. Edited by É. Picard.

Herschel, J.F.W. 1820. *A Collection of Examples of the Applications of the Calculus of Finite Differences.* Cambridge: Cambridge University Press.

Hewitt, E., and Hewitt, R. E. 1980. The Gibbs-Wilbraham phenomenon. *Archive Hist. Exact Sci.*, **21**, 129–160.

Hickerson, D. 1988. A proof of the mock theta conjectures. *Inventiones Math.*, **94**, 639–660.

Hilbert, D. 1893. Über die Transzendenz der Zahlen $e$ und $\pi$. *Gött. Nachr.*, 113–116.

Hilbert, D. 1897. Zum Gedächtnis an Karl Weierstrass. *Gött. Nach.*, 60–69.

Hilbert, D. 1906. Grundzüge einer allgemeinen Theorie der linearen Integralgleichungen, vierte Mitteilung. *Gött. Nach.*, 157–227.

Hilbert, D. 1970. *Gesammelte Abhandlungen.* Berlin: Springer.

Hilbert, D. 1978. *Hilbert's Invariant Theory Papers.* Brookline, MA: Math. Sci. Press. Translated by M. Ackerman with Commentary by R. Hermann.

Hilbert, D. 1993. *Theory of Algebraic Invariants.* New York: Cambridge University Press. Translated by R. C. Laubenbacher.

Hindenburg, C.F. 1778. *Methodus nova et facilis serierum infinitarum exhibendi dignitates exponentis indeterminati.* Leipzig: Langenhem.

Hobson, E.W. 1957a. *The Theory of Functions of a Real Variable.* New York: Dover.

Hobson, E.W. 1957b. *A Treatise on Plane and Advanced Trigonometry.* New York: Dover.

Hoe, J. 2007. *A Study of the Jade Mirror of the Four Unknowns.* Christchurch, N.Z.: Mingming Bookroom.

Hofmann, J.E. 1974. *Leibniz in Paris.* Cambridge: Cambridge University Press. Translated by A. Prag and D.T. Whiteside.

Hofmann, J.E. 1990. *Ausgewählte Schriften.* Zürich: Georg Olms Verlag. Edited by C. Scriba.

Hofmann, J.E. 2003. *Classical Mathematics.* New York: Barnes and Noble.

Hölder, O. 1889. Über einen Mittelwertsatz. *Gött. Nach.*, 38–47.

Horiuchi, A.T. 1994. The *Tetsujutsu sankei* (1722), an 18th century treatise on the methods of investigation in mathematics. Pages 149–164 of: Sasaki, C., Sugiura, M., and Dauben, J.W. (eds), *The Intersection of History and Mathematics.* Basel: Birkhäuser.

Horowitz, D. 1978. A further refinement for coefficient estimates of univalent functions. *Proc. A.M.S.*, **71**, 217–221.

Horowitz, E. 1969. *Algorithms for Symbolic Integration of Rational Functions.* PhD thesis, University of Wisconsin, Madison.

Hua, L.K. 1981. *Starting with the Unit Circle.* New York: Springer-Verlag. Translated by K. Weltin.

Hua, L.K., and Vandiver, H.S. 1948. On the existence of solutions of certain equations in a finite field. *Proc. Nat. Acad. Sci. USA*, **34**, 258–263.

Hua, L.K., and Vandiver, H.S. 1949. On the existence of solutions of certain equations in a finite field. *Proc. Nat. Acad. Sci. USA*, **35**, 94–99.

Hutton, C. 1812. *Tracts on mathematical and philosophical subjects.* London: Rivington.

Ireland, K., and Rosen, M. 1982. *A Classical Introduction to Modern Number Theory.* New York: Springer-Verlag.

Ivory, J. 1796. A new series for the rectification of the ellipses. *Trans. Roy. Soc. Edinburgh*, **4**, 177–190.

Ivory, J. 1812. On the attractions of an extensive class of spheroids. *Phil. Trans. Roy. Soc. London*, **102**, 46–82.

Ivory, J. 1824. On the figure requisite to maintain the equilibrium of a homogeneous fluid mass that revolves upon an axis. *Phil. Trans. Roy. Soc. London*, **114**, 85–150.

Ivory, J., and Jacobi, C.G.J. 1837. Sur le développement de $(1 - 2xz + z^2)^{-1/2}$. *J. Math. Pures App.*, **2**, 105–106.

Jackson, F.H. 1910. On $q$-definite integrals. *Quart. J. Pure App. Math.*, **41**, 193–203.

Jacobi, C.G.J. 1826. Ueber Gauss' neue Methode, die Werthe der Integrale näherungsweise zu finden. *J. Reine Angew. Math.*, **1**, 301–308. Reprinted in Jacobi (1969) vol. 6, pp. 3–11.

Jacobi, C.G.J. 1833. Demonstratio Formulae. *J. Reine Angew. Math.*, **11**, 307. Reprinted in Jacobi (1969) vol. 6, pp. 62–63.

Jacobi, C.G.J. 1834. De usu legitimo formulae summatoriae Maclaurinianae. *J. Reine Angew. Math.*, **12**, 263–272. Reprinted in Jacobi (1969) vol. 6, pp. 64–75.

Jacobi, C.G.J. 1837. Über die Kreistheilung und ihre Anwendung auf die Zahlentheorie. *Monats. Akad. Wiss. Berlin*, 127–136. Reprinted in 1846 in Crelle's Journal; in Jacobi's Werke vol. 6, pp. 254–274; French translation published in 1856 in *Nouv. Ann. Math.*

Jacobi, C.G.J. 1841. De determinantibus functionalibus. *J. Reine Angew. Math.*, **22**, 319–359. Reprinted in Jacobi (1969) vol. 3, pp. 393–438.

Jacobi, C.G.J. 1846. Über einige der Binomialreihe Analoge Reihen. *J. Reine Angew. Math*, **32**, 197–204. Reprinted in Jacobi (1969) vol. 6, pp. 163–173.

Jacobi, C.G.J. 1847. De seriebus ac differentiis observatiunculae. *J. Reine Angew. Math.*, **36**, 135–142. Republished in Jacobi (1969) vol. 4, pp. 174–182.

Jacobi, C.G.J. 1859. Untersuchungen über die Differentialgleichung der hypergeometrische Reihe. *J. Reine Angew. Math*, **56**, 149–165. Reprinted in Jacobi (1969) vol. 6, pp. 184–202.

Jacobi, C.G.J. 1969. *Mathematische Werke*. New York: Chelsea.

Jahnke, H.N. 1993. Algebraic analysis in Germany, 1780–1849: Some mathematical and philosophical issues. *Hist. Math.*, **20**, 265–284.

James, I.M. 1999. *History of Topology*. Amsterdam: Elsevier.

Jensen, J. 1899. Sur un nouvel et important théorèm de la théorie des fonctions. *Acta Math.*, **22**, 359–364.

Jensen, J. 1906. Sur les fonctions convexes et les inégalités entre les valeurs moyennes. *Acta Math.*, **30**, 175–193.

Johnson, W.P. 2002. The curious history of Faà di Bruno's formula. *Am. Math. Monthly*, **109**, 217–234.

Johnson, W.P. 2007. The Pfaff/Cauchy derivative and Hurwitz type extensions. *Ramanujan J. Math.*, **13**, 167–201.

Jordan, C. 1979. *Calculus of Finite Differences*. New York: Chelsea.

Jyesthadeva. 2009. *Ganita-Yukti-Bhasa of Jyesthadeva, 2 volumes*. New Delhi: Hindustan Book Agency and Springer. Jyesthadeva's text translated from Malayalam into English by K. V. Sarma, with notes by editors K. Ramasùbramanian, M.D. Srinivas, and M.S. Sriram.

Kac, M. 1936–1937. Une remarque sur les équations fonctionnelles. *Comment. Math. Helv.*, **9**, 170–171.

Kac, M. 1965. A Remark on Wiener's Tauberian Theorem. *Proc. A.M.S.*, **16**, 1155–1157.

Kac, M. 1979. *Selected Papers*. Cambridge, MA: MIT Press.

Kac, M. 1987. *Enigmas of Chance: An Autobiography*. Berkeley: U. Calif. Press.

Kalman, D. 2009. *Polynomia and Related Realms*. Washington, D.C.: M.A.A.

Karamata, J. 1930. Über die Hardy-Littlewoodschen Umkehrungen des Abelschen Stetigkeitssatzes. *Math. Zeit.*, **32**, 319–320.

Katz, N.M. 1976. An overview of Deligne's proof of the Riemann hypothesis for varieties over finite fields. Pages 275–305 of: Browder, F.E. (ed), *Mathematical Developments Arising from Hilbert Problems*. Providence: A.M.S.

Katz, V.J. 1979. The history of Stokes' theorem. *Math. Mag.*, **52**, 146–156.

Katz, V.J. 1982. Change of variables in multiple integrals: Euler to Cartan. *Math. Mag.*, **55**, 3–11.

Katz, V.J. 1985. Differential forms—Cartan to de Rham. *Archive Hist. Exact Sci.*, **33**, 307–319.

Katz, V.J. 1987. The calculus of the trigonometric functions. *Hist. Math.*, **14**, 311–324.

Katz, V.J. 1995. Ideas of calculus in Islam and India. *Math. Mag.*, **3**(3), 163–174. Reprinted in Anderson, Katz, and Wilson (2004), pp. 122–130.

Katz, V.J. 1998. *A History of Mathematics: An Introduction*. Reading, MA: Addison-Wesley.

Khinchin, A.Y. 1998. *Three Pearls of Number Theory*. New York: Dover.

Khrushchev, S. 2008. *Orthogonal Polynomials and Continued Fractions*. Cambridge: Cambridge University Press.

Kichenassamy, S. 2010. Brahmagupta's derivation of the area of a cyclic quadrilateral. *Hist. Math.*, **37**(1), 28–61.

Kikuchi, D. (ed). 1891. *Memoirs on Infinite Series*. Tokyo: Tokio Math. Phys. Soc. Translation by D. Kikuchi.

Kinkelin, H. 1861–1862. Allgemeine Theorie der harmonischen Reihen, mit Anwendung auf die Zahlentheorie. *Programm der Gewerbeschule Basel*, 1–32.

Klein, F. 1911. *Lectures on Mathematics*. New York: Macmillan.

Klein, F. 1933. *Vorlesungen über die hypergeometrische Funktion*. Berlin: Springer.

Klein, F. 1979. *Development of Mathematics in the 19th Century*. Brookline, MA: Math. Sci. Press. Translated by M. Ackerman.

Knoebel, A., Laubenbacher, R., Lodder, J., and Pengelley, D. 2007. *Mathematical Masterpieces*. New York: Springer.

Knopp, K. 1990. *Theory and Application of Infinite Series*. New York: Dover.

Knuth, D. 1992. Two notes on notation. *Am. Math. Monthly*, **99**, 403–422. Reprinted as chapter 2 in Knuth (2003). Addendum 'Stirling Numbers' published in the *Monthly*, vol. 102, p. 562.

Knuth, D. 1993. Johann Faulhaber and Sums of Powers. *Math. of Computation*, **61**, 277–294.

Knuth, D. 2003. *Selected Papers*. Stanford, CA: Center for the Study of Language and Information (CSLI),.

Knuth, D. 2011. *The Art of Computer Programming, vol. 4A: Combinatorial Algorithms, part 1*. New York: Addison-Wesley.

Koebe, P. 1907–1908. Über die Uniformisierung beliebiger analytischer Kurven. *Gött. Nach.*, 1907: 191–210, 633–649; 1908: 337–358.

Kolmogorov, A.N. 1923. Une série de Fourier-Lebesgue divergente presque partout. *Fund. Math.*, **4**, 324–328.

Kolmogorov, A.N., and Yushkevich, A.P. (eds). 1998. *Mathematics of the 19th Century: Vol. III: Function Theory According to Chebyshev; Ordinary Differential Equations; Calculus of Variations; Theory of Finite Differences*. Basel: Birkhäuser.

Koppelman, E. 1971. The calculus of operations and the rise of abstract algebra. *Archive Hist. Exact Sci.*, **8**, 155–242.

Korevaar, J. 2004. *Tauberian Theory*. New York: Springer.

Kramp, C. 1796. Coefficient des allgemeinen Gliedes jeder willkürlichen Potenz eines Infinitinomiums ... Pages 91–122 of: Hindenburg, C. (ed), *Der polynomische Lehrsatz*. Leipzig: Fleischer.

Kronecker, L. 1968. *Mathematische Werke*. New York: Chelsea. Edited by K. Hensel.

Kubota, K.K. 1977. Linear functional equations and algebraic independence. Pages 227–229 of: Baker, A., and Masser, D.W. (eds), *Transcendence Theory*. New York: Academic Press.

Kummer, E.E. 1836. Über die hypergeometrische Reihe. *J. Reine Angew. Math.*, **15**, 39–83, 127–172.

Kummer, E.E. 1840. Über die Transcendenten, welche aus wiederholten Integrationen rationaler Formeln entstehen. *J. Reine Angew. Math.*, **21**, 74–90, 193–225, 328–371.

Kummer, E.E. 1847. Beitrag zur Theorie der Function $\Gamma(x)$. *J. Reine Angew. Math.*, **35**, 1–4.

Kummer, E.E. 1975. *Collected Papers*. Berlin: Springer-Verlag. Edited by A. Weil.

Kung, J. (ed). 1995. *Gian-Carlo Rota on Combinatorics*. Boston: Birkhäuser.

Kuzmin, R. 1930. In Russian: On a new class of transcendental numbers. *Izvestia Akad. Nauk. SSSR*, **3**, 583–597.

Kuzmin, R. 1938. On the transcendental numbers of Goldbach. *Trudy Leningrad Indust. Inst.*, **1–5**, 28–32.

Lacroix, S., Babbage, C., Herschel, J., and Peacock, G. 1816. *An elementary treatise on the Differential and Integral Calculus.* Cambridge: Cambridge University Press.

Lacroix, S.F. 1800. *Traité des différences et des séries.* Paris: Duprat.

Lacroix, S.F. 1819. *Traité du calcul différentiel et du calcul intégral.* Vol. 3. Paris: Courcier.

Lagrange, J.L. 1770–1771. Réflexions sur la résolution algébraic des équations. *Nouv. Mém. Acad. Roy. Sci. Berlin,* 1770 [1772]: 134–215; 1771 [1773]: 138–253. Reprinted in Lagrange's *Oeuvres,* vol. 3, pp. 205–421.

Lagrange, J.L. 1771. Démonstration d'un théorèm nouveau concernant les nombres premiers. *Nouv. Mém. Acad. Roy. Sci. Belles-Let., 1771,* 125–137. Reprinted in *Oeuvres,* vol. 3, pp. 425–438.

Lagrange, J.L. 1772. Sur une nouvelle espèce de calcul relatif à la différentiation et à l'intégration des quantités variables. *Nouv. Mém. Acad. Roy. Sci. Belle-lettres Berlin,* 441–476.

Lagrange, J.L. 1781. Sur la construction des cartes géographique. *Nouv. Mém. Acad. Roy. Sci. Berlin,* 161–210. Reprinted in Lagrange (1867–1892) vol. 4, pp. 637–692.

Lagrange, J.L. 1797. *Théorie des fonctions analytiques.* Paris: Imprim. de la Répub.

Lagrange, J.L. 1867–1892. *Oeuvres.* Paris: Gauthier-Villars. Edited by J. Serret.

Laguerre, E. 1882. Sur quelques équations transcendantes. *Comptes Rendus,* **94,** 160–163.

Laguerre, E. 1972. *Oeuvres.* New York: Chelsea. Edited by C. Hermite, H. Poincaré, and E. Rouché.

Lambert, J.H. 1768. Mémoire sur quelques propriétés remarquables des quantités transcendentes circulaires et logarithmiques. *Hist. Akad. Berlin, 1761,* **17,** 265–322. See Lambert's *Opera mathematica* (1946, 1948), Zurich: Füssli, vol. I, pp. 194–212 and vol. 2, pp. 112–159.

Landau, E. 1903. Neuer Beweis des Primzahlsatzes und Beweis des Primidealsatzes. *Math. Ann.,* **56,** 645–670.

Landau, E. 1904. Über eine Verallgemeinerung des Picardschen Satzes. *S.B. Preuss. Akad. Wiss.,* **38,** 1118–1133.

Landau, E. 1906. Euler und die Funktionalgleichung der Riemannschen Zetafunktion. *Bib. Math.,* **7,** 69–79.

Landau, E. 1907a. Sur quelques généralisations du théorème de M. Picard. *Ann. École Norm. Sup.,* **24,** 179–201.

Landau, E. 1907b. Über die Konvergenz einiger Klassen von unendlichen Reihen am Rande des Konvergenzgebietes. *Monatsh. Math. Phys.,* **18,** 8–28.

Landau, E. 1907c. Über einen Konvergenzsatz. *Gött. Nach.,* **8,** 25–27.

Landau, E. 1910. Über die Bedeutung einiger neuen Grentzwertsätze der Herren Hardy und Axer. *Prace Mat.-Fiz.,* **21,** 97–177.

Landen, J. 1758. *A Discourse Concerning The Residual Analysis.* London: Nourse.

Landen, J. 1760. A new method of computing the sums of certain series. *Phil. Trans. Roy. Soc. London,* **51,** 553–565.

Landen, J. 1771. A Disquisition concerning Certain Fluents, which are Assignable by the Arcs of the Conic Sections ... *Phil. Trans.,* **61,** 298–309.

Landen, J. 1775. An Investigation of a General Theorem for Finding the Length of Any Arc of Any Conic Hyperbola ... *Phil. Trans.,* **65,** 283–289.

Landis, E.M. 1993. About mathematics at Moscow State University in the late 1940s and early 1950s. Pages 55–73 of: Zdravkovska, S., and Duren, P. (eds), *Golden Years of Moscow Mathematics.* Providence: A.M.S.

Lanzewizky, I.L. 1941. Über die orthogonalität der Fejér-Szegőschen polynome. *D. R. Dokl. Acad. Sci. URSS,* **31,** 199–200.

Laplace, P.S. 1782. Mémoire sur les suites. *Mém. Acad. Roy. Sci. Paris,* 207–309.

Laplace, P.S. 1812. *Théorie analytique des probabilités.* Paris: Courcier.

Laplace, P.S. 1814. *Théorie analytique des probabilités, seconde édition, revue et augmentée par l'auteur.* Paris: Courcier.

Lascoux, A. 2003. *Symmetric Functions and Combinatorial Operators on Polynomials.* Providence: A.M.S.

Laudal, O.A., and Piene, R. (eds). 2002. *The Legacy of Niels Henrik Abel.* Berlin: Springer.

Laugwitz, D. 1999. *Bernhard Riemann 1826–1866.* Boston: Birkhäuser. Translated by A. Shenitzer.

Lebedev, N.A., and Milin, I.M. 1965. In Russian: An inequality. *Vestnik Leningrad University*, **20**, 157–158.

Lebesgue, H. 1902. Intégrale, longueur, aire. *Annali Mat. Pura. App.*, **7**, 231–359.

Lebesgue, H. 1906. *Leçons sur les séries trigonométriques*. Paris: Gauthier-Villars.

Lebesgue, V.A. 1837. Researches sur les nombres. *J. Math Pures Appl.*, **2**, 253–292.

Lebesgue, V.A. 1838. Researches sur les nombres. *J. Math Pures Appl.*, **3**, 113–144.

Legendre, A.M. 1794. *Éléments de Géométrie avec des notes*. Paris: Didot.

Legendre, A.M. 1811–1817. *Exercices de calcul intégral*. Paris: Courcier.

Legendre, A.M. 1825–1828. *Traité des fonctions elliptiques*. Paris: Huzard-Courcier.

Legendre, A.M. 1830. *Théorie des nombres*. Paris: Didot.

Legendre, A.M. 1885. Recherches sur l'attraction des sphéroïdes homogènes. *Mém. Math. Phys. prés. à Acad. Roy. Sci.*, **10**, 411–435.

Leibniz, G.W. 1684. Nova methodus pro maximis et minimis, itemque tangentibus, ... *Acta Erud.*, **3**, 467–473. Reprinted in Leibniz (1971) vol. 5, pp. 220–225.

Leibniz, G.W. 1713. Epistola ad V. Cl. Christianium Wolfium. *Acta Erud.*, **Supp. 5, 1713**, 264–270.

Leibniz, G.W. 1920. *The Early Mathematical Manuscripts of Leibniz*. Chicago: Open Court. Translated with extensive notes by J.M. Child.

Leibniz, G.W. 1971. *Mathematische Schriften*. Hildesheim, Germany: Georg Olms Verlag. edited by K. Gerhardt.

Leibniz, G.W., and Bernoulli, Joh. 1745. *Commercium philosophicum et mathematicum*. Geneva: Bosquet.

Leibniz, G.W., and Knobloch, E. 1993. *De quadratura arithmetica circuli ellipseos et hyperbolae cujus corollarium est trigonometria sine tabulis*. Göttingen: Vandenhoeck and Ruprecht. Presented with commentary by E. Knobloch; translation into German by O. Hamborg.

Lemmermeyer, F. 2000. *Reciprocity Laws*. New York: Springer-Verlag.

Lewin, L. 1981. *Polylogarithms and Associated Functions*. Amsterdam: Elsevier.

Li, Y., and Du, S. 1987. *Chinese mathematics: A concise history*. Oxford: Clarendon. Translated by J. Crossley and A. Lun.

Lindelöf, E. 1902. Mémoire sur la théorie des fonctions entières de genre fini. *Acta Soc. Sci. Fennicae*, **31**, 1–79.

Lindemann, F. 1882. Über die Zahl $\pi$. *Math. Ann.*, **20**, 213–225.

Liouville, J. 1832. Sur quelques questions de géométrie et de mécanique, et sur un nouveau genre de calcul pour résoudre ces questions. *J. École Polytech.*, **13**, 1–69.

Liouville, J. 1837a. Note sur le développement de $(1 - 2xz + z^2)^{-1/2}$. *J. Math. Pures App.*, **2**, 135–139.

Liouville, J. 1837b. Sur la sommation d'une série. *J. Math. Pures Appl.*, **2**, 107–108.

Liouville, J. 1839. Note sur quelques intégrales définies. *J. Math. Pures Appl*, **4**, 225–235.

Liouville, J. 1841. Remarques nouvelles sur l'équation de Riccati. *J. Math. Pures Appl.*, **6**, 1–13.

Liouville, J. 1851. Sur des classes très étendues de quantités dont la valeur n'est ni algébrique, ni même réductible à des irrationnelles algébriques. *J. Math. Pures Appl.*, **16**, 133–142.

Liouville, J. 1857. Sur l'expression $\phi(n)$, qui marque combien la suite $1, 2, 3, \ldots, n$ contient de nombres premiers à $n$. *J. Math. Pures Appl.*, **2**, 110–112.

Liouville, J. 1880. Leçons sur les fonctions doublement périodiques. *J. Reine Angew. Math.*, **88**, 277–310.

Lipschitz, R. 1889. Untersuchungen der Eigenschaften einer Gattung von unendlichen Riehen. *J. Reine Angew. Math.*, **105**, 127–156.

Littlewood, J.E. 1911. The converse of Abel's theorem on power series. *Proc. London Math. Soc.*, **9**, 434–448.

Littlewood, J.E. 1925. On inequalities in the theory of functions. *Proc. London Math. Soc.*, **23**, 481–519. Reprinted in Littlewood (1982) vol. 2, pp. 963–1004.

Littlewood, J.E. 1982. *Collected Papers*. Oxford: Clarendon.

Littlewood, J.E. 1986. *Littlewood's Miscellany*. Cambridge: Cambridge University Press.

Littlewood, J.E., and Paley, R. 1932. A proof that an odd schlicht function has bounded coefficients. *J. London Math. Soc.*, **7**, 167–169. Reprinted in Littlewood (1982) vol. 2, pp. 1046–1048.

Loewner, C. 1988. *Collected Papers*. Boston: Birkhäuser. Edited by L. Bers.

Lototskii, A.V. 1943. Sur l'irrationalité d'un produit infini. *Mat. Sb.*, **12**, 262–272.

Lusin, N. 1913. Sur la convergence des séries trigonométriques de Fourier. *Comptes Rendus.*, **156**, 1655–1658.

Lützen, J. 1990. *Joseph Liouville*. New York: Springer-Verlag.

Maclaurin, C. 1729. A second letter to Martin Folkes, Esq.: Concerning the roots of equations, with the demonstration of other rules in algebra. *Phil. Trans. Roy. Soc.*, **36**, 59–96.

Maclaurin, C. 1742. *A Treatise of Fluxions*. Edinburgh: Ruddimans.

Maclaurin, C. 1748. *A Treatise of Algebra*. London: Millar and Nourse.

Macmahon, P. 1915–1916. *Combinatory Analysis*. Cambridge: Cambridge University Press.

MacMahon, P. A. 1978. *Collected Papers*. Cambridge, MA: MIT Press. Edited by G. Andrews.

Mahlburg, K. 2005. Partition congruences and the Andrews-Garvan-Dyson crank. *Proc. Nat. Acad. Sci. USA*, **102 (43)**, 15373–15376.

Mahler, K. 1982. Fifty years as a mathematician. *J. Number Theory*, **14**, 121–155. Corrected version on the online Kurt Mahler Archive.

Mahnke, D. 1912–1913. Leibniz auf der Suche nach einer allgemeinen Primzahlgleichung. *Bib. Math.*, **13**, 29–61.

Mahoney, M.S. 1973. *The Mathematical Career of Pierre de Fermat (1601–1665)*. Princeton: Princeton University Press.

Malet, A. 1993. James Gregorie on tangents and the Taylor rule. *Archive Hist. Exact Sci*, **46**, 97–138.

Malmsten, C. J. 1849. De integralibus quibusdam definitis. *J. Reine Angew. Math.*, **38**, 1–39.

Manders, K. 2006. Algebra in Roth, Faulhaber and Descartes. *Hist. Math.*, **33**, 184–209.

Manning, K. R. 1975. The emergence of the Weierstrassian approach to complex analysis. *Archive Hist. Exact Sci.*, **14**, 297–383.

Maor, E. 1998. *Tirgonometric Delights*. Princeton: Princeton University Press.

Martzloff, J.C. 1997. *A History of Chinese Mathematics*. New York: Springer.

Masani, P.R. 1990. *Norbert Wiener*. Basel: Birkhäuser.

Mathews, G.B. 1961. *Theory of Numbers*. New York: Chelsea.

Maxwell, J.C. 1873. *A Treatise on Electricity and Magnetism*. Oxford: Clarendon.

Maz'ya, V., and Shaposhnikova, T. 1998. *Jacques Hadamard*. Providence: A.M.S., London Math. Soc. Translated by P. Basarab-Horwath.

McClintock, E. 1881. On the remainder of Laplace's series. *Amer. J. Math.*, **4**, 96–97.

Mclarty, C. 2012. Hilbert on Theology and its Discontents: The Origin of Myth in Modern Mathematics. Pages 105–129 of: Doxiadis, A., and Mazur, B. (eds), *Circles Disturbed*. Princeton: Princeton University Press.

Mehta, M.L., and Dyson, F.J. 1963. Statistical theory of energy levels of complex systems: V. *J. Math. Phys.*, **4**, 713–719.

Meijering, E. 2002. A chronology of interpolation: from ancient astronomy to modern signal and image processing. *Proc. I.E.E.E.*, **90**, 319–342.

Meixner, J. 1934. Orthogonale Polynomsysteme mit einer besonderen Gestalt der erzeugenden Funktion. *J. London Math. Soc.*, **9**, 6–13.

Mengoli, P. 1650. *Novae quadraturae arithmeticae seu de additione fractionum*. Bologna: Montij.

Méray, C. 1869. Remarques sur la nature des quantités définies par la condition de servir de limites à des variables données. *Revue des Soc. Savantes*, **4**, 280–289.

Méray, C. 1888. Valeur de l'intégrale définie $\int_0^\infty e^{-x^2} dx$ déduite de la formule de Wallis. *Bull. Sci. Math.*, **12**, 174–176.

Mercator, N. 1668. *Logarithmotechnia*. Godbid.

Mertens, F. 1874a. Ein Beitrag zur analytischen Zahlentheorie. *J. Reine Angew. Math.*, **78**, 46–62.

Mertens, F. 1874b. Ueber einige asymptotische Gesetze der Zahlentheorie. *J. Reine Angew. Math.*, **77**, 289–338.

Mertens, F. 1875. Über die Multiplicationsregel für zwei unendliche Reihen. *J. Reine Angew. Math.*, **79**, 182–184.

Mertens, F. 1895. Über das nichtverschwinden Dirichletscher Reihen mit reellen Gliedern. *S.B. Kais. Akad. Wiss. Wien*, **104**(Abt. 2a), 1158–1166.

Mertens, F. 1897. Über Dirichlet's Beweis des Satzes, dass jede unbegrenzte ganzzahlige arithmetische Progression, deren Differentz zu ihren Gliedern teilerfreund ist, unendliche viele Primzahlen darstellt. *S.B. Kais. Akad. Wiss. Wien*, **106**, 254–286.

Merzbach, U. 2018. *Dirichlet: A Mathematical Biography*. Cham: Birkhäuser.

Meschkowski, H. 1964. *Ways of Thought of Great Mathematicians*. San Francisco: Holden-Day. Translated by J. Dyer-Bennett.

Mikami, Y. 1914. On the Japanese theory of determinants. *Isis*, **2**, 9–36.

Mikami, Y. 1974. *The Development of Mathematics in China and Japan*. New York: Chelsea.

Milin, I.M. 1964. The area method in the theory of univalent functions, English translation. *Soviet Math. Dokl.*, **5**, 78–81.

Milin, I.M. 1986. Comments on the Proof of the Conjecture on Logarithmic Coefficients. Pages 109–112 of: Baernstein, A. (ed), *The Bieberbach Conjecture*. Providence: A.M.S.

Milne-Thomson, L.M. 1981. *The calculus of finite differences*. New York: Chelsea.

Mittag-Leffler, G. 1923. An introduction to the theory of elliptic functions. *Ann. of Math.*, **24**, 271–351.

Miyake, K. 1994. The establishment of the Takagi-Artin class field theory. Pages 109–128 of: Sasaki, C., Sugiura, M., and Dauben, J.W. (eds), *The Intersection of History and Mathematics*. Basel: Birkhäuser.

Möbius, A.F. 1832. Über eine besondere Art von Umkehrung der Reihen. *J. Reine Angew. Math.*, **9**, 105–123.

Moll, V. 2002. The evaluation of integrals: A personal story. *Notices A.M.S.*, 311–317.

Monsky, P. 1994. Simplifying the proof of Dirichlet's theorem. *Am. Math. Monthly*, **100**, 861–862.

Montmort, P.R. de. 1717. De seriebus infinitis tractatus. *Phil. Trans. Roy. Soc.*, **30**, 633–675.

Moore, G.H. 1982. *Zermelo's Axiom of Choice*. New York: Springer-Verlag.

Morrison, P., and Morrison, E. (eds). 1961. *Charles Babbage and His Calculating Engines: Selected Writings by Charles Babbage*. New York: Dover.

Muir, T. 1960. *The Theory of Determinants in the Historical Order of Development*. New York: Dover.

Mukhopadhyay, A. 1889a. The geometric interpretation of Monge's differential equation to all conics. *J. Asiatic Soc. of Bengal*, **58**, 181–185.

Mukhopadhyay, A. 1889b. On a curve of aberrancy. *J. Asiatic Soc. of Bengal*, **59**, 61–63.

Mukhopadhyay, S. 1909. New methods in the geometry of the plane arc. *Bull. Calcutta Math. Soc.*, **1**, 31–37.

Mukhopadhyaya (Mookerjee), A. 1998. *A Diary of Asutosh Mookerjee*. Calcutta: Mitra and Ghosh.

Mullin, R., and Rota, G.-C. 1970. On the Foundations of Combinatorial Theory: III. Theory of Binomial Enumeration. Pages 167–213 of: Harris, B. (ed), *Graph Theory and its Applications*. New York: Acad. Press. Reprinted in 1995 in *Gian-Carlo Rota on Combinatorics*, Ed. J. Kung, Boston: Birkhäuser, pp. 118–147.

Murphy, R. 1833a. *Elementary Principles of the Theories of Electricity, Heat, and Molecular Actions, Part I*. Cambridge: Cambridge University Press.

Murphy, R. 1833b. On the inverse method of definite integrals, with physical applications. *Trans. Cambridge Phil. Soc.*, **4**, 353–408.

Murphy, R. 1833c. Resolution of Algebraical Equations. *Trans. Cambridge Phil. Soc.*, **4**, 125–153.

Murphy, R. 1835. Second memoir on the inverse method of definite integrals. *Trans. Cambridge Phil. Soc.*, **5**, 113–148.

Murphy, R. 1837. First memoir on the theory of analytic operations. *Phil. Trans. Roy. Soc. London*, **127**, 179–210.

Murphy, R. 1839. *A Treatise on the Theory of Algebraical Equations*. London: Society for Diffusion of Useful Knowledge.

Mustafy, A. K. 1966. A new representation of Riemann's zeta function and some of its consequences. *Norske Vid. Selsk. Forh.*, **39**, 96–100.

Mustafy, A.K. 1972. On a criterion for a point to be a zero of the Riemann zeta function. *J. London Math. Soc. (2)*, **5**, 285–288.

Narasimhan, R. 1991. The coming of age of mathematics in India. Pages 235–258 of: Hilton, P., Hirzebruch, F., and Remmert, R. (eds), *Miscellanea Mathematica*. New York: Springer.

Narayana. 2001. The *Ganita Kamudi* of Narayana Pandit, Chapter 13. *Ganita Bharati*, **23**, 18–82. English translation with notes by Parmanand Singh.

Narkiewicz, W. 2000. *The Development of Prime Number Theory*. New York: Springer.

Needham, J. 1959. *Science and Civilization in China, vol. 3: Mathematics and the Sciences of the Heavens and the Earth*. New York: Cambridge University Press.

Nesterenko, Yu. V. 2006. Hilbert's seventh problem. Pages 269–282 of: Bolibruch, A. A., Osipov, Yu. S., and Sinai, Ya. G. (eds), *Mathematical Events of the Twentieth Century*. Berlin: Springer. Translated by L.P. Kotova.

Neuenschwander, E. 1978. The Casorati–Weierstrass theorem. *Hist. Math.*, **5**, 139–166.

Neumann, O. 2007a. Cyclotomy: From Euler through Vandermonde to Gauss. Pages 323–362 of: Bradley, R.E., and Sandifer, C.E. (eds), *Leonhard Euler: Life, Work and Legacy*. Amsterdam: Elsevier.

Neumann, O. 2007b. The *Disquisitiones Arithmeticae* and the theory of equations. Pages 107–128 of: Goldstein, C., Schappacher, N., and Schwermer, J. (eds), *The Shaping of Arithmetic after C. F. Gauss's Disquisitiones Arithmeticae*. New York: Springer.

Neumann, P. (ed). 2011. *The mathematical writings of Évariste Galois*. Zurich: European Math. Soc. Translated with commentary by P. Neumann.

Nevai, P. 1990. *Orthogonal Polynomials: Theory and Practice*. Dordrecht: Kluwer.

Nevanlinna, R. 1925. Zur Theorie der meromorphen Funktionen. *Acta Math.*, **46**, 1–99.

Nevanlinna, R. 1974. *Le théorème de Picard–Borel et la théorie des fonctions méromorphes*. New York: Chelsea.

Nevanlinna, R., and Paatero, V. 1982. *Introduction to Complex Analysis, 2nd edition*. Providence: A.M.S. Chelsea. Translated by T. Kövari and G.S. Goodman.

Newman, F.W. 1848. On $\Gamma(a)$, especially when $a$ is negative. *Cambridge and Dublin Math. J.*, **3**, 57–63.

Newton, I. 1959–1960. *The Correspondence of Isaac Newton*. Cambridge: Cambridge University Press. Edited by H.W. Turnbull.

Newton, I. 1964–1967. *The Mathematical Works of Isaac Newton*. New York: Johnson Reprint. Edited with introduction by D.T. Whiteside.

Newton, I. 1967–1981. *The Mathematical Papers of Isaac Newton*. Cambridge: Cambridge University Press. Edited by D.T. Whiteside.

Nicole, F. 1717. Traité du calcul des différences finies. *Hist. Acad. Roy. Sci.*, 7–21.

Nicole, F. 1727. Méthode pour sommer une infinité de suites nouvelles, dont on ne peut trouver les sommes par les Méthodes connues. *Mém. Acad. Roy. Sci. Paris*, 257–268.

Nikolić, A. 2009. The story of majorizability as Karamata's condition of convergence for Abel summable series. *Hist. Math.*, **36**, 405–419.

Nilakantha. 1977. *Tantrasangraha of Nilakantha Somayaji, with Yuktidipika and Laghu-vivrti*. Hoshiarpur, India: Punjab University Edited by K.V. Sarma.

Nörlund, N.E. 1923. Mémoire sur le calcul aux différences finies. *Acta Math.*, **44**, 71–212.

Nörlund, N.E. 1924. *Vorlesungen über Differenzenrechnung*. Berlin: Springer.

Ogawa, T., and Morimoto, M. 2018. *Mathematics of Takebe Katahiro and History of Mathematics in East Asia*. Tokyo: Math. Soc. Japan.

Olver, P.J. 1999. *Classical Invariant Theory*. Cambridge: Cambridge University Press.

Ore, Ø. 1974. *Niels Henrik Abel: Mathematician Extraordinary*. Providence: A.M.S.

Ostrogradsky, M.V. 1845. De l'integration des fractions rationelles. *Bull. Physico-Math. Acad. Sci St. Pétersbourg*, **4**, 145–167 and 268–300.

Ozhigova, E.P. 2007. The part played by the Petersburg Academy of Sciences (the Academy of Sciences of the USSR) in the publication of Euler's collected works. Pages 53–74 of: Bogolyubov, N.N., Mikhaĭlov, G.K., and Yushkevich, A.P. (eds), *Euler and Modern Science*. Washington, D.C.: M.A.A.

Papperitz, E. 1889. Ueber die Darstellung der hypergeometrischen Transcendenten durch eindeutige Functionen. *Math. Ann.*, **34**, 247–296.

Parameswaran, S. 1983. Madhava of Sangamagramma. *J. Kerala Studies*, **10**, 185–217.

Patterson, S.J. 2007. Gauss sums. Pages 505–528 of: Goldstein, C., Schappacher, N., and Schwermer, J. (eds), *The Shaping of Arithmetic after C. F. Gauss's Disquisitiones Arithmeticae*. New York: Springer.

Peacock, G. 1843–1845. *Treatise on Algebra*. Cambridge: Cambridge University Press.

Peano, G. 1973. *Selected Works*. London: Allen and Unwin. Edited by H.C. Kennedy.

Peirce, B. 1881. Linear Associative Algebra, with Notes and Addenda by C.S. Peirce. *Amer. J. Math.*, **4**, 97–229.

Pepper, J.V. 1968. Harriot's calculation of the meridional parts as logarithmic tangents. *Archive Hist. Exact Sci.*, **4**, 359–413.

Perron, O. 1929. *Die Lehre von den Kettenbrüchen, second edition*. Leipzig: Teubner.

Peters, C.A. (ed). 1860–1865. *Briefwechsel zwischen C.F. Gauss und H.C. Schumacher, vols. 1–5*. Altona: Esch.

Petrovski, I.G. 1966. *Ordinary Differential Equations*. Englewood Cliffs, N.J.: Prentice-Hall. Translated by R.A. Silverman.

Pfaff, J. 1797a. *Disquisitiones analyticae*. Helmstadii: Fleckheisen.

Pfaff, J. 1797b. Observationes analyticae ad L. Euleri Institutiones Calculi Integralis. *Supplement IV, Historie de 1793, Nova Acta Acad. Sci. Petropolitanae*, **XI**, 38–57.

Picard, É. 1879. Sur une propriété des fonctions entières. *Comptes Rendus*, **88**, 1024–1027.

Pick, G. 1915. Über eine Eigenschaft der konformen Abbildung kreisförmiger Bereiche. *Math. Ann.*, **77**, 1–6.

Pieper, H. (ed). 1998. *Korrespondenz zwischen Legendre und Jacobi*. Leipzig: Teubner.

Pieper, H. 2007. A network of scientific philanthropy: Humboldt's relations with number theorists. Pages 201–234 of: Goldstein, C., Schappacher, N., and Schwermer, J. (eds), *The Shaping of Arithmetic after C. F. Gauss's Disquisitiones Arithmeticae*. New York: Springer.

Pierpoint, W. S. 1997. Edward Stone (1702–1768) and Edmund Stone (1700–1768): Confused Identities Resolved. *Notes and Records Roy. Soc. London*, **51**, 211–217.

Pierpont, J. 1904. The history of mathematics in the nineteenth century. *Bull. A.M.S.*, **11**, 136–159. Reprinted in 2000 in the *Bulletin*, vol. 37, pp. 9–24.

Pietsch, A. 2007. *History of Banach Spaces and Linear Operators*. Boston: Birkhäuser.

Pingree, D. 1970–1994. Census of the exact sciences in Sanskrit. *Amer. Phil. Soc.*, **81, 86, 111, 146, 213**.

Pitt, H.R. 1938. General Tauberian theorems. *Proc. London Math. Soc.*, **44**, 243–288.

Plofker, K. 2009. *Mathematics in India*. Princeton: Princeton University Press.

Poincaré, H. 1883. Sur les fonctions entières. *Bull. Soc. Math. France*, **11**, 136–144.

Poincaré, H. 1886. Sur les intégrales irrégulières des équations linéaires. *Acta Math.*, **8**, 295–344.

Poincaré, H. 1907. Sur l'uniformisation des fonctions analytiques. *Acta Math.*, **31**, 1–64.

Poincaré, H. 1985. *Papers on Fuchsian Functions*. New York: Springer-Verlag. Translated by J. Stillwell.

Poisson, S.D. 1808. Mémoire sur la propagation de la chaleur dans les corps solides (extrait). *Nouv. Bull. Sci. Soc. Philom. Paris*, **1**, 112–116.

Poisson, S.D. 1823. Suite du mémoire sur les intégrales définies et sur la sommation des séries. *J. École Poly.*, **12**, 404–509.

Poisson, S.D. 1826. Sur le calcul numérique des intégrales définies. *Mém. Acad. Sci. France*, **6**, 571–602.

Polignac, A. de. 1857. Recherches sur les nombres premiers. *Comptes Rendus.*, **45**, 575–580.

Pólya, G. 1974. *Collected Papers*. Cambridge, MA: MIT Press. Edited by Ralph Boas.

Pommerenke, C. 1985. The Bieberbach conjecture. *Math. Intelligencer*, **7**(2), 23–25; 32.

Popoff, A. 1861. Sur le reste de la série de Lagrange. *Comptes Rendus*, **53**, 795–798.

Prasad, G. 1931. *Six Lectures on the Mean-Value Theorem of the Differential Calculus*. Calcutta: Calcutta University Press.

Prasad, G. 1933. *Some Great Mathematicians of the Nineteenth Century*. Benares, India: Benares Mathematical Society.

Pringsheim, A. 1900. Zur Geschichte des Taylorschen Lehrsatzes. *Bibliotheca Math.*, **3**, 433–479.

Probst, S. 2015. Leibniz as reader and second inventor: the cases of Barrow and Mengoli. Pages 111–134 of: Goethe, N., Beeley, P., and Rabouin, D. (eds), *G.W. Leibniz, Interrelation between Mathematics and Philosophy*. Dordrecht: Springer.

Purkert, W., and Ilgauds, H.J. 1985. *Georg Cantor*. Leipzig: Vieweg and Teubner.

Rabuel, C. 1730. *Commentaires sur la géométrie de M. Descartes*. Lyon: M. Duplain.

Rajagopal, C.T. 1949. A neglected chapter of Hindu mathematics. *Scripta Math.*, **15**, 201–209.

Rajagopal, C.T., and Rangachari, M.S. 1977. On the untapped source of medieval Keralese mathematics. *Archive Hist. Exact Sci.*, **18**, 89–102.

Rajagopal, C.T., and Rangachari, M.S. 1986. On medieval Keralese mathematics. *Archive Hist. Exact Sci.*, **35**, 91–99.

Rajagopal, C.T., and Vedamurtha Aiyar, T.V. 1951. On the Hindu proof of Gregory's series. *Scripta Math.*, **17**, 65–74.

Rajagopal, C.T., and Venkataraman, A. 1949. The sine and cosine power series in Hindu mathematics. *J. Roy. Asiatic Soc. Bengal, Sci.*, **15**, 1–13.

Ramanujan, S. 1911. Some properties of Bernoulli numbers. *J. Indian Math. Soc.*, **3**, 219–234.

Ramanujan, S. 1917. A series for Euler's constant $\gamma$. *Messenger Math.*, **46**, 73–80.

Ramanujan, S. 1919a. Proof of certain identities in combinatory analysis. *Proc. Camb. Phil. Soc.*, **19**, 214–216. Reprinted in Ramanujan (2000), pp. 214–215.

Ramanujan, S. 1919b. Some properties of $p(n)$, the number of partitions of $n$. *Proc. Camb. Phil. Soc.*, **19**, 207–210. Reprinted in Ramanujan (2000), pp. 210–213.

Ramanujan, S. 1988. *The Lost Notebook and Other Unpublished Papers*. Delhi: Narosa Publishing House. Introduction by G. Andrews.

Ramanujan, S. 2000. *Collected Papers*. Providence: A.M.S. Chelsea. Edited by G.H. Hardy, P.V. Seshu Aiyar, B. M. Wilson, with extensive commentary by B. Berndt.

Rashed, R. 1970. al-Karaji. Pages 240–246 of: *Dictionary of scientific biography, volume 7*. New York: Scribners.

Rashed, R. 1980. Ibn al-Haytham et le théorème de Wilson. *Archive Hist. Exact Sci.*, **22**, 305–321.

Rassias, T.M., Srivastava, H.M., and Yanushauskas, A. 1993. *Topics in Polynomials of One and Several Variables and Their Applications*. Singapore: World Scientific.

Raussen, M., and Skau, C. 2010. Interview with Mikhail Gromov. *Notices A.M.S.*, **57**, 391–403.

Remmert, R. 1991. *Theory of Complex Functions*. New York: Springer-Verlag. Translated by R. Burckel.

Remmert, R. 1996. Wielandt's theorem about the $\Gamma$-function. *Am. Math. Monthly*, **103**, 214–220.

Remmert, R. 1998. *Classical Topics in Complex Function Theory*. New York: Springer-Verlag. Translated by Leslie Kay.

Riccati, J. 1724. Animadversationes in aequationes differentiales. *Acta Erud.*, **Supp. 8, 1724**, 66–73.

Riemann, B. 1859. Ueber die Anzahl der Primzahlen unter einer gegebenen Grösse. *Monatsber. Berlin. Akad.*, 671–680.

Riemann, B. 1899. *Elliptische Functionen. Vorlesungen von Bernhard Riemann. Mit Zusätzen herausgegeben von Hermann Stahl*. Leipzig: Teubner. Edited by H. Stahl.

Riemann, B. 1990. *Gesammelte Mathematische Werke*. New York: Springer-Verlag. Edited by R. Dedekind, H. Weber, R. Narasimham, and E. Neuenschwander.

Riesz, F. 1909. Sur les opérations fonctionnelles linéaires. *Comptes Rendus*, **149**, 974–977.

Riesz, F. 1910. Untersuchungen über Systeme integrierbarer Funktionen. *Math. Ann.*, **69**, 449–497. Reprinted in Riesz (1960) vol. 1, pp. 441–489.

Riesz, F. 1913. *Les systèmes d'équations linéaires a une infinité d'inconnues*. Paris: Gauthier-Villars.

Riesz, F. 1960. *Oeuvres complètes*. Budapest: Académie des Sciences de Hongrie. Edited by Á. Császár.

Riesz, M. 1928. Sur les fonctions conjugées. *Math. Z.*, **27**, 218–44.

Rigaud, S.P. (ed). 1841. *Correspondence of Scientific Men of the Seventeenth Century*. Oxford: Oxford University Press.

Robertson, M.S. 1936. A remark on the odd schlicht functions. *Bull. A.M.S.*, **42**, 366–370.

Robins, B. 1727. A demonstration of the 11th proposition of Sir Isaac Newton's treatise on Quadrature. *Phil. Trans. Roy. Soc.*, **34**, 230–236.

Rodrigues, O. 1816. Mémoire sur l'attraction des sphéroids. *Correspondance École Poly.*, **3**, 361–385.

Rodrigues, O. 1839. Note sur les inversions, ou dérangements produits dans les permutations. *J. Math. Pures Appl.*, **4**, 236–240.

Rogers, L.J. 1888. An extension of a certain theorem in inequalities. *Messenger Math.*, **17**, 145–150.

Rogers, L.J. 1893a. Note on the transformation of an Heinean series. *Messenger Math.*, **23**, 28–31.

Rogers, L.J. 1893b. On a three-fold symmetry in the elements of Heine's series. *Proc. London Math. Soc.*, **24**, 171–179.

Rogers, L.J. 1893c. On the expansion of some infinite products. *Proc. London Math. Soc.*, **24**, 337–352.

Rogers, L.J. 1894. Second memoir on the expansion of certain infinite products. *Proc. London Math. Soc.*, **25**, 318–343.

Rogers, L.J. 1895. Third memoir on the expansion of certain infinite products. *Proc. London Math. Soc.*, **26**, 15–32.

Rogers, L.J. 1907. On function sum theorems connected with the series $\sum_{n=1}^{\infty} \frac{x^n}{n^2}$. *Proc. London Math. Soc.*, **4**, 169–189.

Rogers, L.J. 1917. On two theorems of combinatory analysis and some allied identities. *Proc. London Math. Soc.*, **16**, 321–327.

Rolle, M. 1690. *Traité d'algebre*. Paris: Michallet.

Rolle, M. 1691. *Démonstration d'une Méthode pour résoudre les égalitez de tous les degréz*. Paris: Cusson.

Roquette, P. 2002. The Riemann hypothesis in characteristic $p$, its origin and development. Part I. The formation of the zeta-functions of Artin and of F. K. Schmidt. *Mitt. Math. Ges. Hamburg*, **21**, 79–157.

Roquette, P. 2004. The Riemann hypothesis in characteristic $p$, its origin and development. Part II. The first steps by Davenport and Hasse. *Mitt. Math. Ges. Hamburg*, **23**, 5–74.

Roquette, P. 2018. *The Riemann Hypothesis in Characteristic p in Historical Perspective*. Switzerland: Springer Nature.

Rosen, M. 2002. *Number Theory in Function Fields*. New York: Springer.

Rothe, H.A. 1811. *Systematisches Lehrbuch der Arithmetik*. Erlangen: Barth.

Rowe, D.E., and McCleary, J. 1989. *The History of Modern Mathematics*. Boston: Academic Press.

Roy, R. 1990. The discovery of the series formula for $\pi$ by Leibniz, Gregory and Nilakantha. *Math. Mag.*, **63**(5), 291–306. Reprinted in Anderson, Katz, and Wilson (2004), pp. 111–121.

Roy, R. 1993. The Work of Chebyshev on Orthogonal Polynomials. Pages 495–512 of: Rassias, T., Srivastava, H., and Yanushauskas, A. (eds), *Topics in Polynomials of one and Several Variables and their Applications*. Singapore: World Scientific.

Roy, R. 2017. *Elliptic and Modular Functions*. Cambridge: Cambridge University Press.

Ru, M. 2001. *Nevanlinna Theory and its Relation to Diophantine Approximation*. Singapore: World Scientific.

Rudin, W. 1966. *Real and Complex Analysis*. New York: McGraw-Hill.

Saalschütz, L. 1890. Eine Summationsformel. *Zeit. Math. Phys.*, **35**, 186–188.

Salmon, G. 1879. *A Treatise on the Higher Plane Curves, third edition*. Dublin: Hodges, Foster, and Figgis.

Sandifer, C.E. 2007. *The Early Mathematics of Leonhard Euler*. Washington, D.C.: M.A.A.

Sarma, K.V. 1972. *A History Of The Kerala School Of Hindu Astronomy*. Hoshiarpur, India: Punjab University

Sarma, K.V., and Hariharan, S. 1991. Yuktibhasa of Jyesthadeva. *Indian J. Hist. Sci.*, **26**(2), 185–207.

Sasaki, C. 1994. The adoption of Western mathematics in Meiji Japan, 1853–1903. Pages 165–186 of: Sasaki, C., Sugiura, M., and Dauben, J.W. (eds), *The Intersection of History and Mathematics*. Basel: Birkhäuser.

Sasaki, C., Sugiura, M., and Dauben, J. W. (eds). 1994. *The Intersection of History and Mathematics*. Basel: Birkhäuser.

Schaar, M. 1850. Recherches sur la théorie des résidues quadratiques. *Mém. couronnés et mém. savants étrangers Acad. Roy. Sci. Belgique*, **25**, 1–20.

Scharlau, W., and Opolka, H. 1984. *From Fermat to Minkowski: Lectures on the Theory of Numbers and its Historical Development*. New York: Springer.

Schellbach, K. 1854. Die einfachsten periodischen Functionen. *J. Reine Angew. Math.*, **48**, 207–236.

Schlömilch, O. 1843. Einiges über die Eulerischen Integrale der zweiten Art. *Archiv Math. Phys.*, **4**, 167–174.

Schlömilch, O. 1847. *Handbuch der Differenzial- und Integralrechnung*. Greifswald, Germany: Otte.

Schlömilch, O. 1849. Uebungsaufgaben für Schüler, Lehrsatz von dem Herrn. Prof. Dr. Schlömilch. *Archiv Math. Phys.*, **12**, 415.

Schlömilch, O. 1858. Ueber eine Eigenschaft gewisser Reihen. *Zeit. Math. Phys.*, **3**, 130–132.

Schneider, I. 1968. Der Mathematiker Abraham de Moivre (1667–1754). *Archive Hist. Exact Sci*, **5**, 177–317.

Schneider, I. 1983. Potenzsummenformeln im 17. Jahrhundert. *Hist. Math.*, **10**, 286–296.

Schneider, T. 1934. Transzendenzuntersuchungen periodischer Funktionen: I. Transzendenz von Potenzen; II. Transzendenzeigenschaften elliptischer Funktionen. *J. Reine Angew. Math.*, **172**, 65–74.

Schoenberg, I.J. 1988. *Selected Papers*. Boston: Birkhäuser.

Schönemann, T. 1845. Grundzüge einer allgemeinen Theorie der höheren Congruenzen, deren Modul eine reelle Primzahl ist. *J. Reine Angew. Math.*, **31**, 269–325.

Schur, I. 1917. Ein Beitrag zur additiven Zahlentheorie der Kettenbrüche. *S.B. Preuss. Akad. Wiss. Phys.-Math.*, 302–321.

Schur, I. 1929. Einige Sätze über Primzahlen mit Anwendung auf Irreduzibilitätsfragen. *S-B Akad. Wiss. Berlin Phys. Math. Klasse*, 125–136.

Schwarz, H.A. 1885. Über ein die Flächen kleinsten Flächeninhalts betreffends Problem der Variationsrechnung. *Acta Soc. Scient. Fenn.*, **15**, 315–362. Reprinted in Schwarz (1972) vol. 1, pp. 223–269.

Schwarz, H.A. 1893. *Formeln und Lehrsätze zum Gebrauche der elliptischen Funktionen*. Berlin: Springer.

Schwarz, H.A. 1972. *Abhandlungen*. New York: Chelsea.

Schweins, F. 1820. *Analysis*. Heidelberg: Mohr und Winter.

Schwering, K. 1899. Zur Theorie der Bernoulli'schen Zahlen. *Math. Ann.*, **52**, 171–173.

Scriba, C.J. 1961. Zur Lösung des 2. Debeauneschen Problems durch Descartes. *Archive Hist. Exact Sci.*, **1**, 406–419.

Scriba, C.J. 1964. The inverse method of tangents. *Archive Hist. Exact Sci.*, **2**, 113–137.

Seal, H.L. 1949. The historical development of the use of generating functions in probability theory. *Bull. Assoc. Actuaires Suisses*, **49**, 209–228.

Segal, S.L. 1978. Riemann's example of a continuous "nondifferentiable" function. *Math. Intelligencer*, **1**, 81–82.

Seidel, P.L. 1847. Note über eine Eigenschaft der Reihen, welche discontinuirliche Functionen darstellen. *Abhand. Math. Phys. Klasse der Kgl. Bayrischen Akad. Wiss.*, **5**, 381–394.

Seki, T. 1974. *Takakazu Seki's Collected Works Edited with Explanations*. Osaka: Kyoiku Tosho. Edited by A. Hirayama, K. Shimodaira and H. Hirose; translated by Jun Sudo.

Selberg, A. 1941. Über einen Satz von A. Gelfond. *Arch. Math. Naturvidenskab*, **44**, 159–170. Reprinted in Selberg's *Collected Papers* vol. 1, pp. 62–73.

Selberg, A. 1944. Bemerkninger om et multipelt integral. *Norsk. Mat. Tidskr.*, **26**, 71–78. Republished in Selberg (1989) vol. 1, pp. 204–211.

Selberg, A. 1949. An elementary proof of the prime number theorem. *Ann. of Math.*, **50**, 305–313.

Selberg, A. 1989. *Collected Papers*. New York: Springer-Verlag.

Sen Gupta, D.P. 2000. Sir Asutosh Mookerjee—educationist, leader and institution-builder. *Current Sci.*, **78**, 1566–1573.

Serret, J.A. 1854. *Cours d'algèbre supérieure*. Paris: Mallet-Bachelier.

Serret, J.A. 1868. *Cours de calcul différentiel et intégral*. Paris: Gauthier-Villars.

Serret, J.A. 1877. *Cours d'algèbre supérieure, 4 é.* Paris: Gauthier-Villars.

Shah, S.M. 1948. A note on uniqueness sets for entire functions. *Proc. Indian Acad. Sci., Sect. A*, **28**, 519–526.

Shidlovskii, A.B. 1989. *Transcendental Numbers*. Berlin: de Gruyter. Translated by N. Koblitz.

Shimura, G. 2007. *Elementary Dirichlet Series and Modular Forms*. New York: Springer.

Shimura, G. 2008. *The Map of My Life*. New York: Springer.

Siegel, C.L. 1932. Über die Perioden elliptischer Funktionen. *J. Reine Angew. Math.*, **167**, 62–69.

Siegel, C.L. 1949. *Transcendental Numbers*. Princeton: Princeton University Press.

Siegel, C.L. 1969. *Topics In Complex Function Theory*. New York: Wiley.

Simmons, G.F. 1992. *Calculus Gems*. New York: McGraw-Hill.

Simon, B. 2005. OPUC on One Foot. *Bull. A.M.S.*, **42**, 431–460.

Simpson, T. 1743. *Mathematical Dissertations*. London: Woodward.

Simpson, T. 1750. *The Doctrine and Application of Fluxions*. London: Nourse.

Simpson, T. 1759. The invention of a general method for determining the sum of every second, third, fourth, or fifth, etc. term of a series, taken in order; the sum of the whole being known. *Phil. Trans. Roy. Soc.*, **50**, 757–769.

Simpson, T. 1800. *A Treatise of Algebra*. London: Wingrave.

Sluse, R.F. 1672. An extract of a letter ... concerning his short and easie method of drawing tangents to all geometical curves. *Phil. Trans. Roy. Soc.*, **7**, 5143–5147.

Smith, D.E. (ed). 1959. *A Source Book in Mathematics*. New York: Dover.

Smith, D.E., and Mikami, Y. 1914. *A History of Japanese Mathematics*. Chicago: Open Court.

Smith, H.J.S. 1875. On the Integration of Discontinuous Functions. *Proc. London Math. Soc.*, **6**, 140–153. Reprinted in Smith (1965) vol. 2, pp. 86–100.

Smith, H.J.S. 1965a. *Collected Mathematical Papers*. New York: Chelsea. Edited by J.W.L. Glaisher. Volume 1 includes Smith's *Report on the Theory of Numbers*.

Smith, H.J.S. 1965b. *Report on the Theory of Numbers*. New York: Chelsea.

Snow, J.E. 2003. Views on the real numbers and the continuum. *Rev. Mod. Logic*, **9**, 95–113.

Somayaji, Putumana. 2018. *Karanapaddhati of Putumana Somayaji*. Singapore: Springer Nature. Edited and translated with commentary by V. Pai, K. Ramasubramanian, M.S. Sriram and M.D. Srinivas.

Spence, W. 1809. *An Essay on the Theory of the Various Orders of Logarithmic Transcendents*. London: Murray.

Spence, W. 1819. *Mathematical Essays*. London: Whittaker. Edited by J.F.W. Herschel.

Spiess, O. (ed). 1955. *Der Briefwechsel von Johann Bernoulli*. Basel: Birkhäuser.

Spiridonov, V. 2013. Elliptic hypergeometric functions. Pages 577–606 of: *Special Functions, Russian edition*. Moscow: Cambridge University Press and MCCME. This article was written in Russian as an additional complementary chapter to the Russian edition of *Special Functions*, by Andrews, Askey, Roy.

Sridharan, R. 2005. Sanskrit prosody, Pingala sutras and binary arithmetic. Pages 33–62 of: Emch, G.G. et al. (ed), *Contributions to the history of Indian mathematics*. New Delhi: Hindustan Book Agency and Springer.

Srinivasiengar, C.N. 1967. *The History of Ancient Indian Mathematics*. Calcutta: World Press.

Stäckel, P. 1908. Eine vergessen Abhandlung Leonhard Eulers über die Summe der reziproken Quadrate der natürlichen Zalen. *Biblio. Math.*, **8**, 37–60.

Stäckel, P., and Ahrens, W. (eds). 1908. *Briefwechsel zwischen C.G.J. Jacobi und P.H. Fuss.* Leipzig: Teubner.

Stark, H.M. 1967. A complete determination of the complex quadratic fields with class-number one. *Michigan Math. J.*, **14**, 1–27.

Stark, H.M. 1969. On the 'gap' in a theorem of Heegner. *J. Number Theory*, **1**, 16–27.

Stedall, J. 2000. Catching proteus: The collaborations of Wallis and Brouncker, I and II. *Notes and Records Roy. Soc. London*, **54**, 293–331.

Steele, J.M. 2004. *The Cauchy-Schwarz Master Class.* Cambridge: Cambridge University Press.

Steffens, K.-G. 2006. *The History of Approximation Theory.* Boston: Birkhäuser.

Stephens, L., and Lee, S. (eds). 1908. *Dictionary of National Biography, 22 volumes.* New York: Macmillan.

Stickelberger, L. 1890. Über eine Verallgemeinerung von der Kreistheilung. *Math. Ann.*, **37**, 321–367.

Stieltjes, T.J. 1885a. Sur certains polynômes qui vérifient une équation différentielle linéaire du second ordre et sur la théorie des fonctions de Lamé. *Acta Math.*, **6**, 321–326. Reprinted in Stieltjes's 1993 *Collected Papers* vol. 1, pp. 522–527.

Stieltjes, T.J. 1885b. Sur les polynômes de Jacobi. *Comptes Rendus*, **100**, 620–622. Reprinted in Stieltjes' *Collected Papers* vol. 1, pp. 530–532.

Stieltjes, T.J. 1886a. Recherches sur quelques séries semi-convergentes. *Ann. Sci. École Norm.*, **3**, 201–258. Reprinted in Stieltjes's *Collected Papers* vol. 2, pp. 2–58.

Stieltjes, T.J. 1886b. Sur les racines de l'équation $X_n = 0$. *Acta Math.*, **9**, 385–400. Reprinted in Stieltjes's 1993 *Collected Papers* vol. 2, pp. 77–92.

Stieltjes, T.J. 1889. Sur le développement de $\log\Gamma(a)$. *J. Math. Pures Appl.*, **[4] 5**, 425–444. Reprinted in Stieltjes's *Collected Papers* vol. 2, pp. 215–234.

Stieltjes, T.J. 1890. Note sur l'intégrale $\int_0^\infty e^{-u^2}du$. *Nouv. Ann. Math. Paris*, **9**, 479–480. Reprinted in Stieltjes's Oeuvres Complètes vol. 2, pp. 263–264.

Stieltjes, T.J. 1993. *Collected Papers.* New York: Springer-Verlag. Edited by G. van Dijk.

Stirling, J. 1717. *Lineae Tertii Ordinis Neutonianae.* Oxford: Whistler.

Stirling, J. 1719. Methodus differentialis Newtoniana illustrata. *Phil. Trans.*, **30**, 1050–1070.

Stirling, J. 1730. *Methodus differentialis.* London: Strahan.

Stirling, J., and Tweddle, I. 2003. *James Stirling's Methodus differentialis, An Annotated Translation of Stirling's Text.* London: Springer.

Stokes, G.G. 1849. On the critical values of the sums of periodic series. *Trans. Cambridge Phil. Soc.*, **8**, 533–583.

Stone, E. 1730. *The Method of Fluxions both Direct and Inverse.* London: W. Innys.

Stone, M. 1932. *Linear transformations in Hilbert Spaces and their applications to analysis.* Providence: A.M.S.

Strichartz, R.S. 1995. *The Way of Analysis.* London: Jones and Bartlett.

Struik, D.J. 1969. *A Source Book in Mathematics.* Cambridge, MA: Harvard University Press.

Struik, D.J. 1987. *A Concise History of Mathematics.* New York: Dover.

Stubhaug, A. 2000. *Niels Henrik Abel and His Times.* New York: Springer.

Sturm, C. 1829. Analyse d'un mémoire sur la résolution des équations numériques. *Bull. Sci. Férussac*, **11**, 419.

Sturmfels, B. 2008. *Algorithms on Invariant Theory.* Wien: Springer.

Sylvester, J.J. 1852. On the principles of the calculus of forms. *Cambridge and Dublin Math. J.*, **7**, 52–97, 179–217. Reprinted in Sylvester (1973) vol. 1, pp. 284–327, 328–363.

Sylvester, J.J. 1853a. On a theory of the syzygetic relations of two rational integral functions ... *Phil. Trans.*, **143**, 407–548. Reprinted in Sylvester (1973) vol. 1, pp. 429–586.

Sylvester, J.J. 1853b. On Mr. Cayley's impromptu demonstration of the rule for determining at sight the degree ... *Phil. Mag.*, **5**, 199–202. Reprinted in Sylvester (1973) vol. 1, pp. 595–598.

Sylvester, J.J. 1865. On an elementary proof and generalization of Sir Isaac Newton's hitherto undemonstrated rule for the discovery of imaginary roots. *Proc. London Math. Soc.*, **1**, 1–16. Reprinted in Sylvester's *Collected Mathematical Papers* (1773) vol. 2, pp. 498–513.

Sylvester, J.J. 1869. On a new continued fraction applicable to the quadrature of the circle. *Phil. Magazine*, **37**, 373–375. Republished in Sylvester's *Collected Mathematical Papers* vol. 2, pp. 691–693.

Sylvester, J.J. 1878. Proof of the hitherto undemonstrated theorem of invariants. *Phil. Mag.*, **5**, 178–188. Reprinted in Sylvester (1973) vol. 3, pp. 117–126.

Sylvester, J.J. 1882. A constructive theory of partitions, arranged in three acts, an interact, and an exodion. *Amer. J. Math*, **5**, 251–330. Reprinted in Sylvester (1973) vol. 4, pp. 1–83.

Sylvester, J.J. 1973. *Mathematical Papers.* New York: Chelsea. Edited by H. Baker.

Szegő, G. 1926. Ein Beitrag zur Theorie der Thetafunktionen. *S.B. Preuss. Akad. Wiss. Phys-Math.*, 242–252. Reprinted in Szegő (1982) vol. 1, pp. 795–805.

Szegő, G. 1975. *Orthogonal Polynomials.* Providence: A.M.S.

Szegő, G. 1982. *The Collected Papers of Gabor Szegő.* Boston: Birkhäuser. Edited by R. Askey.

Szekeres, G. 1968. A combinatorial interpretation of Ramanujan's continued fraction. *Canadian Math. Bull.*, **11**, 405–408.

Takagi, T. 1990. *Collected Papers.* Tokyo: Springer-Verlag. Edited by S. Iyanaga, K. Iwasawa, K. Kodaira, and K. Yosida.

Takase, M. 1994. Three aspects of the theory of complex multiplication. Pages 91–108 of: Sasaki, C., Sugiura, M., and Dauben, J.W. (eds), *The Intersection of History and Mathematics.* Basel: Birkhäuser.

Tannery, J., and Molk, J. 1972. *Éléments de la théorie des fonctions elliptiques, 4 vols.* New York: Chelsea.

Tauber, A. 1897. Ein Satz aus der Theorie der unendlichen Reihen. *Monats. Math. und Phys.*, **8**, 273–277.

Taylor, B. 1715. *Methodus incrementorum directa et inversa.* London: W. Innys.

Taylor, B., and Feigenbaum, L. 1981. *Brook Taylor's "Methodus incrementorum": A Translation with Mathematical and Historical Commentary.* Ph.D. thesis, Yale University.

Thomae, J. 1869. Beiträge zur Theorie der durch die Heinsche Reihe ... *J. Reine Angew. Math.*, **70**, 258–281.

Thomson, W., and Tait, P.G. 1890. *Treatise on Natural Philosophy.* Cambridge: Cambridge University Press.

Tignol, J.-P. 1988. *Galois' Theory of Algebraic Equations.* New York: Wiley.

Titchmarsh, E.C., and Heath-Brown, D.R. 1986. *The Theory of the Riemann-Zeta Function.* Oxford: Oxford University Press.

Truesdell, C. 1960. *The Rational Mechanics of Flexible or Elastic Bodies, 1638–1788. Introduction to Vols. 10 and 11, Second Series of Euler's Opera omnia.* Zurich: Orell Füssli Turici.

Truesdell, C. 1984. *An Idiot's Fugitive Essays on Science.* New York: Springer-Verlag.

Tucciarone, J. 1973. The development of the theory of summable divergent series from 1880 to 1925. *Archive Hist. Exact Sci.*, **10**, 1–40.

Turán, P. 1990. *Collected Papers.* Budapest: Akadémiai Kiadó. Edited by P. Erdős.

Turnbull, H.W. 1933. James Gregory: A study in the early history of interpolation. *Proc. Edinburgh Math. Soc.*, **3**, 151–172.

Turnbull, H.W. (ed). 1939. *James Gregory Tercentenary Memorial Volume.* London: Bell.

Tweddle, I. 1984. Approximating $n!$ Historical origins and error analysis. *Amer. J. Phys.*, **52**, 487–488.

Tweddle, I. 1988. *James Stirling: "This about series and such things."* Edinburgh: Scottish Acad. Press.

Tweedie, C. 1917–1918. Nicole's contributions to the foundations of the calculus of finite differences. *Proc. Edinburgh Math. Soc.*, **36**, 22–39.

Tweedie, C. 1922. *James Stirling: A Sketch of His Life and Works along with His Scientific Correspondence.* Oxford: Oxford University Press.

Valiron, G. 1913. Sur les fonctions entières d'ordre fini et d'ordre nul, et en particulier les fonctions à correspondence régulière. *Ann. Fac. Sci. Toulouse*, **5**, 117–257.

Valiron, G. 1949. *Lectures on the General Theory of Integral Functions*. New York: Chelsea.

Vallée-Poussin, C.J. de la. 1896a. Demonstration simplifée du théorème de Dirichlet sur la progression arithmétique. *Mém. Acad. Roy. Soc. Bruxelles*, **53**, 6–8.

Vallée-Poussin, C.J. de la. 1896b. Recherches analytiques sur la théorie des nombres premiers, I–III. *Ann. Soc. Sci. Bruxelles*, **20**, 183–256, 281–362, 363–397.

Van Brummelen, G. 2009. *The Mathematics of the Heavens and the Earth*. Princeton: Princeton University Press.

Van Brummelen, G., and Kinyon, M. (eds). 2005. *Mathematics and the Historian's Craft*. New York: Springer.

Van Maanen, J.A. 1984. Hendrick van Heuraet (1634–1660?): His life and work. *Centaurus*, **27**, 218–279.

van Rootselaar, B. 1964. Bolzano's Theory of Real Numbers. *Archive Hist. Exact Sci.*, **2**, 168–180.

Vandermonde, T.A. 1772. Mémoire sur des irrationnelles de différents ordres avec une application au cercle. *Hist. Acad. Roy. Sci. Paris pour 1772*, 489–498.

Varadarajan, V.S. 2006. *Euler Through Time*. Providence: A.M.S.

Venkatachaliengar, K., and Cooper, S. 2011. *Development of Elliptic Functions According to Ramanujan*. Singapore: World Scientific.

Viète, F. 1593. *Variorum de rebus mathematicis responsorum, Liber. VIII*. Turonis: Iamettium Mettayer.

Viète, F. 1983. *The Analytic Art*. Kent, OH: Kent State University Press. Translated by T.R. Witmer.

Vlǎduţ, S.G. 1991. *Kronecker's Jugendtraum and Modular functions*. New York: Gordon and Breach. Translated by M. Tsfasman.

von Kowalevsky, S. (Kovalevskaya). 1875. Zur Theorie der partiellen Differentialgleichungen. *J. Reine Angew. Math.*, **80**, 1–32.

von Staudt, K.G.C. 1840. Beweis eines Lehrsatzes, die Bernoullischen Zahlen betreffend. *J. Reine Angew. Math.*, **21**, 372–374.

Vorselman de Heer, P. 1833. *Specimen inaugurale de fractionibus continuis*. PhD thesis, Utrecht.

Wagstaff, S. S. 1981. Ramanujan's paper on Bernoulli numbers. *J. Indian Math. Soc.*, **45**, 49–65.

Wali, K.C. 1991. *Chandra: A Biography of S. Chandrasekhar*. Chicago: University Chicago Press.

Walker, J.J. 1891. On the influence of applied on the progress of pure mathematics. *Proc. London Math. Soc.*, **22**, 1–18.

Wallis, J. 1656. *Arithmetica Infinitorum*. Oxford: Lichfield.

Wallis, J. 1659. *Tractatus duo, prior, de cycloide ...* Oxford: Lichfield.

Wallis, J. 1668. Logarithmotechnia Nicolai Mercatoris. *Phil. Trans.*, **3**, 753–759.

Wallis, J. 1685. *A Treatise of Algebra both Historical and Practical*. London: Oxford University

Wallis, J. 1693–1699. *Opera Mathematica, 3 vols*. Oxford: T. Sheldoniano.

Wallis, J., and Stedall, J. 2004. *The Arithmetic of Infinitesimals*. New York: Springer. Translation with notes of Wallis's *Arithmetica Infinitorum* by J.A. Stedall.

Waring, E. 1779. Problems concerning interpolations. *Phil. Trans. Roy. Soc. London*, **69**, 59–67.

Waring, E. 1991. *Meditationes Algebraicae*. Providence: A.M.S. Translated by D. Weeks.

Watson, G.N. 1933. The marquis and the land-agent. *Math. Gazette*, **17**, 5–17.

Watson, G.N. 1938. Ramanujans Vermutung über Zerfällungsanzahlen. *J. Reine Angew. Math.*, **179**, 97–128.

Weber, H. 1895. *Lehrbuch der Algebra*. Braunschweig: Vieweg.

Weierstrass, K. 1856. Über die Theorie der analytischen Facultäten. *J. Reine Angew. Math*, **51**, 1–60. Reprinted in Weierstrass (1894–1927) vol. 1, pp. 153–221.

Weierstrass, K. 1885. Zu Lindemann's Abhandlung "Über die Ludolphsche Zahl." *S.B. Preuss. Akad. Wiss.*, 1067–1085. Reprinted in Weierstrass's *Werke*, vol. 2, pp. 341–361.

Weierstrass, K. 1894–1927. *Mathematische Werke*. Berlin: Mayer and Müller.

Weil, A. 1946. *Foundations of Algebraic Geometry*. Providence: A.M.S.

Weil, A. 1949. Numbers of solutions of equations in finite fields. *Bull. A.M.S.*, **55**, 497–508.

Weil, A. 1974. Two lectures on number theory, past and present. *Enseign. Math.*, **20**, 87–110.

Weil, A. 1976. *Elliptic Functions According to Eisenstein and Kronecker*. New York: Springer.

Weil, A. 1979. *Collected Papers*. New York: Springer-Verlag.

Weil, A. 1984. *Number Theory: An Approach through History from Hammurapi to Legendre*. Boston: Birkhäuser.

Weil, A. 1989a. On Eisenstein's copy of the *Disquisitiones*. *Adv. Studies Pure Math.*, **17**, 463–469.

Weil, A. 1989b. Prehistory of the zeta function. Pages 1–9 of: Aubert, K.E., Bombieri, E., and Goldfeld, D. (eds), *Number Theory, Trace Formulas, and Discrete Groups*. Boston: Academic Press.

Weil, A. 1992. *The Apprenticeship of a Mathematician*. Boston: Birkhäuser.

Westfall, R. 1980. *Never at Rest*. Cambridge: Cambridge University Press.

Whiteside, D.T. 1961a. Henry Briggs: The binomial theorem anticipated. *Math. Gazette*, **45**, 9–12.

Whiteside, D.T. 1961b. Patterns of mathematical thought in the later seventeenth century. *Archive Hist. Exact Sci.*, **1**, 179–388.

Whittaker, E.T., and Robinson, G. 1949. *The Calculus of Observations*. London: Blackie and Son.

Whittaker, E.T., and Watson, G.N. 1927. *A Course of Modern Analysis*. Cambridge: Cambridge University Press.

Wiener, N. 1958. *The Fourier Integral and Certain of Its Applications*. New York: Dover.

Wiener, N. 1979. *Collected Works*. Cambridge, MA: MIT Press. Edited by P. Masani.

Wilbraham, H. 1848. On a certain periodic function. *Cambridge and Dublin Math. J.*, **3**, 198–201.

Wilf, H.S. 2001. The number-theoretic content of the Jacobi triple product identity. Pages 227–230 of: Foata, D., and Han, G.-N. (eds), *The Andrews Festschrift: Seventeen Papers on Classical Number Theory and Combinatorics*. New York: Springer.

Wilson, K. 1962. Proof of a conjecture of Dyson. *J. Math. Physics*, **3**, 1040–1043.

Wintner, A. 1929. *Spektral Theorie der unendlichen Matrizen*. Leipzig: Hirzel.

Woodhouse, R. 1803. *The Principles of Analytical Calculation*. Cambridge: Cambridge University Press.

Yadegari, M. 1980. The binomial theorem: A widespread concept in medieval Islamic mathematics. *Hist. Math.*, **7**, 401–406.

Yandell, B.H. 2002. *The Honors Class*. Natick, MA: Peters.

Young, G.C., and Young, W.H. 1909. On derivatives and the theorem of the mean. *Quart. J. Pure Appl. Math.*, **40**, 1–26.

Young, G.C., and Young, W.H. 2000. *Selected Papers*. Lausanne, Switzerland: Presses Polytechniques. Edited by S.D. Chatterji and H. Wefelscheid.

Young, W.H. 1909. A note on a trigonometrical series. *Messenger Math.*, **38**, 44–48.

Yushkevich, A.P. 1964. *Geschichte der Mathematik in Mittelalter*. Leipzig: Teubner.

Yushkevich, A.P. 1971. The concept of function up to the middle of the 19th century. *Archive Hist. Exact Sci.*, **16**, 37–85.

Zdravkovska, S., and Duren, P. 1993. *Golden Years of Moscow Mathematics*. Providence: A.M.S.

Zolotarev, E. 1876. Sur la série de Lagrange. *Nouvelles Ann. Math.*, **15**, 422–423.

# Index

Printed in the United States
by Baker & Taylor Publisher Services